W9-CLK-476

Lecture Notes in Economics and Mathematical Systems

Managing Editors: M. Beckmann and H. P. Künzi

Control Theory

107

International Conference on ...

Control Theory, Numerical Methods and Computer Systems Modelling

International Symposium, Rocquencourt, June 17–21, 1974

IRIA LABORIA

Institut de Recherche d'Informatique et d'Automatique

Edited by A. Bensoussan and J. L. Lions

Springer-Verlag

Berlin · Heidelberg · New York 1975

Editorial Board

H. Albach · A. V. Balakrishnan · M. Beckmann (Managing Editor) · P. Dhrymes
J. Green · W. Hildenbrand · W. Krelle · H. P. Künzi (Managing Editor) · K. Ritter
R. Sato · H. Schelbert · P. Schönfeld

Managing Editors

Prof. Dr. M. Beckmann
Brown University
Providence, RI 02912/USA

Prof. Dr. H. P. Künzi
Universität Zürich
8090 Zürich/Schweiz

Editors

Dr. A. Bensoussan
Dr. J. L. Lions
IRIA LABORIA
Domaine de Voluceau – Rocquencourt
F-78150 Le Chesnay/France

Library of Congress Cataloging in Publication Data

International Conference on Control Theory, Numerical Methods and Computer Systems Modelling, Rocquencourt, France, 1974.
Control theory, numerical methods, and computer systems modelling.

(Control theory) (Lecture notes in economics and mathematical systems ; 107)
"Sponsored by the International Federation for Information Processing (IFIP) and by the European Institute for Advanced Studies in Management."
English or French.
Bibliography: p.
Includes index.
1. Control theory--Congresses. 2. Numerical analysis--Congresses. 3. Computer simulation--Congresses. I. Bensoussan, Alain, ed. II. Lions, Jacques Louis, ed. III. International Federation for Information Processing. IV. European Institute for Advanced Studies in Management. V. Title. VI. Series: Control theory (Berlin) VII. Series: Lecture notes in economics and mathematical systems ; 107.
QA402.3.I48 1974 629.8'312 74-28484

AMS Subject Classifications (1970): 49-02, 60Gxx, 60Hxx, 60Jxx, 60K25, 60K30, 93Bxx, 93Cxx, 93Exx

---

ISBN 3-540-07020-6 Springer-Verlag Berlin · Heidelberg · New York
ISBN 0-387-07020-6 Springer-Verlag New York · Heidelberg · Berlin

This work is subject to copyright. All rights are reserved, whether the whole or part of the material is concerned, specifically those of translation, reprinting, re-use of illustrations, broadcasting, reproduction by photocopying machine or similar means, and storage in data banks.

Under § 54 of the German Copyright Law where copies are made for other than private use, a fee is payable to the publisher, the amount of the fee to be determined by agreement with the publisher.

© by Springer-Verlag Berlin · Heidelberg 1975. Printed in Germany

Offsetdruck: Julius Beltz, Hemsbach/Bergstr.

Copy 2 Math Sci
Sep

MATH.-SCI. 1417594
QA
402.3
.I48
1974

FOREWORD

IRIA-LABORIA[+] has organized, this year, an International Conference on Control Theory, Numerical Methods and Computer Systems Modelling.

This meeting which was sponsored by the International Federation for Information Processing (IFIP) and by the European Institute for Advanced Studies in Management, took place in June (17-21) with the participation of more than 200 specialists among which 55 participants were representing 12 different countries.

This volume of the Springer-Verlag Series "Lecture Notes" contains the lectures presented during the meeting and demonstrates the high interest of the research which is actually carried out in these fields.

We specially wish to thank Monsieur DANZIN, Director of IRIA, for the interest he has shown for this Symposium, Professor BALAKRISHNAN who has arranged for IFIP to sponsor our meeting and Professor GRAVES, Director of the European Institute for Advanced Studies in Management for his support.

The IRIA Public Relations has been of a great assistance to the Organization Committee and we wish to thank Mademoiselle BRICHETEAU and her staff for their contribution.

At last we express our gratitude to the Sessions Chairmen and all the speakers for the very interesting discussions they have directed.

A. BENSOUSSAN and J.L. LIONS

[+] Institut de Recherche d'Informatique et d'Automatique
Laboratoire de Recherche de l'IRIA

PREFACE

L'IRIA-LABORIA[+] a organisé cette année une Conférence Internationale sur la Théorie du Contrôle, les Méthodes Numériques et la Modélisation des Systèmes Informatiques.

Cette rencontre, placée sous le patronage de l'I. F. I. P. et de l'INSTITUT EUROPEEN D'ETUDES SUPERIEURES EN MANAGEMENT de Bruxelles, s'est déroulée du 17 au 21 Juin 1974 et a réuni plus de deux cents participants dont cinquante-cinq personnalités étrangères représentant douze pays.

Ce volume, rassemblant les différents travaux présentés à ce Colloque dans la Série "Lecture Notes" de Springer-Verlag, témoigne de la diversité et de l'intérêt des recherches entreprises actuellement en ces domaines.

Nous tenons tout particulièrement à remercier Monsieur DANZIN, Directeur de l'IRIA, de l'intérêt qu'il a marqué pour cette manifestation, Monsieur le Professeur BALAKRISHNAN, qui a apporté le patronage de l'I. F. I. P. et Monsieur le Professeur GRAVES, celui de l'INSTITUT EUROPEEN de Bruxelles.

Le Service des Relations Extérieures de l'IRIA a grandement facilité la tâche du Comité d'Organisation ; nous remercions très vivement Mademoiselle BRICHETEAU et ses collaboratrices.

Enfin, que les Présidents des différentes sessions et tous les Conférenciers trouvent ici l'expression de notre gratitude pour les très intéressantes et vivantes discussions qu'ils ont suscitées.

A. BENSOUSSAN et J.L. Lions

[+] Institut de Recherche d'Informatique et d'Automatique
Laboratoire de Recherche de l'IRIA

TABLE DES MATIERES
TABLE OF CONTENTS

THEORIE DU FILTRAGE
FILTERING THEORY

Filtering for linear stochastic hereditary differential systems
S. Mitter, R.B. Vinter 1

A Kalman-Bucy filtering theory for affine hereditary differential equations
R. F. Curtain 22

Linear least-squares estimation of discrete-time stationary processes by means of backward innovations
A. Lindquist 44

Filtrage numérique récursif non-linéaire: résolution du problème mathématique et applications
F. Levieux 64

THEORIE DES JEUX
GAME THEORY

A tale of four information structures
Y.C. Ho, I. Blau, T. Basar 85

Stochastic differential games and alternate play
R.J. Elliott 97

Estimation du saut de dualité en optimisation non convexe
J.-P. Aubin 107

Contrainte d'états dans les jeux différentiels
P. Bernhard, J.F. Abramatic 119

Some general properties of non-cooperative games
I. Ekeland 134

Rationalité et formation des coalitions dans un jeu régulier à n joueurs
H. Moulin 140

CONTROLE DES SYSTEMES DISTRIBUES STOCHASTIQUES ET DES CHAMPS ALEATOIRES
CONTROL OF STOCHASTIC DISTRIBUTED PARAMETER SYSTEMS AND RANDOM FIELDS

Identification and stochastic control of a class of distributed systems with Boundary Noise
A.V. Balakrishnan 163

Distributed parameter stochastic systems in population biology
W.H. Fleming 179

On optimization of random functionals
Yu. A. Rozanov 192

Recursive filtering and detection for two-dimensional random fields
E. Wong 207

Stochastic state space representation of images
S. Attasi 218

Contrôle par feedback d'un système stochastique distribué
M. Robin 231

CONTROLE STOCHASTIQUE
STOCHASTIC CONTROL

A homotopy method for proving convexity in certain optimal stochastic control problems
V.E. Beneš 243

On a class of stochastic bang-bang control problems
J. Ruzicka 250

Some stochastic systems on manifolds
T.E. Duncan 262

Problèmes de contrôle stochastique à trajectoires discontinues
F. Brodeau 271

Théorie du potentiel et contrôle des diffusions markoviennes
J.M. Bismut 283

Contrôle stationnaire asymptotique
J.M. Lasry 296

PROBLEMES EN TEMPS DISCRET ET METHODES NUMERIQUES
DISCRETE TIME PROBLEMS AND NUMERICAL METHODS

On the equivalence of multistage recourse models in stochastic optimization
R.T. Rockafellar 314

The instrinsic model for discrete stochastic control : some open problems
H.S. Witsenhausen 322

Finite difference methods for the weak solutions of the Kolmogorov equations for the density of diffusions and conditional diffusions
H.J. Kushner 336

On the relation between stochastic and deterministic optimization
R. J.-B. Wets 350

Solution numérique de l'équation différentielle de RICCATI rencontrée en théorie de la commande optimale des systèmes héréditaires linéaires
M. Delfour 362

Reduction of the operator RICCATI equation
L. Ljung, J. Casti 384

Algorithme d'identification récursive utilisant le concept de positivité
I.D. Landau 397

CONTROLE DES SYSTEMES A PARAMETRES DISTRIBUES
CONTROL OF DISTRIBUTED PARAMETER SYSTEMS

Problèmes de contrôle des coefficients dans des équations aux dérivées partielles
L. Tartar 420

Etude de la méthode de "boucle ouverte adaptée" pour le contrôle des systèmes distribués
J.P. Yvon 427

Estimation des perméabilités relatives et de la pression capillaire dans un écoulement diphasique
G. Chavent, P. Lemonnier 440

Une méthode d'optimisation de forme de domaine. Application à l'écoulement stationnaire à travers une digue poreuse
P. Morice 454

CONTROLE DE PROCESSUS DE SAUT ET APPLICATIONS AUX MODELES DE SYSTEMES INFORMATIQUES
CONTROL OF JUMP PROCESSES AND COMPUTER MODEL APPLICATIONS

Filtering for systems excited by POISSON white noise
H. Kwakernaak 468

A minimum principle for controlled jump processes
R. Rishel 493

Filtering and control of jump processes
P. Varaiya 509

The martingale theory of point processes over the real half line admitting an intensity
P. Bremaud 519

Response time of a fixed-head disk to variable-length transfers
E. Gelenbe, J. Lenfant, D. Potier 543

PROBLEMES DE FRONTIERE LIBRE ET THEORIE DU CONTROLE
FREE BOUNDARY VALUE PROBLEMS AND CONTROL THEORY

Stopping time problems and the shape of the domain of continuation
A. Friedman 559

Problèmes de temps d'arrêt optimaux et de perturbations singulières dans les inéquations variationnelles
A. Bensoussan, J.L. Lions 567

Méthodes de résolution numérique des inéquations quasi-variationnelles
M. Goursat, S. Maurin 585

Optimisation de structure : application à la mécanique des fluides
O. Pironneau 610

Remarques sur les inéquations quasi-variationnelles
J.L. Joly, U. Mosco 625

Perturbations singulières dans un problème de contrôle optimal intervenant en biomathématique
C.M. Brauner, P. Penel 643

APPLICATIONS DE LA THEORIE DU CONTROLE EN ECONOMIE, EN CONTROLE DE PROCESSUS INDUSTRIEL ET EN RECONNAISSANCE DE FORMES

APPLICATIONS OF CONTROL THEORY IN ECONOMICS, PROCESS CONTROL AND PATTERN RECOGNITION

Theory and applications of self-tuning regulators
K.J. Aström 669

Supply and demand relationships in fisheries management
C.W. Clark 681

Commande stochastique d'un système de stockage
G. Bornard, J.F. Cavassilas 692

Etudes d'automatique sur une unité pilote d'absorption et son mélangeur
Y. Sevely 704

Application du contrôle stochastique à la gestion des centrales thermiques et hydrauliques
C. Leguay, A. Breton 728

Automatic Sequential Clustering
E. Diday 745

FILTERING FOR LINEAR STOCHASTIC HEREDITARY DIFFERENTIAL SYSTEMS*

by

Sanjoy K. Mitter[(1)] and Richard B. Vinter[(2)]

(1) Electrical Engineering Department and Electronic Systems Laboratory, M.I.T., Cambridge, Mass. 02139, U.S.A. and Consiglio Nazionale delle Ricerche, Laboratorio per Ricerche di Dinamica dei Sistemi e di Elettronica Biomedica, Italy.

(2) Electronic Systems Laboratory, M.I.T., Cambridge, Mass. 02139,USA.

* The research of the first author was supported by NSF Grant GK-41647, AFOSR Grant 72-2273 and NASA Grant NGL-22-009-124, all at the Electronic Systems Laboratory, M.I.T., Cambridge, Mass. 02139 and by the Consiglio Nazionale delle Ricerche, Italy. The research of the second author was supported by the Commonwealth Fund (Harkness Fellowship).

§1. Introduction

The theorem on the separation of control and filtering (Wonham 1) for linear optimal stochastic control problems with Gaussian noise processes and a quadratic cost function constitutes one of the central results of stochastic control theory. If the linear system is autonomous, then under appropriate stabilizability, observability, detectability and reachability hypotheses it can be shown that the cascade combination of the Kalman filter and the regulator defines an asymptotically stable closed-loop system.

In previous papers (cf. Delfour-Mitter 1 , Delfour-McCalla-Mitter 1) it has been shown that the infinite-time quadratic cost problem for a general class of hereditary systems can be satisfactory solved. In this paper the filtering problem for infinite dimensional systems described by integral equations involving evolution operators is first studied. It is then shown how these results may be specialized to solve the filtering problem for linear stochastic hereditary differential systems.

This paper may be considered to be an application of the work of (Bensoussan 1) to a class of problems somewhat larger than that considered by him in his book. The unbounded linear operator involved in hereditary systems does not satisfy any coercivity conditions nor is it a generator of a contraction semi-group. Nevertheless, it is possible to exploit the structure of the operator (in particular its spectral properties) to obtain reasonably complete results for the filtering problem. For details regarding the properties of this operator the reader is referred to (Delfour 1 and Vinter 4).

The point of view taken in this paper is that for control purposes it is necessary to estimate the "state" of the system.

For hereditary systems, this means that it is first necessary to set up the stochastic evolution equation corresponding to the stochastic functional differential equation describing the evolution of the state of the system, study its properties and prove that the stochastic evolution equation is an equivalent description of the system. This is a key-step and requires the use of the detailed structure of the evolution operator, its generator and the adjoint of the generator. Once this result is at hand, it is possible to use the general theory developed in earlier sections to obtain the Kalman filter.

The question of filter stability is then studied. It is shown that under an appropriate detectability hypothesis (which is verifiable) the covariance of the estimate, is bounded. In the general case it has not been possible (so far) to give reasonably weak conditions under which the operator defining the Kalman filter is asymptotically stable. Nevertheless if the original system is exponentially stable and the "forcing terms" in the filter are not too large then the asymptotic stability of the filter is assured. This result is significant in generating at least a non-trivial class of stable filters.

The filtering problem for hereditary systems has previously been considered by(Kwakernaak 1 and Lindquist). The approach used here (in particular the manner in which the dual control problem is used) and the emphasis on filter stability appear to be new. Moreover in this framework using the ideas of(Bensoussan-Viot 1) the separation theorem for hereditary systems can be proved (cf. also Lindquist 3).

Thus the linear-quadratic-gaussian problem for hereditary systems is almost as complete as that for linear ordinary differential equations.

§2 Some Preliminary Definitions

Take $X, \mathcal{U}$ real separable Hilbert spaces, $(\Omega, \mathcal{S}, \mu)$ a complete probability space.

§§2.1 Separable Hilbert space-valued random variables

The reader is referred to (Bensoussan 1 , ch.3), (Grenander 1 , ch.6) or (Barucha-Reid 1 , ch.1) for more detailed exposition of this material.

$x : \Omega \to X$ is called an X-valued random variable (r.v.) if it is a (weakly) measurable map. The linear space of X-valued r.v.'s is denoted $Mes(\Omega, \mu; X)$.

An X-valued stochastic process is a map $x(\cdot) : \mathbb{R}^+ \to Mes(\Omega, \mu; X)$. $x(\cdot)$ is a measurable process if the map $(t, \omega) \mapsto x(t, \omega)$ is measurable w.r.t. $\overline{\mu_{\mathcal{L}} \times \mu}$ ($\mu_{\mathcal{L}}$ denotes Lebesgue measure on $\mathbb{R}^+$).

$x \in Mes(\Omega, \mu; X)$ is first order if $x \in L^1[\Omega, \mu; X]$ and second order if $x \in L^2[\Omega, \mu; X]$. For a first order r.v. $x(\omega)$ we define the mean $E\{x(\omega)\}$, $(\bar{x})$

$$E\{x(\omega)\} = \int_\Omega x(\omega)\, d\mu \qquad \text{(Bochner Integral)}$$

For a second order r.v. $x(\omega)$, $(h, \bar{h}) \mapsto E\{\langle x(\omega) - E\{x(\omega)\}, h\rangle \langle x(\omega) - E\{x(\omega)\}, \bar{h}\rangle\}$ is a continuous, symmetric bilinear form which has unique representation through $Q \in \mathcal{L}(X)$, $Q \geq 0$, $Q^* = Q$ as $(h, \bar{h}) \mapsto \langle Qh, \bar{h}\rangle$. Q is the covariance of $x(\omega)$. The covariance of a second order random variable $x(\omega)$ is necessarily nuclear (Grenander 1 p. 129).

Given two X-valued second order r.v.'s $x(\omega), y(\omega)$, $(h, \bar{h}) \mapsto E\{\langle x(\omega) - \bar{x}, h\rangle \langle y(\omega) - \bar{y}, \bar{h}\rangle\}$ has unique representation $(h, \bar{h}) \mapsto \langle Rh, \bar{h}\rangle$, $R \in \mathcal{L}(X)$. R is called the covariance of $x(\omega), y(\omega)$ and is written $\operatorname{cov}\{x(\omega), y(\omega)\}$.

$x(\omega), y(\omega) \in Mes(\Omega, \mu; X)$ are independent if $\langle h, x(\omega)\rangle, \langle \bar{h}, y(\omega)\rangle$ are independent for all $h, \bar{h} \in X$. $x \in L^2[\Omega, \mu; X]$ is Gaussian if $\langle x(\omega), h\rangle$ is normally distributed for each $h \in X$.

§§2.2 Wiener Processes

The $\mathcal{U}$-valued stochastic process $W(t, \omega)$ is a Wiener process if (i) for finite collections $\{t_i\} \in \mathbb{R}^+$, $\{e_j\} \in \mathcal{U}$, $(W(t_i, \omega), e_j)$ is a family of real-valued gaussian r.v.'s (ii) $W(t, \omega)$ is second order for each $t \geq 0$ and there exists some nuclear $Q \in \mathcal{L}(X)$ s.t.

$$E\{\langle W(t_1, \omega), h\rangle \langle W(t_2, \omega), \bar{h}\rangle\} = \langle Qh, \bar{h}\rangle \min\{t_1, t_2\}$$

each $t_1, t_2 \geq 0$, $h, \bar{h} \in \mathcal{U}$ (iii) $E\{W(t, \omega)\} = 0$ each $t \geq 0$. See (Bensoussan 1 , p. 167 et seq.) for properties of $W(t, \omega)$.

Notice that since Q is nuclear, $Q \geq 0$, $Q^* = Q$

$$Q(\cdot) = \sum_i \lambda_i e_i \langle e_i, \cdot \rangle$$

for some $\{\lambda_i\}$, $\lambda_i \geq 0$ with $\sum_i \lambda_i < \infty$ *, some orthonormal sequence $\{e_i\}$ in $\mathcal{U}$. We shall make use of the property (Curtain 1) that the Wiener process $W(t, \omega)$ has unique representation

$$W(t, \omega) = \lim_{N \to \infty} \sum_{i=1}^{N} \beta_i(t, \omega) e_i \quad (\text{limit in } L^2[\Omega, \mu; X])$$

with the β_i's independent real valued Wiener processes.

---

* (Gelfand and Vilenkin 1)

§§2.3 The Wiener Integral

Suppose $b: \mathbb{R}^+ \to X$ is locally essentially bounded, measurable and that $\beta(t,\omega)$ is a real-valued Wiener process. Then the Wiener Integral

$$\int_0^T b(t)\, d\beta(t,\omega)$$

is defined in the usual manner as a limit in $L^2[\Omega,\mu;X]$ through a sequence of simple functions approximating $b(t)$ in $L^2[0,T;X]$. Now suppose that $B(\cdot): \mathbb{R}^+ \to \mathcal{L}(\mathcal{U},X)$ satisfies (i) $\|B(\cdot)\|$ is locally essentially bounded, measurable (ii) $t \mapsto B(t)x$ is measurable for each $x \in X$. The Wiener Integral

$$\int_0^T B(t)\, dW(t,\omega)$$

is defined in this case as

$$\lim_{N\to\infty} \sum_{i=1}^{N} \int_0^T B(t) e_i \, d\beta_i(t,\omega) \quad (\text{limit in } L^2[0,T;X])$$

where each element in the sequence is evaluated as above. ($e_i, \beta_i(t,\omega)$ $i=1,2,..$ as in §§2.2). For $B(\cdot)$ measurable w.r.t. the uniform operator topology this definition* coincides essentially with that in (Bensoussan 1, p. 180 et seq.). Notice that the Wiener Integral is defined modulo null-functions in $L^2[\Omega,\mu;X]$.

§§2.4 Properties of the Wiener Integral

Suppose $B(\cdot): \mathbb{R}^+ \to \mathcal{L}(\mathcal{U},X)$ is such that $\|B(\cdot)\|$ is measurable and locally essentially bounded and such that $B(\cdot)x$ is measurable for each $x \in X$. We have the following easily derived properties

(i) $\langle x, \int_0^T B(t)\, dW(t,\omega)\rangle = \int_0^T \langle B^*(t)x, \cdot\rangle\, dW(t,\omega)$ w.p. 1

(ii) $E\{\|\int_0^T B(t)\,dW(t,\omega)\|^2\} \leq tr\{Q\} \int_0^T \|B(t)\|^2 dt$

(iii) $E\{\int_0^T \langle b_1(t),\cdot\rangle\, dW(t,\omega) \int_0^T \langle b_2(t),\cdot\rangle\, dW(t,\omega)\} = 0$

for $W(t,\omega), V(t,\omega)$ independent Wiener processes, $b_1(\cdot), b_2(\cdot) \in L^\infty_{loc}[\mathbb{R}^+;\mathcal{U}]$.

(iv) $E\{\int_0^{t_1} \langle b_1(t),\cdot\rangle\, dW(t,\omega) \int_0^{t_2} \langle b_2(t),\cdot\rangle\, dV(t,\omega)\} = \int_0^{\min\{t_1,t_2\}} \langle b_1(t), Q\, b_2(t)\rangle dt$

(b_1, b_2 as above).

Finally suppose that $\Phi(\cdot,\cdot): \mathbb{R}^+ \times \mathbb{R}^+ \to \mathcal{L}(\mathcal{U},X)$ is locally essentially bounded and such that $\Phi(\cdot,\cdot)x$ is $\mu_{\mathcal{L}} \times \mu_{\mathcal{L}}$ - measurable for each $x \in \mathcal{U}$. Then**
(v) there exists a measurable version of

$$z(s) = \int_0^T \Phi(s,t)\, dW(t,\omega)$$

and (vi) (integration by parts)

$$\int_0^T [\int_0^T \Phi(s,t)\, ds]\, dW(t,\omega) = \int_0^T [\int_0^T \Phi(s,t)\, dW(t,\omega)]\, ds \qquad \text{w.p. 1}$$

(a measurable version of the integrand being chosen on the right).

§§ 2.5 Perturbed Evolution Equations

Take X a real Hilbert space, $\mathcal{P} = \{(t,s) \in \mathbb{R}^2 \mid t \geq s \geq 0\}$. We say the map $T(\cdot,\cdot): \mathcal{P} \to \mathcal{L}(X)$ is a mild evolution operator if (a) $T(\cdot,\cdot): \mathcal{P} \to \mathcal{L}_s(X)$ is continuous (subscript s denotes "w.r.t. the strong operator topology")

*Suppose $T(\cdot)$ is a C_0 semigroup then $T(\cdot)$ is not strongly measurable w.r.t. the uniform operator topology (except in the trivial situation where $T(\cdot)$ is uniformly continuous). It is precisely because we wish to attach meaning to such integrals as $\int_0^T T(t)\,dW(t,\omega)$ that we must adopt the definition of Wiener integral given here.

** See (Vinter 1)

and (b) $T(t,r)T(r,s) = T(t,s)$, $T(t,t) = I$ (identify in $\mathcal{L}(X)$) $t \geq r \geq s$.

Definition 2.1 Let $T(\cdot,\cdot)$ be a mild evolution operator and suppose that $B: \mathbb{R}^+ \to \mathcal{L}(X)$ is essentially locally bounded and has the property: given any $f \in C[\mathbb{R}^+_s, X_s]$, $t \mapsto B(t)f(t)$ is strongly measurable. Then the B-perturbed mild evolution operator $T_B(\cdot,\cdot)$ (corresponding to $T(\cdot,\cdot)$) is that unique mild evolution operator $\mathcal{U}(\cdot,\cdot)$ such that $\mathcal{U}(t,s)x_0 = T(t,s)x_0 + \int_s^t T(t,\sigma)B(\sigma)\mathcal{U}(\sigma,s)x_0 d\sigma$ all $t \geq s$ for each $s \in \mathbb{R}^+$, each $x_0 \in X$.

It is shown in e.g. (Vinter 2) that $T_B(\cdot,\cdot)$ is well-defined.

§3. System Description

Let the X-valued and $\mathbb{R}^k$-valued processes $\{x(t) | t \geq 0\}$, $\{z(t) | t \geq 0\}$ be defined respectively by

$$x(t) = T(t,0)x_0 + \int_0^t T(t,\sigma)B(\sigma)dw(\sigma) \qquad \text{(process)} \qquad (3.1)$$

$$z(t) = \int_0^t C(\tau)x(\tau)d\tau + \int_0^t F(\tau)dv(\tau) \qquad \text{(observation)} \qquad (3.2)$$

Here,

$X, \mathcal{U}$ are real separable Hilbert spaces .

$\{T(t,s) \in \mathcal{L}(X) | (t,s) \in P\}$ is a mild evolution operator.

$\{w(t) | t \geq 0\}$, $\mathcal{U}$-valued Wiener process with covariance operator W, W some nuclear operator.

$\{z(t) | t \geq 0\}$, $\mathbb{R}^k$-valued Wiener process with covariance operator V . We assume (strongly) sample continuous versions of $w(t)$, $v(t)$ to have been chosen. This is possible (see Bensoussan 1 , p. 179).

$$B(\cdot) \in L^\infty_{loc}[\mathbb{R}^+; \mathcal{L}(\mathcal{U}, X)], \; C(\cdot) \in L^\infty_{loc}[\mathbb{R}^+; \mathcal{L}(X, \mathbb{R}^k)], \; F(\cdot) \in L^\infty_{loc}[\mathbb{R}^+; \mathcal{L}(\mathbb{R}^k, \mathbb{R}^k)] .$$

x_0 is a X-valued Gaussian random variable with zero mean and covariance P_0 .

All random variables are defined w.r.t. the same complete probability space. $dw(t)$, $dv(t)$, x_0 are assumed independent.

In (3.1) $\int_0^t T(t,\sigma)B(\sigma)dW(\sigma)$ is a Wiener integral. It is known that there exists a measurable version of $x(\cdot)$ with summable sample paths; such a version is used in evaluating the observation $z(\cdot)$.

Assumption 3.1 There exists $\nu > 0$ such that

$$\langle F(t)VF^*(t)h,h \rangle \geq \nu \|h\|^2 \quad h \in \mathbb{R}^k \qquad \text{(uniformly in } t \in \mathbb{R}^+\text{)} \blacksquare$$

Assumption 3.2 The map $T^*(\cdot,\cdot): P \to \mathcal{L}(X)$ is strongly continuous. ∎

We remark that Assumption (3.2) is immediately satisfied if $T(\cdot,\cdot)$ is "time-invariant". ∎

From the properties of the stochastic integral, the local essential boundedness of $\|B(\cdot)\|$ and the independence of x_0 and $w(t)$ we deduce the following properties of the process

(i) $x(t)$ is a gaussian r.v. for each $t \geq 0$

(ii) $E\{x(t)\} = 0$

(iii) writing $\Lambda(t,s) = \text{cov}\{x(t), x(s)\}$ we have

$$\langle \Lambda(t,s)h,\bar{h}\rangle = \langle T(t,o)P_o T^*(t,o)h,\bar{h}\rangle + \int_0^{\min\{t,s\}} \langle B^*(\tau)T^*(s,\tau)h, WB^*(\tau)T^*(t,\tau)\bar{h}\rangle d\tau ,$$

$h,\bar{h} \in X$. We have $\Lambda(t,s) \in C_{loc}[\mathbb{R}^+, \mathcal{L}_s(X)]$, $\Lambda^*(t,s) = \Lambda(s,t)$ and $\|\Lambda(t,s)\|$ is locally bounded on $\mathcal{P}$.

§4. The Filtering Problem

We remark that for every $k \in L^2[0,t;\mathbb{R}^k]$

$$\int_0^t \langle k(s),\cdot\rangle dz(s,\omega) \triangleq \int_0^t \langle k(s), \mathcal{C}(s)x(s,\omega)\rangle ds + \int_0^t \langle k(s), \mathcal{F}(s)(\cdot)\rangle dv(s,\omega)$$

is a well-defined second order random variable. Define $Y_t \triangleq \{y \in L^2[\Omega,\mathbb{R}] \mid y = \int_0^t k(s)dz(s,\omega)$ for some $k \in L^2[0,t;\mathbb{R}^k]\}$. For each $t \geq 0$, the mapping

$$X \to L^2[0,t;\mathbb{R}^k], \quad h \mapsto k_h(t,\cdot) \in L^2[0,t;\mathbb{R}^k]$$

is called a <u>mild solution to the filtering problem</u> (at time t) if, writing

$$\hat{x}_h(t,\omega) = \int_0^t \langle k_h(t,s),\cdot\rangle dz(s,\omega) \qquad \text{each } h \in X , \tag{4.1}$$

we have

$$E\{|\langle h,x(t,\omega)\rangle - \hat{x}_h(t,\omega)|^2\} \leq E\{|\langle h,x(t,\omega)\rangle - y|^2\} \tag{4.2}$$

all $y \in Y_t$.

Suppose $k_h(t,\cdot)$ is a mild solution to the filtering problem. Then $\hat{x}_h(t,\omega)$ defined by (4.1) is called an <u>optimal linear estimate</u> of $\langle x(t,\omega),h\rangle$. Further, if $k_h(t,\cdot)$ has representation

$$k_h(t,\cdot) = K^*(t,\cdot)h \quad \text{all } h \in X \quad \text{a.e. on } [0,t]$$

where $K(t,\cdot) \in L^2[0,t;\mathcal{L}(\mathbb{R}^k;X)]$, $K(t,\cdot)$ is called a <u>(strong) solution to the filtering problem</u> and the X-valued random variable $\hat{x}(t,\omega) = \int_0^t K(t,s)dz(s,\omega)$ is called an <u>optimal linear estimate of $x(t,\omega)$</u>. We remark that if $\hat{x}(t,\omega)$ is an optimal linear estimate of $x(t,\omega)$ then, in particular,

$$E\{\|x(t,\omega) - \hat{x}(t,\omega)\|^2\} \leq E\{\|x(t,\omega) - \int_0^t K(t,s)dz(s,\omega)\|^2\}$$

for all $K(t,\cdot) \in L^2[0,t;\mathcal{L}(\mathbb{R}^k,X)]$.

§5. The Mild Wiener-Hopf Equation

<u>Lemma 5.1</u> Given $k_h(t,\cdot) \in L^2[0,t;\mathbb{R}^k]$, $h \in X$ define

$$\hat{x}_h(t) = \int_0^t \langle k_h(t,s),\cdot\rangle dz(s), \quad \tilde{x}_h(t) = \langle x(t),h\rangle - \hat{x}_h(t)$$

Then,

$\hat{x}_h(t)$ is an optimal linear estimate of $\langle x(t),h\rangle \Leftrightarrow E\{\tilde{x}_h(t)\langle z(\tau),p\rangle\}=0$, all $\tau\in[0,t], p\in\mathbb{R}^k$

Proof This is a simple consequence of the projection theorem. See (Vinter and Mitter 1) for details. c.f. Theorem 2.1 of (Curtain 2).❙

Mild solutions to the filtering problem will be characterized through the Mild Wiener-Hopf Equation :

$$\int_0^t \mathcal{C}(\sigma)\Lambda(\sigma,s)\mathcal{C}^*(s)k_h(t,s)ds + \mathcal{F}(\sigma)N\mathcal{F}^*(\sigma)k_h(t,\sigma) = \mathcal{C}(\sigma)\Lambda(\sigma,t)h \quad \text{a.e. } \sigma\in[0,t]$$
$$k_h(t,\cdot)\in L^2[0,t;\mathbb{R}^k],\ h\in X. \tag{5.1}$$

Lemma 5.2 Within the class of $L^2[0,t;\mathbb{R}^k]$ functions, the mild Wiener Hopf equation (5.1) has a unique solution $k_h(t,\cdot)$. We have

$$\text{ess. sup. }\{|k_h(t,\cdot)|\ |\ [0,t]\} \le \text{ constant } \|h\| \tag{5.2}$$

(constant independent of h).❙

Proof Taking note of Assumption 3.1, it is easily shown that the map $L^2[0,t;\mathbb{R}^k]\to L^2[0,t;\mathbb{R}^k]$ defined by $k_h(t,\cdot)\mapsto$ {L.H.S. of Wiener-Hopf equation} has a bounded inverse. This gives uniqueness.

We also find that this map, when restricted to L^∞ takes values in L^∞, and the restriction viewed as a map $L^\infty\to L^\infty$ has a bounded inverse. (5.2) readily follows. See Vinter and Mitter 1 for details.❙

Lemma 5.3 For given $t\ge 0, h\in X$, let $k_h(t,\cdot)\in L^2[0,t;\mathbb{R}^k]$ satisfy the Mild Wiener-Hopf equation (5.1). Then $k_h(t,\cdot)$ is a mild solution to the filtering problem.❙

Proof Let $k_h(t,\cdot)\in L^2[0,t;\mathbb{R}^k]$ satisfy the mild Wiener-Hopf equation. Defining

$$\tilde{x}_h(t) = \langle x(t),h\rangle - \int_0^t \langle k_h(t,s),\cdot\rangle dz(s)$$

we show by direct expansion that

$$E\{\tilde{x}_h(t)\langle z(\tau),p\rangle\} = 0 \qquad \text{each } \tau\in[0,t],\ \text{all } p\in\mathbb{R}^k$$

(see Vinter and Mitter 1 for details.) The assertion now follows from Lemma 5.1.❙

Proposition 5.1 For each $t\ge 0, h\in X$ let $k_h(t,\cdot)$ be the unique solution to the mild Wiener-Hopf equation (5.1). Let $K(t,\cdot)$ be the unique element in $L^\infty[0,t;\mathcal{L}(\mathbb{R}^k,X)]$ such that

$$K^*(t,\cdot)h = k_h(t,\cdot) \quad \text{a.e. on } [0,t],\ \text{every } h\in X \tag{5.3}$$

Then $K(t,\cdot)$ is the unique solution to the filtering problem.❙

Proof Again see (Vinter and Mitter 1) for details. That (5.3) well-defines $K(t,\cdot)$ as a $L^\infty[0,t;\mathcal{L}(\mathbb{R}^k,X)]$ element is easily deduced from (5.2). In view of Lemmas (5.2), (5.3) it remains to show that the filtering problem has a unique mild solution. But the optimal linear estimate $\hat{x}_h(t)$ of $\langle h,x(t)\rangle$ is unique being a projection. We have merely to show therefore that $\hat{x}_h(t)$ has a unique representation

$$\hat{x}_h(t) = \int_0^t \langle k(s),\cdot\rangle dz(s) \qquad k(\cdot)\in L^2[0,t;\mathbb{R}^k]$$

But this follows simply from the coercivity of $\mathcal{F}N\mathcal{F}^*$ hypothesized in Assumption (3.1).❙

§6. Global Optimality of the Filter

Take a fixed time interval $[0,T]$. Then it can be shown that $\tilde{z}:\Omega \to$ {maps from $[0,T]$ into $\mathbb{R}^k$} defined by

$$\tilde{z}(\omega) = \int_0^{(\cdot)} \mathcal{C}(\tau)\,x(\tau,\omega)\,d\tau + \mathcal{F}(\cdot)\,v(\cdot,\omega)$$

takes values almost surely in $L^2[0,T;\mathbb{R}^k]$ and $\omega \mapsto \tilde{z}(\omega)$ defines an $L^2[0,T;\mathbb{R}^k]$ -valued gaussian random variable.

Define $\mathcal{H} = L^2[0,T;\mathbb{R}^k]$ and let σ be the measure induced on the Borel fields of $\mathcal{H}$ under $\tilde{z}$. Define further the closed subspace $\tilde{\mathcal{H}}$ of $L^2[\Omega,\mu;X]$ as

$$\tilde{\mathcal{H}} = \{\varphi \in L^2[\Omega,\mu;X] \mid \varphi(\omega) = g(\tilde{z}(\omega)),\ \text{some } g \in L^2[\mathcal{H},\sigma;X]\}$$

Any element in $\tilde{\mathcal{H}}$ is called an <u>estimate</u> of $x(T,\omega)$. The projection of $x(T,\omega)$ onto $\mathcal{H}$ is called the <u>optimal non-linear estimate of $x(T,\omega)$</u>.

Now in (Benssonssan 1, Ch. 6), where filtering of systems for which the generator of $T(\cdot,\cdot)$ satisfies certain coercivity conditions is treated, it is shown that the optimal linear estimate $\hat{x}(T,\omega)$ of §4 is the optimal <u>non-linear</u> estimate of $x(T,\omega)$ also. The argument carries over almost unaltered to assure that $\hat{x}(T,\omega)$ is still the optimal non-linear estimate of $x(T,\omega)$ in the more general setting under consideration here.* In outline :

(a) Let $x_1(\omega), x_2(\omega)$ be X_1, X_2 -valued zero-mean gaussian random variables respectively (X_1, X_2 real, separable Hilbert spaces), let $\bar{\sigma}$ be the measure induced on the Borel sets of X_2 under $x_2(\omega)$, let $G = L^2[X_2,\bar{\sigma};X_1]$ and

$$\tilde{G} = \{x \in L^2[\Omega,\mu;X_2] \mid x(\omega) = g(x_2(\omega)),\ \text{some } g \in G\}$$

Then a familiar finite-dimensional argument can be patterned (see Bensoussan p. 117-119) to show that if $\tilde{\Lambda} \in \mathcal{L}(X_2,X_1)$ satisfies

$$E\{\langle x_1(\omega) - \tilde{\Lambda}x_2(\omega), h_1\rangle \langle x_2(\omega), h_2\rangle\} = 0, \quad \text{every } h_1 \in X_1,\ h_2 \in X_2$$

then $\tilde{\Lambda}x_2(\omega)$ is the projection of $x_1(\omega)$ onto $\tilde{G}$, i.e. the optimal linear estimate (if it exists) coincides with the optimal non-linear estimate.

(b) $$\omega \mapsto \int_0 \Gamma(t)\,dz(t,\omega)$$

defines an element in $\tilde{\mathcal{H}}$ for any $\Gamma \in L^\infty[0,T;\mathcal{L}(\mathbb{R}^k,X)]$ and in particular for $\Gamma(t) = K(T,t)$ ($K(T,t)$ as in Proposition 5.1)(Bensoussan 1, pp. 230-231). Thus the estimate $\hat{x}(T,\omega)$ of §4 is a linear estimate of $x(T,\omega)$ given $\tilde{z}(\omega)$.

(c) Finally, for arbitrary $h \in X$, we have that

$$E\{\langle x(T,\omega) - \int_0^T K(T,t)\,dz(t,\omega), h\rangle \int_0^T \langle z(t,\omega), f(t)\rangle\,dt\} = 0$$

for $f(t)$ a simple $\mathbb{R}^k$ -valued function by Lemma 5.1. A limiting argument gives the relation for general $f \in L^2[0,T;\mathbb{R}^k]$. Since $x(T,\omega)$ and $\tilde{z}(\omega)$ are zero-mean gaussian r.v.'s we deduce that $\hat{x}(T,\omega)$ is the optimal nonlinear estimate from (a) and (b).

---

*Further, for any bounded linear map $Q: X \to Y$, Y a separable, real Hilbert space, $Q\hat{x}(T,\omega)$ is the optimal non-linear estimate of $Qx(T,\omega)$.

§7. Duality of Estimation and Control

We introduce

Control Problem For given $t \geq 0, h \in X$

$$\begin{cases} \text{minimize } J_t(u,h) \\ \text{subject to } u \in L^2[0,t;\mathbb{R}^k] \text{ and} \\ p(\tau) = T^*(t,\tau)h + \int_\tau^t T^*(s,\tau)\mathcal{C}^*(s)u(s)\,ds \end{cases}$$

Here,

$$J_t(u,h) = \int_0^t [\langle p(\tau), B(\tau)WB^*(\tau)p(\tau)\rangle + \langle u(\tau), \mathcal{F}(\tau)V\mathcal{F}^*(\tau)u(\tau)\rangle]\,d\tau + \langle p(0), P_0\,p(0)\rangle$$

A minimizing u is called an optimal control (for given $t \geq 0, h \in X$).

Now it has been observed (Bensoussan 1, p. 165) that there is in some sense a "duality" between such a control problem and the filtering problem. In the present framework it is convenient to change the emphasis somewhat; rather than merely view this duality as a noteworthy structural property, we exploit it in the actual dynamical filter construction in a fundamental way :

Theorem 7.1 (Duality of Estimation and Control).
For given $t \geq 0, h \in X$, $k_h(t,\cdot) \in L^2[0,t;\mathbb{R}^k]$ is a mild solution to the filtering problem if and only if $-k_h(t,\cdot)$ is an optimal control. ∎

Proof We refer to Vinter and Mitter 1 for details. An outline is the following. For given $t \geq 0, h \in X$ the "cost function" $J_t(u,h)$ is expressed as

$$J_t(u,h) = \pi(u,u) - 2\lambda(u) + \{\text{terms independent of } u\}$$

where $\pi(\cdot,\cdot): L^2 \times L^2 \to \mathbb{R}$ is continuous, bilinear, symmetric, coercive and $\lambda(\cdot): L^2 \to \mathbb{R}$ is bounded, linear.

By standard results concerning minimization of quadratic forms (see e.g. Lions 1), there is a unique optimal control u uniquely characterized through the variational inequality

$$\pi(u,v) = \lambda(v), \quad \text{all} \quad v \in L^2 \tag{7.1}$$

Next it is shown that if $k_h(t,\cdot) \in L^2$ is a mild solution to the filtering problem (for given $h \in X, t \geq 0$) then $u(\cdot) = -k_h(t,\cdot)$ satisfies (7.1) ; the duality theorem follows from the unique characterization of u through (7.1) and the uniqueness of $k_h(t,\cdot)$ established in §5. ∎

§8. Solution to the Control Problem

For convenience we write $N(\tau) = B(\tau)WB^*(\tau)$, $R(\tau) = \mathcal{F}(\tau)V\mathcal{F}^*(\tau)$. The control problem becomes

$$\begin{cases} \text{Minimize } J_t(u,h) \\ \text{subject to } p(\tau) = T^*(t,\tau)h + \int_\tau^t T^*(s,\tau)\mathcal{C}^*(s)u(s)\,ds \end{cases}$$

(for given $t \geq 0, h \in X$), with

$$J_t(u,h) = \int_0^t [\langle p(\tau), N(\tau)p(\tau)\rangle + \langle u(\tau), R(\tau)u(\tau)\rangle]\, d\tau$$
$$+ \langle p\;, P_0\, p(0)\rangle$$

Let us define the family of maps $\{T_t^{\dagger}(r,s) \mid (r,s) \in \mathcal{P}[0,t]\}$ $(\mathcal{P}[0,t] = \{(r,s) \in \mathbb{R}^2 \mid t \geq r \geq s \geq 0\})$ by

$$T_t^{\dagger}(r,s) = T^*(t-s, t-r)$$

It is a straigthforward matter to show the following (see Vinter and Mitter 1 for details)

Lemma 8.1 Under the assumption that $T^*(\cdot,\cdot)$ is strongly continuous on $\mathcal{P}[0,t]$ then $\{T_t^{\dagger}(r,s) \mid (r,s) \in \mathcal{P}[0,t]\}$ is a mild evolution operator. ∎

We now change variables $\tau \to t - \bar{\tau}$ to give

$$\begin{cases} \text{Minimize} \quad \bar{J}_t(\bar{u}, h) \\ \text{subject to} \quad \bar{p}(\bar{\tau}) = T_t^{\dagger}(\bar{\tau},0)h + \int_0^{\bar{\tau}} T_t^{\dagger}(\bar{\tau},s)\bar{\mathcal{C}}^*(s)\bar{u}(s)\, ds \end{cases}$$

with

$$\bar{J}_t(\bar{u},h) = \int_0^t [\langle \bar{p}(\bar{\tau}), \bar{N}(\bar{\tau})\bar{p}(\bar{\tau})\rangle + \langle \bar{u}(\bar{\tau}), \bar{R}(\bar{\tau})\bar{u}(\bar{\tau})\rangle]\, d\bar{\tau} + \langle \bar{p}(t), P_0\, \bar{p}(t)\rangle$$

Here, $p(\tau) = \bar{p}(t-\tau)$ etc.

But this is precisely the control problem studied in (Bensoussan, Delfour and Mitter 1). Indeed by Lemma 8.1, $T_t^{\dagger}(\cdot,\cdot)$ is a mild evolution operator, $\bar{N}(\cdot) \in L^{\infty}[0,t;\mathcal{L}(X)]$, $\bar{R}(\cdot) \in L^{\infty}[0,t;\mathcal{L}(\mathbb{R}^k)]$, $P_0 \in \mathcal{L}(X)$, $\mathcal{C}^*(\cdot) \in L^{\infty}[0,t;\mathcal{L}(\mathbb{R}^k, X)]$; $\bar{N}(t)$, $\bar{R}(t)$ are self-adjoint, non-negative for each t and $\bar{R}(t)$ is coercive.

Given $\bar{P} \in C[0,t;\mathcal{L}_s(X)]$, $\mathcal{Y}_{\bar{P}}^{\dagger}(r,s)$ is taken to be the $(-\bar{\mathcal{C}}^*\bar{R}^{-1}\bar{\mathcal{C}}\bar{P})$-perturbed mild evolution operator corresponding to $T_t^{\dagger}(r,s)$ (see §2).

We know* (Bensoussan, Delfour and Mitter 1) that there exists a unique $\bar{P} \in C[0,t;\mathcal{L}_s(X)]$ such that

$$\langle \bar{P}(\tau)h, \bar{h}\rangle = \langle \mathcal{Y}_{\bar{P}}^{\dagger}(t,\tau)h, P_0 \mathcal{Y}_{\bar{P}}^{\dagger}(t,\tau)\bar{h}\rangle$$
$$+ \int_{\tau}^{t} \langle \mathcal{Y}_{\bar{P}}^{\dagger}(\sigma,\tau)h, [\bar{R}(\sigma) + \bar{P}(\sigma)\bar{\mathcal{C}}^*(\sigma)\bar{R}^{-1}(\sigma)\bar{\mathcal{C}}(\sigma)]\mathcal{Y}_{\bar{P}}^{\dagger}(\sigma,\tau)\bar{h}\rangle\, d\sigma \tag{8.1}$$

and the unique optimal control u is given by

$$\bar{u}(\tau) = -\bar{R}^{-1}(\tau)\bar{\mathcal{C}}(\tau)\bar{P}(\tau)\mathcal{Y}_{\bar{P}}^{\dagger}(\tau,0)h \tag{8.2}$$

§9. The Kalman Bucy Filter

Given $P \in C[0,t;\mathcal{L}_s(X)]$, let us define $\mathcal{Y}_P(\cdot,\cdot)$ to be the $(-P\mathcal{C}^*R^{-1}\mathcal{C})$-perturbed evolution operator corresponding to $T(\cdot,\cdot)$. We shall require the following technical result, proved in(Vinter and Mitter 1):

Lemma 9.1 For each $t \geq \tau \geq \sigma \geq 0$, $P \in C[0,t;\mathcal{L}_s(X)]$, we have

$$[\mathcal{Y}_{\bar{P}}^{\dagger}(t-\sigma, t-\tau)]^* = \mathcal{Y}_P(\tau,\sigma) \;∎$$

---

* Actually in this reference it is only shown that $\bar{P}$ is weakly continuous. However it is simple matter to exploit the property that the adjoint of $T_t^{\dagger}(r,s)$ is strongly continuous to show that $\bar{P}$ is strongly continuous also.

Proposition 9.1 (Integral Riccati Equation)

There exists a unique $P(\cdot) \in C[0,t\,;\mathcal{L}_s(X)]$ such that, all $h, \bar{h} \in X$

$$\langle P(\tau)h,\bar{h}\rangle = \langle \mathcal{Y}_p^*(\tau,0)h, P_0\mathcal{Y}_p^*(\tau,0)\bar{h}\rangle$$
$$+ \int_0^\tau \langle \mathcal{Y}_p^*(\tau,\sigma)h, [B(\sigma)WB^*(\sigma) + P(\sigma)\mathcal{C}^*(\sigma)[\mathcal{F}(\sigma)V\mathcal{F}^*(\sigma)]^{-1}\mathcal{C}(\sigma)P(\sigma)]\mathcal{Y}_p^*(\tau,\sigma)\bar{h}\rangle d\sigma \quad (9.1)$$ ▌

Proof It is clear from Lemma (9.1) that $P(\cdot)$ is a solution to (9.1) if and only if $\bar{P}(\cdot)=P(t-(\cdot))$ is a solution to 8.1. Existence and uniqueness of solutions to (9.1) follow from existence and uniqueness of solutions to (8.1).

The field is now set for construction of the dynamical filter:

Theorem 9.1 (Kalman-Bucy filter)

Let $P(\cdot) \in C[0,t\,;\mathcal{L}_s(X)]$ be the unique solution to (9.1). Then for each $t \geq 0, h \in X$ the optimal non-linear estimate of $x(t)$ is given by

$$\hat{x}(t,\omega) = \int_0^t \mathcal{Y}_p(t,\tau)P(\tau)\mathcal{C}^*(\tau)[\mathcal{F}(\tau)V\mathcal{F}^*(\tau)]^{-1}dz(t,\omega) \quad (9.2)$$

and the process $\hat{x}(t,\omega)$ satisfies

$$\hat{x}(t,\omega) = \int_0^t T(t,\tau)P(\tau)\mathcal{C}^*(\tau)[\mathcal{F}(\tau)V\mathcal{F}^*(\tau)]^{-1}[dz(\tau,\omega) - \mathcal{C}(\tau)\hat{x}(\tau,\omega)\,d\tau] \quad ▌ \quad (9.3)$$

Proof By (8.2) and Theorem (7.1) we have, for each $h \in X$

$$\hat{x}(t,\omega) = \int_0^t \langle k_h(t,s),\cdot\rangle dz(s,\omega)$$

is a mild solution to the filtering problem where

$$k_h(t,s) = [\mathcal{F}(s)V\mathcal{F}^*(s)]^{-1}\mathcal{C}(s)\bar{P}(t-s)\mathcal{Y}_p^{\dagger}(t-\tau,0)h$$

and $\bar{P}(\cdot) \in C[0,t\,;\mathcal{L}_s(X)]$ solves (8.1). We may write

$$\hat{x}_h(t,\omega) = \int_0^t \langle h, \mathcal{Y}_p(t,s)P(s)\mathcal{C}^*(s)[\mathcal{F}(s)V\mathcal{F}^*(s)]^{-1}(\cdot)\rangle dz(s,\omega) \quad (9.4)$$

where $P(\cdot)$ is the unique solution to (9.1) in $C[0,t\,;\mathcal{L}_s(X)]$; we have used Lemma 9.1. In view of Proposition 5.1, we conclude (9.2). (See Vinter and Mitter 1 for details). Finally, (9.3) is deduced from (9.4) by using the defining property for perturbed evolution operators :

$$\mathcal{Y}_p(t,s)\bar{h} = T(t,s)\bar{h} - \int_s^t T(t,\sigma)P(\sigma)\mathcal{C}^*(\sigma)[\mathcal{F}(\sigma)V\mathcal{F}^*(\sigma)]^{-1}\mathcal{C}(\sigma)\mathcal{Y}_p(\sigma,s)\bar{h}\,d\sigma$$

for each $\bar{h} \in X$.

Again we refer to (Vinter and Mitter 1). ▌

§10. Filter Stability

The Generalized Kalman filter for the filtering problem of §4 is taken to be the map $\mathcal{K}$ from $\mathbb{R}^k$-valued measurable processes $\{z(t,\omega)\,|\,t \geq 0\}$ into X-valued measurable processes $\{\hat{x}(t,\omega)\,|\,t\geq 0\}$ defined by

$$\hat{x}(t,\omega) = \int_0^t \mathcal{Y}_p(t,s)K(s)\,dz(s,\omega) \qquad \text{each } t \geq 0 \quad (10.1)$$

$(K(s) = P(s)\mathcal{C}^*(s)[\mathcal{F}(s)V\mathcal{F}^*(s)]^{-1})$, $\mathcal{Y}_p$, P as in §9. The domain of $\mathcal{K}$ comprises all processes $z(t,\omega)$ for which the R.H.S. of (10.1) is defined (for each t) and corresponding to which a measurable version of $\int_0^t \mathcal{Y}_p(t,s)K(s)dz(s,\omega)$

exists.

We note in particular that the process $z^*(t,\omega)$ lies in the domain of $\mathcal{K}$ where

$$\left.\begin{aligned} x^*(t,\omega) &= T(t,0)x_0^*(\omega) + \int_0^t T(t,s)B(s)\,dW(s,\omega) \\ z^*(t,\omega) &= \int_0^t \mathcal{C}(\tau)x^*(\tau,\omega)\,d\tau + \int_0^t \mathcal{F}(\tau)\,dv(\tau,\omega) \end{aligned}\right\} \tag{10.2}$$

(10.2) is identical to (3.1) <u>except</u> that $x_0^*(\omega)$ is now allowed to be an arbitrary X-valued random variable independent of

Of course the process $z^*(t,\omega)$ is interpreted as

$$z^*(t,\omega) = \int_0^t \mathcal{C}(\tau)T(\tau,0)x_0^*(\omega)\,d\tau + \int_0^t \mathcal{C}(\tau)\int_0^\tau T(\tau,\sigma)B(\sigma)\,dW(\sigma,\omega)\,d\sigma + \int_0^t \mathcal{F}(\tau)\,dv(\tau,\omega)$$

Because $x_0^*(\omega)$ is no longer assumed Gaussian, $\hat{x}^*(t,\omega)$ given by

$$\hat{x}^*(t,\omega) = \int_0^t \mathcal{Y}_p(t,s)K(s)\,dz^*(s,\omega)$$

need not be gaussian or even second order.

Now a highly desirable property of the filter is that the error process $\hat{x}(t,\omega) - x(t,\omega)$ be insensitive to errors in the modelling of $x_0(\omega)$. We make this precise in the following definition :

<u>Definition 10.1</u> The filter $\mathcal{K}$ is <u>stable</u> if there exists some measure μ_∞ on the Borel sets of X, such that writing μ_t^* for the measure induced on the Borel sets of X by the X-valued r.v. $\hat{x}^*(t,\omega) - x^*(t,\omega)$ we have

$$\mu_t^* \to \mu_\infty \quad (\text{weakly})$$

for x_0^* an <u>arbitrary</u> X-valued r.v. ▌

Here by weak convergence of μ_t^* we mean the following: for every continuous function $f: X \to \mathbb{R}$ taking values in some bounded set we have

$$\int_X f(x)\,d\mu_t^*(x) \to \int_X f(x)\,d\mu_\infty(x), \quad t \to \infty$$

(Gikhman and Skorokhod 1).

Let us first take notice of the following representation of the "error" process

<u>Proposition 10.1</u> Let the processes $x^*(t,\omega), z^*(t,\omega)$ be as defined in (10.2). Take

$$\hat{x}^*(t,\omega) = \int_0^t \mathcal{Y}_p(t,s)K(s)\,dz^*(s,\omega)$$

$(K(s) = P(s)\mathcal{C}^*(s)[\mathcal{F}(s)V\mathcal{F}^*(s)]^{-1})$. Then

$$\hat{x}^*(t,\omega) - x^*(t,\omega) = \mathcal{Y}_p(t,0)x_0^*(\omega) + \int_0^t \mathcal{Y}_p(t,\sigma)[B(\sigma)\,dW(\sigma,\omega) + K(\sigma)\mathcal{F}(\sigma)\,dv(\sigma,\omega)] \tag{10.3}$$

(Here $P(\cdot)$ and $\mathcal{Y}_p(\cdot,\cdot)$ are taken as in Theorem 9.1) ▌

<u>Proof</u> See (Vinter and Mitter 1) ▌

The following corollary, interpreting $P(t)$ as the "error covariance" in the case when $x_0^* = x_0$ is almost immediate from the results in §2 :

<u>Corollary 10.1</u> Take the processes $\hat{x}(t,\omega)$, $x(t,\omega)$ as in Theorem 9.1. Then

$$\operatorname{cov}\{\hat{x}(t,\omega) - x(t,\omega)\} = P(t) \qquad \text{each } t \geq 0 \text{ ▌}$$

It would appear that minimal assumptions ensuring filter stability are the following

Assumption 10.1

(i) $\|\mathcal{Y}_p(t,0)\| \to 0$ as $t \to \infty$

(ii) There exists some nuclear operator $P_\infty \in \mathcal{L}(X)$ such that

$$\lim_{t\to\infty} \int_0^t \langle \mathcal{Y}_p^*(t,\sigma)h, [B(\sigma)WB^*(\sigma) + K(\sigma)\mathcal{C}(\sigma)P(\sigma)]\mathcal{Y}_p^*(t,\sigma)\bar{h}\rangle d\sigma = \langle P_\infty h, \bar{h}\rangle$$

for each $h, \bar{h} \in X$. ▎

Assumption 10.1 is of an unsatisfactory ad hoc nature; the following lemma supplies sufficient conditions for Assumption 10.1 to be met which, although excessively strong, are at least open to direct verification. The basic idea is that if the process $x(t,\omega)$ is exponentially stable and the maps defining the process and observations are sufficiently "small" then $\mathcal{Y}_p(t,0)$ is exponentially stable.

Lemma 10.1 Suppose that there exist $\omega_0 > 0, M \geq$ such that $\|T(t,\tau)\| < Me^{-\omega_0(t-\tau)}$ and that

$$\{\|P_0\| + \sup_{t\geq 0}[B(t)WB^*(t)]\} \sup_{t\geq 0}\{\|\mathcal{C}^*(t)[F(t)VF^*(t)]^{-1}\mathcal{C}(t)\|\} < \omega_0/M^3$$

then there exists some $\varepsilon_0 > 0$ such that

$$\|\mathcal{Y}_p^*(t,\sigma)\| < Me^{-\varepsilon_0(t-\sigma)} \tag{11.1}$$

and Assumption 10.1 is satisfied. ▎

Proof See (Vinter and Mitter 1) . We note part (i) of Assumption 10.1 is immediate from 11.1. Part (ii) follows essentially from properties (ii) and (iv) of the Wiener Integral given in §§2.4 and the completeness of the trace class operators* in the trace norm. ▎

Proposition 10.2 Suppose that Assumption 10.1 holds (for example, if the conditions of Lemma 10.1 are met). Then the filter $\mathcal{H}$ above is stable and the limiting measure μ_∞ is gaussian with zero mean and covariance operator P_∞ . We have in addition

(i) if $x_0^*(\omega)$ is first order, then mean $\{\mu_t^*\} \to$ mean $\{\mu_\infty\}$ (strongly)

(ii) if $x_0^+(\omega)$ is second order, then

$$\mathrm{cov}\{\mu_t^+\} \to \mathrm{cov}\{\mu_\infty\} \qquad \text{(in the weak } \mathcal{L}(X) \text{ topology)} ▎$$

Proof This makes use of results in (Grenander 1) and patterns arguments in (Vinter 3). ▎

What does not appear obtainable at the present level of generality are conditions for filter stability in terms of (appropriately defined) "controllability" and "detectability" of the system under consideration.

We are able to conclude boundedness of the error covariance however from the appropriate detectability assumption; this is significant because, in applications to delay systems, the detectability hypothesis is directly verifiable.

For simplicity we specialize to time invariant systems.

Definition 10.2 Suppose that (3.1) and (3.2) are time-invariant, i.e. $T(t,\tau) = T(t-\tau)$, $B(t) = B$ etc. Then the system (3.1) - (3.2) is detectable if there exists some bounded linear map $D: X \to \mathbb{R}^k$ such that the semigroup $T_{A^*-\mathcal{C}^*D}(\cdot)$ generated by $A^* - \mathcal{C}^*D$ (A the generator of $T(\cdot)$) is L^2-stable. ▎

Proposition 10.3 Suppose that the system (3.1)-(3.2) is time invariant and detectable. Let $\hat{x}(t,\omega)$ be as given in Theorem 9.1. Write μ_t for the measure induced on the Borel sets of X by $\hat{x}(t,\omega) - x(t,\omega)$ and write

*See (Gelfand and Vilenkin 1)

for $\mathrm{cov}\{\hat{x}(t,\omega)-x(t,\omega)\}$. Then
(i) $\{P(t)\}_{t\geq 0}$ is bounded (in trace class) and $\{\mu_t\}_{t\geq 0}$ is weakly compact*
(ii) when additionally we assume $P_0 = 0$, there exists some gaussian measure μ_∞ with zero mean and covariance P_∞ where P_∞ satisfies

$$\langle P_\infty A^* h, \bar{h}\rangle + \langle h, P_\infty A^* \bar{h}\rangle + \langle BWB^* h, \bar{h}\rangle - \langle P_\infty \mathcal{C}^*[\mathcal{F}\mathcal{N}\mathcal{F}^*]^{-1}\mathcal{C} P_\infty h, \bar{h}\rangle = 0$$

for all $h, \bar{h} \in \mathcal{D}\{A^*\}$ such that

$$\mu_t \longrightarrow \mu_\infty \quad \text{(weakly) and}$$

$$P(t) \longrightarrow P_\infty \quad \text{(in trace class).} \;\rule{1pt}{8pt}$$

<u>Proof</u> See (Vinter and Mitter 1). ▌

Notice that Proposition 10.3 makes no assertion about filter stability in the sense of Definition 10.1. Indeed the conclusions only apply for $x_0^*(\omega)$ the initial condition for which the filter was constructed.

§11. Specialization to Stochastic Delay Systems

We now turn attention to the stochastic delay system :

$$\begin{cases} dx(t,\omega) = \mathcal{L}x_t(\omega)dt + B\,dw(t,\omega) & \text{(process)} \\ x(0,\omega) = [h(\omega)](0) & t \geq 0 \end{cases} \qquad (11.1)$$

$$\begin{cases} dz(t,\omega) = \mathcal{C}x_t(\omega)\,dt + F\,dv(t,\omega) & \text{(observations)} \\ z(0,\omega) = 0 \end{cases} \qquad (11.2)$$

Here

$$\mathcal{L}x_t = \sum_{i=0}^{m} A_i \begin{Bmatrix} x(t+\theta_i) & t+\theta_i \geq 0 \\ h(t+\theta_i) & t+\theta_i < 0 \end{Bmatrix} + \int_{-b}^{0} A(\theta) \begin{Bmatrix} x(t+\theta) & t+\theta \geq 0 \\ h(t+\theta) & t+\theta < 0 \end{Bmatrix} d\theta$$

and

$$\mathcal{C}x_t = C_0 x(t) + \int_{-b}^{0} C(\theta) \begin{Bmatrix} x(t+\theta) & t+\theta \geq 0 \\ h(t+\theta) & t+\theta < 0 \end{Bmatrix} d\theta$$

We suppose

$\mathcal{U}, X$ real, separable, Hilbert spaces

$-b < \theta_m < \theta_{m-1} < \ldots < \theta_0 = 0$; $A_0, A_1, \ldots, A_m \in \mathcal{L}(X)$

$A \in L^\infty[-b,0;\mathcal{L}(X)]$, $B \in \mathcal{L}(\mathcal{U}, X)$, $C_0 \in \mathcal{L}(X;\mathbb{R}^k)$

$C \in L^\infty[-b,0;\mathcal{L}(X,\mathbb{R}^k)]$, $F \in \mathcal{L}(\mathbb{R}^k)$

$w(t,\omega)$ is a measurable, almost surely strongly sample continuous $\mathcal{U}$-valued Wiener process with nuclear covariance W. $v(t,\omega)$ is a measurable, almost surely strongly sample continuous $\mathbb{R}^k$-valued Wiener process with covariance $N \in \mathcal{L}(\mathbb{R}^k, \mathbb{R}^k)$. $h(\omega)$ is an M^2-valued** gaussian r.v.

---

* (Gikhman and Skorokhod 1)

** M^2 denotes the Hilbert space $L^2[-b,0;X] \times X$. Elements in M^2 are written $(h(\theta)\ -b \leq \theta < 0;\ h(0))$

with zero mean and covariance P_0 .

It is assumed that $dw(t,\omega), dv(t,\omega), h(\omega)$ are independent and that $FVF^* > 0$.

The process (11.1) is interpreted as follows : the X-valued stochastic process $\{x(t,\omega) \mid t \geq 0\}$ is a <u>solution</u> to (11.1) if (a) it is almost surely (strongly) sample continuous, and (b) for each $t \geq 0$,

$$x(t,\omega) = [h(\omega)](0) + \int_0^t [\mathcal{L} x_s(\omega)]\, ds + B W(t,\omega)$$

A standard Picard iteration argument gives the solution as unique, if it exists (c.f. Lindquist 1)

Now we take the <u>stochastic abstract evolution equation</u> corresponding to (11.1) to be

$$\tilde{x}(t,\omega) = T(t)\, h(\omega) + \int_0^t T(t-s)\, \tilde{B}\, dW(s,\omega) \qquad (11.3)$$

where $\tilde{X} = M^2, \{T(t) \in \mathcal{L}(\tilde{X}) \mid t \geq 0\}$ is the semigroup describing evolution of "trajectory segments" of the homogeneous equation (Delfour and Mitter 1) and $\tilde{B} : \mathcal{U} \to \tilde{X}$ is defined as

$$[\tilde{B} u](\theta) = \left\{ \begin{matrix} Bu & \theta = 0 \\ 0 & \theta < 0 \end{matrix} \right\} \qquad (\, B \text{ as above})$$

We take the <u>abstract observation process</u> $\tilde{z}(t,\omega)$ to be

$$\tilde{z}(t,\omega) = \mathcal{C}\, \tilde{x}(t,\omega) + F dv(t,\omega)$$

Now for the theory developed in the previous sections to be applicable here, we require that (11.3) and (11.1) be equivalent in an appropriate sense. The necessary result is embodied in the following proposition :

<u>Proposition 11.1</u> There exists an almost surely (strongly) sample continuous version of (11.3) which we also write as $\tilde{x}(t,\omega)$. Defining the projection $H : \tilde{X} \to X$ by $Hh = h(0)$ we have

$$x(t,\omega) \triangleq H\tilde{x}(t,\omega) \quad \text{all } t \geq 0, \text{ each } \omega \in \Omega$$

is the unique solution of (11.1) (to within uniform stochastic equivalence) and for each $t \geq 0$

$$[\tilde{x}(t,\omega)](\theta) = \begin{matrix} x(t+\theta, \omega) & t+\theta \geq 0 \\ [h(\omega)](t+\theta) & t+\theta < 0 \end{matrix}$$

for almost all $\theta \in [-b, 0]$, w.p. 1. ▌

<u>Proof</u> The somewhat lenghtly proof in (Vinter 1) is not reproduced here. The result is crucial in the present development however and the essential steps are therefore outlined.

(a) Let $\tilde{A}$ be the infinitesimal generator of the C_0 semi-group $T(t)$ above. We first need to show that $q(t,\omega) = \int_0^t T(t-s)\tilde{B}\, dW(s,\omega)$ (and thence $\tilde{x}(t,\omega) = T(t)h(\omega) + \int_0^t T(t-s)\tilde{B}\, dW(s,\omega)$) has an almost surely (strongly) sample continuous version. A basic difficultly* is that range$\{\tilde{B}\} \not\subset \mathcal{D}\{\tilde{A}\}$, and

---

* This difficulty seems to preclude development of a satisfactory general theory of stochastic processes described by $dx(t,\omega) = Ax(t,\omega)dt + B dw(t,\omega)$, A the generator of a C_0 semigroup and B a general bounded linear map. The difficulty can be circumvented by relaxing the underlying probability measure to be merely <u>finitely</u> additive and defining a "Wiener process" with absolutely continuous sample paths (Balakrishnan 1).

it does not appear that for $T(t)$ a general C_0 semigroup, $\tilde{\mathcal{B}}$ a general bounded linear map, a sample continuous version of $q(t,\omega)$ necessarily exists. However, exploiting the very special structure of $\tilde{\mathcal{B}}$ and $T(t)$ under consideration, we can show that $q(t,\omega)$ satisfies some generalized Kolmogorov condition and then make use of a result in (Bensoussan 1 , p. 175) to establish the desired sample continuity.

(b) We next show that the (almost surely) strongly sample continuous version of $\tilde{x}(t,\omega)$ uniquely satisfies

$$\langle \tilde{x}(t,\omega),y\rangle = \langle x_0(\omega),y\rangle + \int_0^t \langle \tilde{A}^* y, x(s,\omega)\rangle ds + \langle y, \tilde{\mathcal{B}} W(t,\omega)\rangle \quad \text{w.p.1} \qquad (11.4)$$

for each $y \in \mathcal{D}\{\tilde{A}^*\}$, each $t \geq 0$ within the class of (almost surely) weakly sample continuous processes.

(c) Finally by consideration of the detailed structure of $\tilde{A}^*$ given in (Vinter 4), we show that all sample functions $\hat{x}(t,\omega)$ which are strongly continuous and satisfy (11.4) for all $t \geq 0$ (such sample functions have full measure) have representation as "segments" of some strongly continuous function $\varphi : [-b,\infty) \to X$ with $\varphi(\theta) = h(\theta)$ for $-b \leq \theta \leq 0$ which satisfies (11.1).

It is clear from Proposition 11.1 that the " X-part" of $\tilde{x}(t,\omega)$ is precisely the solution of (11.1), the " L^2-part" of $\tilde{x}(t,\omega)$ describes evolution of "trajectory segments" of the stochastic delay equation (11.1) and that the abstract observation process $\tilde{z}(t,\omega)$ is precisely $z(t,\omega)$. ∎

Comment $\mathcal{C}$ defines a bounded linear operator $M^2 \to \mathbb{R}^k$. Note that in the present framework we cannot treat point delays in the observations, i.e.

$$\mathcal{C} x_t = \sum_{i=0}^{m} C_i x(t+\theta_i) + \int_{-b}^{0} C(\theta) x(t+\theta) d\theta$$

for in this case $\mathcal{C}$ becomes an unbounded operator and the theory above is no longer applicable. Of course point delays in the observations can be approximated by "distributed" delays in an obvious manner. ∎

Now let H be the projection operator introduced in Proposition 11.1. Noting that assumptions (3.1) and (3.2) are met we have from previous sections :

Proposition 11.2 There exists a unique $P(\cdot) \in C[0,t ; \mathcal{L}_s(\tilde{X})]$ such that

$$\langle P(\tau)h,\bar{h}\rangle = \langle \mathcal{Y}_p^*(\tau,0)h, P_0 \mathcal{Y}_p^*(\tau,0)\bar{h}\rangle$$
$$+ \int_0^\tau \langle \mathcal{Y}_p^*(\tau,\sigma)h, [\tilde{\mathcal{B}} W \tilde{\mathcal{B}}^* + P(\sigma)\mathcal{C}^*[FVF^*]^{-1}\mathcal{C} P(\sigma)]\mathcal{Y}_p^*(\tau,\sigma)\bar{h}\rangle d\sigma \qquad (11.5)$$

and the optimal non-linear estimate $\hat{x}(t,\omega)$ of $x(t,\omega)$ is $H\hat{\tilde{x}}(t,\omega)$ where $\hat{\tilde{x}}(t,\omega)$ is given by

$$\hat{\tilde{x}}(t,\omega) = \int_0^t \mathcal{Y}_p(t,s)P(s)\mathcal{C}^*[FVF^*]^{-1} dz(s,\omega)$$

(the evolution operator $\mathcal{Y}_p(t,s)$ interpreted as in §9) and satisfies

$$\hat{\tilde{x}}(t,\omega) = \int_0^t T(t-\tau)P(\tau)\mathcal{C}^*[FVF^*]^{-1}[dz(\tau,\omega) - \mathcal{C}\hat{x}(\tau,\omega)d\tau] \quad ∎$$

We can exploit the fact that $T(\cdot)$ has generator $\tilde{A}$ to further characterize the covariance operator $P(t)$.

Proposition 11.3 Let $P(t) \in C[0,t ; \mathcal{L}_s(X)]$ be the unique solution to (11.5) Then $\langle P(t)h,\bar{h}\rangle$ is absolutely continuous and

$$\begin{cases} d/dt \langle P(t)h,\bar{h}\rangle = \langle P(t)h, A^*\bar{h}\rangle + \langle A^*h, P(t)h\rangle \\ \qquad + \langle h, BWB^*\bar{h}\rangle - \langle h, P\mathcal{C}^*[FVF^*]^{-1}\mathcal{C}P\bar{h}\rangle \qquad (11.6) \\ P(t_0) = P_0 \end{cases}$$

for every $h,\bar{h} \in \mathcal{D}\{A^*\}$ ▮

Proposition 11.4 Suppose that (i) $C_0 = 0$ (ii) $C(\cdot) \in W^{1,2}$ (i.e. $\mathcal{C}x_t = \int_{-b}^{0} C(\theta)x(t+\theta)d\theta$ with $C(\cdot)$ sufficiently smooth) then $P(t)$ is the unique solution to (11.6) within the class of maps $\varphi \in C[0,\infty;\mathcal{L}_w(\bar{X})]$ such that $\langle \varphi h,\bar{h}\rangle$ is absolutely continuous for each $h,\bar{h} \in \mathcal{D}\{A^*\}$. ▮

Proposition 11.3 is proved in (Vinter 2). Proposition 11.4 also follows from results in (Vinter 2) noting that the additional hypotheses on $\mathcal{C}$ ensure that $\text{range}\{\mathcal{C}^*\} \subset \mathcal{D}\{A^*\}$ (see Vinter 4). ▮

We now turn to specialization of the filter stability results of §10.

Even with the additional structure here present it does not appear possible to obtain satisfactory stability conditions in terms of suitably defined controllability and detectability of appropriate operator pairs.

We remark though that some results are now available characterizing the growth of $T(t)$ in terms of the maps $A_0, A_1, ..$ so that the hypotheses of Lemma 11.1 may be verified. We can therefore generate at least a class of non-trivial delay systems to which correspond stable optimal filters.

It is possible however to give a fairly simple characterization of detectable delay systems in the sense of Definition 11.2 and hence (in view of Proposition 10.3) of systems for which the error covariance is bounded (in trace norm). First take note of :

Lemma 11.1 Take $\tilde{A}$ the generator of $T(t)$. Then $\tilde{A}^*$ has compact resolvent and the generalized eigenmanifold $m^{\oplus}$ corresponding to the spectral set comprising all poles of $\tilde{A}^*$ with non-negative real parts is finite dimensional. ▮

Proof This follows simply from the property that $\tilde{A}$ has compact resolvent implies $\tilde{A}^*$ has compact resolvent and the known spectral properties of $\tilde{A}$. (See Vinter 4). ▮

Using notions in (Hale 1 , p. 98 et seq.) concerning decomposition of the state space of time-invariant delay systems into generalized eigenmanifolds of the generator it is not difficult to show

Proposition 11.5 Let $m^{\oplus}$ be as in Lemma 11.1. Then the delay system (11.1), (11.2) is detectable if and only if

$$m^{\oplus} \subset \text{range}\{\mathcal{C}^*\} \quad ▮$$

§12. A comment on time varying stochastic delay systems

Attention has been restricted to time-invariant delay systems in §11. Now consider the time-varying situation. Write $T(t,s)$ for the evolution operator describing evolution of trajectory segments of the homogeneous time-varying delay system and write $\tilde{A}(t)$ for its generator (we make suitable assumptions on $A_i(t)$, $A(t,\theta)$ as time functions, see Vinter 4). The following serious difficulties arise :

(i) it is not clear in this case that $T^*(\cdot,\cdot)$ is strongly continuous, a crucial hypothesis in development of the filter.
(ii) it is not apparent how we would prove "equivalence" of $[\tilde{x}(t,\omega)](0)$ and $x(t,\omega)$ in this situation; without this crucial step the whole development is of questionable value.
(iii) there is no hope in general of the error covariance $P(t)$ being uniquely characterized through a differential Riccati equation.

Difficilties (ii) and (iii) arise in connection with the following awkward property of the adjoint $\tilde{A}^*(t)$ of the generator $\tilde{A}(t)$: in general*

$$\bigcap_{t\geq 0} \mathcal{D}\{\tilde{A}^*(t)\} \quad \text{is not dense in } \tilde{X}$$

See (Vinter 4).

---

* Indeed for the scalar system $dx/dt = t\,x(t-1)$,

$$\bigcap_{t\geq 0} \mathcal{D}\{A^*(t)\} \subset \{h \in M^2 \mid h(0) = 0\}$$

References

A.V. Balakrishnan 1 , to appear

A.T. Bharucha-Reid 1 , "Random Integral Equations" Academic Press, New York, 1972

A. Bensoussan 1 , "Filtrage Optimal des Systèmes Lineaires" Dunod, Paris, 1971

A. Bensoussan, M.C. Delfour and S.K. Mitter 1 , "Notes on Infinite Dimensional Systems" Monograph (to appear)

A. Bensoussan and M. Viot , "Optimal Control of Stochastic Linear Distributed Parameter Systems" IRIA Technical Report, 1974

R.T. Curtain 1 , "Stochastic Differential Equations in a Hilbert Space" J. Diff. Eqns., 10, 1971, pp. 412-431

R.T. Curtain 2 , "Infinite Dimensional Filtering" SIAM J. Control (to appear)

M.C. Delfour , "State Theory of Linear Hereditary Differential Systems" to appear

M.C. Delfour and S.K. Mitter 1 , "Controllability, Observability and Optimal Feedback Control of Hereditary Differential Systems" SIAM J. Control, 10, 1972, pp. 298-328

M.C. Delfour, C. McCalla and S.K. Mitter , "Stability and the Infinite-Time Quadratic Cost Problem for Linear Hereditary Differential Systems" SIAM J. Control, vol. 13, 1, 1975 (to appear)

I.M. Gelfand and N.Y. Vilenkin 1 , "Generalized Functions" vol. 4 Academic Press, New York, 1964

I.I. Gikhman and A.V. Skorokhod 1 , "Introduction to the Theory of Random Processes" Saunders, London, 1969

U. Grenander 1 , "Probabilities on Algebraic Structures" Wiley, New York, 1966

J. Hale 1 , "Functional Differential Equations" Springer Verlag, New York, 1971

H. Kwakernaak 1 , "Optimal Filtering in Linear Systems with Time Delay" IEEE Trans. Aut. Con., 12,2,1967, pp. 169-173

A. Lindquist 1 , "A Theorem on Duality between Estimation and Control for Linear Stochastic Systems with Time Delay" J. Math. Analysis and Appl., 37,2,1972 pp. 516-536

A. Lindquist 2 , "Optimal Control of Linear Stochastic Systems with Applications to Time Lag Systems" Information Sciences, 5,1973 pp. 81-126

A. Lindquist , "On Feedback Control of Linear Stochastic Systems" SIAM J. of Control, vol. 11,2,May 1973 pp. 323-343

J.L. Lions 1 , "Control Optimal de Systems Gouvernés par des Equations aux Dérivées Partielles" Dunod, Paris, 1968

R.B. Vinter 1 , "Stochastic Delay Equations Formulated as Stochastic Evolution Equations" MIT Technical Report 1974

R.B. Vinter 2 , "Some Results Concerning Perturbed Evolution Equations with Applications to Delay Systems"
Technical Report Electronic Systems Laboratory, M.I.T.1974

R.B. Vinter 3 , "Invariant Measures Induced by Stochastic Evolution Equations"
Technical Report Electronic Systems Laboratory, MIT 1974

R.B. Vinter 4 , "On the Evolution of the State of Linear Differential Delay Equations in M^2 : Properties of the Generator"
Technical Report Electronic Systems Laboratory, MIT 1974

R.B. Vinter and S.K. Mitter 1 , "Filtering of Stochastic Evolution Equations"
Technical Report Electronic Systems Laboratory, MIT,1974

W.M. Wonham , "Random Differential Equations in Control Theory" in "Probabilistic Methods in Applied Mathematics" Editor Bharucha-Reid, Academic Press, New York, 1970

A KALMAN-BUCY FILTERING THEORY FOR AFFINE HEREDITARY DIFFERENTIAL EQUATIONS

Ruth F. Curtain
Control Theory Centre
University of Warwick
Coventry CV4 7AL
England

INTRODUCTION

There have been a number of different approaches to the filtering problem for delay systems, for example, Kwakernaak [12], Kushner and Barnea [11], Kailath [9] and Lindquist [13]. The main theoretical contribution to the problem is by Lindquist, who proves a duality theorem between estimation and control for stochastic systems with time delay, using the (nonrandom) theory of linear functional differential equations as expounded by Halanay, Hale, Banks et al neatly avoiding the Riccati equation which occurs in the Kalman-Bucy theory. This paper incorporates a more direct approach and generalizes the Kalman-Bucy filtering theory for a class of linear delay equations, along the lines of Kwakernaak in [12]. This is done by formulating the problem as one in the abstract Hilbert space $\mathcal{M}^2$ introduced by Delfour and Mitter in their theory of affine hereditary differential equations in [7], and using a similar approach to that in "Infinite Dimensional Filtering" [4].

1. PRELIMINARIES ON AFFINE HEREDITARY DIFFERENTIAL SYSTEMS

The following formulation is that of Delfour and Mitter in [7] and [8].

(1.1) The affine hereditary differential system considered is

$$\frac{dx(t)}{dt} = A_{00}(t)x(t) + \sum_{i=1}^{N} A_i \begin{cases} x(t+\theta_i) & ;\ t+\theta_i \geq 0 \\ h(t+\theta_i) & ;\ t+\theta_i < 0 \end{cases} +$$

$$\int_{-b}^{0} A_{01}(t,\theta) \begin{cases} x(t+\theta) & ;\ t+\theta \geq 0 \\ h(t+\theta) & ;\ t+\theta < 0 \end{cases} d\theta + B(t)u(t) + f(t)$$

$$x(0) = h(0)$$

where $t \in [0,T] = T$, a real, finite time interval, $A_{00} \in L_\infty(T; \mathcal{L}(R^n))$,

$A_i \in \mathcal{L}(R^n)$, $A_{01} \in L_\infty(T\times[-b,0]; \mathcal{L}(R^n))$, $f \in L_2(T; R^n)$,

$B \in L_\infty(T; \mathcal{L}(R^m,R^n))$, $u \in L_2(T; R^m)$ and

$-b < -\theta_N < \dots < -\theta_0 = 0.$

We remark that this is not the most general system considered by Delfour and Mitter in [7] and [8], where $A_i(t)$ depend on t. This restriction is necessary to ensure that $\mathcal{D}(\mathcal{A}^*(t))$ is time invariant (see §1.9 and [14]).

(1.2) <u>Definition of spaces $\mathcal{M}^2$, and $\mathcal{A}\mathcal{C}^2$</u>

Consider $\mathcal{L}^2(-b,0;\mathbb{R}^n)$, the space of maps: $[-b,0] \to \mathbb{R}^n$

under the seminorm $\|y\|_{\mathcal{M}^2} = \left(|y(0)|^2 + \int_{-b}^{0} |y(\theta)|^2 d\theta \right)^{1/2}$

where $|.|$ is a Euclidean norm on $\mathbb{R}^n$.

Then $\mathcal{M}^2(-b,0;\mathbb{R}^n)$ is the quotient space of $\mathcal{L}^2(-b,0;\mathbb{R}^n)$ generated by equivalence classes under $\|\cdot\|_{\mathcal{M}^2}$. $\mathcal{M}^2$ is then a Hilbert space and is isometrically isomorphic to $\mathbb{R}^n \times L_2(-b,0;\mathbb{R}^n)$.

$\mathcal{A}\mathcal{C}^2(t_0,t;\mathbb{R}^n)$ is the space of absolutely continuous maps: $[t_0,t] \to \mathbb{R}^n$ with a derivative in $L_2(t_0,t;\mathbb{R}^n)$ under the norm

$$\|x\|_{\mathcal{A}\mathcal{C}^2} = \left(|x(t_0)|^2 + \int_{t_0}^{t} \left|\frac{dx(s)}{ds}\right|^2 ds \right)^{1/2}$$

(1.3) <u>Existence Theorem for (1.1)</u>

Consider the homogeneous form of (1.1) with $u = 0 = f$ and the initial datum

$x(s) = h(0)$, $h \in \mathcal{D} = \mathcal{M}^2(-b,0;\mathbb{R}^n) \cap \mathcal{A}\mathcal{C}^2(s,T;\mathbb{R}^n)$.

Then (1.1) has the unique solution $\phi_s(\cdot\,;h) \in \mathcal{A}\mathcal{C}^2(s,T;\mathbb{R}^n)$.

The map $(t,s) \to \phi_s(t\,;h)$ generates the 2-parameter semigroup $\Phi(t,s)$ or evolution operator satisfying:

(1.4) (i) $\Phi(t,s) \in \mathcal{L}(\mathcal{M}^2)$; $t \geq s \geq 0$.

(ii) $\Phi(t,r)\Phi(r,s) = \Phi(t,s)$; $t \geq r \geq s \geq 0$.

(iii) $\Phi(t,t) = \mathcal{I}$, the identity operator in $\mathcal{L}(\mathcal{M}^2)$.

(iv) $\Phi(t,s)$ is strongly continuous in s and t for $s \leq t$.

(v) $\Phi(t,s) : \mathcal{D} \to \mathcal{D} = \mathcal{M}^2(-b,0;R^n) \cap \mathcal{A}\mathcal{C}^2(-b,0;R^n)$,

and $\frac{\partial}{\partial t}(\Phi(t,s)y) = \mathcal{A}(t)\Phi(t,s)y \quad \forall\, y \in \mathcal{D}$.

where $\mathcal{A}(t)$ is a closed operator on $\mathcal{M}^2$ with domain $\mathcal{D}$ and is defined by:

$$(\mathcal{A}(t)h)(\theta) = \begin{cases} \mathcal{A}^0(t)h & , \theta = 0 \\ (\mathcal{A}^1 h)(\theta) & , \theta \neq 0. \end{cases}$$

where $\mathcal{A}^0(t) : \mathcal{D} \to R^n$ and $\mathcal{A}^1 : \mathcal{D} \to L_2(-b,0;R^n)$ are given by

$$\mathcal{A}^0(t)h = A_{00}(t)h(0) + \sum_{i=1}^{N} A_i h(\theta_i) + \int_{-b}^{0} A_{01}(t,\theta)h(\theta)d\theta$$

and $(\mathcal{A}^1 h)(\theta) = \frac{dh(\theta)}{d\theta}$

(vi) For $t-s \geq b$, $\Phi(t,s) : \mathcal{M}^2 \to \mathcal{M}^2$ is compact.

(1.5) Abstract Evolution Equation Description

So we see that the homogeneous form of (1.1) may be expressed as an abstract evolution equation on $\mathcal{M}^2(-b,0;R^n)$.

$$(1.5) \qquad \begin{cases} \dot{\phi}(t) = \mathcal{A}(t)\phi(t) \\ \phi(0) = h \quad \in \mathcal{D} \end{cases}$$

where $\mathcal{A}(t)$ is an unbounded operator generating the evolution operator $\Phi(t,s)$ and this is the most convenient form to consider the filtering problem.

The inhomogeneous form of (1.1) may be expressed as

$$(1.6) \qquad \begin{cases} \dot{\phi}(t) = \mathcal{A}(t)\phi(t) + \tilde{\mathcal{B}}(t)u(t) + \tilde{f}(t) \\ \phi(0) = h \in \mathcal{D} \end{cases}$$

where $\tilde{\mathcal{B}} \in L_\infty(T;\mathcal{L}(R^m,\mathcal{M}^2))$, $\tilde{f} \in L_2(T;\mathcal{M}^2)$, and $u \in L_2(T;R^m)$. are given by

$$(1.7) \qquad (\tilde{\mathcal{B}}(t)v)(\theta) = \begin{cases} B(t)v & , \theta = 0 \\ 0 & , \theta \neq 0 \end{cases} \quad \text{for} \quad v \in R^m$$

(1.8) $$(\tilde{f}(t))(\theta) = \begin{cases} f(t) & , \theta = 0 \\ 0 & , \theta \neq 0 \end{cases}$$

Delfour and Mitter in [7] show that under the stated conditions, (1.6) has a unique solution in $\mathcal{M}^2$ given by

$$\phi(t) = \Phi(t,0)\,h + \int_0^t \Phi(t,s)\,\vec{B}(s)\,u(s)\,ds + \int_0^t \Phi(t,s)\,\tilde{f}(s)\,ds$$

and the solutions of (1.1) and (1.6) are equivalent.

In the filtering problem we also use the following characterization of the $\mathcal{M}^2$- adjoint of $\mathcal{A}(t)$ from Vinter [14].

(1.9) $\mathcal{A}^*(t) \in \mathcal{L}(\mathcal{M}^2)$ is given by

$$(\mathcal{A}^*(t)\,h)(\theta) = \begin{cases} A^*(t,\theta)\,h(\theta) - \dfrac{dh(\theta)}{d\theta} & , -b \leq \theta < 0 \\ A_0^*(t)\,h(0) + \lim\limits_{\varepsilon \to 0^+} h(-\varepsilon) & , \theta = 0 \end{cases}$$

for $h \in \mathcal{D}(\mathcal{A}^*(t)) = \mathcal{D}^*$, which is given by

$$\mathcal{D}^* = \left\{ \begin{array}{l} h \in \mathcal{M}^2 : h \text{ is differentiable on } [\theta_i, \theta_{i-1}) \text{ with} \\ \displaystyle\int_{\theta_i}^{\theta_{i-1}} \left\| \frac{dh(\theta)}{d\theta} \right\|^2 d\theta < \infty \\ \qquad\qquad \text{and predetermined jumps} \\ h(\theta_i) - \lim\limits_{\varepsilon \to 0^+} h(\theta_i - \varepsilon) = A_i^*\,h(0) \quad , i = 1, \ldots, N-1 \\ h(-b) = 0 \end{array} \right\}$$

2. PRELIMINARIES ON ABSTRACT EVOLUTION EQUATIONS AND THE INFINITE DIMENSIONAL RICCATI EQUATION

Here we just state the definition of evolution operators and the existence and uniqueness theorem for the Riccati equation from [5].

(2.1) Consider the abstract evolution equation on a Banach space $\underline{X}$:

$$(2.1)\quad \begin{cases} \dfrac{dz(t)}{dt} = \mathcal{A}(t)z(t) \\ z(s) = z_0 \end{cases} \qquad 0 \le s \le T < \infty$$

where $\mathcal{A}(t)$ is a closed operator on $\underline{X}$ for each $t \in T$. Then under rather general conditions on $\mathcal{A}(t)$, there exists a unique solution $z(t) = \mathcal{U}(t,s)z_0$ for all $z_0 \in \mathcal{D}(\mathcal{A}(s))$, where the evolution operator $\mathcal{U}(t,s)$ has the properties

(2.2)

(i) $\mathcal{U}(t,t) = \mathcal{I}$, the identity operator on $\underline{X}$.

(ii) $\mathcal{U}(t,r)\,\mathcal{U}(r,s) = \mathcal{U}(t,s)$ for $0 \le s \le r \le t$

(iii) $\mathcal{U}(t,s) : \mathcal{D}(\mathcal{A}(s)) \to \mathcal{D}(\mathcal{A}(t))$

(iv) $\frac{\partial}{\partial t}(\mathcal{U}(t,s)z_0) = \mathcal{A}(t)\,\mathcal{U}(t,s)\,z_0$ for $z_0 \in \mathcal{D}(\mathcal{A}(s))$.

(v) $\mathcal{U}(t,s)$ is strongly continuous in s and t for $0 \le s \le t \le T$.

(2.3) <u>The Infinite Dimensional Riccati equation</u>

The key to the Kalman-Bucy filtering theory is a Riccati equation and here we state results from [5], which are formulated to cover the Riccati equations which arise in the quadratic cost control problem and in the filtering theory of affine hereditary differential systems.(1)

Let $\mathcal{A}(t)$ be a closed operator on a Hilbert space $\mathcal{H}$ which generates an evolution operator $\mathcal{U}(t,s)$ satisfying (2.2). Let $\mathcal{V}$ be another Hilbert space and $\mathcal{B}(\cdot) \in L_\infty(T; \mathcal{L}(\mathcal{V},\mathcal{H}))$, $\mathcal{W}(\cdot) \in L_\infty(T; \mathcal{L}(\mathcal{H}))$, $\mathcal{R}(\cdot)$, $\mathcal{R}(\cdot)^{-1} \in L_\infty(T; \mathcal{L}(\mathcal{V}))$, $\mathcal{G} \in \mathcal{L}(\mathcal{H})$ and suppose $\mathcal{W}(\cdot)$ and $\mathcal{G}$ are self adjoint, positive semidefinite

---

(1) In the case where (2.1) represents a distributed parameter system, a different set of sufficient conditions to ensure the existence of a solution of the infinite dimensional Riccati equation (2.4) are given in [6].)

and $R(\cdot)$, $R(\cdot)^{-1}$ self adjoint and positive definite. Then if

(1) $\mathcal{B}(s) : \mathcal{V} \to \mathcal{D}(\mathcal{A}(t)) \quad \forall\, s < t \in T$ and $\sup_{s<t\in T} \| \mathcal{A}(t)\mathcal{B}(s)x \|$ exists for all $x \in \mathcal{V}$.

(2) $\sup_{t\in T} \| \mathcal{A}(t) y \| = \alpha_y < \infty$ exists for each $y \in \bigcap_{t\in T} \mathcal{D}(\mathcal{A}(t))$.

the following inner product Riccati equation has a unique strongly continuous solution $\mathcal{Q}_\infty(t) \in \mathcal{L}(\mathcal{H})$

$$\left\langle \left[\frac{d\mathcal{Q}_\infty(t)}{dt} + \mathcal{A}^*(t)\mathcal{Q}_\infty(t) + \mathcal{Q}_\infty(t)\mathcal{A}(t) + \mathcal{W}(t) - \mathcal{Q}_\infty(t)\, \mathcal{B}(t) R^{-1}(t) \mathcal{B}^*(t) \mathcal{Q}_\infty(t) \right] y, x \right\rangle = 0$$

(2.4) where $x, y \in \bigcap_{t\in T} \mathcal{D}(\mathcal{A}(t))$ and $\mathcal{Q}_\infty(T) = \mathcal{G}$.

3. PRELIMINARIES ON ABSTRACT PROBABILITY THEORY

Let $(\Omega, \mathcal{S}, \mu)$ be our basic probability space which is assumed complete. Let $T = [0,T]$ be a real, finite interval and $\mathcal{H}, \mathcal{K}$ real Hilbert spaces. Then an $\mathcal{H}$-valued random variable is a function $u(\cdot) ; \Omega \to \mathcal{H}$ which is measurable with respect to the μ-measure. If $u(\cdot)$ is also integrable then we define the expectation $E\{u(\cdot)\} = \int_\Omega u(\omega)\, d\mu(\omega)$. An $\mathcal{H}$-valued stochastic process is a function $u(\cdot,\cdot) : T \times \Omega \to \mathcal{H}$ which is measurable in the pair (t,ω), using Lebesgue measure on T. We also recall the definition of an $\mathcal{H}$-valued Wiener process from [2] or [3].

(3.1) <u>$w(t)$ is a Wiener process on $\mathcal{H}$</u> if it is an $\mathcal{H}$-valued stochastic process with the following properties:

(i) $E\{w(t) - w(s)\} = 0 \quad \forall\, s,t \in T.$

(ii) $w(t)$ is continuous in t on T with probability one (w.p.1.).

(iii) $E\{ (w(t) - w(s)) \circ (w(t) - w(s)) \} = (t-s)\, \mathcal{W} \quad \forall\, s < t \in T.$

where $\mathcal{W} \in \mathcal{L}(\mathcal{H})$ and is a positive, nuclear operator with eigenvalues $\{\lambda_i\}$ and orthonormal eigenvectors $\{e_i\}$ and is called the covariance operator of $w(\cdot)$

[$u \circ v \in \mathcal{L}(\mathcal{H})$ is defined for all $u, v \in \mathcal{H}$

by $u \circ v\,(h) = u \langle v, h\rangle \quad \forall\, h \in \mathcal{H}$]

(iv) $E\{\|w(t)-w(s)\|^2\} < \infty \quad \forall\, s, t \in T \qquad \langle w(t_2)-w(t_1), e_i\rangle$, $\langle w(t_4)-w(t_3), e_i\rangle$ are independent real random variables for $t_1 < t_2 \leq t_3 < t_4$ and all eigenvectors e_i of $\mathcal{W}$.

Then it is shown that $w(t)$ has the unique representation :

(3.2) $\quad w(t) = \sum_{i=0}^{\infty} \beta_i(t,\omega) e_i \quad (t,\omega)$ almost everywhere, where $\beta_i(t,\omega)$ are mutually orthogonal real Wiener processes and $\{e_i\}$ is the orthornormal basis of $\mathcal{H}$ generated by the eigenvectors of $\mathcal{W}$.

(3.3) Generalize the $\circ$ operation between 2 different Hilbert spaces as follows:

$$u \circ v(x) = u \langle v, x\rangle$$

for fixed $u \in \mathcal{H}$, $v \in \mathcal{K}$ and $\forall\; x \in \mathcal{K}$.

Then $\quad u \circ v \in \mathcal{L}(\mathcal{K}, \mathcal{H})$.

We also need some properties of $\mathcal{H}$-valued Wiener stochastic integrals,

$\int_T B(t)\, dw(t,\omega)$ where $B(t) \in \mathcal{L}(\mathcal{H},\mathcal{K})$ and $\int_T \|B(t)\|^2 dt < \infty$.

(3.4) <u>Properties of the Wiener integral : $\int_T B(t)\, dw(t)$</u> ($B(t)$ nonrandom). (see [2], [3]).

(i) $\quad E\{\int_T B(t)\, dw(t)\} = 0$

(ii) $\quad E\{\|\int_T B(t)\, dw(t)\|^2\} \leq \operatorname{trace} \mathcal{W} \int_T \|B(t)\|^2 dt$

(iii) $\quad \int_T B(t)\, dw(t) = \sum_{i=0}^{\infty} \int_T B(t) e_i \, d\beta_i(t)$

where $\{\beta_i(t)\}$, $\{e_i\}$ are as in (3.2).

(iv) If $W(t)$ and $V(t)$ are independent Wiener processes on $\mathcal{H}, \mathcal{K}$ respectively, then

$$E\left\{\int_{t_1}^{t_2} B_1(t)\, dw(t) \circ \int_{s_1}^{s_2} B_2(t)\, dv(t)\right\} = 0$$

for nonrandom $B_1(\cdot) \in L_2(T; \mathcal{L}(\mathcal{H},\mathcal{K}))$ and $B_2(\cdot) \in L_2(T; \mathcal{L}(\mathcal{K},\mathcal{K}_1))$ respectively and (t_1, t_2), (s_1, s_2) are any intervals contained in T.

(v) $$E\left\{\int_0^{s_1} \mathcal{B}_1(t)\,dw(t) \circ \int_0^{s_2} \mathcal{B}_2(t)\,dw(t)\right\} = \int_0^{\min(s_1,s_2)} \mathcal{B}_1(t)\, W\, \mathcal{B}_2^*(t)\,dt$$

for $\mathcal{B}_1(\cdot) \in L_2(T; \mathcal{L}(\mathcal{H},\mathcal{K}_1))$ and $\mathcal{B}_2(\cdot) \in L_2(T; \mathcal{L}(\mathcal{H},\mathcal{K}_2))$.

(3.5) <u>The Itô Differential</u>

If $z_0 \in \mathcal{K}$, $a(\cdot) \in L_1(T; \mathcal{K})$, $\mathcal{B}(\cdot) \in L_2(T; \mathcal{L}(\mathcal{K},\mathcal{H}))$ and $w(t)$ is an $\mathcal{H}$-valued Wiener process, then

$z(t) = z_0 + \int_0^t a(s)ds + \int_0^t \mathcal{B}(s)\,dw(s)$ is

a well-defined $\mathcal{K}$-valued stochastic process with $E\{\|z(t)\|^2\} < \infty$ and we write it in differential notation.

$$\begin{cases} dz(t) = a(t)\,dt + \mathcal{B}(t)\,dw(t) \\ z(0) = z_0 \end{cases}$$

(3.6) <u>The Stochastic Analogue of the Abstract Evolution Equation</u>

Consider the linear stochastic evolution equation on

$$\begin{cases} du(t,\omega) = \mathcal{A}(t)u(t,\omega)\,dt + f(t)\,dt + \mathcal{B}(t)\,dw(t,\omega) \\ u(0) = u_0 \end{cases} \tag{3.6}$$

where $\mathcal{A}(t)$ generates an evolution operator $\mathcal{U}(t,s)$ satisfying properties (2.2), $u_0 \in \mathcal{H}$, $f \in L_2(T; \mathcal{H})$, $w(t,\omega)$ is a $\mathcal{K}$-valued Wiener process and $\mathcal{B}(\cdot) \in L_\infty(T; \mathcal{L}(\mathcal{K},\mathcal{H}))$. Then we define the mild solution of (3.6) to be

$$u(t,\omega) = \mathcal{U}(t,0)u_0 + \int_0^t \mathcal{U}(t,s)f(s)\,ds + \int_0^t \mathcal{U}(t,s)\mathcal{B}(s)\,dw(s,\omega) \tag{3.7}$$

which is an $\mathcal{H}$-valued stochastic process with $E\{\|u(t)\|^2\} < \infty$ uniformly on T.

(3.7) is a strong solution of (3.6), that is it satisfies (3.6) w.p.l. and is strongly continuous in t w.p.l, provided $\mathcal{A}(t)$ and $\mathcal{B}(t)$ satisfy some extra conditions (see [2]). The concept of mild solution was introduced by Pritchard and Curtain in [6] and is similar to the concept of weak solutions of partial differential equations. In fact the solution of the stochastic evolution equations in this paper turn out to be strong solutions, because of the special nature of the delay equation.

4. <u>FILTERING FOR AFFINE HEREDITARY DIFFERENTIAL SYSTEMS</u>

We now give a model for the filtering problem for affine hereditary differential systems based on the ideas in § 1 and § 3. Consider the following linear stochastic evolution equation on $\mathcal{M}^2$

(4.1)
$$\begin{cases} d\phi(t,\omega) = \mathcal{A}(t)\phi(t,\omega)\,dt + \tilde{\mathcal{B}}(t)dw(t,\omega) \\ \phi(0) = \phi_0 \qquad\qquad t \in T \end{cases}$$

where $\mathcal{A}(t)$ is as in (1.4) (v), $w(t,\omega)$ is an m-dimensional Wiener process with covariance matrix $\mathcal{W}$ and $\tilde{\mathcal{B}}(\cdot) \in L_\infty(T;\mathcal{L}(R^m,\mathcal{M}^2))$.

is defined as in (1.7). (4.1) always has the mild solution:

(4.2)
$$\phi(t,\omega) = \Phi(t,0)\phi_0 + \int_0^t \Phi(t,s)\tilde{\mathcal{B}}(s)\,dw(s,\omega)$$

from (3.7). (For additional conditions on $\tilde{\mathcal{B}}$ and $w(\cdot)$ to ensure that $\phi(t,\omega) = \phi(t)$ is a strong solution see [2] .)

The observation process is assumed to have the following form

(4.3)
$$\begin{cases} dz(t,\omega) = \mathcal{C}(t)\,\phi(t,\omega)\,dt + \mathcal{F}(t)dv(t,\omega) \\ z(0,\omega) = z_0(\omega) \qquad\qquad t \in T \end{cases}$$

where $\mathcal{F}(\cdot)$, $\mathcal{F}(\cdot)^{-1} \in L_\infty(T;\mathcal{L}(R^k))$, $v(t,\omega)$ is a k- dimensional Wiener process with covariance matrix $\mathcal{V}$. v and w are assumed independent and ϕ_0 is an $\mathcal{M}^2$-valued random variable independent of v and w with zero expectation and covariance operator $P_0 = E\{\phi_0 \circ \phi_0\}$. $\mathcal{C}(\cdot) \in L_\infty(T;\mathcal{L}(\mathcal{M}^2,R^k))$ which we assume to have the following form

(4.4)
$$\mathcal{C}(t)u = \int_{-b}^0 \mathcal{C}(t,\theta)u(\theta)\,d\theta \quad \text{for} \quad u \in \mathcal{M}^2$$

where $\mathcal{C}, \frac{\partial \mathcal{C}}{\partial\theta} \in L_\infty(T\times[-b,0];\mathcal{L}(R^n,R^k))$ and $\mathcal{C}(t,0) = 0 = \mathcal{C}(t,-b)$.

Discussion of the filtering model

(4.1) represents a very general class of affine hereditary systems in the literature, but (4.3) does not explicitly allow for delays (i.e. for terms like $c_i\,x(t+\theta_i)$) as does for example Kwakernaak in [12]. This is because the known results in infinite dimensional filtering theory do not apply for $\mathcal{C}$ unbounded and the most general bounded $\mathcal{C}: \mathcal{M}^2 \to R^k$ has the form

$$\mathcal{C}(t)u = \mathcal{C}_0(t)u(0) + \int_{-b}^0 \mathcal{C}(t,\theta)u(\theta)\,d\theta \quad , \quad \text{where}$$

$\mathcal{C}_0 \in L_\infty(T;\mathcal{L}(R^n,R^k))$ and $\mathcal{C}(\cdot) \in L_\infty(T\times[-b,0];\mathcal{L}(R^n,R^k))$.

which from the point of view of the model is no better than (4.4). (The restrictions in (4.4) are due to technicalities in lemmas 2 and 3). However, we can allow for delays in the observation in the following way:

Choose $\ell(t,\theta)$ such that $\int_{-b}^{0} \ell(t,\theta)\,u(\theta)\,d\theta \simeq u(\theta_i)$.

So the formulation is not as restrictive as it may appear at first and so does represent a mathematically rigorous version of Kwakernaak's model in [12].

We now propose the filtering problem to be to find the best estimate of $\phi(t,\omega)$ based on the observations $z(t,\omega)$, $0 \le s \le t$, which has the form

$$\hat{\phi}(t,\omega) = \int_0^t K(t,s)\,dz(s)$$

where $K(t,\cdot) \in L_2(T;\mathcal{L}(\mathbb{R}^K,\mathcal{M}^2))$ for almost all t and which minimizes $E\{\|\phi(t)-\hat{\phi}(t)\|^2\}$ $\forall\ t \in T$.

First we need some simple technical lemmas.

<u>Lemma 1</u>

$\sup_{t\in T} \|\mathcal{A}^*(t)\,y\|_{\mathcal{M}^2} = \alpha_y$ exists for each $y \in \mathcal{D}^*$, the domain of $\mathcal{A}^*(t)$, the $\mathcal{M}^2$- adjoint of $\mathcal{A}(t)$.

<u>Proof</u>

From (1.9), we have that

$$(\mathcal{A}^*(t)\,y)(\theta) = \begin{cases} A^*(t,\theta)\,y(\theta) - \dfrac{dy(\theta)}{d\theta}, & -b \le \theta < 0 \\ A_0^*(t)\,y(0) + \lim\limits_{\varepsilon\to 0^+} h(-\varepsilon), & \theta = 0. \end{cases}$$

for $y \in \mathcal{D}^*$, which is independent of t.

$$\begin{aligned} \therefore \|\mathcal{A}^*(t)\,y\|^2_{\mathcal{M}^2} &= |A_0^*(t)\,y(0) + \lim_{\varepsilon\to 0^+} h(-\varepsilon)|^2 + \int_{-b}^{0} |A_0^*(t,\theta)y(\theta) - \frac{dy(\theta)}{d\theta}|^2 d\theta \\ &\le |A_0^*(t)\,y(0)|^2 + \lim_{\varepsilon\to 0^+} |h(-\varepsilon)|^2 + \int_{-b}^{0} |A_0^*(t,\theta)\,y(\theta)|^2 d\theta \\ &\quad + \sum \int_{\theta_i}^{\theta_{i-1}} |\frac{dy(\theta)}{d\theta}|^2 d\theta \end{aligned}$$

$<$ constant independent of t.

since $A_0(t)$ and $A(t,\theta)$ are uniformly bounded in t and $y \in \mathcal{D}^*$ ensures $\int_{\theta_{i-1}}^{\theta_i} |\frac{dy(\theta)}{d\theta}|^2 d\theta < \infty$.

Lemma 2

$\mathcal{b}(s)\,\mathcal{A}(t)$ is a bounded operator from $\mathcal{M}^2$ to $\mathbb{R}^K$ for each $s,t \in T$ and $\sup_{s,t\in T} \| \mathcal{b}(s)\,\mathcal{A}(t) \| < \infty$.

Proof

$$\mathcal{b}(s)\,\mathcal{A}(t)\,y = \int_{-b}^{0} \mathcal{b}(s,\theta)\,\frac{dy}{d\theta}\,d\theta \qquad \text{by (1.4) (v) and (4.4)}$$

$$= \left[\mathcal{b}(s,\theta)\,y(\theta)\right]_{-b}^{0} - \int_{-b}^{0} y(\theta)\,\frac{\partial \mathcal{b}(s,\theta)}{\partial\theta}\,d\theta$$

$$= -\int_{-b}^{0} \frac{\partial \mathcal{b}}{\partial\theta}(s,\theta)\,y(\theta)\,d\theta \qquad \text{by (4.4)}$$

$$\therefore |\mathcal{b}(s)\mathcal{A}(t)y|^2 \le \int_{-b}^{0} \left|\frac{\partial \mathcal{b}}{\partial\theta}(s,\theta)\right|^2 d\theta \int_{-b}^{0} |y(\theta)|^2\,d\theta \qquad \text{by Schwarz inequality}$$

$$\le \text{const.}\ \|y\|^2_{\mathcal{M}^2}\ , \qquad \text{by assumptions (4.4) on } \mathcal{b}.$$

So $\mathcal{b}(s)\,\mathcal{A}(t)$ can be extended to all $y \in \mathcal{M}^2$, with

$\sup_{s,t\in T} \| \mathcal{b}(s)\,\mathcal{A}(t) \| < \infty$.

Lemma 3

Under the assumptions in (4.1), (4.3) and (4.4) the following inner product version of the infinite dimensional Riccati equation has a unique solution $P(t) \in \mathcal{L}(\mathcal{M}^2)$.

(4.5)
$$\left\langle \left[\frac{dP(t)}{dt} - \mathcal{A}(t)P(t) - P(t)\mathcal{A}^*(t) - \tilde{B}(t)W\tilde{B}^*(t) + P(t)\mathcal{b}^*(t)\left(\mathcal{F}(t)V\mathcal{F}^*(t)\right)^{-1}\mathcal{b}(t)P(t) \right] x, y \right\rangle = 0$$

where $x, y \in \mathcal{D}^*$, $P(0) = P_0$.

Proof

Let $t' = T-t$ and $\mathcal{A}_1^*(t') = \mathcal{A}(T-t)$, $\mathcal{b}_1(t') = \mathcal{b}^*(T-t)$, $P_1(t') = P(T-t)$, $B_1(t') = \tilde{B}(T-t)$, $\mathcal{F}_1(t') = \mathcal{F}(T-t)$.

Then (4.5) has the form of (2.4). The conditions we need to verify are now

(1) $\sup \| \mathcal{A}^*(T-t)\,y \|$ exists for each $y \in \mathcal{D}^*$

and

(2) $\mathcal{B}^*(T-t): \quad R^K \to \mathcal{D}^*$

and $\quad \| \mathcal{A}^*(T-t)\mathcal{B}^*(T-s)x \| \leq g(s) \in L_1(T) \quad \forall \; x \in R^K.$

(1) follows from lemma 1

(2) holds since $\mathcal{A}^*(T-t)\mathcal{B}^*(T-s)$ is a linear map from R^K to $\mathcal{M}^2$ and hence bounded and $\sup\limits_{s,t\in T} \| \mathcal{B}^*(s)\tilde{\mathcal{A}}^*(t) \| = \sup\limits_{s,t\in T} \| \mathcal{A}(t)\mathcal{B}(s) \| < \infty$ by lemma 2.

With these preliminaries we can develop the filtering theory as in [4].

Theorem 1

$\hat{\phi}(t) = \int_0^t K(t,s)\,dz(s)$ is a solution of the filtering problem if and only if $E\{\tilde{\phi}(t) \circ (z(\sigma) - z(\tau))\} = 0$ for all σ, τ such that $0 \leq \tau \leq \sigma \leq T$, where $\tilde{\phi}(t) = \phi(t) - \hat{\phi}(t)$.

Proof

Let h be fixed and define the Hilbert space $\overline{X}(h) = \{ <u,h>$, where u is an $\mathcal{H}$-valued random variable with $E\{\|u\|^2\} < \infty \}$.
The inner product is $[<u,h>,<v,h>] = E\{<u,h><v,h>\}$.

(We note that $E\{u \circ v\} = 0$ iff $[<u,h>,<v,h>] = 0 \quad \forall \; h \in \mathcal{H}$).

We also define for fixed t, the subspace

$$\overline{X}_t(h) = \left\{ \begin{array}{ll} <y_t, h>, \text{ where } & y_t = \int_0^t R(t,s)\,dz(s) \\ & \\ \text{and } R(t,\cdot) \in L_2(T; \mathcal{L}(\mathcal{K},\mathcal{H})) & \text{for almost all } t \end{array} \right\}$$

Now $<\tilde{\phi}(t), h> \in \overline{X}_t(h)$ and we wish to minimize $\tilde{\phi}(t) = \phi(t) - \hat{\phi}(t)$ in the $\overline{X}(h)$ norm for all h. By the orthogonal projection lemma, this is equivalent to requiring $<\tilde{\phi}(t), h> \perp \overline{X}_t(h)$ in $\overline{X}(h)$ for all $h \in \mathcal{H}$.

i.e. $E\{<h, \tilde{\phi}(t)> <h, y_t>\} = 0 \quad \forall \; <h, y_t> \in \overline{X}_t(h), \; \forall \; h \in \mathcal{H}.$

i.e. iff $<h, E\{\tilde{\phi}(t) \circ y_t\} h> = 0$

by definition of $\circ$. So we need only establish that $E\{\tilde{\phi}(t) \circ y_t\} = 0$ iff $E\{\tilde{\phi}(t) \circ (z(\sigma) - z(\tau))\} = 0$ for $0 \leq \tau \leq \sigma \leq T$.

Suppose first that $E\{\tilde{\phi}(t) \circ (z(\sigma) - z(\tau))\} = 0$

Consider $y_t = \int_0^t R(t,s)\,dz(s)$, where $R(t,s)$ is a step function in s.

Then $E\{\tilde{\phi}(t) \circ y_t\} = \sum\limits_j E\{\tilde{\phi}(t) \circ R_j \Delta z_j\}$

(with the obvious incremental notation) $= \sum_j E\{\tilde{\phi}(t) \circ \Delta z_j\} R_j^*$

(since R is nonrandom)

$= 0$ (by assumption).

For general $R(t,s)$, we can approximate by a sequence $\{R_n(t,s)\}$ of step functions $\ni \int_0^t \|R(t,s) - R_n(t,s)\|^2 ds \to 0$ as $n \to \infty$

Then

$$E\{\|\int_0^t [R(t,s) - R_n(t,s)] dz(s)\|^2\}$$

$$\leq 2E\{\|\int_0^t R(t,s) - R_n(t,s)\, b(s)\phi(s) ds\|^2\} + 2E\{\|\int_0^t (R - R_n) \mathcal{F} dv\|^2\}$$

$$\leq 2 \int_0^t \|R - R_n\|^2 ds \sup_{s \in T}\{\|b(s)\|^2\} \cdot \int_0^t E\{\|\phi(s)\|^2\} ds \qquad \text{by (2.2)}$$

$$+ 2 \operatorname{trace} \mathcal{V} \sup_{s \in T}\{\|\mathcal{F}(s)\|^2\} \cdot \int_0^t \|R - R_n\|^2 ds$$

by Schwarz inequality and property (3.4) (ii)

$\leq \text{Const} \cdot \int_0^t \|R - R_n\|^2 ds$ since $b(\cdot)$, $\mathcal{F}(\cdot)$ are uniformly bounded in norm and using (3.7)

$\to 0$ as $n \to \infty$

But $\|E\{\tilde{\phi}(t) \circ y_t\}\|^2 \leq (E\{\|\tilde{\phi}(t) \circ y_t\|\})^2$

$= (E\{\|\tilde{\phi}(t)\| \|y_t\|\})^2$ by definition of $\circ$

$\leq E\{\|\tilde{\phi}(t)\|^2\} E\{\|y_t\|^2\}$ by Schwarz inequality

So by approximating y_t by $y_t^n = \int_0^t R_n(t,s) dz(s)$, we see that $E\{\tilde{\phi}(t) \circ y_t\} = 0$ for all $y_t \ni \langle y_t, h\rangle \in \underline{X}_t(h)$.

Clearly the argument is independent of h.

Suppose conversely that $E\{\tilde{\phi}(t) \circ (z(\sigma) - z(\tau))\} \neq 0$ for some σ, τ.

Then define

$$R(t,s) = \begin{cases} E\{\tilde{\phi}(t) \circ (z(\sigma) - z(\tau))\} & \text{for } \tau \leq s \leq \sigma \\ 0 & \text{otherwise} \end{cases}$$

Then $\int_0^t \|R(t,s)\|^2 ds \le \int_\tau^6 E\{\|\tilde{\phi}(t)\|^2\} ds \cdot \int_\tau^6 E\{\|z(6)-z(\tau)\|^2\} ds$

by the usual inequality arguments

$< \infty$

So $y_t = \int_0^t R(t,s) dz(s)$ is such that $\langle y_t, h \rangle \in \bar{X}_t(h)$.

Now $\langle h, E\{\tilde{\phi}(t) \circ y_t\} h \rangle = \langle h, E\{\tilde{\phi}(t) \circ (z(6)-z(\tau))\} E\{\tilde{\phi}(t) \circ (z(6)-z(\tau))\}^* h \rangle$

$= \| E\{\tilde{\phi}(t) \circ (z(6)-z(\tau))\}^* h \|^2$

$\neq 0$ for some h.

So $E\{\tilde{\phi}(t) \circ y_t\} \neq 0$.

<u>Lemma 4</u>

Let $\Lambda(t,s) = E\{\phi(t) \circ \phi(s)\}$, then

(4.6)

(a) $\Lambda(t,s) = \Phi(t,0) P_0 \Phi(s,0) + \int_0^{\min(t,s)} \Phi(t,\tau) \tilde{B}(\tau) W \tilde{B}^*(\tau) \Phi(s,\tau) d\tau = \Lambda^*(s,t)$

(b) $\langle [\frac{\partial}{\partial t} \Lambda^*(t,s) - \Lambda^*(t,s) \mathcal{A}^*(t)] x, y \rangle = 0$ for $x, y \in \mathcal{D}^*$.

(c) $\langle [\frac{\partial}{\partial t} \Lambda^*(t,t) - \Lambda^*(t,t) \mathcal{A}^*(t) - \mathcal{A}(t) \Lambda^*(t,t) - \tilde{B}(t) W \tilde{B}^*(t)] x, y \rangle = 0$

for $x, y \in \mathcal{D}^*$.

<u>Proof</u>

(a) Follow by direct calculation using (3.4) (iv) and (v)

(b) $\frac{\partial}{\partial t} \Phi^*(t,\tau) x = \Phi^*(t,\tau) \mathcal{A}^*(t) x$ for $x \in \mathcal{D}^*$

follows from (1.4) (v), differentiating under the integral sign is justified by lemma 1 (a) and since all other operators are L_∞ on T.

(c) Similar to (b).

<u>Lemma 5</u>

Let $K(t,\cdot) \in L_2(T; \mathcal{L}(R^K, \mathcal{H}^2))$ and $\phi(t)$ the solution of (4.1)

Then $E\{\int_0^t K(t,s) dz(s) \circ \phi(6)\} = \int_0^t K(t,s) b(s) \Lambda(s,6) ds$.

<u>Proof</u>

Direct verification using (3.4) (iv).

Theorem 2

Under the assumptions in (4.1) and (4.3) only, the following integral equation has a unique solution $\mathcal{K}(t,\cdot) \in L_\infty(T; \mathcal{L}(R^k, \mathcal{H}^2))$ for each fixed $t \in T$.

(4.7) $\int_0^t \mathcal{K}(t,s)\mathcal{B}(s)\Lambda(s,\sigma)\mathcal{B}^*(\sigma)ds + \mathcal{K}(t,\sigma)\mathcal{F}(\sigma)V\mathcal{F}^*(\sigma) = \Lambda(t,\sigma)\mathcal{B}^*(\sigma)$.

Proof (follows [1]).

Let t be fixed and define the operators $\mathcal{Q}_1, \mathcal{Q}_2$ by

$$\mathcal{Q}_1 = \mathcal{B}(\tau)\Lambda(\tau,t)$$

$$\mathcal{Q}_2 f(\sigma) = \mathcal{F}(\sigma)V\mathcal{F}(\sigma)^* f(\sigma) + \int_0^t \mathcal{B}(\sigma)\Lambda(\sigma,s)\mathcal{B}^*(s)f(s)ds$$

Then $\mathcal{Q}_1 : \mathcal{H} \to L_\infty([0,t); \mathcal{K})$

$$\mathcal{Q}_2 : L_2([0,t]; \mathcal{K}) \to L_2([0,t]; \mathcal{K})$$

and $$\langle\langle \mathcal{Q}_2 f, f \rangle\rangle = \int_0^t \langle \mathcal{F}(\sigma)V\mathcal{F}^*(\sigma)f(\sigma), f(\sigma)\rangle d\sigma + \int_0^t \langle \int_0^t \mathcal{B}(\sigma)\Lambda(\sigma,s)\mathcal{B}^*(s)f(\sigma)ds, f(\sigma)\rangle d\sigma$$

$$> 0 \quad \forall f \in L_2([0,t]; \mathcal{K})$$

from the form of $\Lambda(\cdot,\cdot)$ in (4.6) (a) and since $\mathcal{F}(\sigma)V\mathcal{F}(\sigma)^*$ is invertible. Hence $\mathcal{Q}_2^{-1}$ exists.

Define $\mathcal{K}(t,\sigma) = (\mathcal{Q}_2^{-1}\mathcal{Q}_1)^*$ and let $k(\sigma) = \mathcal{K}(t,\sigma)^* h$ for some $h \in \mathcal{H}$.

Then $\mathcal{Q}_2 k(\sigma) = \mathcal{Q}_1 h$

or $\mathcal{F}(\sigma)V\mathcal{F}(\sigma)^* k(\sigma) + \int_0^t \mathcal{B}(\sigma)\Lambda(\sigma,s)\mathcal{B}^*(s)k(s)ds = \mathcal{B}(\sigma)\Lambda(\sigma,t)h$

and $k(\cdot) \in L_2([0,t]; \mathcal{H})$.

But $k(\sigma) = \mathcal{K}(t,\sigma)^* h$ and we see that $\mathcal{K}(t,\cdot) \in L_\infty([0,t]; \mathcal{L}(\mathcal{K},\mathcal{H}))$

$\mathcal{K}(t,\cdot)$ satisfies the integral equation (4.7) substituting $k(\sigma) = \mathcal{K}(t,\sigma)^* h$ in (4.8) and taking adjoints. So $\mathcal{K}(t,\sigma) = (\mathcal{Q}_2^{-1}\mathcal{Q}_1)^*$ is a solution of (4.7).

The uniqueness is similarly proved using the linearity of (4.7)

Theorem 3

There is a solution $\hat{\phi}(t) = \int_0^t K(t,s)\,dz(s)$ to the filtering problem if and only if (4.7) has a solution.

Proof

Suppose firstly that there is a solution $\hat{\phi}(t) = \int_0^t K(t,s)\,dz(s)$ to the filtering problem. Let t be fixed and define for $0 \le \sigma < t$,

$$y(\sigma) = \int_0^\sigma b(s)\phi(s)\,ds = z(\sigma) - z(0) - \int_0^\sigma f(s)\,dv(s)$$

Now
$$\frac{d}{d\sigma} E\{\tilde{\phi}(t) \circ y(\sigma)\} = E\{\tilde{\phi}(t) \circ b(\sigma)\phi(\sigma)\}, \quad \text{since } \sigma < t$$
$$= E\{\phi(t) \circ \phi(\sigma)\} b^*(\sigma) - E\{\int_0^t K(t,s)\,dz(s) \circ \phi(\sigma)\} b^*(\sigma)$$
$$= \Lambda(t,\sigma) b^*(\sigma) - \int_0^t K(t,s) b(s) \Lambda(s,\sigma) b^*(\sigma)\,ds \tag{4.9}$$

by lemmas 4 and 5.

But
$$E\{\tilde{\phi}(t) \circ y(\sigma)\} = E\{\tilde{\phi}(t) \circ -\int_0^\sigma f(s)\,dv(s)\} \quad \text{applying theorem 1}$$
$$= E\{\int_0^t K(t,s) f(s)\,dv(s) \circ \int_0^\sigma f(s)\,dv(s)\}$$

expanding $\phi(\cdot)$ and $\hat{\phi}(\cdot)$, since $v(\cdot)$ is independent of ϕ_0 and $w(\cdot)$ and using property (3.4) (iv)

$$= \int_0^\sigma K(t,s) f(s) V f(s)^*\,ds \quad \text{by (3.4) (v)}$$

$$\therefore \frac{d}{d\sigma} E\{\tilde{\phi}(t) \circ y(\sigma)\} = K(t,\sigma) f(\sigma) V f(\sigma)^* \tag{4.10}$$

Equating (4.9) and 4.10) gives us the integral equation (4.7).

Suppose now that $K(t,s)$ is the unique solution of (4.7).

To prove that $\hat{\phi}(t) = \int_0^t K(t,s)\,dz(s)$ satisfies theorem 1.

i.e. $E\{\tilde{\phi}(t) \circ (z(\sigma) - z(\tau))\} = 0$ for all $0 \le \tau \le \sigma \le T$.

From the linearity, we may let $\tau = 0$.

Now
$$E\{\tilde{\phi}(t) \circ z(\sigma) - z(0)\} = E\{\tilde{\phi}(t) \circ y(\sigma)\} + E\{\tilde{\phi}(t) \circ \int_0^\sigma f(s)\,dv(s)\}$$

from the definition of $y(\sigma)$

$$= \int_0^\sigma \Big[\Lambda(t,\alpha) b^*(\alpha) - \int_0^t K(t,s) b(s) \Lambda(s,\alpha) b^*(\alpha)\,ds\Big] d\alpha$$
$$- \int_0^\sigma K(t,s) f(s) V f^*(s)\,ds$$

from (4.9) and (3.4) (v)

$$= 0$$

since $\mathcal{K}(t,s)$ is a solution of (4.7).

Corollary

Under the assumptions of our theorem, there exists a unique solution to the filtering problem. We remark that this result is independent of the special form of $\tilde{\mathcal{B}}(t)$ and $\tilde{\mathcal{b}}(t)$ and w uld hold for arbitrary $\tilde{\mathcal{B}} \in L_\infty(T; \mathcal{L}(R^m, \mathcal{M}^2))$ and $\mathcal{b}(\cdot) \in L_\infty(T; \mathcal{L}(\mathcal{M}^2, R^k))$. However, to obtain the Kalman-Bucy filtering equations we do need the special form of $\mathcal{b}$ (though not of $\tilde{\mathcal{B}}$)

Lemma 6

If the solution $\mathcal{K}(t,\cdot)$ of (4.7) satisfies the additional regularity properties (4.11)

(i) $\mathcal{K}(t,s)$ is strongly differentiable in t on T for $t > s$ and

$$\left\| \frac{\partial \mathcal{K}^*(t,s)}{\partial t} y \right\| \leq C \quad \text{for} \quad y \in \mathcal{D}^*$$

(ii) $\| \mathcal{K}^*(t,s) \mathcal{A}^*(t) y \| \leq C$, for $y \in \mathcal{D}^*$ then

$$\left[\mathcal{A}(t)\mathcal{K}(t,s) - \mathcal{K}(t,t)\mathcal{b}(t)\mathcal{K}(t,s) - \frac{\partial \mathcal{K}}{\partial t}(t,s)\right] x = 0$$

for x in its domain of definition.

Proof

Differentiating the following with respect to t for $t > \sigma$

$$< \left[\int_0^t \mathcal{b}(\sigma)\Lambda^*(s,\sigma)\mathcal{b}^*(s)\mathcal{K}^*(t,s)\,ds + \mathcal{J}(\sigma)\mathcal{V}\mathcal{J}^*(\sigma)\mathcal{K}^*(t,\sigma) - \mathcal{b}(\sigma)\Lambda^*(t,\sigma)\right] x, y > = 0$$

for $x, y \in \mathcal{D}^*$.

using Lemma 4 (b), (4.7) yields

$$< \left[\int_0^t \mathcal{b}(\sigma)\Lambda^*(s,\sigma)\mathcal{b}^*(\sigma)\Delta^*(t,s)\,ds + \mathcal{J}(\sigma)\mathcal{V}\mathcal{J}^*(\sigma)\Delta^*(t,\sigma)\right] x, y > = 0$$

where $\Delta(t,s) = \mathcal{A}(t)\mathcal{K}(t,s) - \mathcal{K}(t,t)\mathcal{b}(t)\mathcal{K}(t,s) - \frac{\partial \mathcal{K}}{\partial t}(t,s)$.

Conditions (i) and (ii) were necessary to justify differentiating under the integral sign.

$$\therefore \int_0^t < \mathcal{b}(\sigma)\Lambda(\sigma,s)\mathcal{b}^*(s)x, \Delta(t,s)y > ds + < \mathcal{J}(\sigma)\mathcal{V}\mathcal{J}^*(\sigma)x, \Delta(t,\sigma)y > = 0$$

or $\quad < \mathcal{Q}_2 x, \Delta(t,\sigma) y > = 0$

where $\mathcal{Q}_2 : L_2(T; R^k) \to L_2([0,t]; R^k)$ is given by

$$\mathcal{Q}_2 f(\sigma) = \mathcal{F}(\sigma) V \mathcal{F}^*(\sigma) f(\sigma) + \int_0^t \mathcal{C}(\sigma) \Lambda(\sigma, s) \mathcal{C}^*(s) f(s)\, ds$$

and is strictly positive. $\therefore\ \Delta(t,\sigma) = 0$ on its domain (see theorem 2).

<u>Theorem 4</u> <u>Generalized Kalman-Bucy Filtering Theorem</u>

If $P(t)$ is the unique solution of (4.5), let $\mathcal{K}(t) = P(t)\mathcal{C}^*(t)(\mathcal{F}(t)V\mathcal{F}^*(t))^{-1}$ and $\mathcal{Y}(t,s)$ the evolution operator generated by $\mathcal{A}(t) - \mathcal{K}(t)\mathcal{C}(t)$. Then $\hat{\phi}(t) = \int_0^t \mathcal{Y}(t,s)\mathcal{K}(s)\,dz(s)$ is the solution to the filtering problem and satisfies the following 'Kalman-Bucy' type equation.

$$\text{(4.12)} \quad \begin{cases} d\hat{\phi}(t) = (\mathcal{A}(t) - \mathcal{K}(t)\mathcal{C}(t))\hat{\phi}(t)\,dt + \mathcal{K}(t)\mathcal{C}(t)\phi(t)\,dt + \mathcal{K}(t)\mathcal{F}(t)\,dv(t) \\ \qquad\quad = (\mathcal{A}(t) - \mathcal{K}(t)\mathcal{C}(t))\hat{\phi}(t)\,dt + \mathcal{K}(t)\,dz(t) \\ \hat{\phi}(0) = 0 \end{cases}$$

<u>Proof</u>

(a) $\mathcal{K}(t,s) = \mathcal{Y}(t,s)\mathcal{K}(s) = \mathcal{Y}(t,s)P(s)\mathcal{C}^*(s)(\mathcal{F}(s)V\mathcal{F}^*(s))^{-1}$ satisfies the assumptions in lemma 6 since $\mathcal{Y}(t,s)$ is an evolution operator satisfying (2.2) with generator $\mathcal{A}(t) - \mathcal{K}(t)\mathcal{C}(t)$ (see [5] for details on perturbation theory for evolution operators).

So $\mathcal{K}(t,s)$ satisfies the differentiated version of (4.7)

i.e. $\int_0^t \mathcal{Y}(t,s)\mathcal{K}(s)\mathcal{C}(s)\Lambda(s,\sigma)\mathcal{C}^*(\sigma)\,ds - \Lambda(t,\sigma)\mathcal{C}^*(\sigma) + \mathcal{Y}(t,\sigma)\mathcal{K}(\sigma)(\mathcal{F}(\sigma)V\mathcal{F}^*(\sigma))^{-1}$

$= R(\sigma)$ some operator valued function of σ independent of t. We now show that $R(\sigma) = 0$ and so $\mathcal{Y}(t,s)\mathcal{K}(s)$ satisfies (4.7). It is sufficient to show this at $\sigma = t$ i.e. To show that

$$\text{(4.13)} \quad P(\sigma) = \Lambda(\sigma,\sigma) - \int_0^\sigma \mathcal{Y}(\sigma,s)P(s)\mathcal{C}^*(s)(\mathcal{F}(s)V\mathcal{F}^*(s))^{-1}\mathcal{C}(s)\Lambda(s,\sigma)\,ds$$

But this is just an integrated version of the Riccati equation (4.5), which is easily verified by differentiation of $\langle \mathcal{N}^*(\sigma)x, y\rangle$, where $x, y \in \mathcal{D}^*$ and $\mathcal{N}(\sigma)$ is the left hand side of (4.13). (Use properties (2.2) for $\mathcal{Y}(t,s)$ and lemma 4). Differentiation under the integral is again justified since all bounded operators are L_∞ and $\mathcal{A}(t)$ satisfies (a) of lemma 1. Since (4.5) has a unique solution, $P(\sigma) \equiv \mathcal{N}(\sigma)$.

(b) (4.12) is a linear stochastic evolution equation of the form (3.6) and so has the mild solution

$$\hat{\phi}(t) = \mathcal{Y}(t,0)\,0 + \int_0^t \mathcal{Y}(t,s)\,\mathcal{K}(s)\,\mathcal{b}(s)\,\phi(s)\,ds + \int_0^t \mathcal{Y}(t,s)\,\mathcal{K}(s)\,\mathcal{J}(s)\,dV(s)$$

$$= \int_0^t \mathcal{K}(t,s)\,dz(s)$$

For completeness, we show that $P(t)$ is actually the covariance operator for the error $\tilde{\phi}(t) = \phi(t) - \hat{\phi}(t)$

Lemma 7

$$P(t) = \mathrm{Cov}\,[\phi(t) - \hat{\phi}(t)]$$

$$= E\,\{(\phi(t) - \hat{\phi}(t)) \circ (\phi(t) - \hat{\phi}(t))\}$$

Proof

Let $R(t)$ denote the covariance operator.

Then $R(t) = E\{(\phi(t)-\hat{\phi}(t)) \circ (\phi(t)-\hat{\phi}(t))\}$ since the expectations of $\phi(t)$ and $\hat{\phi}(t)$ are zero using the expression (4.2) for ϕ and

$$\hat{\phi}(t) = \int_0^t \mathcal{K}(t,s)\,dz(s)$$

$$= \int_0^t \mathcal{K}(t,s)\,\mathcal{b}(s)\,\phi(s)\,ds + \int_0^t \mathcal{K}(t,s)\,\mathcal{J}(s)\,dV(s) \qquad (4.14)$$

and property (3.4) (i) of the stochastic integral.

$$\therefore R(t) = \Lambda(t,t) - 2\int_0^t \mathcal{K}(t,s)\,\mathcal{b}(s)\,\Lambda(s,t)\,ds + E\{\hat{\phi}(t) \circ \hat{\phi}(t)\}$$

by lemmas 4 (a) and 5

Now $$\hat{\phi}(t) = \int_0^t \mathcal{K}(t,s)\mathcal{b}(s)\Phi(s,0)\phi_0\,ds + \int_0^t \mathcal{K}(t,s)\mathcal{b}(s)\int_0^s \Phi(s,\alpha)\,\tilde{B}(\alpha)\,dW(\alpha)\,ds$$
$$+ \int_0^t \mathcal{K}(t,s)\,\mathcal{b}(s)\,\mathcal{J}(s)\,dV(s)$$

using (4.14) and expression (4.2) for ϕ, and using the independence of ϕ_0, $W(\cdot)$ and $V(\cdot)$ and repeated application of (3.4) (iv) and (v), one obtains that

$$E\{\hat{\phi}(t) \circ \hat{\phi}(t)\} = \int_0^t \mathcal{K}(t,s)\mathcal{J}(s)\,\mathcal{V}\,\mathcal{J}^*(s)\,\mathcal{K}^*(t,s)\,ds$$
$$+ \left(\int_0^t \mathcal{K}(t,s)\mathcal{b}(s)\Phi(s,0)\,ds\right) P_0 \left(\int_0^t \mathcal{K}(t,s)\mathcal{b}(s)\Phi(s,0)\,ds\right)^*$$

$$+ \int_0^t\left(\int_0^t \mathcal{K}(t,s)\mathcal{b}(s)\left[\int_0^{\min(t,s)} \Phi(s,\alpha)\,\tilde{B}(\alpha)\,\mathcal{W}\,\tilde{B}^*(\alpha)\,\Phi^*(r,\alpha)\,d\alpha\right]\mathcal{b}^*(r)\,\mathcal{K}^*(t,r)\,dr\right) ds$$

$$\therefore E\{\hat{\phi}(t)\circ\hat{\phi}(t)\} = \int_0^t \mathcal{K}(t,r)\mathcal{F}(r)V\mathcal{F}^*(r)\mathcal{K}^*(t,r)\,ds + \int_0^t\int_0^t \mathcal{K}(t,s)\mathcal{b}(s)\Lambda(s,r)\mathcal{b}^*(r)\mathcal{K}^*(t,r)\,dr\,ds$$

by lemma 4

$$-\int_0^t \Lambda(t,r)\mathcal{b}^*(r)\mathcal{K}^*(t,r)\,dr$$ since $\mathcal{K}(t,s)$ satisfies (4.7)

$$= \int_0^t \mathcal{K}(t,r)\mathcal{b}(r)\Lambda(r,t)\,dr$$ since $a_0 d$ is self adjoint

$$\therefore \mathcal{R}(t) = \Lambda(t,t) - \int_0^t \mathcal{K}(t,r)\mathcal{b}(r)\Lambda(r,t)\,dr$$

$$= \mathcal{P}(t)$$

by (4.13), since $\mathcal{K}(t,r) = \mathcal{Y}(t,r)\mathcal{P}(r)\mathcal{b}^*(r)(\mathcal{F}(r)V\mathcal{F}^*(r))^{-1}$.

Acknowledgements

I would like to thank Richard Vinter for clarification of the nature of the adjoint $\mathcal{A}^*(t)$ which is a key element in the theory.

REFERENCES

1. Bensoussan, A. "Filtrage Optimal des Systems Lineaires," Dunod 1971.

2. Curtain R.F. and Falb P.L. Stochastic Differential Equations in a Hilbert Space J. Diff Eqns 10 (1971) 412 - 430.

3. Curtain R.F. On the Itô Stochastic Integral in a Hilbert Space. Control Theory Centre Report No. 11, University of Warwick, England.

4. Curtain R.F. Infinite Dimensional Filtering. Siam. J. Control 13 (1975).

5. Curtain R.F. The Infinite Dimensional Riccati Equation with Applications to Affine Hereditary Differential Systems. Control Theory Centre Report No 24. University of Warwick, England. (Submitted to Siam J. Control).

6. Curtain R.F. and Pritchard A.J. The Infinite Dimensional Riccati Equation. J. Math. Anal. Appl. 46

7. Delfour M.C. and Mitter S.K. Hereditary Differential Systems with Constant Delays I - General Case (J. Diff Eqns 12 (1972), 213-235) II-A Class of Affine Systems and the Adjoint Problem. J. Diff Eqns.

8. Delfour M.C. and Mitter S.K. Controllability, Observability and Optimal Feedback Control of Hereditary Differential Systems, SIAM J Control, 10 (1972), 298-328.

9. Kailath T. An innovations approach to least squares estimation, Part I: Linear Filtering in additive white noise, IEEE Trans. Aut. Control AC (13) (1968) (646-655)

10. Kalman R.E. and Bucy R.S. New results in linear filtering and prediction theory. J. Basic Eng. ASME 83 (1961) pp 95-108

11. Kushner H.J. and Barnea D.I. On the control of a linear functional - differential equation with quadratic cost SIAM J Control 8 (1970) pp 257-272

12. Kwakernaak H. Optimal Filtering in Linear Systems with Time delays IEEE Trans. AC - 12 No 2 1967 (p 169-173)

13. Lindquist A. A Theorem on Duality Between Estimation and Control for Linear Stochastic Systems with Time delay. J. Math Anal and Appl. Vol 37 No 2 1972 (p 516 - 536)

14. Vinter, R. On the Evolution of the state of linear differential delay equations in $\mathcal{M}^2$. Properties of the generator. Report E SL - R - 541, Electronic Systems Laboratory, M.I.T.

Linear Least-Squares Estimation of Discrete-Time Stationary Processes by Means of Backward Innovations

Anders Lindquist[†]

1. Introduction

There has lately been a considerable interest in fast algorithms for recursive linear least squares estimation. This is clearly witnessed by a series of recent papers by Casti, Kalaba and Murthy [1], Rissanen [2], Casti and Tse [3], Kailath [4,5,6], and Lindquist [7,8,9], to just mention some contributions related to the work presented in this paper. Among these [1,3,4,5,8,9] concern stochastic processes in continuous time, while [2,6,7,9] deal with discrete-time processes. For an account on the relation between these papers, among which [4,5,6,7,8] and to a certain extent [3] are concerned with Kalman-Bucy filtering,we refer the reader to [9] where we also try to clarify the connections between these recent results and some classical results in filtering [10,11,12], the theory of polynomials orthogonal on the unit circle [13,14,15] and the theory of Fredholm integral equations [16,17,18,19].

In this paper we shall consider the algorithm for the discrete-time Kalman-Bucy gain first presented in [7]. In the important case when the number of outputs are much fewer that the dimension n of the system, this algorithm, which holds for stationary systems,requires a number of scalar equations of order n rather than n^2 as with the conventional method based on the Riccati equation.

The continuous-time counterpart of this algorithm was first derived by Kailath [4] by means of a decomposition of the Riccati equation due to Bucy [20]. Since this decomposition holds for all constant Kalman-Bucy models, Kailath has been able to obtain similar equations [5] also for certain nonstationary systems, although the computional advantage of these algorithms rely heavily on the possible low rank of a certain n x n - matrix—a condition which is automatically fulfilled in the stationary case. However, in the sequel only stationary models will be considered.

Our discrete-time result [7] was obtained independently by a method based on the work [10,11,12,13, 14,15], and somewhat suprisingly these equations are more complicated than their continuous-time counterparts. In a subsequent paper [6] Kailath et al have demonstrated that the algorithm [7] can also be derived from the discrete-time Riccati-equation by a decomposition akin to that of [4], which of course is only to be expected. However, now the decomposition is no longer unique so that several versions of the algorithm emerge. Unfortunately, the Riccati - approach [6] gives very little insight into the relation between these versions.

This paper will be devoted to a more thorough study of the discrete - time algorithm [7]. By proceeding from basic principles we shall be able to present a number of different versions *and* explain the relation between them. This will be done in Sections 3 and 4 by an essentially *deterministic* technique, the

[†]Department of Mathematics, University of Kentucky, Lexington, Kentucky 40506, U.S.A.

basic feature of which is a certain reversed-time operation previously used in [9]. In Section 5 we provide a *stochastic* interpretation of this approach in terms of the *forward and backward innovation processes.* This method, which suggested the investigation of Sections 3 and 4, was first used in [8] and is based on the simple observation that the available data can be orthogonalized either in the forward or the backward direction, thereby providing two different innovation processes.

As we did in [7,8,9], we shall find it convenient to develop our results in a smoothing context, although our primary interest is the one-step predictor and the pure filter. This is of course no coincidence, because forward and backward recursions do play a central role in the theory of linear smoothing. In particular our approach reminds one of the interpretation, due to Mayne and Fraser, of the optimal smoother as a combination of a forward and backward filter [21,22,23,24]. However, in contrast to the basic idea of the Mayne-Fraser technique we use the backward innovation for the filtering part and the forward for the smoothing part. In fact this is the key idea of our method.

Although our main interest is in Kalman-Bucy filtering, we shall deliberately introduce the Gauss-Markov condition as late in our development as possible. The reason for this is to pinpoint what properties of the algorithms depend on this assumption and what properties hold for wide sense stationary processes in general.

Finally, to explain why our discrete-time results are considerably more complex than the corresponding results in continuous time, in Section 6 we shall briefly discuss the analogous continuous-time approach.

2. Preliminaries

Let $\{x(t); t = \ldots -2, -1, 0, 1, 2, \ldots\}$ be an n-dimensional vector process with zero mean and covariance function

$$\Gamma_{ts} = E\{x(t)\, x(s)'\} \tag{2.1}$$

and consider the linear least-squares estimate $\hat{x}(t|r)$ of $x(t)$ given $z(0), z(1), \ldots, z(r)$, where the m-dimensional process z is defined by

$$z(t) = H\, x(t) + w(t). \tag{2.2}$$

Here H is an m x n - matrix and w is a zero mean white noise sequence with covariance

$$E\{w(t)\, w(s)'\} = I\, \delta_{ts}. \tag{2.3}$$

We assume that x and w are uncorrelated.

Now, denoting the estimation error

$$\tilde{x}(t|r) = x(t) - \hat{x}(t|r), \tag{2.4}$$

we define the error covariance function

$$P_r(t,s) = E\{\tilde{x}(t|r)\,\tilde{x}(s|r)'\} \tag{2.5}$$

which provides us with the weighting function of the estimate:

Proposition 2.1:
$$\hat{x}(t|r) = \sum_{s=0}^{r} P_r(t,s)\,H'z(s) \tag{2.6}$$

Proof: The projection theorem implies that

$$E\{\tilde{x}(t|r)\,z(s)'\} = E\{\tilde{x}(t|r)\,x(s)'\}H' - E\{\hat{x}(t|r)\,w(s)'\} \tag{2.7}$$

is zero for all $s = 0,1,\ldots,r$. Therefore inserting

$$\hat{x}(t|r) = \sum_{s=0}^{r} N_s z(s)$$

in (2.7), we have

$$N_s = E\{\tilde{x}(t|r)\,x(s)'\}H'$$

from which our assertion follows.■

However note that (2.5) defines $P_r(t,s)$ for arguments $(s<0,\ s>r)$ which are not needed in the representation (2.6). The function P has the following properties:

Proposition 2.1: *The function* P *satisfies*

$$P_r(s,t) = P_r(t,s)', \tag{2.8}$$

and is the unique solution of the system of linear equations

$$P_r(t,s) + \sum_{i=0}^{r} \Gamma_{ti}\,H'H\,P_r(i,s) = \Gamma_{ts}. \tag{2.9}$$

Moreover it satisfies the relation

$$P_{r+1}(t,s) = P_r(t,s) - P_{r+1}(t,r+1)H'HP_r(r+1,s) \tag{2.10}$$

Proof: Relation (2.8) follows directly from the definition. To obtain (2.9) insert (2.6) into

$$P_r(t,s) = E\{x(t)\,\tilde{x}(s|r)'\}.$$

Since the matrix T formed by the blocks

$$T_{ij} = I\,\delta_{ij} + \Gamma_{ij}H'H$$

is nonsingular (it is the sum of an identity matrix and the product of two non-negative matrices), (2.9) has a unique solution. Finally, to prove (2.10) first note that for $t = 0,1,\ldots,r$ we can write (2.9) as

$$\sum_{i=0}^{r} T_{ti}\,P_r(i,s) = \Gamma_{ts},$$

and for the same values of t we also have

$$\sum_{i=0}^{r} T_{ti} P_{r+1}(i,s) + \Gamma_{t,r+1} H'HP_{r+1}(r+1,s) = \Gamma_{ts}.$$

Upon eliminating Γ between these two equations we obtain

$$\sum_{i=0}^{r} T_{ti} [P_{r+1}(i,s) - P_r(i,s) + P_r(i,r+1)H'HP_{r+1}(r+1,s)] = 0 \qquad (2.11)$$

and since T is nonsingular the quantities within the square brackets must be zero. Hence we can exchange T_{ti} for $\Gamma_{ti}H'H$ in (2.11), where t is now arbitrary, and apply (2.9) to cancel all sums. Then take the transpose and invoke (2.8) to obtain (2.10). ■

We may regard (2.9) as the discrete-time analog of the Fredholm integral equation occurring in the continuous-time theory and (2.10) as the corresponding *Bellman-Krein equation* [17,18]. (Also see [9].)

In Section 5 we shall need the innovation process

$$\nu(t) = z(t) - H\hat{x}(t|t-1). \qquad (2.12)$$

It is well-known and easy to show (see e.g. [25]) that ν is a white noise process

$$E\{\nu(s)\nu(t)'\} = R_t\,\delta_{ts}, \qquad (2.13)$$

where

$$R_t = I + HP_{t-1}(t,t)H'. \qquad (2.14)$$

In the same way it can be shown that

$$\mu(t) = z(t) - H\hat{x}(t|t) \qquad (2.15)$$

is also a white noise process:

$$E\{\mu(s)\,\mu(t)'\} = \bar{R}_t\;\delta_{ts}, \qquad (2.16)$$

where

$$\bar{R}_t = I - HP_t(t,t)H', \qquad (2.17)$$

but we shall postpone the discussion of this until Section 5, for the moment defining R_t and $\bar{R}_t$ by (2.14) and (2.17) respectively.

We shall be interested in the *gain functions*

$$Q_t = P_{t-1}(t,t)H' \qquad (2.18)$$

$$K_t = P_t(t,t)H' \qquad (2.19)$$

in terms of which we can write

$$R_t = I + HQ_t \tag{2.20}$$

$$\bar{R}_t = I - HK_t . \tag{2.21}$$

Then defining the *feedback function*

$$\tilde{F}_t = I - K_t H, \tag{2.22}$$

we can list the following useful relations:

Proposition 2.2: *The functions defined above are related in the following way:*

$$P_t(t,s) = \tilde{F}_t P_{t-1}(t,s) \tag{2.23}$$

$$H\tilde{F}_t = \bar{R}_t H \tag{2.24}$$

$$K_t = Q_t \bar{R}_t \tag{2.25}$$

$$\bar{R}_t = R_t^{-1} \tag{2.26}$$

$$\tilde{F}_t^{-1} = I + Q_t H \tag{2.27}$$

Proof: Putting $r = t - 1$ in (2.10) we have

$$P_t(t,s) = [I - P_t(t,t)H'H]\ P_{t-1}(t,s)$$

which is precisely (2.23). Relation (2.24) follows immediately from (2.21) and (2.22), and (2.25) is then a consequence of (2.23) and (2.24):

$$\begin{aligned} P_t(t,t)H' &= P_{t-1}(t,t)\ \tilde{F}_t'\ H' \\ &= P_{t-1}(t,t)H'\bar{R}_t \end{aligned}$$

where we have also used (2.8). To see that (2.26) holds, note that by (2.21) and (2.25)

$$\bar{R}_t = I - HQ_t\bar{R}_t$$

or

$$(I + HQ_t)\ \bar{R}_t = I$$

which together with (2.20) yields the desired result. Finally, to prove (2.27) first observe that $(I + Q_tH)$ is nonsingular, Q_tH being the product of the two nonnegative matrices $P_{t-1}(t,t)$ and $H'H$. Then apply the "matrix inversion lemma" to obtain

$$(I + Q_tH)^{-1} = I - Q_t(I + HQ_t)^{-1}H.$$

Therefore, by successively applying (2.20), (2.26) and (2.25), we have

$$(I + Q_tH)^{-1} = I - K_tH,$$

which in view of (2.22) is the same as (2.27). ∎

The reason for our interest in the functions K_t and $\tilde{F}_t$ will be made clear presently upon applying (2.23) to the representation (2.6) which yields

$$\hat{x}(t|t) = \tilde{F}_t\hat{x}(t|t-1) + K_tz(t) \tag{2.28}$$

This formula, which of course holds without any special assumptions on the x-process, constitutes the *measurement update* of the Kalman filter. (See e.g. [26].) In order to obtain the *time update* part of the filter we need to impose a Markov structure on the covariance functions of x, i.e.

$$\Gamma_{t+1,s} = F\Gamma_{ts} \quad \text{for } t \geqslant s, \tag{2.29}$$

where F is an n x n matrix. Indeed, by applying this condition to (2.9) we have

$$P_r(t+1,s) = FP_r(t,s) \tag{2.30}$$

whenever $t \geqslant s,r$, which together with (2.6) gives the time update formula

$$\hat{x}(t+1|t) = F\hat{x}(t|t). \tag{2.31}$$

Hence we can combine (2.28) and (2.31) to obtain

$$\hat{x}(t+1|t) = F\tilde{F}_t\hat{x}(t|t-1) + FK_tz(t) \tag{2.32}$$

which is the Kalman filter formula for the one-step predictor. Of course we may instead want the pure filtering formula

$$\hat{x}(t|t) = \tilde{F}_tF\,\hat{x}(t-1|t-1) + K_tz(t). \tag{2.33}$$

In any case, the application of these recursive filtering formulas requires determining the gain function K_t, which is usually done by solving a matrix Riccati equation. In this paper, however, we shall take a different course and, for the case when x is stationary, develop a different set of equations for K_t from basic principles. This will be done in the following sections.

Finally, we should point out that of course the results of this section do not require that the matrices H and F be constant as our notations suggest. However we have deliberately left out the time arguments since the main result of this paper concerns stationary processes.

3. Reversed-time estimation of stationary processes

In the sequel we shall assume that the covariance function (2.1) is given by

$$\Gamma_{ts} = C_{t-s}, \tag{3.1}$$

i.e. the process x is wide sense stationary. Clearly the sequence C_t must satisfy

$$C_{-t} = C_t' \,. \tag{3.2}$$

The error covariance function P is uniquely determined by C, being the unique solution of (2.9) with Γ given by (3.1). To remind ourselves of this fact, we may write P[C] although we shall refrain from this whenever there is no reason for misunderstanding.

Much of what follows will heavily rely on the simple observation that P* defined by

$$P_r^*(t,s) = P_r(r-t, r-s) \tag{3.3}$$

is also an error covariance function of type (2.5) and that consequently we can define "starred" versions of the quantities defined in Section 2 by merely exchanging P for P* everywhere. In fact, we have

Proposition 3.1: $$P_r^*[C] = P_r[C'] \tag{3.4}$$

Proof: Insert (3.1) into (2.9), make a simple change in the order of summation, and observe (3.2) to see that P* is the unique solution of (2.9) with $\Gamma_{ts} = C_{t-s}'$. (This new Γ is clearly a covariance function.) ■

Hence the new functions K*, Q*, $\overline{R}$*, $\tilde{F}$* etc. have precise meanings. We note that the starred version of the important relation (2.10) is unchanged since it does not depend on C, and that the star operation applied twice gives us the original quantity back (for P** = P). Also note that the star operation degenerates for m = 1, so that the starred quantities are equal to the unstarred.

As explained in Section 2 we shall be interested in obtaining equations for the gain function K_t. Since (2.25), (2.26) and (2.20) provide us with the relation

$$K_t = Q_t(I + HQ_t)^{-1}, \tag{3.5}$$

equations in Q_t will also serve our purpose. Hence we shall develop several sets of equations in both K_t and Q_t and explain the relation between them.

Our basic tool is the Bellman-Krein type equation (2.10) presented in the previous section, i.e.

$$P_{r+1}(t,s) = P_r(t,s) - P_{r+1}(t,r+1)H'HP_r(r+1,s) \tag{3.6}$$

In order to determine K_t and Q_t, we need $P_t(t,t)$ and $P_{t-1}(t,t)$ respectively. Equ. (3.6) only provides us with a recursion in the first (index) argument, but by introducing the Markov structure (2.29) we can readily derive a recursion updating all three arguments. This will lead to the *Riccati equation*, the non-linear term of which is supplied by (3.6). We shall however proceed in a different direction:

First note that

$$P_{t-1}(t,t) = P_{t-1}^*(-1,-1)$$

and that

$$P_t(t,t) = P_t^*(0,0) \,.$$

Therefore the *starred version* of (3.6) will immediately provide us with a recursion of desired type, for only the index argument need to be updated. Indeed,

$$P_t^*(-1,-1) = P_{t-1}^*(-1,-1) - P_t^*(-1,t)H'HP_{t-1}^*(t,-1) \tag{3.7}$$

and

$$P_{t+1}^*(0,0) = P_t^*(0,0) - P_{t+1}^*(0,t+1)H'HP_t^*(t+1,0) \tag{3.8}$$

Since, in view of (3.3) and (2.8), the last terms of (3.7) and (3.8) can be written as

$$P_t(t+1,0)H'HP_{t-1}(t,-1)'$$

and

$$P_{t+1}(t+1,0)H'HP_t(t,-1)'$$

respectively, this suggests introducing the following *auxiliary* functions formed in analogy with Q_t and K_t,

$$U_t = P_{t-1}(t,-1)H' \tag{3.9}$$

$$V_t = P_t(t+1,0)H' \tag{3.10}$$

$$X_t = P_t(t,-1)H' \tag{3.11}$$

$$Y_t = P_{t+1}(t+1,0)H', \tag{3.12}$$

the physical interpretation of which will be made clear in Section 5 upon introducing the *backward innovation processes.* Therefore, postmultiplying (3.7) and (3.8) by H', we obtain

$$Q_{t+1} = Q_t - V_tU_t'H' \tag{3.13}$$

$$K_{t+1} = K_t - Y_tX_t'H' \tag{3.14}$$

and, in view of (2.20) and (2.21),

$$R_{t+1} = R_t - HV_tU_t'H' \tag{3.15}$$

$$\overline{R}_{t+1} = \overline{R}_t + HY_tX_t'H'. \tag{3.16}$$

Consequently, it remains to determine the functions U_t, V_t, X_t and Y_t, and to this end we shall first investigate the relation between them:

Lemma 3.2: *The functions* U_t, V_t, X_t *and* Y_t *are related in the following way:*

$$X_t = \tilde{F}_tU_t \tag{3.17}$$

$$Y_t = \tilde{F}_{t+1}V_t \tag{3.18}$$

$$V_t = U_t\overline{R}_t^* \tag{3.19}$$

$$Y_t = X_t\overline{R}_{t+1}^* \; . \tag{3.20}$$

Proof: Relations (3.17) and (3.18) follow directly from (2.23). To show that (3.19) holds, observe that V_t can be written

$$P_t^*(-1,t)H'$$

and therefore we can invoke the starred version of (2.23) and (2.24) together with (2.8) to obtain

$$V_t = P_{t-1}^*(-1,t)\tilde{F}_t^{*\prime}H'$$

$$= P_{t-1}^*(-1,t)H'\bar{R}_t^*$$

which is equal to the right member of (3.19). In the same way we prove (3.20). ■

Lemma 3.2 provides us with a means to obtain any one of the quantities U_t, V_t, X_t and Y_t in terms of any of the others by a linear transformation, for both $\tilde{F}_t$ and $\bar{R}_t^*$ are clearly invertible, $\tilde{F}_t^{-1}$ being given by (2.27) and $\bar{R}_t^*$ being equal to $(R_t^*)^{-1}$ by † (2.26)*. These transformations can now be used to reformulate (3.13)–(3.16) so that only one auxiliary function is needed. This can be done in several obvious ways and we shall return to this in the next sections. In this context we may also note that (3.17) and (3.18) together with (2.24) give us

$$HX_t = \bar{R}_t HU_t \tag{3.21}$$

$$HY_t = \bar{R}_{t+1} HV_t, \tag{3.22}$$

where we should remember that, by (2.26), $\bar{R}_t$ equals R_t^{-1}. However, while $\tilde{F}_t$ and R_t together with their inverses can be determined via the recursions developed sofar, R_t^* is as yet an unknown quantity.

To determine R_t^* and $\bar{R}_t^*$ we need starred versions of (3.15) and (3.16). We can obtain these by the simple observation that

$$HU_t^* = (HU_t)' \tag{3.23}$$

$$HV_t^* = (HX_t)' \tag{3.24}$$

$$HX_t^* = (HV_t)' \tag{3.25}$$

$$HY_t^* = (HY_t)', \tag{3.26}$$

which is an immediate consequence of the definitions and (3.3). Therefore, the appropriate modifications of (3.15) and (3.16) yield

$$R_{t+1}^* = R_t^* - X_t'H'HU_t \tag{3.27}$$

$$\bar{R}_{t+1}^* = \bar{R}_t^* + Y_t'H'HV_t \tag{3.28}$$

We should however point out that, in view of (2.14)* and (2.17)*, we have

$$R_t^* = I + HP_{t-1}(-1,-1)H'$$

and

$$\bar{R}_t^* = I - HP_t(0,0)H',$$

---

† (a)* means "the starred version of (a)"

so that in fact (3.31) and (3.32) can be obtained directly from the *unstarred* Bellman-Krein type equation (3.6), the stationarity assumption being unnecessary in this case. Equations (3.27) and (3.28) can also be reformulated using the transformations of Lemma 3.2, but we shall postpone the discussion of this to the next section.

The above results leave us with the problem to find recursions for the auxiliary functions U_t, V_t, X_t and Y_t. The transformations of Lemma 3.2 only provide *static* relations between these functions in that they relate quantities with the same time index. To obtain *dynamic* relations we introduce the Markov condition (2.29) which in our present (stationary) setup reads

$$C_t = F^t C_0 \quad \text{for } t \geqslant 0, \tag{3.29}$$

where F and C_0 are constant n x n - matrices. (For $t > 0$, C_t is defined through (3.2).) Then relation (2.30) holds, and, in view of the definitions (3.9)-(3.12), we have:

$$U_{t+1} = FX_t \tag{3.30}$$

$$V_{t+1} = FY_t, \tag{3.31}$$

which together with (3.17) and (3.18) provide us with recursions for the auxiliary functions. We shall return to this in the next section.

However, let us first summarize the results of this section in the following theorem:

Theorem 3.3: *Assume that* x *is wide sense stationary. Then the gain functions* K_t *and* Q_t *defined by* (2.18) *and* (2.19) *and related through* (3.5), *satisfy recursion* (3.13) *and* (3.14) *respectively. The auxiliary functions* U_t, V_t, X_t *and* Y_t *are related by the linear* static *transformations* (3.17) - (3.22), *where* $\tilde{F}_t$ *and* $\tilde{F}_t^{-1}$ *are given by* (2.22) *and* (2.27), R_t^* *and its inverse* $\bar{R}_t^*$ *satisfy the recursions* (3.27) *and* (3.28), *and* R_t *and its inverse* $\bar{R}_t$ *can be determined either from* (2.20) *and* (2.21) *or by the recursions* (3.15) *and* (3.16). *Also given the Markov condition* (3.29), *we have the* dynamic *relations* (3.30) *and* (3.31) *for the auxiliary functions, which can then also be determined recursively.*

4. Algorithms for the Kalman-Bucy gain

Equipped with the results of the previous section we are now in a position to formulate algorithms for the gain function K_t of the optimal filters (2.32) and (2.33). Of course we assume that x is (wide sense) stationary with covariance function (3.29). Since K_t can be determined from Q_t by means of (3.5) we shall be interested in equations in Q_t also.

The form of the algorithm will primarily depend on our choice of auxiliary functions. So, for example, using U_t we have the following algorithm for Q_t:

$$Q_{t+1} = Q_t - U_t \bar{R}_t^* U_t' H' \tag{4.1}$$

$$U_{t+1} = F\tilde{F}_t U_t \tag{4.2}$$

$$\bar{R}_{t+1}^* = \bar{R}_t^* + \bar{R}_t^* U_t' H' \bar{R}_{t+1}^{-1} H U_t \bar{R}_t^*, \tag{4.3}$$

where $\tilde{F}_t$ and R_t stand for

$$\tilde{F}_t = I - Q_t(I + HQ_t)^{-1} \tag{4.4}$$

and

$$R_t = I + HQ_t. \tag{4.5}$$

To obtain (4.1) and (4.2) we have substituted (3.19) into (3.13) and (3.17) into (3.30) respectively. To see that (4.3) holds, first insert (3.22) into (3.28) which gives us

$$\bar{R}^*_{t+1} = \bar{R}^*_t + V'_t H' R^{-1}_{t+1} HV_t. \tag{4.6}$$

Then use transformation (3.19) to obtain (4.3). Relations (4.4) and (4.5) follow immediately from (2.22), (3.5) and (2.20). Instead of (4.3) we could use

$$R^*_{t+1} = R^*_t - U'_t H' R^{-1}_t HU_t, \tag{4.7}$$

obtained by plugging (3.21) into (3.27), but then we would have to invert R^*_t in each step to get the inverse $\bar{R}^*_t$. (In fact, (4.3) can be determined from (4.7) by using the matrix inversion lemma.) Since $P_{-1}(0,0)$ and $P_{-1}(-1,-1)$ both equal C_0 we have the following initial conditions:

$$Q_0 = C_0 H' \tag{4.8}$$

$$U_0 = FC_0 H' \tag{4.9}$$

$$R^*_0 = I + HC_0 H', \tag{4.10}$$

$\bar{R}^*_0$ being simply the inverse of R^*_0. In determining (4.9) we have also used (2.30).

Now, with (4.4) and (4.5) properly inserted, (4.1)–(4.3) provide us with $2mn + \frac{1}{2}m(m+1)$ scalar equations to determine Q_t, for Q_t and U_t are $m \times n$ - matrices and $\bar{R}^*_t$ is a symmetric $m \times m$ - matrix. When $m = 1$ the star operation (3.3) degenerates so that R^*_t is given by (4.5). Then the $\bar{R}^*_t$- equation (4.3) becomes superfluous and only $2n$ scalar equations are needed. This should be compared with the $\frac{1}{2}n(n+1)$ scalar equations of the Riccati equation ; a much larger number whenever, as often is the case, $m \ll n$.

Other versions of the above algorithm can now be constructed by instead using a different auxiliary function. From (3.13), (3.19), (3.31) and (3.18) we have

$$Q_{t+1} = Q_t - V_t R^*_t V'_t H' \tag{4.11}$$

$$V_{t+1} = F\tilde{F}_{t+1} V_t \tag{4.12}$$

$$R^*_{t+1} = R^*_t - R^*_t V'_t H' R^{-1}_t HV_t R^*_t, \tag{4.13}$$

the last equation of which can be obtained by applying (3.19) to (4.7). Again we have used the short-hand notation (4.4) and (4.5), and the initial condition V_0 is provided by (3.19)

$$V_0 = FC_0 H'(I + HC_0 H)^{-1}. \tag{4.14}$$

We could use (4.6) instead of (4.13) but then again we would have to take the inverse in each step.

Yet another version is provided by

$$Q_{t+1} = Q_t - FX_{t-1}(R_t^*)^{-1}X_{t-1}'F'H' \tag{4.15}$$

$$X_t = \widetilde{F}_t FX_{t-1} \tag{4.16}$$

$$R_{t+1}^* = R_t^* - X_t'H'HFX_{t-1}. \tag{4.17}$$

These equations are simply obtained by applying the transformation (3.30) to (4.1), (3.17) and (3.27) respectively. Equations (3.17) and (4.9) give us the initial condition

$$X_0 = \widetilde{F}_0 FC_0 H'. \tag{4.18}$$

Note that the essential difference between this algorithm and (4.1) - (4.3) is the equation for R_t^*, and we can get still other versions by instead using the symmetric form

$$R_{t+1}^* = R_t^* - X_t'H'R_t HX_t \tag{4.19}$$

derived from (3.21) and (3.27), or the inverse recursion

$$\overline{R}_{t+1}^* = \overline{R}_t^* + \overline{R}_t^* X_t'H'R_t R_{t+1}^{-1} R_t HX_t \overline{R}_t^*, \tag{4.20}$$

obtained by applying transformation (3.21) to (4.3).

Proceeding exactly analogously we also have

$$Q_{t+1} = Q_t - FY_{t-1}(\overline{R}_t^*)^{-1}Y_{t-1}'F'H' \tag{4.21}$$

$$Y_t = \widetilde{F}_{t+1} FY_{t-1} \tag{4.22}$$

$$\overline{R}_{t+1}^* = \overline{R}_t^* + Y_t'H'HFY_{t-1}, \tag{4.23}$$

with the alternative of using

$$\overline{R}_{t+1}^* = \overline{R}_t^* + Y_t'H'R_{t+1}HY_t \tag{4.24}$$

or

$$R_{t+1}^* = R_t^* - R_t^* Y_t'H'R_{t+1}R_t^{-1}R_{t+1}HY_t R_t^* \tag{4.25}$$

in place of (4.23). The initial condition of (4.17) is

$$Y_0 = \widetilde{F}_1 FC_0 H'(I + HC_0 H')^{-1}. \tag{4.26}$$

We can also formulate algorithms directly in terms of the gain K_t. We have, for example,

$$K_{t+1} = K_t - X_t(R_{t+1}^*)^{-1}X_t'H' \tag{4.27}$$

$$X_{t+1} = \widetilde{F}_{t+1} FX_t \tag{4.28}$$

$$R_{t+1}^* = R_t^* - X_t'H'(\overline{R}_t)^{-1}HX_t, \tag{4.29}$$

into which we should substitute $\tilde{F}_t$ and $\bar{R}_t$ as originally defined by (2.22) and (2.21):

$$\tilde{F}_t = I - K_t H \tag{4.30}$$

$$\bar{R}_t = I - HK_t. \tag{4.31}$$

Equation (4.27) is immediately obtained from (3.14) and (3.20), and (4.28) and (4.29) are identical to (4.16) and (4.19) respectively. Instead of (4.29) we could use (4.17), thereby avoiding one inversion. Equ. (4.20) on the other hand could *not* be used, since we need K_{t+1} to determine R^*_{t+1} which in turn is needed to determine K_{t+1}. The initial condition

$$K_0 = C_0 H'(I + HC_0 H')^{-1} \tag{4.32}$$

is obtained from (3.5) and (4.8).

Similarily, we can also express K_t in terms of Y_t:

$$K_{t+1} = K_t - Y_t R^*_{t+1} Y_t' H', \tag{4.33}$$

but this equation is unfortunately unusable since the corresponding auxiliary equation (4.22) requires knowledge about K_{t+1}.

We should point that from a computional point of view a separate equation for R_t or $\bar{R}_t$ is usually to be preferred. However, in view of (4.5) and (4.31), such a recursion is immediately obtained by pre-multiplying the appropriate Q_t or K_t equation by H.

The various sets of equations presented above are of course essentially different versions the same algorithm first presented in [7], where we gave the first version (4.1)-(4.3). In the subsequent paper [6] by Kailath et. al., both (4.1)-(4.3) and (4.11) – (4.13) were obtained by two different decompositions of the Riccati equation, but the relation between them was not clearly explained. The other versions, among which the ones in K_t are of particular interest, seem to appear here for the first time.

The purpose of this section has been to demonstrate that the continuous-time algorithm [4,5,8], to which we shall briefly return in Section 6, have *many* counterparts in discrete time (*some* of which we have presented in this section) and to determine the relation between them. We are not at this time prepared to comment on the computional properties of the different versions.

5. Forward and backward innovations

We shall now give a stochastic interpretation of the results presented above. To this end let us first summarize some useful facts about the innovation processes (2.12) and (2.15):

Proposition 5.1: *The processes ν and μ, defined by* (2.12) *and* (2.15) *respectively, satisfy* (2.13), (2.16) *and*

$$E\{\mu(t)\,\nu(s)'\} = I\,\delta_{ts}. \tag{5.1}$$

Moreover we have the innovations representations

$$\hat{x}(t|r) = \sum_{s=0}^{r} P_s(t,s)H'\nu(s) \tag{5.2}$$

and

$$\hat{x}(t|r) = \sum_{s=0}^{r} P_{s-1}(t,s)H'\mu(s), \tag{5.3}$$

where P *is defined by* (2.5).

Proof: The proof of (2.13) is straight-forward using orthogonality arguments and we refer the reader to [25] for it. We can prove (2.16) and (5.1) along the same lines, but instead we shall use relation (2.28) to write

$$\mu(t) = z(t) - H\tilde{F}_t\hat{x}(t|t-1) - HK_t\, z(t).$$

Therefore we can invoke (2.24) and (2.21) to obtain

$$\mu(t) = \bar{R}_t\nu(t), \tag{5.4}$$

which, in view of (2.13) and (2.26), gives us (2.16) and (5.1). It is easy to see [25] that z(t) can be expressed as a linear function of $\nu(0), \nu(1), \ldots, \nu(t)$ so that

$$\hat{x}(t|r) = \sum_{i=0}^{r} \Lambda_i\, \nu(i)$$

for some weighting function Λ. Since

$$E\{\tilde{x}(t|r)\mu(s)'\} = 0 \qquad (s = 0, 1, \ldots, r),$$

using (5.1), we have

$$E\{x(t)\,\tilde{x}(s|s)'\}H' - \Lambda_s = 0,$$

and therefore (5.2) holds. To prove (5.3) we proceed analogously, first noting that, in view of (5.4), z(t) is also a linear function of $\mu(0), \mu(1), \ldots, \mu(t)$. ■

Remark 5:2: We are now in a position to give an alternative (innovations) proof of the Bellman-Krein type equation (2.10). In fact, using representation (5.2) for $\hat{x}(t|r)$ and (5.3) for $\hat{x}(s|r)$, we have

$$E\{\hat{x}(t|r)\,\hat{x}(s|r)'\} = \sum_{i=0}^{r} P_i(t,i)H'HP_{i-1}(i,s) \tag{5.5}$$

where we have also used (5.1) and (2.8). Then (2.10) is obtained by inserting (5.5) into

$$P_r(t,s) = E\{x(t)x(s)'\} - E\{\hat{x}(t|r)\,\hat{x}(s|r)'\} \tag{5.6}$$

and making the appropriate reformulation. Likewise by using *one* of the representations (5.2) and (5.3) for *both* $\hat{x}(t|r)$ and $\hat{x}(s|r)$, we can also derive two symmetric versions of (2.10), both of which can also be obtained directly from (2.10) by using (2.23) and (2.24). ■

We shall now proceed to the main result of this section: Consider the string of data

$$\{z(0), z(1), z(2), \ldots, z(T)\} \tag{5.7}$$

where z, of course, is defined by (2.2). The innovation approach amounts to orthogonalizing the data (5.7) and to express the estimate in terms of the so constructed innovation process. This orthogonalization can be performed in two obvious ways: either start with z(0) and proceed in the forward direction up to z(T) or begin with z(T) and go backwards. The former procedure will provide us with a (forward) innovation process such as (2.12) or (2.15), while the latter will give us a *backward innovation process.*

To formalize this idea we define the processes $x_T(t)$ and $w_T(t)$ to be equal to $x(T-t)$ and $w(T-t)$ respectively. Then, defining z_T analogously, we have the following counterpart of (2.2)

$$z_T(t) = Hx_T(t) + w_T(t). \tag{5.8}$$

Hence we shall consider the linear least squares estimate $\hat{x}_T(t|r)$ of $x_T(t)$ given the data

$$\{z_T(0), z_T(1), z_T(2), \dots, z_T(r)\}, \tag{5.9}$$

and the corresponding estimation error

$$\tilde{x}_T(t|r) = x_T(t) - \hat{x}_T(t|r). \tag{5.10}$$

We can now define the backward innovation processes

$$\nu_T(t) = z_T(t) - H\hat{x}_T(t|t-1) \tag{5.11}$$

and

$$\mu_T(t) = z_T(t) - H\hat{x}_T(t|t), \tag{5.12}$$

for which we have the following result:

Theorem 5.3: *Assume that* x *is wide sense stationary. Then, for all* T, *the error covariance*

$$E\{\tilde{x}_T(t|r)\tilde{x}_T(s|r)'\} = P_r^*(t,s), \tag{5.13}$$

where P^* *is defined by* (3.3) *and* (2.5). *The innovation processes* (5.11) *and* (5.12) *satisfy*

$$E\{\nu_T(t)\,\nu_T(s)'\} = R_t^*\,\delta_{ts}, \tag{5.14}$$

$$E\{\mu_T(t)\,\mu_T(s)'\} = \bar{R}_t^*\,\delta_{ts}, \tag{5.15}$$

and

$$E\{\mu_T(t)\,\nu_T(s)'\} = I\,\delta_{ts}, \tag{5.16}$$

where R_t^* *and its inverse* $\bar{R}_t^*$ *are defined as in Section* 3. *Moreover, we have the following representations for* $\hat{x}(t|r)$:

$$\hat{x}(t|r) = \sum_{s=0}^{r} P_s(s+t-r, 0)\,H'\nu_r(s) \tag{5.17}$$

and

$$\hat{x}(t|r) = \sum_{s=0}^{r} P_{s-1}(s+t-r-1, -1)\,H'\mu_r(s), \tag{5.18}$$

where P *is defined by* (2.5).

Proof: Since w_T has the same covariance function (2.3) as w, the only difference between the problem to determine $\hat{x}_T(t|r)$ and the estimation problem considered earlier in this paper is that we now have x_T where we previously had x. However, given that x has the covariance function (3.1), we can use (3.2) to see that

$$E\{x_T(t)x_T(s)'\} = C'_{t-s},$$

which does not depend on T. Therefore (5.13) immediately follows from Proposition 3.1. Consequently, (5.14), (5.15) and (5.16) are merely starred versions of (2.13), (2.16) and (5.1). Also,

$$\hat{x}_r(t|r) = \sum_{s=0}^{r} P_s^*(t,s)H'\nu_r(s) \tag{5.19}$$

is the starred version of (5.2) taking T to be r. Since the data (5.7) and (5.9) coincide for T = r, we have

$$\hat{x}(t|r) = \hat{x}_r(r-t|r), \tag{5.20}$$

and therefore, in view of (3.3), (5.19) is the same as (5.17). The proof of (5.17) is analogous. ■

Remark 5.4: If we wish to make the above proof independent of any result in Sections 2 and 3, we may simply note that the stationarity of x implies that the left member of (5.13) is independent of T so that we can put T = r. Then we can use (5.20) to see that this error covariance is $P_r(r-t, r-s)$, which is precisely equal to the right member of (5.13). ■

However, our prime interest is in the one-step predictor and the pure filter. Therefore we shall now invoke the representations (5.17) and (5.18) to obtain

$$\hat{x}(t+1|t) = \sum_{s=0}^{t} U_s \mu_t(s) \tag{5.21}$$

$$= \sum_{s=0}^{t} V_s \nu_t(s) \tag{5.22}$$

and

$$\hat{x}(t|t) = \sum_{s=0}^{t} X_{s-1}\mu_t(s) \tag{5.23}$$

$$= \sum_{s=0}^{t} Y_{s-1}\nu_t(s), \tag{5.24}$$

where U,V,X and Y are defined by (3.9) - (3.12), hence providing the previously mentioned stochastic interpretation of the auxiliary functions introduced in Section 3. We can now use these representations to derive the equations for Q_t and K_t presented in Sections 3 and 4. In fact, inserting (5.21) and (5.22) in appropriate combinations into

$$Q_t = C_0H' - E\{\hat{x}(t|t-1)\,\hat{x}(t|t-1)'\}H'$$

(obtained from (5.6)), and observing (5.14), (5.15) and (5.16), we obtain (4.1), (4.11) and (3.13) respectively. Likewise we derive equations (4.27), (4.33) and (3.14) by substituting (5.23) and (5.24) into

$$K_t = C_0 H' - E\{\hat{x}(t|t)\,\hat{x}(t|t)'\}H'.$$

To derive equations for R_t^* and its inverse $\overline{R}_t^*$, we shall need *forward* innovation representations for $\hat{x}_t(t+1|t)$ and $\hat{x}_t(t|t)$. Therefore, invoking (5.2), (5.3), (2.8) and (5.20), we have

$$H\hat{x}_t(t+1|t) = \sum_{s=0}^{t} U_s' H'\mu(s) \tag{5.25}$$

$$= \sum_{s=0}^{t} X_s' H'\nu(s) \tag{5.26}$$

and

$$H\hat{x}_t(t|t) = \sum_{s=0}^{t} Y_{s-1}' H'\nu(s) \tag{5.27}$$

$$= \sum_{s=0}^{t} V_{s-1}' H'\mu(s), \tag{5.28}$$

which plugged into

$$R_t^* = I + HE\{\hat{x}_{t-1}(t|t-1)\hat{x}_{t-1}(t|t-1)'\}H' \tag{5.29}$$

and

$$\overline{R}_t^* = I - HE\{\hat{x}_t(t|t)\hat{x}_t(t|t)'\}H' \tag{5.30}$$

in different combinations, yield (3.27), (3.28), (4.6), (4.7), (4.19) and (4.24). Here (5.29) and (5.30) have been obtained by inserting (5.13) with appropriate choice of T and r into the starred versions of (2.14) and (2.17) respectively.

In this section we have so far made no use of the Markov condition (3.29), the need of which enters upon deriving the auxiliary equations (4.2), (4.12), (4.16) and (4.22). In our present stochastic setting the most suitable way to introduce this condition is to assume (as usual) that x is generated by the stochastic difference equation

$$x(t+1) = Fx(t) + v(t)$$

where v is a white sequence. It is then well-known that $\hat{x}(t+1|t)$ is given by (2.32) and consequently

$$\tilde{x}(t+1|t) = F\tilde{F}_t\tilde{x}(t|t-1) + v(t) - FK_t w(t)$$

which inserted into

$$U_t = E\{\tilde{x}(t|t-1)\,x(-1)'\}H'$$

gives us (4.2) (for $x(-1)$ is uncorrelated with v and w). Similar arguments give us the other auxiliary equations (4.12), (4.16) and (4.22). However for X_t and Y_t we must use the pure filtering formula (2.33) instead.

The main purpose of this section has been to provide a stochastic framework for the previous, essentially nonrandom, development. However, despite frequent references to results in Sections 2–4, the presentation has been essentially self-contained.

6. A remark on the continuous-time result

For the sake of comparison we shall *briefly* outline the continuous-time analog of the development in the previous sections. The forward-backward innovation method has already been described in continuous time in [8], so we shall only have to consider the analog of the method of Sections 3 and 4.

The two gain functions K and Q defined in Section 2 have only one counterpart in continuous time, namely

$$K(t) = P_t(t,t)H'. \tag{6.1}$$

Likewise, there is only one auxiliary function

$$Y(t) = P_t(t,0)H', \tag{6.2}$$

corresponding to the functions U,V,X and Y defined by (3.9)-(3.12).

We have the following *Bellman-Krein* equation:

$$\frac{\partial P_r}{\partial r}(t,s) = -P_r(t,r)H'HP_r(r,s) \tag{6.3}$$

to replace (3.10), the starred version of which gives us

$$\frac{\partial P_t^*}{\partial t}(0,0) = -P_t^*(0,t)H'HP_t^*(t,0),$$

which, in view of (3.3) and "(2.8)", provides us with the counterpart of (3.13) and (3.14), namely

$$\dot{K}(t) = -Y(t)\, Y(t)'H'. \tag{6.4}$$

To determine an equation for Y, first note that (6,3) implies

$$\left[\frac{\partial P_t}{\partial t}(s,0)H'\right]_{s=t} = -K(t)HY(t). \tag{6.5}$$

Then, to proceed we must introduce the counterpart of (2.30): the Markov condition

$$\frac{\partial P_r}{\partial t}(t,s) = FP_r(t,s) \quad \text{for } t > r,s, \tag{6.6}$$

which with a simple limit argument gives us

$$\left[\frac{\partial P_t}{\partial s}(s,0)H'\right]_{s=t} = FY(t) \tag{6.7}$$

Hence (6.5) and (6.7) provide us with the counterpart of (4.2), (4.12), (4.16) and (4.22), namely

$$\dot{Y}(t) = [F - K(t)H]\, Y(t)\ . \tag{6.8}$$

Equations (6.4) and (6.8) together with their initial conditions $K(0) = Y(0) = C_0H'$ constitute the continuous-time analog of *all* algorithms of Section 4. These matrix differential equations, which contain 2n scalar equations, were first obtained (independently of our own work [7]) by Kailath [4] who derived them from the Riccati equation.

References

1. J. Casti, R. Kalaba and V.K. Murthy, A new initial value method for on–line filtering and estimation, IEEE Trans. on Information Theory, July 1972, pp. 515–517.

2. J. Rissanen, A fast algorithm for optimum linear prediction, IEEE Transactions Automatic Control, vol. AC–18, p.555, Oct. 1973.

3. J. Casti and E. Tse, Optimal linear filtering theory and radiative transfer: Comparisons and interconnections, J. Math. Anal. and Appl., vol. 40, October 1972, pp. 45–54.

4. T. Kailath, Some Chandrasekhar–type algorithms for quadratic regulators, Proc. IEEE Decision and Control Conference, Dec. 1972.

5. T. Kailath, Some new algorithms for recursive estimation in constant linear systems, IEEE Trans. Information Theory, Vol. IT–19, November 1973, pp. 750–760.

6. T. Kailath, M. Morf and S. Sidhu, Some new algorithms for recursive estimation in constant discrete–time linear systems, Proc. Seventh Princeton Symp. on Information and System Science, March 1973.

7. A. Lindquist, A new algorithm for optimal filtering of discrete–time stationary processes, SIAM J. Control (to appear).

8. A Lindquist, Optimal filtering of continuous–time stationary processes by means of the backward innovation process, SIAM J. Control (to appear).

9. A. Lindquist, On Fredholm Integral Equations, Toeplitz Equations and Kalman–Bucy Filtering, International J. Appl. Math. and Optimization (to appear).

10. N. Levinson, The Wiener RMS (root mean square) error in filter design and prediction, Appendix B in N. Wiener, Extrapolation, Interpolation and Smoothing of Stationary Time Series, M.I.T. Press, Cambridge, Mass., 1942.

11. P. Whittle, On the fitting of multivariate autoregressions and the approximate canonical factorization of a spectral density matrix, Biometrica, vol. 50, 1963, pp. 129–134.

12. R.A. Wiggins and E.A. Robinson, Recursive solutions to the multichannel filtering problem, J. Geophys. Research 70, pp. 1885–1891 (1965).

13. L. Ya. Geronimus, Orthogonal Polynomials, Consultant Bureau, New York, 1961.

14. N.I. Akiezer, The Classical Moment Problem, Hafner Publishing Company, New York 1965.

15. U. Grenander and G. Szego, Toeplitz Forms and their Applications, University of California Press, Berkeley and Los Angeles, 1958.

16. S. Chandrasekhar, Radiative Transfer, Oxford University Press, 1950.

17. R. Bellman, Functional equations in the theory of dynamic programming–VII: A partial differential equation for the Fredholm resolvent, Proc. Amer. Math. Soc., 8, 1957, pp. 435–440.

18. M. Krein, On a new method for solving linear integral equations of the first and second kinds, Dokl. Akad. Nauk.SSSR, 100, 1955, pp. 413–416.

19. M. Krein, Continuous analogs of propositions concerning polynomials orthogonal on the unit circle, Dokl. Akad. Nauk. SSSR, 105, 1955, pp.637–640.

20. R.S. Bucy and P.D. Joseph, Filtering for Stochastic Processes with Applications to Guidance, Interscience Publishers, New York 1968.

21. D.Q. Mayne, A solution of the smoothing problem for linear dynamic systems, Automatica 4, 1966, pp. 73–92.

22. D.C. Fraser, On the application of optimal linear smoothing techniques to linear and nonlinear dynamic systems, PhD Thesis, M.I.T., Cambridge, Mass., 1967.

23. D.C. Fraser and J.E. Potter, The optimum linear smoother as a combination of two optimum linear filters, IEEE Trans. AC, August 1969.

24. R.K. Mehra, On optimal and suboptimal linear smoothing, Proc. Nat. Electr. Conf. XXIV, 1968.

25. T. Kailath, An innovations approach to linear least squares estimation, Part I: Linear filtering in additive white noise, IEEE Trans. AC–13, 1968, pp. 646–655.

26. P.G. Kaminski, A.E. Bryson and S.F. Schmidt, Discrete square root filtering: A survey of current techniques, IEEE Trans. AC–16, 1971, pp. 727–736.

FILTRAGE NUMERIQUE RECURSIF NON LINEAIRE : RESOLUTION DU PROBLEME MATHEMATIQUE ET APPLICATIONS

F. LEVIEUX

LABORIA-IRIA

(FRANCE)

Résumé

Cette étude se propose de poursuivre deux objectifs nettement différenciés. D'abord donner un cadre mathématique précis au modèle utilisé classiquement en filtrage récursif non-linéaire, en étudiant la solution d'une équation parabolique non-linéaire introduite formellement par H. KUSHNER pour décrire l'évolution de la densité de probabilité conditionnelle d'un processus de diffusion observé à l'aide d'un processus de ITO. Ensuite faire l'analyse numérique du problème pour définir un algorithme numériquement stable, qui fournisse une approximation de cette densité de probabilité.

I. Introduction.

Le problème du filtrage numérique récursif non-linéaire est généralement présenté comme une extension du cas linéaire et des solutions récursives proposées par R.E. KALMAN et R.S. BUCY.

Le modèle choisi est le suivant : Soit Ω un espace d'épreuves ω, μ une mesure sur Ω, $X(t)$ un processus vectoriel markovien, de dimension n, défini par l'équation différentielle stochastique de ITO :

$$(1.1) \qquad d\,X(t) = f_X(X(t),t)\,dt + g_X\,(X(t),t)\,d\,b_1(t)$$

où f_X est une fonction vectorielle de dimension n, g_X une fonction matricielle de dimension $n \times n$, b_1 un mouvement brownien vectoriel unitaire de dimension n. L'équation (1.1) décrit l'évolution de l'état d'un système dynamique perturbé par un bruit. Le processus $X(t)$ est observé par un appareillage que l'on décrit par une autre équation de ITO :

$$(1.2) \qquad dz(t) = H(X(t),t)\,dt + d\,b_2(t)$$

où $b_2(t)$ est aussi un mouvement brownien unitaire, scalaire et indépendant de b_1, et $H(X(t),t)$ une fonction scalaire.

Comme l'a montré formellement H. KUSHNER (voir [1]), le problème du filtrage non-linéaire est résolu si l'on peut résoudre en p, l'équation aux dérivées partielles suivantes :

$$(1.3)\quad d\, p(x,t) + A(t)\, p(x,t)\, dt = p(x,t)\, \Big[H(x,t) - \int_{R^n} H(\alpha,t)\, p(\alpha,t)\, d\alpha\Big] \Big[dz(t) - dt \int_{R^n} H(\alpha,t)\, p(\alpha,t)\, d\alpha\Big]$$

où $A(t)$ est l'opérateur de Fokker-Planck associé à l'équation (1.1), et décrit par la relation :

$$(1.4)\quad A(t). = \sum_{i=1}^{n} \frac{\partial\, [f_i\ .]}{\partial\, x_i} - \frac{1}{2} \sum_{i,j=1}^{n} \frac{\partial^2 [g_{ij}\ .]}{\partial\, x_i\, \partial\, x_j} ,$$

où les f_i sont les composantes de f_x et g_{ij} les composantes de $g_x\, g'_x$.

L'étude du problème mathématique requiert deux étapes : une étude d'un cas simplifié déterministe, qui est l'objet du deuxième paragraphe, et l'étude de l'équation stochastique complète utilisant les résultats de l'étape précédente.

II. Résultats préliminaires dans un cadre déterministe.

Le résultat suivant précise dans un cadre déterministe les propriétés des trajectoires mesurées du processus. Il s'énonce de la façon suivante.

Théorème 1

Sous les hypothèses :

(H1) : les coefficients $f_i(x,t)$ et $g_{ij}(x,t)$, pour $i,j = 1,\dots,n$, de f_x et $g_x\, g'_x$ respectivement, sont éléments de $C^{n_1}\{(0,T)\ ;\ C^{n_0}\ (R^n)\}$ avec $n_0 = \text{Entier}\ [\frac{n}{2} + 1]$ et $n_1 = \text{Entier}\ [\frac{n_0}{2} + 1]$ ∎

(H2) les fonctions f_i et g_{ij} sont bornées sur $R^n \times (0,T)$, et vérifient une condition de Hölder sur $R^n \times (0,T)$ ∎

(H3) il existe $\lambda,\ \gamma > 0$ tels que, pour tout $\varphi \in L^2\{(0,T)\ ;\ V\}$

$$< A(t)\, \varphi,\ \varphi >_{VV'} + \lambda||\varphi||^2_H > \gamma||\varphi||^2_V ,$$

où $V \equiv H^1(R^n)$, $H \equiv H' \equiv L^2(R^n)$, $V' \equiv H^{-1}(R^n)$ ■

(H4) $H(x,t)$ est élément de $L^\infty[R^n \times (0,T)] \cap C^o\,[R^n \times (0,T)]$ et $\dfrac{\partial H}{\partial x}$, $\dfrac{\partial H}{\partial t}$ sont éléments de $L^2[R^n \times (0,T)]$ ■

(H5) $p_0(x)$ est élément de $H^{n_o}(R^n)$ ■

(H6) Il existe un prolongement $A^{**}(t)$ de l'opérateur $A(t)$ pour $t < 0$ vérifiant (H1) à (H3) tel que : Soit $\tau_0 \geq 0$ et $\rho_{\tau_0}(x,t)$ la fonction définie par :

$$\rho_{\tau_0}(x, -\tau_0) = \delta(x)$$

$$(2.1) \qquad \frac{d\rho_{\tau_0}}{dt}(x,t) + A^{**}(t)\,\rho_{\tau_0}(x,t) = 0 ,$$

alors il existe τ_0 fini et $\delta > 0$ tels que $p_0(x)$ vérifie les relations :

$$0 < p_0(x) < \delta\,\rho_{\tau_0}(x,0)$$

$$\int_{R^n} p_0(x)\,dx = 1 \quad ■$$

Alors, si $u(t)$ est un élément de $L^2(0,T)$, l'équation :

$$p(x,0) = p_0(x)$$

$$(2.2)\ \frac{dp}{dt}(x,t) + A(t)p(x,t) = p(x,t)\left[H(x,t) - \int_{R^n} H(\alpha,t)p(\alpha,t)d\alpha\right] u(t)$$

admet une solution unique dans $L^1[R^n \times (0,T)] \cap L^\infty[R^n \times (0,T)]$.

Cette solution vérifie les propriétés suivantes :

$$p(x,t) \text{ est élément de } C^o\{(0,T)\ ;\ H^{n_o}(R^n)\} ,$$

$$p(x,t) > 0 ,$$

$$\int_{R^n} p(x,t)\,dx = 1.$$

De plus l'application $\{p_0(x), u(t)\} \rightarrow p(x,t)$ est continue de $L^2(R^n) \times L^2(0,T)$ dans $L^2\{R^n \times (0,T)\}$. ■

Démonstration.

Nous donnons ici les principales étapes de la démonstration. Un exposé plus détaillé est fait au chapitre II de la référence [2].

Il faut d'abord montrer l'équivalence de l'équation (2.2) et de l'équation intégrale suivante :

(2.3) $$p(x,t) = K_0[p_0(x)] + K_1[p(x,t)]$$

où K_0 désigne l'opérateur :

(2.4) $$K_0[p_0(x)] = \int_{R^n} q(0,y\ ;\ t,x)\ p_0(y)\ dy\ ,$$

et K_1 désigne l'opérateur :

(2.5) $$K_1[p(x,t)] = \int_0^t d\sigma \int_{R^n} q(\sigma,y\ ;\ t,x)\ p(y,\sigma)[H(y,\sigma) - \int_{R^n} H(\alpha,\sigma)p(\alpha,\sigma)d\alpha]u(\sigma)\ dy,$$

La fonction q étant la solution des équations

(2.6) $$q(\sigma,y\ ;\ \sigma,x) = \delta(x - y)$$

(2.7) $$\frac{dq}{dt}\ (\sigma,y\ ;\ t,x) + A(t)\ q(\sigma,y\ ;\ t,x) = 0.$$

Par application de l'opérateur $\frac{d}{dt} + A(t)$ aux deux membres de l'équation (2.5), puis par utilisation de la dérivation sous le signe somme et de la condition (2.6), on peut vérifier formellement l'identité des relations (2.2) et (2.3). On peut ensuite justifier ces calculs en se servant de la grande régularité de la fonction q vis à vis de t et de x sur tout intervalle de la forme $[\varepsilon, T] \times R^n$ avec $\varepsilon > 0$ arbitraire. Cette propriété d'équivalence sera d'une grande utilité lors de la définition des schémas d'approximation de la solution de (2.2).
On utilise ensuite le résultat partiel suivant :

Lemme 2.1.
Sous les hypothèses (H1) à (H6) il existe un intervalle de mesure non nulle $[0, \tau]$ inclus dans $[0, T]$ sur lequel l'équation (2.2) admet une solution unique. Cette solution vérifie de plus les propriétés énoncées à la fin du théorème 1. ∎

Démonstration.
On choisi une fonction $\rho(x,t)$ vérifiant les propriétés suivantes :

- $\rho(x,t)$ est élément de $C^0[(0,T)\ ;\ H^{\infty}(R^n)]$

- $\rho(x,t) > 0$ sur tout intervalle de $(0,T) \times R^n$
- $||\rho(x,t)||_{L^1(R^n)} = 1.$
- $\rho(x,t)$ vérifie la relation suivante :

$$\int_{R^n} q(\sigma,y\ ;\ t,x)\ \rho(y,\sigma)\ dy = \rho(x,t) \tag{2.8}$$

Enfin il existe $\delta > 0$ tel que :

$$p_0(x) < \delta\ \ \rho(x,0) \tag{2.9}$$

quelque soit x dans R^n. Une telle fonction existe d'après l'hypothèse (H6). Définissons l'espace $E_{\rho,I}$ des fonctions $p(x,t)$ mesurables sur $R^n \times (0,T)$ et vérifiant la propriété suivante :

$$\exists M > 0, \quad \sup_{x \in R^n\ t \in [0,\tau]} \left| \frac{p(x,t)}{\rho(x,t)} \right| < M \tag{2.10}$$

et désignons par $||p(x,t)||_{\rho,\tau}$ la norme, induite de $||\ \ ||_\infty$, associée à l'espace $E_{\rho,\tau}$ par la relation (2.10).
Par majorations systématiques des différents termes des opérateurs K_0 et K_1, on peut établir les relations :

$$||K_1[p(x,t)]||_{\rho,\tau} < \alpha(\tau)\ ||\ p(x,t)||_{\rho,\tau} + \beta(\tau)||p(x,t)||^2_{\rho,\tau} \tag{2.11}$$

$$||K_1[p_1(x,t)] - K_1[p_2(x,t)]||_{\rho,\tau} < [C + \gamma(\tau)]||p_1(x,t) - p_2(x,t)||_{\rho,\tau} \tag{2.12}$$

où $\alpha(\tau)$, $\beta(\tau)$, $\gamma(\tau)$ sont des fonctions, qui tendent vers zéro avec τ. Désignons par $K_{1,L}$ et $K_{1,NL}$ respectivement les parties linéaires en p et non-linéaire en p de l'opérateur K_1. On défini dans $E_{\rho,\tau}$ la récurrence :

$$p_1(x,t) = K_0\ [p_0(x)] \tag{2.13}$$

$$p_{m+1}(x,t) - K_{1,L}[p_{m+1}(x,t)] = K_{1,NL}[p_m(x,t)] + K_0[p_0(x)] \tag{2.14}$$

et on montre que si τ est suffisamment petit, l'opération $p_m \to p_{m+1}$ est une contraction de $E_{\rho,\tau}$ ce qui prouve le lemme 1. La vérification des propriétés vérifiées par la solution est immédiate.

Fin de la démonstration du théorème 1

Elle est basée sur la constatation suivante. Si, après application du lemme 1, on étudie les propriétés de la fonction $p(x,\tau)$ obtenue à l'extrémité de l'intervalle $[0,\tau]$, on constate qu'elle vérifie les hypothèses (H1) à (H6) si τ est différent de T. On peut donc définir un nouvel intervalle $[\tau,\tau']$ sur lequel on applique le lemme 2.1. Il reste à prouver que cette procédure est finie, ce qui est aisé, les fonctions $\alpha(\tau)$, $\beta(\tau)$ et $\gamma(\tau)$, qui conditionnent l'application du lemme 2.1., n'étant dépendantes, que de la norme dans $L^2[0,\tau]$ de la fonction $u(t)$. Celle-ci étant élément de $L^2[0,\tau]$, la procédure peut se poursuivre en un nombre fini d'étapes, au moins jusqu'à l'instant T.

III. Résultat principal dans le cas stochastique.

Nous allons maintenant montrer l'existence et l'unicité de la solution de l'équation (1.3) lorsque $z(t)$ est un processus de ITO généré par les relations (1.1) et (1.2). Soit $b(t)$ l'innovation associée à $z(t)$ et définie par la relation.

$$(3.1) \qquad d\,b(t) = dz(t) - dt \int_{R^n} H(\alpha,t)\, p(\alpha,t)\, d\alpha$$

Le processus $b(t)$ ne dépend que des réalisations du processus $z(t)$, bien que la relation (3.1) puisse faire croire, qu'il dépend aussi de p. $b(t)$ est un mouvement brownien isonome à $b_2(t)$, tel que, toute martingale de carré sommable mesurable sur la famille croissante de σ-algebres $B(z_0^t)$ construite à partir des réalisations de $z(t)$, puisse être définie comme intégrale stochastique par rapport à $b(t)$. La réciproque n'est pas vraie, contrairement au cas linéaire (voir FUSIJAKI, KALLIANPUR, KUNITA [5]). On peut donc réécrire l'équation (1.3) sous la forme :

$$(3.2) \qquad p(x,t) + \int_0^t A(s)\, p(x,s)\, ds = p_0(x)$$
$$+ \int_0^t p(x,s)\, [H(x,s) - \int_{R^n} H(\alpha,s)\, p(\alpha,s)\, d\alpha]\, d\,b(s).$$

Remarquons que la relation (3.2) est une écriture sous forme intégrale de la relation (1.3) dans laquelle on met en évidence l'intégrale stochastique du deuxième membre. Enonçons, maintenant, le résultat principal.

Théorème 2

Soit $z(t)$ un processus de ITO défini par les relations (1.1), (1.2) et $b(t)$ l'innovation associée. Sous les hypothèses (H1) à (H6), il existe une solution unique de l'équation (3.2) dans l'espace $\mathcal{H} = L^2\{\Omega,\mu;L^2[(0,T)\;;\;H]\}$.
De plus, cette solution appartient à l'espace $L^2\{\Omega,\mu\;;\;C^o[(0,T)\;;\;H^{n_o}(R^n)]\}$ ■

Démonstration

Comme pour le théorème 1, nous donnons ici les principales étapes de cette démonstration, dont on trouvera un exposé détaillé dans la référence [2]. Elle nécessite la démonstration préalable de deux lemmes :

Lemme 3.1.

Soit $u_m(t)$ une suite de fonctions dans $L^2[\Omega,\mu\ ;\ L^2(0,T)]$ telle que, pour toute fonction $\varphi(t,\omega)$ dans $L^2[\Omega,\mu\ ;\ L^2(0,T)]$, mesurable pour la famille $B(z_0^t)$, l'expression

$$(3.3)\qquad \int_0^t \varphi(t,\omega)\, u_m(t,\omega)\, dt$$

converge uniformément en moyenne quadratique vers l'intégrale stochastique :

$$(3/4)\qquad \int_0^t \varphi(t,\omega)\, d\, b(t,\omega)$$

Soit $p_m(x,t)$ la solution de l'équation (2.2) associée à l'élément $u_m(t)$ de la suite, alors la suite p_m est uniformément bornée dans $\mathcal{H}$ ∎

Démonstration

Choisissons la séquence $u_m(t,\omega)$ sous la forme :

$$(3.5)\qquad u_m(t,\omega) = \sum_{i=1}^{m} \frac{1}{\Delta t}\,[b(t_{i+1},\omega) - b(t_i,\omega)]\, \Pi_i(t)$$

où t_i, $i = 1 \dots m$ désigne une subdivision de l'intervalle $(0,T)$, de pas constant égal à $\Delta t = T/m$, et Π_i la fonction caractéristique de l'intervalle $[t_{i-1}\ ,\ t_i]$. Alors, pour toute fonction φ élément de $L^2[\Omega,\mu\ ;\ L^2(0,T)]$ mesurable sur $B(b_0^t)$, la relation suivante est satisfaite :

$$(3.6)\qquad E\{\int_0^T \varphi(s,\omega)\, u_m(s,\omega)\, ds\} = 0$$

Notons également u_m, la sous-suite infinie issue de la précédente, des fonctions telles que pour tout φ, la relation suivante est satisfaite.

$$(3.7)\qquad E\{||\int_0^T \varphi(t,\omega)\, u_m(t,\omega)\, dt||_H^2\} < \eta\ E\{\int_0^T ||\varphi||_H^2\ dt\}$$

où η est un réel positif. L'existence d'une telle sous-séquence est une conséquence directe du théorème de Baire-Hausdorff (voir YOSHIDA [8] p 69).

Soit p_m la solution de (2.2) associée à u_m.
La fonction p_m appartient aux espaces $\mathcal{H}$ et $\mathcal{V} = L^2\{\Omega,\mu\ ;\ L^2[(0,T)\ ;\ V]\}$, comme fonction de x,t et ω, puisque l'opérateur $u_m(t) \to p_m(x,t)$ est continu d'après le théorème 1, et que cette continuité implique la mesurabilité de $p(x,t;\omega)$ par rapport à la mesure de base μ.

Soit $\chi(p)$ l'expression :

$$(3.8)\qquad \chi(p) = p(x,t;\omega)[H(x,t) - \int_{R^n} H(\alpha,t)p(\alpha,t;\omega)\, d\alpha]$$

Alors, si V' désigne le dual fort de V, on a la relation suivante :

$$(3.9)\qquad \frac{dp_m}{dt}(x,t;\omega) + A(t)p_m(x,t;\omega) = \chi[p_m(x,t;\omega)]\ u_m(t,\omega)$$

de (3.9) on déduit :

$$(3.10)\quad \langle \frac{d}{dt}\, p_m,\, p_m \rangle_{VV'} + \langle A(t)\, p_m,\, p_m \rangle_{VV'} = \langle \chi[p_m]\, u_m,\, p_m \rangle_H$$

où $\langle\ ,\ \rangle_{VV'}$ désigne le produit de dualité entre V et V'et $\langle\ ,\ \rangle_H$ désigne le produit scalaire dans H. En intégrant la relation (3.10) on obtient

$$(3.11)\quad \langle p_m,\, p_m \rangle_H + 2\int_0^t \langle A(s)p_m,\, p_m\rangle_{VV'}\, ds = \langle p_0,\, p_0\rangle_H + 2\int_0^t \langle \chi(p_m)u_m, p_m\rangle_H\, ds$$

En utilisant l'hypothèse (H3) on obtient la majoration :

$$(3.12)\ ||p_m||_H^2 + 2\,\gamma\int_0^t ||p_m||_V^2\, ds < ||p_0||_H^2 + 2\,\lambda\int_0^t ||p_m||_H^2\, ds + 2\int_0^t \langle\, \chi(p_m)\, u_m,\, p_m\rangle_H\ ds$$

en prenant l'espérance mathématique des deux membres de (3.12), on obtient :

$$(3.13)\ E\{||p_m||_H^2\} < ||p_0||_H^2 + 2\,\lambda \int_0^t E\{||p_m||^2\}\ ds + 2\ E\{\int_0^t \langle\chi(p_m)u_m,\, p_m\rangle_H\ ds\}$$

Il faut maintenant trouver une majoration du dernier terme du second membre de (3.13) soit :

$$(3.14)\ E\{\int_0^t u_m(s,\omega)ds \int_{R^n} p_m^2(x,s;\omega)dx\ [H(x,s) - \int_{R^n} H(\alpha,s)\ p_m(\alpha,s)\ d\alpha]\}$$

L'expression (3.14) n'est pas égale à zéro, malgré la relation (3.6), car la fonction $\chi(p_m)$ n'est pas mesurable par rapport à $B(z_0^t)$. Pour éliminer cette difficulté, on considère une subdivision $t_0 \ldots t_i \ldots t_N$ de $[0,T]$. Supposons que s appartienne à l'intervalle $[t_i, t_{i+1}]$ et soit $\Delta p_m(x,s;\omega)$ l'expression

$$(3.15) \qquad \Delta p_m(x,s;\omega) = p_m(x,s;\omega) - p_m(x,t_i;\omega)$$

En intégrant la relation (2.2), nous obtenons la relation suivante :

$$(3.16) \quad \Delta p_m = \int_{t_i}^{s} p_m(x,\sigma;\omega)[H(x,\sigma) - \int_{R^n} H(\alpha,\sigma)p_m(\alpha,\sigma;\omega)\, d\alpha] \times u_m(\sigma)\, d\sigma - \int_{t_i}^{s} A(\sigma)\, p_m(x,\sigma;\omega)\, d\sigma$$

En introduisant dans l'expression (3.14) le développement (3.16), puis en isolant les termes fonction de l'instant t_i des termes infiniment petits avec $t_{i+1} - t_i$, on obtient la relation suivante :

$$\begin{aligned}(3.17) \quad & E\{\int_0^t u_m(s,\omega)\, ds \int_{R^n} p_m^2\, (x,s;\omega)\, dx\, [H(x,s) - \int_{R^n} H(\alpha,s)p_m(\alpha,s;\omega)\, d\alpha]\} \\ & = E\{\int_0^t u_m(s,\omega)\, ds \int_{R^n} p_m^2\, (x,t_i;\omega)\, dx\, [H(x,t_i) - \int_{R^n} H(\alpha,t_i)p_m(\alpha,t_i;\omega)\, d\alpha]\} \\ & + 2\, E\{\int_0^t u_m(s,\omega)[b(s)-b(t_i)]ds \int_{R^n} p_m^2(x,t_i;\omega)dx[H(x,t_i) - \int_{R^n} H(\alpha,t_i)p_m(\alpha,t_i;\omega)d\alpha]^2 \\ & - E\{\int_0^t u_m(s)[b(s)-b(t_i)]ds \int_{R^n} p_m^2\, (x,t_i;\omega)\, dx \int_{R^n} H(\alpha,t_i)\, p_m(\alpha,t_i;\omega)\, d\alpha \\ & [H(x,t_i) - \int_{R^n} H(\alpha,t_i)\, p_m(\alpha,t_i;\omega)\, d\alpha] \\ & + \mathcal{O}_m\end{aligned}$$

où t_i désigne l'élément de la subdivision qui précède immédiatement la variable s, et $\mathcal{O}_m$ une fonction qui tend vers zéro en moyenne quadratique, quand m tend vers l'infini.

Le premier terme du second membre de (3.17) est nul d'après la relation (3.6). Les deuxièmes et troisièmes termes sont des intégrales de Stielges et l'on peut réécrire (3.17) sous la forme :

$$(3.18);\quad (3.14) = \int_0^t ds\, E\{ \int_{R^n} p_m^2 \, dx\, [H - \int_{R^n} Hp\, d\alpha]^2$$

$$- \int_0^t ds\, E\{ \int_{R^n} p_m^2 \, dx \int_{R^n} H\, p_m\, d\alpha [H - \int_{R^n} H\, p_m\, d\alpha]\}$$

En majorant le second membre de (3.18) par les relations :

$$(3.19) \qquad | H(x,s) | < H_{max}$$

$$(3.20) \qquad | \int_{R^n} H(\alpha,s)\, p_m(\alpha,s;\omega)\, d\alpha | < H_{max}$$

on obtient :

$$(3.21) \qquad | E \{ \int_0^t u_m(s,\omega) ds \int_{R^n} p_m^2(x,s;\omega)\, dx [H(x,s) - \int H(\alpha,s) p_m(\alpha,s;\omega) d\alpha]\}|$$

$$< 6\, H_{max}^2 \int_0^t ds \quad E\{ \int_{R^n} p_m^2\, (x,s;\omega)\, dx\}$$

De (3.21) et (3.13), nous déduisons :

$$(3.22) \quad E\{||p_m||_H^2\} < ||p_0||^2 + [2\lambda + 6H_{max}^2] \int_0^t ds\, E\{||p_m||_H^2\}$$

La relation (3.22) est une inégalité de Gronwall ce qui implique l'existence de deux réels positifs k_1 et k_2 tels que

$$(3.23) \qquad ||p_m||_{\mathcal{H}}^2 < k_1\, e^{k_2 T}$$

ce qui démontre le lemme. On en déduit qu'on peut extraire de la suite p_m une sous-suite infinie qui converge faiblement dans $\mathcal{H}$ vers un élément $p(x,t;\omega)$ de cet espace.

Lemme 3.2.

Soit $\varphi(x,t;\omega)$ un élément de l'espace $\mathcal{D}[(0,T) \times R^n] \otimes L^\infty(\Omega,\mu;R)$. Alors l'expression :

$$(3.24) \quad E\{ \int_0^t \int_{R^n} \varphi(x,t;\omega)\, p_m(x,t;\omega)[H(x,t) - \int_{R^n} H(\alpha,t) p_m(\alpha,t;\omega)\, d\alpha]\, u_m(t,\omega)\, dx\, dt\}$$

converge vers l'expression :

$$(3.25)\quad E\{\int_0^t \int_{R^n} \varphi(x,t;\omega)\ p(x,t;\omega)\ [H(x,t) - \int_{R^n} H(\alpha,t)\ p_m(\alpha,t;\omega)\ d\alpha]\ dx\ db(t,\omega)\}\ \blacksquare$$

Démonstration

Il suffit de prouver la convergence pour les fonctions φ du type :

$$(3.26)\qquad \varphi(x,t;\omega) = \lambda(x,t)\ \nu(\omega)$$

où λ appartient à $\mathcal{D}((0,T) \times R^n)$ et ν à $L^\infty(\Omega,\mu;R)$.

On effectue ensuite la différence entre les expressions (3.24) et (3.25). On remarque alors qu'il suffit d'appliquer la propriété de convergence d'une expression du type (3.4) vers une expression du type (3.5) d'une part, et de se servir de la convergence faible de p_m vers p d'autre part pour achever la démonstration. ■

Fin de la démonstration du théorème 2.

Le lemme (3.2) fournit un résultat de convergence d'un second membre d'une relation du type (2.2) vers le second membre d'une relation du type (3.2).

Prenons donc une fonction φ de l'espace $\mathcal{D}[(0,T) \times R^n) \otimes L^\infty(\Omega,\mu;R)$.

Par définition p_m et u_m sont liés par la relation

$$(3.27)\qquad \frac{dp_m}{dt}(x,t;\omega) + A(t)\ p_m(x,t;\omega) = \chi(p_m)\ u_m(t,\omega)$$

Donc, on a :

$$(3.28)\qquad \langle \frac{dp_m}{dt}, \varphi \rangle_H + \langle A(t)\ p_m, \varphi\rangle_{VV'} = \langle \chi(p_m)u_m, \varphi\rangle_H$$

Désignant par A* l'opérateur adjoint de A, on obtient :

$$(3.29)\qquad - \langle p_m, \frac{d\varphi}{dt} \rangle_H + \langle p_m, A^*\varphi\rangle_H = \langle \chi(p_m)u_m, \varphi\rangle_H$$

Quand m tend vers l'infini, appliquons la convergence faible de p_m vers p, puis le lemme 3.2 ; on obtient :

$$(3.30)\ E\{\int_{R^n} \varphi(x,s;\omega)\ dx\ [dp(x,s;\omega) + A(s)\ p(x,s;\omega)\ ds - \chi(p)\ db(s,\omega)] = 0$$

Cette relation étant vraie pour toute fonction φ cela montre que p est solution de l'équation (3.2). Reste à prouver l'unicité de la solution : Soit p_1 et p_2 deux solutions de (3.2). Alors on peut écrire :

$$(3.31)\quad p_1 - p_2 + \int_0^t A(s)\,[p_1 - p_2]\,ds = \int_0^t db(s)\{[p_1 - p_2]\,[H - \int_{R^n} H\,p_1\,d\alpha] - p_2 \int_{R^n} H(p_1 - p_2)\,d\alpha\}$$

D'où l'on déduit :

$$(3.32)\quad \langle p_1 - p_2\,,\, dp_1 - dp_2\rangle_{VV'} + \langle p_1 - p_2,\, A(t)[p_1 - p_2]\rangle_{VV'}$$
$$= \langle p_1 - p_2,\, [p_1 - p_2][H - \int_{R^n} H\,p_1\,d\alpha] - p_2 \int_{R^n} H\,(p_1 - p_2)\,d\alpha\rangle_H \times db(t)$$

En appliquant le calcul différentiel de ITO au premier terme du premier membre de (3.32) et en intégrant, on obtient :

$$(3.33)\quad ||p_1 - p_2||_H^2 + 2\int_0^t \langle p_1 - p_2,\, A(s)(p_1 - p_2)\rangle_{VV'}\,ds$$
$$= 2\int_0^t \langle p_1 - p_2,\, \Phi(p_1 - p_2)\rangle_H\,db(s) + \int_0^t ||\Phi(p_1 - p_2)||_H^2\,ds$$

où Φ désigne le facteur de $(p_1 - p_2)$ et de db(t) dans le second membre de (3.32). En prenant les espérances mathématiques des deux membres de (3.33) et en utilisant la majorations (3.19) et (3.20) on obtient :

$$(3.34)\quad E\{\,||p_1 - p_2||_H^2\} < [2\,\lambda + 2\,H_{max} + k_1\,e^{k_2 H_{max}}] \times \int_0^t E\{||p_1 - p_2||_H^2\,\}\,ds.$$

C'est une inéquation de Gronwall sans terme constant, ce qui prouve l'unicité de la solution de (3.2).

Signalons enfin, un résultat auxiliaire :

<u>Théorème 3</u>

Désignons par $\mathcal{V}$ l'espace $L^2\{\Omega,\mu\ ;\ L^2[(0,T)\ ;\ V]\}$. La fonction p solution de (3.2) vérifie l'inégalité de l'énergie suivante :

$$(3.35)\quad E\{||p||_H^2\} + 2\,\gamma\,||p||_{\mathcal{V}}^2 < ||p_0||_H^2 + k\,||p||_{\mathcal{H}}^2$$

où γ et k sont des constantes positives ∎

IV. Liaison avec le problème du filtrage non-linéaire.

La liaison entre le théorème 2 et l'évolution de la densité de probabilité conditionnelle utilise les résultats de KUSHNER [1], ZAKAI [4] et FUSIJAKI, KALLIANPUR, KUNITA [5]. Soit $\Psi(x,t)$ un élément de $\mathcal{D}[R^n \times (0,T)]$. On sait que l'on a la relation suivante :

$$(4.1)\qquad E\{\Psi(X(t),t) \mid B(z_0^t)\} + \int_0^t E\{A^* \Psi(X(s),s) \mid B(z_0^s)\}\, ds$$

$$= E\{\Psi(0),0\} + \int_0^t [E\{\Psi(X(s),s)\, H(X(s),s) \mid B(z_0^s)\}$$

$$- E\{\Psi(X(s),s) \mid B(z_0^s)\}\, E\{H(X(s),s) \mid B(z_0^s)\}]\, d\, b(s)$$

Soit $\tilde{p}$ la densité de probabilité de $X(s)$ conditionnée par z_0^s. Cette fonction vérifie dans l'espace $\mathcal{D}'[(0,T) \times R^n)$ des distributions, une relation du type (3.2) puisque :

$$(4.2)\qquad E\{\Psi(X(s),s) \mid B(z_0^s)\} = \int_{R^n} \Psi(x,s)\, \tilde{p}(x,s;\omega)\, dx$$

D'un autre côté le théorème de représentation (BUCY [6], BENSOUSSAN [7]) prouve que $\tilde{p}$ est élément de $\mathcal{H}$.

D'après le théorème 2, il ne peut exister deux fonctions de $\mathcal{H}$ distinctes, qui induisent sur l'espace $\mathcal{D}'$ la même relation (3.2). Donc $\tilde{p}$ et p sont identiques et la relation (3.2) décrit bien dans $\mathcal{H}$ l'évolution de la densité conditionnelle.

V. Algorithme d'approximation de la fonction p.

Examinons maintenant le point de vue pratique.

Supposons que nous connaissions un enregistrement du processus z, en temps réel ou différé. Nous désirons construire une approximation de la densité de probabilité conditionnelle. Il faut donc définir un schéma de discrétisation de l'équation (2.2) puisque nous sommes dans des conditions déterministes. Il y a trois possibilités différentes pour construire un tel schéma.

α) Le schéma explicite :

$$(5.1)\qquad \frac{1}{h}(p^+ - p) + A(h)\, p = \chi(p,u)$$

où h désigne le pas de discrétisation en temps, p la fonction connue à un instant t, p^+ la fonction inconnue à l'instant t + h, A(h) l'opérateur discret, déduit de l'opérateur A(t) à l'instant t, enfin χ le second membre discrétisé de l'équation :

$$(5.2)\quad \frac{dp}{dt} + A(t)p = p[H(x,t) - \int_{R^n} H(\alpha,t)p\, d\alpha][u(t) - \int_{R^n} H(\alpha,t)\, p\, d\alpha]$$

où u désigne la mesure physique associée à $\frac{dz}{dt}(t)$.

β) Le schéma mixte qui diffère du précédent par la relation :

$$(5.3)\quad \frac{1}{h}(p^+ - p) + A(h)[\theta p^+ + (1-\theta)\, p] = \chi(p,u)$$

où θ appartient à $]0,1[$.

γ) Le schéma implicite :

$$(5.4)\quad \frac{1}{h}(p^+ - p) + A(h)\, p^+ = \chi(p,u)$$

Il faut décider entre ces trois possibilités, laquelle conduit à un algorithme convergent. Le schéma explicite correspond à une économie maximale de calcul, mais est généralement instable. Les schémas mixtes et implicites sont généralement plus stables, mais nécessitent l'inversion numérique d'une matrice de dimension importante. Pour définir la meilleure structure, nous allons utiliser la démonstration du théorème 1. On introduit une équation de type intégral :

$$(5.5)\qquad p(x,t) = K_{0,h}[p_0(x)] + K_{1,h}[p(x,t)]$$

où $K_{i,h}$ désigne le discrétisé de l'opérateur K_i introduit dans la démonstration du théorème 1.

Par exemple :

$$(5.6)\qquad K_{1,h} = \sum_{l=1}^{k} h \int_{R^n} q_h(k+1,x;l,y)\, \chi[p(y,l),\, u(l)]\, dy$$

où q_h est solution du système d'équation :

$$(5.7)\qquad q_h(k,x;k,y) = \delta(x - y)$$

$$(5.8)\qquad \frac{1}{h}[q_h(k+1,x;l,y) - q_h(k,x;l,y)] + A_h(k)\, . = 0$$

l'opérande de $A_h(k)$ étant $q_h(k,x;l,y)$ si l'on choisit le schéma explicite et

$q_h(k+1,x;l,y)$ si l'on choisit le schéma implicite. Dans le cas du schéma implicite l'équation (5.8) est très fortement régularisante et la présence d'une impulsion de Dirac dans la relation (5.7) n'a aucune importance. Au contraire, dans tous les autres cas, cette condition initiale diffuse au cours du temps des fonctions de type Dirac, qui provoquent l'instabilité de l'algorithme.
Dans le cas du schéma implicite, si à l'instant kh la fonction q_h est élément de l'espace de Sobolev H^j , à l'instant (k+1)h cette même fonction est élément de l'espace H^{j+2}. Ceci permet d'utiliser telle quelle la démonstration du théorème 1 pour prouver la convergence forte de la fonction calculée p_h vers la fonction continue p dans l'espace $L^2[(0,T) ; H]$ lorsque h tend vers zéro. Le détail de cette étude est donné dans la référence [2].

VI. Expériences numériques.

Un algorithme numérique est déduit de la relation (5.4) par utilisation d'une discrétisation de l'espace R^n selon la méthode des éléments finis. L'espace R^n n'étant pas un ouvert borné, il faut utiliser le fait que la fonction inconnue p^+ est une densité de probabilité, donc une fonction de norme L^1 égale à 1, pour se retrouver dans un cadre classique. L'algorithme a ensuite été testé sur un système linéaire simulé, où il est aisé de vérifier les résultats à l'aide d'un filtre de Kalman.
Les dimensions de l'espace d'état considérées sont un et deux, bien qu'il soit envisageable de traiter des problèmes dont l'état est tridimensionnel. La précision de calculs décroit lorsque la dimension d'état augmente. Elle est de deux à trois pour cent pour un système monodimensionnel et de six à sept pour cent pour un système bidimensionnel. Les résultats sont meilleurs pour l'estimée du filtre optimal, l'intégration effectuée pour obtenir le filtre ayant un effet de lissage sur les incertitudes du calcul de la densité de probabilité.
On peut vérifier sur la figure 1, le petit nombre de biais constatés sur un système monodimensionnel.

Il est indispensable, pour des raisons théoriques de choisir un schéma implicite de discrétisation. On peut constater sur la figure 2,la très rapide divergence d'un schéma explicite utilisé parallèlement à un schéma implicite sur un système linéaire.

Enfin, on peut constater figure 3 et 4 l'intérêt d'une telle procédure dans la détection d'hypothèses a priori multiples, que l'on modélise par la position initiale de l'état d'un système dynamique faiblement bruité. L'observation est faite à l'aide d'une loi sinusoïdale fortement bruitée (problème de la détection de phase initiale d'un phénomène oscillatoire perturbé).

VII. Principales notations.

Ω, ω	Espace de probabilité et épreuve
μ	Mesure sur Ω
R	Ensemble des réels
x, α, y	Eléments de R^n
t, σ, s, T	Eléments de R
$L^2[(0,T) ; V]$	Ensemble des fonctions de carré sommable sur (0,T) à valeur dans V
V'	Dual fort de V
$< , >_H$	Produit scalaire dans H
$< , >_{VV'}$	Produit de dualité entre éléments de V' et de V
$\Vert \quad \Vert_H$	Norme hilbertienne dans H
$B(z_0^t)$	Plus petite σ algèbre contenant tous les événements du type $\alpha < z(s) < \beta$, α, β appartenant à R, s à (0,t).

REFERENCES.

1 Kushner H., "Dynamical equations for optimal non-linear filtering", Journal of differential equation, Vol. 3, 1967, p. 179-190.

2 Levieux F., "Filtrage non-linéaire et analyse fonctionnelle", Rapport LABORIA, Available by writing to IRIA 78150 Rocquencourt, France.

3 Levieux F., "Functional analysis approach of the partial differential equation arising from non-linear filtering theory", Third symposium on non-linear estimation theory and its applications, San Diego, 1972, p. 142-147.

4 Zakai M., "On the optimal filtering of diffusion processes", Z. Wahrscheinlichkeitstheorie verw. Geb. , Vol. 11, 1969, p. 230-243.

5 Fusijaki, Kallianpur, Kunita, "Stochastic differential equation for the non-linear filtering problem", Osaka Journal on Mathematics, Vol. 9, 1972, p. 19-40.

6 Bucy J.S., "Non-linear filtering theory", correspondence IEEE-TAC, Vol. AC-10, 1965, p. 198.

7 Bensoussan A., "Problèmes d'estimation statistique dans les espaces de Hilbert : Application au théorème de représentation de Bucy", Trans. of the 5 th symposium on information theory, statistical decision functions and random processes, Prague, 1971.

8 Yoshida K., "Functional Analysis", Springer Verlag - 1965.

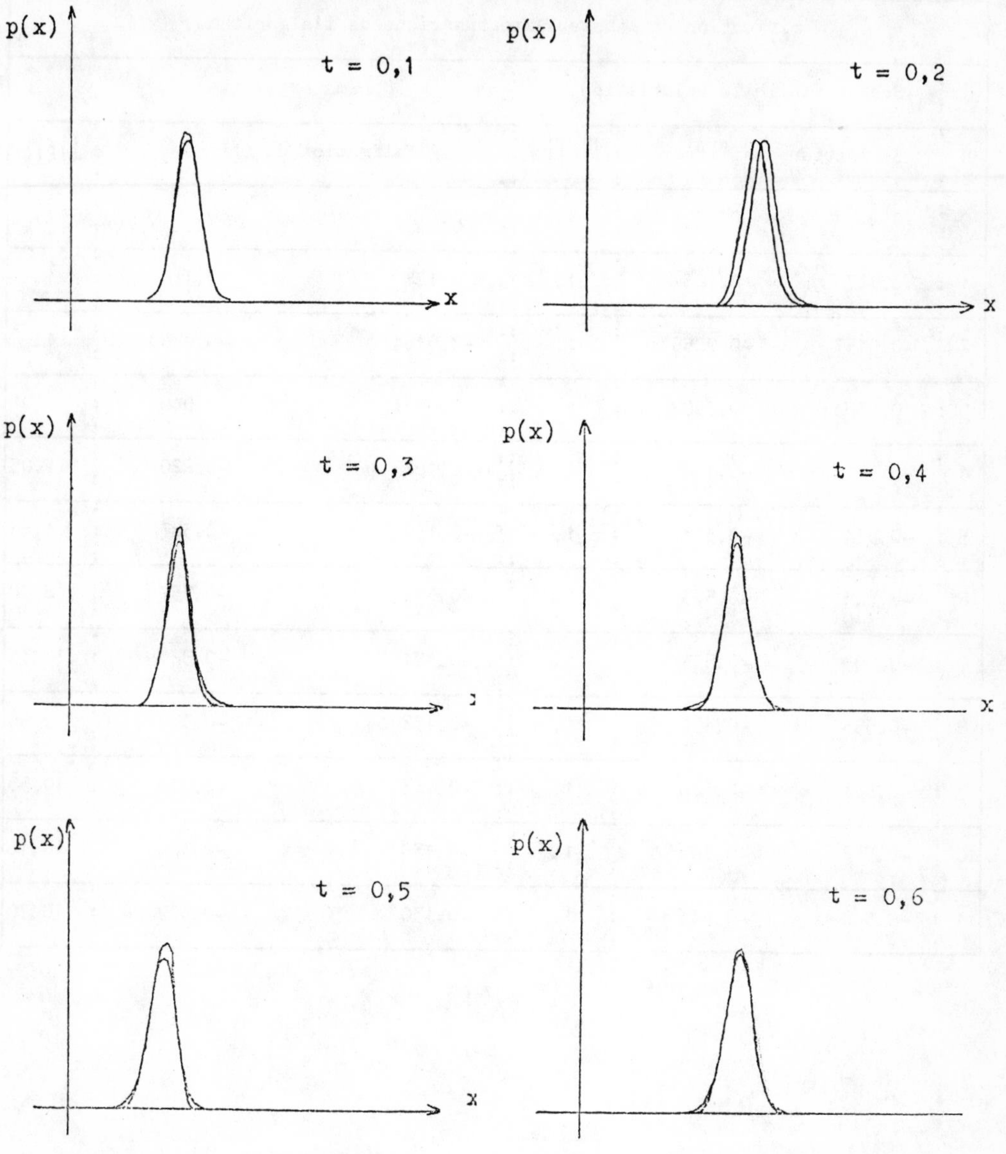

p(x)
t = 0,1
x
p(x)
t = 0,2
x
p(x)
t = 0,3
p(x)
t = 0,4
x
p(x)
t = 0,5
x
p(x)
t = 0,6

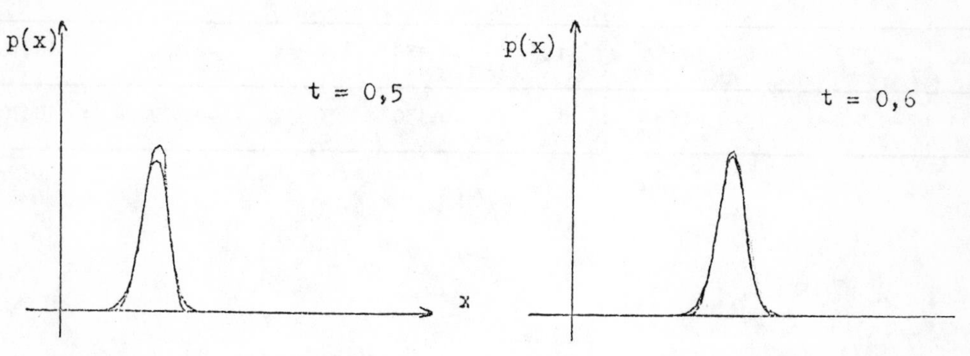

Figure 1 : équations du modèle : $dx = x\ dt + 0{,}6\ d\ b_1(t)$

$dz = x\ dt + 0{,}1\ d\ b_2(t)$

conditions initiales : $p_o(x) = \delta(x)$

Vérification de la stabilité numérique de l'algorithme						
	Schéma implicite (simplifié)			Schéma explicite		
N	Estimation du filtre		$\|\|p\|\|_1$	Estimation du filtre		$\|\|p\|\|_1$
0	0.	0.	1.	0.	0.	1.
1	0.127	0.126	1.	0.128	0.128	1.
2	-0.0157	-0.0255	1.	-0.0164	-0.0269	1.
3	0.0583	0.0502	1.	0.0584	0.048	1.01
4	-0.171	-0.184	1.	-0.182	-0.220	1.05
5	-0.314	-0.313	1.	-0.159	-0.327	1.66
6	-0.525	-0.503	1.	-0.136	-0.357	8.08
7	-0.593	-0.533	1.	-0.124	-0.321	10.35
8	-0.555	-0.451	1.	-0.132	-0.339	9.96
9	-0.661	-0.528	1.	-0.126	-0.316	10.58
10	-0.562	-0.383	1.	-0.130	-0.324	10.16
11	-0.542	-0.333	1.	-0.130	-0.309	10.70

Figure 2

Calcul du filtre d'un même système linéaire d'état bidimensionnel par un schéma implicite puis explicite.

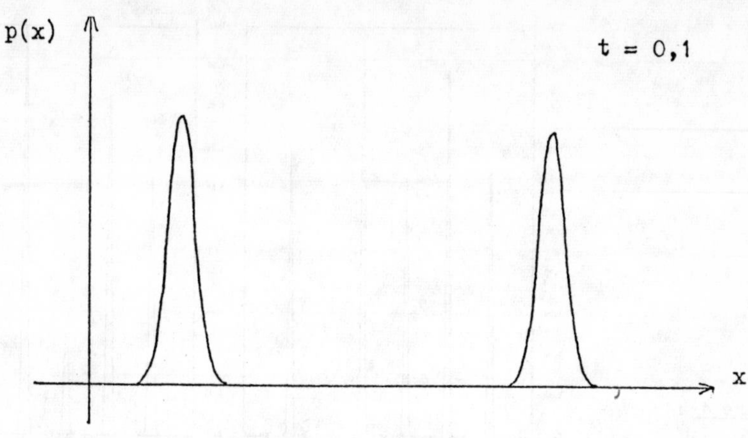

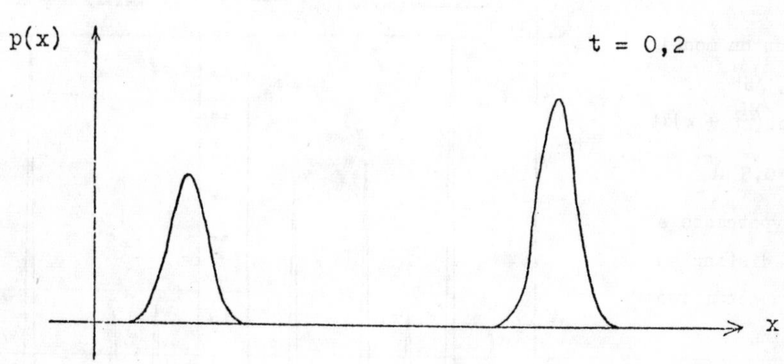

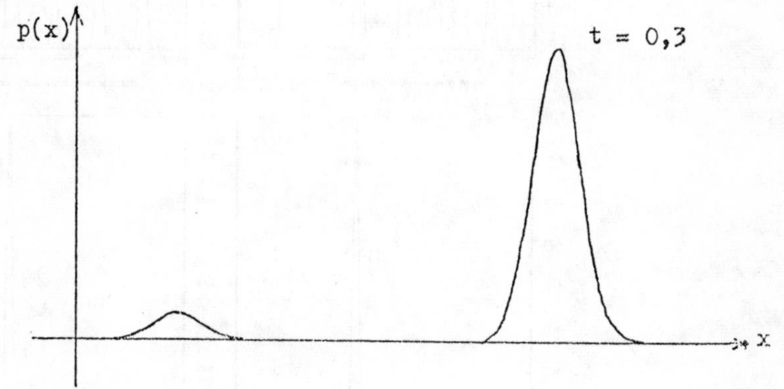

figure 3

équations du modèle : $dx = 0{,}1 \; d\, b_1(t)$

$dz = \text{Arctg}\,(x)\, dt + d\, b_2(t)$

conditions initiales : $p_o(x) = 0{,}5\ \delta(x-1) + 0{,}5\ \delta(x-5)$

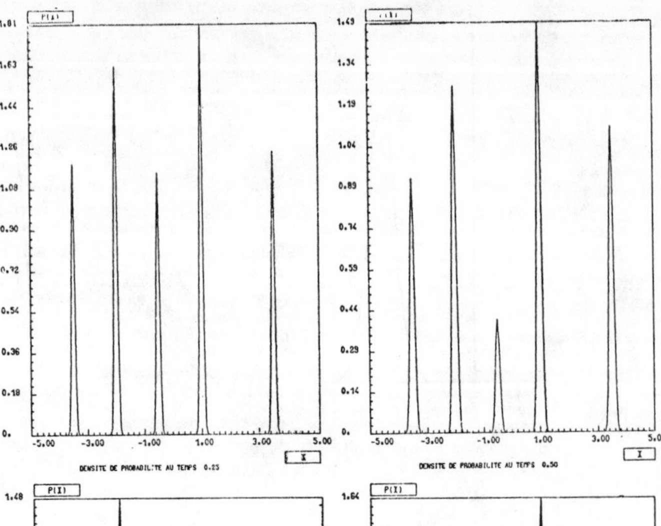

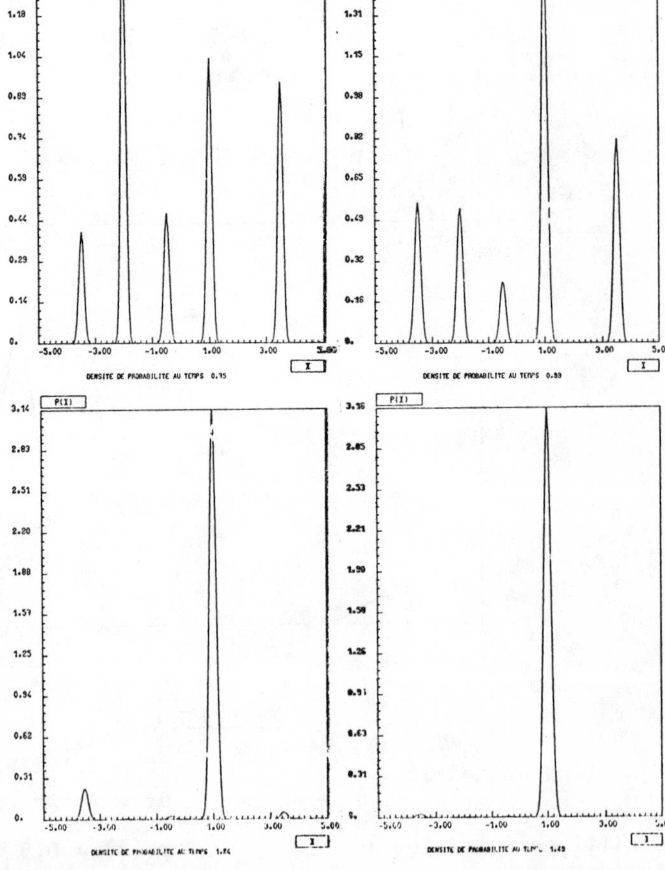

Figure 4 :

équation du modèle

$dx = 0{,}1\ db_1$

$dz = \cos(\frac{\Pi}{2}t + x)dt + 0{,}5\ db_2$

cinq hypothèses a priori distinctes pour la condition initiale.

A TALE OF FOUR INFORMATION STRUCTURES

by

Y. C. Ho
Harvard University

I. Blau
Massachusetts Institute of Technology

and

T. Basar
Marmara Scientific and Industrial Research Institute, Turkey

I. Introduction

From a decision-theoretic viewpoint, the value of information in a stochastic optimization problem is roughly characterized as

Value of Information = the Best the decision maker can do with the information − the best the decision maker can do without the information.

More precisely, let $(\Omega, \mathscr{B}, \mathscr{P})$ be a probability space and ξ with probability density function $p(\xi)$ be a vector random variable defined on this probability space. The decision variable (vector) $u \in U$ is given an observation $z \in Z$. The information structure η of the problem is then defined by the probability density function $p(z/\xi)$. Equivalently, when physical motivations permit, the information structure η can also be specified by defining

(1) $$z = h(\xi, \varepsilon)$$

where ε is another vector random variable on Ω and we are given the joint density function $p(\xi, \varepsilon)$. In this case, ε often plays the role of measurement noise. The strategy of the decision maker is a map γ: $Z \to U$ and the payoff is the expected loss (utility) function

(2) $$J = E[L(\xi, u = \gamma(z))]$$

Under this setup, we have the

(3) $$-\text{ Value of Information Structure } \eta = \min_{\gamma \in \Gamma} E[L(\xi, u = \gamma(z))] - \min_{\gamma \in \Gamma_c} E[L(\xi, u = \gamma(z))]$$

where

Γ - the class of z-measurable maps γ.

Γ_c - class of constant maps. (i.e. $\gamma \in \Gamma_c$ is not dependent on z.)

When there are more than one DM involved, we can extend the idea of the value of information structure in a natural way. However, since in game theory

---

The research reported in this paper was made possible through support extended by the Division of Engineering and Applied Physics, Harvard University, the U.S. Office of Naval Research under the Joint Services Electronics Program by Contracts N00014-67-A-0298-0006 and by the National Science Foundation under Grant GK 31511.

solution concept prolificate, we must be more specific in the definition of "best". In this paper, we consider a two person nonzero sum stochastic optimization problem and solve for the Nash equilibrium under four different information structures. The change in the Nash costs for the two DMs under the different η's will show some surprising (at least at first glance) results.

II. Problem Statement

Let $u, v \in \mathbb{R}$ be the two decision variables controlled by DM1 and DM2 and $\xi \in \mathbb{R}$ be a scalar gaussian random variable N(0,1) representing the state of the world. The cost function for the two DMs are respectively

(4)
$$J_1 = E[(\xi + u + v)^2 + D_1 u^2] \quad D_1 > 0$$
$$J_2 = E[(\xi + u + v)^2 + D_2 v^2] \quad D_2 > 0$$

The information structure of u, η_1, is given by

(5) $$z_1 = \xi + \varepsilon_1 \qquad \varepsilon_1 \sim N(0, s_1) \text{ independent of } \xi$$

We now consider four different information structures for v as follows:

Case A

(6) $$\eta_2: \; z_2 = \xi + \varepsilon_2 \qquad \varepsilon_2 \sim N(0, s_2) \text{ independent of } \xi \text{ and } \varepsilon_1$$

Case B

(7) $$\eta_2: \begin{cases} \xi + \varepsilon_2 = z_2 \\ z_1 \end{cases}$$

i.e. DM2 knows the observation of DM1, z_1, in addition to his own observation.

Case C

(8) $$\eta_2: \begin{cases} z_2, z_1 \\ u \end{cases}$$

i.e. DM2 knows in addition to z_1 the actual decision of DM1, u.

We interpret this information structure as the one that leads to player 1 acting first in the decision problem.

Case D

(9) $$\eta_2: \begin{cases} z_2 \\ \xi + \varepsilon_3 \end{cases} \qquad \varepsilon_3 \sim N(0, s_3) \text{ independent of } \xi, \varepsilon_1, \text{ and } \varepsilon_2.$$

This case can be viewed equivalently as Case A but with a smaller s_2.

In each of the cases, we wish to find a strategy pair $(\gamma_1^*: \eta_1 \to u, \gamma_2^*: \eta_2 \to v)$

such that

$$J_1(\gamma_1^*,\gamma_2^*) \le J_1(\gamma_1,\gamma_2^*) \qquad \forall\, \gamma_1 \in \Gamma_1$$

(10)

$$J_2(\gamma_1^*,\gamma_2^*) \le J_2(\gamma_1^*,\gamma_2) \qquad \forall\, \gamma_1 \in \Gamma_2$$

where Γ_1, Γ_2 are the appropriate class of admissible (measurable w.r.t. the observations) strategies. We denote J_1^*, J_2^* as the Nash cost.

It should be noted that no generality is gained if we consider instead cost functions of the type $J_1 = E[(\alpha_1\xi + \alpha_2 u + \alpha_3 v)^2 + D_1 u^2]$, $J_2 = E[(\alpha_1\xi + \alpha_2 u + \alpha_3 v)^2 + D_2 v^2]$ or by letting $\xi \sim N(0,\sigma^2)$ instead of $N(0,1)$. By redefining u, v, D_1, and D_2, we can always reduce the problem to the form of (4). Also, assuming nonzero mean for the random variables merely complicates notation without adding anything conceptually new.

III. Existence and Uniqueness of Nash Equilibrium Strategies

We shall prove that a unique Nash strategy pair exists for all four cases. This is significant and important because uniqueness of Nash strategies are by no means obvious even in deterministic linear-quadratic nonzero sum problems. Without uniqueness little credence can be attached to the Nash solution, and consequently, the values of information structure to be derived.

Cases A, B, and D will be treated first since the information structure under each case is static [4] (i.e. observation depends only on the state of the world and not on the decisions of other DMs). We have

Lemma 1: (γ_1^*,γ_2^*) is a Nash equilibrium pair for (4) if and only if it is also the person-by-person optimal strategy pair for the team payoff

(11) $$J = E[(\xi + u + v)^2 + D_1 u^2 + D_2 v^2]$$

Proof: Consider $\tilde{J}_1 \equiv J_1 + E[D_2 v^2] = \tilde{J}_2 \equiv J_2 + E[D_1 u^2] = J$ since information is static, addition of $E[D_2\gamma_2^2]$ and $E[D_1\gamma_1^2]$ to J_1 and J_2 respectively does not effect the inequality in (10). ■

Now by a well known theorem of Radner [5], we know that for a Linear-Quadratic-Gaussian team problem with strictly convex cost function, we have a unique affine person-by person optimal strategy for all the DMs which is also globally optimal. Consequently, we have

Theorem 1. The Nash equilibrium strategy (γ_1^*,γ_2^*) for Cases A, B, D are unique and affine in the observations.

Proof: Immediate via Lemma 1 and Radner's Theorem.

The problem of Case C is slightly different since the information structure is dynamic. (DM2's information depends on what DM1 has done). We note, however, that in this case DM2 has a unique permanently optimal strategy γ_2^* against any strategy choice of u. This is derived by considering

$$\underset{\gamma_2 \varepsilon \Gamma_2}{\text{Min}} \; E[(\xi + \gamma_1(z_1) + \gamma_2(z_1, z_2, u))^2 + D_2\gamma_2^2]$$

$$= E_{\eta_2} \underset{v}{\text{Min}} \; E\{[(\xi + \gamma_1(z_1) + v)^2 + D_2 v^2] | \eta_2\}$$

$$(12) = E_{\eta_2} \underset{v}{\text{Min}} \; [(1 + D_2)v^2 + 2uE(\xi|\eta_2) + 2v[u + E|\xi|\eta_2)] + u^2 + E(\xi^2|\eta_2)]$$

in view of the fact that $E/\eta_2 \, (\gamma_1) = u$, $E/\eta_2(v) = v$. The minimization problem in (12) is strictly convex in v and has a unique solution

$$(13) \qquad v = -\frac{1}{(1+D_2)} [u + E(\xi|\eta_2)] \equiv \beta_1 z_1 + \beta_2 z_2 + \beta_3 u$$

Now substituting (13) into J_1, we have

$$(14) \qquad \underset{\gamma_1 \varepsilon \Gamma_1}{\text{Min}} \; J_1 = \underset{\gamma_1 \varepsilon \Gamma_1}{\text{Min}} \; E[(\xi + (1+\beta_3)u + \beta_1 z_1 + \beta_2 z_2)^2 + D_1 u^2]$$

This is a well defined strictly convex one person decision problem which has a unique solution for u.

$$(15) \qquad u = -\frac{(1+\beta_3)}{[(1+\beta_3)^2+D_1]} [\beta_1 z_1 + \beta_2 E(z_2/z_1) + E(\xi/z_1)] \equiv az_1$$

Eqs. (13) and (14) then constitute the unique Nash strategy pair (γ_1^*, γ_2^*) for Case C.

One final remark should be stressed here is that under Case C we are precluding situations where DM2 can make a pre-game announcement of committing irrevocably (and believably to I) to a particular strategy before DM1 acts. Otherwise, DM2 can always announce that he will ignore u, and/or z_1 and thus guarantee his cost under Case A. Other complications may also arise [1]. The purpose of this analysis is to display certain phenomenon. How to change the "rules" so as to eliminate or secure the phenomenon is important but not within the scope of this paper.

III. Comparison of Costs for the Four Information Structures

By virtue of the discussions in section II, we can solve for the unique affine Nash strategies for u and v in the form of

$$(16) \qquad \begin{aligned} u &= az_1 \\ v &= \beta_1 z_1 + \beta_2 z_2 + \beta_3 u + \beta_4 z_3 \end{aligned}$$

where β_1, β_3, or β_4 may be zero depending on whether or not it is Case B, C, or D. The specific formulas for a and β_i's in terms of s_1, s_2, D_1, D_2 are given in the appendix. Substituting (16) into (4) will then yield the Nash Equilibrium Cost

J_1 and J_2[†]. As a simple example to illustrate the variation of Nash cost in Cases A-D, we chose the specific symmetric case $s_1 = s_2 = D_1 = D_2 = \frac{1}{2}$.

It is clear that the Nash costs J_1^*, J_2^* are equal under Case A, i.e.

(17) $$J_{1A}^* = J_{2A}^*$$

what is perhaps surprising (at least at first glance) is that

(18)
$$J_{1A}^* > J_{1B}^*$$
$$J_{2A}^* < J_{2B}^*$$

i.e. increasing information entails higher cost for the receiver of information and lower cost for the giver. Even more surprising is that the situation accentuates

(19)
$$J_{1B}^* > J_{1C}^*$$
$$J_{2B}^* < J_{2C}^*$$

Finally, we have

(20)
$$J_{1A}^* > J_{1D}^*$$
$$J_{2A}^* > J_{2D}^* \quad \text{BUT} \quad J_{1D}^* < J_{2D}^*$$

i.e. availability of z_3 (or equivalently improving DM2's independent information) to DM2 helps DM1 more than DM2.

The inequalities of (18-20) can be dismissed as curiosities if they hold only for the particular values of s_i, D_j in question. However, we find that (19-20) holds for all values of all parameters and (18) holds over wide range or parameter value (about two orders of magnitude around the norminal value of $s_1 = s_2 = D_1 = D_2 = \frac{1}{2}$). The reasons behind these counter-intuitive results are explained below.

IV. Some Partial Explanations

Let us consider $L_1(u,v,\xi)$ and $L_2(u,v,\xi)$ both strictly convex in u and v. Let η_1, and η_2 be the information available to DM1 and DM2 respectively. Assume there exists a unique Nash equilibrium strategy pair $u = \gamma_1^*(\eta_1)$, $\gamma_1 \varepsilon \Gamma_1$ and $v = \gamma_2^*(\eta_2)$, $\gamma_2 \varepsilon \Gamma_2$ for the costs $J_1 = E[L_1(u,v,\xi)]$ $J_2 = E[L_2(u,v,\xi)]$ where Γ_1, Γ_2

[†] The resultant expressions are extremely complicated and requires computer program with symbolic manipulations capabilities to handle it. It should be emphasized that the theorems and results in section IV would not have been suspected if we did not have the help of these symbolic manipulative tools.

are respectively the classes of η_1, η_2 measurable strategies. Now we furnish DM2 with the additional information as to the value of u, i.e. $\eta_2' = (\eta_2, u)$. Let us assume again that there exists a unique Nash strategy pair $u = \gamma_1^o(\eta_1)$, $\gamma_1 \varepsilon \Gamma_1$ and $v = \gamma_2^o(\eta_2')$, $\gamma_2 \varepsilon \Gamma_2'$ for the costs where Γ_2' is similarly defined as the class of η_2'-measurable strategies. We then have

Theorem 2. $J_1(\gamma_1^*, \gamma_2^*) \geq J_1(\gamma_1^o, \gamma_2^o)$ if $\eta_2 \supseteq \eta_1$

In other words, so long as the player v knows as much as u, then it always pays for u to reveal his action to v.

Proof: By definition of (γ_1^o, γ_2^o) we have

(21) $$J_1(\gamma_1^o, \gamma_2^o) \leq J_1(\gamma_1^*, \gamma_2^o)$$

Now consider $\underset{\gamma_2 \varepsilon \Gamma_2'}{\text{Min}} J_2(\gamma_1^*, \gamma_2)$

$$= \underset{\gamma_2 \varepsilon \Gamma_2'}{\text{Min}} E_{\eta_2'} E[L_2(u = \gamma_1^*(\eta_1), v, \xi)/\eta_2']$$

$$= E_{\eta_2'} \underset{v}{\text{Min}} E[L_2(u = \gamma_1^*(\eta_1), v, \xi)/\eta_2'] \Rightarrow v = \gamma_2^o(\eta_2')$$

where $E(\gamma_1/\eta_2') = u$ regardless of the strategy γ_1 employed by DM1. In short, γ_2^o is a permanently optimal strategy in the sense of Von Neumann. On the other hand, since $\eta_2 \supseteq \eta_1$, and $E[\gamma_1^* (\eta_1)/\eta_2] = \gamma_1^*(\eta_1) = E[\gamma_1^*(\eta_1)/\eta_2'] = u$

$$\underset{\gamma_2 \varepsilon \Gamma_2}{\text{Min}} J_2(\gamma_1^*, \gamma_2) \Leftrightarrow \underset{v}{\text{Min}} E[L(u,v,\xi)/\eta_2']$$

or

$$\gamma_2^o(\eta_2'; \gamma_1^*) = \gamma_2^*(\eta_2; \gamma_1^*)$$

Consequently, we have

(22) $$J_1(\gamma_1^*, \gamma_2^o) = J_1(\gamma_1^*, \gamma_2^*)$$

which in connection with (21) is the desired result. ■

Theorem 2 explains the first inequality of (19) completely. The second inequality can be explained as follows:

First of all note that, for our problem, the information available to DM2 i.e. z_1, z_2, could equally well be expressed by $\hat{\xi} \equiv E(\xi/z_1)$, and $e_2 \equiv z_2 - \hat{\xi}_1$

Now consider cases B' and C' where the information structures are

Case B' $\quad \eta_1: \hat{\xi}_1$

$\quad \eta_2: \hat{\xi}_1$

Case C' η_1: $\hat{\xi}_1$

η_2: $(\hat{\xi}_1, u)$

By virtue of the results of § III, we know that there exists a unique affine Nash strategy pair for each case with

$$u = a_{B'}\hat{\xi}_1$$

(23) for Case B'

$$v = b_{B'}\hat{\xi}_1$$

$$u = a_{C'}\hat{\xi}_1$$

(24) for Case C'

$$v = \beta_{1C'}\hat{\xi}_1 + \beta_{3C'}u$$

Lemma 2. If we express the Nash strategy pair under Cases B and C as

$$u = a_B\hat{\xi}_1$$

(25)

$$v = \beta_{1B}\hat{\xi}_1 + \beta_{2B}e_2$$

$$u = a_C\hat{\xi}_1$$

(26)

$$v = \beta_{1C}\hat{\xi}_1 + \beta_{2C}e_2 + \beta_{3C}u$$

then

$$a_B = a_{B'} \qquad a_C = a_{C'}$$

$$\beta_{2B} = \beta_{2C}$$

$$\beta_{1B} = \beta_{1B'} \qquad \beta_{3C} = \beta_{3C'}$$

Proof: Let us first consider information structures B and B'. For either case, the Nash strategies satisfy

(27a) $$u = -\frac{1}{1+D_1}E[\xi + \gamma_2(\eta_2)|\eta_1]$$

(27b) $$v = -\frac{1}{1+D_2}E[\xi + \gamma_1(\eta_1)|\eta_2]$$

with η_1 and η_2 interpreted accordingly for each case. Now, substituting (27b) into (27a) we have

$$u = -\frac{1}{1+D_1} E[\xi|\eta_1] + \frac{1}{1+D_1} \cdot \frac{1}{1+D_2} E\{E[\xi + \gamma_1(\eta_1)|\eta_2]|\eta_1\}$$

$$= -\frac{1}{1+D_1} E[\xi|\eta_1] + \frac{1}{1+D_1} \cdot \frac{1}{1+D_2} E\{E[\xi|\eta_2]|\eta_1\}$$

$$(28) \qquad + \frac{1}{1+D_1} \cdot \frac{1}{1+D_2} E\{E[\gamma_1(\eta_1)|\eta_2]|\eta_1\}.$$

Since $\eta_2-\eta_1$ is statistically independent of η_1 for both cases and since $\gamma_1(.)$ is η_1-measurable the last expression in (28) can be written as

$$(29) \qquad \frac{1}{1+D_1} \cdot \frac{1}{1+D_2} \gamma_1(\eta_1)$$

and it follows from (28) that $a_B = a_B'$. Now using this fact in (27b) gives

$$(30) \qquad v = -\frac{1}{1+D_2} E[\xi|\eta_2] - \frac{1}{1+D_2} a_B\hat{\xi}_1$$

for both information structures. It should be obvious from (30) that independence of $\hat{\xi}_1$ and e_2 imply $\beta_{1B} = \beta_{1B'}$.

Now let us consider information structures C and C'. Interpreting these as information structures giving rise to player 1 acting first, the permanently optimal policy for player 2 is

$$(31) \qquad v = -\frac{1}{1+D_2} [E[\xi|\eta_2] + u]$$

from which it follows that $\beta_{3C} = \beta_{3C'}$. Comparing (31) with (30) imply $\beta_{2B} = \beta_{2C}$. Now, substituting (31) into J_1, optimizing over η_1-measurable strategies for player 1 and using the fact $E\{E[\xi|\eta_2]|\eta_1\} = E[\xi|\eta_1]$ for both information structures (because $\eta_1 \subseteq \eta_2$), it follows that the optimal policy of player 1 is the same regardless of which which information structure is used. Hence $a_C = a_{C'}$. ■

The significance of Lemma 2 is that for the purpose of investigating the inequalities of (19), we might as well be dealing with B' and C' instead of B and C. We can see this easily if we substitute into the costs the respective strategies (23-26). Thus

<u>Lemma 3</u>. $J^*_{1B} - J^*_{1C} = J^*_{1B'} - J^*_{1C'}$

$$J^*_{2B} - J^*_{2C} = J^*_{2B'} - J^*_{2C'}$$

Proof: Follows from the equality $\beta_{2B} = \beta_{2C}$ and independence of e_2 and $\hat{\xi}_1$.

Now (27) under Case B' is a pair of simultaneous linear equations in u and v parameterized by $\hat{\xi}_1$. A typical case is illustrated in Figure 1 for the case $D_1 = D_2 = 1$ $\hat{\xi}_1 = 1$. The Nash strategy pair is the point N_B

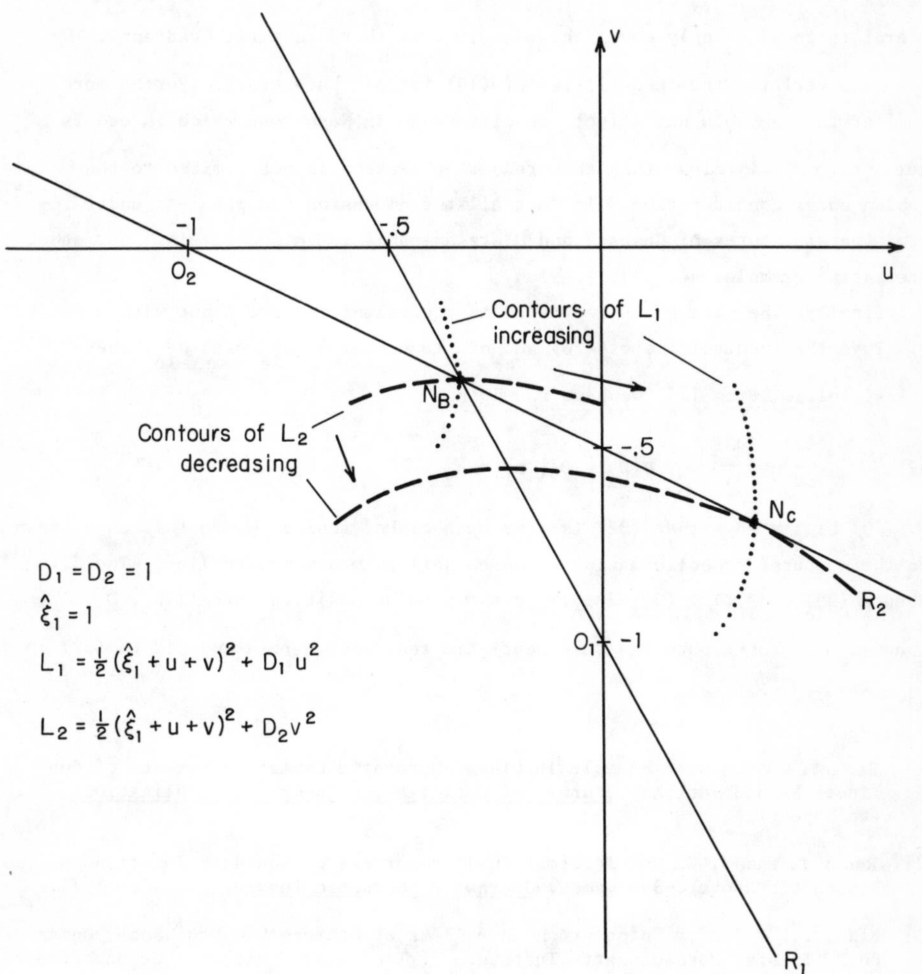

R_1 and R_2 are the reaction curves of u and v respectively (equivalent to the two linear equations defined by (27)). The contour of constant $(\hat{\xi}_1 + u + v)^2 + D_1u^2$ and $(\hat{\xi}_1 + u + v)^2 + D_2v^2$ are ellipses with center at point 0_1 and 0_2. Now under Case C', DM1 knows that DM2 will react according to R_2. Thus he must choose v such that the least cost contour $(\hat{\xi}_1 + u + v)^2 + D_1 u^2$ is tangent to R_2. This implies that N_C will be located Southeast of N_B; consequently, at a point of higher cost for $(\hat{\xi}_1 + u + v)^2 + D_2 v^2$. This reduces the explanation of the second (as well as the first) inequality of (19) to a simple graphical inspection problem as illustrated in Figure 1. Since D_1, $D_2 > 0$ and s_1 effects only $\hat{\xi}_1$, Figure 1 in fact is a representation of the Case B vs. Case C situation in general (sign of $\hat{\xi}_1$ only moves the picture from third to first quadrant). We have thus verified the inequalities of (19) for all parameters. Furthermore, it it clear that s_2 does not effect the difference in Nash cost which indeed is true. Also, it is clear that this reduction process is not limited to the problem under consideration. In fact all two dimension LQG problems under information structures of Cases B and C are amenable to this analysis. For other interesting examples see [3].

Finally, the Case D can be viewed as equivalent to Case A but with a smaller s_2. Thys the inequalities of (20) amounts to comparing J^*_{iA} with $J^*_{iA'}$ where $s_2 > s'_2$ i=1,2. From [2], we have shown that

$$(32) \qquad \frac{\partial J_1^*}{\partial s_2} < \frac{\partial J_2^*}{\partial s_2} \qquad \forall\, D_1 = D_2,\ s_1 = s_2$$

Eq. (20) simply says that (32) is true on a global scale. We do this once again via the computer symbolic manipulation capability mentioned earlier. The difference $(J^*_{1A} - J^*_{1A'}) - (J^*_{2A} - J^*_{2A'})$ is shown to be positive for all D_1, D_2, s_1 and s_2. We attribute the same heuristic reasoning here as we did to (32) in [2].

References

[1] Basar,T., A Counter Example in Linear-Quadratic Games: Existence of Nonlinear Nash Solutions, Journal of Optimization Theory and Applications, (to appear).

[2] Basar,T. and Y. C. Ho, Informational Properties of the Nash Solutions of Two Stochastic Nonzero-Sum Games, Journal of Economic Theory, 7, April 1974.

[3] Blau, I., Value of Information in a Class of Nonzero Sum Stochastic Games, Ph.D. Thesis, Massachusetts Institute of Technology, Mathematics Department, June 1974.

[4] Ho, Y. C. and K. C. Chu, Information Structure in Dynamic Multi-Person Control Problems, Automatica, to appear July 1974.

[5] Radner, R., Team Decision Problems, Annals of Mathematical Statistics, Vol. 33, No. 3, pp. 857-881, 1962.

Appendix

Here we provide the formulas for the Nash strategies for various cases.

Case A

$$u = az_1$$

$$v = \beta_1 z_1 + \beta_2 Z_2 + \beta_3 u + \beta_4 z_3$$

where

$$\beta_1 = \beta_3 = \beta_4 = 0$$

$$a = -\frac{1}{1+D_1}\left(\frac{D_2}{1+D_2} + s_2\right)/K$$

$$\beta_2 = -\frac{1}{1+D_2}\left(\frac{D_1}{1+D_1} + s_1\right)/K$$

$$K = [(1+s_1)(1+s_2) - \frac{1}{1+D_1} \cdot \frac{1}{1+D_2}]$$

Case B

$$u = az_1$$

$$v = \beta_1 z_1 + \beta_2 z_2 + \beta_3 u + \beta_4 z_3$$

where

$$\beta_3 = \beta_4 = 0$$

$$a = \frac{-D_2}{(D_2+D_1+D_1D_2)(1+s_1)}$$

$$\beta_1 = [\frac{s_2}{(s_1s_2+s_1+s_2)} - \frac{D_2}{(D_1+D_2+D_1D_2)(1+s_1)}](-\frac{1}{1+D_2})$$

$$\beta_2 = (-\frac{1}{1+D_2})\frac{s_1}{(s_1s_2+s_1+s_2)}$$

Case C

$$u = az_1$$

$$v = \beta_1 z_1 + \beta_2 z_2 + \beta_3 u + \beta_4 z_3$$

where

$$\beta_4 = 0$$

$$a = \frac{-(1+\chi_1)}{[(1+\chi_1)^2+D_1]}[\frac{1}{1+s_1} + \chi_2 + \frac{\chi_3}{1+s_1}]$$

$$\chi_1 = -\frac{1}{1+D_2} \qquad \chi_2 = \frac{\chi_1 s_2}{\Sigma} \qquad \chi_3 = \frac{\chi_1 s_1}{\Sigma} \qquad \Sigma = s_1+s_2+s_1s_2$$

$$\beta_1 = -\frac{1}{1+D_2}\left[\frac{s_2}{s_1 s_2 + s_1 + s_2}\right]$$

$$\beta_3 = -\frac{1}{1+D_2}$$

$$\beta_2 = -\frac{1}{1+D_2}\left[\frac{s_1}{s_1 s_2 + s_1 + s_2}\right]$$

<u>Case D</u>

Same as Case A with $s_2'(<s_2)$ replacing s_2 in the above expressions.

STOCHASTIC DIFFERENTIAL GAMES AND ALTERNATE PLAY

ROBERT J. ELLIOTT,

UNIVERSITY OF HULL, ENGLAND

1. INTRODUCTION

Consider a static two-person zero sum competitive situation where a continuous real valued function $\phi(y,z)$ is given. A player, or controller, J_1 is to choose y from a compact space Y with the object of maximising ϕ , and a second player J_2 is at the same time to choose z from a similar space Z with the object of minimising ϕ . If Y and Z are finite the situation is that of the classic zero sum matrix game.

A saddle point is a pair of controls (y^*,z^*) such that

$$\phi(y,z^*) \leq \phi(y^*,z^*) \leq \phi(y^*,z)$$

for all other pairs $(y,z) \in YxZ$.

If a saddle point exists $\phi(y^*,z^*)$ is called the value of the game; it is the best that either player can force. In general, however, there is not a saddle point unless ϕ satisfies convexity conditions. In the matrix game situation von Neumann introduced mixed strategies, that is probabilities over the finite sets Y and Z. The outcome of a pair of mixed strategies is defined to be the expected value. In this more general situation the outcome is linear in the mixed strategies and so a saddle point value exists; this is the conclusion of von Neumann's celebrated minimax theorem.

By introducing the space of probability measures $\Lambda(Y)$ and $\Lambda(Z)$ over Y and Z, respectively, Wald [7] extended von Neumann's result to the situation described in the first paragraph. For $\sigma \in \Lambda(Y)$ and $\tau \in \Lambda(Z)$ the outcome $\phi(\sigma,\tau)$ is defined to be $\int_Y\int_Z \phi(y,z)\, d\sigma(y)\, d\tau(z)$.

$\phi(\sigma,\tau)$ is again linear in σ and τ, with the weak-* topologies $\Lambda(Y)$ and $\Lambda(Z)$ are compact, and so a saddle point (σ^*,τ^*) exists.

Clearly the outcome $\phi(\sigma,\tau)$ can be considered as the limit of averages of repeated plays. However, in this paper we wish to consider a different method of averaging repeated plays in which the two players make choices of controls alternately rather than simultaneously as in the above model.

By introducing related stochastic games we then consider two person zero sum differential games which are played in this manner. We show their value exists and is approximated by the solution of a non-linear parabolic equation. The results generalize those announced by Danskin [1], but the methods are different. The work described here was done in co-operation with A. Friedman and N. Kalton and a full account will appear in the Journal of Differential Equations.

2. ALTERNATE PLAY

ϕ, Y, Z, J_1 and J_2 are as in paragraph 1. For $0 \leq \sigma \leq 1$ write:

$$V_n^\sigma(z) = \max_y \min_\zeta \left(\sigma\phi(y,z) + (1-\sigma)\phi(y,\zeta) + V_{n-1}^\sigma(\zeta)\right),$$

$$V_0^\sigma(z) = 0$$

$$V_n^\sigma = \min_z V_n(z)$$

V_n^σ can be considered as the value of a two-person zero sum game with $2n+1$ alternate moves and payoff $\int\phi(y,z)dt$. Minimizing player J_2, choosing elements of Z, makes an initial choice of z at time 0 and then choices at times $j+\sigma$, $j = 0,\ldots,n-1$. Maximising player J_1, who chooses elements of Y, makes choices at times $j = 0,\ldots,n-1$. Write $B = \max_y \max_z |\phi(y,z)|$. Expanding V_{m+n}^σ we see

$$V_{m+n}^\sigma = \min_{z_0} \max_{y_1} \min_{z_1} \ldots \max_{y_{m+n}} \min_{z_{m+n}} \left(\sigma\phi(y_1,z_0) + \ldots + (1-\sigma)\phi(y_{m+n},z_{m+n})\right)$$

and so $$|V_{m+n}^\sigma - V_m^\sigma - V_n^\sigma| \le 4B. \quad (2.1)$$

Consider the related sequence $r_n^\sigma = V_n^\sigma/n$, $n = 1, 2, \ldots$.

Then r_n^σ can be considered as the value of a two-person zero sum game played over the interval $[0,1]$, where J_2 makes choices of z at times 0 and $(j+\sigma)/n$ and J_1 makes choices of y at times j/n, $j = 0,\ldots n-1$.

By using (2.1) and considering r_{mn} it is easily established that

$$|r_{mn}^\sigma - r_m^\sigma| \le 4B/m,$$

and $$|r_{mn}^\sigma - r_n^\sigma| \le 4B/n.$$

Consequently

$$|r_m^\sigma - r_n^\sigma| \le 8B(1/m + 1/n).$$

Thus r_n^σ is a Cauchy sequence. Writing $\phi_\sigma = \lim_{n\to\infty} r_n^\sigma$ we see that ϕ_σ is the limit of average values of the $\phi(y,z)$ game when the moves of the two players overlap in the ratio $\sigma : (1-\sigma)$.

Clearly $\phi_0 = \max_y \min_z \phi(y,z)$

$\phi_1 = \min_z \max_y \phi(y,z)$

and $\phi_0 \le \phi_\sigma \le \phi_1$.

Note that in the games used to define V_n^σ and r_n^σ the players used constant controls at each step. It is interesting that if they can use measurable functions in general a different result is obtained.

DEFINITION 2.1. A function $m(t)$ from $[a,b]$ into a compact space M is said to be measurable if the composite function $f(m(t))$ is Lebesgue measurable for all continuous real valued functions f on M.

Write $M_1([a,b])$ (resp. $M_2([a,b])$) for the set of measurable functions on $[a,b]$ with the values in Y (resp. Z).

For $0 \le \sigma \le 1$ define

$$W_n(z) = \sup_{y \in M_1([0,1])} \inf_{\zeta \in M_2([\sigma,1+\sigma])} \left(\int_0^\sigma \phi(y(t),z(t))dt + \int_\sigma^1 \phi(y(t),\zeta(t))dt + W_{n-1}^\sigma(\zeta) \right),$$

$$W_n^\sigma(z) = 0 \ ,$$

and $W_n^\sigma(z) = \inf_{z \in M_2([0,\sigma])} W_n^\sigma(z)$.

Because J_1 and J_2 are now playing functions they can anticipate their opponent's choice of control and

$$W_n = n(\sigma\phi_1 + (1-\sigma)\phi_0) \ .$$

So if $\quad s_n = W_n^\sigma/n \quad$ we see

$$\lim_{n\to\infty} s_n = \tilde{\phi}_\sigma = \sigma\phi_1 + (1-\sigma)\phi_0 \ .$$

3. DIFFERENTIAL GAMES

Consider a differential game $G(0,0)$ with dynamics

$$\frac{dx}{dt} = f(t,x,y,z)$$

and initial condition $x(0) = 0 \in R^m$. Here $x \in R^m$, $t \in [0,1]$ and Y and Z are compact metric spaces. $f : [0,1] \times R^m \times Y \times Z \to R^m$ is a continuous function satisfying a uniform Lipschitz condition in x and t:

$$\|f(t_1,x_1,y,z) - f(t_2,x_2,y,z)\| \le K(\|x_1 - x_2\| + |t_1 - t_2|) \ .$$

NOTATION: K will denote the Lipschitz constant in all cases.

The payoff is

$$P = g(x(1)) + \int_0^1 h(t,x,y,z)dt$$

but to simplify the exposition we suppose that P has the form

$$P = g(x(1))$$

where g is twice continuously differentiable and its derivatives $\frac{\partial g}{\partial t}$, $\frac{\partial g}{\partial x_i}$, $\frac{\partial^2 g}{\partial x_i \partial x_j}$ are uniformly Lipschitzian in (t,x). More general payoffs can then be discussed by approximation as in [2].

As in [2] section 9 we may suppose f and g vanish outside compact sets. Write B for an upper bound for $\|f\|$ and $|g|$.

The game is zero sum so the player J_1 controlling the y variable is trying to maximise P while J_2 controlling z is trying to minimize P. Consider for any integer N a partition of $[0,1]$ into N subintervals

$$I_1 = [0,t_1],\ I_j = [t_{j-1},t_j],\ 0 \le j \le N$$

where $t_j = j\delta$ and $\delta = 1/N$.

Previously Fleming [5] and Friedman [6] have approximated the method of play in differential games by considering for each N an "upper" and a "lower" game. In the "upper" games J_2 plays first on each interval I_j and then, with the knowledge of the control chosen by J_2, J_1 plays second on I_j . In the "lower" game J_1 played first and J_2 second on each I_j .

Similarly to that used in the definition of r_n^σ consider an alternate method of play: for $0 \le \sigma \le 1$ and $0 \le j \le N-1$ write $t_{j+\sigma} = j+\sigma/N$. At $t_0 = 0$, J_2 chooses $z_0 \in Z$. With the knowledge of z_0 , J_1 chooses $y_1 \in Y$. Then at $t_{j+\sigma}$, with the knowledge of $z_0,\ldots z_j$ and $y_1 \ldots y_{j+1}$, J_2 chooses $z_{j+1} \in Z$, and at t_j with a knowledge of z_0, ... z_j and $y_1 \ldots y_j$, J_1 chooses $y_{j+1} \in Y$. These choices determine piecewise constant control functions and so a trajectory and payoff.

Define

$$VC^N(\sigma,0,0) = \min_{z_0} \max_{y_1} \ldots \max_{y_n} \min_{z_n} g(x(1)) .$$

This is the value of the approximation $2N+1$ stage differential game $GC^N(\sigma,0,0)$ (C for piecewise constant controls) in which J_1 and J_2 have reaction times $(1-\sigma)/N$ and σ/N respectively.

However, below we also consider a related alternate move game $GF^N(\sigma,0,0)$, (F for functions), in which the players choose their controls at the same times as in $GC^N(\sigma,0,0)$ but they are now allowed to play measurable functions. They, therefore, choose a family of functions $z_0(t),y_1(t),z_1(t)$, ... $y_n(t),z_n(t)$ which substituting into the dynamics determines a trajectory and payoff.

Define

$$VF^N(\sigma,0,0) = \inf_{z_0(t)} \sup_{y_1(t)} \ldots \sup_{y_n(t)} \inf_{z_n(t)} g(x(1))$$

4. STOCHASTIC GAMES

NOTATION

To obtain the approximation results for GC games it is necessary to consider sub-partitions of $[0,1]$ into MN equal subintervals. This sub-partitioning is not necessary for GF games, but they will be discussed this way so that both games can be treated simultaneously.

For $0 \le p < M$ and $0 \le q < N$ write $t_{(p,q)} = (pN+q)/MN$

and $t_{(p,q+\sigma)} = (pN + q + \sigma)/MN$ for $0 \le \sigma \le 1$.

Write $t_{(p,0)} = t_p$ and $t_{(p,N)} = t_{(p+1,0)}$.

We now wish to consider alternate move games $GC^{MN}(\sigma,t_j,\xi)$ and $GF^{MN}(\sigma,t_j,\xi)$ which start at the intermediate time t_j and position $\xi \in R^M$. Corresponding to the MN partition the method of alternate play is as described in section 3: J_2 chooses control values $z_{p,q+1} \in Z$ in GC, (resp. control functions $z_{p,q+1}(t)$ in GF), at times $t_{(p,q+\sigma)}$ and J_1 chooses control values $y_{p,q+1} \in Y$ in GC (resp. control functions $y_{p,q+1}(t)$ in GF) at times $t_{(p,q)}$.

A strategy α for J_1 in $GC^{MN}(\sigma,t_j,\xi)$ (resp. $GF^{MN}(\sigma,t_j,\xi)$) consists of functions:

$$\begin{array}{llll} \alpha_{j,1} & \alpha_{j,2} & \cdots & \alpha_{j,N} \\ \alpha_{j+1,1} & \alpha_{j+1,2} & \cdots & \alpha_{j+1,N} \\ \vdots & & & \\ \alpha_{M-1,1} & \alpha_{M-1,2} & \cdots & \alpha_{M-1,N} \end{array}$$

such that $\alpha_{j+p,q}$ is a function of $z_{j,0}, y_{j,1} \dots y_{j+p,q-1}, z_{j+p,q-1}$ with values in Y (resp. a function of $z_{j,0}(t), y_{j,1}(t), \dots y_{j+p,q-1}(t), z_{j+p,q-1}(t)$ with values in $M_1([t_{(j+p,q)}, t_{(j+p,q+1)}]))$. Write $AC(j)$ for the set of all such strategies for J_1 in GC and $AF(j)$ for the set of such strategies in GF.

A strategy β for J_2 in $GC^{MN}(\sigma,t_j,\xi)$ (resp. $GF^{MN}(\sigma,t_j,\xi)$) consists of a control value $\beta_{j,0} = z_{j,0} \in Z$ (resp. a function $\beta_{j,0} = z_{j,0}(t) \in M_2([t_j, t_{(j,\sigma)}]))$ together with functions

$$\begin{array}{llll} \beta_{j,1} & \beta_{j,2} & \cdots & \beta_{j,N} \\ \beta_{j+1,1} & \beta_{j+1,2} & \cdots & \beta_{j+1,N} \\ \vdots & & & \\ \beta_{M-1,1} & \cdots\cdots\cdots & & \beta_{M-1,N} \end{array}$$

such that $\beta_{j+p,q}$ is a function of $z_{j,0}, y_{j,1} \dots, z_{j+p,q-1}, y_{j+p,q}$, (resp. a function of $z_{j,0}(t), y_{j,1}(t), \dots z_{j+p,q-1}(t), y_{j+p,q}(t)$). Write $BC(j)$ for the set of all such strategies in GC and $BF(j)$ for the set of such strategies in GF.

Given a pair of strategies $\alpha \in AC(j)$, $\beta \in BC(j)$ piecewise constant control functions for J_1 and J_2 are determined step by step. Substituting in the dynamics a trajectory $x(t)$ and payoff $g(x(1))$ are obtained and the value of the game $GC^{MN}(\sigma,t_j,\xi)$ is defined to be

$$VC^{MN}(\sigma,t_j,\xi) = \sup_{\alpha\in AC(j)} \inf_{\beta\in BC(j)} g(x(1)).$$

Similarly strategies $\alpha \in AF(j)$, $\beta\in BF(j)$ determine step-by-step measurable control functions and the value of $GF^{MN}(\sigma,t_j,\xi)$ is

$$VF^{MN}(\sigma,t_j,\xi) = \sup_{\alpha\in AF(j)} \inf_{\beta\in BF(j)} g(x(1)) \quad .$$

STOCHASTIC GAMES

We now introduce $GC_\varepsilon^{MN}(\sigma,t_j,\xi)$ and $GF_\varepsilon^{MN}(\sigma,t_j,\xi)$, stochastic versions of GC^{MN} and GF^{MN} in which an amount ε of "noise" is introduced into the dynamics at the times t_{j+p}, $p=1,\ldots, M-j$.

Suppose $\{\eta_{ij}, i=1,2,\ldots,N,\ j=1,2,\ldots,m\}$ is a collection of independent random variables each taking the values ∓ 1 with probability $1/2$. Write η_i for the vector $(\eta_{i1},\ldots\eta_{im}) \in R^M$ and $\eta = (\eta_1,\ldots,\eta_N)$.

In $GC_\varepsilon^{MN}(\sigma,t_j,\xi)$ (resp. $GF_\varepsilon^{MN}(\sigma,t_j,\xi)$) piecewise constant controls (resp. measurable functions) are chosen by J_1 and J_2 at the same times as in $GC^{MN}(\sigma,t_j,\xi)$ or $GF^{MN}(\sigma,t_j,\xi)$. A strategy α^ε for J_1 in $GC_\varepsilon^{MN}(\sigma,t_j,\xi)$ (resp. $GF_\varepsilon^{MN}(\sigma,t_j,\xi)$) consists of functions $\{\alpha^\varepsilon_{j+p,q}\}$ as before, but now $\alpha^\varepsilon_{j+p,q}$ is a function of $z_{j,0}$, $y_{j,1} \ldots y_{j+p,q-1}$, $z_{j+p,q-1}$, and $\eta_{j+1},\ldots \eta_{j+p}$. Write $AC^\varepsilon(j)$ (resp. $AF^\varepsilon(j)$) for the set of all such strategies.

A strategy β^ε for J_2 in $GC_\varepsilon^{MN}(\sigma,t_j,\xi)$ (resp. $GF_\varepsilon^{MN}(\sigma,t_j,\xi)$) consists of $\beta^\varepsilon_{j,0} = z_{j,0} \in Z$ (resp. $\beta_{j,0} = z_{j,0}(t) \in M_2([t_j,t_{j+\sigma}])$) and functions $\{\beta^\varepsilon_{j+p,q}\}$, $q \geq 1$, where $\beta^\varepsilon_{j+p,q}$ is a function of $z_{j,0}$, $y_{j,1} \ldots z_{j+p,q-1}$, $y_{j+p,q}$ and $\eta_{j+1}, \ldots , \eta_{j+p}$. Write $BC^\varepsilon(j)$ (resp. $BF^\varepsilon(j)$) for the set of all such strategies.

NOTATION

Below we wish to discuss both piecewise constant alternate move games $GC^{MN}(\sigma,t_j,\xi)$, measurable control function alternate move games $GF^{MN}(\sigma,t_j,\xi)$ and the stochastic versions of them GC_ε^{MN} , GF_ε^{MN} . We have seen that their definitions and the definitions of their strategies and values are identical, except that one is referring to piecewise constant controls and the other to measurable controls. Therefore below G^{MN} will denote either GC^{MN} or GF^{MN} and $A(j)$ will denote either $AC(j)$ or $AF(j)$, according to context. $B(j)$, $A^\varepsilon(j)$ and $B^\varepsilon(j)$ should be interpreted similarly.

STOCHASTIC TRAJECTORIES

Given $\alpha^{\varepsilon} \in A\ (j)$, $\beta^{\varepsilon} \in B\ (j)$ and $\eta = (\eta_1,\dots,\eta_N)$ a pair of control functions $y(t)$, $z(t)$ are determined by the components of α^{ε} and β^{ε}. The trajectory is defined to be the discontinuous solution of the equation

$$x_{\eta}(t) = \xi + \int f(t,x_{\eta}(t),y(t)z(t))dt + \varepsilon\delta^{1/2} \sum_{t_j \le t_k < t} \eta_k ,$$

where $\delta = M^{-1}$.

The corresponding payoff is

$$P_{\eta}^{j}(\sigma,\xi,\ \alpha^{\varepsilon},\beta^{\varepsilon}) = g(x_{\eta}(1))$$

and with E denoting expectation

$$P^{j}(\sigma,\xi,\ \alpha^{\varepsilon},\beta^{\varepsilon}) = Eg(x_{\eta}(1))$$

The value of α^{ε} to J_1 is defined to be:

$$U_j(\sigma,\xi,\alpha^{\varepsilon}) = \inf_{\beta^{\varepsilon} \in B^{\varepsilon}(j)} P^{j}(\sigma,\xi,\alpha^{\varepsilon},\beta^{\varepsilon})$$

and the σ value of the game $G^{MN}(\sigma,t_j,\xi)$ is:

$$V^{MN}(\sigma,t_j,\xi) = \sup_{\alpha^{\varepsilon} \in A^{\varepsilon}(j)} U_j(\sigma,\xi,\alpha^{\varepsilon}) .$$

5. APPROXIMATIONS TO THE VALUE

Recall that B denotes an upper bound for $\|f\|$ and $|g|$ and K the general Lipschitz constant. By establishing Lipschitz continuity properties of V^{MN} in t and ξ the following result can be established.

THEOREM 5.1. If $s < N$

$$|V^{MN+s}(\sigma,0,0) - V^{MN}(\sigma,0,0)| \le Qs/MN{+}s$$

where

$$Q = (B + K + 3BK)e^{K} .$$

By adapting methods of [4] the following relation between the values of the deterministic and stochastic games is established:

THEOREM 5.2.

$$|V_{\varepsilon}^{MN}(\sigma,t_j,\xi) - V^{MN}(\sigma,t_j,\xi)| \le 2m\ Ke^{K}\varepsilon .$$

Consider now the function of $y \in Y$ and $z \in Z$:

$$H(t,x,p;y,z) = p.f(t,x,y,z)$$

for fixed $t \in [0,1]$, $x \in R^{m}$ and $p \in R^{m}$. Similar to the sequence r_n^{σ} discussed in section 2 for the *GC* games we introduce the following Hamiltonians: for $0 \le \sigma \le 1$ and $n = 1,2,\ldots$ consider the sequence of functions:

$$H_n^C(t,x,p) = \min_{z_0} \max_{y_1} \ldots \max_{y_n} \min_{z_n} n^{-1}(\sigma H(t,x,p;y_1,z_0) + (1-\sigma)H(t,x,p;y_1,z_1) + \ldots +$$
$$+ \sigma H(t,x,p;y_n,z_{n-1}) + (1-\sigma)H(t,x,p;y_n,z_n)) \quad .$$

In section 2 we showed the functions H_n^C converge to a limit function $H^C(t,x,p)$.

For the *GF* games we introduce the Hamiltonians

$$H_\sigma^F(t,x,p) = \sigma \min_z \max_y H(f,x,p;y,z) + (1-\sigma) \max_y \min_z H(t,x,p;y,z) \quad .$$

NOTATION: H_σ below will denote either H_σ^C or H_σ^F according to context.

From Friedman [6] we quote:

THEOREM 5.3. For $\varepsilon > 0$ there is a unique solution ϕ of the equation

$$(\varepsilon^2/2)\,\nabla^2\phi + \partial\phi/\partial t + H_\sigma(t,x,\nabla\phi) = 0$$

satisfying

$$\phi(1,\xi) = g(\xi) \quad .$$

Further, ϕ has the property that $\partial\phi/\partial t$ and $\partial^2\phi/\partial x_i\partial x_j$ satisfying Hölder conditions of the form:

$$|\psi(t,x) - \psi(t',x')| \le Q(|t-t'|^{\gamma/2} + \|x-x'\|^\gamma) \text{ for } 0 < \gamma < 1 \quad .$$

By adapting the method of [4] and working step-by-step we can then relate the value of the stochastic game with the above solution as follows:

THEOREM 5.4. For $\varepsilon > 0$ and any time t_j

$$\sup_{\xi\in R^m} |\phi(t_j,\xi) - V^{MN}(t_j,\xi)| \le CM^{-\gamma/2} + DN^{-1}$$

where C and D are independent of M,N,j and ε. Here V_ε^{MN} denotes either VC_ε^{MN} or VF_ε^{MN}.

In particular

$$\sup_{\xi\in R^m} |\phi(0,0) - V_\varepsilon^{MN}(\sigma,0,0)| \le CM^{-\gamma/2} + DN^{-1} \quad .$$

6. THE EXISTENCE OF VALUE

By comparing the value $V^N(\sigma,0,0)$ with the value of the related stochastic game, as in Theorem 5.2, and the value of the stochastic game with the solution of the parabolic equation, as in Theorem 5.4, it can be shown that $\lim_{N\to\infty} V^N(\sigma,0,0)$ exists, that is the game played in the alternate manner described in section 2 has a limiting value as the size of the partition decreases.

THEOREM 6.1. $V^N(\sigma,0,0)$ is a Cauchy sequence, so $\lim_{N\to\infty} V^N(\sigma,0,0) = V(\sigma)$ exists.

PROOF. Choose $\nu > 0$. With the notation of Theorems 5.1 and 5.4 choose M_0 such that

$$Q/M < \nu/6 \text{ if } M \geq M_0$$

$$Q/M^{\gamma/2} < \nu/6 \text{ if } M \geq M_0 \quad .$$

Choose N_0 such that

$$D/M < \nu/6 \text{ if } N \geq N_0$$

Then if n_1, n_2 are greater than

$$n_0 = M_0 N_0$$

$$n_1 = M_1 N_0 + s_1 \text{ with } M_1 \geq M_0 \text{ and } s_1 < N_0$$

$$n_2 = M_2 N_0 + s_2 \text{ with } M_2 \geq M_0 \text{ and } s_2 < N_0 \quad .$$

Consequently, by Theorem 5.1

$$|V^{n_1}(\sigma,0,0) - V^{M_1N_0}(\sigma,0,0)| \leq Qs_1/M_1N_0 + s_1 < Q/M_1 < \nu/6 \ ,$$

and $$|V^{n_2}(\sigma,0,0) - V^{M_2N_0}(\sigma,0,0)| < \nu/6 \ .$$

By Theorem 5.2, for any ε

$$|V^{M_1N_0}(\sigma,0,0) - V_\varepsilon^{M_1N_0}(\sigma,0,0)| \leq 2mKe^K\varepsilon$$

and $$|V^{M_2N_0}(\sigma,0,0) - V_\varepsilon^{M_2N_0}(\sigma,0,0)| \leq 2mKe^K\varepsilon$$

and from Theorem 5.4:

$$|\phi(0,0) - V_\varepsilon^{M_1N_2}(\sigma,0,0)| \leq C/M_1^{\gamma/2} + D/N_0 < \nu/3 \ ,$$

and $$|\phi(0,0) - V^{M_2N_0}(\sigma,0,0)| \leq \nu/3 \quad .$$

As ε is arbitrary, if n_1,n_2 are greater than n_0

$$|V^{n_1}(\sigma,0,0) - V^{n_2}(\sigma,0,0)| < \nu \quad .$$

Consequently $V^N(\sigma,0,0)$ is Cauchy and $V(\sigma,0,0) = \lim_{N\to\infty} V^N(\sigma,0,0)$ exists.

Finally, similarly to techniques used in [3] it can be shown that the values satisfy the following intermediate Isaacs' equations.

THEOREM 6.2. At points of differentiability $VC(\sigma,t,\xi)$ satisfies the differential equation

$$\frac{\partial V}{\partial t} + H_\sigma^C(t,x,\nabla V) = 0$$

and $VF(\sigma,t,\xi)$ satisfies

$$\frac{\partial V}{\partial t} + H_\sigma^F(t,x,\nabla V) = 0 \ .$$

REFERENCES

1. Danskin, J. 'Values in differential games'. To appear in Bull. Amer. Math. Soc.

2. Elliott, R. J. and Kalton, N. J. 'The existence of value in differential games', Memoir of the American Math. Soc. 126, Providence, R.1. (1972).

3. Elliott, R. J. and Kalton, N. J., 'Cauchy Problems for certain Isaacs-Bellman Equations and games of survival', Trans. Amer. Math. Soc. 1974, to appear.

4. Elliott, R. J. and Kalton, N. J., 'Upper values of differential games', J. Diff. Equations, 14(1973), 89-100.

5. Fleming, W. H., 'The convergence problem for differential games', J. Math. Analysis and Appl. 3 (1961), 102-116.

6. Friedman, A., 'Partial Differential Equations of Parabolic Type', Prentice Hall, Englewood Cliffs, N.J., (1964).

7. Wald, A., 'Statistical Decision Functions', John Wiley & Sons, New York, London. (1950).

ESTIMATION DU SAUT DE DUALITE EN OPTIMISATION NON CONVEXE

Jean-Pierre AUBIN

Introduction.

Les résultats suivants sont dûs à Ivar Ekeland et l'auteur. Ils ont pour but d'illustrer le fait que les fonctions non convexes vérifient des propriétés des fonctions convexes avec une erreur. Cette erreur peut être déterminée en fonction des modules de non convexité $\rho(f)$ des fonctions non convexes utilisées. Ce module est défini par

$$\rho(f) = \sup_{\substack{\text{Combinaisons} \\ \text{convexes}}} \left(f\left(\sum \alpha_i x_i\right) - \sum \alpha_i f(x_i)\right)$$

Nous allons montrer essentiellement que si g est convexe,

$$\inf_{x \in \mathbb{R}^m} (f(x) + g(x)) \leqslant - \inf_{p \in \mathbb{R}^{m*}} (f^*(p) + g^*(-p)) + \rho(f)$$

lorsque des hypothèses topologiques convenables sont satisfaites.

On retrouve le théorème de dualité de Fenchel lorsque f est aussi convexe (c'est-à-dire $\rho(f) = 0$).

On précise en fait ce résultat lorsque la fonction f s'écrit

$$f(x) = \frac{1}{T} \sum_{t=1}^{T} f_t(x_t), \quad f_t : \mathbb{R}^{m_t} \longrightarrow]-\infty, +\infty].$$

On montre alors que si $T \gg m$

$$\inf_{x_t \in \mathbb{R}^m} \left[\frac{1}{T} \sum_{t=1}^{T} f_t(x_t) + g\left(\frac{1}{T} \sum_{t=1}^{T} x_t\right)\right] \leqslant$$

$$- \inf_{p \in \mathbb{R}^{m*}} \left[\frac{1}{T} \sum_{t=1}^{T} f_t^*(p) + g^*(-p)\right] + \frac{m}{T} \sup_t \rho(f_t)$$

En d'autres termes, ce résultat montre que la somme $f(x) = \frac{1}{T}\sum_{t=1}^{T} f_t(x_t)$ d'un grand nombre de fonctions f_t est approximativement convexe.

On utilise pour cela le théorème de Shapley-Folkman, exposé dans [1], p. 392 et utilisé par Starr (voir [5]).

Après avoir défini le module de non convexité d'une fonction, nous allons énoncer les résultats qui seront démontrés à la fin de cet exposé.

1 - Module de non convexité d'une fonction.

Considérons

(1-1) Un sous-ensemble convexe X d'un espace vectoriel V.

On introduit

(1-2) $\left\{\begin{array}{l} \text{l'ensemble } M(X) \text{ des mesures discrètes de probabilité } m = \sum_{i=1}^{k} \alpha_i\, \delta(x_i) \text{ où} \\ \delta(x_i) \text{ désigne la mesure de Dirac en } x_i \text{, où } \alpha_i \geq 0,\ \sum_{i=1}^{k} \alpha_i = 1 \end{array}\right.$

ainsi que

(1-3) $\left\{\begin{array}{l} \text{l'opérateur barycentrique } \beta : M(X) \longrightarrow X \text{ associant à tout } m = \sum_{i=1}^{k} \alpha_i \delta(x_i) \\ \text{le barycentre } \beta m = \sum_{i=1}^{k} \alpha_i\, x_i \,. \end{array}\right.$

Définition.

Soit f : X ⟶ R une fonction numérique définie sur X. On appelle "module de non-convexité" $\rho(f)$ de f le scalaire

(1-4) $$\rho(f) = \sup_{m \in M(X)} \left(f(\beta m) - \langle m,f \rangle\right)$$

où $\langle m,f \rangle = \sum_{i=1}^{k} \alpha_i\, f(x_i)$ *si* $m = \sum_{i=1}^{k} \alpha_i\, \delta(x_i)$.

Il est clair qu'une fonction f est convexe si et seulement si son module de non-convexité $\rho(f)$ est nul.

De façon imagée, on peut dire que plus $\rho(f)$ est grand, moins f est convexe.

Nous allons utiliser ce module de non convexité pour majorer le "saut de dualité" du problème d'optimisation ci-dessous.

2 - Enoncé des résultats.

Considérons

deux fonctions semi-continues inférieurement

(2-1) $$\begin{cases} f : \mathbb{R}^m \longrightarrow]-\infty,+\infty] \\ g : \mathbb{R}^n \longrightarrow]-\infty,+\infty] \end{cases}$$

et

(2-2) un opérateur linéaire $A \in \mathcal{L}(\mathbb{R}^m,\mathbb{R}^n)$.

On se propose d'évaluer

(2-3) $$v = \inf_{x \in \mathbb{R}^m} (f(x) + g(Ax))$$

en fonction du problème dual

(2-4) $$v^* = - \inf_{p \in \mathbb{R}^n} (f^*(A^*p) + g^*(-p))$$

où

(2-5) $$\begin{cases} \text{i) } A^* \in \mathcal{L}(\mathbb{R}^{n*}, \mathbb{R}^{m*}) \text{ désigne le transposé de A} \\ \text{ii) } f^*(q) = \sup\limits_{x \in \mathbb{R}^m} (\langle q,x \rangle - f(x)) \;\; ; \; q \in \mathbb{R}^{m*} \\ \text{iii) } g^*(p) = \sup\limits_{y \in \mathbb{R}^n} [\langle p,y \rangle - g(y)] \; ; \; p \in \mathbb{R}^{n*} \end{cases}$$

On sait que l'on a toujours

(2-6) $$v^* \leqslant v$$

On rappelle que l'hypothèse

(H) 0 appartient à l'intérieur de Dom g - A dom f

et l'hypothèse de convexité

(C^2) les fonctions f et g sont convexes

impliquent l'existence de $\bar{p} \in \mathbb{R}^{n*}$ tel que

(2-7) $\quad v = v^* = -(f^*(A^*\bar{p}) + g^*(-\bar{p}))$

(voir [4] , §31 par exemple).

Nous allons montrer le résultat suivant.

Théorème 1.

Supposons que l'hypothèse (H) soit satisfaite et que

$$(C)\begin{cases} \text{i) seule la fonction g soit convexe.} \\ \text{ii) le domaine de f est convexe.} \end{cases}$$

Alors il existe $\bar{\bar{p}} \in \mathbb{R}^n$ tel que

(2-8) $\quad v \leq -(\overset{*}{g}(-\bar{p}) - f^*(A^*\bar{p})) + \rho(f)$

En fait, nous allons préciser ce résultat et celui de [2] dans le cas où

$$(2\text{-}9)\begin{cases} \text{i)}\ \mathbb{R}^m = \prod_{t=1}^{T} \mathbb{R}^{m_t} \\ \text{ii)}\ f(x) = \sum_{t=1}^{T} f_t(x_t)\ ;\ f_t : \mathbb{R}^{m_t} \longrightarrow]-\infty,+\infty] \\ \text{iii)}\ Ax = \sum_{t=1}^{T} A_t\, x_t\ ;\ A_t \in \mathcal{L}(\mathbb{R}^{m_t}, \mathbb{R}^n) \end{cases}$$

On pourrait remplacer $\rho(f)$ par $\sum_{t=1}^{T} \rho(f_t)$ dans (2-8) puisque $\rho(f) \leq \sum_{t=1}^{T} \rho(f_t)$.

Grâce à un théorème de Shapley-Folkman, nous allons démontrer le résultat suivant.

Théorème 2.

Supposons que les hypothèses (2-9), (H) et (C) soient satisfaites. Alors, il existe $\bar{\bar{p}} \in \mathbb{R}^{n*}$ tel que

$$(2\text{-}11)\quad \left\{ \begin{array}{l} \inf\limits_{x_t \in \mathbb{R}^{m_t}} \left(\frac{1}{T} \sum\limits_{t=1}^{T} f_t(x_t) + g\left(\frac{1}{T} \sum\limits_{t=1}^{T} A_t x_t\right)\right) \leqslant \\ \qquad - \left(g^*(-\bar{p}) + \frac{1}{T} \sum\limits_{t=1}^{T} f_t^*(A_t^*\bar{p})\right) + \frac{n+1}{T} \sup\limits_{1 \leqslant t \leqslant T} \rho(f_t) \end{array} \right.$$

Ce théorème est susceptible de nombreuses applications (voir par exemple [3]). Mentionnons seulement le théorème d'approximation suivant.

Posons

$$(2\text{-}12)\qquad f_T(u) = \inf_{\frac{1}{T}\sum\limits_{t=1}^{T} x_t = u} \frac{1}{T} \sum_{t=1}^{T} f(x_t)$$

Il est clair que

$$(2\text{-}13)\qquad f^{**}(u) \leqslant f_T(u) \leqslant f(u).$$

On déduit tout d'abord que

$$(2\text{-}14)\qquad f_T(u) - f^{**}(u) \leqslant \frac{m+1}{T} \rho(f).$$

Ensuite, nous obtenons le

<u>Corollaire</u> 3.

<u>Supposons que les hypothèses</u> (H) <u>et</u> (C) <u>soient satisfaites</u>. <u>Alors, pour tout</u> T, <u>il existe</u> $\bar{p} \in \mathbb{R}^n$ <u>tel que</u>

$$(2\text{-}15)\quad \left\{ \inf_{u \in \mathbb{R}^m} (f_T(u) + g(Au)) \leqslant - (g^*(-\bar{p}) + f^*(A^*\bar{p})) + \frac{n+1}{T} \rho(f). \right.$$

Rappelons que le sous-différentiel $\partial_\alpha f(x)$ à α près de f au point x est l'ensemble

$$(2\text{-}16) \qquad \partial_\alpha f(x) = \left\{ q \in \mathbb{R}^{n*} \text{ tels que } f(x) + f^*(q) - \langle q,x \rangle \leq \alpha \right\}$$

et que le sous-différentiel de f au point x est l'ensemble

$$(2\text{-}17) \qquad \partial f(x) = \left\{ q \in \mathbb{R}^{n*} \text{ tels que } f(X) + f^*(q) - \langle q,x \rangle = 0 \right\}$$

Dans le cadre du théorème 3, les relations d'extrémalité deviennent comme suit.

Théorème 4

Faisons les hypothèses (2.9), (H) et (C) . Considérons un élément $\bar{p}$ vérifiant (2.11).

Si $\bar{x} = (\bar{x}_t)_{1 \leq t \leq T}$ minimise

$$g\left(\frac{1}{T} \sum_{t=1}^{T} A_t x_t\right) + \frac{1}{T} \sum_{t=1}^{T} f_t(x_t) \quad \text{sur } \mathbb{R}^m = \prod_{t=1}^{T} \mathbb{R}^{m_t} ,$$

Alors

$$(2.19) \begin{cases} \text{i/} \quad -\bar{p} \in \partial_{\frac{n+1}{T}\rho}\, g\left(\frac{1}{T} \sum_{t=1}^{T} A_t \bar{x}_t\right) \\ \text{ii/} \quad A_t^* \bar{p} \in \partial_{(n+1)\rho} f_t(\bar{x}_t) \qquad 1 \leq t \leq T \end{cases}$$

(n+1)ρ

où

$$\rho = \sup_{1 \leq t \leq T} \rho(f_t)$$

Inversement si

$$(2.20) \begin{cases} \text{i/} \quad \frac{1}{T} \sum_{t=1}^{T} A_t \hat{x}_t \in \partial g^*(-\bar{p}) \\ \text{ii/} \quad \forall t,\ \hat{x}_t \text{ minimise } f_t(x_t) - \langle A_t^* \bar{p}, x_t \rangle \end{cases}$$

alors

$$(2.21) \begin{cases} g\left(\frac{1}{T} \sum_{t=1}^{T} A_t \bar{x}_t\right) + \frac{1}{T} \sum_{t=1}^{T} f_t(\bar{x}_t) \\ -g\left(\frac{1}{T} \sum_{t=1}^{T} A_t \hat{x}_t\right) - \frac{1}{T} \sum_{t=1}^{T} f_t(\hat{x}_t) \leq \frac{n+1}{T} \rho \end{cases} .$$

3. Démonstrations

Nous allons commencer par la

démonstration du théorème 2.

On introduit

$$(3.1)\quad \begin{cases} \text{i/ l'espace } \mathbb{R}\times\mathbb{R}^n \\ \text{ii/ le vecteur } \theta = (1,0) \in \mathbb{R}\times\mathbb{R}^n \\ \text{iii/ les cônes } P = [0,+\infty[\times\{0\} \text{ et } \mathring{P} =]0,\infty[\times\{0\} \\ \quad \text{de } \mathbb{R}\times\mathbb{R}^n \end{cases}$$

On pose

$$(3.2)\quad X = \operatorname{Dom} f \subset \mathbb{R}^m \;;\quad Y = \operatorname{Dom} g \subset \mathbb{R}^n$$

et l'on définit l'opérateur $\phi : X \times Y \longmapsto \mathbb{R}\times\mathbb{R}^n$ par

$$(3.3)\quad \phi(x,y) = (f(x) + g(y)\ ,\ y - Ax) \in \mathbb{R}\times\mathbb{R}^n \ .$$

On considère le scalaire

$$(3.4)\quad w = \inf_{m\in M(X)} [\langle m,f\rangle + g(A\beta m)]$$

Tout d'abord, on vérifie que

$$(3.5)\quad \begin{cases} \text{i/ } w\theta \notin \operatorname{co}[\phi(X\times Y)] + \mathring{P} \\ \\ \text{ii/ } \forall \varepsilon > 0\ ,\ (w+\varepsilon)\theta \in \operatorname{co}[\phi(X\times Y)] + P \end{cases}$$

Preuve de (3.5) i/

En effet, si $w\theta \in \operatorname{co}[\phi(X\times Y)] + \mathring{P}$, il existerait des $x_i \in X (1\leq i\leq k)$, des $y_i \in Y(1\leq i\leq k)$, des $\alpha_i \geqslant 0$ $(1\leq i\leq k)$ tels que $\sum_{i=1}^k \alpha_i = 1$ vérifiant

$$(3.6)\quad \begin{cases} w > \sum_{i=1}^k \alpha_i\ f(x_i) + \sum_{i=1}^k \alpha_i\ g(y_i) \geqslant \\ \\ \sum_{i=1}^k \alpha_i\ f(x_i) + g(\sum_{i=1}^k \alpha_i\ y_i) \text{ (puisque } g \text{ est convexe)} \end{cases}$$

et

$$(3.7)\quad 0 = \sum_{i=1}^k \alpha_i\ y_i - \sum_{k=1}^k \alpha_i\ A\ x_i$$

En posant $m = \sum_{i=1} \alpha_i\ \delta\ (x_i) \in M(X)$, on obtient

$$(3.8)\quad \begin{cases} w > \sum_{i=1}^{k} \alpha_i f(x_i) + g(A(\sum_{i=1}^{k} \alpha_i x_i)) = \\ \langle m,f \rangle + g(A\beta m) \geq w . \end{cases}$$

Ceci est impossible et par suite, $w\theta \notin co[\phi(X \times Y)] + \mathring{P}$.

Preuve de (3.5) ii/

D'autre part, la définition de w implique que pour tout $\varepsilon > 0$, il existe $m = \sum_{i=1}^{k} \alpha_i \delta(x_i)$ tel que

$$(3.9)\quad \begin{cases} \langle m,f \rangle + g(A\beta m) = \sum_{i=1}^{k} \alpha_i (f(x_i) + g(y)) \\ \leq w + \varepsilon \end{cases}$$

où l'on a posé $y = A\beta m$, c'est-à-dire

$$(3.10)\quad 0 = \sum_{i=1}^{k} \alpha_i (y - Ax_i) .$$

Par suite, $(w+\varepsilon)\theta$ appartient bien à $co[\phi(X \times Y)] + P$.

Conséquence de (3.5) i/

Ceci étant, on déduit de (3.5) i/ et du théorème de séparation l'existence d'une forme linéaire non nulle $(c,p) \in \mathbb{R}^* \times \mathbb{R}^{n*}$ telle que

$$(3.11)\quad \begin{cases} \langle (c,p), w\theta \rangle = c.w \leq \\ \inf_{\substack{x \in X \\ y \in Y \\ a > 0}} (c(f(x) + g(y)) + \langle p, y \rangle - Ax + ca) \end{cases}$$

En premier lieu, on déduit que $\inf_{a \geq 0} ca$ est fini, et donc que $c \geq 0$ et que $\inf_{a \geq 0} ca = 0$. En fait, l'hypothèse (H) implique que $c > 0$. Sinon, $c = 0$ et on déduirait que

$$0 \leq \inf_{\substack{x \in X \\ y \in Y}} \langle p, y - Ax \rangle = \inf_{z \in Y - A(X)} \langle p,z \rangle$$

Mais puisque 0 est intérieur à $Y - A(X)$, on sait que $\inf_{z \in Y - A(X)} \langle p,z \rangle < 0$ sauf si $p = 0$.

Par suite, $p = 0$, ce qui contredit le fait que la forme linéaire (c,p) est non nulle.

Donc, en divisant par $c > 0$ et en posant $\bar{p} = \frac{p}{c}$, on déduit de (3.11) que

(3.12) $$\begin{cases} w \leq \inf\limits_{\substack{x \in X \\ y \in Y}} [f(x) + g(y) + \langle \bar{p}, y \rangle - \langle A^{*} \bar{p}, x \rangle] \\ = -(f^{*}(A^{*}\bar{p}) + g^{*}(-\bar{p})) \leq v^{*} \end{cases}$$

Conséquence de (3.5) ii/

D'autre part, on va déduire de (3.5) ii/ une minoration de w. Pour cela, on écrit

(3.13) $$\mathrm{co}\,[\phi(X \times Y)] + P = G + \mathrm{co} \sum_{t=1}^{T} F_t$$

où

(3.14) $$\begin{cases} G = \{(g(y), y)\}_{y \in Y} + P \text{ est l'épigraphe de } g, \\ \text{qui est convexe d'après (C)} \end{cases}$$

et où

(3.15) $$F_t = \{(f_t(x_t), -A_t x_t)\}_{x_t \in X_t} \; ; \; X_t = \mathrm{Dom}\, f_t \, .$$

Par suite, on peut écrire

(3.16) $$\begin{cases} (w + \varepsilon) = \gamma + \varphi \text{ ou } \gamma = (g(y) + a, y) \in G, \\ \text{et où } \varphi \in \mathrm{co}(\sum_{t=1}^{T} F_t) \subset \mathbb{R} \times \mathbb{R}^n . \end{cases}$$

Or, le théorème de Shapley-Folkman (voir [1], p. 392) implique l'existence d'un ensemble S d'au plus (n+1) indices tel que

(3.17) $$\varphi = \sum_{t \notin S} \varphi_t + \sum_{s \in S} \varphi_s$$

où

(3.18) $$\begin{aligned} \varphi_t &= (f_t(x_t), -A_t x_t) \in F_t \quad \text{si } t \notin S \\ \varphi_s &= (\sum_{i=1}^{k} \alpha_i^s \, (f_s(x_s^i), -A_s(x_s^i)) \\ &= (f_s(x_s), -A_s x_s) - (f_s(\beta m_s) - \langle m_s, f_s \rangle, 0 \in \mathrm{co}(F_s) \quad \text{si } s \in S \end{aligned}$$

Cet élément s'écrit donc :

(3.19) $$\varphi = \sum_{t=1}^{T} (f_t(x_t), -A_t x_t) - \sum_{s \in S} (f_s(\beta w_s) - \langle m_s, f_s \rangle), 0)$$

Ainsi, on obtient

$$(3.20)\quad \begin{cases} w + \mathcal{E} = g(y) + a + \sum_{t=1}^{T} f_t(x_t) - \sum_{s \in S} (f_s(\beta m_s) - \langle m_s, f_s \rangle) , \\ 0 = y - \sum_{t=1}^{T} A_t \; x_t \end{cases}$$

On en déduit que

$$w + \mathcal{E} = g(\sum_{t=1}^{T} A_t \; x_t) + a + \sum_{t=1}^{T} f_t(x_t)$$
$$- \sum_{s \in S} (f_s(\beta m_s) - \langle m_s, f_s \rangle) \geqslant$$
$$v - \sum_{s \in S} \rho(f_s) \geqslant v - (n+1) \sup_s \rho(f_s)$$

Puisque $a \geqslant 0$ et puisque S contient au plus $(n+1)$ indices.

On a donc prouvé que

$$(3.21)\qquad v \leqslant w + (n+1) \sup_t \rho(f_t)$$

Les inégalités (3.12) et (3.21) impliquent le théorème 2.

Démonstration du théorème 1.

Lorsque $T = 1$, on déduit de (3.13) que

$$(3.22)\qquad (w + \mathcal{E})\theta = \gamma + \varphi \quad \text{où } \gamma \in G \quad \text{et où } \varphi \in co(F) ,$$

et par suite, que l'on peut écrire

$$(3.23)\quad \begin{cases} w + \mathcal{E} = g(Ax) + a + f(x) - (f(\beta m) - \langle m, f \rangle) \\ \qquad \geqslant v - \rho(f) \; . \end{cases}$$

On établit ainsi le théorème 1.

Démonstration du théorème 3.

On obtient le théorème 3 en utilisant le théorème 2 lorsque g est remplacée par la fonction $y \to Tg(\frac{y}{T})$, dont la fonction conjuguée est $p \to Tg^*(p)$.

On obtient alors

$$\inf_{x_t} \; (Tg(\frac{1}{T} A (\sum_{t=1}^{T} x_t) + \sum_{t=1}^{T} f_t(x_t))$$
$$\leqslant - (Tg^*)(-\bar{p}) + \sum_{t=1}^{T} f_t^*(A_t^* \bar{p})) + (n+1) \sup_t \rho(f_t) \; .$$

En divisant par T , on obtient le théorème 3.

Démonstration du corollaire 3.

En posant $A_t = A$ et $f_t = f$, on obtient le corollaire 3.

Démonstration de l'inégalité (2.14)

En prenant $R^m = R^n$, $A_t = 1$, $f_t = f$ et $g(x) = \Psi_{\{u\}}(x)$ la fonction indicatrice du point u , on déduit alors que

$$(3.24)\quad \left\{ \begin{array}{l} \inf\limits_{\frac{1}{T}\sum_{t=1}^{T} x_t = u} \frac{1}{T}\sum f(x_t) \leqslant \langle \bar{p},u\rangle - f^*(\bar{p}) + \frac{(m+1)}{T}\rho(f) \\ \qquad\qquad \leqslant f^{**}(u) + \frac{m+1}{T}\rho(f) . \end{array} \right.$$

Démonstration du théorème 4.

L'inégalité (2.11) peut s'écrire :

$$(3.25)\quad \left\{ \begin{array}{l} \frac{1}{T}\sum_{t=1}^{T} (f_t(\bar{x}_t) + f_t^*(A_t^*\bar{p}) - \langle A_t^*\bar{p},\bar{x}_t\rangle) \\ + (g(\frac{1}{T}\sum_{t=1}^{T} A_t\bar{x}_t) + g^*(-\bar{p}) - \langle -\bar{p}, \frac{1}{T}\sum_{t=1}^{T} A_t\bar{x}_t\rangle) \\ \leqslant \frac{n+1}{T}\rho . \end{array} \right.$$

Puisque chacun des termes de cette somme sont positifs, on en déduit que

$$(3.26)\quad \left\{ \begin{array}{l} g(\frac{1}{T}\sum_{t=1}^{T} A_t\bar{x}_t) + g^*(-\bar{p}) - \langle -\bar{p}, \frac{1}{T}\sum_{t=1}^{T} A_t\bar{x}_t\rangle \leqslant \frac{n+1}{T}\rho \\ f_t(\bar{x}_t) + f_t^*(A_t^*p) - \langle A_t^*\bar{p}, \bar{x}_t\rangle \leqslant (n+1)\rho \end{array} \right.$$

c'est-à-dire, les relations (2.19) du théorème 4. Inversement, les hypothèses(2.20) du théorème impliquent que

$$(3.27)\quad g^*(-\bar{p}) + g(\frac{1}{T}\sum_{t=1}^{T} A_t\hat{x}_t) = -\langle \bar{p}, \frac{1}{T}\sum_{t=1}^{T} A_t\hat{x}_t\rangle$$

et que

$$(3.28)\quad f_t^*(A_t^*\bar{p}) + f_t(\hat{x}_t) = \langle \bar{p}, A_t\hat{x}_t\rangle .$$

Par suite, on obtient

$$(3.29) \quad -g^{*}(-\bar{p}) - \frac{1}{T} \sum_{t=1}^{T} f_t^{*}(A_t^{*} \bar{p}) = g(\frac{1}{T} \sum_{t=1}^{T} A_t \hat{x}_t) + \frac{1}{T} \sum_{t=1}^{T} f_t(\hat{x}_t) .$$

Donc, l'inégalité (2.21) du théorème 4 résulte des inégalités (2.11) et (3.29) .

BIBLIOGRAPHIE

[1] ARROW-HAHN "General competitive analysis". Holden Day 1971.

[2] EKELAND "Une estimation à priori en programmation non convexe. Cahiers de Mathématiques de la Décision - Université de Paris 9 - 1974.

[3] EKELAND & TEMAN Analyse convexe et problèmes variationnels. Dunod - Gauthier Villars - 1974.

[4] ROCKAFFELAR Convex Analysis - Princeton University Press 1970.

[5] STARR Quasi equilibria in markets with non convex preferences. Econometrica 37 (1967) pp. 25-38.

CONTRAINTE D'ETATS DANS LES JEUX DIFFERENTIELS

P. BERNHARD
Centre d'Automatique
de l'E.N.S.M.P.
et I.R.I.A.

J.F. ABRAMATIC
Centre d'Automatique
de l'E.N.S.M.P.

Introduction.

Nous nous intéressons à des jeux différentiels à deux joueurs et somme nulle. L'étude de jeux particuliers a très tôt fait intervenir des contraintes d'états, Breakwell, notamment, en a mis plusieurs en évidence : [1], [2], [3]. Le cas le plus courant est celui où la cible comporte une partie "non utile". C'est-à-dire que le joueur qui y a intérêt peut toujours éviter de pénétrer cette partie de la cible. Cela introduit une contrainte d'états pour ce joueur. D'une manière plus générale, une contrainte d'états sera, pour nous, une variété S de dimension (n-1) de l'espace d'états dont un des deux joueurs peut et doit éviter que l'état ne la traverse. On voit que c'est une contrainte asymétrique. On dira qu'elle est sous la responsabilité de ce joueur.

La solution à de tels problèmes comporte en général des arcs de trajectoire "saturant" la contrainte, c'est-à-dire situés dans S. La façon classique de les construire, dans le contexte de la théorie d'Hamilton-Jacobi-Isaacs,consiste à résoudre un jeu "réduit" ou jeu "contraint". On limite les commandes du joueur qui a la responsabilité de la contrainte à être telles que l'état reste sur S. Ceci définit une dynamique sur cette variété, donc de dimension n-1. Par ailleurs, le champ de trajectoires construit depuis la cible définit une sous variété B de dimension n-2 sur S, d'où la trajectoire optimale quitte S et est connue.

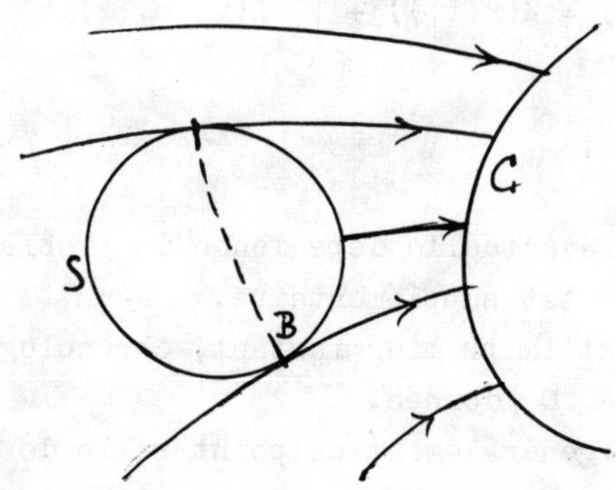

Ainsi, on connaît la valeur du jeu sur B, et cette valeur sert de coût final et B de variété finale pour résoudre le jeu réduit sur S. Ceci permet de connaître la valeur du jeu en tout point de S, et de traiter alors S comme variété finale, avec cette valeur pour coût final, pour construire les trajectoires qui rejoignent S.

Cette méthode n'est pas totalement satisfaisante, comme un exemple le montrera. Nous nous proposons, dans cet article, de faire une étude systématique de cette question.

Après avoir précisé la formulation du problème et les notations, nous démontrerons un théorème très simple qui précise la nature de la solution cherchée. Puis nous proposerons une méthode pour construire cette solution, et nous étudierons en particulier une difficulté technique qu'elle soulève. Enfin, nous présenterons l'exemple annoncé, où la méthode antérieure échoue, mais que nous avons pu résoudre de la façon proposée ici.

Formulation.

Soit :

$$\dot{x} = f(x, u, v) \tag{1}$$

la dynamique du jeu. x est le vecteur d'état, $x(t) \in R^n$, u est la commande du joueur u, $u(t) \in R^m$, v est la commande du joueur v, $v(t) \in R^p$. La fonction f(., ., .) est supposée être de classe C^1 dans $R^n \times R^m \times R^p$.

Un objectif est fixé par le critère J du jeu :

$$\max_v \min_u J(x, u, v) = V(x) , \quad u(t) \in U, \quad v(t) \in V . \tag{2}$$

$$J(x_o, u, v) = K(x(t_1)) + \int_0^{t_1} L(x, u, v)dt \quad ,$$

$$x(0) = x_o \quad .$$

On a éliminé une éventuelle dépendence du problème en t à l'aide d'une variable d'état supplémentaire.

L'instant t_1 est le premier instant, chronologiquement, où l'état pénètre une cible C donnée.

On cherche généralement un point selle de J. Cependant, un

tel point selle n'existe généralement pas du fait de la contrainte que nous allons étudier. On admettra alors qu'on recherche un maximin. C'est-à-dire que u pourra, pour ne pas violer la contrainte, connaître à chaque instant la commande $v(t)$, en plus de l'état $x(t)$ qui constitue l'information habituelle. Breakwell [2] a montré comment la valeur du jeu que nous allons calculer sous ces hypothèses peut être approchée arbitrairement près si u ne connaît pas $v(t)$. (Ou si les états de S font partie des états non admissibles).

On suppose que la variété S est définie par :

$$x_n = \text{constante} = c \quad ,$$

et l'ensemble des états admissibles par :

$$x_n \geq c \quad ,$$

où x_n est la dernière coordonnée du vecteur d'état. Les hypothèses que nous ferons reviennent à nous limiter à une contrainte du premier ordre en u :

$$\frac{\partial f_n}{\partial u} \neq 0 \quad . \tag{3}$$

Nous noterons $\hat{x}$ le vecteur des $(n-1)$ premières coordonnées de x. $\hat{x}$ paramétrise S. De même pour $\hat{f}$, et les autres variables que nous aurons à introduire.

Pour tout $x \in S$ l'équation :

$$f_n(x, u, v) = 0 \tag{4}$$

définit une variété $\Sigma(\hat{x})$ de dimension $m+p-1$ dans l'espace $R^m \times R^p$ des commandes (u, v). Cette variété divise cet espace en une région "admissible" :

$$(u, v) \in R_{ad}(x) \Leftrightarrow f_n(x, u, v) \geq 0 \quad , \tag{5}$$

et son complémentaire ou région non admissible. Pour tout v, ceci définit dans U une variété $W(\hat{x}, v)$ de dimension $m-1$, limitant une région $U_{ad}(x, v)$:

$$U_{ad}(x, v) = \{u | (u, v) \in R_{ad}(x)\} \tag{6}$$

$$W(\hat{x}, v) = \{u \mid (u, v) \in \Sigma(\hat{x})\} \quad . \tag{7}$$

Nous complétons la définition de U_{ad} par :

$$U_{ad}(x, v) = U \qquad \forall x \in S, \quad x_n > c \ .$$

Nous faisons l'hypothèse qu'en tout point de S, U_{ad} est non vide, quel que soit v (d'où (3)). En effet, s'il n'en était pas ainsi le jeu aurait une solution triviale pour un état initial situé sur S. Le joueur γ choisirait une commande telle que U_{ad} soit vide. (On serait dans la "partie utile" de la cible). Par continuité de f, cela implique que pour tout v, tel que U_{ad} soit différent de U, il existe $u \in W(x, v)$. Nous utiliserons l'hamiltonien :

$$H(x, \lambda, u, v) = L(x, u, v) + \lambda' f(x, u, v) \quad (*)$$

et noterons $u^*(x, \lambda)$, $v^*(x, \lambda)$ les arguments du point selle de H en (u, v), dont nous supposons l'existence pour tout x et tout λ. Nous noterons aussi :

$$\hat{H}(x, \hat{\lambda}, u, v) = L(x, u, v) + \hat{\lambda}' \hat{f}(x, u, v)$$

de sorte que :

$$H(x, \lambda, u, v) = \hat{H}(x, \hat{\lambda}, u, v) + \lambda_n f_n(x, u, v) \quad . \tag{8}$$

En fait, $\hat{H}$ sera utilisé pour des états $x \in S$, de sorte que $x_n = c$ et que $\hat{H}$ ne dépendra que de $\hat{x}$. Chaque fois que cela ne sera pas ambigu, nous omettrons tout ou partie des arguments dans les quantités que nous avons définies.

Toute l'analyse proposée ci-après sera basée sur la recherche d'une solution à l'équation d'Hamilton-Jacobi-Isaacs [4] modifiée :

$$\max_{v \in V} \min_{u \in U_{ad}(v)} H(x, \frac{\partial V}{\partial x}, u, v) = 0 \ . \tag{9}$$

Il est en effet immédiat de généraliser la théorie classique, établie pour la recherche d'un point selle, à ce cas, et de démontrer que (9),

(*) L'accent sur un vecteur désigne la transposition. Ceci évite la notation $\langle \lambda, f \rangle$.

avec la condition aux limites $V(x) = K(x)$ sur C, est une condition suffisante pour résoudre (2), avec la structure d'information décrite.

Un théorème sur le raccordement.

Nous allons prouver le résultat suivant :

THEOREME 1. Si l'hamiltonien a, pour tout λ, un point selle unique strict, alors la contrainte est rejointe tangentiellement par les trajectoires optimales.

DEMONSTRATION. En un point de S, la trajectoire incidente est engendrée par les commandes (u^*, v^*). Supposons que $(u^*, v^*) \notin R_{ad}$, de sorte qu'il doive y avoir discontinuité des commandes à la jonction.

Alors, $u^* \notin U_{ad}(v^*)$. Et on a :

$$\forall u \in U_{ad}(v^*), \quad H(x, \lambda, u, v^*) > 0 \quad ,$$

et a fortiori :

$$\max_{v} \min_{u \in U_{ad}(v)} H(x, \lambda, u, v) > 0 \ , \quad \forall \lambda \ ,$$

ce qui exclut l'existence d'une solution à (9).

Au contraire, si $(u^*, v^*) \in R_{ad}$, mais pas à Σ, alors la trajectoire engendrée quitte S, et constitue la trajectoire optimale depuis x, contredisant l'hypothèse de travail qu'il existe un arc saturé.

Donc un arc saturé ne peut être rejoint que par une trajectoire tangente à S, le théorème est démontré.

Ce théorème, extrêmement simple améliore considérablement celui présenté dans [1]. Il a deux intérêts. D'une part, il établit la condition pour qu'une contrainte soit rejointe avec une discontinuité des commandes. D'autre part, quand cette discontinuité est exclue, il précise la nature de la solution de (9) recherchée, comme nous allons le voir.

Le problème réduit.

Pratiquement, nous voulons construire le champ de trajectoires optimales sur S, en travaillant en $\hat{x}$. La solution classique consistait à chercher à résoudre :

$$\max_{v\in V}\ \min_{u\in W(v)} \hat{H}(\hat{x}, \hat{\lambda}, u, v) = 0 \quad , \qquad (10)$$

par la méthode des équations d'Euler-Lagrange. La justification est bien sûr que pour $U \in W(v)$, $\hat{H} = H$. Toutefois, l'hypothèse implicite ici est que le max min de (9), qui est atteint par hypothèse en un point de Σ, est aussi un max min de $\hat{H}$. Alors la solution de (10) sera celle de (9). Le cas difficile est bien entendu celui où celà ne se produit pas. Il faut alors distinguer, dans la maximisation en v, les valeurs de v pour lesquelles u doit choisir u dans W(v), et celle pour lesquelles il est libre de choisir u^*.

Soient donc V_1 et V_2 deux régions de V définies par :

$$V_1(\hat{x}, \lambda) = \{v \,|\, (u^*(x, \lambda), v) \notin \overset{o}{R}_{ad}\} \quad , \qquad (11)$$

où $\overset{o}{R}_{ad}$ désigne l'intérieur de R_{ad}, et V_2 est le complémentaire de V_1 dans V.

Nous proposons de considérer le "jeu réduit" suivant :

$$\max_{v\in V_1}\ \min_{u\in W(v)} \hat{H}(\hat{x}, \frac{\partial V}{\partial \hat{x}}, u, v) = 0 \quad , \qquad (12)$$

où λ, qui est nécessaire pour définir V_1 à l'aide de (11), est calculé à l'aide de :

$$\lambda' = (\frac{\partial V}{\partial \hat{x}}, \lambda_n) \qquad (13)$$

et λ_n choisi tel que :

$$H(x, \lambda, u^*(x, \lambda), v^*(x, \lambda)) = 0 \quad . \qquad (14)$$

Dans (14), seul λ_n est inconnu. Nous supposerons que cette équation le fixe, en fonction de $\hat{\lambda}$. Nous énonçons les résultats suivants :

LEMME. <u>Si</u> $\hat{H}(\hat{x}, \hat{\lambda}, u, v)$ <u>a un maximin unique strict, alors une solution de</u> (12) (13) (14) <u>est telle que le max min dans</u> (12) <u>est obtenu en</u> (u^*, v^*).

PREUVE. Soit $\hat{u}$, $\hat{v}$ l'argument du max min en (12). Si $\hat{v} \neq v^*$, alors, par définition de $\hat{v}$:

$$\min_{u \in W(v^*)} \hat{H}(\hat{x}, \hat{\lambda}, u\ v^*) < \min_{u \in W(\hat{v})} \hat{H}(\hat{x}, \hat{\lambda}, u, \hat{v}) = 0 \quad .$$

Or, pour $u \in W(v^*)$, et $v = v^*$, $H = \hat{H}$ quel que soit λ_n. De plus $W(v^*) \subset U_{ad}(v^*)$. Donc :

$$\min_{u \in U_{ad}(v^*)} H(x, \lambda, u, v^*) < \min_{u \in W(v^*)} H(x, \lambda, u, v^*) < 0 \quad ,$$

ce qui est en contradiction avec :

$$\min_{u \in U} H(x, \lambda, u, v^*) = 0 \quad ,$$

qui est impliqué par la relation (14). Le lemme est démontré.

THEOREME. <u>Si on a résolu le problème réduit</u> (12, (13), (14), <u>on a, dans le voisinage de</u> S, <u>une solution de</u> (9).

DEMONSTRATION. Notons d'abord que si $x \notin S$, la solution de (14) peut être prolongée par des techniques classiques, et comme alors $U_{ad} = U$, on a bien une solution de (9).

Si $x \in S$, il faut distinguer la région de V où se trouve v. Si $v \in V_1$, mais $v \neq v^*$, alors u^* n'est pas admissible. u doit choisir u dans $U_{ad}(v)$. Comme $v^* = \hat{v}$, on sait que :

$$\min_{u \in W(v)} H(x, \lambda, u, v) = \min_{u \in W(v)} \hat{H}(x, \hat{\lambda}, u, v) < 0 \quad .$$

A fortiori, le minimum pour $u \in U_{ad}(v)$ est négatif.
Si $v = v^*$, alors :

$$\min_{u \in U} H(x, \lambda, u, v^*) = 0 = \min_{u \in U_{ad}(v^*)} H(x, \lambda, u, v^*) \quad ,$$

la première égalité à cause de (14), la seconde parce que u^* appartient à $W(v^*)$, donc à $U_{ad}(v^*)$.
Finalement, si $v \in V_2$, alors u^* est admissible, et

$$\min_{u \in U_{ad}(v)} H(x, \lambda, u, v) \leq H(x, \lambda, u^*, v) < 0 \quad .$$

Donc v^* donne bien le plus grand minimum, et celui-ci est nul. Le théorème est démontré.

REMARQUE. Le problème (12) a l'interprétation suivante, en termes de jeu. Le joueur v cherche à faire aussi bien que possible, tout en s'assurant qu'il "gêne" u, c'est-à-dire qu'il ne lui permet pas de quitter la contrainte tout en jouant sa stratégie optimale en l'absence de cette contrainte : u^*. D'où l'obligation qu'il s'impose de choisir v dans V_1.

Equations d'Euler-Lagrange.

On souhaite résoudre (12), (14) à l'aide des équations des caractéristiques, qui pour l'équation d'Hamilton-Jacobi habituelle sont les équations d'Euler-Lagrange. Pour pouvoir pousser l'analyse plus loin, nous devons faire l'hypothèse que toutes les fonctions en cause sont assez régulières pour que les dérivées que nous utilisons existent. La théorie classique permet encore d'établir que les caractéristiques sont solution de :

$$\frac{d\hat{x}'}{dt} = \frac{\partial \hat{H}^*}{\partial \hat{\lambda}} \tag{15}$$

$$\frac{d\hat{\lambda}'}{dt} = - \frac{\partial \hat{H}^*}{\partial \hat{x}} \tag{16}$$

où $\hat{H}^*$ est une fonction de $\hat{x}$ et $\hat{\lambda}$ seulement :

$$\hat{H}^*(\hat{x}, \hat{\lambda}) = \max_{v \in V_1(\hat{\lambda})} \min_{u \in W(v)} \hat{H}(\hat{x}, \hat{\lambda}, u, v) \quad .$$

V_1 dépend de λ, mais (13), (14) lient λ à $\hat{\lambda}$, de sorte que V_1 peut être considéré comme fonction de $\hat{\lambda}$.

Nous devons analyser l'équation (15) plus en détail. Soient :

$$\hat{u}(\hat{x}, \hat{\lambda}, v) = \operatorname{Arg} \min_{u \in W(v)} \hat{H}(\hat{x}, \hat{\lambda}, u, v) \quad ,$$

$$\hat{v}(\hat{x}, \hat{\lambda}) = \operatorname{Arg}\max_{v \in V_1(\hat{\lambda})} \hat{H}(\hat{x}, \hat{\lambda}, \hat{u}(\hat{x}, \hat{\lambda}, v), v) .$$

L'équation (15) s'écrit :

$$\frac{d\hat{x}'}{dt} = \hat{f}'(\hat{x}, \hat{u}, \hat{v}) + \frac{\partial\hat{H}}{\partial u}\frac{\partial\hat{u}}{\partial\hat{\lambda}} + \left[\frac{\partial\hat{H}}{\partial u}\frac{\partial\hat{u}}{\partial v} + \frac{\partial\hat{H}}{\partial v}\right]\frac{\partial\hat{v}}{\partial\hat{\lambda}} .$$

Le terme en $\partial\hat{u}/\partial\lambda$ fait l'objet d'une analyse classique. W est définie par :

$$f_n(\hat{x}, u, v) = 0 , \tag{17}$$

et d'après le théorème de Lagrange, il existe un vecteur de multiplicateurs, μ, tel que :

$$\frac{\partial\hat{H}}{\partial u} + \mu'\frac{\partial f_n}{\partial u} = 0 .$$

(17) est satisfaite identiquement par $\hat{u}(\hat{x}, \hat{\lambda}, v)$. Dérivant en $\hat{\lambda}$ il vient :

$$\frac{\partial f_n}{\partial u}\frac{\partial\hat{u}}{\partial\hat{\lambda}} = 0 .$$

En comparant les deux dernières relations, on en déduit :

$$\frac{\partial\hat{H}}{\partial u}\frac{\partial\hat{u}}{\partial\hat{\lambda}} = 0 .$$

(Si U est limité par des inégalités $g_i(x, u) \leq 0$, ce même raisonnement s'étend. Voir [5]).

Toutefois, pour le terme en $\partial v/\partial\hat{\lambda}$, l'équation définissant la frontière de V_1, sur laquelle nous savons que se trouve $\hat{v}$, dépend explicitement de $\hat{\lambda}$. Soit

$$\varphi(\hat{x}, \hat{\lambda}, v) = 0$$

cette équation. On trouverait qu'il existe un vecteur ν tel que

$$\left(\frac{\partial\hat{H}}{\partial u}\frac{\partial\hat{u}}{\partial v} + \frac{\partial\hat{H}}{\partial v}\right)\frac{\partial\hat{v}}{\partial\hat{\lambda}} = \nu'\frac{\partial\varphi}{\partial\hat{\lambda}}$$

La conclusion serait qu'en général, les caractéristiques (15) (16) de l'équation d'Hamilton-Jacobi-Isaacs ne sont plus supportées par les trajectoires optimales. Elles permettraient néanmoins de calculer $V(x)$, u^* et v^* en tout point de S. Mais il resterait à intégrer les trajectoires.

En fait, la situation est plus simple. Notons que

$$\frac{\partial H}{\partial u} = \frac{\partial \hat{H}}{\partial u} + \lambda_n \frac{\partial f_n}{\partial u} , \quad \frac{\partial H}{\partial v} = \frac{\partial \hat{H}}{\partial v} + \lambda_n \frac{\partial f_n}{\partial v} .$$

On peut dériver (17), en y remplaçant u par $\hat{u}$, en $\hat{\lambda}$ et en v :

$$\frac{\partial f_n}{\partial u} \frac{\partial \hat{u}}{\partial \hat{\lambda}} = 0 \quad , \quad \frac{\partial f_n}{\partial u} \frac{\partial \hat{u}}{\partial v} + \frac{\partial f_n}{\partial v} = 0 \quad .$$

De ces deux relations, on déduit :

$$\frac{\partial \hat{H}}{\partial u} \frac{\partial \hat{u}}{\partial \hat{\lambda}} = \frac{\partial H}{\partial u} \frac{\partial \hat{u}}{\partial \hat{\lambda}} , \quad \text{et} \quad \frac{\partial \hat{H}}{\partial u} \frac{\partial \hat{u}}{\partial v} + \frac{\partial \hat{H}}{\partial v} = \frac{\partial H}{\partial u} \frac{\partial \hat{u}}{\partial v} + \frac{\partial H}{\partial v} .$$

Comme la première de ces deux quantités est nulle, on peut écrire :

$$\left[\frac{\partial \hat{H}}{\partial u} \cdot \frac{\partial \hat{u}}{\partial v} + \frac{\partial \hat{H}}{\partial v}\right] \frac{\partial \hat{v}}{\partial \hat{\lambda}} = \frac{\partial H}{\partial u} \frac{\partial \hat{u}}{\partial \hat{\lambda}} + \left[\frac{\partial H}{\partial u} \frac{\partial \hat{u}}{\partial v} + \frac{\partial H}{\partial v}\right] \frac{\partial \hat{v}}{\partial \hat{\lambda}} \quad .$$

Maintenant, nous savons que :

$$\hat{u}(\hat{x}, \hat{\lambda}, \hat{v}(\hat{x}, \hat{\lambda})) = u^*[(\hat{x}, c), (\hat{\lambda}, \lambda_n(\hat{\lambda}))] \qquad \forall(\hat{x}, \hat{\lambda}) ,$$

$$\hat{v}(\hat{x}, \hat{\lambda}) = v^*[(\hat{x}, c), (\hat{\lambda}, \lambda_n(\hat{\lambda}))] \qquad \forall(\hat{x}, \hat{\lambda}) .$$

On peut dériver ces relations en $\hat{\lambda}$, et il vient :

$$\frac{\partial \hat{u}}{\partial \hat{\lambda}} + \frac{\partial \hat{u}}{\partial v} \frac{\partial \hat{v}}{\partial \hat{\lambda}} = \frac{\partial u^*}{\partial \hat{\lambda}} + \frac{\partial u^*}{\partial \lambda_n} \frac{d\lambda_n}{d\hat{\lambda}}$$

$$\frac{\partial \hat{v}}{\partial \hat{\lambda}} = \frac{\partial v^*}{\partial \hat{\lambda}} + \frac{\partial v^*}{\partial \lambda_n} \frac{d\lambda_n}{d\hat{\lambda}}$$

et en reportant il vient :

$$\left[\frac{\partial \hat{H}}{\partial u} \frac{\partial \hat{u}}{\partial v} + \frac{\partial \hat{H}}{\partial v}\right] \frac{\partial \hat{v}}{\partial \hat{\lambda}} = \frac{\partial H}{\partial u}\left[\frac{\partial u^*}{\partial \hat{\lambda}} + \frac{\partial u^*}{\partial \lambda_n} \frac{d\lambda_n}{d\hat{\lambda}}\right] + \frac{\partial H}{\partial v}\left[\frac{\partial v^*}{\partial \hat{\lambda}} + \frac{\partial v^*}{\partial \lambda_n} \frac{d\lambda_n}{d\lambda}\right]$$

Cependant, la théorie habituelle montre que :

$$\frac{\partial H}{\partial u}\frac{\partial u^*}{\partial \lambda} = [\frac{\partial H}{\partial u}\frac{\partial u^*}{\partial \lambda}, \frac{\partial H}{\partial u}\frac{\partial u^*}{\partial \lambda_n}] = 0 \quad ,$$

et de même pour v. On en conclut que le terme étudié est nul, et que les caractéristiques restent les trajectoires optimales.

Faisons une remarque qui a un intérêt numérique certain. Dans la résolution du problème (12), si on connaît explicitement la forme de $u^*(x, \lambda)$, $v^*(x, \lambda)$, on peut remplacer (14) par :

$$f_n(x, u^*(x, \lambda), v^*(x, \lambda)) = 0 \quad . \tag{18}$$

Ceci assure que :

$$H(x, \lambda, u^*, v^*) = \hat{H}(x, \lambda, u^*, v^*) \quad .$$

On prend alors systématiquement u^*, v^*, pour faire progresser les équations (15), (16), assurant ainsi que :

$$\hat{H}(x, \lambda, u^*, v^*) = 0 \quad .$$

Il reste à vérifier qu'on soit bien aux maximin de $\hat{H}$. Un calcul semblable au précédent permet de montrer qu'on est à un point stationnaire. Il reste donc seulement à vérifier soit le signe des dérivées secondes, soit la valeur de $\hat{H}$ en certains autres points.

L'avantage de cette méthode est le suivant : l'équation (14) ne peut pas être résolue en λ_n numériquement par une méthode de Newton, et sera très difficile à résoudre par toute méthode. En effet, la racine est double, comme le montre la relation :

$$\frac{\partial H^*}{\partial \lambda_n} = \frac{dx_n}{dt} = 0 \quad .$$

Par contre, (18) peut être numériquement mieux conditionnée.

Conclusion : un exemple.

Nous présentons un exemple où u et v sont scalaires, de sorte que $W(v)$ se réduit à une commande $u = \varphi(v)$ unique, simplifiant considérablement l'analyse mathématique.

Dans le jeu des gendarmes et des voleurs, un voleur $\mathcal{U}$, courant à la vitesse $b > 1$, veut atteindre la prison : un cercle C de rayon $l > 1$ centré à l'origine, afin d'y délivrer ses camarades.

Un gendarme v veut empêcher u d'atteindre C en le capturant, c'est-à-dire en passant à une distance inférieure à 1 de lui.
On cherche à résoudre le problème où le critère à maximinimiser est le temps mis par u pour atteindre C. On voit que le seul moyen qu'à v d'agir sur ce temps est d'user de la contrainte d'état qu'il impose à u de rester à une distance $d \geq 1$ de lui.

Pour décrire le problème avec le plus petit nombre possible de variables, on choisit des axes ox, oy centrés au centre de C, ox aligné sur u. L'état est constitué des coordonnées x, y de v dans ce repère et par l'abscisse z de u.

La cible est C : $z < 1$.
La contrainte d'états : $(x - z)^2 + y^2 \geq 1$.
Pour être dans la situation de la théorie ci-dessus, on utilisera aussi les coordonnées (r, α, z) définies par :

$$r = \sqrt{(x - z)^2 + y^2}$$

$$x = z + r \cos \alpha$$

$$y = r \sin \alpha \ .$$

Les commandes sont les directions des vitesses respectives des joueurs, représentées par l'angle qu'elles font avec l'axe ox. Les équations de la dynamique sont :

$$\overset{\circ}{x} = \cos v + b \frac{y}{z} \sin u$$

$$\dot{y} = \sin v - b \frac{x}{z} \sin u$$

$$\dot{z} = b \cos u$$

ou en coordonnées r, α, z :

$$\dot{r} = \cos(\alpha - v) - b \cos (\alpha - u)$$

$$\dot{\alpha} = \frac{1}{r}[- \sin(\alpha - v) + b \sin(\alpha - u)] - \frac{b}{z} \sin u \ .$$

La variété S est le cylindre oblique $r = 1$, et $W(v)$ se réduit aux commandes u vérifiant :

$$\dot{r} = \cos(\alpha - v) - b\cos(\alpha - u) = 0 \quad .$$

Les trajectoires construites depuis la cible en remontant le temps correspondent, dans l'espace physique, à une course rectiligne pour u, la stratégie de v n'étant pas définie quand il n'a plus espoir de gêner u dans le futur. Ces trajectoires rencontrent le cylindre S le long de la génératrice :

$$\alpha = \alpha_o \quad : \quad \cos\alpha_o = \frac{1}{b}$$

qui, avec l'arc $z = 1$, $\alpha > \alpha_o$ joue le rôle de variété B. (En fait, il faut considérer $\alpha < \alpha_1$, où la valeur de α_1 est donnée par $\sin\alpha_1 = 1/b$, et l'arc symétrique par rapport à ox).

La solution classique consisterait à poser $u = \varphi(v)$:

$$\cos u = \frac{1}{b}[\cos\alpha\cos(\alpha - v) - \sin\alpha\sqrt{b^2 - \cos^2(\alpha - v)}]$$

$$\sin u = \frac{1}{b}[\sin\alpha\sin(\alpha - v) + \cos\alpha\sqrt{b^2 - \cos^2(\alpha - v)}] \quad .$$

La non unicité de u dans $W(u)$ est vite éliminée par des arguments de continuité en B, ou de simple logique. La dynamique du jeu réduit est alors :

$$\dot{\alpha} = [-\sin(\alpha-v) - \sqrt{b^2-\cos^2(\alpha-v)}] - \frac{1}{z}[\cos\alpha\sqrt{b^2-\cos^2(\alpha-v)}+\sin\alpha\cos(\alpha-v)]$$

$$\dot{z} = \cos\alpha\cos(\alpha-v) - \sin\alpha\sqrt{b^2-\cos^2(\alpha-v)}$$

et la commande $\hat{v}$ en un point de la génératrice α_o, se trouverait être donnée par :

$$\cos\hat{v} = +\frac{1}{b}$$

$$\sin\hat{v} = +\frac{\sqrt{b^2-1}}{b}$$

Et on aurait une commande semblable dans un voisinage, $\alpha > \alpha_o$. C'est-à-dire que v se comportant comme si u était tenu de se maintenir sur S s'éloigne de C pour en éloigner u. C'est bien sûr un comportement absurde. Si v joue cette commande, u l'ignorera et ira directement vers C. Au contraire, si on applique la théorie plus fine que nous avons développée, en α_o on trouve que V_1 se réduit à v^* opposée à la commande que nous venons de donner. Et dans un voisinage, $\alpha > \alpha_0$, la commande sera peu différente. γ se dirigeant vers u pour le gêner.

On peut discuter la différence entre les deux méthodes sur le graphique en u, v, ou nous représentons Σ, u^* et v^*.

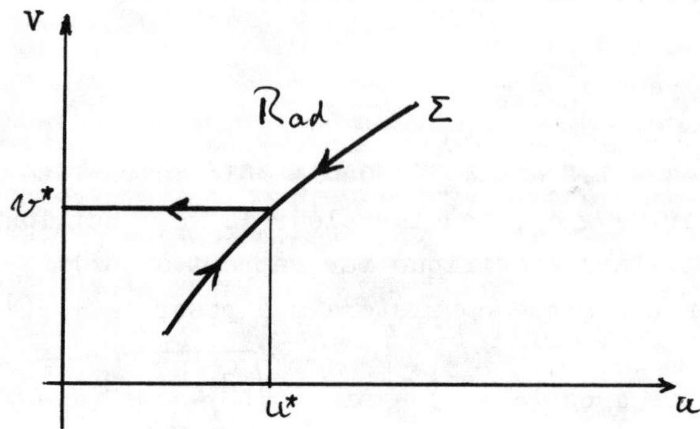

Dans l'ancienne méthode, on recherche un minimum sur Σ en u maximum en v. Soit un maximin le long des courbes indiquées en gras sur la figure où H doit croître quand on les parcourt dans le sens indiqué par les flèches. Il faut au contraire rechercher ce maximin le long de la courbe indiquée en gras sur la figure ci-dessous :

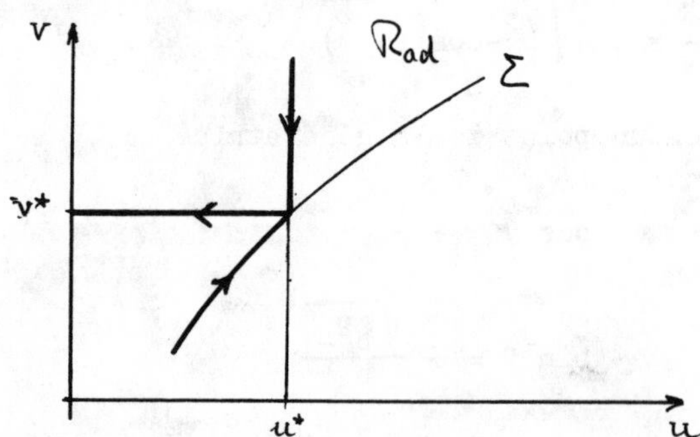

Cette démarche aboutit, malgré des difficultés numériques assez sérieuses, dues au fait que (18) est discontinue pour l'argument que nous cherchons, et nulle seulement dans sa limite à gauche.

Références

[1] P. BERNHARD : "Linear Differential Games and the Isotropic Rocket", PhD dissertation, SUDAAR 413, Stanford University, 1970.

[2] J.V. BREAKWELL : "The Dolichobrachistochrome Game", Unpublished lecture notes, Stanford University, 1969, and Report C-18/2, Centre d'Automatique de l'Ecole Nationale Supérieure des Mines de Paris, Fontainebleau, 1971.

[3] J.V. BREAKWELL and A. MERZ : "Towards a complete solution of the homicidal chauffeur Game", Proceedings of the First International Conference on the Theory and Applications of Differential Games. Amherst, 1970.

[4] R. ISAACS : "Differential Games", Willey, New York, 1965.

[5] R.E. KALMAN : "The Calculus of Variations and Optimal Control Theory", in Mathematical Optimization Techniques, R. Bellman ed., University of California Press, Berkeley, 1963.

SOME GENERAL PROPERTIES OF NON-COOPERATIVE GAMES

Ivar EKELAND
U.E.R. MATHEMATIQUES DE LA DECISION
UNIVERSITE PARIS 9 DAUPHINE
75775 - Parix CEDEX 16

Let there be given two strategy sets Σ_1 and Σ_2, and two payoff functions f_1 and f_2 on $\Sigma_1 \times \Sigma_2$. This is the usual setting of two-persons-non-cooperative game theory. We are looking for situations $(\bar\sigma_1, \bar\sigma_2) \in \Sigma_1 \times \Sigma_2$ which exhibit some kind of stability when each player strives to maximize his own payoff. Such situations will be called "equilibria" or "solutions" of the game, and many different kinds have been discovered since the beginning of game theory. For instance:

(a) <u>the von Neumann equilibria</u>, certainly the best-known of all. A situation $(\bar\sigma_1, \bar\sigma_2) \in \Sigma_1 \times \Sigma_2$ is a von Neumann equilibrium iff :

$$f_1(\bar\sigma_1, \bar\sigma_2) \geq f_1(\sigma_1, \sigma_2) \qquad \forall \sigma_1 \in \Sigma_1$$

$$f_2(\bar\sigma_1, \bar\sigma_2) \geq f_2(\bar\sigma_1, \sigma_2) \qquad \forall \sigma_2 \in \Sigma_2$$

Such equilibria are very satisfying from a conceptual point of view but do not always exist, even when the strategy sets are assumed to be compact and the payoff functions continuous. Other stringent assumptions are needed, such as convexity of the strategy sets and payoff functions. ■

b) <u>the Stackelberg equilibria</u>. Let M denote the set of situations $(\sigma_1, \sigma_2) \in \Sigma_1 \times \Sigma_2$ such that :

$(\sigma_1, \sigma_2) \in M \longleftrightarrow f_2(\sigma_1, \sigma_2) \geq f_2(\sigma_1, \tau_2) \quad \forall \tau_2 \in \Sigma_1$

A situation $(\bar{\sigma}_1, \bar{\sigma}_2)$ is a Stackelberg equilibrium iff :

$f_1(\bar{\sigma}_1, \bar{\sigma}_2) \geq f_1(\sigma_1, \sigma_2) \quad \forall (\sigma_1, \sigma_2) \in M.$

Such equilibria always exist if the strategy sets are assumed to be compact and the payoff functions continuous. Note that the first player is better off with a Stackelberg equilibrium than with a von Neumann equilibrium. Indeed, if (τ_1, τ_2) is a von Neumann equilibrium and (σ_1, σ_2) a Stackelberg equilibrium, then $(\tau_1, \tau_2) \in M$ and hence $f_1(\sigma_1, \sigma_2) \geq f_1(\tau_1, \tau_2)$. ■

Up to now, game theory has been set up in a linear analysis framework, the main question to be answered being usually the existence of von Neumann equilibria. In this lecture, we will use instead a differential-topologic setting. That means we will drop the usual convexity assumptions in favour of regulaty assumptions : Σ_1 and Σ_2 will be smooth manifolds (without boundary), and f_1 and f_2 will be C^2-functions That will enable us to raise sharper questions.

<u>Question 1</u> : <u>are the von Neumann equilibria finite in number, and stable with respect to small perturbations of the payoff functions</u> ?

This question of stability is important with respect to applications. Indeed, if the payoff functions f_1 and f_2 are intended to describe a concrete situation, we know them but with a certain degree of accuracy, i.e. a small error is necessarily made on the payoff functions. For the concept of equilibrium to be relevant, that should mean a small error only on the corresponding equilibria, i.e. stability is expected.

The answer to this question is "almost always" yes. Indeed, we have the theorem :

Theorem 1 *There is an open and dense subset Ω of $\mathcal{C}^2(\Sigma_1 \times \Sigma_2) \times \mathcal{C}^2(\Sigma_1 \times \Sigma_2)$ with the following property : on every connected component w of Ω there exist a finite number of $\mathcal{C}^1$ mappings :*

$$\gamma_i : \omega \longrightarrow \Sigma_1 \times \Sigma_2 \ , \ i \in I_\omega$$

such that the set $\{\gamma_i(f_1, f_2)\}$, $i \in I_\omega$, is the set of all von Neumann equilibria of the game with payoff functions f_1 and f_2.

Let us point out that an open and dense subset of a complete metric space such as $\mathcal{C}^2 \times \mathcal{C}^2$ is very large. Indeed, its complement is nowhere dense, and moreover the Baire category theorem holds ; for instance, every denumerable intersection of open dense subsets will still be dense.

Question 2 : *can a Stackelberg equilibrium $(\bar{\sigma}_1, \bar{\sigma}_2)$ be approximated by a sequence $(\sigma_1^n, \sigma_2^n) \in M$ such that σ_2^n is the only maximum of the payoff $\tau_2 \mapsto f_2(\sigma_1^n, \tau_2)$ on Σ_2 ?*

To understand this question, let us recall the conventional interpretation of the Stackelberg equilibrium. The first player is the leader : he chooses first strategy $\tau_1 \in \Sigma_1$, and announces it. The second player is the follower : he answers by picking some strategy $\tau_2 \in \Sigma_2$ which maximizes his payoff $f_2(\tau_1, \tau_2)$. That means (τ_1, τ_2) will always belong to M, and the leader will try to reach the point $(\bar{\sigma}_1, \bar{\sigma}_2)$ of M where his payoff f_i is maximal, i.e. the Stackelberg equilibrium.

So far so good. Suppose now the leader plays σ_1, in the hope that he will be answered by σ_2. If $\bar{\sigma}_2$ is the only maximum of the payoff $f_2(\bar{\sigma}_1, \sigma_2)$, it will indeed be picked by the follower, and so the Stackelberg equilibrium will be reached automatically. But it is easily seen that usually the function $f_2(\bar{\sigma}_1, \bar{\sigma}_2)$ will have several maxima.

In that case, the follower may choose any of them ; he might pick $\bar{\sigma}_2$, and he might pick quite another one. So there is no guaranty that the Stackelberg equilibrium will be reached.

But if the answer to question 2 is positive, the leader will not play σ_1 directly. He will reach $\bar{\sigma}_1$ only by going through all the σ_1^n, so that the follower will be taken through all the σ_2^n , and will finally play $\bar{\sigma}_2$ by continuity.

We could also say that, if the answer to question 2 is positive, although the leader cannot ensure that the precise Stackelberg equilibrium will be reached, he can approximate it up to any degree of accuracy. Indeed, if the leader picks one of the σ_1^n close enough to $\bar{\sigma}_1$, the follower has no choice but σ_2^n, which will be close to $\bar{\sigma}_2$.

Once again, the answer to question 2 will be "almost always" yes. As a by-product, we shall also answer the question about the stability of Stackelberg equilibria with respect to perturbations of the payoff functions.

Theorem 2 There exists in $C^\infty(\Sigma_1 \times \Sigma_2)$ and open and dense subset $\mathcal{U}_2$ with the following property : for every $f_2 \in \mathcal{U}_2$, each point $(\sigma_1, \sigma_2) \in M$ is the limit of a sequence $(\sigma_1^n, \sigma_2^n) \in M$ such that σ_2^n is the only maximum of $\tau_2 \longrightarrow f_2(\sigma_1^n, \tau_2)$ on Σ_2.

Corollary If $f_2 \in \mathcal{U}_2$, there exists in $C^2(\Sigma_1 \times \Sigma_2)$ an open and dense subjet $\mathcal{U}_1$ and a C^1 mapping $s : \mathcal{U}_1 \longrightarrow M$ such that $s(f_1)$ is the only Stackelberg equilibrium of the game with payoff functions f_1 and f_2.

Let us dwell a little more on theorem 2. The set M is the graph of the upper semi-continuous compact-valued mapping $m : \Sigma_1 \longrightarrow \Sigma_2$ defined by :

$$m(\sigma_1) = \{\sigma_2 \mid f_2(\sigma_1, \sigma_2) \geq f_2(\sigma_1, \tau_2) \ \forall \tau_2 \in \Sigma_2\}.$$

The proof of theorem 2 goes to show that, for every $f_2 \in \mathcal{U}_2$, the multi-valued mapping m contains at most (dimension Σ_1+1) points. Moreover, the strategy set Σ_1 can be partitioned into subsets $\Sigma_1^i, 1 \leq i \leq 1+ \dim \Sigma_1$:

$$\Sigma_1^i : \{\sigma_1 \mid f_2(\sigma_1, .) \text{ has i maxima on } \Sigma_2\}.$$

It is shown that Σ_1^i is a regular (non-connected) submanifold of Σ_1, of dimension $\dim \Sigma_1 - (i-1)$, i.e. of codimension (i-1). If $j > i$, then Σ_1^j is contained in the boundary of Σ_1^i, and satisfies certain regularity conditions.

For instance, Σ_1^1 is an open and dense subset of Σ_1. If $\sigma_1 \in \Sigma_1^1$, the function $f_2(\sigma_1,.)$ has a unique maximum $\varphi(\sigma_1) \in \Sigma_2$. It is shown that this mapping $\varphi : \Sigma_1^1 \longrightarrow \Sigma_2$ is C^∞, and that M is the closure of the graph of φ in $\Sigma_1 \times \Sigma_2$.

Any component of Σ_1^2 separates two components of Σ_1^1. If $\sigma_1 \in \Sigma_1^2$, the function $f_2(\sigma_1,.)$ has two maxima σ_2 and τ_2. Now σ_1 can be approximated from both sides of Σ_1^2. On one side, we find a sequence $\sigma_1^n \longrightarrow \sigma_1$ with $\sigma_1^n \in \Sigma_1^1$ such that the corresponding σ_2^n go to σ_2. On the other side, we find a sequence $\tau_1^n \longrightarrow \sigma_1$ with $\tau_1^n \in \Sigma_1^1$ such that the corresponding τ_2^n go to τ_2.

For further information and details the reader is referred to "Topologie différentielle et théorie des jeux", to be published in Topology, 1974.

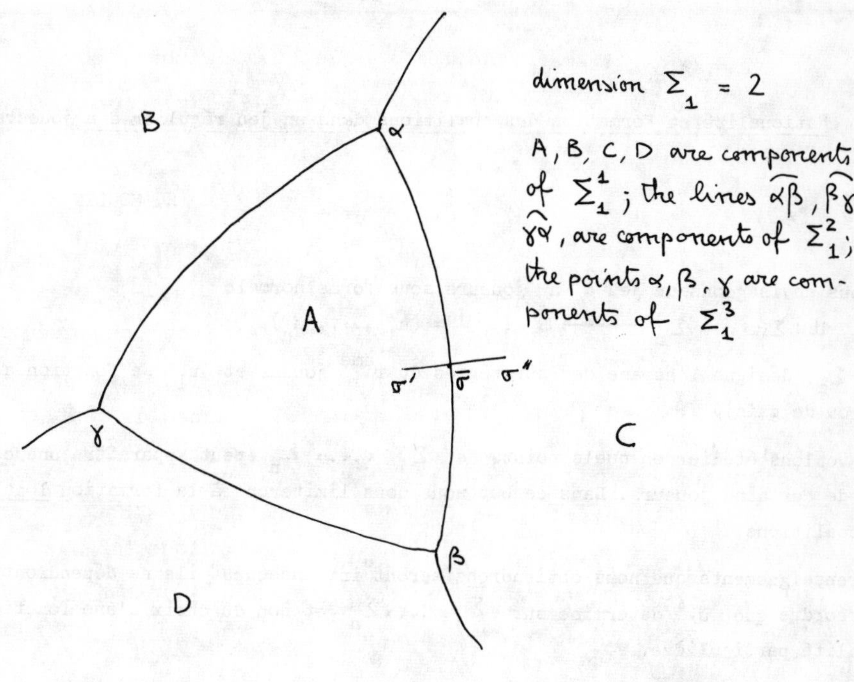

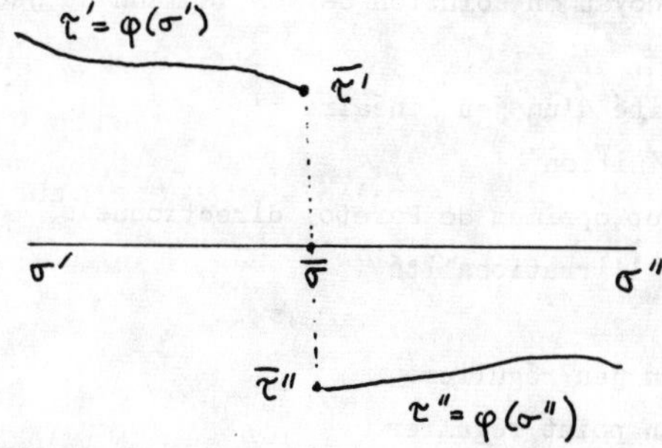

diagram of M in $\Sigma_1 \times \Sigma_2$ at a crossing point $\bar{\sigma}$ of Σ_1^2; note that m is double-valued at $\bar{\sigma}$ and single valued otherwise.

Rationalité et Formation des coalitions dans un jeu régulier à n joueurs

H. MOULIN

Nous envisageons un jeu à n joueurs sous forme normale

(0) $\mathcal{U} : \Sigma_1 \times \dots \times \Sigma_n \longrightarrow \mathbb{R}^n$; $\mathcal{U} = (u_1, \dots, u_n)$

(où Σ_i désigne l'espace des stratégies du $i^{\text{ème}}$ joueur et u_i sa fonction d'utilité ou de gain).

Nous voulons étudier en quels points de $\Sigma_1 \times \dots \times \Sigma_n$ peut apparaître une <u>coalition</u> de certains joueurs. Dans ce but nous nous limiterons à la formation <u>locale</u> des coalitions.

Les renseignements que nous obtiendrons seront "intrinsèques": ils ne dépendront que du préordre que u_i détermine sur $\Sigma_1 \times \dots \times \Sigma_n$ et non du choix d'une fonction d'utilité particulière.

Dans la partie I on étudie la rationalité d'un jeu linéaire et on applique ces résultats à la différentielle de l'application $\mathcal{U}$ dans la partie II. Puis on illustre les résultats par des exemples issus de la theorie économique. Enfin on étudie d'un point de vue local, le noyau ou solution de Von Neumann du jeu (0) .

I Rationalité et irrationalité d'un jeu linéaire

1. Irrationalité, définition
2. Equilibre de Nash et optimum de Pareto directionnels
3. Caractérisation de l'irrationalité

II Rationalité locale dans un jeu régulier

1. Irrationalité en un point régulier
2. Rationalité et surface de Pareto
3. Exemples économiques
4. Le noyau local d'un jeu régulier

I Rationalité et irrationalité d'un jeu linéaire

On se donne n espaces de Hilbert Σ_i et n formes linéaires continues u_i définies sur $\Sigma_1 \times \dots \times \Sigma_n$. On appelle jeu dont les ensembles de stratégies sont Σ_i et les fonctions d'utilité u_i la donnée de l'application $\mathcal{U}$

$$(I.1.) \quad \mathcal{U} : \Sigma_1 \times \dots \times \Sigma_n \longrightarrow \mathbb{R}^n \; ; \; \mathcal{U}(\sigma_1, \dots, \sigma_n) = (u_1(\sigma_1, \dots, \sigma_n), \dots, u_n(\sigma_1, \dots, \sigma_n))$$

1 Irrationalité; définition

Définition (I.1.)

On dit que le jeu (I.1.) *est irrationnel si il existe un vecteur* $(\sigma_1, \dots, \sigma_n)$ *dans* $\Sigma_1 \times \dots \times \Sigma_n$ *tel que*

$$(I.2.) \qquad \forall i = 1, \dots, n \qquad u_i(0, \dots, 0, \sigma_i, 0 \dots 0) > 0$$

$$(I.3.) \qquad \forall i = 1, \dots, n \qquad u_i(\sigma_1, \dots, \sigma_n) < 0$$

Notons $\langle , \rangle_i$ le produit scalaire du Hilbert Σ_i. Il existe donc des vecteurs α_i^j dans Σ_j tels que :

$$(I.4.) \qquad \forall i = 1, \dots, n \qquad u_i(\sigma_1, \dots, \sigma_n) = \sum_{j=1}^{n} \langle \alpha_i^j , \sigma_j \rangle_j$$

Proposition (I.1.) *Supposons que les vecteurs* α_i^i *sont non nuls. Le jeu* (I.1.) *est irrationnel si et seulement si il n'existe pas de vecteurs* (p,q) *tels que* :

$$(I.5.) \qquad p \in \mathbb{R}_+^n - \{0\} \; ; \; q \in \mathbb{R}_+^n \; ; \; \forall j = 1, \dots, n \quad \sum_{i=1}^{n} p_i \alpha_i^j = q_j \alpha_j^j$$

Démonstration

Si le jeu (I.1.) est irrationnel, il existe $\sigma = (\sigma_1, \dots, \sigma_n)$ dans $\Sigma_1 \times \dots \times \Sigma_n$ tel que :

$$(I.6.) \qquad \forall i = 1, \dots, n \; \langle \alpha_i^i , \sigma_i \rangle > 0 \quad \text{et} \quad \mathcal{U}(\sigma) \in \overset{\circ}{\mathbb{R}}{}_-^n$$

Supposons qu'il existe aussi une solution (p,q) au système (I.5.). Alors on aura :

(I.7.) $\langle {}^t\mathcal{U} p, \sigma \rangle = \sum_{j=1}^{n} \langle ({}^t\mathcal{U}p)_j, \sigma_j \rangle_j = \sum_{j=1}^{n} q_j \langle \alpha_j^j, \sigma_j \rangle \geq 0$

et d'autre part

(I.8.) $\langle {}^t\mathcal{U} p, \sigma \rangle = \langle p, \mathcal{U}(\sigma) \rangle < 0$

Donc si (I.1.) est irrationnel, le système (I.5.) n'a pas de solution.

Supposons maintenant que (I.1.) est rationnel et définissons l'application linéaire S :

(I.9.) $S : \Sigma_1 \times \ldots \times \Sigma_n \longrightarrow \mathbb{R}^{2n} : S(\sigma_1, \ldots, \sigma_n) = (u_1(\sigma_1 \ldots \sigma_n), \ldots, u_n(\sigma_1 \ldots \sigma_n), - u_1(\sigma_1), \ldots, -u_n(\sigma_n))$

Alors il est clair que la rationalité de (I.1.) s'écrit :

(I.10.) $S(\Sigma_1 \times \ldots \times \Sigma_n) \cap \mathring{\mathbb{R}}^{2n}_- = \emptyset$

D'après le théorème de Hahn Banach il existe donc un couple (p,q) de $\mathbb{R}^n \times \mathbb{R}^n$ tel que :

(I.11.) $(p,q) \in \mathbb{R}^{2n}_+ - \{0\} \; ; \; \forall (\sigma_1, \ldots, \sigma_n) \in \Sigma_1 \times \ldots \times \Sigma_n \; \langle (p,q), S(\sigma_1, \ldots, \sigma_n) \rangle \geq 0$

c'est à dire que pour tout $(\sigma_1, \ldots, \sigma_n)$ on a :

(I.12.) $\langle p, \mathcal{U}(\sigma_1, \ldots, \sigma_n) \rangle - \sum_{i=1}^{n} q_i u_i(\sigma_i) \geq 0$

ou encore :

(I.13.) $\sum_{i=1}^{n} \langle ({}^t\mathcal{U}p)_i \; , \sigma_i \rangle_i - \langle q_i \alpha_i^i, \sigma_i \rangle_i \geq 0$

Cette relation étant vraie pour tout $(\sigma_1, \ldots, \sigma_n)$ il vient :

(I.14.) $\forall i = 1, \ldots, n \; ({}^t\mathcal{U}p)_i = q_i \alpha_i^i$

Or d'après (I.11.) (p,q) est dans $\mathbb{R}^{2n}_+ - \{0\}$. Donc si p était nul, et puisque les vecteurs α_i^i sont non nuls, on obtiendrait une contradiction. Ceci achève de prouver que (p,q) vérifie le système (I.5.).

Nous allons maintenant interpréter la rationalité d'un jeu linéaire.

2. Equilibre de Nash et optimums de Pareto directionnels

Nous notons $P_+(\Sigma_i)$ le quotient de $\Sigma_i - \{0\}$ par la relation de collinéarité

(I.15.) $x \sim y$ signifie $\exists \lambda > 0 \quad x = \lambda y$

Soit $\rho = (\rho_1, \ldots, \rho_n)$ un élément de $\mathring{\mathbb{R}}^n_+$, nous notons $\Sigma_i(\rho_i)$ la boule de rayon ρ_i dans Σ_i et $\mathcal{U}(\rho)$ la restriction de $\mathcal{U}$ à ces boules :

(I.16.) $$\mathcal{U}(\rho) : \Sigma_1(\rho_1) \times \ldots \times \Sigma_n(\rho_n) \longrightarrow \mathbb{R}^n$$

Proposition (I.2.)

a/ L'ensemble des optimums de Pareto faibles du jeu $\mathcal{U}(\rho)$ (I.16.) a une image dans $\prod_{i=1}^{n} P_+(\Sigma_i)$ qui ne dépend pas de ρ. On l'appelle l'ensemble des optimums de Pareto directionnels et on la note $\dot{\mathrm{Par}}(\mathcal{U})$.

b/ De plus, si les vecteurs α_i^i sont tous non nuls, l'unique équilibre de Nash du jeu $\mathcal{U}(\rho)$ (I.16.) a une image dans $\prod_{i=1}^{n} P_+(\Sigma_i)$ qui ne dépend pas de ρ. On l'appelle l'équilibre de Nash directionnel et on le note $\acute{\mathrm{Na}}(\mathcal{U})$.

Démonstration

Un point $(\bar{\sigma}_1, \ldots, \bar{\sigma}_n)$ de $\Sigma_1(\rho_1) \times \ldots \times \Sigma_n(\rho_n)$ est un optimum de Pareto faible du jeu $\mathcal{U}(\rho)$ si et seulement si il existe p dans $\mathbb{R}^n_+ - \{0\}$ tel que :

(I.17.) $$< p, \mathcal{U}(\bar{\sigma}_1, \ldots, \bar{\sigma}_n)> = \max_{\substack{\|\sigma_i\|_i \leq \rho_i \\ i=1,\ldots,n}} < p, \mathcal{U}(\sigma_1, \ldots, \sigma_n)>$$

Ce problème d'optimisation se décentralise en :

(I.18.) $$\forall i = 1, \ldots, n \qquad <({}^t\mathcal{U} p)_i, \bar{\sigma}_i>_i = \max_{\|\sigma_i\|_i \leq \rho_i} < ({}^t\mathcal{U} p)_i, \sigma_i>_i$$

Et il est clair que pour un p fixe les solutions de (I.18.) ont une image dans $\prod_{i=1}^{n} P_+(\Sigma_i)$ qui ne dépend pas de ρ. D'où la partie a/ de la proposition Pour la partie b/ on constate que l'équilibre de Nash (unique) du jeu $\mathcal{U}(\rho)$(I.16.) est :

(I.19.) $$i = 1, \ldots, n \qquad \bar{\sigma}_i = \frac{\rho_i}{\|\alpha_i^i\|_i} \alpha_i^i$$

Théorème (I.2.)

Supposons que toutes les restrictions $\mathcal{U}_{\{i\}}$ de $\mathcal{U}$ sont surjectives :

(I.20.) $\mathcal{U}_{\{i\}} : \Sigma_i \longrightarrow \mathbb{R} \quad \mathcal{U}_{\{i\}}(\sigma_i) = u_i(0,\ldots,0,\sigma_i,0\ldots0)$

Alors il est équivalent de dire que le jeu (I.1.) est rationnel ou que l'équilibre de Nash directionnel est un optimum de Pareto directionnel :

(I.21.) $\quad N\acute{a}(\mathcal{U}) \in P\acute{a}r(\mathcal{U})$

Démonstration

L'hypothèse de surjectivité des $\mathcal{U}_{\{i\}}$ est exactement la non nullité des vecteurs α^i_i (ce qui d'après la proposition (I.2.) donne un sens à $N\acute{a}(\mathcal{U})$) . Supposons donc (I.21.) et choisissons $\rho_i = \|\alpha^i_i\|_i$. Dans le jeu $\mathcal{U}(\rho)$ (I.16.) associé, la relation (I.21.) s'écrit :

(I.22.) $\quad (\alpha^1_1,\ldots,\alpha^n_n)$ est optimum de Pareto faible du jeu $\mathcal{U}(\rho)$

Autrement dit il existe p dans $\mathbb{R}^n_+ - \{0\}$ tel que :

(I.23.) $\quad \forall i = 1,\ldots,n \quad < ({}^t\mathcal{U}p)_i, \alpha^i_i > = \max\limits_{\|\sigma_i\|_i \le \|\alpha^i_i\|_i} < ({}^t\mathcal{U}p)_i , \sigma_i >$

Posons $q = (q_1,\ldots,q_n)$ avec ;

(I.24.) $\quad \forall i = 1,\ldots,n \quad q_i = 0$ si $({}^t\mathcal{U}p)_i = 0$; $q_i = \dfrac{\|({}^t\mathcal{U}p)_i\|_i}{\|\alpha^i_i\|_i}$ sinon

Alors d'après (I.23.) il vient :

(I.25.) $\quad \forall i = 1,\ldots,n \quad ({}^t\mathcal{U}p)_i = q_i\alpha^i_i$

Donc (p,q) vérifie le système (I.5.) de la proposition (I.1.) et d'après cette même proposition le jeu (I.1.) est rationnel.

Réciproquement supposons que le jeu (I.1.) est rationnel et utilisons encore la proposition (I.1.) : il existe une solution (p,q) du système (I.5.) :

(I.26.) $\quad p \in \mathbb{R}^n_+ - \{0\} \quad q \in \mathbb{R}^n_+ \quad \forall i = 1,\ldots,n \; ({}^t\mathcal{U}p)_i = q_i\alpha^i_i$

On en déduit clairement que p vérifie (I.23.) (si $({}^t\mathcal{U}p)_i = 0$ c'est trivial) et donc que $(\alpha^1_1,\ldots,\alpha^n_n)$ est optimum de Pareto faible du jeu $\mathcal{U}(\rho)$ avec $\rho_i = \|\alpha^i_i\|$. Autrement dit

(I.27.) $\quad Na(\mathcal{U}(\rho)) \in Par(\mathcal{U}(\rho))$

ce qui entraine (I.21.) et achève la démonstration.
Le théorème (I.2.) permet d'interpréter la rationalité du jeu (I.1.) : dans tout jeu restreint $\mathcal{U}(\rho)$ (I.16.) l'équilibre non coopératif (de Nash) fait partie des équilibres coopératifs (de Pareto). Il est naturel de dire alors que la coalition de tous les joueurs ne peut pas se former.

3. Caractérisation de l'irrationalité

Nous supposons dans ce paragraphe que les espaces Σ_i sont de dimension 1 et identifions donc le jeu (I.1.) avec une matrice (n-n) :

(I.28.) $$\mathcal{U} = \left[\alpha_i^j\right]_{i=1,\dots,n}^{j=1,\dots,n}$$

Si A désigne une matrice (n-n) on note A(i,j) la sous matrice obtenue en omettant la $i^{ème}$ ligne et la $j^{ème}$ colonne. On désigne par $\mathcal{P}_n$ le sous-ensemble de l'espace des matrices (n-n) défini par la récurrence suivante :

(I.29.) $\quad [a] \in \mathcal{P}_1 \quad : a < 0$

(I.30.) $\quad A \in \mathcal{P}_n \quad$:- ou bien A est inversible et les lignes de A^{-1} sont dans $\mathbb{R}_-^n$
- ou bien il existe j tel que pour tout i (i = 1,...,n) on ait : $A(i,j) \in \mathcal{P}_{n-1}$

<u>Proposition (I.3.)</u>

<u>Supposons que toutes les sous matrices de $\mathcal{U}$ sont de rang maximal.</u>

<u>Alors le jeu $\mathcal{U}$(I.1.) est irrationnel si et seulement si la matrice</u> A <u>suivante est dans</u> $\mathcal{P}_n$:

(I.31.) $$A = \left[a_i^j\right] \qquad a_i^j = \alpha_i^j \, \alpha_j^j$$

On constate donc que la condition d'irrationalité de $\mathcal{U}$ est un ensemble d'inéquations portant sur les coefficients de $\mathcal{U}$. Par exemple si n = 2 ce sont les inéquations :

(I.32.) $$\alpha_1^1 \alpha_2^1 < 0 \ , \ \alpha_1^2 \alpha_2^2 < 0 \ ; \ \alpha_1^1 \alpha_2^2 \left[\alpha_1^1 \alpha_2^2 - \alpha_1^2 \alpha_2^1\right] < 0$$

Démonstration

Il est clair d'après la définition que l'irrationalité de $\mathcal{U}$ équivaut à :

(I.33.) $A(\mathring{\mathbb{R}}^n_+) \cap \mathring{\mathbb{R}}^n_- \neq \emptyset$

Nous allons montrer que cette relation équivaut à

(I.34.) $A \in \mathcal{S}_n$

C'est clair pour $n = 1$, et nous allons le montrer par récurrence. Soit donc A une matrice dont toutes les sous matrices sont de rang maximal et qui vérifie (I.33.). Alors 2 cas sont possibles :

1er cas : $A(\mathring{\mathbb{R}}^n_+) \supset \mathring{\mathbb{R}}^n_-$

Comme A est inversible ce cas équivaut à

(I.35.) $A^{-1}(\mathring{\mathbb{R}}^n_+) \subset \mathring{\mathbb{R}}^n_-$

c'est à dire que les lignes de A sont dans $\mathbb{R}^n_-$.

2ème cas : $A(\mathring{\mathbb{R}}^n_+) \not\supset \mathring{\mathbb{R}}^n_-$

Alors il existe x dans $\mathring{\mathbb{R}}^n_-$ qui appartient à la frontière (au sens topologique ou au sens de la convexité) de $A(\mathring{\mathbb{R}}^n_+)$. Autrement dit x est une combinaison linéaire positive de $(n-1)$ seulement parmi les n vecteurs colonnes de la matrice A . Comme $\mathring{\mathbb{R}}^n_-$ est ouvert on peut même supposer que les coefficients de cette combinaison sont positifs stricts. Nous venons de montrer que si $A_{\check{j}}$ désigne la $(n,(n-1))$ matrice obtenue en otant la $j^{\text{ème}}$ colonne de A il existe j tel que :

(I.36.) $A_{\check{j}}(\mathring{\mathbb{R}}^{n-1}_+) \cap \mathring{\mathbb{R}}^n_- \neq \emptyset$

Ceci suffit à assurer que pour tout i on a :

(I.37.) $A(i,j)(\mathring{\mathbb{R}}^{n-1}_+) \cap \mathring{\mathbb{R}}^{n-1}_- \neq \emptyset$

Autrement dit (hypothèse de récurrence) $A(i,j)$ est dans $\mathcal{S}_{n-1}$. D'après la définition (I.30.) , A est donc un élément de $\mathcal{S}_n$. Nous avons montré que dans tous les cas - si $\mathcal{U}$ est irrationnel alors A est dans $\mathcal{S}_n$.
Réciproquement soit A la matrice associée au jeu $\mathcal{U}$ et supposons que A est dans $\mathcal{S}_n$. Deux cas sont possibles : ou bien A^{-1} a tout ses coefficients négatifs et on a (I.35.) . Donc $\mathcal{U}$ est irrationnel. Ou bien toute sous matrice $A(i,n)$ de

$A_{\check{n}}$ (par exemple) est dans $\mathcal{P}_{n-1}$. Nous allons montrer qu'alors :

(I.38.) $\qquad A(\mathring{\mathbb{R}}_{+}^{n-1}) \cap \mathring{\mathbb{R}}_{-}^{n} \neq \emptyset$

Ceci suffit à établir (I.33.) et donc l'irrationalité de $\mathcal{U}$. Supposons que (I.39.) n'est pas vraie, le théorème de Hahn-Banach et l'inversibilité de A montre qu'il existe p dans $\mathbb{R}_{+}^{n} - \{0\}$ tel que :

(I.39.) $\qquad {}^{t}A(p) \in \mathbb{R}_{+}^{n-1} - \{0\}$

Si $W_1, \ldots, W_n$ désignent les lignes de la matrice $A_{\check{n}}$ on a donc :

(I.40.) $\qquad x = p_1 W_1 + \ldots + p_n W_n \in \mathbb{R}_{+}^{n-1} - \{0\}$

Si W_1 appartient au cône engendré par $W_2, \ldots, W_n$ on en déduit immédiatement que ce cône coupe $\mathbb{R}_{+}^{n-1}$ autrement dit que :

(I.41.) $\qquad {}^{t}A(1,n)\ (\mathbb{R}_{+}^{n-1} - \{0\}) \cap \mathbb{R}_{+}^{n-1} \neq \emptyset$

ce qui met en défaut la relation :

(I.42.) $\qquad A(1,n)\ (\mathring{\mathbb{R}}_{+}^{n-1}) \cap \mathring{\mathbb{R}}_{-}^{n-1} \neq \emptyset$

Or on sait que $A(1,n)$ est dans $\mathcal{P}_{n-1}$ et, par hypothèse de récurrence, il vérifie donc (I.42.)

Si maintenant W_1 n'appartient pas au cône engendré par $W_2, \ldots, W_n$ écrivons

(I.43.) $\qquad x = \dfrac{p_1}{1+p_1} \left[(1+p_1) W_1 \right] + \dfrac{1}{1+p_1} \left[\sum_{i \geq 2} (1+p_1) p_i W_i \right]$

et posons

(I.44.) $\qquad a = (1+p_1) W_1 \; ; \; b = \sum_{i \geq 2} (1+p_1) p_i W_i$

Sur le segment $[a,b]$ il existe donc un point de la frontière du cône $A(1,n)\ (\mathbb{R}_{+}^{n-1})$ Puisque la matrice $A(1,n)$ est inversible, il existe donc c sur le segment $[a,b]$ qui est une combinaison linéaire positive de $(n-2)$ par mi les vecteurs $\{W_2, \ldots, W_n\}$. Donc x, qui est soit dans $[a,c]$ soit dans $[c,b]$ est de toutes façons combinaison linéaire positive de $(n-1)$ seulement parmi les vecteurs W_i. Donc il existe i tel que :

(I.45.) $\qquad A(i,n)\ (\mathbb{R}_{+}^{n-1}) \cap \mathbb{R}_{+}^{n-1} - \{0\} \neq \emptyset$

ce qui de la même façon, contredit le fait que $A(i,n)$ est dans $\mathcal{P}_{n-1}$.

II Rationalité locale dans un jeu régulier

On se donne n variétés Σ_i de classe C^2, modelées sur des espaces de Hilbert et n fonctions de classe C^2 u_i définies sur $\Sigma_1 \times \ldots \times \Sigma_n$.
On appelle jeu régulier dont les espaces de stratégies sont Σ_i et les fonctions d'utilité u_i, la donnée de l'application $\mathcal{U}$:

$$\text{(II.1.)} \quad \mathcal{U} : \Sigma_1 \times \ldots \times \Sigma_n \longrightarrow \mathbb{R}^n, \ \mathcal{U}(\sigma_1, \ldots, \sigma_n) = (u_1(\sigma_1, \ldots, \sigma_n), \ldots, u_n(\sigma_1, \ldots, \sigma_n))$$

1. Irrationalité en un point régulier

Nous appelons point régulier du jeu (II.1.) un point $\sigma = (\sigma_1, \ldots, \sigma_n)$ tel que

$$\text{(II.2.)} \quad \forall i = 1, \ldots, n \quad \frac{\partial u_i}{\partial \sigma_i}(\sigma) \neq 0 \quad \text{dans} \quad T^*_{\sigma_i}(\Sigma_i).$$

D'après [1] de tels points constituent génériquement un ouvert dense de $\Sigma_1 \times \ldots \times \Sigma_n$.
Nous appelons chemin régulier partant du point σ une application $\tilde{\sigma}$ de classe C^1 :

$$\text{(II.3.)} \quad \tilde{\sigma} : [0,1] \longrightarrow \Sigma_1 \times \ldots \times \Sigma_n \ ; \ \tilde{\sigma}(t) = (\sigma_1(t), \ldots, \sigma_n(t)) \ ; \ \tilde{\sigma}(0) = \sigma$$

telle que :

$$\text{(II.4.)} \quad \forall i < \frac{\partial u_i}{\partial \sigma_i}(\sigma), \frac{d\tilde{\sigma}_i}{dt}(0) >_i \neq 0 \ ; \ \sum_{j=1}^{n} < \frac{\partial u_i}{\partial \sigma_j}(\sigma), \frac{d\tilde{\sigma}_j}{dt}(0) >_j \neq 0$$

Définition (II.1.)

Le jeu (II.1.) est dit irrationnel au point régulier σ si il existe un chemin régulier $\tilde{\sigma}$ partant de σ tel que

$$\text{(II.5.)} \quad \forall i = 1, \ldots, n \ \forall t \in]0,1[\quad u_i(\sigma_1, \ldots, \sigma_{i-1}, \tilde{\sigma}_i(t), \sigma_{i+1} \ldots \sigma_n) > u_i(\sigma_1, \ldots, \sigma_n)$$

et

$$\text{(II.6.)} \quad \forall i = 1, \ldots, n \ \forall t \in]0,1[\quad u_i(\tilde{\sigma}_1(t), \ldots, \tilde{\sigma}_n(t)) < u_i(\sigma_1, \ldots, \sigma_n)$$

Le jeu est donc irrationnel au point σ si il existe un déplacement $\tilde{\sigma}_i(t)$ de chacun des joueurs, qui est avantageux pour chaque joueur si il est le seul à se déplacer (II.5.), mais désavantageux pour tous si ils se déplacent tous ensemble (II.6.).

La définition (II.1.) montre d'autre part que l'irrationalité d'un point est une notion "intrinséque" c'est à dire qu'elle ne dépend pas des fonctions d'utilité choisies mais seulement des préordres qu'elles définissent sur $\Sigma_1 \times \ldots \times \Sigma_n$ (on aurait pu définir seulement n préordres et les supposer "différentiables" voir [2]).

Théorème (II.1.)

Soit σ un point régulier du jeu régulier (II.1.) . Alors le jeu est irrationnel au point σ si et seulement si le jeu "tangent" suivant :

(II.7.) $T_\sigma \mathcal{U} : T_{\sigma_1} \Sigma_1 \times \dots \times T_{\sigma_n} \Sigma_n \longrightarrow \mathbb{R}^n$

où $T_\sigma \mathcal{U}$ désigne la différentielle de $\mathcal{U}$ au point σ , est irrationnel au sens de la définition (I.1.)

Démonstration

Si le jeu (II.1.) est irrationnel au point σ , il existe un chemin régulier $\tilde{\sigma}$ tel que :

(II.8.) $t \in]0,1] : h_i(t) = u_i(\sigma_1,\dots,\sigma_{i-1}, \tilde{\sigma}_i(t), \sigma_{i+1}, \dots, \sigma_n) > h_i(0) = u_i(\sigma)$.

Puisque h_i est de classe C^1 et de dérivée non nulle (régularité de $\tilde{\sigma}$) sa dérivée à l'origine est positive :

(II.9.) $\forall i = 1, \dots, n < \dfrac{\partial u_i}{\partial \sigma_i} , \dfrac{d\tilde{\sigma}_i}{dt}(0) >_i > 0$

On montre de même que :

(II.10.) $\forall i = 1, \dots, n \quad \sum_{j=1}^{n} < \dfrac{\partial u_i}{\partial \sigma_j}(\sigma) , \dfrac{d\tilde{\sigma}_j}{dt}(0) >_j < 0$

La définition (I.1.) montre alors que le jeu linéaire (II.7.) est irrationnel.
Réciproquement supposons que le jeu (II.7.) est irrationnel. Alors il existe n vecteurs v_i ($v_i \in T_{\sigma_i} \Sigma_i$) tels que :

(II.11.) $\forall i = 1, \dots, n < \dfrac{\partial u_i}{\partial \sigma_i}(\sigma) , v_i >_i > 0$

(II.12.) $\forall i = 1, \dots, n \quad \sum_{j=1}^{n} < \dfrac{\partial u_i}{\partial \sigma_j}(\sigma) , v_j >_j < 0$

Il ne reste plus qu'à construire un chemin régulier $\tilde{\sigma}$ dont la dérivée à l'origine soit $(v_1, \dots, v_n)$ et de constater que $\tilde{\sigma}$ vérifie (II.5.) et (II.6.) dans un voisinage de $t = 0$.
Le théorème (II.1.) nous permet d'utiliser tous les résultats de la partie I. Par exemple le théorème (I.2.) suggère d'imposer des bornes $\rho_1, \dots, \rho_n$ au module de la vitesse de déplacement de chaque joueur

(II.13.) $\forall i = 1, \dots, n \quad \| \dfrac{d\tilde{\sigma}_i}{dt}(0) \|_i \leq \rho_i$

et de borner en conséquence les ensembles de stratégies du jeu tangent (II.7.) . On obtient ainsi un jeu "tangent borné" $T_{\sigma}\mathcal{U}(\rho)$ dont l'équilibre de Nash est unique : s'il est en plus Pareto optimal le jeu est rationnel et nous dirons que la coalition de tous les joueurs ne peut pas se former; s'il n'est pas Pareto optimal, le jeu est irrationnel et il est possible que les joueurs jouent de façon coopérative : la coalition de tous les joueurs peut se former.

La proposition (I.1.) nous donne d'autres renseignements sur la rationalité locale. Appelons p_i $(i = 1, \ldots, n)$ la dimension (éventuellement infinie) de la variété Σ_i On constate que si les dimensions vérifient :

$$\text{(II.14.)} \qquad \sum_{i=1}^{n} p_i \geq 2n$$

Alors la rationalité du jeu (II.1.) en un point σ exige au moins une égalité liant les composantes des dérivées partielles des fonctions u_i au point σ. Tandis que si la relation (II.14.) est fausse, la rationalité du jeu (II.1.) au point σ se traduit par un ensemble d'inégalités (cf. par exemple I. §3) : autrement dit, si les dimensions des espaces de stratégies sont trop grandes, la rationalité locale est "exceptionnelle".

2. Rationalité et surface de Pareto.

Nous supposons dans ce paragraphe que chaque variété Σ_1 est de dimension 1 $(p_i=1)$ Utilisant l'esprit de [3] nous nous plaçons au voisinage d'un optimum de Pareto local : un point est un optimum de Pareto local s'il n'existe pas de chemin $\tilde{\sigma}$ de classe C^1 partant de σ et tel que pour tout i :

$$\text{(II.15.)} \qquad \frac{d}{dt} u_i(\tilde{\sigma}(t)) > 0 \quad \text{presque surement en } t .$$

Nous notons $T_{\sigma}\mathcal{U} = [\alpha_i^j(\sigma)]$ la matrice de l'application tangente au point σ à $\mathcal{U}$ $[A_i^j(\sigma)]$ la comatrice de la précédente et $D(\sigma)$ le déterminant de $T_{\sigma}\mathcal{U}$

Lemme (II.1.)

Un point σ de $\Sigma_1 \times \ldots \times \Sigma_n$ vérifiant

$$\text{(I.16.)} \qquad D(\sigma) = 0$$

$$\text{(I.17.)} \qquad \forall i = 1, \ldots, n \quad A_1^1(\sigma) \cdot A_i^1(\sigma) > 0$$

$$\text{(II.18.)} \qquad \sum_{i=1}^{n} A_1^i(\sigma) \frac{\partial}{\partial \sigma_i}(D)(\sigma) < 0$$

<u>est un optimum de Pareto local au voisinage duquel l'équation (II-16) est celle de la variété des optimums de Pareto locaux</u>

<u>Démonstration.</u>

L'équation (II-16) et les inéquations (II-17) sont exactement les conditions de premier ordre de [3].

Puisque $A_i^1(\sigma)$ n'est pas nul (II-17) posons :

$$\text{(II-19)}\quad \lambda_i = \frac{A_i^1(\sigma)}{A_1^1(\sigma)} = \frac{A_i^j(\sigma)}{A_1^j(\sigma)} \quad i,j = 1,\dots, n\ ;\ \lambda_i > 0.$$

On a donc :

$$\text{(II-20)}\quad \text{gradient}\ \Big(\sum_{i=1}^{n} \lambda_i u_i\Big)(\sigma) = 0.$$

Supposons qu'il existe un chemin C^1, $\tilde{\sigma}$ partant de σ et vérifiant (II-15) :

$$\text{(II-21)}\qquad \tilde{\sigma} : [0,1] \longrightarrow \Sigma_1 \times \dots \times \Sigma_n \quad \tilde{\sigma}(0) = \sigma.$$

Alors on a :

$$\text{(II-22)}\qquad \forall i = 1,\dots n \quad u_i(\tilde{\sigma}(t)) \text{ est croissante en } t \text{ (strictement).}$$

Et pour presque tout t dans un voisinage de 0, la dérivée de la fonction $t \longrightarrow u_i(\tilde{\sigma}(t))$ est positive stricte et on a donc :

$$\text{(II-23)}\qquad \forall i = 1,\dots n \quad \sum_{j=1}^{n} \alpha_i^j(\tilde{\sigma}(t))\, \frac{d}{dt}\, \tilde{\sigma}_j(t) > 0 \quad \text{p.s. en } t\,.$$

Puisque la relation (II-17) se conserve dans un voisinage de σ on en déduit que dans un voisinage de 0, presque surement en t :

$$\text{(II-24)}\qquad D(\tilde{\sigma}(t)) \neq 0.$$

Et nous pouvons en ces points inverser la matrice $T_\sigma \mathcal{U}$:

$$\text{(II-25)}\qquad (T_\sigma \mathcal{U})^{-1}(\tilde{\sigma}) = \frac{1}{D(\tilde{\sigma})}\, [A_j^i(\tilde{\sigma})].$$

Donc, en ces points :

$$\text{(II-26)} \qquad \frac{d\tilde{\sigma}_i}{dt}(t) = \frac{1}{D(\tilde{\sigma}(t))} \left[\sum_{j=1}^{n} A_j^i(\tilde{\sigma}(t)).\ \frac{du_j}{dt}(\tilde{\sigma}(t)) \right].$$

Calculons :

$$\text{(II-27)} \quad \frac{dD}{dt}(\tilde{\sigma}(t)) = \sum_{i=1}^{n} \frac{\partial D}{\partial \sigma_i}(\tilde{\sigma}(t)) \frac{d\tilde{\sigma}_i}{dt}(t) = \frac{1}{D(\tilde{\sigma})} \left[\sum_{j=1}^{n} \left(\sum_{i=1}^{n} A_j^i(\tilde{\sigma}) \frac{\partial D}{\partial \sigma_i}(\tilde{\sigma}) \right) \frac{du_j}{dt}(\tilde{\sigma}) \right]$$

Soit d'aprés (II-19)

$$\text{(II-28)} \quad \frac{1}{2}\frac{d}{dt}\left[D^2(\tilde{\sigma})\right] = \left[\sum_{j=1}^{n} \lambda_j(\tilde{\sigma}) \frac{du_j}{dt}(\tilde{\sigma}) \right] \left[\sum_{i=1}^{n} A_1^i(\tilde{\sigma}) \frac{\partial D}{\partial \sigma_i}(\tilde{\sigma}) \right].$$

Donc d'après (II-18), pour presque tout t dans un voisinage de 0 :

$$\text{(II-29)} \qquad \frac{1}{2}\frac{d}{dt}\left[D^2(\tilde{\sigma}(t))\right] < 0$$

Or $D^2(\tilde{\sigma}(0)) = 0$ d'après (II-16). Nous obtenons la contradiction désirée.

Il reste à remarquer qu'au voisinage de σ les inégalités (II-17) et (II-18) restent vraies, donc que l'équation (II-16) constitue bien l'équation de la sous-variété des optimums de Pareto locaux.

Nous allons maintenant étudier la rationalité du jeu au voisinage d'un tel optimum de Pareto.

Théorème (II-2)

Soit σ un optimum de Pareto du jeu (II-1) vérifiant les hypothèses du lemme (II-1) et régulier (II-2).

i) Il existe un voisinage de σ dans lequel la sous-variété des optimums de Pareto locaux, d'équation (II-16) partage $\Sigma_1 \times \ldots \times \Sigma_n$ en 2 demi-espaces dont l'un est formé de points où le jeu est rationnel.

ii) Si de plus les cofacteurs $A_i^j(\sigma)$ sont tous de même signe, le jeu est irrationnel dans l'autre demi-espace.

L'interprétation du cas ii) est la suivante : si les jou eurs suivent un chemin $t \longrightarrow \tilde{\sigma}(t)$ le long duquel ils ont un comportement non coopératif :

$$\text{(II-30)} \qquad \forall i \quad \forall t \qquad \frac{d}{dt}\,\tilde{\sigma}_i(t) \text{ a le même signe que } \frac{\partial u_i}{\partial \sigma_i}(\tilde{\sigma}(t))$$

alors, au moment où un tel chemin traverse la variété des optimums de Pareto, leur interêt commum est justement d'adopter un comportement coopératif et de ne pas la traverser.

On constate que si $n = 2$ on se trouve toujours dans le cas ii).

<u>Démonstration du théorème.</u>

Soit σ_o un optimum de Pareto vérifiant (II-16) (II-17) et (II-18). Alors le vecteurs $V_i(\sigma_o)$:

$$\text{(II-31)} \qquad V_i(\sigma_o) = (A_i^j(\sigma_o))_{j=1,..n}$$

sont tous paralléles et de même sens, et il existe donc un vecteur W de $\mathring{\mathbb{R}}_+^n$, ou bien $\mathring{\mathbb{R}}_-^n$, tel que :

$$\text{(II-32)} \qquad \forall i=1,\ldots n \qquad < W, V_i(\sigma_o) >> 0$$

Nous choisissons un voisinage de σ_o où cette relation (II-32) est encore vraie.

Comme $D(\sigma)$ a un gradient non nul au point σ_o(II-18) nous pouvons restreindre ce voisinage de sorte que la surface de Pareto ($D(\sigma) = 0$) y sépare $\Sigma_1 \times \ldots \times \Sigma_n$ en 2 demi-espaces ($D(\sigma) > 0$ et $D(\sigma) < 0$).

Si par exemple $W \in \mathring{\mathbb{R}}_+^n$ nous choisissons σ tel que

$$\text{(II-33)} \qquad D(\sigma) > 0$$

On a alors, d'après (II-32)

$$\text{(II-34)} \qquad \frac{1}{D(\sigma)}\left[A_i^j(\sigma)\right]\left[W\right] \in \mathring{\mathbb{R}}_+^n$$

Donc :

$$\text{(II-35)} \qquad {}^t(T\mathcal{U}_\sigma)^{-1}(W) \in \mathring{\mathbb{R}}_+^n$$

D'après la proposition (I-1), et sans réserve qu'on ait d'abord orienté Σ_i de sorte que $\frac{\partial u_i}{\partial \sigma_i}(\sigma_o) > 0$ (σ_o est régulier), nous en déduisons qu'au point σ le jeu (II-1) est rationnel.

Pour demontrer ii) il suffit de remarquer que lorsque les $A_i^j(\sigma)$ sont tous positifs (par exemple) alors sur le demi-espace $D(\sigma) < 0$ on a :

(II-36) $$^t(T\,u_\sigma)^{-1}(\mathbb{R}^n_+ - \{0\}) \subset \mathbb{R}^n_- - \{0\}$$

ce qui est contradictoire avec :

(II-37) $$^t(T\,u_\sigma)\,(\mathbb{R}^n_+ - \{0\}) \cap \mathbb{R}^n_+ - \{0\} \neq \emptyset$$

Donc, d'après la proposition (I-1) le jeu est irrationnel au point σ.

3 - Exemples économiques.

a) Duopole : Les deux fontaines

C'est un exemple type de duopole. Chaque joueur possède une fontaine et controle à sa guise la quantité d'eau σ_i ($i=1,2$; $0 < \sigma_i < 1$) qu'il offre sur le marché de l'eau. Nous supposons que le prix que détermine alors la loi du marché est $2 - \sigma_1 - \sigma_2$. Comme le coût de production de l'eau est nul, les recettes u_1 et u_2 des 2 joueurs déterminent le jeu :

(II-38) $$(\sigma_1, \sigma_2) \in]0,1[\qquad u_i(\sigma_1, \sigma_2) = (2 - \sigma_1 - \sigma_2)\,\sigma_i \quad (i=1,2)$$

On constate, en utilisant les relations (II-16) à (II-18) et (I-32), que les optimums de Pareto sont les point du segment :

(II-39) $$\sigma_1 + \sigma_2 = 1$$

et les points où le jeu est irrationnel (où les 2 joueurs peuvent se coaliser) ceux où :

(II-40) $$\forall i=1,2 \qquad 1 - \sigma_i > \sigma_1 + \sigma_2 - 1 > 0$$

Sur la figure (II-1) on a indiqué : la zone où le jeu est irrationnel, l'équilibre de Nash directionnel en chaque point, ainsi que l'équilibre de Nash $(\frac{2}{3}, \frac{2}{3})$.

On constate, comme l'indique le théorème (II-2) que la traversée de la surface de Pareto n'a pas lieu ou a lieu selon que les 2 joueurs ont un comportement coopératif ou non.

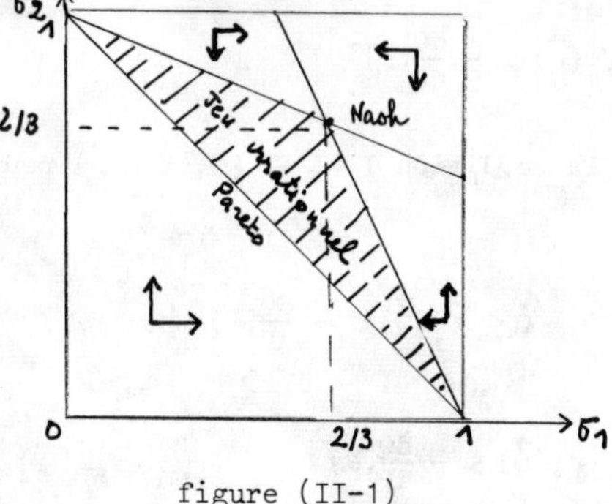

figure (II-1)

b) Cas général : oligopole de Cournot.

C'est la généralisation de l'exemple précédent. Le $i^{\text{ème}}$ producteur controle la quantité Q_i du bien envisagé qu'il produit au coût $C_i(Q_i)$. Lorsque les n producteurs ont choisi un "coup" $\vec{Q} = (Q_1, \ldots, Q_n)$ il en résulte une production totale $Q = Q_1 + \ldots + Q_n$ et un prix de vente $P(Q)$. Nous devons donc étudier le jeu:

(II-41) $\forall i=1,\ldots n \qquad u_i(\vec{Q}) = Q_i . P(Q) - C_i(Q_i).$

La seule hypothèse que nous faisons est la décroissance de la fonction $Q \rightarrow P(Q)$. A l'aide du lemme (III-1) on calcule les optimums de Pareto. Posons :

(II-42) $\forall i=1,\ldots n \qquad \gamma i(\vec{Q}) = \dfrac{Q_i}{P(Q) - \frac{dCi}{dQ_i}(Q_i)}$.

Alors les optimums de Pareto réguliers (c'est-à-dire vérifiant (II-16) à (II-18) sont les points où :

(II-43) $\forall i=1,\ldots n \qquad 0 < \gamma_i(\vec{Q}) < - \frac{dQ}{dP}(P)$

et

(II-44) $$\sum_{i=1}^{n} \gamma_i(\vec{Q}) = -\frac{dQ}{dP}(P)$$

Les points où le jeu est irrationnel sont ceux où (II-43) est vraie ainsi que

(II-45) $$\sum_{i=1}^{n} \gamma_i(\vec{Q}) > -\frac{dQ}{dP}(P)$$

Enfin, les points où la coalition I ($I \subset \{1,\dots n\}$) peut se former sont ceux où :

(II-46) $$\forall i \in I \quad 0 < \gamma_i(\vec{Q}) < -\frac{dQ}{dP}(P)$$

et

(II-47) $$\sum_{i \in I} \gamma_i(\vec{Q}) > -\frac{dQ}{dP}(P)$$

Par conséquent, au voisinage de la surface des optimums de Pareto, seule la coalition de tous les joueurs peut se former.

c) Le jeu du monopole-monopsone.

Le marché d'un bien comporte un producteur, qui contrôle le prix de vente P et un consommateur, qui contrôle sa consommation Q. Le producteur doit payer C(Q) pour produire la quantité Q et le consommateur a une fonction d'utilité $u(\rho, Q)$ dépendant de sa richesse ρ et de sa consommation Q du bien envisagé (on a agrégé tous les autres biens dans la richesse ρ). Sa richesse initiale est R. Le jeu décrivant ce marché est donc :

(II-48) $$\begin{cases} u_1(P,Q) = PQ - C(Q) \\ u_2(P,Q) = u(R - PQ, Q). \end{cases}$$

Les quantités économiquement significatives sont les 3 prix P, prix de vente réel, $\frac{dC}{dQ}(Q)$ coût marginal de production (noté $C_m(Q)$) et P_u le prix subjectif du consommateur :

(II-49) $$P_u(\rho, Q) = \frac{\frac{\partial u}{\partial Q}(\rho, Q)}{\frac{\partial u}{\partial \rho}(\rho, Q)}$$

On remarquera que le prix subjectif, qui est la pente à la courbe d'indifférence en un point donné, ne dépend que du préordre des préférences du consommateur et non de sa fonction d'utilité cardinale. On constate que les optimums de Pareto sont les points où

(II-50) $$P_u = C_m$$

et que les points où le jeu est irrationnel sont de 2 types : ceux où on a :

(II-51) $$P < P_u < C_m$$

et ceux où on a

(II-52) $$C_m < P_u < P$$

Dans le premier cas l'équilibre de Nash directionnel suggère au consommateur de consommer davantage et au producteur d'élever le prix de vente.

Mais le coût marginal de production est si élevé que le producteur acceptera de baisser ses prix si le consommateur accepte de consommer moins.

Dans le second cas l'équilibre de Nash directionnel (le comportement "égoïste") suggère une baisse de consommation et une hausse des prix.

La coopération entre les 2 joueurs permettra une augmentation de consommation et une baisse des prix.

d) L'Oligopsone.

De façon analogue au b) on peut traiter le cas de l'oligopsone, c'est à dire, le cas de n consommateurs d'une même bien. Si $Q_i (i=1,\ldots n)$ désigne la consommation du $i^{\text{ème}}$ joueur, alors, avec les notations du b) le prix de vente sera $P(Q)$ et le jeu a pour fonctions de gain

(II-53) $$\forall i=1,\ldots n \quad u_i(\vec{Q}) = a_i(R_i - P(Q).Q_i, Q_i)$$

où R_i désigne la richesse du $i^{\text{ème}}$ joueur et a_i sa fonction d'utilité "agrégée" (comme en c)).

Si P_i désigne le prix subjectif du $i^{\text{ème}}$ joueur alors on obtient exactement les mêmes formules qu'en b) en remplaçant γ_i par γ'_i :

(II-54) $$\gamma'_i(\vec{Q}) = \frac{Q_i}{P(Q) - P_i(\vec{Q})}$$

4 - Le noyau local d'un jeu régulier.

Nous allons donner une définition locale du concept de solution de Von Neumann d'un jeu sous forme normale ous dirons le noyau de ce jeu.

Nous devons pour cela définir les chemins efficaces pour une coalition :

Définition (II-2)

Soit I une coalition, c'est-à-dire un sous-ensemble de $\{1,..,n\}$. On appelle chemin I-efficace un chemin régulier $\tilde{\sigma}^I$ tel que :

(II-55) $$\tilde{\sigma}^I : [0,1] \longrightarrow \Sigma_1 \times \ldots \times \Sigma_n$$

(II-56) $\forall i \notin I \quad t \longrightarrow \tilde{\sigma}^I_i(t)$ est une application constante.

(II-57) $\forall i \in I \quad t \longrightarrow u_i(\tilde{\sigma}^I(t))$ est strictement croissante sur $[0,1]$.

Définition (II-3)

On dit qu'un point σ d'une sous-variété N de $\Sigma_1 \times \ldots \times \Sigma_n$ est localement stable dans N si il n'existe pas de chemin I-efficace partant de σ et à valeurs dans N.

Notons Θ l'ensemble des optimums de Pareto au sens du lemme (II-1). Nous dirons qu'un élément σ de Θ est localement absorbant si pour tout voisinage U de σ dans Θ, il existe un voisinage W de σ dans $\Sigma_1 \times \ldots \times \Sigma_n$ tel que pour tout τ dans W et tout chemin $\tilde{\sigma}^{\{1,..n\}}$, $\{1,..n\}$- efficace partant de τ et se finissant dans Θ, on ait:

(II-58) $\tilde{\sigma}^{\{1,..n\}}(1) \in U$

Conformément à SMALE (qui appelle ces points les optimums de Pareto stables : [3] p.6) nous notons Θ_S le sous-ensemble de Θ formé des optimums localement absorbants. D'après SMALE c'est génériquement une sous-variété de dimension (n-1) de $\Sigma_1 \times \ldots \times \Sigma_n$ (au sens de la proposition p.13 de [3]).

Définition (II.4)

Le noyau local du jeu (II-1) est le sous-ensemble de Θ_S formé des points localement stables dans Θ_S.

Nous supposons maintenant que toutes les variétés Σ_i sont de dimension 1. Si I désigne une coalition nous notons $U_I(\sigma)$ la sous matrice de $T\,\mathcal{U}_\sigma$ obtenue en conservant les éléments dont la ligne et la colonne sont dans I ; si h est un vecteur de $\mathbb{R}^N$, h_I désignera sa projection canonique sur $\mathbb{R}^I$.

Proposition (II-1)

Soit un point σ où toutes les matrices U_I sont de rang maximal et où toutes les composantes du gradient $\nabla D(\sigma)$ sont non nulles.

Si σ est dans l'intérieur de Θ_S, alors il est dans le noyau local si et seulement si

(II-59) $$\forall I \subset \{1,\ldots n\} \quad [{}^tU_I(\sigma)]^{-1} [\nabla D(\sigma)]_I \in \mathbb{R}^I_+ \cup \mathbb{R}^I_- .$$

Démonstration.

Si σ est dans l'intérieur de Θ_S c'est qu'au voisinage de σ, Θ_S a pour équation :

(II-60) $$D(\sigma) = 0$$

D'après la définition (II-2) si $\tilde{\sigma}^I$ désigne un chemin I-efficace partant de σ, et si $\frac{d}{dt}(\tilde{\sigma}^I(0))$ est noté $(h_1, \ldots, h_n)$ alors on a

(II-61) $$\sum_{i \in I} \frac{\partial D}{\partial \sigma_i}(\sigma)\ h_i = 0$$

et

(II-62) $$U_I(\sigma)\ (h_I) \in \mathring{\mathbb{R}}_+^I$$

Si on pose $k = ({}^tU_I(\sigma))^{-1}\ (\nabla D(\sigma))_I$ il vient :

(II-63) $$< U_I(\sigma)(h_I), k > \ = \ < h_I\ , {}^tU_I(\sigma)(k) > \ = \ < h_I, (\nabla D(\sigma))_I > \ = 0$$

ce qui contredit la relation :

(II-64) $$k \in (\mathbb{R}_+^I \cup \mathbb{R}_-^I) - \{0\}$$

Réciproquement, si la relation (II-59) est fausse il existe I tel que par exemple le vecteur : $k = ({}^tU_I(\sigma))^{-1}[\nabla D(\sigma)]_I$ ait sa première composante positive et sa seconde négative. Donc il existe un vecteur ℓ de $\mathbb{R}^I$ tel que :

(II-65) $$\ell \in \mathring{\mathbb{R}}_+^I \quad \text{et} \quad < \ell, k > \ = 0$$

Soit $h_I = U_I^{-1}(\sigma)\ (\ell)$ alors on a :

(II-66) $$U_I(\sigma)\ (h_I) \in \mathring{\mathbb{R}}_+^I \quad \text{et} \quad < h_I\ ,\ (\nabla D(\sigma))_I > \ = 0$$

Il existe donc un chemin $\tilde{\sigma}^I$ partant de σ tel que

(II-67) $$\forall i \notin I \qquad \tilde{\sigma}_i^I(t) \text{ est constant en } t$$

(II-68) $$\forall i \in I \qquad \frac{d}{dt}\tilde{\sigma}_i^I(0) = (h_I)_i$$

Ce chemin contredit la stabilité locale de σ dans Θ_S.

Exemple (II-4). Il y a 3 joueurs.

On suppose que $\Sigma_i = \mathbb{R}$ (i=1,2,3) et on donne 3 points a_1, a_2, a_3 de $\mathbb{R}^3$. On pose $\sigma = (\sigma_1, \sigma_2, \sigma_3)$ et

(II-69) $\quad i=1,2,3 \quad u_i(\sigma) = - \| \sigma - a_i \|^2$ (norme euclidienne)

On constate aisément que Θ_S (qui est égal à Θ) est le triangle $\{a_1, a_2, a_3\}$.

Pour chercher le noyau local, il est clair qu'il suffit de s'interesser aux coalitions formées de 2 joueurs.

Soit p un vecteur normal au plan $\{a_1, a_2, a_3\}$ et soit $\{e_1, e_2, e_3\}$ la base canonique de $\mathbb{R}^3$.

La relation (II-59) pour la coalition $\{i,j\}$, et si k désigne le joueur restant, équivaut à :

(II-70) $\quad < \sigma - a_i \; , \; p \wedge e_k > . < \sigma - a_j \; , \; p \wedge e_k > \; < \; 0$

et cette relation délimite un sous-triangle de Θ_S.

L'intersection de ces 3 triangles peut être vide ou non, autrement dit le noyau local est un sous-triangle de Θ_S éventuellement vide.

Notons T_k le sous-triangle défini par la relation (II-70), les 2 cas de figure, dans le plan du triangle a_1, a_2, a_3 sont les suivants :

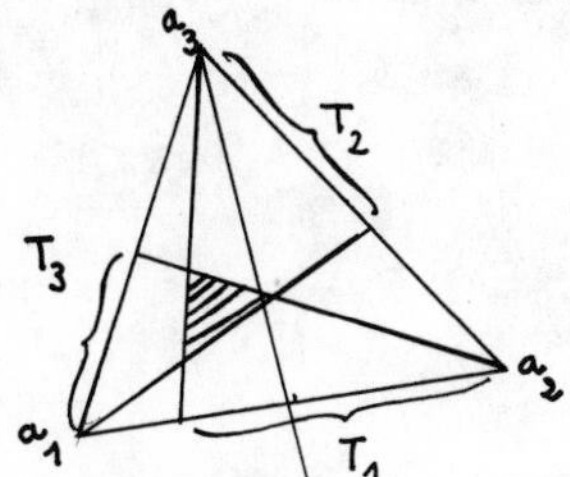

hachuré : le noyau local

figure (II-2)

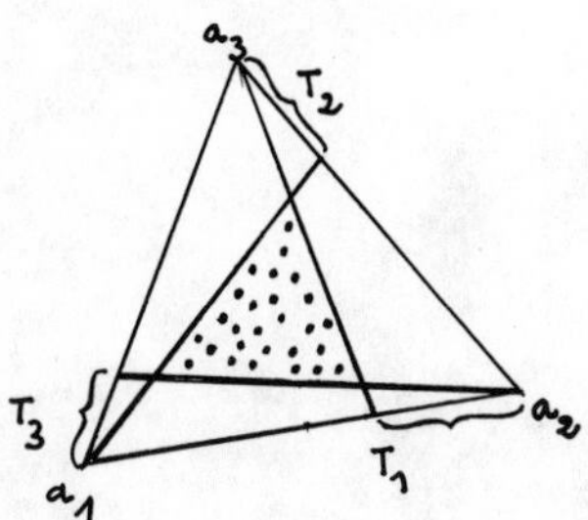

Le noyau local est vide.

La zone pointillée est globalement stable, mais pas localement stable.

figure (II-3)

Interpétation : la zone T_i s'interprète comme la zone où la coalition des 2 autres joueurs ne peut pas se former, étant donné qu'ils jouent un jeu dont l'espace des états est Θ_S au lieu de $\Sigma_1 \times \ldots \times \Sigma_n$. Le noyau $\bigcap_{i=1}^{3} T_i$ est alors le coeur de ce nouveau jeu.

REFERENCES

[1] THOM, R. Morphogenése et Stabilité Structurelle.Benjamin 1973

[2] MOULIN, H. Préordres différentiables. Cahiers de Mathématiques de la Décision n° 7218.

[3] SMALE, S. Global Analysis and Economics. Pareto optim um and a generalization of Mors e theory 1973.

IDENTIFICATION AND STOCHASTIC CONTROL OF A CLASS OF DISTRIBUTED SYSTEMS WITH BOUNDARY NOISE

A. V. Balakrishnan*
Professor, School of Engineering and Applied Science
University of California, Los Angeles

Introduction

We study the problem of identification and stochastic boundary control of a class of linear partial differential equations characterized by random noise on the boundary. Such a problem does not have an analogue in the case of ordinary differential equations and the phenomena are quite different. We are mainly concerned with an abstract-theoretic formulation of the problem and computational aspects are not discussed. In the formulation, we use a white-noise theory based on Gauss measure in a Hilbert Space (in contrast to the usual Wiener process theory) initiated in [1,2]. Even though the measure is only finitely additive we have the advantage of retaining the original topology introduced for the solution of the p.d.e. Moreover, as noted in [3] the white-noise interpretation is essential to provide a meaningful model where noise on the observation is involved (as here). Another feature is the use of semigroup theory (as opposed to the Lions-Magenes variational theory as in [4]) which although limited to time-invariant systems, helps to separate out those aspects of the problem that are of a more general (if abstract) nature and not specifically tied to properties of the p.d.e. involved, and in this way also to pin-point the latter.

The first exhaustive study of stochastic filtering and control problems for distributed system is due to Bensoussan [4] following the Lions-Magenes [6] work for deterministic systems. He does not consider boundary input problems as here and the approach in [2] to both filtering and control problems is quite different, the results also being more general. The basic estimation theory for linear random variables is essentially the same in [2] as in Bensoussan [4], although the extension to causal operators here is different. The R-N derivative formula allowing

* Research supported in part under AFOSR Grant No. 73-2492, USAF, Applied Math. Div.

for white-noise on the observation also appears here for the first time, the result for distributed inputs being given in [5]. References to earlier mathematical work may be found in [4] as well as in [2].

2. Preliminary Notions: White Noise

We begin with a brief description of our notion of white noise (see [2] for more details). Let H be a real separable Hilbert space and let μ be the Gauss measure thereon. This is a finitely additive measure on the ring of cylinder sets, countably additive on cylinder sets with base in the same finite-dimensional sub-space. Let ω denote points in H. Then for any h in H

$$\int_H e^{i[\omega,h]}\, d\mu = \exp\left(- ||h||^2/2\right)$$

which characterizes the cylinder measure completely. Let

$$W = L_2\ [\ [0,1];\ H].$$

Then W is also as separable Hilbert Space and if we now consider the Gauss measure on W, and denote the elements in W by ω, and define

$$n(t,\omega) = \omega(t) \quad \text{a.e. } 0 \leq t \leq 1,$$

then for each ω

$$n(t,\omega) \qquad 0 \leq t \leq 1$$

is a 'white noise' wave form. Let L be a linear bounded transformation mapping W into another Hilbert Space. Then the measure induced on the sets of the form

$$[\omega \mid L\omega \in B]$$

where B is a Borel set, is countably additive if and only if L is Hilbert Schmidt and in that case

$$E||L\omega||^2 = \text{Tr } L^*L = ||L||^2$$

Further if $L + L^*$ is nuclear, then a similar property holds for

$$[L\omega,\omega]$$

in the sense that if P_n is any sequences of projection operators converging to the Identity strongly, then the inverse images of the map

$$[LP_n\omega,P_n\omega]$$

are cylinder sets on which μ is countably additive and thus we have an ordinary random variable and

$$E\,(e^{it[LP_n\omega,P_n\omega]}) = C_n(t)$$

yields a sequence of characteristic functions converging to a unique characteristic function and defines the distribution of the variable $[L\omega,\omega]$.

For details on estimation theory for linear random variables (that is of the form $L\omega$, where L is linear bounded although not necessarily Hilbert-Schmidt) see [2] and [4]; see also [2] for stochastic differential equations.

3. The Filtering Problem

We begin with the filtering or state estimation problem since it plays an essential role in both identification and control.

Let Ω be a bounded open set in R_n with boundary Γ. (Typical domains for us will be the sphere or the box, the latter typifying the case where the boundary is not C^∞). The observed process is described by

$$v(t) = C\, f(t,\cdot) + n_o(t) \qquad 0 \le t \le 1$$

where

$$\left.\begin{aligned} &\frac{\partial f}{\partial t} = \Delta^2 f \text{ in } \Omega \\ &f(t,\cdot)\big|_\Gamma = n_s(t) \\ &f(0,\cdot) = 0\,. \end{aligned}\right\} \tag{3.1}$$

Here $n_s(t)$ and $n_o(t)$ are white noise processes, mutually independent, the specific spaces to be made precise in a moment. We specify the differential operator to be the Laplacian in order to have a concrete case. The particular spatial operator involved will not be essential, as we chall see below. C is bounded linear transformation (not necessarily finite dimensional, although that would be a case of interest in practice) mapping the (real) Hilbert Space $L_2(\Omega) = H_s$ into another Hilbert Space (real, separable) H_o, sufficiently 'smooth' in a manner to be specified below.

We first obtain an abstract formulation of the boundary input problem (3.1). Let $f(\cdot)$ be a function continuous in $\bar{\Omega}$. Then the 'boundary value' of f, denoted $f|_\Gamma$ is clearly well defined as the values of $f(\cdot)$ on Γ. Let $g(\cdot)$ be a function defined and continuous on the boundary Γ. Then we assume now that Ω is such that the Dirichlet problem

$$\left.\begin{aligned} \Delta^2 f &= 0 \text{ in } \Omega \\ f|_\Gamma &= g \end{aligned}\right\} \tag{3.2}$$

has a unique solution in $C^2\,[\Omega]$, continuous on $\bar{\Omega}$. Moreover the corresponding linear transformation, denoted D, will be assumed to be such that

$$||f|| = ||Dg|| \leq \text{const. } ||g||$$

where $||f||$ denotes the norm of $f(\cdot)$ in H_s and $||g||$ denotes the norm in $H_b = L_2(\Gamma)$. For any g in H_b, Dg will be taken as the generalized solution of (3.2) as an element of H_s.

Next let A denote the "zero boundary-value" restriction of Δ^2; more precisely: A is the restriction of τ, the distributional Laplacian:

$$\mathcal{D}(A) = H_0^1\,(\Omega) \cap \mathcal{D}(\tau)$$

$$Af = \tau f$$

where $H_0^1\,(\Omega)$ is the closure of functions in $C_0^\infty\,(\Omega)$ with inner product:

$$[f,h]_1 = [f,h] + \sum_1^n\,[D_i f,\, D_i h]$$

D_i denoting first-order derivatives.

Then A is the infinitesimal generator of a strongly continuous semigroup of Hilbert Schmidt operators, and what is crucial, the semigroup is analytic. Let us denote the semigroup by $S(t)$. Then for any x in H_s,

$$S(t)\,x \in \mathcal{D}(A) \text{ for } t > 0.$$

In our particular case, A is of course self-adjoint, although this is not essential to the theory developed here, and will not be used.

Let the boundary function $n_s(t,\cdot)$, denoted $n_s(t)$, be such that $n_s(\cdot) \in W_b = L_2[(0,1);\, H_b]$. Assume moreover that $n_s(t)$ has continuous second derivatives. Then we can invoke the technique of Fattorini [7] to obtain a solution to (3.1) at first formally. We seek a solution in the form:

$$x(t) = x_o(t) + D\, n_s(t)$$

$$\dot{x}_o(t) = Ax_0(t) + z(t)$$

where z(t) is to be determined. We have, differentiating,

$$\dot{x}(t) = \tau(x(t) - D\, n_s(t)) + z(t) + D\, \dot{n}_s(t)$$

so that we need only choose

$$z(t) = -D\, \dot{n}_s(t)$$

and thus the solution is claimed to be:

$$x(t) = S(t)\ (x(0) - D\, u(0)) - \int_0^t S(t-\sigma)\, D\, \dot{n}_s\, (\sigma)\, d\sigma$$
$$+ D\, n_s(t). \tag{3.3}$$

Note that

$$x(t) - D\, n_s(t)\ \varepsilon\ \mathscr{D}(A) \text{ for } t > 0.$$

This solution can be verified to satisfy

$$\dot{x}(t) = \tau\, x(t) \qquad t > 0$$

$$x(t)|_\Gamma = n_s(t)$$

$$||x(t) - x(0)|| \to 0 \text{ as } t \to 0$$

and unique as well for the class of (smooth) inputs $n_s(t)$ prescribed. For our purposes, the appearance of the derivative in (3.3) is awkward and we wish now to get rid of it, and this can be done by an integration by parts leading to

$$x(t) = -\int_0^t A\, S(t-\sigma)\, D\, n_s(\sigma)\, d\sigma \qquad \text{a.e. in } t \quad 0 \leq t \leq 1$$

for any $n_s(\cdot)$ in W_b.

It is important to note the fact that x(t) is defined inly a.e. in t. However, $x(\cdot)$, thus defined, is an element of $W_s = L_2\ [(0,1);H_s]$ and moreover, what is crucial, if we denote the mapping on W_b into W_s by L, we have that

$$||L\, n_s|| \leq \text{const. } ||n_s||$$

or, L is linear bounded, and hence (3.4) is defined to be the generalized solution of (3.1). The proof that (3.4) is defined a.e and that L is linear bounded is given in [8] and can be proved directly for the present case by a detailed consideration of the eigenvalues of A, or generally by an argument in [9] invoking merely the analyticity of the semigroup.

We thus have the abstract formulation of the problem (with the control set equal to zero):

$$v(t) = C\, x(t) + n_o(t) \qquad \text{a.e.} \qquad 0 \le t \le 1$$

$$x(t) = -\int_0^t A\, S(t-\sigma)\, D\, n_s(\sigma)\, d\sigma \qquad \text{a.e.} \quad 0 \le t \le 1$$

where we have assumed the initial condition $x(0) = 0$.

We now make precise the formulation for the stochastic case. Thus let

$$H = H_s \times H_o.$$

Let ω denote the elements in

$$W = L_2[(0,1);\, H_s \times H_o]$$

and let

$$n(t,\omega) = \begin{bmatrix} n_s(t,\omega) \\ n_o(t,\omega) \end{bmatrix} = \omega(t) \qquad\qquad 0 \le t \le 1$$

With Gauss measure on W, $n(t,\omega)$ represents white noise and let us introduce the trivial coordinate maps:

$$F\, n = n_s$$

$$G\, n = n_o$$

$$FG^* = 0$$

Then our system becomes:

$$v(t,\omega) = C\, x(t,\omega) + G\, n(t,\omega) \quad \text{a.e.} \quad 0 \le t \le 1$$

$$x(t,\omega) = -\int_0^t A\, S(t-\sigma)\, DG\, n(\sigma,\omega)\, d\sigma \ \text{a.e.} \quad 0 \le t \le 1.$$

Note that

$$x(\cdot,\omega) = L\, G\, \omega$$

where LG is a linear bounded transformation mapping W into W_s.

The "smoothness" assumption on C is that C^* maps into $\mathcal{D}(A^*)$ so that $A^* C^*$ is bounded. For example if

$$C\, x(t,\omega) = [C, x(t,\omega)]$$

where C is an element of H_s, then the smoothness assumption is that

$$C \in \mathcal{D}(A^*).$$

The similarity to the finite dimensional case or the standard case [2] is apparent except for the appearance of the operator A in the integral. Since the range of D is not contained in the domain of A, we cannot write

$$A\, S(t-\sigma)\, D\, n_s(\sigma)$$

as

$$S(t-\sigma)\, AD\, n_s(\sigma).$$

If we went ahead anyway and wrote the Riccati equation corresponding to 'filtering error' formally, we would have

$$\dot{P} = AP + PA^* - PC^*CP + (AD)^*(D^*A^*)$$

It is true D^*A^* is well-defined, but is an unbounded, unclosable operator, so that even writing 'in some sense'

$$[(AD)^*\ (D^*A^*)\, x,y] = [(D^*A^*)\, x,\ D^*A^*y\,]$$

we are in trouble. Actually since x(t) is only defined a.e. in t, and is not continuous in t, we cannot expect P (t) to be defined for every t anyway.

Let us now see how to resolve these difficulties.

First of all let us look at the definition of $x(t,\omega)$. Since $x(t\cdot,\omega)$ is only defined a.e. in t for each ω, clearly the estimate cannot be defined for each t either. Hence we proceed in the following way. Let L(t,s) be a bounded linear transformation mapping H_o into H_s, strongly continuous in $0 \le s \le t \le 1$, such that

$$\int_0^1\int_0^t ||L(t,s)||^2\, ds\, dt < \infty.$$

Define L mapping W_o into W_s by

$$Lf = g\ ;\quad g(t) = \int_0^t L(t,s)\, f(s)\, ds \qquad 0 \le t \le 1$$

Let ℓ denote this class of operators. For any L in ℓ,

$$Lv\ (\cdot,\omega)$$

defines a 'physically realizable' or 'non-anticipating' or 'causal' linear transformation of the process $v(t,\omega)$. For any h in W_s, we now seek to minimize

$$E([x(\cdot,\omega) - Lv(\cdot,\omega), h]^2)$$

over ℓ. The minimum need not be attained in ℓ. We shall be satisfied if there is a sequence L_n in ℓ such that L_n converges strongly to L_o such that

$$\operatorname*{Inf}_{L\in\ell} E\left([x(\cdot,\omega) - Lv(\cdot,\omega), h]^2\right)$$

$$= \lim_n E\left([x(\cdot,\omega) - L_n v(\cdot,\omega), h]^2\right)$$

for every h in W_s.

In particular we shall define

$$\hat{x}(\cdot,\omega) = L_o v(\cdot,\omega).$$

This definition coincides with the definition in [2] when specialized. The difficulty in general is to show existence of L_o. Here we can exploit the fact that we can express $x(t,\omega)$ as:

$$x(t,\omega) = A\, y(t,\omega) \qquad \text{a.e. } 0 \le t \le 1$$

where

$$y(t,\omega) = -\int_0^t S(t-\sigma)\; DFn(\sigma,\omega)$$

and $y(t,\omega)$ satisifies

$$\dot{y}(t,\omega) = Ay(t,\omega) - DFn\,(t,\omega);\; y(0,\omega) = 0. \tag{3.5}$$

Also

$$v(t,\omega) = C\,Ay(t,\omega) + GN(t,\omega) \quad \text{a.e.}$$

But since C is smooth enough so that

$$\text{Range of } C^* \subset \mathcal{D}(A^*)$$

we have that

$$(CA)x = (A^*C^*)^*x, \quad x \in \mathcal{D}(A)$$

and of course (CA) can be extended to be linear bounded. Let

$$A^*C^* = Q.$$

Then we can apply the results of [2] to obtain that

$$\dot{\hat{y}}(t,\omega) = A\,\hat{y}(t,\omega) + P(t) \quad Q[v(t,\omega) - Q^* y(t,\omega)].$$

For x,y in $\mathscr{D}(A^*)$,

$$[\dot{P}(t)x,y] = [P(t)A^* x,y] + [P(t)x,A^*\hat{y}]$$

$$+ [D\,D^* x,y] - [P(t)\,Q\,Q^*\;P(t)\,x,y]$$

$$P(0) = 0.$$

Moreover for any L in ℓ, and h_1, h_2 in W_s

$$E\,([y(\cdot,\omega) - \hat{y}\,(\cdot,\omega),\,h_1]\;\;[Lv(\cdot,\omega),\,h_2]) = 0.$$

Now from

$$\hat{y}(t,\omega) = \int_0^t S(t-\sigma)\;P(\sigma)\;Q\;v(\sigma,\omega)\;d\sigma$$

$$-\int_0^t S(t-\sigma)\;P(\sigma)\;Q\;Q^*\hat{y}(\sigma,\omega)\;d\sigma$$

it follows that if we use the notation

$$\hat{y}(t,\omega) = \int_0^t L_y(t,s)\;v(s;\omega)\;ds,$$

then

$$\hat{y}(t,\omega) \in \mathscr{D}(A) \quad \text{a.e., and}$$

$$A\hat{y}\;(t,\omega) = \int_0^t A\;L_y(t,s)\;v(s,\omega)\;ds \text{ a.e. } 0 \le t \le 1$$

or,

$$A\hat{y}\;(\cdot,\omega) = A\;L_y v(\cdot,\omega)$$

where

$L_y v\;(\cdot,\omega)$ is defined as $\int_0^t L_y(t,s)\;v(s,\omega)\;ds,\;0 \le t \le 1.$

Next we shall prove that

$$\hat{x}(\cdot,\omega) = A\;\hat{y}(\cdot,\omega) = A\;L_y v(\cdot,\omega)$$

or,

$$L_o = A\;L_y$$

This follows from the fact

$$E\ (([x(\cdot,\omega) - Lv(\cdot,\omega),\ h]^2)$$

$$= E([A\ y(\cdot,\omega) - A\hat{y}\ (\cdot,\omega),\ h]^2$$

$$+ [(A\ L_y - L)\ v(\cdot,\omega),\ h]^2$$

$$+ 2\ [A\ y(\cdot,\omega) - A\ \hat{y}\ (\cdot,\omega)\ ,\ h][(A\ L_y - L)\ v(\cdot,\omega),\ h]).$$

Let h be such that

$$h(t) \in \mathscr{D}(A^*) \quad \text{a.e.}$$

and

$$A^* h \in W_s.$$

Then

$$E([A\ \hat{y}(\cdot,\omega) - A\ y(\cdot,\omega),h]\ [(A\ L_y - L)\ v(\cdot,\omega),\ h])$$

$$= E\ ([\hat{y}(\cdot,\omega) - y(\cdot,\omega),\ A^*h]\ [(A\ L_y - L)\ v(\cdot,\omega),\ h])$$

$$= 0.$$

Since such $h(\cdot)$ are dense in W_s, it follows that

$$L_o = A\ L_y\ .$$

We cannot of course talk about

$$E\ ([x(t,\omega) - \hat{x}(t,\omega),\ h]^2) \text{ for } h \text{ in } H_s$$

but we can obtain

$$E\ ([x(\cdot,\omega), - \hat{x}(\cdot,\omega),h])^2$$

for h in W_s. For $h(\cdot)$ such that

$$h(t) \in \mathscr{D}(A^*) \quad \text{a.e.}$$

$$A^* h(t) \in W_s$$

we have that

$$E\ ([x(\cdot,\omega) - \hat{x}\ (\cdot,\omega),h]^2$$

$$= E\ ([y\ (\cdot,\omega) - \hat{y}\ (\cdot,\omega),\ A^*h]^2)$$

$$= [RA^*h,\ A^*h]$$

where from [2] we know that R is the operator defined by

$$Rf = g\ ;\ g(t) = \int_0^1 R(t,s)\ f(s)\ ds$$

where

$$R(t,\sigma) = S(t-\sigma)\, P(\sigma) \qquad t \geq \sigma$$

$$= R\,(\sigma,t)^* \qquad \sigma \geq t.$$

Now from the fact that

$$P(t)\, x = \int_0^t S(t-\sigma)\, D\, D^* S^*(t-\sigma)\, x\, d\sigma$$

$$+ \int_0^t S(t-\sigma)\, P\,(\sigma)\, Q\, Q^* P(\sigma) S(t-\sigma)^*\, x\, d\sigma$$

it follows that

$$P(t)\, x \in \mathcal{D}(A) \qquad \text{a.e. in } t$$

and that

$$\int_0^1 ||A\, P(t)\, x\, ||^2\, dt < \infty .$$

Let

$$\tilde{P}(t) = AP(t) \qquad \text{a.e. in } t.$$

Then it follows that

$$P(t)A^* = \tilde{P}(t)^* \qquad \text{on } \mathcal{D}(A^*)$$

can be extended to be linear bounded a.e., and hence we have:

$$E([x(\cdot,\omega) - \hat{x}\,(\cdot,\omega,\, h]^2)$$

$$= [R_x h, h]$$

where R_x is defined by

$$g = R_x f\ ; \qquad g(t) = \int_0^t AS(t-\sigma)\, \tilde{P}(\sigma)\, h(\sigma)\, d\sigma$$

$$+ \int_t^1 \tilde{P}(t)^*\, (AS(\sigma-t))^*\, h(\sigma)\, d\sigma$$

or, the corresponding kernel is

$$A\, S(t-\sigma)\, \tilde{P}(\sigma) \qquad t > \sigma$$

$$\tilde{P}(t)^* (AS(\sigma-t))^* \qquad t < \sigma$$

4. Control Problem

Of the many stochastic control problems that may be formulated in this context we study only the simplest — the quadratic-regular problem but with control on the boundary.

In formulating the problem we shall take advantage of the abstract set-up we already have developed. Let H denote a real separable Hilbert Space and let B denote a Hilbert Schmidt transformation mapping H into $H_b = L_2(\Gamma)$. With u(t) denoting the control with range in H, the dynamic equations corresponding to boundary control can be written

$$\begin{aligned} v(t,\omega) &= C\,x(t,\omega) + n_o(t\cdot,\omega) \\ \dot{y}(t,\omega) &= A\,y(t,\omega) + D(n_s(t;\omega) + Bu(t,\omega)) \\ x(t,\omega) &= -A(t,\omega) \qquad \text{a.e.} \end{aligned} \tag{4.1}$$

The control u(t) will be required to be determined by a causal operator on the data. Let R denote a Hilbert Schmidt operator mapping H_s into H where H is a separable Hilbert Space, such that it is again "smooth" in the sense that

$$\left.\begin{aligned} &\text{Range of } R^* \subset \mathscr{D}(A^*) \\ \text{and} \quad &A^*R^* \text{ is Hilbert Schmidt} \end{aligned}\right\} \tag{4.2}$$

We shall only consider controls of the form:

$$u(t,\omega) = \int_0^t L(t,s)\; v(s,\omega)\; ds \tag{4.3}$$

where furthermore, representing this by:

$$u(\cdot,\omega) = K\omega.$$

K is Hilbert Schmidt also. The regulator problem is that of determining the operator kernel in (4.3) such that

$$E\;[Rx(\cdot,\omega),\; Rx(\cdot,\omega)] + \lambda\; E[u(\cdot,\omega),\; u(\cdot,\omega)], \quad \lambda > 0 \tag{4.4}$$

is minimized. First of all it should be noted that the first term in (4.4) is finite because of our assumption (4.2). Since

$$R\;x(\cdot,\omega) = RA\;y(\cdot,\omega) \tag{4.5}$$

and denoting this element by $g(\cdot,\omega)$, we have

$$g(t,\omega) = -\int_0^t RA\, S(t-\sigma)\, D\, (n_s(\sigma,\omega) + Bu(\sigma,\omega))\, d\sigma$$

$$= -(A^*R)^* \int_0^t S(t-\sigma)\, D(n_s(\sigma,\omega) + Bu(\sigma,\omega))\, d\sigma$$

and hence by (4.2)

$$R\, x(\cdot;\omega) = M\omega$$

where M is Hilbert Schmidt.

Exploiting (4.5), we can follow now the development in [2] and deduce the solution to the optimal boundary control problem as:

$$u(t,\omega) = \frac{1}{\lambda}\, (DB)^*\, P_c(t)\, \hat{y}\, (t,\omega)$$

$$\dot{\hat{y}}\, (t,\omega) = A\, \hat{y}\, (t;\omega) - \frac{1}{\lambda}\, (DB)\, (DB)^*\, P_c(t)\, \hat{y}(t,\omega)$$

$$+ P(t)\, Q(v(t,\omega) - Q^*\hat{y}\, (t,\omega))$$

where $P_c(t)$ is the solution of (for x, y $\varepsilon\mathscr{D}(A)$):

$$\frac{d}{dt}\, [P_c(t)x,y] + P_c(t)x,Ay] + [P_c(t)Ax,y] + \left[(A^*R)(A^*R)^*\, x,y\right]$$

$$- \frac{1}{\lambda}\left[P_c(t)\, (DB)\, (DB)^*\, P_c(t)\, x,y\right] = 0$$

$$P_c(1) = 0,$$

(a solution for which is established as in [2]).

5. Identification Problem

As with stochastic control, a wide variety of identification problems for distributed systems with boundary inputs can be formulated. Here we shall consider one special class. Rather than return to the partial differential equation, we shall make use of the abstract formulation already developed. Thus we shall assume that we have a parametrized family of infinitesimal generators $A(\theta)$ where θ has its range in a metric space. Our observation is

$$v(t,\omega) = C\, x(t,\omega) + n_o(t,\omega) \tag{5.1}$$

$$x(t,\omega) = A(\theta)y(t,\omega) \qquad \text{a.e.} \tag{5.2}$$

$$\dot{y}(t,\omega) = A(\theta)\, y(t,\omega) - D(\theta)\, (n_s(t,\omega) + Bu(t)) \tag{5.3}$$

where $D(\theta)$ is the Dirichlet operator depending on the parameter θ, B is a Hilbert-Schmidt operator as in Section 4 and u(t) is now a known (given) input function.

Our problem is to 'estimate' θ given the observation in $0 \leq t \leq 1$. In our approach, we assume that there is a true value θ_o, which lies on a known open set S. The criterion used in formulating the problem is the method of maximum likelihood, see [5]. More precisely, we note first of all that the covariance operator of the observation process, say R, where

$$E([v(\cdot,\omega)f]\ [v(\cdot,\omega),g]) = [R_v f,g]$$

has the form

$$R_v = I + R_x$$

where R_x is a trace-class operator, provided we now assume that

$$A^*C^* \tag{5.4}$$

is Hilbert-Schmidt. This is trivially true if in the observation the functional C has the form

$$C\ x(t,\omega) = \{[C_i,\ x(t,\omega)]\} \qquad i = 1,\ldots m,$$

$$C_i \in \mathscr{D}(A^*).$$

Moreover under the assumption (5.4) we note that measure induced by $v(\cdot,\omega)$ is absolutely continuous with respect to the measure induced by the white Gaussian noise $n_o(\cdot,\omega)$, and the Radon-Nikodym derivative has the form:

$$p(v;\theta) = \exp -1/2\ \{[C\hat{x}(\cdot,\omega),\ C\hat{x}(\cdot,\omega)] - 2\ [C\hat{x}(\cdot,\omega),\ v(\cdot,\omega)]$$

$$+ 2\ \mathrm{Tr.}\int_0^1 (A^*C^*)^*\ P(t)(A^*C^*)\ dt\ \} \tag{5.5}$$

where P(t) is given by (3.6), and

$$\hat{x}(\cdot,\omega) = A(\theta)\hat{y}(\cdot,\omega)$$

$$\dot{\hat{y}}(t,\omega) = A(\theta)\hat{y}(t,\omega) - D(\theta)Bu(t) + P(t)Q(v(t,\omega) - Q^*\hat{y}(t,\omega)).$$

The identification algorithm is to seek a root of the 'gradient' of the logarithm of (5.5) in a sufficiently small neighborhood of a nominal value. The definition of the gradient is immediate if the parameter set is finite dimensional; we can then also utilize the following form of the Newton-Raphson type algorithm:

$$\theta_{n+1} = \theta_n - R(\theta_n)^{-1}\ \nabla_{\theta_n}\ q(\theta_n)$$

where

$$q(\theta_n) = \text{Log } p(v;\theta_n)$$

$R(\theta)$ is the matrix with components:

$$\left[\frac{\partial}{\partial\alpha_i} C\hat{x}(\cdot,\omega), \frac{\partial}{\partial\alpha_j} C\hat{x}(\cdot,\omega)\right]$$

where α_i denote the components of θ.

Let $R_T(\theta)$ denote the matrix based on an observation of length T. Then the 'identifiability' condition is that the matrix

$$\lim_{T\to\infty} \quad R_T(\theta_o)/T$$

is non-singular. See [5] for more on this. The estimation technique is quite similar to the linearization technique in the so-called "Inverse" Problems as for example in Lavrentiev et al., [10] except that our stochastic formulation gives a rationale for the procedures in [10]. For more on this see [11]. See also [11] for the generalization to the case where the parmeter is not finite dimensional.

References

1. Balakrishnan, A. V.: Introduction to Optimization Theory in a Hilbert Space, Lecture Notes No. 42, Springer-Verlag 1970.

2. Balakrishnan, A. V.: Stochastic Optimization Theory in Hilbert Spaces - I, Journal of Applied Mathematics and Optimization, Vol. 1, No. 2, 1974.

3. Balakrishnan, A. V.: Approximation of Ito integrals by band-limited processes, SIAM Journal on Control, 1974.

4. Bensoussan, A.: Filtrage Optimale des Systemes Lineares, Dunod, Paris, 1970.

5. Balakrishnan, A. V.: Identification of Systems Subject to Random State Disturbance, in Lecture Notes in Computer Science No. 3, Springer-Verlag, 1974.

6. Lions, J. L.: Optimal Control of Systems Governed by Partial Differential Equations, Springer-Verlag, 1971.

7. Fattorini, H.: Boundary Control Systems, SIAM Journal on Control, Vol. 6, 1968, p. 349-385.

8. Balakrishnan, A. V.: Boundary Control of the Diffusion Equation: A semigroup theoretic approach: to appear.

9. De Simon, Luciano: Un' applicazione della teoria degli integrali singulari allo studio delle equazioni differenziale lineari astratte del primo ordine, Rendicotti del Seminario Matematic della Universita di Padua, 1964.

10. Laurentiev, Romanov, Vasiliev: Multidimensional Inverse Problems for Differential Equations, Springer-Verlag, Lecture Notes in Mathematics, No. 167, 1970.

11. Balakrishnan: Identification and Inverse Problems — a Stochastic Formulation, Proceedings of the 6th IFIP Conference on Optimization, Novosibirsk, 1974 (to be published by Springer-Verlag).

Distributed Parameter Stochastic Systems in Population Biology*

Wendell H. Fleming
Brown University
Providence, R.I. 02912

1. Introduction. Various models have been used to study changes over time in the numbers of individuals in a natural population, or in the relative frequencies of different types in a population with more than one type of individual. We shall consider in this paper populations with a geographic structure, there being continual movement of individuals within the habitat where the population is situated. We suppose that the population is continuously distributed over a habitat R, contained in r dimensional space ($r = 1,2,$ or 3). This is a distributed parameter model in engineering terminology. An alternative, which we do not consider, is to lump the population into discrete colonies (or niches) with certain rates of exchange of individuals among the colonies.

We suppose that movement of individuals within R is of a local character, described by a linear elliptic second order partial differential operator (the operator L_o in §3,6). This kind of movement is often called dispersal in the population biology literature. If stochastic effects are neglected, then the changes in population densities obey a parabolic equation ((3.1) below) or perhaps a system of parabolic equations [20]. A similar example is Fisher's equation for gene frequencies [8], [21].

The purpose of the present paper is to discuss some stochastic distributed parameter models of population growth and dispersal (§'s 3,4), and a population genetics model (§6). The latter model, which

* This research was supported by the Air Force Office of Scientific Research under grant AF-AFOSR 71-2078B.

we first formulate in discrete time, is related to a model of Malécot [14]. We then pass from discrete to continuous time, indicating some difficulties involved.

The author wishes to thank M. Clegg, E. Pardoux, and S. Sawyer for helpful suggestions.

2. A simple stochastic model of population growth. Let us begin with a population having only one type of individual and no geographic structure. Let $N(t)$ denote the number of individuals at time t. A very simple model for the change in population size is:

$$\frac{dN}{dt} = f(N). \tag{2.1}$$

If $f(N) = kN$, then the population growth is malthusian (i.e. exponential). If $f(N) = kN - \ell N^2$, then the growth is logistic.

Now suppose that the population growth rate fluctuates randomly with time, due for instance to environmental fluctuations. Instead of (2.1) let us consider the stochastic model:

$$\frac{dN}{dt} = f(N) + \sigma N \frac{dw_1}{dt}, \tag{2.2}$$

where $\sigma > 0$ and w_1 is a 1 dimensional Wiener process (or brownian motion). In the malthusian case, with $f(N) = kN$, we see that $N(t)$ has a lognormal distribution as follows. By the Ito stochastic differential rule,

$$d \log N = N^{-1} dN + \frac{1}{2}\sigma^2 N^2 (\log N)'' dt,$$

$$\log N_o^{-1} N = (k - \frac{1}{2}\sigma^2)t + \sigma w_1(t),$$

where N_o is the initial population size and $w_1(0) = 0$. Hence $\log N_o^{-1} N$ is normal with mean $(k - \frac{1}{2}\sigma^2)t$ and variance $\sigma^2 t$. For a discussion of this model see [5], [13]. The more complicated case when f is logistic is treated in [11, Chap. 3]. We have interpreted (2.2) as a stochastic differential equation in the Ito sense. The solution of (2.2) is then a Markov diffusion process with generator $\frac{1}{2}\sigma^2 N^2 d^2/dN^2 + f(N)d/dN$ [10]. In [11], the equation corresponding to (2.2) is implicitly taken in a sense such that $d \log N = N^{-1} dN$. This formula holds if the Stratonovich stochastic calculus is used [22]. However, with the Stratonovich interpretation the generator would be $\frac{1}{2}\sigma^2 N^2 d^2/dN^2 + (f(N) + \frac{1}{2}\sigma^2 N)d/dN$.

The simple model above can be elaborated in several ways. In [11] models of several interacting populations are considered in which (2.2) is replaced by a system of stochastic differential equations. The terms corresponding to $f(N)$ have there a special (quadratic) form, similar to the classical Lotka-Volterra preditor-prey model. In the next section, we consider a different extension of (2.2), with one kind of individual distributed over a habitat R.

3. <u>A stochastic model for geographically structed populations</u>. Let $N(t,x)$ denote the density of individuals at time t and place x in the habitat R. If stochastic effects are ignored, then N is assumed to evolve according to:

$$\frac{\partial N}{\partial t} = L_o N + f(N), \tag{3.1}$$

with $f(N)$ as in (2.1) and L_o an operator describing the dispersal of individuals within R. It is assumed that L_o is linear, elliptic of second order. The example usually treated in the literature is L_o a constant times the Laplace operator, corresponding to a normal law of dispersal.

Now suppose that the growth term is subject to random fluctuations in time, as in §2. Instead of (3.1) let us take the model:

$$\frac{\partial N}{\partial t} = L_o N + f(N) + \sigma N \frac{dw_1}{dt} . \tag{3.2}$$

Let us consider the stochastic partial differential equation (3.2) in a bounded habitat R, with either Dirichlet or Neumann conditions on the boundary of R. We shall not justify the boundary conditions here. See however [19], and for the case of zero normal derivative boundary conditions in a population genetics model [9, §6], also end of §6 below.

This problem can be treated rigorously using the thesis of Pardoux [17]. If $f(N)$ is Lipschitz, then his results give existence and uniqueness of a solution N such that $N(t,\cdot)$ regarded as a stochastic process in the Hilbert space $L^2(R)$ has sample paths almost surely continuous with

$$E\int_0^T \int_R (N^2 + |N_x|^2)\,dx\,dt < \infty, \quad \text{any} \quad T > 0.$$

Here N_x is the gradient in the habitat variables $x = (x_1,\ldots,x_r)$.

In (3.2) random environmental fluctuations were allowed in time but not in space. More generally, one could allow fluctuations both in time and in space:

$$\frac{\partial N}{\partial t} = L_o N + f(N) + \sigma N \frac{\partial w}{\partial t}, \tag{3.3}$$

where $w(t,x)$ is a process with independent increments in time, but correlated in space. For w let us take a Wiener process in the Hilbert space $L^2(R)$, whose covariance operator has a kernel $Q(x,y)$ (see [2, Chap. 3.3] or [3, p. 99] for definitions). Thus, if $\Delta w(t,x) = w(t+h,x) - w(t,x)$,

$$E(\Delta w(t,x)) = 0$$

$$E(\Delta w(t,x)\Delta w(t,y)) = Q(x,y)h + o(h).$$

The method of Pardoux again gives existence and uniqueness for (3.3) with the boundary conditions. For the malthusian case, $f(N) = kN$, the mean and covariance of population densities obey partial differential equations, as follows. Let

$$M(t,x) = EN(t,x)$$

$$V(t,x,y) = \operatorname{cov}[N(t,x),N(t,y)].$$

The equations for M and V are

$$\frac{\partial M}{\partial t} = L_o M + kM, \tag{3.4}$$

$$\frac{\partial V}{\partial t} = (L_{ox}+L_{oy})V + 2kV + k^2\sigma^2 Q(M(t,x)M(t,y) + V), \tag{3.5}$$

where the notation L_{ox}, L_{oy} indicates that L_o acts respectively in the variables x and y.

If the spatial correlations of increments are weak except for x near y, one can try (as an idealization) to replace the process w in (3.3) by a process W whose time increments are also uncorrelated in space, i.e. with $Q(x,y)$ replaced by $\delta(x-y)$ where δ is the Dirac function in r dimensions (product of 1 dimensional Dirac functions). The process W can be defined [2, p. 90], [7, p. 288], but not as a process in $L^2(R)$. Formally, $\partial W/\partial t$ is a space-time white noise. If R is a finite 1 dimensional interval, then Dawson showed [7, Theorem 2, p. 300] that the equation

$$\frac{\partial N}{\partial t} = LN + g(N)\frac{\partial W}{\partial t}, \tag{3.6}$$

with L linear elliptic, has a unique solution if g is Lipschitz. However, Dawson's condition (8), [7, p. 291], does not hold in dimension $r > 1$. If we took, in particular, $L = L_o + k$, $g(N) = \sigma N$, for a solution of (3.6) the covariance V should obey (3.5) with $Q(x,y)$ replaced by $\delta(x-y)$. However, one will then not have a process $N(t,x)$ with finite variance $V(t,x,x)$ when $r > 1$. Whether there is a solution of (3.6) in some weaker sense is not clear.

4. <u>A branching diffusion process model</u>. In §3 dispersal was treated deterministically through the elliptic operator L_o. A population undergoing dispersal could also be modelled in the following way. Intuitively, each individual wanders according to a Markov diffusion in r dimensional space, independently of all other individuals. Moreover, in each small time interval h the individual dies with probability $\mu h + o(h)$, and gives birth to a new individual at the same place with probability $\lambda h + o(h)$. This intuitive idea can be made precise as a branching diffusion, in the sense of [12]. See also [18]. The operator L_o corresponds to the adjoint of the generator of this spatial diffusion; and the malthusian constant is $k = \lambda - \mu$

Bailey [1] considered a corresponding model for a population lumped into discrete colonies. He found differential equations for the changes over time of the mean population number per colony and the covariance of population numbers. Dawson [7, Appendix 1] considered a model with continuous 1 dimensional habitat corresponding to Bailey's. In that model the population density $N(t,x)$ obeys

$$\frac{\partial N}{\partial t} = \frac{\partial^2 N}{\partial x^2} + cN^{\frac{1}{2}} \frac{\partial W}{\partial t} .$$

(Existence and uniqueness of a solution are not yet known, since $N^{\frac{1}{2}}$ is not Lipschitz.) For both Dawson's model and the branching diffusion model, the mean population density M obeys (3.4). Instead of the last term of (3.5), Dawson's covariance equation [7, (A8)] has the term $c^2 M\delta(x-y)$. On the other hand, S. Sawyer pointed out that the corresponding partial differential equation for the covariance is different in case of the branching diffusion model; in fact, that equation involves the derivative of a δ function. Thus, these two models are not equivalent extensions of Bailey's.

5. A stochastic population genetics model. To motivate section 6, we first recall a simpler model with no geographic structure. Consider a population with a fixed number N of individuals. Each individual has, at a given gene locus on two homologous chromosomes, a pair of genes of the possible types (alleles) A_1 or A_2. Let $P_j = (2N)^{-1}$ times the number of A_1 genes be the frequency of type A_1 in a given generation $j = 0,1,2,\ldots$. In the simplest case, the $2N$ genes which occur in the next generation $j + 1$ are chosen according to a binomial distribution from a population in which A_1, A_2 have frequencies P_j, $1 - P_j$. The Markov chain $P_o, P_1, P_2, \ldots$ is approximated by a 1 dimensional Markov diffusion process p as follows. Make the time change $t = N^{-1}j$; and for large N replace P_j by $p(N^{-1}j)$. The generator of the Markov process p is $\mathscr{G} = \frac{1}{4}\, p(1-p) d^2/dp^2$. The name random genetic drift is given to the fluctuations of gene frequencies occurring in this way through random sampling.

If reversible mutations between A_1 and A_2, and selective advantages among the three possible genotypes A_1A_1 A_1A_2 A_2A_2, are allowed in addition to random genetic drift, then the generator of the diffusion approximation p becomes

$$\mathscr{G} = \frac{1}{4}\, p(1-p)\, \frac{d^2}{dp^2} + g(p)\, \frac{d}{dp}$$

with $g(p)$ a cubic polynomial of a certain form. See [6, Chaps. 8,9] Such a 1 dimensional diffusion p can be represented as the solution of the stochastic differential equation.

$$\frac{dp}{dt} = g(p) + \left[\frac{p(1-p)}{2}\right]^{\frac{1}{2}} \frac{dw_1}{dt} ,$$

with w_1 a 1 dimensional Wiener process.

In the next section, we shall formulate a discrete generation stochastic model with geographic structure. We would then like to approximate the discrete time gene frequency process by a diffusion, whose state space is some space of functions on the habitat R. As we shall indicate, some difficulties are encountered in making this last step.

6. <u>A stochastic population genetics model for geographically structured populations</u>. A well known model of this type is the stepping stone model, in which the population is subdivided into discrete colonies [4], [15], [23]. As before we suppose instead that the population is continuously distributed over a habitat R. We begin with a discrete generation model. It is rather similar to a model of Malécot [14] who considered, however, nonreversible mutations each of which produces a new gene type. For simplicity, we exclude mutation and selective advantages. Fluctuations in gene frequencies are then due entirely to dispersal and random genetic drift. Let $P_j(x)$ be the frequency of type A_1 in generation $j = 0,1,2,...$ and place x in R. The (j+1)st generation is formed in two stages. In the first stage, individuals are mated, offspring are produced, and among them individuals are randomly selected to form the new generation. This is to occur independently at each place x in R; and the population density D is to remain constant over all x in R and all generations. Formally, the density $P_j^{\prime}(x)$ after the first stage should have mean $P_j(x)$, conditioned on $P_j(\cdot)$. The conditional covariance of frequencies at two places x and y should be $[(2D)^{-1}P_j(x)(1-P_j(x)]^{1/2}\delta(x-y)$. This is suggested by lumping the population into a large but finite number of colonies, each located in a subdivision of R, and performing stage 1 independently in each colony; compare with §5. In the second stage the new individuals disperse, so that

$$P_{j+1}(x) = \int_R G(x,\xi)P_j^{\prime}(\xi)d\xi, \qquad \int_R G(x,\xi)d\xi = 1.$$

Here $G(x,\xi)$ is the density of individuals at x which migrated from ξ. The equation for $P_{j+1}(x)$ can be given a precise meaning as follows:

$$P_{j+1}(x) = \int_R G(x,\xi)P_j(\xi)d\xi + \int_R G(x,\xi)\left[\frac{P_j(\xi)(1-P_j(\xi))}{2D}\right]^{\frac{1}{2}} \zeta_j(\xi), \qquad (6.1)$$

where $\zeta_o, \zeta_1, \zeta_2, \ldots$ are independent, r dimensional white noise processes. Assuming that the kernel G is continuous, nonnegative, and P_o is continuous, it can be shown that (6.1) recursively defines $P_1, P_2, \ldots$ and each P_j is continuous on R with probability 1.

Let us now attempt to pass from discrete to continuous time. For this purpose, we introduce a new time scale $t = D^{-1}j$. For population density D large (but $D < \infty$) we wish to approximate $P_j(x)$ by $p(D^{-1}j,x)$, where $p(t,x)$ is some continuous time Markov diffusion process whose states $p(t,\cdot)$ lie in some space of functions on R. Let $h = D^{-1}$, so that h corresponds to one generation on the old time scale and h is small for large D. Suppose that, for smooth enough functions $p(x)$,

$$\int_R G(x,\xi)p(\xi)d\xi = p(x) + hL_o p(x) + o(h), \qquad (6.2)$$

with L_o a linear elliptic partial differential operator. Now formally replace $P_j(\cdot)$ by $p(t,\cdot)$, and apply (6.2) to the first term on the right side of (6.1). For the second term we use (somewhat arbitrarily) the following approximation. Let us replace there $P_j(\xi)$ by $p(t,x)$, and $h^{\frac{1}{2}}\zeta_j(\xi)$ by $W(t+h,\xi) - W(t,\xi)$ with W as in §3. Then (6.1) is replaced by the stochastic partial differential equation

$$\begin{aligned} \frac{\partial p}{\partial t} &= L_o p + \left[\frac{p(1-p)_+}{2}\right]^{\frac{1}{2}} \frac{\partial w}{\partial t}, \\ w(t,x) &= \int_R G(x,\xi)W(t,\xi)d\xi, \end{aligned} \qquad (6.3)$$

and $p(1-p)_+ = p(1-p)$, if $0 \leq p \leq 1$, $= 0$ otherwise. A reasonable boundary condition for p is $\partial p/\partial n = 0$, as we indicate in the example below. The process w is a Wiener process on $L^2(R)$ with covariance kernel (compare with §3)

$$Q(x,y) = \int_R G(x,\xi)G(y,\xi)d\xi.$$

Existence and uniqueness would follow from [17] if we replaced $[p(1-p)_+]^{\frac{1}{2}}$ by a Lipschitz function of p. The solution might not, however, satisfy $0 \leq p(t,x) \leq 1$. The quantity

$$m(x) = \int_R |\xi-x|^2 G(x,\xi)d\xi$$

is of order $h = D^{-1}$; $m(x)$ is the mean square migration distance per generation for individuals arriving at x. Since h is small, instead of (6.3) one might consider

$$\frac{\partial p}{\partial t} = L_o p + \left[\frac{p(1-p)_+}{2}\right]^{\frac{1}{2}} \frac{\partial W}{\partial t}, \tag{6.4}$$

with the same boundary conditions. This perhaps has a solution, satisfying $0 \leq p(t,x) \leq 1$, if R is a finite 1 dimensional interval. For $L_o = \frac{1}{2} d^2/dx^2$ the corresponding covariance equation was solved in [9]. Any solution of (6.4) satisfying $0 \leq p(t,x) \leq 1$ must clearly have finite variance. As noted at the end of §3 this cannot happen for habitats of dimension $r > 1$. For this reason we have considered (6.3) instead of (6.4)

Example. Consider a normal migration law

$$\Gamma(x,\xi) = (2\pi k)^{-\frac{r}{2}} \exp[-(2k)^{-1}|x-\xi|^2].$$

Since Γ is symmetric, it can be interpreted both as the density of destinations x of genes with given origin ξ, and as the density of origins ξ of genes with given destination x. Now suppose that genes from an origin ξ in R which migrate outside R are lost, and there is no migration into R from outside. Set

$$G(x,\xi) = \Gamma(x,\xi)\left[\int_R \Gamma(x,\eta)\,d\eta\right]^{-1}, \quad x,\xi \text{ in } R.$$

Suppose that the mean square migration distance per generation $m = rk$ satisfies $m = Ch$, $h = D^{-1}$, for some constant C. For x in R, not on the boundary of R, the denominator above is nearly 1. From (6.2), $L_o = \frac{C}{2r}\Delta$ with Δ the Laplace operator in $x = (x_1,\ldots,x_r)$. Let us obtain boundary conditions for the nonstochastic equation $\partial p/\partial t = L_o p$. It is plausible that the same boundary conditions are appropriate for (6.3). At least the corresponding boundary conditions must be satisfied by the means and covariances of $p(t,x)$, $p(t,y)$. See [9, §6]. In the nonstochastic case

$$P_{j+1}(x) - P_j(x) = \int_R G(x,\xi)[P_j(\xi) - P_j(x)]\,d\xi.$$

Replace $P_{j+1}(x)$ by $p(t+h,x)$, and $P_j(x), P_j(\xi)$ by $p(t,x), p(t,\xi)$. For x a boundary point of R, the left side is of order h, while the right side is of order $h^{\frac{1}{2}}$ unless $\partial p/\partial n = 0$ at x. Thus the boundary condition is zero normal derivative.

References

[1] Bailey, N.T.J., Stochastic birth, death and migration processes for spatially distributed populations, Biometrika 55 (1968), 189-198.

[2] Bensoussan, A., Filtrage Optimal des Systèms Linéaires, Dunod, Paris, 1971.

[3] Bensoussan, A. and R. Temam, Équations aux derivées partielles stochastiques nonlinéaires, Israel J. Math. 11 (1972), 95-129.

[4] Bodmer, W. F. and L. L. Cavalli-Sforza, A migration matrix model for the study of random genetic drift, Genetics 59 (1968), 565-592.

[5] Capocelli, R. M. and L. M. Ricciardi, A diffusion model for population growth in a random environment, Theo. Popn. Biol. 5 (1974), 28-41.

[6] Crow, J. F. and M. Kimura, An Introduction to Population Genetics Theory, Harper and Row, New York, 1970.

[7] Dawson, D. A., Stochastic evolution equations, Math. Biosci. 15 (1972), 287-316.

[8] Fisher, R. A., The wave of advance of advantageous genes, Ann. Eugen. 7 (1937), 355-369.

[9] Fleming, W. H. and C. H. Su, Some one dimensional migration models in population genetics theory, Theo. Popn. Biol. 5, June 1974.

[10] Gikhman, I. I. and A. V. Skorokhod, Introduction to the Theory of Random Processes, Saunders, Philadelphia, 1969.

[11] Goel, N. S., S. C. Maitra, and E. W. Montroll, Nonlinear Models of Interacting Populations, Academic Press, New York, 1971.

[12] Ikeda, N., M. Nagasawa, and S. Watanabe, Branching Markov processes I, II, III, J. Math. Kyoto Univ. 8 (1968), 233-278, 365-410 and 9 (1969), 95-160.

[13] Lewontin, R. C. and D. Cohen, On population growth in a randomly varying environment, Proc. Nat. Acad. Sci. 62 (1969), 1056-1060.

[14] Malécot, G., Identical loci and relationship, Proc. Fifth Berkeley Symp. Math. Stat. Prob. IV: 317-332, Univ. of Calif. Press, Berkeley, 1967.

[15] Maruyama, T., Stepping stone models of finite length, Adv. Appl. Prob. 2 (1970), 229-258.

[16] Moran, P. A. P., The Statistical Processes of Evolutionary Theory, The Clarendon Press, Oxford, 1962.

[17] Pardoux, E., Thesis, Paris, 1974.

[18] Sawyer, S., A formula for semigroups, with an application to branching diffusion processes, Trans. Amer. Math. Soc. 152 (1970), 1-38.

[19] Skellam, J. G., Random dispersal in theoretical populations, Biometrika 38 (1951), 196-218.

[20] Skellam, J. G. , The formulation and interpretation of mathematical models of diffusionary processes in population biology pp. 63-85 of The Mathematical Theory of the Dynamics of Biological Populations, eds. M.S. Bartlett and R.W. Hiorns, Academic Press, 1973.

[21] Slatkin, M., Gene flow and selection in a cline, Genetics 75 (1973), 733-756.

[22] Stratonovich, R. L., A new representation for stochastic integrals and equations, SIAM J. Control 4 (1966), 362-371.

[23] Weiss, G. K. and M. Kimura, A mathematical analysis of the stepping stone model of genetic correlation, J. Appl. Prob. 2 (1965), 129-149.

On Optimization of Random Functionals

Yu. A. Rozanov
IIASA

Let $f(\alpha,x)$ be a functional of a variable $x\varepsilon X$ where is some "unobservable" random parameter with a probability distribution P. Suppose we have to choose some point $x^{o}\varepsilon X$ and we like to optimize this procedure in some sense of minimization of $f(\alpha,x)$, x X, with unknown parameter .

For example, $f(\alpha,x)$ may be a cost function of some economical model concerning future time, say

$$f(\alpha,x) = \sum_{1}^{n} \alpha_j x_j \quad , \qquad x = (x \ ,\cdots,x_n)\varepsilon X \tag{1}$$

where X is a given convex set in n-dimensional vector space formed with inequalities

$$\sum_{1}^{n} \alpha_j x_j \geq b_i \quad ; \qquad i = 1,\cdots,m$$

(including $x_j \geq 0$; $j = 1,\cdots,n$), and $\alpha = (\alpha_1,\cdots,\alpha_n)$ is a vector of "cost coefficients" which are expected to take values with some probability distribution $P(\cdot|\delta)$ under conditions of some given data δ.

Sometimes one uses a criterion based on minimization of mean value $Ef(\alpha,x)$, $x\varepsilon X$, and considers x^{o} as the optimal point if

$$Ef(\alpha,x^{o}) = \min_{x\varepsilon X} Ef(\alpha,x) \quad . \tag{3}$$

This criterion looks quite reasonable if one is going to deal with a big number N of similar models and the total cost function can be approximately described (according to central limit theorem) as

$$\sum_{k=1}^{N} f(\alpha_k,x) \approx [Ef(\alpha,x)]\cdot N + \Theta\sqrt{N} \quad ,$$

where Θ is a random (normal) variable with mean zero and variance $\sigma^2(x) = Df(\alpha,x)$. But if you have to put in a big investment only once then mean value criterion may not work well; moreover the minimum point x^o of mean value function $Ef(\alpha,x)$, $x\varepsilon X$, can be the maximum point of the cost function $f(\alpha,x)$, $x\varepsilon X$, with a great probability.

In order to make this obvious remark clearer, let us mention a model of a non-symmetric coin-game with two outcomes: $\alpha = \alpha_1 , \alpha_2$ which takes place with corresponding probabilities $p_1 , p_2 = 1 - p_1$ and cost function is $f(\alpha,x)$ with $x = x_1 , x_2$. One has to pay $f_{ij} = f(\alpha_i,x_j)$ under the outcome α_i if he chooses in advance the strategy x_j $(i,j = 1,2)$. Suppose $f_{ij} = C$ $(i\neq j)$ where C is the all gambler capital (so he will loose this capital C under the strategy x_j if it be the outcome α_i, $i\neq j$) and $f_{ii} = -M_i C$ (he will increase the initial capital C in M_i times). The mean value function is

$$Ef(\alpha,x) = \begin{cases} C(-M_1p_1 + p_2) & \text{if } x=x_1 \\ C(p_1-M_2p_2) & \text{if } x=x_2 \end{cases} .$$

Suppose the outcome α_1 takes place with a great probability p_1 (say $p_1 = 0.999$) and M_2 is so big that

$$p_1-M_2p_2 < -M_1p_1 + p_2 \quad .$$

Using mean value criterion we obtain $x^0 = x_2$ as the optimal point but obviously this is a very foolish strategy except in the case when one should very much like to loose his capital (because it will be with the great probability 0.999). Another similar example: suppose the cost function is

$$f(\alpha,x) = \begin{cases} \alpha_{10} + \alpha_{11}x & \text{with probability } p_1 \\ \alpha_{20} + \alpha_{21}x & \text{with probability } p_2 = 1 - p_1 \end{cases}$$

(say $p_1 = 0.999$, $p_2 = 0.001$) where $0 \leq x \leq 1$ and the cost coefficients α_{11},α_{21} are such that $\alpha_{11} > 0$; $\alpha_{11}p_1 + \alpha_{21}p_2 < 0$.

Using mean value criterion we have to choose $x^0 = 1$ though with the great probability p_1 ($p_1 = 0.999$) it will be the <u>maximum</u> point (see Fig. 1) of the actual cost function $f(\alpha,x)$, $0 \leq x \leq 1$.

Concerning the mean value type criterion, we wish to say some other things. It is very easy to realize that one may prefer a random variable $\eta_1 = f(\alpha,x_1)$ in comparison to

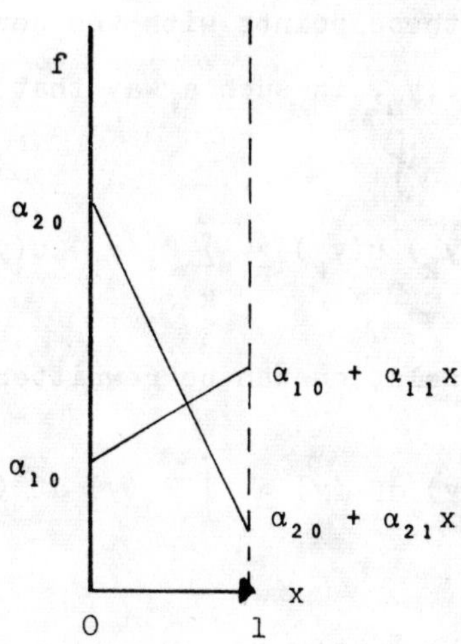

Figure 1

another random variable $\eta_2 = f(\alpha, x_2)$ if for some crucial point y

$$F_1(y) = P\{\eta_1 \leq y\} \geq \{\eta_2 \leq y\} = F_2(y) \quad .$$

Of course, there may be a few, in some sense, crucial points $y = y_1, \ldots, y_n$. Suppose it is possible to estimate "an importance" of these points with the corresponding values $u(y)$, $y = y_1, \ldots, y_n$, in such a way that one prefers η_1 (as compared to η_2) if

$$\sum_k F_1(y_k)\, u(y_k) \geq \sum_k F_2(y_k)\, u(y_k) \quad .$$

The preference relation can be rewritten in the form

$$\int F_1(y)\, dU\,(y) \geq \int F_2(y)\, dU\,(y) \quad .$$

where

$$U(y) = \sum_{y_k \leq y} u(y_k) \,, \qquad -\infty < y < \infty \quad .$$

Because for any distribution function $F(y)(F(-\infty) = 0,\ F(\infty) = 1)$ we have

$$\int F(y)\, dU\,(y) = - \int U(y)\, dF(y) + U(\infty)$$

the preference criterion can be represented in the form

$$EU\,(\xi_1) \leq EU\,(\xi_2) \tag{4}$$

where $E(\cdot)$ is the corresponding mean value.

One can consider (4) for arbitrary distribution type function $U(y)$, $-\infty < y < \infty$, as the general mean value criterion. Obviously, if the corresponding density $u(y)$, $-\infty < y < \infty$, is positive, then $U(y)$, $-\infty < y < \infty$, is monotone increasing function. Besides, if for any $y_1 \leq y_2$ on some interval we consider y_1 as "more important" in comparison with y_2, more precisely if

$$u(y_1) \geq u(y_2) \quad , \quad y_1 \leq y_2 \quad ,$$

i.e. the density $u(y)$, $x \varepsilon I$, is monotone decreasing function on the interval I, then the preference function $U(y)$, $y \varepsilon I$, is convex (see Fig. 2).

We are going to suggest below a few other types of criteria of optimization for random cost functions.

1. Let $f(\alpha,x)$, $x \varepsilon X$, be a cost function which depends on a random parameter α. Suppose for some acceptable cost value C we can neglect a probability that the actual cost will exceed C. Suppose that minimal (random) cost

$$C(\alpha) = \min_{x \varepsilon X} f(\alpha,x)$$

has a probability distribution with a rather small range and corresponding minimum point $\xi \varepsilon X$:

$$f(\alpha,\xi) = \min_{x \varepsilon X} f(\alpha,x)$$

has a discrete distribution (may be with a very big dispersion).

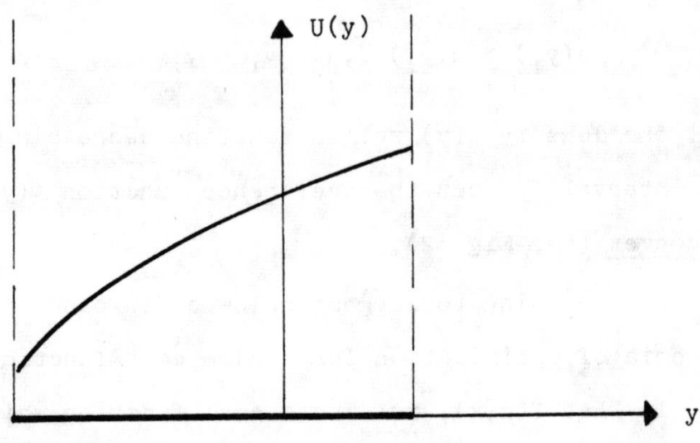

Figure 2

It seems quite reasonable to take a risk to choose such point $x^o \varepsilon X$ for which

$$P\{f(\alpha\ x^o) = C(\alpha)\} = \max_{x \varepsilon X} P\{f(\alpha,x) = C(\alpha)\} \ . \qquad (5)$$

Note that if the probability in the relation (5) equals to 1, in other words, there is a point $x^o \varepsilon X$ for which

$$f(\alpha,x^o) = \min_{x \varepsilon X} f(\alpha,x) \text{ with probability } 1,$$

then our criterion gives the usual minimum of cost function.

Let us consider the linear cost function

$$f(\alpha,x) = \sum_1^n \alpha_j\ x_j$$

of $x = (x_1,\ldots,x_n) \varepsilon X$ where $\alpha = (\alpha_1,\ldots,\alpha_n)$ is the random vector with a given probability distribution P and X is a simplex in n-dimensional vector space of the type (2):

$$\sum_1^n a_{ij} x_j \geq b_i \quad ; \qquad i = 1,\ldots,m \ .$$

Denote $x^1,\ldots,x^N$ extreme points of simplex X. As well known a minimum point $\xi \varepsilon X$ (ξ depends of α) can be chosen among $x^1,\ldots,x^N$, so $x^o = x^1,\ldots,x^N$ is the optimal point in the sense of the criterion (5) if

$$P\{\xi = x^o\} = \max_{1 \leq k \leq N} P\{\xi = x^k\} \ . \qquad (6)$$

Thus, the problem is to find all probabilities*

$$P_k = P\{\xi = x^k\} \quad ; \qquad k = 1,\dots,N$$

and to choose the optimal x^o as the point among x^k; $k = 1,\dots,N$, with the greatest probability P_k; $k = 1,\dots,N$.

We have $P_k = P(Y^k)$ where Y^k is the set of all vectors $y = (y_1,\dots,y_n)$ for which the corresponding linear function

$$f(y,x) = \sum_1^n y_j x_j \quad , \ x \varepsilon X$$

has x^k as the minimum point:

$$f(y,x^k) = \min_{x \varepsilon X} f(y,x) \quad .$$

In order to make our elementary consideration more clear let us shift x^k to the origin point $x = 0$. Obviously the extreme point $x^k = 0$ gives a minimum of $f(y,x)$, $x \varepsilon X$, iff

$$\sum_1^n y_j x_j \geq 0 \quad \text{for all } x \varepsilon X \quad ,$$

in other words, iff the vector $y = (y_1,\dots,y_n)$ belongs to so-called <u>polar cone</u>.

Let us take <u>all</u> hyperplains

$$\sum_1^n a_{ij} x_j = b_i \quad , \ i \varepsilon I_k \quad , \tag{7}$$

---

* Note the events $\{\xi = x^k\}$; $k = 1,\dots,N$, generally are not disjoined and $\sum_1^N P_k$ not necessary equals to 1.

(see (2)) containing the extreme point x^k (in the case $x^k = 0$ we have $b_i = 0$, $i \varepsilon I_k$). Let us introduce a cone

$$X^k = \bigcap_{i \varepsilon I_k} \{x : \sum_1^n a_{ij} x_j \geq 0\} \quad .$$

The corresponding polar cone is exactly the set Y_0^k of all vectors $y = (y_1, \ldots, y_n)$ such that $\sum_1^n y_j x_j \geq 0$, $x \varepsilon X^k$ (see Fig. 3). This polar cone Y_0^k is formed by all linear combinations

$$y = \sum_{i \varepsilon I_k} \lambda_i a_i \quad ; \quad \lambda_i \geq 0 \quad , \qquad (8)$$

of the vectors $a_i = (a_{i1}, \ldots, a_{in})$, $i \varepsilon I_k$, because a dual polar cone for the set of all vectors (8) coincides with X^k: obviously

$$\sum y_j x_j = \sum_{i \varepsilon I_k} \lambda_i (\sum a_{ij} x_j) \geq 0$$

for all $\lambda_i \geq 0$, iff $x \varepsilon X^k$ (see, for example, duality theorem in [1]). Thus, Y^k is the set of all vectors

$$y = x^k + \sum_{i \varepsilon I_k} \lambda_i a_i \quad , \quad \lambda_i \geq 0 \quad , \qquad (9)$$

where $a_i = (a_{i1}, \ldots, a_{in})$ are all vectors such that for $x = x^k$ at the relations (2) we have strict equalities, and the optimal point can be found among x^k, $k = 1, \ldots, N$, as a point with maximum probability

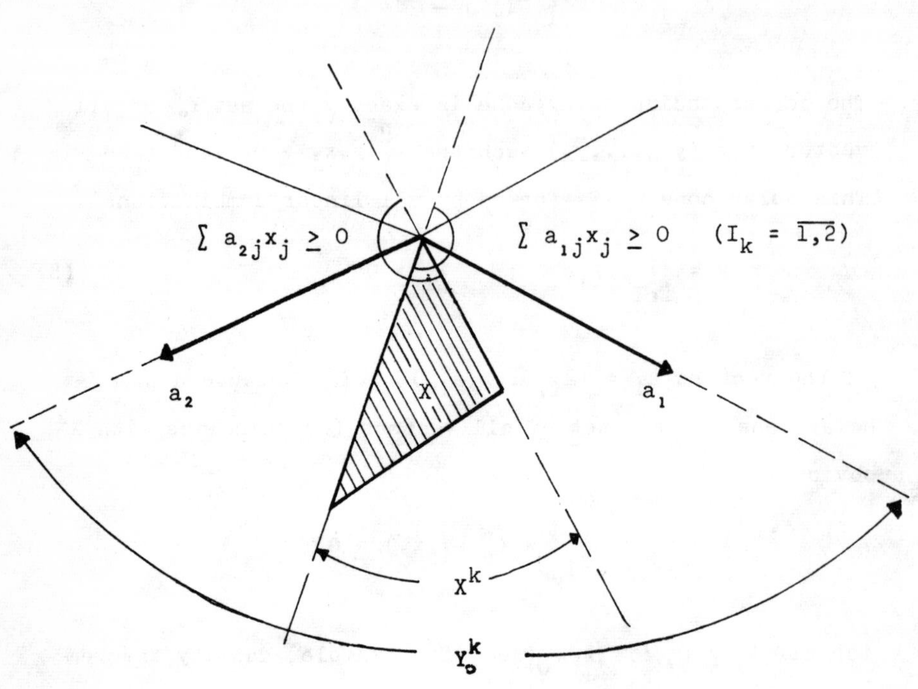

$\sum a_{2j}x_j \geq 0$
$\sum a_{1j}x_j \geq 0 \quad (I_k = \overline{1,2})$
a_2
X
a_1
X^k
Y_o^k

Figure 3

$$P(Y_k) = P\{\alpha \varepsilon Y_k\} \quad ; \quad k = 1,\ldots,N \ . \tag{10}$$

2. Suppose as above there is the acceptable cost which can be exceeded only with a corresponding small probability but the situation is different in the sense that the range of the minimum cost distribution is considerably big. (For example, the minimum point $\xi = x^1, x^2$ can be distributed with almost equal probabilities $P_1 > P_2$ but corresponding cost values are such that $f(\alpha,x^1) >> f(\alpha,x^2)$ so there is no reason to choose the point x^1 with the greatest probability P_1 as optimum.)

Suppose that one is going to risk in order to make the cost value less than some level C_o. (Probability $P\{C(\alpha) \leq C_o\}$ has to be considerably big.) Then one can choose optimal point $x^o \varepsilon X$ in the sense that

$$P\{f(\alpha,x_n^o) \leq C_o\} = \max_{x\varepsilon X} P\{f(\alpha,x) \leq C_o\} \ . \tag{11}$$

This criterion is of mean value type (4) concerning a new cost function $EU(f(\alpha,x))$, $x\varepsilon X$, where

$$U(y) = \begin{cases} 1 & \text{if } y \leq C_o \\ 0 & \text{if } y > C_o \ , \end{cases}$$

namely

$$EU(f(\alpha,x^o)) = \min_{x\varepsilon X} EU(f(\alpha,X)) \ . \tag{12}$$

(Note it is impossible to restrict "y" in order to deal with the convex function $U(y)$, $y\varepsilon I$.)

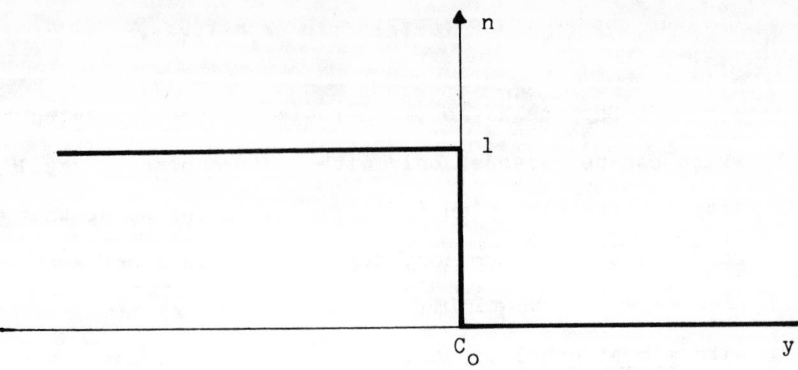

Figure 4

3. Suppose now there is a good deal of risk to pay a big amount if we use "extreme strategy" x^o of the types (5) or (11) because with considerably big probability cost value $f(\alpha,x^o)$ may be too much. Suppose one should like to prevent a danger of dealing with the "almost worst" outcome α and the problem is to find optimal strategy against "very clever random enemy." In this situation it seems quite reasonable the following criterion (similar to the <u>minimax principal</u> of game theory).

Namely suppose one agrees (roughly speaking) to risk only with a small probability $\varepsilon \geq 0$. Let $C(x)$ be the "ε-quantil" for the random variable $f(\alpha,x)$:

$$C(x) = \min C \mid \ P\{f(\alpha,x) \leq C\} \geq 1-\varepsilon \quad . \tag{13}$$

One can choose the point $x^o \varepsilon X$ which is optimal in the sense that

$$C(x^o) = \min_{x\varepsilon X} C(x) \quad . \tag{14}$$

In the case of $\varepsilon = 0$ our criterion of optimality coincides with well known minimax principal of the game theory which was mentioned above because if $\varepsilon = 0$ then

$$C(x) = \sup_{\alpha} f(\alpha,x)$$

(we mean so-called essential sup $f(\alpha,x)$ concerning the probability distribution P of the random variable α).

For the linear cost function (1) with the coefficients $\alpha = (\alpha_1,\dots,\alpha_n)$ which are weakly dependent one can expect the random variable $f(\alpha,x) = \sum_1^n \alpha_j x_j$ is normally distributed (due to the central limit theorem) with a mean value

$$(c,x) = \sum_1^n c_j x_j$$

and variance

$$\|\sigma^{\frac{1}{2}} x\|^2 = \sum_1^n \sum_1^n \sigma_{ij} x_i x_j$$

$(c_i = E\alpha_i; \ \sigma_{ij} = E(\alpha_i - c_j)(\alpha_j - c_j); \ i,j = 1,\dots,n)$.

If it holds true then

$$C(x) = \sum_1^n c_i x_i + y_\varepsilon (\sum_1^n \sum_1^n \sigma_{ij} x_i x_j)^{\frac{1}{2}}, \ x\varepsilon X \quad ,$$

where y_ε denotes ε-quantil for the standard normal distribution:

$$\frac{1}{\sqrt{2\pi}} \int_{y_\varepsilon}^{\infty} e^{-y^2/2}\, dy = \varepsilon \quad .$$

This function

$$C(x) = (c,x) + y_\varepsilon \, \|\sigma^{\frac{1}{2}} x\| \quad , \; x \varepsilon X \quad ,$$

(where $\sigma^{\frac{1}{2}}$ means the square root of the positive matrix $\{\sigma_{ij}\}$) for $y_\varepsilon > 0$ is <u>concave</u> because

$$\left\|\sigma^{\frac{1}{2}} \frac{x_1 + x_2}{2}\right\| \leq \frac{1}{2}\left(\|\sigma^{\frac{1}{2}} x_1\| + \|\sigma^{\frac{1}{2}} x_2\|\right)$$

and besides has a property

$$C(\lambda x) = \lambda C(x)$$

for all $\lambda > 0$. For functions of this type G. Daytzig [2] suggested a generalization of his well known simplex-method which gives a good method of computing the minimum point x^0 for our function $C(x)$.

References

[1] Karlin S., Mathematical methods and theory in games, programming, and economics, vol. 1, 1959.

[2] Dantzig G., "A Generalized Programming Solution to a Convex Programming Problem with a Hemogeneous Objective". See IIASA research report No. 73-21, December 1973.

RECURSIVE FILTERING AND DETECTION FOR TWO-DIMENSIONAL RANDOM FIELDS

Eugene Wong
Department of Electrical Engineering and Computer Sciences
and the Electronics Research Laboratory
University of California, Berkeley, California 94720

1. Introduction

By a two-dimensional random field we mean a family of random variables parameterized by points in a two-dimensional space. In this paper we shall consider the detection and filtering problems of a random signal in additive white Gaussian noise when both the signal and noise are two-dimensional random fields. To be more specific, let the observed field be

$$\xi_t = Z_t + \eta_t \quad , \quad t \in T = (0,a_1) \times (0,a_2) \tag{1.1}$$

where the signal $\{Z_t, t \in T\}$ is a random field and $\{\eta_t, t \in R^2\}$ is a Gaussian random function satisfying

$$E\eta_t = 0 \quad , \quad E\eta_t\eta_{t'} = N_0\, \delta(t_1-t_1')\, \delta(t_2-t_2') \tag{1.2}$$

Our principal results will be the following:

(a) a recursive formula relating the likelihood ratio of ξ_t (with respect to the measure induced by η_t above) to the first and second moments of Z_t given $\{\xi_\tau, \tau \in (0,t_1) \times (0,t_2)\}$.

(b) a recursive filtering formula for the case where Z_t is a Markov process (in a sense to be defined).

The counterparts of these results in one-dimension are both well known and important.

2. Two-Parameter Wiener Process

As in one dimension, we define a two-parameter white noise $\{\eta_t, t \in R^2\}$ as a zero mean random field with

$$E\eta_t\eta_{t'} = \delta(t-t') = \delta(t_1-t_1')\, \delta(t_2-t_2') \tag{2.1}$$

Just as in the one dimensional case, the calculus of white noise is made precise by the introduction of a random field

$$W_t = \int_0^{t_1}\int_0^{t_2} \eta_{(s_1,s_2)}\, ds_1 ds_2, \quad t \in R_+^2 = [0,\infty)^2 \tag{2.2}$$

which is a zero-mean random field with

$$EW_tW_{t'} = \min(t_1,t_1')\, \min(t_2,t_2') \tag{2.3}$$

If η is Gaussian, we shall call W a Wiener process.

Actually, it is convenient to introduce a random set function

---

Research sponsored by the Army Research Office - Durham, Grant DAHC04-67-C-0046, the National Science Foundation, Grant GK-10656X3, and the Air Force Office of Scientific Research, Contract F44620-71-C-0087.

$$(2.4) \qquad W(A) = \int_A \eta_s \, ds \quad , \quad A \subset R^2$$

which has a covariance property

$$(2.5) \qquad EW(A)W(B) = \int_{A \cap B} ds$$

If η is a Gaussian white noise then W(A) is an additive process (or a process with independent areas). That is, if $A_1, A_2, \ldots, A_m$ are disjoint sets then $W(A_1), W(A_2), \ldots, W(A_m)$ are independent random variables.

For points in R^2, we define a partial ordering $\succ$ by

$$t \succ t' \Leftrightarrow t_1 \geq t_1', \ t_2 \geq t_2'$$

With respect to $\succ$, the Wiener process is a martingale, i.e.,

$$(2.6) \qquad E(W_t | W_s, \ s \prec t') = W_{t'}, \ t' \prec t$$

To prove (2.6), let A_t denote the rectangle $[0,t_1] \times [0,t_2]$ and $\bar{A}_t$ denote its complement. Then, for $t \succ t'$ we have

$$W_t = W(A_t) = W(A_{t'}) + W(A_t \cap \bar{A}_{t'})$$

Since $A_{t'}$ and $A_t \cap \bar{A}_{t'}$ are disjoint, $W(A_{t'})$ and $W(A_t \cap \bar{A}_{t'})$ are independent so that

$$E(W_t | W_s, \ s \prec t') = W(A_{t'}) + EW(A_t \cap \bar{A}_{t'}) = W_{t'}$$

Multiparameter Wiener processes are not new, and have been studied by a number of authors [1,2,3]. It provides a natural framework for dealing with white Gaussian noise in a precise way. For example, (1.1) can now be rewritten as

$$X_t = \int_{A_t} Z_\tau \, d\tau + W_t$$

where W_t is a Wiener process.

3. Stochastic Integrals

Experience in the one dimensional case makes it clear that the cornerstone of analysis involving Gaussian white noise must be a stochastic calculus of the Ito type. For the two dimensional case this was done in [4], and it involved defining stochastic integrals of two types:

$$I_1(\phi) = \int_{R_+^2} \phi_t \, W(dt)$$

and

$$I_2(\psi) = [\int_{R_+^2} \times \int_{R_+^2}] \psi_{t,t'} W(dt)\, W(dt')$$

Let $\{W_t, t \in R_+^2\}$ be a Wiener process and $\{\phi_t, t \in R_+^2\}$ be a random field such that

(3.1) (a) For any $t \in R_+^2$, $\{\phi_s, W_s, s \in A_t\}$ is independent of $\{W(\Delta), \Delta \subset \bar{A}_t\}$.

(b) $\int_{R_+^2} E\phi_t^2\, dt < \infty$

We interpret (a) to mean that "future" increments of W are independent of the "past" of W and ϕ. First, take a rectangle A of finite area and sub-divide it by a sequence of partitions $\Pi_n = \{\Delta_{\eta,n}\}$ such that $\max_{\nu} \text{Area}(\Delta_{\nu,n}) \xrightarrow[n \to \infty]{} 0$. We define

(3.2) $$\int_A \phi_t W(dt) = \lim_{n \to \infty} \text{in q.m.} \sum_{\nu} \phi_{t_{\nu,n}} W(\Delta_{\nu,n})$$

where $t_{\nu,n}$ denotes the lower left corner of $\Delta_{\nu,n}$, i.e.,

$$t_{\nu,n} = (\inf t_1, \inf t_2),\ t \in \Delta_{\nu,n}$$

It is clear that (3.2) makes the stochastic integral a forward increment integral. To complete our definition, set

(3.3) $$\int_{R_+^2} \phi_t\, W(dt) = \lim_{m \to \infty} \text{in q.m.} \int_{[0,m]^2} \phi_t W(dt)$$

The stochastic integral so defined has a number of important properties which are direct generalizations of one-dimensional counterparts.

(3.4) Linearity: $$\int_{R_+^2} (a\phi_t + b\psi_t) W(dt) = a \int_{R_+^2} \phi_t W(dt) + b \int_{R_+^2} \psi_t W(dt)$$

(3.5) Martingale: If $X_t = \int_{A_t} \phi_s W(ds)$, then $\{X_t, t \in R_+^2\}$ is a martingale.

It is sample continuous if a separable version is chosen.

(3.6) If $X_t = \int_{A_t} \phi_s W(ds)$, then $Y_t = X_t^2 - \int_{A_t} \phi_s^2 ds$ is a martingale.

We note that (3.5) implies $EX_t = 0$ and (3.6) implies

$$EX_t^2 = \int_{A_t} \phi_s^2 \, ds$$

which together with linearity imply

$$\text{(3.7)} \qquad E\left(\int_{A_t} \phi_s W(ds)\right)\left(\int_{A_t} \psi_s W(ds)\right) = \int_{A_t} E\phi_s\psi_s \, ds$$

So far, there are no surprises. For a one-dimensional Wiener process W_t, $t \in [0,\infty)$, $W_t^2 - t$ is a martingale which can be expressed (by Ito's differentiation rule) as $W_t^2 - t = 2\int_0^t W_s \, dW_s$. For $t \in R_+^2$, although $W_t^2 - t_1t_2$ is a martingale, it cannot be expressed as a stochastic integral $\int_{A_t} \phi_s W(ds)$ for any ϕ. In this sense the integrals defined by (3.2) and (3.3) are incomplete. We need stochastic integrals of a second type which we shall denote by

$$[\int_{R_+^2} \int_{R_+^2}] \, \psi(t,t') \, W(dt) \, W(dt')$$

where ψ satisfies two conditions similar to (3.1), viz.,

(a) For any $t \in R_+^2$, $\{\psi(s,s'), W(s); s,s' \in A_t\}$ is independent of $\{W(\Delta), \Delta \subset \bar{A}_t\}$.

(b) $\displaystyle\int_{R_+^2} \int_{R_+^2} E\psi^2(t,t') \, dt \, dt' < \infty$

Briefly, the definition is given as follows: Again, first consider a finite rectangle and subdivide it by a sequence of rectangular partitions $\{\Delta_{\nu,n}\}$. Denote the lower-left corner of $\Delta_{\nu,n}$ by $t_{\nu,n}$. We set

$$\text{(3.8)} \qquad [\int_{A} \times \int_{A}] \, \psi(t,t') \, W(dt) \, W(dt')$$

$$= \lim_{n \to \infty} \text{ in q.m. } \sum_{\nu} \sum_{\mu} \psi(t_{\nu,n}, t_{\mu,n}) \, W(\Delta_{\nu,n}) \, W(\Delta_{\mu,n})$$

where the double sum is taken over only those pairs (ν,μ) for which $t_{\nu,n}$ and $t_{\mu,n}$ are unordered. Extending the integral from a finite rectangle A to R_+^2 can be done in the usual manner.

Stochastic integrals of the second kind have a number of interesting properties, some of which are given below. More details can be found in [4].

(3.9) Linearity: $[\int_A \times \int_A] [a\psi(t,t') + b\phi(t,t')] W(dt) W(dt')$

$$= a [\int_A \times \int_A] \psi(t,t') W(dt) W(dt')$$

$$+ b [\int_A \times \int_A] \phi(t,t') W(dt) W(dt')$$

(3.10) Martingale: $X_t = [\int_{A_t} \times \int_{A_t}] \psi(s,s') W(ds) W(ds')$ is a martingale.

(3.11) The integrand $\psi(t,t')$ can be set equal to zero at all ordered pairs (t,t') without changing the integral.

(3.12) The integral is symmetric so that $\psi(t,t')$ can be replaced by $\frac{1}{2} [\psi(t,t') + \psi(t',t)]$ without changing the integral.

It was shown in [4] that any square integrable functional of a two-parameter Wiener process can be represented in terms of stochastic integrables of the first and second types. This was proved by using an important differentiation rule which will be stated below.

Let $\{M_t, t \in R_+^2\}$ be a martingale defined by

$$M_t = \int_{A_t} \phi_s W(ds) \tag{3.13}$$

and denote

$$V_t = \int_{A_t} \phi_s^2 ds \tag{3.14}$$

Suppose that $f(x,t)$ is a function satisfying

$$\frac{1}{2} f''(x,t) \nabla V_t + \nabla f(x,t) = 0 \tag{3.15}$$

where ∇ denotes gradient with respect to t and $f'' = \frac{\partial^2 f}{\partial x^2}$. Then, $f(X_t,t)$ is a martingale and can be expressed as

$$f(M_t,t) - f(0,(t_1,0)) - f(0,(0,t_2)) + f(0,(0,0))$$

$$= \int_{A_t} f'(M_s,s)\phi_s W(ds) + \frac{1}{2} [\int\int_{A_t \times A_t}] f''(M_{s\vee s'}, s\vee s')\phi_s\phi_{s'} W(ds)W(ds')$$

where $f' = \frac{\partial f}{\partial x}$, $f'' = \frac{\partial^2 f}{\partial x^2}$ and $s \vee s' = (\max(s_1,s_1'), \max(x_2,s_2'))$.

Functions satisfying (3.15) may be called harmonic functions. One important example of harmonic functions is $x^2 - V_t$ and (3.16) yields

$$M_t^2 - V_t = 2\int_{A_t} M_s\phi_s W(ds) + [\int\int_{A_t \times A_t}] \phi_s\phi_{s'} W(ds)\, W(ds') \tag{3.17}$$

Another important example is $e^{x - \frac{1}{2}V_t}$. If we set

$$\Lambda_t = e^{M_t - \frac{1}{2}V_t} \tag{3.18}$$

then

$$\Lambda_t = 1 + \int_{A_t} \Lambda_s\phi_s W(ds) + \frac{1}{2}[\int_{A_t} \times \int_{A_t}] \phi_s\phi_{s'}\Lambda_{s\vee s'} W(ds)\, W(ds') \tag{3.19}$$

4. Stochastic Integral Equations and Markovian Fields

Cairoli [5] defined a class of two-parameter processes which are transformations of Wiener processes via stochastic integral equations of the form

$$X_t = X_o + \int_{A_t} m(X_s,s)\, ds + \int_{A_t} \sigma(X_s,s)\, W(ds) \tag{4.1}$$

Existence and uniqueness of a solution X can be established in exactly the same way as in one dimension provided that m and σ satisfy a linear growth condition and a uniform Lipschitz condition [5].

Let Δ be a rectangle $[t_1,t_1+\delta_1] \times [t_2,t_2+\delta_2]$. We can write

$$X_{(t_1+\delta_1,t_2+\delta_2)} - X_{(t_1+\delta_1,t_2)} - X_{(t_1,t_2+\delta_2)} + X_{(t_1,t_2)} \tag{4.2}$$

$$= \int_\Delta m(X_s,s)\, ds + \int_\Delta \sigma(X_s,s)\, W(ds)$$

Now consider a point t and the rectangle A_t. For any point τ outside of A_t, X_τ can be expressed in terms of $\{X_s,\ s \in \partial A_t\}$ and $\{W(\Delta),\ \Delta \subset \bar{A}_t\}$, where ∂A_t is the boundary

$$\partial A_t = \{(s_1,t_2),\ s_1 \le t_1\} \cup \{(t_1,s_2),\ s_2 \le t_2\}$$

and $\bar{A}_t$ is the complement of A_t. To see this, consider three cases:

(1) $\tau \succ t$, i.e., $\tau_1 \ge t_1$, $\tau_2 \ge t_2$

(2) $\tau_1 \ge t_1$, $\tau_2 < t_2$

(3) $\tau_1 < t_1, \tau_2 \geq t_2$

In case (1) we can write from (4.1)

$$X_\tau - X_t = \int_{A_\tau \cap \bar{A}_t} m(X_s,s)\, ds + \int_{A_\tau \cap \bar{A}_t} \sigma(X_s,s)\, W(ds)$$

For case (2) we can write

$$X_\tau - X_{(t_1,\tau_2)} = \int_{A_\tau \cap \bar{A}_t} m(X_s,s)\, ds + \int_{A_\tau \cap \bar{A}_t} \sigma(X_s,s)\, W(ds)$$

Indeed, for all three cases we can denote $t \wedge \tau = (\min(\tau_1,t_1), \min(\tau_2,t_2))$ and write

$$X_\tau - X_{t\wedge\tau} = \int_{A_\tau \cap \bar{A}_t} m(X_s,s)\, ds + \int_{A_\tau \cap \bar{A}_t} \sigma(X_s,s)\, W(ds), \quad \tau \in \bar{A}_t \tag{4.3}$$

Now, take a point $t' \succ t$, and consider (4.3) for $\tau \in A_{t'} \cap \bar{A}_t$. Then, (4.3) can be viewed as a stochastic integral equation for X_τ, $\tau \in A_{t'} \cap \bar{A}_t$, with $\{X_s, s \in \partial A_t\}$ as "initial" condition. Hence, for any $\tau \in A_{t'} \cap \bar{A}_t$, X_τ can be expressed in terms of the values of X on ∂A_t and $\{W(\Delta), \Delta \subset A_{t'} \cap \bar{A}_t\}$. It follows from the independence property of $W(\Delta)$ that for $s \in A_t$ and $\tau \in \bar{A}_t$, X_s and X_τ are conditionally independent given X on ∂A_t. We shall call a random field satisfying this property a Markovian field.

We should note that this definition of a Markovian field is similar, but not identical, to the definition due to Lévy [6] and McKean [7]. The difference is that we require conditional independence of "future" and "past" given the "present," only when the "present" is the boundary of a rectangle, while Lévy and McKean allowed more general boundaries.

The Markov property can also be stated in terms of propagation from boundary to boundary. Suppose $\tau \succ t$ and X is Markov. Then, given $\{X_s, s \in A_t\}$, the distribution of $\{X_s, s \in \partial A_\tau\}$ depends only on $\{X_s, s \in \partial A_t\}$.

5. A Change of Probability

The two problems stated in the introduction can now be stated more precisely. Let $\{X_t, Z_t, t \in T\}$ be a pair of two-parameter processes defined on a probability space $(\Omega, \mathcal{F}, \mathcal{P})$ such that

$$W_t = X_t - \int_{A_t} Z_\tau\, d\tau \tag{5.1}$$

is a Wiener process independent of the Z process. Assume $\int_T Z_\tau^2\, d\tau < \infty$ with probability 1, and $\int_T E|Z_\tau|\, d\tau < \infty$. We wish to consider the following two problems: (For their one-dimensional versions see [8, 9, 10].)

(a) Let $\mathcal{P}_o$ be any probability measure such that X is a Wiener process. We want to derive the likelihood ratio

$$L_t = E_o\left(\frac{d\,\mathcal{P}_x}{d\,\mathcal{P}_{ox}}\,|\mathcal{F}_{xt}\right)$$

where $\mathcal{F}_{xt}$ denotes the σ-field generated by $\{X_\tau,\ \tau \in A_t\}$ and $\mathcal{P}_x$ and $\mathcal{P}_{ox}$ are the restrictions of the respective probability measures to the σ-field generated by the X process.

(b) For $t \succ \tau$ derive the conditional distribution of Z_t given $\{X_s,\ s \in A_\tau\}$.

We begin with an absolutely continuous transformation of $\mathcal{P}$ to get a probability measure which has the property of $\mathcal{P}_o$ in (a).

Theorem 1. Let $(\Omega,\mathcal{F},\mathcal{P})$ be a probability space and let $(Z_t,\ W_t,\ t \in T)$ be a pair of random fields defined on it. We assume that: (a) Z and W are independent processes. (b) W is a Wiener process. (c) With probability 1, $\int_T Z_s^2\, ds < \infty$. Define a measure $\tilde{\mathcal{P}}$ on $(\Omega,\mathcal{F})$ by

$$\frac{d\tilde{\mathcal{P}}}{d\mathcal{P}} = \exp\{-\int_T Z_s\, W(ds) - \frac{1}{2}\int_T Z_s^2\, ds\} \tag{5.2}$$

Then, the following assertions are true:

(1) $\tilde{\mathcal{P}}$ is a probability measure mutually absolutely continuous with respect to $\mathcal{P}$.

(2) (X,Z) have the same distribution under $\tilde{\mathcal{P}}$ as (W,Z) under $\mathcal{P}$.

Proof: To prove that $\tilde{\mathcal{P}}$ is a probability, we must show

$$E\left\{ e^{-\int_T Z_s\, W(ds) - \frac{1}{2}\int_T Z_s^2\, ds} \right\} = 1 \tag{5.3}$$

Since Z and W are independent, given $\{Z_s,\ s \in T\}$, $\int_T Z_s W(ds)$ is a Gaussian random variable with zero mean and variance $\int_T Z_s^2\, ds$. Hence,

$$E\left[e^{-\int_T Z_s\, W(ds)} \middle| Z_s,\ s \in T \right] = e^{\frac{1}{2}\int_T Z_s^2\, ds}$$

and (5.3) is proved.

To prove (2), we can compute the characteristic function

$$\tilde{E}\ e^{i\int_T u(s)\, X(ds) + i\int_T r(s)\, Z_s\, ds}$$

and show that is is the same as $E\ e^{i\int_T u(s)\, W(ds) + i\int_T r(s)\, Z_s\, ds}$. The details are straightforward and will be omitted. [c.f. 10]

Now, it follows that

$$\tilde{E}\left\{e^{\int_T Z_s X(ds) - \frac{1}{2}\int_T Z_s^2 ds}\right\} = 1$$

Since

$$e^{\int_T Z_s X(ds) - \frac{1}{2}\int_T Z_s^2 ds} = \left(\frac{d\tilde{\mathcal{P}}}{d\mathcal{P}}\right)^{-1}$$

this proves equivalence between $\tilde{\mathcal{P}}$ and $\mathcal{P}$, and

(5.4) $$\frac{d\mathcal{P}}{d\tilde{\mathcal{P}}} = e^{\int_T Z_s X(ds) - \frac{1}{2}\int_T Z_s^2 ds}$$

Let Λ_t be defined by

(5.5) $$\Lambda_t = e^{\int_{A_t} Z_s X(ds) - \frac{1}{2}\int_{A_t} Z_s^2 ds}$$

Since under $\tilde{\mathcal{P}}$, X is a Wiener process, (5.5) has the form of (3.18) and (3.19) takes on the form

(5.6) $$\Lambda_t = 1 + \int_{A_t} Z_s \Lambda_s X(ds) + \frac{1}{2}\left[\int_{A_t}\times\int_{A_t}\right] Z_s Z_{s'} \Lambda_{s\vee s'} X(ds) X(ds')$$

Since Λ is to be considered a random field on $(\Lambda, \mathcal{F}, \tilde{\mathcal{P}})$, the integrals in (5.6) are well-defined stochastic integrals.

6. A Recursive Filtering Formula

Let Z, X, W be defined as in theorem 1, and assume that the conditions of theorem 1 are satisfied. Let f be any random variable which is a functional of $\{Z_s, s \in \partial A_t\}$ such that $E|f| = \tilde{E}|f| < \infty$. For $t \succ \tau$ define

(6.1) $$H_{t,\tau} f = E(f | Z_s, s \in A_\tau)$$

If Z is a Markovian field then clearly $H_{t,\tau} f$ is a functional of $\{Z_s, s \in \partial A_\tau\}$. Now, define

(6.2) $$\Pi_t f = \tilde{E}(\Lambda_t f | X_s, s \in A_t)$$

Note that [See e.g., 10, p. 234]

(6.3) $$E(f | X_s, s \in A_t) = \frac{\tilde{E}(\Lambda_t f | X_s, s \in A_t)}{\tilde{E}(\Lambda_t | X_s, s \in A_t)} = \frac{\Pi_t f}{\Pi_t(1)}$$

Therefore, Π_t effectively incorporates the conditional distribution of $\{Z_s, s \in \partial A_t\}$ given $\{X_s, s \in A_t\}$ under $\mathcal{P}$ measure.

Theorem 2. Let (X,Z,W) be as defined in theorem 1, and let the conditions of theorem 1 be satisfied. In addition, let Z be a Markovean field. Let f be a function of $\{Z_s, s \in \partial A_t\}$ such that $E|f| < \infty$. Then

$$(6.4) \qquad \Pi_t f = Ef + \int_{A_t} \Pi_s(Z_s H_{t,s} f)\, X(ds)$$

$$+ \frac{1}{2} \left[\int_{A_t} \times \int_{A_t}\right] \Pi_{s\vee s'}[Z_s Z_{s'} H_{t,s\vee s'}(f)]\, X(ds)\, X(ds')$$

Proof: Equation (6.4) follows from (5.6) by multiplying both sides of (5.6) by f and taking expectation with respect to $\tilde{\mathcal{P}}$ measure. Observe that

$$\tilde{E}\left[\int_{A_t} Z_s \Lambda_s f\, X(ds) \mid X_\tau, \tau \in A_t\right]$$

$$= \int_{A_t} \tilde{E}(Z_s \Lambda_s f \mid X_\tau, \tau \in A_t)\, X(ds)$$

Because under $\tilde{\mathcal{P}}$ X is a Wiener process independent of Z

$$\begin{aligned}
&\tilde{E}(Z_s \Lambda_s f | X_\tau, \tau \in A_t) \\
&\quad = \tilde{E}\{Z_s \Lambda_s E(f | X_\tau, \tau \in A_t, Z_\tau, \tau \in A_s) | X_\tau, \tau \in A_t\} \\
&\quad = \tilde{E}\{Z_s \Lambda_s H_{t,s} f | X_\tau, \tau \in A_t\} \\
&\quad = \tilde{E}\{Z_s \Lambda_s H_{t,s} f | X_\tau, \tau \in A_s\} \\
&\quad = \Pi_s(Z_s H_{t,s} f)
\end{aligned}$$

The last term in (6.4) follows by taking nearly identical steps. ⌑

Equation (6.4) can now be regarded as an equation for the evolution of Π_t, which is an operator on functions of $\{Z_s, s \in \partial A_t\}$.

7. A Likelihood Formula

Observe that on the sample space of the X process the probability measures $\mathcal{P}_o$ and $\tilde{\mathcal{P}}$ referred to in section 5 agree. Therefore,

$$\begin{aligned}
(7.1) \qquad L_t &= E_o\left(\frac{d\mathcal{P}_x}{d\mathcal{P}_{ox}} \mid \mathcal{F}_{xt}\right) \\
&= \tilde{E}\left(\frac{d\mathcal{P}_x}{d\tilde{\mathcal{P}}_x} \mid \mathcal{F}_{xt}\right) \\
&= \tilde{E}(\Lambda_t \mid \mathcal{F}_{xt})
\end{aligned}$$

$$= \Pi_t(1)$$

From (6.4) we have

(7.2) $$L_t = L + \int_{A_t} \Pi_s(Z_s)\, X(ds)$$

$$+ \frac{1}{2} \left[\int_{A_t} \times \int_{A_t}\right] \Pi_{svs'}\, (Z_s Z_{s'})\, X(ds)\, X(ds')$$

If we make use of (6.3), then (7.2) can be rewirtten as

(7.3) $$L_t = 1 + \int_{A_t} L_t\, E(Z_s | \mathcal{F}_{xs})\, X(ds)$$

$$+ \frac{1}{2} \left[\int_{A_t} \times \int_{A_t}\right] L_{svs'}\, E(Z_s Z_{s'} | \mathcal{F}_{x\ svs'})\, X(ds)\, X(ds')$$

which relates L_t to the first and second conditional moments of Z given the observation and represents a generalization of the one-dimensional likelihood ratio expression in terms of the conditional mean.

We note that (6.4) was derived under the assumption that Z is a Markov process under $\mathcal{P}$. Equation (7.3) can be derived without this assumption as is done in [11], and as can be easily seen by tracing the steps in the proof of theorem 2 for the special case of f = 1.

References

[1] K. Ito, "Multiple Wiener integral," J. Math. Soc. Japan 3 (1951), 157-169.
[2] J. Yeh, "Wiener measure in a space of functions of two variables," Amer. Math. Soc. Trans. 95 (1960), 433-450.
[3] W. J. Park, "A multi-parameter Gaussian process," Ann. Math. Statist. 41 (1970), 1582-1595.
[4] E. Wong and M. Zakai, "Martingales and stochastic integrals for processes with a multidimensional parameter," to be published in Zeit. Wahrscheinlichkeits-theorie. verw. Geb.
[5] R. Cairoli, "Sur une équation différentielle stochastique," Compte Rendus Acad. Sc. Paris 274 (June 12, 1972) Ser. A, 1739-1742.
[6] P. Lévy, "A special problem of Brownian motion and a general theory of Gaussian random functions," Proc. 3rd Berkeley Symp. Math. Stat. and Prob., University of California Press, 1956, Vol. II, pp. 133-175.
[7] H. P. McKean, Jr., "Brownian motion with a several dimensional time," Theor. Prob. Appl., Vol. 8, pp. 335-365, 1963.
[8] T. Kailath, "A general likelihood formula for random signals in Gaussian noise," IEEE Trans. Information Theory IT-15 (1969) 351-361.
[9] M. Zakai, "On the optimal filtering of diffusion processes," Zeit. Wahrschein-lichkeitstheorie verw. Geb., 11 (1969) 230-243.
[10] E. Wong, Stochastic Processes in Information and Dynamical Systems, McGraw-Hill Book Co., New York, 1971.
[11] E. Wong, "A likelihood formula for two-dimensional random fields," IEEE Trans. Information Theory, to be published, July, 1974.

Stochastic State Space Representation of Images

par Samer ATTASI

IRIA - LABORIA

I - Preface.

The representation of "time sequences" by recursive models driven by white noise has naturally contributed in the sixties to the development of numerous real time recursive processing algorithms.

The extension of this type of representation to double indexed sequences $\{y_{k,l};\ (k,l) \in Z^2\}$, with evident applications to image processing, has been greatly retarded, in our opinion due both to philosophical (meaning of causality attached to a recursive model) and technical reasons (lack of algebraic tools when dealing with functions of two variables).

In this paper, a class of double indexed sequences built on recursive models driven by white noise is studied, with the spirit of extending and adapting all what can be saved from the "nice" properties of time sequences generated by recursive models driven by white noise : generality (approximation theorems), study of autocorrelation function and spectrum, identification algorithms...

Notation list

$H, F_1, F_2, G, P\ldots$	Matrices
G'	Transpose of matrix G
$P > 0$	The symmetric matrix P is positive definite
i,j,k,l,α,β	Integers
$\mathbb{R}$	Set of real numbers
Z	Set of integers
$\{y_{i,j} ; (i,j) \in Z^2\}$	Double indexed sequence of vectors (random or deterministic)
l_p	Set of p-summable sequences
$\cap$	Intersection
$\delta_{k,l}$	Double indexed Kronecker symbol : $\delta_{k,l} = 1$ if $k = l = 0$ $= 0$ otherwise

II - A class σ of recursive systems.

II-1 Definition

A linear system defined over a two dimensional time set, of the considered class σ, transforms an input sequence $\{u_{k,l}\ ;\ (k,l) \in Z^2\}$ into an "output" sequence $\{y_{k,l}\ ;\ (k,l) \in Z^2\}$, according to the model

$$(1) \qquad y_{k,l} = H X_{k,l}$$

$$(1') \qquad X_{k+1,l+1} = F_1 X_{k,l+1} + F_2 X_{k+1,l} - F_1 F_2 X_{k,l} + G u_{k,l}.$$

where $u_{k,l}$ is the input vector $\in \mathbb{R}^m$, $y_{k,l}$ the output vector $\in \mathbb{R}^p$ and $X_{k,l}$ the "state vector" $\in \mathbb{R}^n$.

Matrices H, F_1, F_2, G have appropriate dimensions and verify the commutativity condition :

$$(2) \qquad F_1 F_2 = F_2 F_1$$

The impulse response $\{A_{i,j}\ ;\ (i,j) \in Z^2\}$ and the transfer function $R(z_1, z_2)$ of this system are respectively given by

$$(3) \qquad \begin{cases} A_{i,j} = H F_1^{i-1} F_2^{j-1} G & i > 0,\ j > 0 \\ A_{i,j} = 0 & \text{otherwise.} \end{cases}$$

$$(4) \qquad R(z_1, z_2) = H(z_1 I - F_1)^{-1} (z_2 I - F_2)^{-1} G$$

II-2 Approximation properties.

Let Γ be the space of double indexed sequences of the form :

$$(5) \qquad \omega = \{\omega_{i,j} = H F_1^i F_2^j G\ ;\ i \geq 0,\ j \geq 0\}$$

where H, F_1, F_2, G are finite dimensional matrices with $F_1 F_2 = F_2 F_1$.

Theorem 2.1. $\Gamma \cap l_p$ is dense in l_p $(p < +\infty)$ where l_p is the space of p-summable sequences $\{\omega_{i,j}\ ;\ i \geq 0,\ j \geq 0\}$.

Principle of proof (see ATTASI [1]):
We show that Γ is a linear space and that we can build by means of nilpotent matrices F_1 and F_2 each one of the elements of the canonical base of the space γ of finite support sequences ; therefore $\gamma \subset \Gamma$. But since γ is dense in l_p, $\Gamma \cap l_p$ is also dense in l_p.

The proof of this theorem gives also an approximation algorithm of any sequence in l_p, by a sequence in Γ.

II-3 Extension of classical properties.

The proofs of the theorems of this paragraph are not indicated, (see ATTASI [1]) ; they are pure extensions of proofs of similar properties for linear discrete dynamical systems, KALMAN [2], SILVERMAN [3].

II-3-1 Controllability.

Definition 2.2. The system σ (or the triple $\{F_1, F_2, G\}$) is said to be controllable if for all $(k,l) \in Z^2$, and all $\delta \in \mathbb{R}^n$, there exists a finite sequence of inputs $\{u_{i,j} \, ; \, i \in (k-N,k-1), \; j \in (l-M,l-1)\}$ such that from the initial conditions $X_{k-N,\cdot} = 0$, $X_{\cdot,l-M} = 0$, equation (1') leads to $X_{k,l} = \delta$.

Theorem 2.2. The following properties are equivalent

i) σ (or $\{F_1, F_2, G\}$) is controllable

ii) the controllability matrix $C_n = [G,\ldots,F_1^i \, F_2^j \, G,\ldots,F_1^{n-1} \, F_2^{n-1} \, G]$ where $0 \leqslant i \leqslant n-1$, $0 \leqslant j \leqslant n-1$, is of full rank.

iii) when matrices F_1 and F_2 are both asymptotically stable (eigenvalues of modulus less than 1), the matrix equation

$$(7) \quad P - F_1 P F_1' - F_2 P F_2' + F_1 F_2 P F_1' F_2' = GG'$$

has a unique positive definite solution P.

Remark : In ATTASI [1], concepts of stability have been introduced, and are characterized by the asymptotic stability of matrices F_1 and F_2. It is noticeable that equation (7) enables us to define for stable σ-systems, a class of quadratic Lyapunov functions, and has been thus called "double indexed Lyapunov equation".

II-3-2 Observability.

<u>Definition 2.3</u>. The system σ(or the triple $\{H, F_1, F_2\}$) is observable if for all $(k,l) \in Z^2$, there exist two positive integers N and M, such that the knowledge of both the output

$$\{y_{u,v} \; ; \; k \leqslant u \leqslant k+N, \; l \leqslant v \leqslant l+M\}$$

and the input

$$\{u_{\alpha,\beta} \; ; \; k \leqslant \alpha \leqslant k+N-1, \; l \leqslant \beta \leqslant l+M-1\}$$

with the initial conditions $X_{k,l} = \delta$, and $X_{i,j} = 0$ if $i < K$ or $j < l$, determines uniquely the state δ.

<u>Theorem 2.3.</u> The triple $\{H, F_1, F_2\}$ is observable if and only if the "dual" triple $\{F_1', F_2', H'\}$ is controllable.

We also define the observability matrix O_n of a triple as the transpose of the controllability matrix of the "dual" triple.

II-3-3 Minimal realizations.

<u>Definition 2.4</u>. The quadruple $\{H, F_1, F_2, G\}_n^*$ with $F_1F_2 = F_2F_1$ is said to be a realization of the sequence $\{A_{i,j}; (i,j) \in Z^2\}$ if

$$(8) \qquad \forall i \geqslant 0, \forall j \geqslant 0 \quad ; \quad A_{i,j} = H F_1^i F_2^j G$$

<u>Definition 2.5</u>. Two realizations $\{H, F_1, F_2, G\}_n$ and $\{\overline{H}, \overline{F}_1, \overline{F}_2, \overline{G}\}_n$ of the same sequence $\{A_{i,j} \; ; \; (i,j) \in Z^2\}$ are said to be equivalent if there exists an invertible matrix T, such that

$$(8') \qquad \overline{F}_1 = TF_1T^{-1}, \; \overline{F}_2 = TF_2T^{-1}, \; \overline{H} = HT^{-1}, \; \overline{G} = TG.$$

The following theorems yield a decomposition of the state space associated to a given realization, and enable us to define the set of realizations of minimal dimension, which turns to be a single equivalence class.

<u>Theorem 2.4.</u> Any realization $\{H, F_1, F_2, G\}_n$ of the sequence $\{A_{i,j} \; ; \; i \geqslant 0, \; j \geqslant 0\}$, where we pose rank $(O_n \; C_n) = n_1 \leqslant n$, is equivalent to a realization $\{\overline{H}, \overline{F}_1, \overline{F}_2, \overline{G}\}_n$ in which :

---

* the subscript n denotes the dimension of matrices F_1 and F_2.

$$(9) \quad \bar{F}_i = \begin{bmatrix} \bar{F}_{11,i} & 0 & \bar{F}_{13,i} \\ \bar{F}_{21,i} & \bar{F}_{22,i} & \bar{F}_{23,i} \\ 0 & 0 & F_{33,i} \end{bmatrix} \quad i \in (1,2) \; ; \; \bar{G} = \begin{bmatrix} \bar{G}_1 \\ \bar{G}_2 \\ 0 \end{bmatrix} \; ; \; \bar{H} = [\bar{H}_1, 0, \bar{H}_2]$$

where $\{\bar{H}_1, \bar{F}_{11,1}, \bar{F}_{11,2}, \bar{G}_1\}_{n_1}$ is a controllable and observable realization of the same sequence $\{A_{i,j} \; ; (i,j) \in Z^2\}$.

Theorem 2.5. The following propositions are equivalent

i) the realization $\{H, F_1, F_2, G\}_n$ is of minimal dimensions.

ii) rank $(O_N \, C_N) = n \quad \forall N \geq n$.

iii) the realization $\{H, F_1, F_2, G\}_n$ is controllable and observable.

Theorem 2.6. The set of minimal realizations (realizations of minimal dimension) is a single equivalence class defined by equation (8').

Note that theorem 2.4. also yields a reduction procedure of any realization of a given sequence, to a minimal realization.

II-4 Algebraic study - Minimal realization algorithm.

In this paragraph we give a characterization of σ-systems in terms of linear space morphisms. This characterization yields naturally to a minimal realization algorithm. For the sake of simplicity we shall consider single input single output σ-systems ($p = m = 1$).

Let us consider the linear spaces

$$\Omega = \{\omega = \{\omega_{i,j}\} \; ; \; i \leq 0, \; j \leq 0, \; \omega_{i,j} \in \mathbb{R}\}$$

$$\Gamma = \{\gamma = \{\gamma_{i,j}\} \; ; \; i > 0, \; j > 0, \; \gamma_{i,j} \in \mathbb{R}\} .$$

To each sequence $\{A_{i,j} \; ; \; i > 0, \; j > 0\}$, we associate a linear application $f : \Omega \to \Gamma$ defined by

$$(10) \qquad f(e_{1-i,1-j}) = \begin{bmatrix} A_{i,j} & A_{i+1,j} & \cdots \\ A_{i,j+1} & \cdots & \\ \vdots & & \\ \vdots & & \end{bmatrix}$$

where $e_{\alpha,\beta}$ denotes the sequence $\omega = \{\omega_{k,l} = \delta_{\alpha-k,\beta-l}\}$

Definition 2.6. The linear application f defined by equation (10) is said to be realizable if and only if there exists a σ-system of impulse response $\{A_{i,j} ; i > 0, j > 0\}$.

It follows then from linearity and homogeneity (extension of stationarity of dynamical systems) of σ-systems that these are entirely described by the corresponding f applications.

Theorem 2.6. The linear application f defined by equation (10) is realizable if and only if it is of finite rank. Moreover we have

rank f = dimension of the minimal realizations.

The proof of this theorem is based on a special factorization of the linear application f [KALMAN[4], RISSANEN[5]], characterized by the following commutative diagram

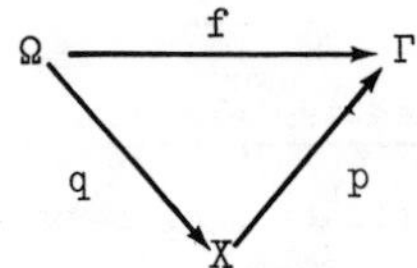

where X is a linear space isomorphic to the range of f in Γ.
p is a linear injection.
q is a linear surjection.

Moreover, when f is of finite rank n, we introduce transition diagrams starting from the above commutative diagram. We then get a minimal realization of dimension n, given by

$$(11) \quad H = \rho \circ p, \; F_1 = p^{-1} \circ \sigma_{\Gamma_1} \circ p, \; F_2 = p^{-1} \circ \sigma_{\Gamma_2} \circ p, \; G = q \circ j$$

where ρ is the canonical projection of Γ onto ℝ
j is the canonical injection of ℝ into Ω
p^{-1} is the inverse application of p, defined on the range of f in Γ.
σ_{Γ_1} and σ_{Γ_2} are translation operators, operating in Γ defined by

$$\forall \{\gamma_{i,j}\} \in \Gamma, \quad \sigma_{\Gamma_1}\{\gamma_{i,j}\} = \{\gamma_{i+1,j}\}$$
$$\sigma_{\Gamma_2}\{\gamma_{i,j}\} = \{\gamma_{i,j+1}\}.$$

Starting from equations (11), an algorithm determining recursively the applications p and q (in a particular basis) is proposed in ATTASI[1]. Owing to the particular form of matrices P and Q representing the applications p and q, the realization $\{H, F_1, F_2, G\}$ is composed of submatrices of P and Q.

III - A class Σ of homogeneous gaussian sequences.

III-1 Définition.

A sequence of random vectors $\{y_{i,j} ; (i,j) \in Z^2\}$ is said to be homogeneous if its correlation function $f(i,j,k,l) \triangleq E[y_{i,j}\, y'_{k,l}] = \Lambda(i-k, j-l)$.

A sequence of homogeneous gaussian vectors $\{y_{i,j} ; (i,j) \in Z^2\}$ taking values in $\mathbb{R}^p$, belongs to the considered class Σ if it admits the following recursive model

$$(12) \qquad \begin{cases} y_{i,j} = H\, X_{i,j} \\ X_{i,j} = F_1 X_{i-1,j} + F_2\, X_{i,j-1} - F_1 F_2 X_{i-1,j-1} + v_{i-1,j-1} \end{cases}$$

where

i) $\{v_{i,j} ; (i,j) \in Z^2\}$ is homogeneous white noise taking values in $\mathbb{R}^n$, with $E[v_{i+k,j+l}\, v'_{i,j}] = Q\, \delta_{k,l} = L\, L'\, \delta_{k,l}$.

ii) $H; F_1, F_2$ are matrices of appropriate dimensions. Moreover, matrices F_1 and F_2 commute and are both asymptotically stable.

The sequence $\{y_{i,j} ; (i,j) \in Z^2\}$ can also be expressed by

$$(13) \qquad y_{i,j} = H \sum_{\substack{k>0 \\ l>0}} F_1^{k-1}\, F_2^{l-1}\, v_{i-k,j-l}.$$

As an immediate consequence of the preceeding results of section II, all Σ-sequences $\{y_{i,j} ; (i,j) \in Z^2\}$ admit models of type (12) with the following properties

(14)

- $\{F_1, F_2, L\}$ is controllable.
- $\{H, F_1, F_2\}$ is observable.

These models should remind the minimal Markov realizations of stationnary time series [FAURRE[6]]. Only these models will be considered from now and on. Moreover, the quadruple $\{H, F_1, F_2, L\}_n$ associated to such a model will be called stochastic realization of the gaussian sequence $\{y_{i,j} ; (i,j) \in Z^2\}$.

III-2 Approximation properties.

We consider the set $\mathfrak{F}$ of all sequences of gaussian vectors $\{y_{i,j} ; (i,j) \in Z^2\}$ taking values in $\mathbb{R}^p$, which can be obtained by

$$(15) \qquad y_{i,j} = \sum_{(k,l) \in Z^2} \beta_{k,l} \; v_{i-k,j-l}$$

where $\{v_{i,j} ; (i,j) \in Z^2\}$ is homogeneous white noise taking values in $\mathbb{R}^m$ and $\{\beta_{k,l} ; (k,l) \in Z^2\}$ is a sequence of matrices of appropriate dimensions.

Theorem 3.1. Any sequence $\{y_{i,j} ; (i,j) \in Z^2\}$ of the set $\mathfrak{F}$ is the uniform limit in quadratic mean of Σ sequences (convergence in the space $l^\infty (Z ; L^2(\Omega ; \mathbb{R}^p))$).

Principle of proof (see ATTASI[7]):

Equation (15) has a meaning provided the sequence of matrices $\{\beta_{k,l} ; (k,l) \in Z^2\}$ is square summable. The sequence $\{y_{i,j} ; (i,j) \in Z^2\}$ given by equation (15) can then be approximated uniformly in the quadratic mean by the sequences

$$(16) \quad y^{(N)}_{i,j} = \sum_{\substack{-N<k<N \\ -N<l<N}} \beta_{k,l} \; v_{i-k,j-l} = \sum_{\substack{0<k<2N \\ 0<l<2N}} \gamma_{k,l} \; w_{i-k,j-l}$$

where
$$\gamma_{k,l} = \beta_{k-N,l-N}$$
$$w_{i,j} = v_{i+N,j+N}$$

It follows then from the proof of theorem (2.1) applied to the finite support sequence $\{\gamma_{k,l}\}$ that the sequences $\{y^{(N)}_{i,j} ; (i,j) \in Z^2\}$ are Σ-sequences.

Corollary 3.1. The autocorrelation function of any sequence $\{y_{i,j} ; (i,j) \in Z^2\}$ belonging to $\mathfrak{F}$ is the uniform limit of autocorrelation functions of Σ sequences.

Note that this concept of convergence is quite adapted to stochastic identification algorithms starting from an estimate of the autocorrelation function of an image .

III-3 Statistical properties.

Starting from equations (13), we get through relatively simple calculations an expression of the autocorrelation function of a Σ-sequence in terms of its stochastic realization (H, F_1, F_2, L). This expression enables us to characterize the set of stochastic realizations associated to a given autocorrelation function.

Theorem 3.2. The set (if not empty) of the stochastic realizations associated to a given function $\Lambda(.,.)$ is characterized by the following relations

$$(17)\qquad \begin{cases} \Lambda(i,j) = H F_1^i F_2^j P H' & \\ \Lambda(-i,j) = H F_2^j P F_1'^{\,i} H' & \forall i \geqslant 0,\ j \geqslant 0 \\ \Lambda(-i,-j) = \Lambda'(i,j) & \forall (i,j) \in Z^2 \\ P > 0 & \\ P - F_1PF_1' - F_2PF_2' + F_1F_2PF_1'F_2' = LL' & \end{cases}$$

with the conditions (14).

Valid almost everywhere in the class Σ (outside a null measure set) the following theorem gives a two step characterization of the set of stochastic realizations, and leads to stochastic identification algorithms based mainly on the algorithms of section II.

Theorem 3.3. The set (if not empty) of the stochastic realizations (H, F_1, F_2, L) associated to a given function $\Lambda(.,.)$ can be obtained in the following manner

i) Search for the minimal dimension quadruples $\{H, F_1, F_2, G\}$ verifying

$$(18)\qquad \forall\, i \geqslant 0,\ j \geqslant 0 \qquad \Lambda(i,j) = H F_1^i F_2^j G$$

ii) for each such quadruple, search for the set of positive definite symmetric matrices P verifying

$$(19)\qquad \begin{cases} PH' = G \\ P - F_1PF_1' - F_2PF_2' + F_1F_2PF_1'F_2' = LL' \geqslant 0 \\ H F_2^j P F_1'^{\,i} H' = \Lambda(-i,j), \ \forall i \geqslant 0,\ j \geqslant 0. \end{cases}$$

Principle of proof (see ATTASI[7]).

Conditions (14) imply that outside a zero measure set, every quadruple (H, F_1, F_2, PH') associated to a stochastic realization belongs to the single equivalence class of quadruples (H, F_1, F_2, G) verifying condition i). The decomposition of the problem in i) and ii) follows. Finally, using the properties of this equivalence class leads to equations (19).

The following theorem, valid almost everywhere in the class Σ, is a consequence of theorem (3.3.)

Theorem 3.4. The set of stochastic realizations (H, F_1, F_2, L) associated to a given function $\Lambda(.,.)$ is a single equivalence class defined by the coordinate transformation

$$(H, F_1, F_2, L) \rightarrow (HT^{-1}, TF_1T^{-1}, TF_2T^{-1}, TL).$$

Principle of proof (see ATTASI[7]).
The change of coordinates T is defined by the set of solutions of equation (18), of minimal dimension. We then prove that for fixed H, F_1, F_2, G, equations (8) have a unique solution L.

The announced stochastic identification procedure goes then as follows :

i) Apply the algorithm proposed in section II, on the sequence $\{\Lambda(i,j) ; i \geq 0, j \geq 0\}$ (where $\Lambda(.,.)$ is an estimate of the autocorrelation function) to get a quadruple of minimal dimensions verifying equation (18).

ii) Simple algebraic calculations, starting from equations (19) yield the matrix L.

These results can be translated into the spectral domain language as follows.

Definition (3.1.) The spectrum of the gaussian homogeneous sequence $\{y_{i,j} ; (i,j) \in Z^2\}$ is the function of two complex variables (z_1, z_2) defined as

$$S(z_1, z_2) = \sum_{(i,j)\in Z^2} \Lambda(i,j)\, z_1^{-i}\, z_2^{-j} \tag{20}$$

where $\Lambda(.,.)$ is the autocorrelation function of the considered sequence.

Theorem 3.5. Under conditions (14), the function $S(z_1, z_2)$ defined by equation (20) admits a "weak" factorization of the form

$$S(z_1, z_2) = \mathcal{H}(z_1, z_2)\, \mathcal{H}'(\tfrac{1}{z_1}, \tfrac{1}{z_2})$$

with

$$\mathcal{H}(z_1, z_2) = H(I-z_1^{-1} F_1)^{-1}(I-z_2^{-1} F_2)^{-1} L$$

if and only if the quadruple (H, F_1, F_2, L) is a stochastic realization associated to the function $\Lambda(.,.)$.

<u>Theorem 3.6.</u> The spectrum of a Σ-sequence, admits almost everywhere in the class Σ, a unique factorization.

Note the fundamental difference with the properties of spectras of stationnary time sequences which admit a whole set of factorizations(weak and strong factorization, FAURRE[6], YOULA[8], ANDERSON[9]).

IV - Conclusion.

Thus, the basis for the stochastic modelization of a large class of double indexed gaussian sequences has been established. We are now investigating applications of this modelization to image processing.

References

1 - Attasi S. ; Systèmes linéaires homogènes à deux indices ; Rapport Laboria n° 31; Septembre 1973.

2 - Kalman R.E., Falb P.L., Arbib K.A. ; Topics in mathematical system theory ; Mc Graw Hill Book Company ; 1969.

3 - Silverman L.M. ; Realization of linear dynamical systems ; IEEE trans. on AC ; Vol. AC16 ; p. 554 ; Dec. 1971.

4 - Kalman R.E. ; Mathematical description of linear dynamical systems ; SIAM. J. Control ; Vol. 1 ; p. 152, 1963.

5 - Rissanen J. ; Recursive identification of linear systems ; SIAM J. Control ; Vol. 9 ; p. 420 ; Aug. 1969.

6 - Faurre P. ; Réalisations Markoviennes de processus stationnaires ; Rapport Laboria n° 13, Mars 1973.

7 - Attasi S. ; Modèles stochastiques de suites gaussiennes à deux indices ; Rapport Laboria n° 56 ; Février 1974.

8 - Youla D.C. ; On the factorization of rational matrices ; IRE Trans. on IT, Vol. IT 7 , p. 172 ; July 1961.

9 - Anderson B.D.O. ; An algebraic solution to the spectral factorization problem ; IEEE Trans. on AC, Vol. 12, p. 410 ; Aug. 1967.

CONTROLE PAR FEEDBACK D'UN SYSTEME STOCHASTIQUE DISTRIBUE

M. ROBIN

IRIA - LABORIA

I - INTRODUCTION

On considère un système d'équations aux dérivées partielles stochastiques décrivant une réaction chimique. On cherche à contrôler ce système par des feedbacks a priori linéaires,

- dans le cas d'une information parfaite sur l'état,
- puis dans le cas d'une information "bruitée".

Dans le premier cas, on étudie l'existence d'une solution pour les équations d'état (en utilisant les travaux de A. BENSOUSSAN R. TEMAM [2] et E. PARDOUX [8]) et l'existence d'un contrôle optimal.
Dans le second cas on calcule formellement une estimation de l'état et on se ramène à un problème en information parfaite sur le système formé par l'état et son estimation. On donne des applications numériques permettant de comparer les deux situations précédentes.

II - LE MODELE PHYSIQUE.

On considère les équations suivantes (écrites formellement par l'instant)

$$(1)\quad \begin{cases} \dfrac{\partial y_1}{\partial t} - \dfrac{\partial^2 y_1}{\partial x^2} + f_1(y_1,y_2) = w & (x,t) \in \,]0,1[\times]0,T[\\ \dfrac{\partial y_2}{\partial t} + f_2(y_1,y_2) = 0 & (x,t) \in [0,1]\times]0,T[\\ \dfrac{\partial y_1}{\partial x}(0,t) = u(t) & \\ \dfrac{\partial y_1}{\partial x}(1,t) = 0 & \\ y_1(x,0) = y_{10} + \zeta\,, \quad y_2(x,0) = y_{20} & \end{cases}$$

Avec

$$(2)\quad \begin{cases} f_1(y_1,y_2) = -\,by_2 \exp\left(-a/|y_1|\right) \\ f_2(y_1,y_2) = cy_2 \exp\left(-a/|y_1|\right) \end{cases}$$

- a,b,c sont des constantes positives,

- ζ,w sont des perturbations aléatoires dont la nature sera précisée au § 3,
- u(t) est un contrôle à préciser également.

Avec des hypothèses convenables sur la chimie de la réaction les équations représentent l'évolution de la température (y_1) et de la concentration de l'un des deux produits de la réaction (y_2) (les variables de (1) sont sans dimension - cf. [6] pour un modèle analogue).

On considère le critère suivant où z_d sera une valeur donnée

$$J(u) = E\{\alpha \int_0^T |y_1(1,t) - z_d|^2 dt + \int_0^T |u|^2 \, dt\} \tag{3}$$

et on envisage les problèmes de contrôle suivant :

Problème 1

Information parfaite

On se restreint à

$$u(t) = u_1 + u_2(y_1(1,t) - z_d) \tag{4}$$

avec

$$(u_1,u_2) \in \mathcal{U}_{ad} = \{v \in \mathbb{R}^2 \ |v_i| \leq M\} \tag{5}$$

et on veut minimiser (3) sur $\mathcal{U}_{ad}$.

Problème 2

Information "bruitée"

On suppose que l'observation est donnée par

$$dz(t) = Cy(t)dt + dv_t \tag{6}$$

où

$$Cy(t) = \begin{cases} y_1(1,t) \\ y_2(0,t) \\ y_2(1,t) \end{cases} \tag{7}$$

v_t sera un processus de wiener à valeur dans $\mathbb{R}^3$ de covariance $\sigma_v^2 I$

On prend alors des contrôles de forme :

$$u(t) = u_1 + u_2(\hat{y}_1(1,t) - z_d) \tag{8}$$

$(u_1,u_2) \in \mathcal{U}_{ad}$ donné par (5), et $\hat{y}_i(x,t)$ sera une estimation (à préciser) de y_i.

Problème 3

Contrôle open loop.

Il est en fait contenu dans la formulation du problème 1 en imposant $u_2=0$.

Remarque 1

On ne prétend aucunement que les problèmes de contrôles considérés ici sont très réalistes. Le but est ici de comparer les résultats sur un exemple.

III - ETUDE DU PROBLEME EN INFORMATION PARFAITE.

3.1. Hypothèses et notations.

On notera $\mathcal{O}$ un ouvert borné de R^n, de frontière Γ régulière ; on note $H=L^2(\mathcal{O})$, $V=H^1(\mathcal{O})$ V' dual de V, H est identifié à son dual.
Soit (Ω,μ) un espace de probabilité, on note X l'espace $L^2(\Omega,\mu;H)$, Y l'espace $L^2(\Omega,\mu;V)$. On note $L^2(X)$ (resp. $L^2(Y)$) l'espace $L^2(0,T;X)$ (resp. $L^2(0,T;Y)$).

$\langle .,. \rangle$ désignera la dualité V,V' ;
$(.,.)$ le produit scalaire dans H, $|.|$, $\|.\|$ les normes dans H et V respectivement.
On notera h un processus de wiener hilbertien à valeur dans H au sens de [1], c'est-à-dire :

$$\forall e, e_1, e_2 \in H, \quad \forall t, t_1, t_2$$

i) $(h(t),e)$ est une variable gaussienne,
ii) $E[(h(t),e)] = 0$

iii) $E[(h(t_1),e_1).(h(t_2),e_2)] = \int_0^{Min(t_1,t_2)} (Q(s)e_1,e_2)ds$

où $Q(t)$ est un opérateur positif autoadjoint et nucléaire et trace $Q(t) \leq M \quad \forall t$.
On donne également

$$y_{10} \in X, \quad y_{20} \in L^\infty(\mathcal{O}) \qquad y_{20} \geq 0$$

y_Γ sera la trace de y sur Γ pour $y \in V$.

On donne enfin un opérateur linéaire continu de $L^2(\Gamma)$ dans lui-même.
On définit alors un opérateur linéaire de V dans V' par

$$\langle Ay,\varphi \rangle = \int_{\mathcal{O}} \frac{\partial y}{\partial x}.\frac{\partial \varphi}{\partial x}\, dx - (By_\Gamma, \varphi_\Gamma)_{L^2(\Gamma)} \qquad \forall \varphi \in V \tag{9}$$

(le cas où B est affine continu, se traite de façon analogue).
On considère maintenant le système

$$y_1(t) - y_{10} + \int_0^t A\, y_1(s)ds + \int_0^t f_1(y_1,y_2)ds = h(t)-h(0) \tag{10}$$

$$\begin{cases} \dfrac{\partial y_2}{\partial t} + f_2(y_1,y_2) = 0 \\ y_2(0) = y_{20} \end{cases} \tag{11}$$

3.2. Existence d'une solution.

On a alors le résultat suivant :

Théorème

Avec les hypothèses et notations du §3.1 il existe y_1, y_2 uniques tels que

$$y_1 \in C(0,T;X) \cap L^2(Y)$$

$$y_2 \in C(0,T;X)$$

$$0 \leq y_2 \leq |y_{20}|_{L^\infty(\mathcal{O})} \quad \text{p.s. } \omega, \text{ pp sur } \mathcal{O} \times]0,T[$$

et y_1 vérifie (10) dans $C(0,T;Y')$

y_2 vérifie (11) dans $L^2(X)$.

On a de plus

$$E|y_1|_H^2 + 2\,E\int_0^t \langle Ay_1,y_1\rangle\, ds + 2E\int_0^t (f_1(y_1,y_2),y_1)ds = E|h(t)|^2 - E|h(0)|^2 + E|y_{10}|^2$$

On indique brièvement les étapes de la démonstration.

a) A est un opérateur coercif sur H^1

i.e. $\exists\ \lambda,\ \alpha > 0$ tels que $\forall\ y \in V$

$$(12) \qquad \langle Ay,y\rangle + \lambda|y|^2 \geq \alpha\|y\|^2$$

c'est une conséquence de (9) et du résultat suivant (cf. [3])

$\forall\ \varepsilon,\ \exists\ C_\varepsilon$ tel que

$$(13) \qquad \forall\ y \in H^1 \quad |y_\Gamma|^2_{L^2(\Gamma)} \leq \varepsilon\, \|y\|^2 + C_\varepsilon |y|^2$$

b) on considère alors le schéma itératif suivant y_1^0 étant donné dans $L^2(X)$. On définit y_2^1 par

$$(14) \qquad \frac{\partial y_2^1}{\partial t} + f_2(y_1^0, y_2^1) = 0$$

qui, d'après (2) est une équation linéaire en y_2^1.

On montre facilement qu'elle a une solution unique vérifiant

$$(15) \qquad 0 \leq y_2^1 \leq |y_{20}|_{L^\infty(\mathcal{O})} \quad \text{p.s. } \omega, \text{ p.p. sur } \mathcal{O}\times]0,T[\ .$$

Puis on définit y_1^1 par

$$(16) \qquad y_1^1(t) - y_{10} + \int_0^t Ay_1^1(s)ds + \int_0^t f_1(y_1^0, y_2^1)ds = h(t) - h(0)$$

qui est une équation stochastique <u>linéaire</u> avec A coercif ; on peut donc appliquer le résultat de [2], c'est-à-dire : il existe y_1^1 unique dans $C(0,T;X) \cap L^2(Y)$ vérifiant (14) (dans $C(0,T;Y')$) avec

$$(17)\quad E|y_1^1(t)|_H^2 + 2E\int_0^t \langle Ay_1^1,y_1^1\rangle ds + 2E\int_0^t (f_1(y_1^o,y_2^1),y_1^1)ds = E|y_{10}|^2+E|h(t)|^2 - E|h(0)|^2$$

On a ainsi définit une application φ de $L^2(X)$ dans lui même $y_1^1 = \varphi(y_1^o)$.

On utilise alors les propriétés de f_1,f_2 (linéaires en y_2, uniformément lipschitziennes et bornées en y_1) ainsi que (14)(15)(16)(17) pour montrer que φ admet une itérée contractante. Un théorème classique montre que cette application a un point fixe unique $\bar{y}_1$ dans $L^2(X)$ et il est clair que si $\bar{y}_2$ est donné par (11) pour $y_1=\bar{y}_1$, alors le couple $(\bar{y}_1,\bar{y}_2)$ vérifie (10)-(11).

3.3. Existence d'un contrôle optimal.

Dans le cas particulier des équations (1), avec le critère (3) et $\mathcal{U}_{ad}$ donné par (5), il suffit de montrer que l'application $(u_1,u_2) \to y_1(1,t)$ est continue de R^2 dans $L^2(\Omega,\mu;L^2(0,T))$.

En notant (y_1,y_2) la solution pour (u_1,u_2) et $(\bar{y}_1,\bar{y}_2)$ la solution pour (v_1,v_2) en posant $z_i=y_i-\bar{y}_i$, on a p.s. ω :

$$(18)\quad \begin{cases} z_1' -\Delta z_1 + f_1(y_1,y_2) - f_1(\bar{y}_1,\bar{y}_2) = 0 \\ z_2' + f_2(y_1,y_2)-f_2(\bar{y}_1,f_2) =0 \\ \dfrac{\partial z_1}{\partial x}(0,t) = (v_1 - v_1) + u_2 z_1(1,t) + (u_2 - v_2)(y_1(1,t)-z_d) \\ \dfrac{\partial z_1}{\partial x}(1,t) = 0 \\ z_i(x,0) = 0 \end{cases}$$

On utilise les estimations habituelles, les propriétés de f_1,f_2 et l'inégalité (13) pour obtenir

$$(19)\quad |z_1(t)|_H^2 + \int_0^t \|z_1(s)\|^2 ds \le C_1\int_0^t \{|z_1(s)|^2 + |z_2(s)|^2\}ds + C_2(|u_1-v_1|^2 + |u_2-v_2|^2) \quad \text{p.s. } \omega$$

et de même

$$(20)\quad |z_2(t)|^2 \le C_3\int_0^t \{|z_1(s)|^2 + |z_2(s)|^2\}ds$$

d'où quand $v_i \to u_i$ $i=1,2$

$$(21)\quad \int_0^T \|z_1(s)\|^2 ds \to 0 \quad \text{p.s. } \omega$$

donc

$$(22) \qquad \int_0^T |z_1(1,s)|^2 ds \to 0 \quad \text{p.s. } \omega$$

le théorème de Lebesgue donne alors le résultat.

Remarque 2

- L'existence d'un contrôle optimal open loop dépendant du temps avec $|u|_{L^2(0,T)} \leq M$ peut être obtenue en utilisant des méthodes de compacité (cf. [3]).
- La démonstration précédente resterait la même si on avait u_1, u_2 dépendant du temps mais restant dans un compact d'un espace métrique mais on ne sait pas démontrer l'existence d'un contrôle optimal dans $\mathcal{U}_{ad}$ borné de $L^\infty(0,T)$ par exemple.

IV - ESTIMATION DE L'ETAT

On considère maintenant le problème 2, l'observation étant donnée par (6)-(7). Pour simplifier les notations, on expose la méthode sur le problème suivant

$$(23) \qquad y(t) + \int_0^t Ayds + \int_0^t f(y(s))ds = u + h(t) + y_0$$

$$(24) \qquad J(u) = E\,\{\alpha \int_0^T |y-z_d|_H^2\, ds + \int_0^T |u|_H^2 ds\,\}$$

On considère a priori l'équation

$$(25) \quad \hat{y}(t) + \int_0^t A\hat{y}(s)ds + \int_0^t f(\hat{y}(s))ds + \int_0^t P(s)C^*R^{-1}(C\hat{y}ds - dz_s) = \hat{y}_0 + u$$

où P sera la solution de l'équation de Riccati (qui correspondrait au système linéarisé déduit de (23))

$$(26) \qquad \frac{dP}{dt} + (A+f'_yI)P+P(A^*+f'^*_yI)+PC^*R^{-1}CP = Q$$

$$P(0) = P_0$$

f'_y dans (26) étant évalué sur une trajectoire arbitraire. On peut obtenir <u>formellement</u> les équations (25)-(26) à partir de la fonctionnelle des moindres carrés correspondant à (23). (cf.[5] par exemple). On remarque que P ne dépend pas du contrôle. Alors si l'on cherche un contrôle de forme

$$(27) \qquad u(t) = u_1 + u_2(\hat{y}(t)-z_d(t))$$

on a ramené le problème à la recherche d'un contrôle sur le système en $(y,\hat{y})$ donné par (23)-(25) et le critère (24). (l'équation (26) étant résolue a priori).

Remarque 3

On peut démontrer, par exemple avec A linéaire coercif de V dans V', f lipschitzien de H dans H, et quand l'observation n'est pas frontière

i) que (26) a une solution positive dans $L^{\infty}(0,T;\mathcal{L}(H,H))$ (cf. [9]).

ii) que (23)(25)(27) a une solution unique et qu'il existe u_1,u_2(réels) minimisant (24).

V - METHODES ET RESULTATS NUMERIQUES

5.1. Simulation des termes aléatoires.

On se restreint dans les applications à un processus h de forme $h=h_1\lambda(t)$ avec $h_1 \in H$ et $\lambda(t)$ wiener réel de variance $\sigma_\lambda^2 t$ et on considère la discrétisation suivante de (23).

$$(28) \qquad \frac{y^{n+1}-y^n}{\Delta t} + Ay^{n+1} + f(y^{n+1}) = h_1 \frac{\lambda^{n+1}-\lambda^n}{\Delta t} + u^n \qquad n=1,N$$

de sorte que $\xi^n = \frac{\lambda^{n+1}-\lambda^n}{\Delta t}$ est une variable gaussienne centrée de variance $\frac{\sigma_\lambda^2}{\Delta t}$ facile à simuler.

On se ramène alors à un problème d'optimisation déterministe en simulant M trajectoires du bruit, c'est-à-dire M suites $(\xi_i^n \ n=1,N)$ $i=1,M$ $(y_i^n \ n=1,N)$ $i=1,M$. Le coût approché est alors

$$(29) \qquad J_{MN} = \frac{1}{M}\sum_{i=1}^{M} \{\sum_n \alpha|y_i^n - z_d^n|^2\Delta t + |u(y_i^n)|^2\Delta t\}$$

l'intérêt essentiel de ce type de méthode est que l'on peut alors calculer un système adjoint et donc un gradient de J_{MN} (ce qui pose des problèmes sous la forme continue(23)).

Remarque 4

On sait montrer, au moins en dimension finie, la convergence de la solution optimale pour J_{MN} vers la solution optimale pour J_N quand le nombre M de simulations tend vers l'infini (cf.[4] pour une étude complète).

5.2. Données et résultats numériques.

- L'inconvénient de la "méthode de simulations" est bien entendu que le temps de calcul croît très vite avec le nombre de simulations. On s'est restreint ici à M=5.
- On a utilisé un algorithme de gradient ordinaire, arrêté quand la variation relative du coût entre deux itérations successives est inférieure à 10^{-3}.
- Les données numériques étaient les suivantes

  $a=34$, $b=1.39\ 10^{11}$, $c=1.25\ 10^{11}$, $y_{10} = 1.25$ $y_{20} = 1$, $\sigma_h=3$, $\sigma_v = 0.2$, $\alpha = 10^4$,

  $z_d = 1.26$, $\sigma_o = 0.02$.
- Le tableau 1 donne les valeurs des paramètres u_1,u_2 calculés, le coût correspondant et les temps de calcul (sur IBM 360/91) pour les cas déterministes, open loop,

feedback en information parfaite, feedback en information bruitée.

- La figure 1, montre un exemple de trajectoire pour les trois derniers cas.
- Les figures 2 et 3 donnent un exemple de trajectoires simulées et des estimations correspondantes.
- On observe que le contrôle obtenu en information imparfaite est sensiblement différent du cas information parfaite. D'autre part, l'estimation suit assez bien la trajectoire simulée.

	u_1	u_2	coût initial	coût final	temps CPU
Déterministe	1.14	0.01		3.2	5 s
Open loop	1.46	0	1.28	26.5	50 s
Information parfaite	0.6	14.5	34.0	10.9	50 s
Information bruitée	1	8.1	56.0	15.1	3 mois

Tableau 1

CONCLUSION

Il nous manque évidemment une théorie du filtrage non linéaire en dimension infinie, et même en information parfaite, on ne dispose pas de beaucoup de méthode pour traiter des problèmes de contrôle stochastiques pour des équations aux dérivées partielles. Cependant au moins sur des systèmes du type considéré ici (dimension 1, en espace nombre réduit d'équations) on peut obtenir une première approximation d'un contrôle optimal, même avec une méthode d'estimation heuristique.

<u>Figure 1</u>

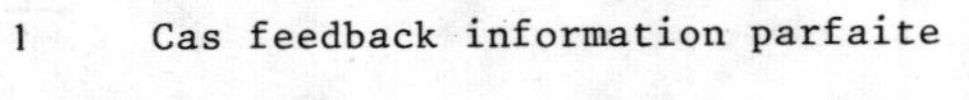
1 Cas feedback information parfaite

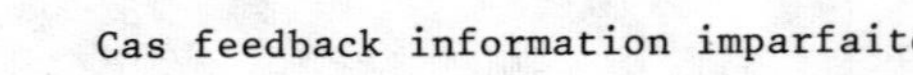
2 Cas feedback information imparfaite

3 Cas openloop

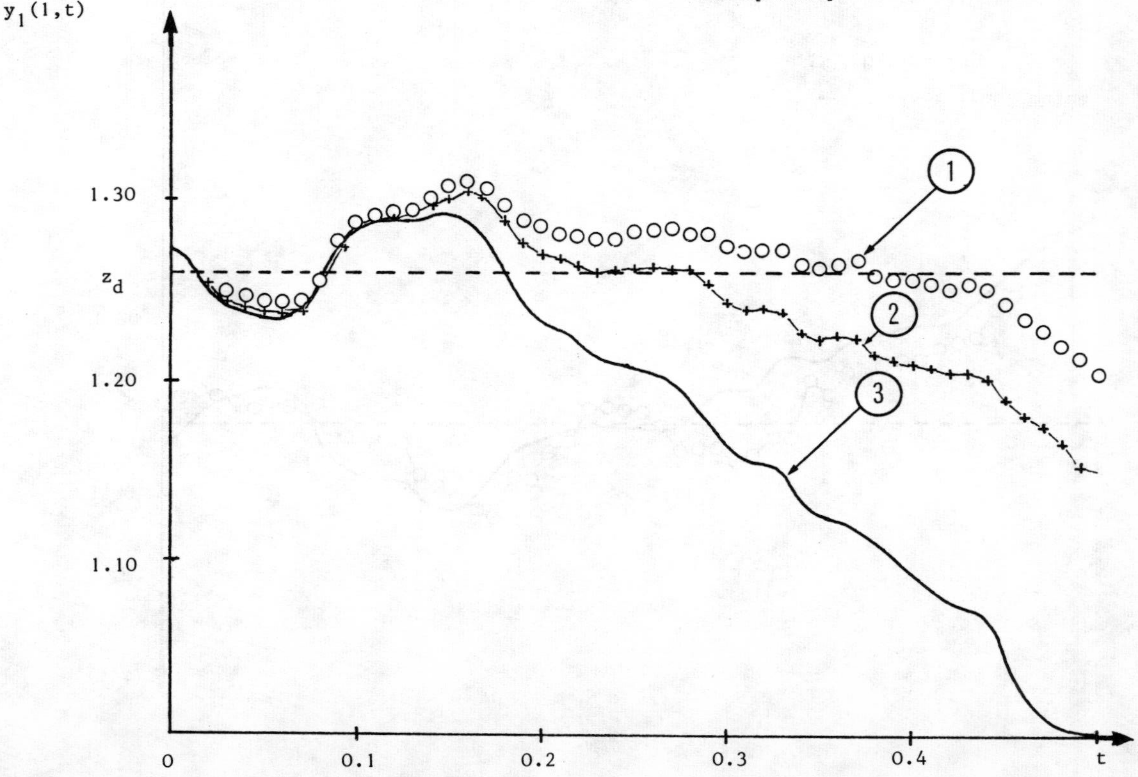

$y_1(1,t)$

1.30

z_d

1.20

1.10

0 0.1 0.2 0.3 0.4 t

① ② ③

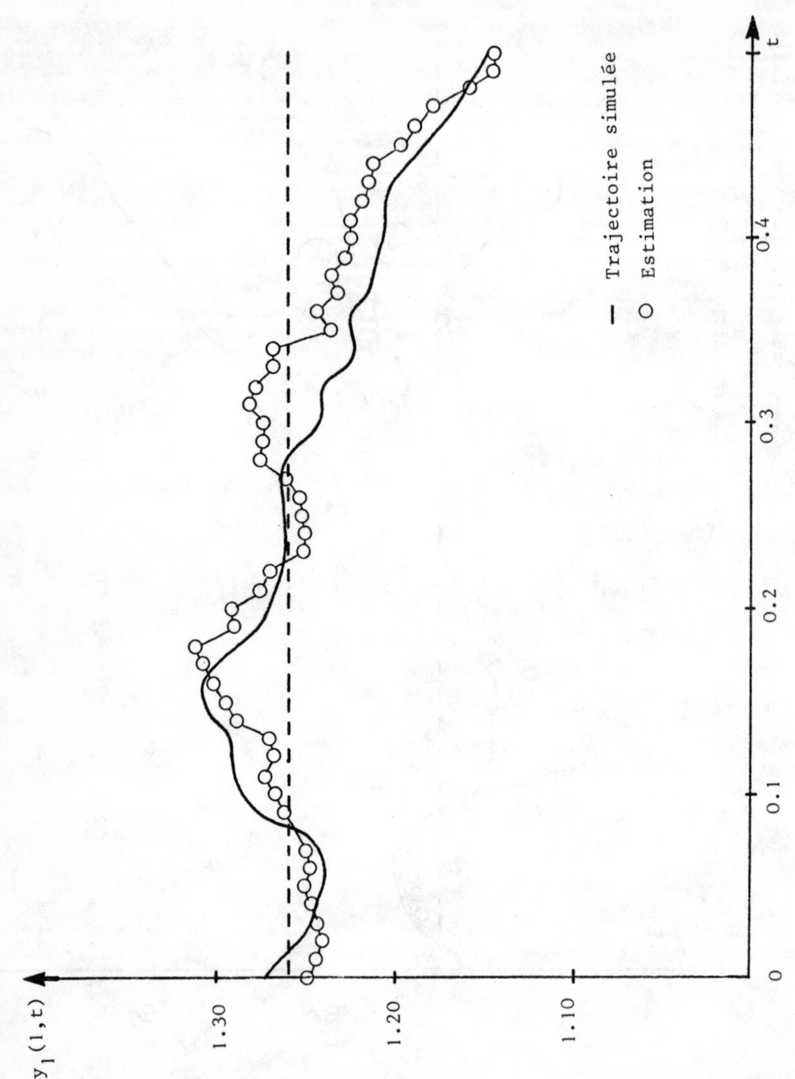
$y_1(1,t)$
1.30
1.20
1.10
0
0.1
0.2
0.3
0.4
t
— Trajectoire simulée
O Estimation

Figure 2

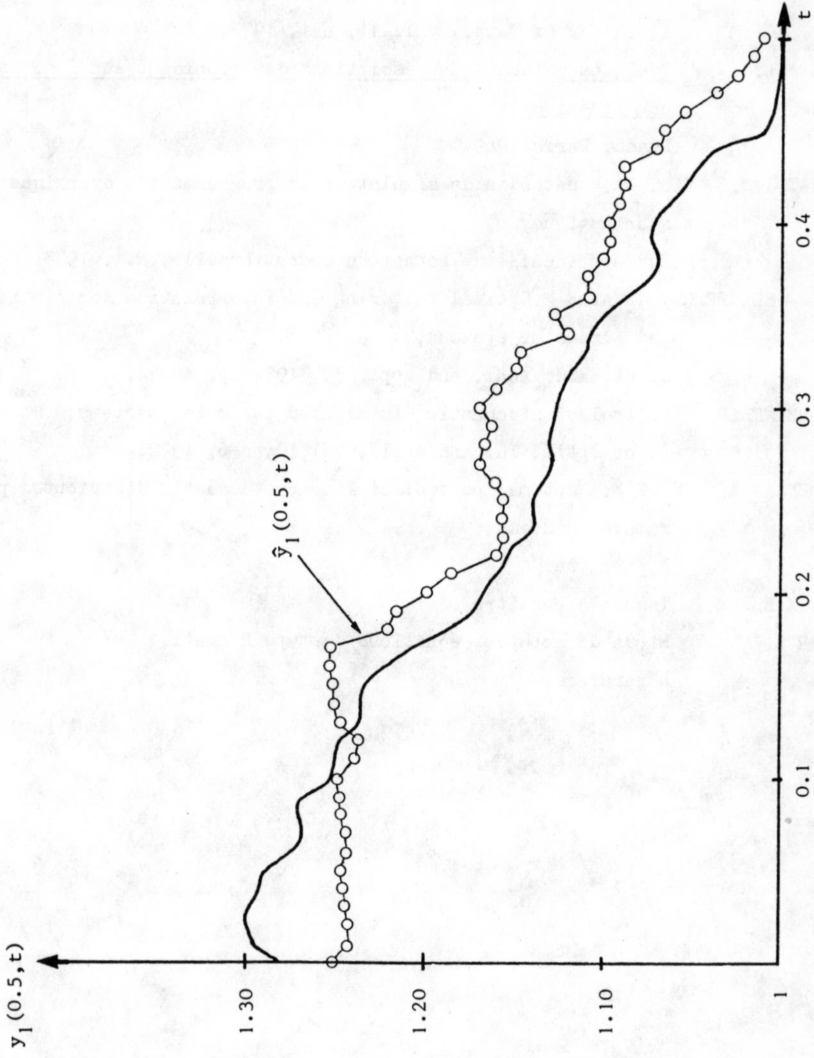

Figure 3

REFERENCES

[1] BENSOUSSAN A. *Filtrage optimal des systèmes linéaires*, Dunod, Paris 1970.

[2] BENSOUSSAN A. - TEMAM R. Equations aux dérivées partielles stochastiques non linéaires, Israël J. of Math., Vol. 11, n°1, 1972.

[3] LIONS J.L. *Quelques méthodes de résolutions des problèmes aux limites non linéaires*, Dunod, Paris 1969.

[4] QUADRAT J.P. - VIOT M. Méthodes de simulation en programmation dynamique stochastique, Revue française de Recherche opérationnelle, R.1. 1973.

[5] HWANG - SEINFELD - GAVALAS Optimal filtering and interpolation for distributed parameter systems, J. of Math. Anal. and Appl. 39, 1972, p. 49-74.

[6] YU - SEINFELD Control of stochastic distributed parameter systems, J. of Optim. Th. and Appl., Vol 10, n°6, 1972.

[7] BENSOUSSAN A. - VIOT M. Optimal control of linear stochastic distributed parameter systems, à paraître.

[8] PARDOUX E. Thèse (à paraître).

[9] TARTAR L. Etude directe des équations de type Riccati, à paraître.

A HOMOTOPY METHOD FOR PROVING CONVEXITY IN CERTAIN OPTIMAL STOCHASTIC CONTROL PROBLEMS

by

V. E. Beneš
Bell Laboratories
Murray Hill, New Jersey

Let $w_{\cdot}$ be a d-dimensional Wiener process and let $k: R \to R^+$ be even and increasing on the half-line. It is known[1] that the solution of the final value stochastic control problem

$$E\, k(|x_1|) = \min, \text{ subject to}$$

$$dx_t = u(t,x_t)dt + dw_t \quad 0 \le t \le 1$$

$$|u| \le 1$$

is given by the control law $u(t,x) = -x/|x|$. For an admissible, smooth control law u the value function $v(\tau,x) = E\{k(|x_1|)|x_{1-\tau} = x\}$ satisfies $v(0,x) = k(|x|)$, and the PDE

$$\mathcal{L}[u]\, v = \left\{ -\frac{\partial}{\partial\tau} + \frac{1}{2}\nabla^2 + u(1-\tau,x)'\nabla \right\} v = 0$$

When the criterion is a convex or "bowl-shaped" function $k(\cdot)$ of $|x|$ it is reasonable to expect that this property is propagated backwards in time along $v(\cdot,\cdot)$ through the action of $\mathcal{L}[u]$. We study some instances of this phenomenon.

In solving[1] the control problem it was natural to consider the subclass of admissible laws that point at (or away from) the origin, i.e., those of the form $u(t,x) = x\, y(t,|x|)$, $|y(t,r)| \leq r^{-1}$. For such laws $v(\tau,x)$ is of the form $\xi(\tau,|x|)$ where ξ satisfies a PDE related to the Bessel process with drift $y(t,r)$ associated with dimension d; in these cases the convexity question can be reduced to questions about ξ. However, here we follow a devious alternative.

Convexity is expressed by the positive definiteness $D_2 v > 0$ of the matrix $D_2 = (\partial^2/\partial x_i \partial x_j)$ applied to v, and thus by the positivity of det $D_2^S v$, where S is any subset of coordinate indices, and D_2^S is the restriction of D_2 to S. The condition det $D_2^S v > 0$ is in turn equivalent to the property that $\nabla^S v$, considered as a map of coordinates in S with other coordinates fixed, has isolated images of index +1.

These facts suggests the following methodological exercise: To prove that $v(\tau,\cdot)$ is convex by showing that the maps $\nabla^S v$ have isolated image points and then using the invariance of topological degree under homotopy.[2]

Fortunately, there is a natural homotopy at hand. Since[1] $v(\tau,x)$ is representable as $v^u(\tau,x) =$

$$E\, k(|x+w_\tau|) \exp \zeta_0^\tau(u(1-\tau+\cdot, x+w_\cdot))$$

where

$$\zeta_0^\tau(u(1-\tau+\cdot, x+w_\cdot)) = \int_0^\tau u(1-\tau+s, x+w_s)dw_s - \frac{1}{2}\int_0^\tau |u(1-\tau+s, x+w_s)|^2 ds$$

we look at the map η: $R^d \times [0,1] \to R^d$ defined (with τ fixed) by

$$\eta(x,\varepsilon) = \nabla E\, k(|x+w_\tau|) \exp \zeta_0^\tau(\varepsilon u(1-\tau+\cdot, x+w_\cdot)) = \nabla v^{\varepsilon u}(\tau,x).$$

Intuitively, this amounts to turning down the gain on the law u until it becomes the law $u \equiv 0$. For at $\varepsilon = 0$ η is the ∇ of $E\, k(|x+w_\tau|)$, which is convex if $k(|\cdot|)$ is; thus ∇v^0 has isolated images of index +1. At $\varepsilon = 1$ we have $\eta(x,1) = \nabla v^u(\tau,x)$. It is easy to show by known methods[3] that η is jointly continuous in x and ε, and so is a homotopy.

Suppose now that all the maps $\nabla v^{\varepsilon u}$, $0 \leq \varepsilon \leq 1$, have isolated images, and fix τ, x. For each ε in $[0,1]$ there is then a closed ball B_ε about x such that

$$q(x,\varepsilon) = \nabla v^{\varepsilon u}(\tau,x) \notin \partial \nabla v^{\varepsilon u}(\tau,B_\varepsilon)$$

i.e., $\nabla v^{\tau u}$ does not map any point on the sphere $S_\epsilon(x)$ of radius ε about x into the same point as it maps x. The degree of $\nabla v^{\varepsilon u}$ at $q(x,\varepsilon)$ with respect to B_ε is defined, and by invariance under homotopy is equal to +1 for each ε. In particular, then, all maps $\nabla^S v^{\varepsilon u}$ have isolated images, and have degree +1 at $q(x,\varepsilon)$ with respect to the closed subset of B_ε obtained by fixing the values of coordinates not in S. Thus

$$\det \nabla^S v^{\varepsilon u} > 0, \quad 0 \leq \varepsilon \leq 1,$$

and so $v^{\varepsilon u}(\tau,.)$ are all convex for $0 \leq \varepsilon \leq 1$.

It remains to find out when the maps $\nabla v^{\varepsilon u}$ have isolated images. This is the difficult part. The arguments to be given will work for any control law u of the form $u(t,x) = xy(t,|x|)$, so we can drop the superscripts on v. It can be seen from the representation of $v(\tau,x)$ that v is invariant under, and thus ∇v commutes with, an orthogonal matrix T whenever u commutes with T. Clearly u of the form above commute with every orthogonal matrix. Indeed in such cases we may say a little imprecisely that ∇v rotates in the direction of T as x does. Thus for x fixed and B a closed ball about x, $y \in \partial B$ implies $\nabla v(\tau,y) \neq \nabla v(\tau,x)$ unless y is collinear with x. Hence, the problem of showing that ∇v has isolated images reduces to showing that the length of ∇v is strictly monotone with $r = |x|$. To do this it is convenient to use the PDE for v, which will lead us back to Bessel processes again.

We use the following result, which, in case it is not intuitively obvious, is proved in the Appendix of Ref. 1: Lemma: Let $f: R^d \to R^d$ be continuous and such that f commutes with every orthogonal matrix. Then for $|x| > 0$

$$f(x) = \frac{|f(x)|}{|x|} x.$$

Thus $\nabla v(\tau,x) = f(\tau,r)x$ and we can use the PDE for v to get one for f. By the lemma, u must have the form $xy(t,r)$, so that the Jacobian matrix $\partial u/\partial x$ is $y_r xx' + yI$. Using this, and the equation

$$\nabla v_\tau = \frac{1}{2}\nabla^2\nabla v + \nabla(u(1-\tau,x)'\nabla v),\ \nabla v(0,x) = k'(r)\frac{x}{r}$$

we find that (with $d \geq 2$ the dimension of x) f satisfies

$$f_\tau = \frac{1}{2} f_{rr} + \frac{d+1}{2r} f_r + ry(1-\tau,r)\, f_r + f(ry_r+2y)$$

$$f(0,r) = r^{-1}k_r.$$

Let r_t be a Bessel process, corresponding to dimension $d + 2$, started at $r > 0$. Then $f(\tau,r)$ can be represented as a Feynman-Kac integral

$$(1)\quad f(\tau,r) = E_r f(0,r_\tau)\exp\left\{\int_0^\tau y(1-\tau+s,r_s)r_s dr_s - \frac{1}{2}\int_0^\tau |y(1-\tau+s,r_s)r_s|^2 ds + \int_0^\tau V(1-\tau+s,r_s)ds\right\}$$

where the "potential" term V is given by

$$V(t,r) = ry_2(1-t,r) + 2y(1-t,r),\ y_2 = \frac{\partial}{\partial r} y$$

Evidently f is positive if k_r is. Further, $\nabla v(\tau,x) = f(\tau,r)x$ so that $|\nabla v| = rf(\tau,r)$. We need then to show that rf is increasing in r when k is strictly convex: $k_{rr} > 0$. To this end note that rf satisfies the equation

$$g_\tau = \frac{1}{2} g_{rr} + \left(\frac{d-1}{2r} + ry\right) g_r + g\left(ry_2 + y - \frac{d-1}{2r^2}\right)$$

$$g(0,r) = k'(r)$$

Note that $|ry| = |u| \leq 1$.

For y smooth enough in r we can differentiate the equation for g to obtain a similar equation for $h = g_2$, viz.

$$h_\tau = \frac{1}{2} h_{rr} + \left(\frac{d-1}{2r} + ry\right) h_r + 2\left(ry_r + y - \frac{d-1}{2r}\right) h + g\left(ry_{rr} + 2y_2 + \frac{d-1}{r^3}\right)$$

$$h(0,r) = k''(r)$$

Denoting by U the potential term $2(ry_r + y - \frac{d-1}{2r})$ and by L the Lagrangian term $ry_{22} + 2y_2 + \frac{d-1}{r^3}$ $(= \frac{1}{2} U_r)$, a Feynman-Kac argument similar to that leading to (1) allows us to represent $h(\tau,r)$ as

$$E\, k''(r_\tau)\, e^{\zeta_0^\tau} + E \int_0^\tau L(1-\tau+s, r_s)\, e^{\zeta_0^\tau}\, ds$$

where ζ is the exponent functional in (1) with V replaced by U. If now k is strictly convex and $L \geq 0$, then $h > 0$ and $r\dot{f}\uparrow$ strictly. The condition $L \geq 0$ is certainly met by the known optimal control law $u(t,x) = -r^{-1}x$, for which $ry \equiv -1$. Alternatively the positivity of h could be proved from $k'' > 0$ and $L \geq 0$ by the maximum principle.

REFERENCES

1. V. E. Beneš, Composition and invariance methods for solving some stochastic control problems, to appear.
2. J. Cronin, Fixed points and topological degree in nonlinear analysis, AMS Math. Surveys, No. 11, Providence, 1964.
3. V. E. Beneš, Full "bang" to reduce predicted miss is optimal, to appear.

ON A CLASS OF STOCHASTIC BANG-BANG CONTROL PROBLEMS

J. RUZICKA
UNIVERSITY OF CALIFORNIA, LOS ANGELES.

1. INTRODUCTION. To begin with, let us recall a well known separation result for a stochastic linear regulator problem [6],[9],[1]. Let (Ω,P) be the underlying probability space and consider the following system:

$$(1.1)\quad \begin{aligned} dx(t,\omega) &= [A(t)x(t,\omega) + B(t)u(t,\omega)]dt + F(t)dW(t,\omega), \\ x(0,\omega) &= x_0(\omega), \\ dy(t,\omega) &= C(t)x(t,\omega)dt + G(t)dW(t,\omega), \quad y(0,\omega)=0, \\ 0 &\leq t \leq t_1, \end{aligned}$$

where x, y, u correspond to state, observation and control processes respectively, $W(t,\omega)$ is a Wiener process, the matrices $A,B,\ldots$ are, say, continuous in t and of appropriate dimensions, x_0 is a r.v. independent of $W(t,\omega)$, $t\geq 0$. For the sake of simplicity let

$$(1.2)\quad F(t)G^*(t) \equiv 0, \quad G(t)G^*(t) \equiv I$$

(the state noise is independent of the observation noise). The class of admissible controls $\mathcal{U}_y$ is specified as follows:

$$(1.3)\quad \begin{cases} u(t,\omega) \text{ is measurable } (t,\omega) \text{ jointly}, \\ u(t,\omega) \text{ is adapted } \mathcal{B}_W(t), \\ \int_0^{t_1} E\{\|u(s,\omega\|^2\}ds < \infty, \end{cases}$$

$$(1.4)\quad u(t,\omega) \text{ is adapted } \mathcal{B}_y(t),$$

Here by $\mathcal{B}_y(t)$ is denoted the σ-field generated by $y(\ ,\omega)$, $s\leq t$. Let

$$(1.5)\quad J(u) = \int_0^{t_1} E\{(Q(s)x(s,\omega), x(s,\omega)) + \lambda\|u(s,\omega)\|^2\}ds$$

be the performance index, $\lambda>0$, $Q(t)\geq 0$ is continuous on $(0,t_1)$. A control $u_0(t,\omega)\in\mathcal{U}_y$ is said to be optimal if

$$(1.6)\quad J(u_0) \leq J(u), \qquad \text{for all } u\in\mathcal{U}_y.$$

In the generality of the condition (1.4) it has been proved, [1], that the optimal control for the above problem has a feedback representation

(1.7) $$u(t,\omega) = -\frac{1}{\lambda} B^*(t) P_c(t) \hat{x}(t,\omega)$$

where $P_c(t)$ is the solution of certain Riccati equation , and the state estimate

(1.8) $$\hat{x}(t,\omega) = E\{x(t,\omega) \mid \mathcal{B}_y(t)\}$$

is updated via the recursive (Kalman) filtering equations. ((1.4) expresses the requirement that the control be dependent on the observation data available.) Explicit as it stands, the above separation result is still the only one of its kind available in the pertinent literature. Generalizations attempts have essentially reduced to the case as stated above, but with the additional requirements on the admissible controls. [9],[2],[3]; in particular

(1.9) $$u \in \mathcal{U}_y^b \equiv \{u \in \mathcal{U}_y : \|u(t,\omega)\| \leq 1 \text{ a.e. jointly}\}$$

In the first of the above mentioned contributions the admissible class (1.9) is actually specialized further in order to fit the tools used (such as Flemming-Nissio existence theorem). This specialization seems to have backfired a bit in the sense that any hints as to the actual structure of optimal controllers are lost in the comlexity of the optimality conditions derived. The main thrust of the Balakrishnan's contribution [2] focuses on certain σ-field equivalence which plays a key role in the separation problems in general.

The aim of this paper is to obtain an optimal control law for the problem (1.1),(1.2),(1.5),1.9) in the spirit of the classical separation result (1.7),(1.8), and in the light of Balakrishnan's contribution. As a final remark let us note that since $u(t,\omega)$ is already bounded, we put

(1.10) $$\lambda = 0$$

(by no means a simplification ,see [9]). Also , we will freely omit the argument ω and/or t unless there is a confussion.

2. EXISTENCE AND OPTIMALITY CONDITIONS.

As indicated in section 1, the problem is as follows:

Problem statement I. Given the system (1.1),(1.2), the performance index

$$J(u) = \int_0^{t_1} E\{[Q(s)x(s), x(s)]\} ds \tag{2.1}$$

and the admissible class $\mathcal{U}_y^b$ (see (1.9)), find $u_0 \in \mathcal{U}_y^b$ such that

$$J(u_0) \leq J(u) , \text{ all } u \in \mathcal{U}_y^b . \tag{2.2}$$

The first goal is to reformulate the problem in a more convenient fashion. To this end define $\tilde{x}(t,\omega)$, $\tilde{y}(t,\omega)$ by

$$\begin{aligned} d\tilde{x} &= A\tilde{x}\,dt + F dW , \quad \tilde{x}(0) = x_0 , \\ d\tilde{y} &= C\tilde{x}\,dt + G dW , \quad \tilde{y}(0) = 0 . \end{aligned} \tag{2.3}$$

Let

$$\hat{\tilde{x}}(t,\omega) = E\{\tilde{x}(t,\omega) \mid \mathcal{B}_{\tilde{y}}(t)\} . \tag{2.4}$$

Then, as is well known

$$Z(t,\omega) = \tilde{y}(t,\omega) - \int_0^t C(s)\hat{\tilde{x}}(s,\omega) ds \tag{2.5}$$

is a Wiener process, and moreover

$$\mathcal{B}_Z(t) \equiv \mathcal{B}_{\tilde{y}}(t),$$

$$\begin{aligned} d\hat{\tilde{x}}(t) &= A(t)\hat{\tilde{x}}(t)\,dt + P(t)C^*(t)\,dZ(t) , \\ \hat{\tilde{x}}(0) &= E x_0 , \end{aligned} \tag{2.6}$$

$$\begin{aligned} \dot{P}(t) &= A(t)P(t) + P(t)A^*(t) + F(t)F^*(t) - P(t)C^*(t)C(t)P(t) , \\ P(0) &= \operatorname{Cov} x_0 . \end{aligned} \tag{2.7}$$

Denote

$$\mathcal{H}_Z \overset{def}{=} \{u(\cdot,\cdot) : (1.3) \text{ holds } \& \ u(t,\omega) \text{ adapted } \mathcal{B}_Z(t)\} \tag{2.8}$$

Then $\mathcal{H}_Z$ is a Hilbert space in the inner product

$$[u_1, u_2]_{\mathcal{H}} = \int_0^{t_1} E\{[u_1(s), u_2(s)]\}\, ds$$

Theorem 2.1. (Balakrishnan separation theorem)

(2.9) $$\mathcal{U}_y^b \equiv \{u \in \mathcal{H}_Z : \|u(t,\omega)\| \leq 1 \;\; a.e.\ (t,\omega)\}$$

For proof we refer to [2],[3].

Using this theorem it is now straightforward to verify that for $u \in \mathcal{U}_y^b$ we also have that

(2.10) $$Z(t,\omega) = y(t,\omega) - \int_0^t C(s)\hat{x}(s,\omega)\, ds$$

where $$d\hat{x}(t,\omega) = [A(t)\hat{x}(t,\omega) + B(t)u(t,\omega)]\, dt + P(t)C^*(t)\, dZ(t,\omega),$$

(2.11) $$\hat{x}(0) = Ex_0 \; ; \quad \hat{x}(t) = E\{x(t) \mid \mathcal{B}_y(t)\} .$$

The Riccati equation (2.7) stays unchanged. By the same token the cost index can be rewritten as

$$J(u) = E\int_0^{t_1} [Q\hat{x}, \hat{x}]\, ds + \int_0^{t_1} Tr(QP)\, ds .$$

Having (2.9) in mind, we arrive at the following equivalent formulation of the problem statement I:

Problem statement II. Given (211),(2.7), the performance index

(2.12) $$\hat{J}(u) = E\int_0^{t_1} [Q\hat{x}, \hat{x}]\, ds$$

and the admissible class

(2.13) $$\mathcal{U}_Z^b \equiv \{u \in \mathcal{H}_Z : \|u(t,\omega)\| \leq 1 \;\; a.e.\}$$

find $u_0 \in \mathcal{U}_Z^b$ such that

(2.14) $$\hat{J}(u_0) \leq \hat{J}(u), \quad all \quad u \in \mathcal{U}_Z^b .$$

Thus, besides separating the filtering and the control problems, the admissible set of controls has been specified in the framework of the Hilbert space $\mathcal{H}_Z$. For the rest of this section the standard Hilbert space techniques are used to obtain the existence and optimality results.

In particular, no recourse to the theory of partial differential equations is required. Define the linear operator $L : \mathcal{H}_Z \to \mathcal{H}'_Z$ by

(2.15) $$(Lu)(t,\omega) \stackrel{def}{=} \int_0^t \Phi(t)\Phi^{-1}(s)B(s)u(s,\omega)ds$$

(2.16) $$\dot{\Phi} = A\Phi , \quad \Phi(0) = I ,$$

($\mathcal{H}'_Z$ is defined the same way as $\mathcal{H}_Z$ except the dimensionality of its elements). Let

(2.17) $$v(t,\omega) \stackrel{def}{=} \Phi(t)\hat{x}(0) + \int_0^t \Phi(t)\Phi^{-1}(s)P(s)C^*(s)dZ(s,\omega) ,$$

(2.18) $$[\sqrt{Q}\,u](t,\omega) \stackrel{def}{=} \sqrt{Q(t)}\,u(t,\omega) .$$

Then

(2.19) $$\hat{x} = Lu + v$$

(2.20) $$\hat{J}(u) = \|\sqrt{Q}\,(Lu + v)\|^2_{\mathcal{H}'_Z}$$

Clearly L is bounded, and it is not difficult to show that $\mathcal{U}^b_Z$ is a closed, convex and bounded subset of $\mathcal{H}_Z$. Hence $\sqrt{Q}(L\mathcal{U}^b_Z + v)$ is closed, bounded and convex subset of $\mathcal{H}'_Z$, and therefore there is an element of minimal norm in this set. Thus we have

<u>Theorem 2.2.</u> <u>There exists an optimal control.</u>

The optimality conditions are obtained by applying the well known variational inequality to characterize the above mentioned minimal element: That is, $\xi_0 \in \sqrt{Q}(L\mathcal{U}^b_Z + v)$ is of minimal norm iff

$$[\xi_0 , \xi_0 - \xi]_{\mathcal{H}'_Z} \le 0 , \quad all \ \ \xi \in \sqrt{Q}(L\mathcal{U}^b_Z + v).$$

In other words

$$[L^*Q(Lu_0 + v), u_0 - u]_{\mathcal{H}_Z} \le 0 , \ \forall u \in \mathcal{U}^b_Z .$$

Since $L^* : \mathcal{H}'_Z \to \mathcal{H}_Z$ is given by

$$(L^*\xi)(t,\omega) = B^*(t)\Phi^{*-1}(t)\int_t^{t_1} \Phi^*(s)E\{\xi(s,\omega)\,|\,\mathcal{B}_Z(t)\}ds ,$$

we have

<u>Theorem 2.3.</u> <u>The control</u> $u_0(t,\omega)$ <u>is optimal iff for all</u> $u \in \mathcal{U}^b_Z$

(2.21) $$[u_0 , B^*\hat{\eta}]_{\mathcal{H}_Z} \le [u , B^*\hat{\eta}]_{\mathcal{H}_Z}$$

where

(2.22) $$\hat{\eta}(t,\omega) = E\{\eta(t,\omega) \mid \mathcal{B}_z(t)\}$$

(2.23) $$\dot{\eta}(t,\omega) = -A^*(t)\,\eta(t,\omega) - Q(t)\,\hat{x}(t,\omega)\,,\quad \eta(t_1,\omega) = 0,$$

(2.24) $$d\hat{x}(t,\omega) = [A(t)\,\hat{x}(t,\omega) + B(t)\,u_o(t,\omega)]dt + P(t)C^*(t)dZ(t,\omega)$$
$$\hat{x}(0) = E x_o\,,$$

($B^*\hat{\eta}$ in (2.21) is defined similarly as (2.18)).

<u>Corollary 2.1.</u> <u>For u_o to be optimal it is sufficient that the following two conditions be satisfied:</u>

(2.25) $$P\{\omega \mid Measure\{t \mid B^*(t)\,\hat{\eta}(t,\omega) = 0\} = 0\} = 1,$$

(2.26) $$[u_o(t,\omega), B^*(t)\hat{\eta}(t,\omega)] = \underset{\|u(t,\omega)\| \le 1}{Min}\, [u(t,\omega), B^*(t)\hat{\eta}(t,\omega)]$$

<u>a.e. jointly.</u>

Note that under the condition of the corollary (2.26) is equivalent to

(2.27) $$u_o(t,\omega) = -\frac{B^*(t)\,\Phi^{*-1}(t)\,E\{\int_t^{t_1}\Phi^*(s)\,Q(s)\,\hat{x}(s,\omega)\,ds \mid \mathcal{B}_z(t)\}}{\|\,\text{———}\,\|}$$

3. SOLVING THE OPTIMIZATION PROBLEM.

Some heuristic considerations based on the corresponding deterministic problems in the absence of singular subarcs, and also on the formula (2.27) suggest the following conjecture regarding the form of the optimal control:

(3.1) $$u_o(t,\omega) = -\frac{S(t)\,\hat{x}(t,\omega)}{\|S(t)\,\hat{x}(t,\omega)\|}\,,$$

for a suitable matrix-function $S(t)$. In this generality,a proof or a counterexample are not yet available. This conjecture can however be verified in some special cases, namely those which reduce to the scalar problem. For the rest of this section we will deal with the scalar, time invariant case:

(3.2) $$\dim x = \dim y = \dim u = 1\,,\ \dim W = 2\,,\ A(t) \equiv a,$$
$$B(t) \equiv 1\,,\ C(t) \equiv c\,,\ P(t) = p(t)\,,\ Q(t) \equiv 1\,,\ FF^* \equiv GG^* \equiv 1\,.$$

A formal examination of (2.27), which can be written in this case as

$$u(t,\omega) = -\operatorname{sgn} E\left\{\int_t^{t_1} e^{-a(t-s)}\hat{x}(s)\,ds \,\middle|\, \mathcal{B}_z(t)\right\}$$

(3.3)

$$= -\operatorname{sgn}\left(\hat{x}(t) + E\left\{\int_t^{t_1} \frac{e^{2a(t_1-s)}-1}{e^{2a(t_1-t)}-1}\, u(s)\,ds \,\middle|\, \mathcal{B}_z(t)\right\}\right),$$

shows that the conjecture (3.1) actually becomes

(3.4) $$u_0(t,\omega) = -\operatorname{sgn}\hat{x}(t,\omega)$$

It will be seen that (3.4) is indeed the optimal control in this case. But for such a statement to have any meaning at all one must first show that the non-Lipschitzian stochastic differential equation which thereby results makes good sense. Let us note at this point that in the case of stochastic control problems without the observation noise one may adopt a different point of view, and phrase the optimization problem in terms of weak solutions (or solution-measures) of the stochastic differential equations. For examples of this kind we refer to V.Beneš [4],[5].

Theorem 3.1. Let $a(\cdot), \zeta(\cdot), \zeta^{-1}(\cdot)$ be elements of $L^\infty(t_0,t_1)$, let $W(t,\omega)$ be a Wiener process on (Ω, P), and let φ_0 be a finite number. Then there exists a unique solution of

$$d\varphi(t,\omega) = \left(a(t)\varphi(t,\omega) - \operatorname{sgn}\varphi(t,\omega)\right)dt + \zeta(t)\,dW(t,\omega)$$

(3.6)

$$\varphi(0) = \varphi_0$$

on $[t_0,t_1]$ with a.s. continuous sample path and adapted to $\mathcal{B}_W(t)$.[1)]

Proof.

Define the families of functions $\{f_n^+\}, \{f_n^-\}$ on $(-\infty,+\infty)$ as follows:

---

1) As pointed out to the present author by Professor S. Watanabe, an existence and uniqueness theorem for the scalar, non-Lipschitzian equations with bounded drifts has also been published earlier this year by A. K. Zvonkin [10]. His proof invokes deep results from partial differential equations theory. It seems that the proof here is more direct and much simpler. McKean's book being the main inspiration, [7].

$$
(3.7)\begin{cases}
\text{(i)} & f_n^{\pm}(\cdot) \subset C^{(1)}(-\infty,\infty) \text{ - the space of continuously differentiable functions on } (-\infty,\infty)\,, \quad n = 1, 2, \cdots \\
\text{(ii)} & f_n^{\pm}(\cdot) \text{ are non-increasing} \\
\text{(iii)} & f_n^{+}(\varphi) \downarrow -\operatorname{sgn}\varphi \quad \text{if } \varphi \neq 0\,, \quad f_n^{+}(0) = 1\,, \\
& f_n^{-}(\varphi) \uparrow -\operatorname{sgn}\varphi \quad \text{if } \varphi \neq 0\,, \quad f_n^{-}(0) = -1\,, \\
& \forall\, n\,, \quad \forall\, \varphi \in R^1 .
\end{cases}
$$

Consider the equations

$$
(3.8)\quad
\begin{aligned}
d\varphi_n^{+} &= (a\varphi_n^{+} + f_n^{+}(\varphi_n^{+}))\,dt + \zeta\,dW\,, \quad \varphi_n^{+}(0) = \varphi_0\,, \\
d\varphi_n^{-} &= (a\varphi_n^{-} + f_n^{-}(\varphi_n^{-}))\,dt + \zeta\,dW\,, \quad \varphi_n^{-}(0) = \varphi_0 .
\end{aligned}
$$

As is well known, (3.8) have unique , $\mathcal{B}_W(t)$-adapted solutions with a.s. continuous sample paths, $\varphi_n^{+}, \varphi_n^{-}$. It can be easily shown that except for an ω-set of zero probability

$$
\varphi_1^{+} \geq \varphi_2^{+} \geq \cdots \geq \varphi_2^{-} \geq \varphi_1^{-}\,,
$$

and further that $\{\varphi_n^{+}(\cdot,\omega)\}, \{\varphi_n^{-}(\cdot,\omega)\}$ are equibounded, equi-continuous families. From here

$$
(3.11)\quad
\begin{aligned}
&\varphi_n^{+}(t,\omega) \uparrow \varphi^{+}(t,\omega)\,, \quad \varphi_n^{-}(t,\omega) \downarrow \varphi^{-}(t,\omega)\,, \\
&\varphi^{-}(t) \leq \varphi^{+}(t)\,,
\end{aligned}
$$

uniformly on (t_0, t_1) for a. a. ω , where φ^{+}, φ^{-} are $\mathcal{B}_W(t)$-adapted processes with a.s. continuous sample paths. Moreover, $\varphi^{+}(t,\omega), \varphi^{-}(t,\omega)$ are equally distributed. To see that, rewrite first the equations (3.8) in the clock-time of $\int_{t_0}^{t} \zeta(s)\,dW(s)$, that is

$$
(3.12)\qquad \tau = v(t) \equiv t_0 + \int_{t_0}^{t} \zeta^2(s)\,ds\,,
$$

and then apply the Girsanov theorem [8]. There results

$$
(3.13)\qquad P\{\varphi_n^{\pm}(t) < \alpha\} = P_n^{\pm}\{\widetilde{W}_{\varphi_0}(v(t)) < \alpha\}
$$

where $\widetilde{W}_{\varphi_0}(\tau)$, $t_0 \leq \tau \leq v(t_1)$ is a new Wiener process on $(\Omega, P, \mathcal{B}_W(t))$ starting with φ_0 , α -a real number, and ($w(\tau)$ is the inverse function of $v(t)$)

$$dP_n^{\pm} = dP \exp\left\{ \int_{t_0}^{v(t_1)} \beta_n^{\pm}(\tau) d\tilde{W}_{\varphi_0}(\tau) - \frac{1}{2} \int_{t_0}^{v(t_1)} \beta_n^{\pm 2}(\tau) d\tau \right\},$$

$$\beta_n^{\pm}(\tau,\omega) = \zeta^{-2}(w(\tau)) \left(a(w(\tau)) \tilde{W}_{\varphi_0}(\tau,\omega) + \varphi_n^{\pm}(\tilde{W}_{\varphi_0}(\tau,\omega)) \right), \quad n = 1, \ldots$$

Next it can be shown that

(3.14) $$P_n^{\pm} \longrightarrow P^{\circ} \quad \text{weakly,}$$

where the probability measure P° is defined by

$$dP^{\circ} = dP \exp\left\{ \int_{t_0}^{v(t_1)} \beta(\tau) d\tilde{W}_{\varphi_0}(\tau) - \frac{1}{2} \int_{t_0}^{v(t_1)} \beta^2(\tau) d\tau \right\},$$

(3.15) $$\beta(\tau,\omega) = \zeta^{-2}(w(\tau)) \left(a(w(\tau)) \tilde{W}_{\varphi_0}(\tau,\omega) - \operatorname{sgn} \tilde{W}_{\varphi_0}(\tau,\omega) \right),$$

$$v(w(\tau)) \equiv \tau .$$

Hence

(3.16) $$P\{\varphi^{\pm}(t,\omega) < \alpha\} = P^{\circ}\{\tilde{W}_{\varphi_0}(v(t)) < \alpha\}$$

as claimed. Therefore by (3.1)

(3.17) $$\varphi^{+}(t,\omega) \overset{a.s}{\equiv} \varphi^{-}(t,\omega) .$$

Denote the common value by $\psi(t,\omega)$ .From (3.16),(3.15)

(3.18) $$P\{ \operatorname{Mes} \{\psi(t,\omega) = 0\} = 0\} = 1 ,$$

since $P_0 \ll P$. With this at hand it is straightforward to verify

$$\psi(t) = \varphi_0 + \int_{t_0}^{t} (a\psi - \operatorname{sgn}\psi) dt + \int_{t_0}^{t} \zeta \, dW ,$$

since the sgn-function is defined a. e. t ,w.p.1. The uniqueness and $\mathcal{B}_W(t)$-measurability are not difficult to obtain. Q.E.D.

Let us now return to the problem statement II from the section 2, which in the present case specializes as follows:

(3.19) $$\hat{J}(u) = E \int_0^{t_1} \hat{x}^2(s) ds \longrightarrow \underset{u \in \mathcal{U}_z^b}{\operatorname{Min}}$$

(3.20) $$d\hat{x}(t) = (a\hat{x}(t) + u(t)) dt + cp(t) dZ(t)$$

$$\hat{x}(0) = E x_0 \overset{def}{=} \xi$$

$$\dot{p} = 2ap + 1 - c^2p^2 ,$$

(3.21) $$p(0) = p_0 = \mathrm{Cov}\, x_0$$

$$0 \leq t \leq t_1$$

Theorem 3.2. The optimal control for the problem (3.19) - (3.21) is given by

(3.22) $$u(t,\omega) = -\,\mathrm{sgn}\, \hat{x}(t,\omega)$$

a.e. jointly.

Proof. We may assume $p_0 > 0$ (the case $p_0 = 0$ will follow by a simple argument). The solution $p(t)$ of the Riccati equation (3.21) has the property

$$p_0 \wedge p_\infty \leq p(t) \leq p_0 \vee p_\infty , \quad 0 \leq t \leq \infty,$$

where $0 < p_\infty = \lim_{t\to\infty} p(t) < \infty$ .Thus by the theorem 3.1 we know that

(3.23) $$d\hat{x}(t) = (a\hat{x}(t) - \mathrm{sgn}\,\hat{x}(t))dt + cp(t)\,dZ(t), \quad \hat{x}(0) = \xi,$$

augmented by (3.21) has a unique continuous solution. For the proof of optimality of (3.22) we use the corollary 2.1. Therefore we must verify

(3.24) $$P\{\mathrm{Meas}\,\{t \mid E\{\int_t^{t_1} e^{-a(t-s)}\hat{x}(s)\,ds \mid \mathcal{B}_Z(t)\} = 0\} = 0\} = 1$$

and

(3.25) $$\mathrm{sgn}\,\hat{x}(t) = \mathrm{sgn}\, E\{\int_t^{t_1} e^{-a(t-s)}\hat{x}(s)\,d \mid \mathcal{B}_Z(t)\}.$$

Applying (3.18) to the present situation, it is easy to see that (3.25) implies (3.24). Hence it is enough to verify (3.25). To this end define the Brownian stopping time

(3.26) $$\text{⧣} = t_1 \wedge \inf[s > t \mid \hat{x}(s;\xi,p_0) = 0]$$

with the corresponding σ-field

(3.27) $$\mathcal{B}_{\text{⧣}} \equiv \sigma\{B \in \mathcal{B}_Z(t_1) \mid B \cap \{\text{⧣} < t\} \in \mathcal{B}_Z(t),\ \text{all } t \in (0,t_1)\}$$

The conditional expectation in (3.25) can be split as

(3.28) $$E\{\int_t^{t_1} e^{-a(t-s)}\hat{x}(s)\,ds \mid \mathcal{B}_Z(t)\} = E\{\int_t^{\text{⧣}} \mid \mathcal{B}_Z(t)\} + E\{\int_{\text{⧣}}^{t_1} \mid \mathcal{B}_Z(t)\}.$$

If one can prove that

(3.29) $$E\{ \int_{\mathbb{t}}^{t_1} e^{-a(t-s)} \hat{x}(s)\, ds \mid \mathcal{B}_z(t) \} = 0 \quad ,$$

then the proof is complete, since, clearly,

$$\operatorname{sgn} E\{ \int_t^{\mathbb{t}} e^{+a(s-t)} \hat{x}(s)\, ds \mid \mathcal{B}_z(t) \} = \operatorname{sgn} \hat{x}(t) \; .$$

In order to prove (3.28), let us note that the pair $\{\hat{x}(t;\xi,p_0), p(t,p_0)\}$ has the strong Markov property:

(3.30) $$E\left\{ \begin{matrix} f(\hat{x}(\mathbb{t}+s;\xi,p_0)) \\ p(\mathbb{t}+s;p_0) \end{matrix} \;\middle|\; \mathcal{B}_{\mathbb{t}} \right\} = E\left\{ \begin{matrix} f(\hat{x}(s;\eta,\pi)) \\ p(s;\pi) \end{matrix} \right\}_{\substack{\eta=\hat{x}(\mathbb{t};\xi,p_0) \\ \pi=p(\mathbb{t};p_0)}}$$

(it can be seen the same way as in the Lipschitzian case). Using this and the fact that

$$t \le \mathbb{t} \;\Rightarrow\; \mathcal{B}_z(t) \subset \mathcal{B}_{\mathbb{t}} \; ,$$

the proof of (3.29) is reduced to showing

(3.31) $$E\{ \operatorname{sgn} \hat{x}(s;0,\pi) \} = 0 \quad \forall \; s \ge 0$$

In order to evaluate the expression above, let us apply the formula (3.16) with a proper identification reflecting the present case. The result is

(3.32) $$E \operatorname{sgn} \hat{x}(s;0,\pi) = E\{ \operatorname{sgn} W(t,\omega)\, e^{\int_0^t \beta(s)dW(s) - \frac{1}{2}\int_0^t \beta^2(s) dW(s)} \}$$

where

$$\beta(t,\omega) = \left(a\,W(t,\omega) - \operatorname{sgn} W(t,\omega) \right) q(t) \; ,$$

and q is a positive bounded function. Now the expectation in (3.32) is certainly finite ; but most importantly, the expression in the curly brackets on the right is anti-symmetric in ω, (the ω-space can be taken as the sample space of $W(t)$). To see that, the only thing which need be done is to show

$$(\int_0^t \beta\, dW)(\omega) = (\int_0^t \beta\, dW)(-\omega) \; .$$

Since

$$P\{ \int_0^{t_1} \beta^2(s,\omega)\, ds < \infty \} = 1$$

it is possible to find simple non-anticipative functions $\beta_n(s)$, n=1,...

such that [7]

(3.33) $$P\left\{ \int_0^{t_1} (\beta - \beta_n)^2 dt \le 2^{-n},\ n \uparrow \infty \right\} = 1$$

and

(3.34) $$\int_0^t \beta\, dW = \lim_{n\to\infty} \int_0^t \beta_n\, dW .$$

Define new simple functions

$$\tilde{\beta}_n(t,\omega) = \tfrac{1}{2}\left(\beta_n(t,\omega) - \beta_n(t,-\omega)\right).$$

$\tilde{\beta}_n$ satisfy (3.33), and since for such functions (3.34) is independent of the particular choice, we have

$$\int_0^t \beta\, dW = \lim_{n\to\infty} \int_0^t \tilde{\beta}_n\, dW .$$

The claim follows from here. Therefore (3.32) is zero, and so the proof of the theorem is complete.

The methods of this section do not generalize directly to the non-scalar case, unless for special cases. In particular, this is true of the proof of the theorem 3.1. The n-dimensional case is the subject of the forthcomming paper by the author.

ACKNOWLEDGEMENT. This reseach was supported in part by the Office of Scientific Reseach, U.S. Air force, Applied Mathematics Division, under Grant no. AFOSR 73-2492.

REFERENCES.

[1] Balakrishnan, A. V. Stochastic Differential Systems I, Springer Verlag, Berlin 1973.
[2] Balakrishnan, A. V. "Stochastic Control: A Function Space Approach" SIAM J. Control, Vol. 10, No. 2,pp.285-297, May 1972.
[3] Balakrishnan, A. V. "A Note on the Strusture of Optimal Stochastic Controls," Inter. J. Appl. Math. Opt., Vol. 1, No. 1, 1974.
[4] Beneš, V. E. "Girsanov Functionals and Optimal Bang-Bang Laws for Final Value Stochastic Control," to be published.
[5] Beneš, V. E. "Existence of Optimal Stochastic Control Laws," SIAM J. Control, Vol.9, No. 3, pp.446-472, August 1971.
[6] Joseph, P. D. and J. T. TOU. "On Linear Control Theory", AIEE Trans. Appl. and Ind., Part II, Vol. 80, pp.193-196, 1961.
[7] McKean,Jr., H. P. Stochastic Integrals, AP ,New York, 1969.
[8] Shiryaev, A.V. and R. S. Lipcer. "On the Absolute Continuity of Measures ...", Izv. Akad. Nauk SSSR, Ser. Mat. Tom 36 (1972), No.4
[9] Wonham, W. M. "Random Differential Equation in Control Theory", in Probabilistic Methods in Appl. Math., Vol.2,pp.131-212, AP 1970.
[10] Zvonkin, A. K. "A Drift Annulating Phase Space Map for a Diffusion" Matematičeskii Sbornik, Tom 93(135), No.1, 1974,(in russian).

SOME STOCHASTIC SYSTEMS ON MANIFOLDS

T. E. Duncan

Department of Applied Mathematics and Statistics

SUNY, Stony Brook, N. Y.

1. Introduction

Some stochastic systems described by stochastic processes with values in a manifold will be considered. The manifolds will be smooth, complete, connected Riemannian manifolds. These manifolds seem to have the properties that are used in $\mathbb{R}^n$ to construct Brownian motion and to study stochastic systems. They have also been the setting for some calculus of variation problems in differential geometry. Furthermore, this family of manifolds forms a large, interesting class for applications. For global differential geometry, principal bundles seem to be the most natural approach. Since the manifolds considered here are Riemannian it is natural to consider the bundle of orthonormal frames.

The bundle of orthonormal frames allows a global description in one respect of the manifold and the parallelism of vectors along curves has a direct global interpretation in the bundle from the notion of the (horizontal) lift of a curve in the manifold to the bundle.

Since Brownian motion and processes related to it have been useful models for stochastic systems in $\mathbb{R}^n$ it is natural to construct Brownian motion in a smooth, connected Riemannian manifold, M. Similarly the notion of stochastic differential equations in TM will be useful as models for stochastic systems. Since parallelism along Brownian paths will be useful, Brownian motion will be constructed in the bundle of orthonormal frames, O(M). Stochastic systems described by stochastic differential equations in TM will describe some stochastic optimal control problems. It will be shown that for a class of stochastic systems optimal controls exist by establishing properties of a family of Radon-Nikodym derivatives used in the tranformation of the measure for M-valued Brownian motion. This result is similar to the result for $\mathbb{R}^n$-valued Brownian motion [2, 4].

For $\mathbb{R}^n$-valued Brownian motion the Hilbert space (or more precisely the Sobolev space), $L_0^{2,1}(\mathbb{R}^n)$, of continuous functions with zero as initial value that are absolutely continuous (with respect to Lebesgue measure) and whose derivative is square integrable plays a fundamental role. In fact in some respects this Hilbert space which has measure zero with respect to Wiener measure is more fundamental than

Research supported by NSF Grant GK32136. Part of the research was done in the summer of 1973 while the author was at IRIA.

the Banach space of $\mathbb{R}^n$-valued continuous functions with zero as initial value that has measure one with respect to Wiener measure (e.g. absolute continuity questions and stochastic integrals). This phenomenon was apparently first abstracted by I. E. Segal [14] with the notion of a canonical normal distribution on a Hilbert space. This cylinder set measure is a rotationally invariant measure on each finite dimensional subspace of the Hilbert space. L. Gross [9], abstracting Wiener's method of proof for Wiener measure, obtained a sufficient condition, which is also necessary [3], on a seminorm on the Hilbert space so that the completion of the Hilbert space with respect to the seminorm enables the extension of the canonical normal distribution to a measure on this Banach space.

Let $L_0^{2,1}(T_aM)$ be the family of continuous functions from [0,1] to the tangent space, T_aM, with zero as initial value that are absolutely continuous and whose derivative is square integrable and let $L_a^{2,1}(M)$ be the family of continuous functions from [0, 1] to M with $a \in M$ as initial value that are absolutely continuous and whose derivative is square integrable (using the Riemannian metric). $L_a^{2,1}(M)$ is a Hilbert manifold by the existence of convex neighborhoods and the compactness of [0, 1] (Eells [6]).

Let M be a Riemannian manifold and let $d(x,y)$ be the distance between x and y, where $x, y \in M$, defined as the infimum of the lengths of all piecewise differentiable curves of class C^1 joining x and y. The distance function, d, defines the same topology as the manifold topology of M. If M is connected and complete, then (M, d) is a locally compact, complete metric space.

Curves from [0, 1] to M with $a \in M$ as initial value can be identified with curves from [0, 1] to T_aM with $0 \in T_aM$ as initial value by means of the development. The development of a curve in M starting at a into the (affine) tangent space T_aM is obtained by the parallel transport of the vectors of the curve along the curve to the (affine) space T_aM and then forming a curve in the tangent space from the vectors. Conversely, given a curve in T_aM starting at 0 there is a curve in M starting at a whose development on T_aM is the given curve. In fact the development defines a diffeomorphism between the Hilbert manifold, $L_a^{2,1}(M)$, and the Hilbert space $L_0^{2,1}(T_aM)$.

It is reasonable to conjecture that a T_aM-valued Brownian motion can be identified with an M-valued Brownian motion by using the development. McKean [11, 12] has used this technique for Lie groups and Eells-Elworthy [7] have reported the result for complete, connected Riemannian manifolds.

The extension of the development to almost all Brownian paths enables the identification of the two Brownian motions but for the identification to be more useful tangent vectors along the T_aM-valued Brownian motion should be identified with tangent vectors along the M-valued Brownian motion and conversely. This identification requires the notion of parallelism along M-valued Brownian paths. This notion for differential forms was apparently first introduced by K. Itô [10]. Parallelism and global differential geometry seem to be most effectively approached using a principal bundle. Since M is Riemannian it is natural to consider the bundle of orthonormal frames. Curves in the manifold M can be lifted to the bundle and there is a unique lift with a fixed starting point in the bundle that has the property that its vectors are in the horizontal subspace of the tangent space of the bundle at each point [13]. The connection form enables the determination of this (horizontal) lift. Parallelism of the vectors along a curve is obtained from the construction of the (horizontal) lift of the curve in M. Therefore for M-valued Brownian motion it is desirable to obtain a (horizontal) lift of it to the bundle of orthonormal frames, O(M) thereby obtaining parallelism of vectors along Brownian paths. From this parallelism real-valued stochastic integrals obtained from the T_aM-valued Brownian motion can be identified with real-valued stochastic integrals obtained from the M-valued Brownian motion.

2. Brownian motion in the bundle of orthonormal frames.

The following result gives the O(M)-valued Brownian motion.

Theorem 1. Let $(B_t)_{t\in[0,1]}$ be an n-dimensional standard Brownian motion on T_aM where M is a smooth, complete, connected, n-dimensional Riemannian manifold. There is an O(M)-valued process, $(F_t)_{t\in[0,1]}$, such that the process $(\pi(F_t))_{t\in[0,1]}$, where $\pi: O(M) \to M$ is the projection, is an M-valued Brownian motion which is developed almost surely on $(B_t)_{t\in[0,1]}$. The process, $(F_t)_{t\in[0,1]}$, is the (horizontal) lift of $(\pi(F_t))_{t\in[0,1]}$ such that $F_o \equiv u$ and $\pi(u) = a$.

The proof of the result above as well as the following results will be given elsewhere. The result above implies that the development can be extended to almost all Brownian paths, i.e. the diffeomorphism of the Hilbert space, $L_o^{2,1}(T_aM)$ and the Hilbert manifold, $L_a^{2,1}(M)$ can be extended to almost all of the T_aM-valued Brownian paths and the M-valued Brownian paths to define an almost sure bijection between these two processes. Similarly from the (horizontal) lift of the M-valued Brownian motion, vectors along the T_aM-valued Brownian motion can be identified with vectors along the M-valued Brownian motion. This identification enables one to relate real-valued Wiener integrals and more generally stochastic integrals for the T_aM-valued Brownian motion with the corresponding objects for for the M-valued Brownian motion. The almost sure bijection of the processes enables the identification of the natural families of sub-σ-algebras associated with the two Brownian motions $(B_t)_{t\in[0,1]}$ and $(\pi(F_t))_{t\in[0,1]}$.

Measurability properties for processes with continuous sample paths in M or for processes in the tangent bundle of the former processes can often easily be determined from the local triviality of the Banach manifold because there is a countable cover of the manifold by patches.

A notion of measurability for TM-valued processes will be defined as well as a family of these measurable processes that will be used as integrands for stochastic integrals with the M-valued Brownian motion.

Definition 1: A function $f\colon [0,1]\times\Omega \to TM$ where Ω is the family of continuous maps from $[0,1]$ to M starting at $a \in M$ is said to be predictable (with respect to the M-valued Brownian motion, $(C_t)_{t\in[0,1]}$ where $\pi(F_t) = C_t$) if

i) $f(t,w) \in T_{C_t(w)} M$ for almost all $w \in \Omega$

ii) The parallelism of f along the Brownian paths to T_aM determines an $\mathbb{R}^n$-valued function that is measurable with respect to the σ-algebra on $[0,1] \times \tilde{\Omega}$ generated by the left continuous $\mathbb{R}^n$-valued processes that are adapted to $(\mathfrak{F}_t)_{t\in[0,1]}$ where the parallel transport of f is expressed as a functional of the T_aM-valued Brownian motion, $\tilde{\Omega}$ is the family of continuous functions from $[0,1]$ to $\mathbb{R}^n$ with initial value zero, and $(\mathfrak{F}_t)_{t\in[0,1]}$ is the augmented family of sub-σ-algebras generated by the T_aM-valued Brownian motion.

Remark: If it is assumed that f has some integrability properties then the notion of measurability described above can be given by the local triviality of the Banach manifold.

Definition 2: A predictable TM-valued process, f, is said to be in $L^2(TM; \langle C,C\rangle)$ where $(C_t)_{t\in[0,1]}$ is the M-valued Brownian motion if

$$\iint (f(t,w), f(t,w))_{C_t(w)} dt\, dP(w) < \infty$$

where $(\cdot,\cdot)_x$ is the inner product in T_xM induced from the Riemannian metric.

With these definitions real-valued stochastic integrals can be defined for the M-valued Brownian motion.

Proposition 1: Let $f \in L^2(TM; \langle C,C\rangle)$. Then there is a version of the process

$$Y_t = \int_0^t (f, dC)_C$$

such that $(Y_t, G_t, P)_{t\in[0,1]}$ is a real-valued continuous square integrable martingale where G_t is the completion of $\sigma(C_u;\ u \leq t)$ and P is the measure for the M-valued Brownian motion.

Since M may not be a linear space the measures that are absolutely continuous with respect to the Wiener measure on the space of M-valued continuous functions are conveniently described by retaining the same process and changing the measure and then describing the process locally in terms of the new measure, specifically the process is described by a stochastic differential equation in TM with respect to the new measure.

The following result is similar to the transformation of measure results for $\mathbb{R}^n$-valued Brownian motion (e.g. Girsanov [8]).

Proposition 2: Let $(\varphi_t)_{t\in[0,1]}$ be a TM-valued predictable process (with respect to the M-valued Brownian motion $(C_t)_{t\in[0,1]}$) such that

$$\int_0^1 (\varphi_t, \varphi_t)_{C_t}\, dt < \infty \qquad \text{a.s. } P$$

Then the real-valued process $(M_t, G_t, P)_{t\in[0,1]}$ defined as

$$M_t = \exp\left[\int_0^t (\varphi_s, dC_s)_{C_s} - \frac{1}{2}\int_0^t (\varphi_s, \varphi_s)_{C_s} ds\right]$$

is a continuous local martingale.

Furthermore, if the process $(\varphi_t)_{t\in[0,1]}$ is uniformly bounded, then $(M_t, G_t, P)_{t\in[o,1]}$ is a continuous martingale. In this case the TM-valued stochastic differential equation

$$dC_t = \varphi_t\, dt + d\tilde{B}_t$$

where the differentials are in $T_{C_t}M$ defines an M-valued process such that $(\tilde{B}_t, G_t, \overline{P})_{t\in[0,1]}$ is a Brownian motion where $d\overline{P} = M dP$.

3. Applications to stochastic systems.

The stochastic control systems that will be considered are described by stochastic differential equations of the following form

$$dX_t = f(t, X, u(t,X))dt + d\tilde{B}(t)$$

where the differentials are in $T_{X_t}M$ and $(\tilde{B}_t)_{t\in[0,1]}$ is an M-valued Brownian motion such that $\tilde{B}(0) \equiv a$. Various conditions will be attached to f subsequently. The control, u, takes values in a complete, connected m-dimensional Riemannian manifold.

The solution of the stochastic control systems will be obtained from the transformation of measures as described in Proposition 2.

Some notation that will be useful later is introduced in the following definition.

Definition 3: a) Let $C_a([0,1];M)$ be the Banach manifold of continuous functions from $[0,1]$ to M with $a \in M$ as initial value.

b) For each $t \in [0,1]$ let S_t be the minimum augmented (with respect to the Wiener measure on C_a) σ-algebra of subsets of C_a that makes the random variables $\{C(u),\ u \le t\}$ measurable.

c) Let $S = S_1$.

Remark: $C_a = C_a([0,1];M)$ is a Banach manifold by the existence of convex neighborhoods and the compactness of $[0,1]$ (Eells [6]). The distance metric on M can be used to define the topology.

The conditions on the TM-valued drift term, f, are described in the following. The measurability properties can be determined from the local triviality of the Banach manifold, C_a.

C1. The drift, f, is measurable with respect to the product σ-algebra $\mathbb{B}\otimes S\otimes\mathbb{B}_U$ where $\mathbb{B}(\mathbb{B}_U)$ is the family of Borel subsets of $[0,1]$ (U).

C2. For $t\in[0,1]$, $f(t,\cdot,\cdot)$ is $S_t\otimes\mathbb{B}_U$ measurable.

C3. For $(t,z)\in[0,1]\times C_a$, $f(t,z,\cdot)\colon U \to T_{z_t}M$ is continuous.

C4. The family, $(f(t,z,u))$, of TM-valued drift terms is uniformly bounded.

C5. For $(t,z)\in[0,1]\times C_a$ the set

$$f(t,z,U) = \{f(t,z,u)\colon u \in U\}$$

is a closed, convex subset of $T_{z_t}M$.

For the optimal control problem it is necessary to define a family of controls that will be used.

Definition 4: a) An admissible control is a map $u:[0,1] \times C_a \to U$ that is $\mathbb{B}\otimes S$ measurable such that (u_t) is adapted to (S_t). $\mathcal{U}$ will denote the family of all admissible controls.

b) The drift corresponding to $u \in \mathcal{U}$ is the TM-valued function, g_u, where $g_u(t,z) = f(t,z,u(t,z))$. $G = \{g_u\colon u \in \mathcal{U}\}$.

Definition 5: A function $\varphi:[0,1] \times C_a \to TM$ is said to be causal if it is $\mathbb{B}\otimes S$ measurable and if $(\varphi_t)_{t\in[0,1]}$ is adapted to $(S_t)_{t\in[0,1]}$. Φ will denote the family of causal TM-valued functions.

The following characterization of causal functions follows from a result of Beneš [1] by using local triviality of the Banach manifold, C_a, and the fact that the control takes values in a connected,complete Riemannian manifold which is therefore a locally compact, complete metric space.

Lemma 1: Let g be a causal TM-valued function. $g \in G$ if and only if $g(t,z) \in f(t,z,U)$ for all $(t,z) \in [0,1] \times C_a$.

As in the results for stochastic systems in a linear space [2,4,5] the following result is fundamental in showing that an optimal control exists.

Theorem 2: Let $D(\Phi)$ be the subset of $L^1(P)$ given by

$$D(\Phi) = \left\{ \exp\left[\int_0^1 (\varphi_s, dC_s)_{C_s} - \frac{1}{2}\int_0^1 (\varphi_s, \varphi_s)_{C_s}\, ds\right] : \varphi \in \Phi \right\}$$

Then $D(\Phi)$ is a closed, convex and uniformly integrable subset of $L^1(P)$.

The results described above can be applied to M-valued stochastic systems obtaining the existence of optimal controls for a large class of control problems as well as the existence of saddle points for certain stochastic differential games. The solution of the stochastic system is defined by the transformation of the Wiener measure as described in Proposition 2.

Theorem 3: Consider the M-valued stochastic control system described by

$$dX(t) = f(t,X,u(t,X))\, dt + d\widetilde{B}(t)$$

where $(\widetilde{B}(t))_{t\in[0,1]}$ is an M-valued Brownian motion such that $\widetilde{B}(0) \equiv a$, $t \in [0,1]$ and $X_0 \equiv a$. Suppose that f satisfies C1-C5. Let $L:C_a \to \mathbb{R}$ be a bounded S-measurable function.

Define

$$J(u) = E\left\{ L(X) \exp\left[\int_0^1 (g_u(s), dC(s))_{C(s)} - \frac{1}{2}\int_0^1 (g_u(s), g_u(s))_{C(s)}\, ds \right] \right\}$$

where g_u is the drift corresponding to $f(\cdot,\cdot,u)$. Then there is an optimal control $u^* \in \mathcal{U}$ such that

$$J(u^*) \leq J(u) \qquad \forall u \in \mathcal{U}.$$

A result for the existence of a saddle point to a stochastic differential game that is similar to Theorem 5 in [4] can also be obtained.

References

1. Beneš, V. E., Existence of optimal strategies based on specified information, for a class of stochastic deeision problems, SIAM J. Control, 8 (1970) 179-188.

2. Beneš, V. E., Existence of optimal stochastic control laws, SIAM J. Control 9 (1971)

3. Dudley, R. M., Feldman, J., LeCam L., On seminorms and probabilities, and abstract Wiener spaces, Ann. of Math. 93 (1971), 390-408.

4. Duncan, T., Varaiya, P., On the solutions of a stochastic control system, SIAM J. Control 9 (1971), 354-371.

5. Duncan, T., Varaiya, P., On the solutions of a stochastic control system II, to appear in SIAM J. Control.

6. Eells, J., On the geometry of functions spaces, Symp. Inter. de Top. Alg. Mexico 1956 (1958), 303-308.

7. Eells, J., Elworthy, K. D., Wiener integration on certain manifolds. Some problems in Non-linear Analysis, C.I.M.E. 1970.

8. Girsanov, I. V., On transforming a certain class of stochastic processes by absolutely continuous substituion of measures, Theor. Probability Appl. 5 (1960), 285-301.

9. Gross, L., Abstract Wiener spaces, Proc. Fifth Berkeley Symposium on Math. Stat. and Prob., University of California Press, 1965.

10. Itô, K., The Brownian motion and tensor fields on Riemannian manifold, Proc. Int. Cong. Math., Stockholm 1963, 536-539.

11. McKean, H. P., Brownian motions on the 3-dimensional rotation group, Mem. Coll. Sci. Kyoto Univ. 33 (1960), 25-38.

12. McKean, H. P., Stochastic Integrals, Academic Press. New York 1969.

13. Kobayashi, S., Nomizu, K., Foundations of Differential Geometry V.I Interscience New York 1963.

14. Segal, I. E., Distributions in Hilbert space and canonical systems of operators, Trans. Amer. Math. Soc. 88 (1958), 12-41.

PROBLEMES DE CONTROLE STOCHASTIQUE A TRAJECTOIRES DISCONTINUES

F. BRODEAU

1. Modèle mathématique.

Dans [1] et [2] nous avons introduit et justifié un modèle mathématique adapté à l'étude de problèmes de contrôle stochastique lorsque le système contrôlé est soumis à des demandes, ou perturbations, présentant, en des instants aléatoires, des discontinuités de première espèce.

Nous utilisons le formalisme suivant, qui fait appel aux notions de mesures aléatoires de Poisson telles qu'elles sont définies dans [3] et [4] .

Tous les éléments aléatoires utilisés sont définis sur un même espace probabilisé $(\Omega, \mathcal{A}, P)$.

Soit T un nombre positif donné $(T>0)$.

On désigne par $\mathcal{B}_0$ la réunion, pour tous les nombres ε tels que $0<\varepsilon<1$, des tribus $\mathcal{B}_\varepsilon$, où, pour chaque ε , $\mathcal{B}_\varepsilon$ est la trace de la tribu de Borel $\mathcal{A}$ de $\mathbb{R}$ sur le sous-ensemble des éléments x de $\mathbb{R}$ vérifiant la condition : $\varepsilon \leq |x| \leq \frac{1}{\varepsilon}$.

Soit π une mesure positive donnée sur $(\mathbb{R}, \mathcal{A})$ prenant des valeurs finies sur l'ensemble des éléments bornés de $\mathcal{A}$ dont l'adhérence ne contient pas l'origine.

On note p la mesure aléatoire de Poisson dans $[0,T] \times \mathbb{R}$ pour laquelle, quels que soient t, t' éléments de $[0,T]$, $t \leq t'$, et A élément de $\mathcal{B}_0$, $p([t,t'] \times A)$ est une variable aléatoire de Poisson de paramètre $(t'-t)\pi(A)$.

La mesure aléatoire q sur $[0,T] \times \mathbb{R}$ utilisée dans le formalisme est définie par la condition :

$$q([t,t'] \times A) = p([t,t'] \times A) - (t'-t)\pi(A) \qquad (0 \leq t \leq t' \leq T, A \in \mathcal{B}_0)$$

$\mathbf{w}$ désigne un processus de Wiener sur $[0,T]$, et X_0 une variable aléatoire donnée.

On suppose que q, w et X_0 sont indépendants, et on note $\mathcal{A}_t$, pour tout élément t de $[0,T]$, la tribu engendrée par X_0 , $w(s) - w(o)$, $q([o,s] \times A)$, quels que soient s élément de $[0,T]$ et A élément de $\mathcal{B}_0$.

La classe C des commandes est l'ensemble des fonctions aléatoires mesurables réelles définies sur $[0,T]$, adaptée à $\{\mathcal{A}_t\}_{t \in [0,T]}$, et à valeurs dans un ensemble compact donné U de $\mathbb{R}$.

A toute commande Y , élément de C , on associe une fonction aléatoire X , représentant l'état d'un système, solution de l'équation :

$$X(t) = X_0 + \int_0^t f[u, X(u), Y(u)]\, du + \int_0^t g[u, X(u)]\, dw(u) + \ldots$$

1.1)

$$\dots + \int_0^t \int_{\mathbb{R}} h[u, X(u), v]\, q(du, dv)$$

où f, g, h sont des fonctions réelles données, supposées mesurables par rapport à leurs arguments.

On minimise sur C l'expression

1.2) $$J_Y = E \int_0^T F[X(t), Y(t), t]\, dt ,$$

considérée comme fonction de Y , X correspondant à Y par 1.1) ; F est une fonction réelle mesurable donnée.

Une condition nécessaire d'optimalité, du type théorème du maximum, est d'abord établie. Cette condition est appliquée pour préciser la nature des stratégies optimales. Enfin des hypothèses assurant l'existence, a priori, d'une solution sont données.

2. Hypothèses.

θ désigne une application de $\mathbb{R}^+$ dans $\mathbb{R}^+$ telle que : $\lim_{h\to 0} \theta(h) = 0$.

H1. X_o admet un moment d'ordre 4.

H2. $\forall t, t' \in [0,T]$, $\forall x, x' \in \mathbb{R}$, $\forall y, y' \in U$, $\exists K > 0$ tel que

$$|f(t,x',y) - f(t,x,y)|^2 + |g(t,x') - g(t,x)|^2 + \int_{\mathbb{R}} |h(t,x',u) - h(t,x,u)|^2 \pi(du)$$
$$\leq K|x-x'|^2$$

$$|f(t,x,y)|^2 + |g(t,x)|^2 + \int_{\mathbb{R}} |h(t,x,u)|^2 \pi(du) \leq K[1+|x|^2]$$

$$\int_{\mathbb{R}} |h(t,x,u) - h(t,x',u)|^4 \pi(du) \leq K|x-x'|^4$$

$$\int_{\mathbb{R}} |h(t,x,u)|^4 \pi(du) \leq K[1+|x|^4]$$

$$|F(x,y,t) - F(x',y',t')| \leq K[|x-x'| + |y-y'| + \theta(|t-t'|)][|x'| + |x| + 1]$$

H3. $f(t,x,y)$, $g(t,x)$, $h(t,x,v)$, $F(x,y,t)$ admettent des dérivées partielles par rapport à x notées y^1, g^1, h^1, F^1, respectivement.

De plus on a : $\forall t \in [0,T]$, $\forall x, x' \in \mathbb{R}$, $\forall y \in U$, $\exists K$ tel que

$$|f(t,x,y) - f(t,x',y) - (x-x') f^1(t,x,y)| \leq K|x-x'|^2$$

La même condition est imposée à g et à F .

Pour h on suppose que

$$h(t,x,u) - h(t,x',u) - (x-x')\, h^1(t,x,u) = (x-x')^2\, k(t,x,x',u)$$

où $$k(t,x,x',.) \in L^2(\mathbb{R}, \mathcal{A}, \pi)$$

Enfin on a $h^1(t,x,.) \in L^2(\mathbb{R},\mathcal{A},\pi) \cap L^4(\mathbb{R},\mathcal{A},\pi)$.

Des théorèmes d'existence de solutions d'équations différentielles stochastiques de [3] on déduit immédiatement, à l'aide de H1 et H2, que, pour tout élément Y de C , l'équation 1.1) admet une solution unique adaptée à $\{\mathcal{a}_t\}_{t\in[0,T]}$.

Il y a unicité si on considère comme équivalentes deux fonctions aléatoires coïncidant pour tout élément de $[0,T]$ presque-sûrement. Dans la classe d'équivalence ainsi définie on peut toujours choisir, ce qui sera fait dans toute la suite, un représentant X presque-sûrement à trajectoires continues à droite, donc mesurable. J_Y est alors défini par 1.2) à l'aide de X , ce qui a un sens, la valeur de J_Y ne dépendant pas, compte tenu de H2, du représentant mesurable choisi dans la classe d'équivalence.

X est une fonction aléatoire d'ordre 4, n'admettant pas, avec probabilité un, de discontinuités de deuxième espèce. De plus il existe une constante K , indépendante de Y , telle que : $\forall t,t' \in [0,T]$

$$E|X(t)|^4 \le K \quad , \quad E|X(t)-X(t')|^2 \le K|t-t'| \quad , \quad E|X(t)-X(t')|^4 \le K|t-t'| \quad .$$

Il faut remarquer que c'est uniquement dans le but de développer un calcul variationnel afin d'obtenir une condition nécessaire d'optimalité, que des hypothèses assurant l'existence d'une solution de 1.1) qui soit d'ordre 4 sont imposées.

3. Condition nécessaire d'optimalité.

Sous les hypothèses du paragraphe 2 on démontre le théorème suivant :

Théorème 3.1. : Soit Y^o un élément de C supposé optimal auquel correspond X^o par 1.1). Pour tout élément t de $[0,T]$ l'équation

$$3.1)\quad \begin{aligned} A(u,t) = 1 &+ \int_t^u A(v,t)\, f^1(v,X^o(v),Y^o(v))\,dv + \int_t^u A(v,t) g^1(v,X^o(v))\,dw(v) \\ &+ \int_t^u \int_{\mathbb{R}} A(v,t)\, h^1(v,X^o(v),s)\,q(dv,ds) \end{aligned}$$

admet une solution unique sur $[t,T]$.

Si $\psi(t)$ est la fonction aléatoire définie sur $[0,T]$ par

$$3.2)\qquad \psi(t) = -\int_t^T A(u,t)\; F^1(X^o(u),Y^o(u),u)\,du$$

on a la condition d'optimalité suivante :

pour presque tout t élément de $[0,T]$

$$3.3)\quad \begin{aligned} &E[F(X^o(t),\pi,t) - \psi(t)\; f(t,X^o(t),\pi \mid \mathcal{a}_t] \ge \qquad \text{p.s.} \\ &E[F(X^o(t),Y^o(t),t) - \psi(t)\; f(t,X^o(t),Y^o(t) \mid \mathcal{a}_t] \end{aligned}$$

<u>pour toute variable aléatoire π mesurable par rapport à $\mathcal{a}_t$ et à valeurs dans U</u> .

<u>Commentaires</u> -

- L'unicité de la solution de 3.1) est obtenue avec les mêmes conventions que dans le cas de 1.1).

- Sous des hypothèses assurant que l'on peut résoudre 3.1) en utilisant la formule de différentiation de Ito [3], on a

$$\psi(t) = -\int_t^T \{\exp[\int_t^u f^1(v,X^o(v),Y^o(v))dv - \frac{1}{2}\int_t^u g^1(v,X^o(v))^2 dv + \int_t^u g^1(v,X^o(v))dw(v)$$

$$+ \int_t^u \int_{\mathbb{R}} \mathrm{Log}\,|1+h^1(v,X^o(v),s)|\,q(dv,ds)]\}\, F^1(X^o(u),Y^o(u),u)du \quad .$$

- La formulation et les résultats obtenus s'étendent de façon immédiate au cas de fonctions aléatoires à valeurs dans des espaces de type $\mathbb{R}^n$.

- Un résultat de même nature est obtenu dans le cas de stratégies adaptées à l'état du système, c'est-à-dire de stratégies Y , telles que pour tout t

$$Y(t) = \phi(t,X(t)) \quad , \quad X \text{ correspondant à } Y \text{ par } 1.1),$$

si l'on impose à ϕ des conditions de la forme

$$|\phi(t,x) - \phi(t,x')| \leq |x-x'| \quad .$$

- Si $h \equiv 0$, le modèle se réduit à un modèle classique et la condition d'optimalité coïncide avec celle obtenue dans [5], où toutefois le modèle est d'autre part enrichi par l'imposition de contraintes sur l'état du système.

- Les techniques récentes de démonstration de conditions nécessaires d'optimalité pour les problèmes de contrôle consistent à interpréter ces problèmes comme des problèmes d'optimisation auxquels on applique des résultats très généraux dûs à Neustadt [6], ainsi dans [7] pour le cas déterministe et [5] pour un cas stochastique. Il nous a semblé difficile d'employer une telle démarche, compte tenu du fait que les trajectoires présentent des discontinuités, et que des conditions peu réalistes semblent devoir être imposées aux stratégies. En conséquence nous avons préféré utiliser une technique directe de démonstration. Cette démonstration étant longue nous nous contentons d'en indiquer les principales étapes. Une rédaction détaillée doit être prochainement publiée.

<u>Indications sur la démonstration du théorème</u> -

Le lemme suivant, adaptation d'un résultat dû à Kushner [5], se révèle essentiel.

<u>Lemme 3.1.</u> -

1. <u>Il existe un sous-ensemble négligeable Θ_1 de $[0,T]$ tel que, pour tout élément t de $[0,T]-\Theta_1$</u>

$$\underline{\lim_{\tau\to 0^+} \frac{1}{\tau} \int_t^{t+\tau} \{f[u,X^o(u),Y^o(u)] - f[t,X^o(t),Y^o(t)]\}\, du = 0 \quad , \text{ p.s.}}$$

2. Il existe un sous-ensemble négligeable Θ_2 de $[0,T]$ tel que, pour tout élément t de $[0,T]-\Theta_2$, pour toute variable aléatoire π à valeurs dans U ,

$$\underline{\lim_{\tau\to 0^+} \frac{1}{\tau} \int_t^{t+\tau} \{f[u,X^o(u),\pi] - f[t,X^o(t),\pi]\}\, du = 0 \quad , \quad \text{p.s.}}$$

L'ensemble négligeable de $[0,T]$ pour lequel la condition 3.3) n'est pas satisfaite, contient, entre autres ensembles négligeables, Θ_1 et Θ_2 .

On définit la stratégie Y^p , à partir de Y^o , de la façon suivante. Soit t_1 élément de $[0,T[$, et $\tau>0$, tel que $t_1+\tau \le T$. Soit, d'autre part, π une variable aléatoire mesurable par rapport à $\mathcal{a}_{t_1}$ et à valeurs dans U . On pose

$$Y^p(t) = \begin{cases} Y^o(t) & , \ t \in ([t_1,t_1+\tau[)^c \\ \pi & , \ t \in [t_1,t_1+\tau[\end{cases} .$$

Le principe de la démonstration consiste à montrer que $\frac{1}{\tau}(J_Y - J_{Y^p})$ admet une limite, nécessairement positive ou nulle, lorsque τ tend vers 0 .

X^p désigne le processus associé à Y^p par 1.1).

Un lemme préliminaire étudie le comportement de $X^o(t) - X^p(t)$.

Lemme 3.2. - Il existe une constante $H>0$ telle que, pour tout élément t de $[t_1,T]$, on ait :

$$\underline{E|X^p(t) - X^o(t)|^4 \le H\tau^4} \quad .$$

Etude de $J_{Y^p} - J_{Y^o}$.

On pose $J_{Y^p} - J_{Y^o} = \alpha(t_1,\tau) + \beta(t_1,\tau)$, où

$$\left|\begin{aligned} \alpha(t_1,\tau) &= E\int_{t_1}^{t_1+\tau} \{F[X^p(t),Y^p(t),t] - F[X^o(t),Y^o(t),t]\}\, dt \\ \beta(t_1,\tau) &= E\int_{t_1+\tau}^{T} \{F[X^p(t),Y^p(t),t] - F[X^o(t),Y^o(t),t]\}\, dt \quad . \end{aligned}\right.$$

Lemme 3.3. - On a

$$\underline{\alpha(t_1,\tau) = \tau E\{F[X^o(t_1),\pi,t_1] - F[X^o(t_1),Y^o(t_1),\pi]\} + R(t_1,\tau)} \quad ,$$

où, pour presque tout t_1 élément de $[0,T[$, et ce indépendamment de π ,

$$\underline{\lim_{\tau\to 0^+} R(t_1,\tau) = 0} \quad .$$

$A(.,.)$ désignant la résolvante définie par 3.1), la linéarisation de l'équation 1.1) conduit au résultat suivant

Lemme 3.4. - Pour tout t, $t \geq t_1 + \tau$, on a

$$X^p(t) - X^o(t) = \tau A(t,t_1)\{f[t_1, X^o(t_1), \pi] - f[t_1, X^o(t_1), Y^o(t_1)]\} + S(t_1,\tau) ,$$

où, pour presque tout t_1, et ce indépendamment de π,

$$\lim_{\tau \to 0^+} \frac{1}{\tau^2} E|S(t_1,\tau)|^2 = 0 .$$

Après introduction de ψ, défini par 3.2) et utilisation du lemme 3.4 on obtient le lemme :

Lemme 3.5. - On a :

$$\beta(t_1,\tau) = -\tau E\{\psi(t_1)[f(t_1, X^o(t_1), \pi) - f(t_1, X^o(t_1), Y^o(t_1)]\} + T(t_1,\tau)$$

où, pour presque tout t_1, et ce indépendamment de π,

$$\lim_{\tau \to 0^+} \frac{1}{\tau} T(t_1,\tau) = 0 .$$

Des lemmes 3.3 et 3.5 on déduit immédiatement la condition nécessaire d'optimalité suivante :

$$E[F(X^o(t_1), \pi, t_1) - \psi(t_1)\, f(t_1, X^o(t_1), \pi)] \geq$$
$$E[F(X^o(t_1), Y^o(t_1), t_1) - \psi(t_1)\, f(t_1, X^o(t_1), Y^o(t_1))]$$

valable pour presque tout t_1.

La condition 3.3) s'obtient alors, de façon classique, par l'absurde en utilisant les propriétés des espérances conditionnelles.

4. Applications du théorème 3.1.

Nous commençons par dégager des conditions de régularité, en moyenne quadratique, des stratégies optimales.

L'inégalité 3.3) se met en fait sous la forme :

$$\begin{aligned} &F[X^o(t), \pi, t] - E[\psi(t)\,|\,\mathcal{A}_t].f[t, X^o(t), \pi] \geq \\ &F[X^o(t), Y^o(t), t] - E[\psi(t)\,|\,\mathcal{A}_t].f[t, X^o(t), Y^o(t)] \end{aligned} \qquad \text{p.s.}$$

Un rôle essentiel est ainsi joué par le processus $V(t) = E[\psi(t)\,|\,\mathcal{A}_t]$ pour lequel nous dégageons des conditions de régularité.

Lemme 4.1. - Il existe une constante $H>0$ telle que, quels que soient t, t' éléments de $[0,T]$,

$$E|\psi(t)-\psi(t')|^2 \leq K|t-t'|^{1/2} .$$

Ce lemme, de démonstration élémentaire, permet d'obtenir la propriété suivante.

Propriété 4.1. - En tout élément t de $]0,T[$, $V(t)$ admet, en moyenne quadratique, $E[\psi(t)|\mathcal{Q}_t-]$ et $E[\psi(t)|\mathcal{Q}_t+]$ pour limites à gauche et à droite, respectivement.

Soit en effet $\{t_n\}_{n\in\mathbb{N}}$ une suite croissante d'éléments de $[0,T]$ convergeant vers t. On a

$$E[\psi(t_n)|\mathcal{Q}_{t_n}] - E[\psi(t)|\mathcal{Q}_t-] = (E[\psi(t_n)|\mathcal{Q}_{t_n}] - E[\psi(t)|\mathcal{Q}_{t_n}])$$
$$+ (E[\psi(t)|\mathcal{Q}_{t_n}] - E[\psi(t)|\mathcal{Q}_t-]) .$$

Or, d'une part

$$E|E\{\psi(t_n)-\psi(t))|\mathcal{Q}_{t_n}\}|^2 \leq E|\psi(t_n)-\psi(t)|^2 \leq K|t_n-t|^{1/2} ,$$

d'après le lemme 4.1., d'autre part, d'après un résultat classique concernant les martingales ([8] par exemple) $E[\psi(t)|\mathcal{Q}_{t_n}] - E[\psi(t)|\mathcal{Q}_t-]$ tend vers 0, avec probabilité un, puisque, pour t fixé, $\{E[\psi(t)|\mathcal{Q}_{t_n}]\}_{n\in\mathbb{N}}$ est une martingale.

Cette martingale étant, de plus, fermée à droite par $E[\psi(t)|\mathcal{Q}_t]$, variable aléatoire du second ordre, un autre résultat classique permet d'affirmer que $E[\psi(t)|\mathcal{Q}_{t_n}]$ converge en moyenne quadratique vers $E[\psi(t)|\mathcal{Q}_t-]$.
Il est alors clair que $\lim_n \text{m.q. } V(t_n) = E[\psi(t)|\mathcal{Q}_t-]$, et le résultat de la propriété concernant la limite à gauche s'en déduit immédiatement ; par des techniques analogues on démontre la résultat concernant la limite à droite. En utilisant un résultat classique de [8] on en déduit la propriété

Propriété 4.2. - Il existe un sous-ensemble T_o de $[0,T]$, au plus dénombrable, tel que, pour tout t élément de $[0,T]-T_o$, $V(t^+) \overset{p.s.}{=} V(t^-) \overset{p.s.}{=} V(t)$.

D'après un autre résultat classique on est d'ailleurs assuré de l'existence d'une modification de $V(t)$ qui est mesurable.
Supposons alors que, quels que soient x, v réels, $t\in[0,T]$, l'expression

$$F[x,y,t] - f[t,x,y]v ,$$

considérée comme fonction de y, admette sur U un minimum unique $\phi(t,x,v)$.

Si, de plus, il existe une constante $K>0$ telle que :

$$\forall t,t' \in [0,T] \quad , \quad x,x',v,v' \in \mathbb{R}$$
$$|\phi(t,x,v) - \phi(t',x',v')| \leq K\{\theta(|t-t'|) + |x-x'| + |v-v'|\} \quad ,$$

on a le résultat, obtenu à l'aide des propriétés précédentes, de la continuité en moyenne quadratique de X^o , et du théorème 3.1. :

<u>Propriété 4.3.</u> - <u>Soit Y^o une stratégie optimale. Pour tout élément t de $]0,T[$, $Y^o(t)$ admet, en moyenne quadratique, une limite à gauche et une limite à droite. Il existe, de plus, un ensemble T_o de $[0,T]$, au plus dénombrable tel que, pour tout élément t de $[0,T] - T_o$,</u>

$$Y^o(t^+) \overset{p.s.}{=} Y^o(t^-) \overset{p.s.}{=} Y^o(t) \quad .$$

Un tel Y^o est bien mesurable.

<u>Remarque</u> : Dans le cas où $h \equiv 0$, les stratégies sont en fait uniquement adaptées au processus de Wiener w que l'on peut toujours supposer standard, au sens de [9]. Il est alors légitime de supposer que la famille $\{\mathcal{Q}_t\}_{t\in T}$ est continue à droite. Dans ce cas, pour tout t élément de $[0,T[$, $Y^o(t)$ est continu, en moyenne quadratique, à droite.

<u>Cas particuliers</u> -

Si les fonctions f,g,h sont de la forme

$$f(t,x,y) = \alpha x + \beta y \quad , \quad g(t,x) = \gamma x \quad , \quad h(t,x,u) = \delta(u)x \quad ,$$

où α, β, γ sont des constantes, et δ un élément de $L^2(\pi) \cap L^4(\pi)$, les hypothèses concernant ces fonctions sont satisfaites.

Toutes les hypothèses formulées sont alors vérifiées si l'on considère des fonctions F de la forme :

$$F(x,y,t) = ax + by \quad , \quad a,b \in \mathbb{R} \quad , \quad \text{ou} \quad F(x,y,t) = ax^2 + by^2 \quad , \quad a,b \in \mathbb{R}^+ \quad .$$

Des types plus généraux peuvent d'ailleurs être envisagés.

Dans le cas où $F(x,y,t) = ax + by$, le théorème 3.1. permet d'obtenir une solution Y^o du problème :

$$Y^o(t) = \begin{cases} \ell & \text{si } b - \beta V(t) > 0 \\ m & \text{si } b - \beta V(t) < 0 \\ \text{indéterminé} & \text{si } b - \beta V(t) = 0 \end{cases} \qquad \text{pour presque tout } t \ .$$

si $U = [\ell, m]$.

Cette solution peut d'ailleurs être obtenue explicitement si l'on impose, de plus, la condition :

$$\forall u \in \mathbb{R} \ , \ \delta(u) \geq k > -1 \ , \quad k \text{ constante donnée.}$$

On a en effet alors

$$\psi(t) = -a \int_t^T \exp[(\alpha - \frac{1}{2}\gamma^2)(u-t) . \exp[\gamma(w(u) - w(t))] \ldots$$

$$\ldots \exp[\int_{\mathbb{R}} \text{Log}(1+\delta(v)) \, q([t,u],dv)] \, du \ ,$$

et

$$V(t) = -a \int_t^T \exp[(\alpha - \frac{1}{2}\gamma^2)(u-t) . E\{\exp[\gamma(w(u) - w(t))]\} \ldots$$

$$\ldots E\{\exp[\int_{\mathbb{R}} \text{Log}(1+\delta(v)) \, q([t,u],dv)]\} \, du \ .$$

$V(t)$ est ainsi une fonction réelle qu'il est possible d'obtenir par intégration : d'une part $w(u) - w(t)$ suit la loi normale $\mathcal{N}(o, u-t)$, ce qui permet le calcul de $E\{\exp[\gamma(w(u) - w(t))]\}$, d'autre part on montre que

$$E\{\exp[\int_{\mathbb{R}} \text{Log}(1+\delta(v)) q([t,u],dv)]\} = \exp[(u-t)\int_{\mathbb{R}} \{\delta(v) - \text{Log}(1+\delta(v))\} \pi(dv) \ .$$

5. Existence d'une solution optimale.

Sous des hypothèses relativement larges on peut conclure à l'existence, a priori, d'une solution optimale.

Supposons que

$$f(t,x,y) = \alpha(t)x + \beta(t)y \ , \ g(t,x) = \gamma(t)x \ , \ h(t,x,u) = \delta(t,u)x \ ,$$

où les fonctions $\alpha, \beta, \gamma, \delta$ sont telles que f, g, h vérifient les hypothèses imposées au paragraphe 2 ; ces hypothèses peuvent d'ailleurs être affaiblies compte tenu du caractère linéaire de l'équation d'évolution.

On suppose que F, outre les conditions imposées au paragraphe 2 est telle que, pour tout t, $F(x,y,t)$ est convexe par rapport à (x,y).

On considère la classe C des stratégies comme un sous-ensemble de l'espace $\mathcal{H}$ des classes de fonctions aléatoires $Y(t)$ définies et mesurables sur $(\Omega \times [0,T], \mathcal{A}\otimes\mathcal{B}_T)$, $\mathcal{B}_T$ tribu de Borel de $[0,T]$, vérifiant la condition :

$$\int_0^T E|Y(t)|^2 \, dt < +\infty \ .$$

J peut être interprété comme une application de C dans $\mathbb{R}$.

Les deux lemmes suivants précisent la nature de C et de J .

Lemme 5.1. - C est faiblement compact.

Lemme 5.2. - J est une application convexe et continue de C dans $\mathbb{R}$, $\mathcal{H}$ étant muni de la topologie forte.

J est ainsi faiblement semi-continue inférieurement et atteint donc son minimum sur C . D'où

Théorème : sous les hypothèses faites, il existe une solution optimale.

Sous des hypothèses supplémentaires assurant que J est strictement convexe on est d'ailleurs assuré de l'unicité, au sens de $\mathcal{H}$, de la solution.

6. Etude des stratégies optimales.

Il peut être jugé peu réaliste d'adapter les stratégies à w et q , et de nombreux auteurs considèrent en fait des stratégies adaptées à l'état du système. Nous montrons que sous certaines hypothèses, à l'aide de résultats déduits du théorème 3.1., les stratégies optimales dans le cas de notre formalisme sont en fait "peu différentes" de stratégies adaptées, en chaque instant, à l'état du système à cet instant.
Les hypothèses sont celles du paragraphe 5, auxquelles on ajoute les conditions assurant au paragraphe 4 la validité de la propriété 4.3.
Y^o désigne une solution optimale à laquelle est associée X^o .
Soit $t_1 \in]0,T[$; on désigne, pour tout $t > t_1$, par $\mathcal{A}_t^1$ la tribu engendrée par $X^o(t_1)$, $w(s) - w(t_1)$, $q([t_1, s] \times A)$, quels que soient s élément de $[t_1, t]$ et A élément de $\mathcal{B}_o$.
Soit $Y^1(u)$ la fonction aléatoire définie par la condition

$$Y^1(u) = \begin{cases} Y^o(u) & \text{si} \quad u < t_1 \\ E[Y^o(u) \mid \mathcal{A}_u^1] & \text{si} \quad u \geq t_1 \end{cases}$$

Lemme 6.1. - Il existe une modification de $Y^1(u)$ qui est élément de C .

La mesurabilité d'une modification de $Y^1(u)$ découle en effet, comme au 4, du fait que, en dehors d'un ensemble de $[0,T]$ au plus dénombrable, Y^o est continu en moyenne quadratique. C'est cette modification que nous utilisons dans la suite. On désigne par $X^1(u)$ la solution de 1.1) qui lui est associée.
Compte tenu du caractère linéaire de 1.1) on peut alors écrire :

$$X^o(t) = X^o(t_1)\ \exp\Big[\int_{t_1}^{t} (\alpha(u) - \tfrac{1}{2}\gamma^2(u))du + \int_{t_1}^{t} \gamma(u)dw(u) + \int_{t_1}^{t}\int_{\mathbb{R}} \mathrm{Log}(1+\delta(u,v))q(du,dv)\Big]$$

6.1)

$$+ \int_{t_1}^{t} \beta(s)Y^o(s)\ \exp\Big[\int_{1}^{t}(\alpha(u)-\tfrac{1}{2}\gamma^2(u))du + \int_{s}^{t}\gamma(u)dw(u) + \int_{s}^{t}\int_{\mathbb{R}} \mathrm{Log}(1+\delta(u,v))q(du,dv)\Big]\,ds$$

Soit $\tilde{X}^1(u)$ la fonction aléatoire définie par la condition

$$\tilde{X}^1(u) = \begin{cases} X^o(u) & \text{si} \quad u < t_1 \\ E[X^o(u)\,|\,\mathcal{A}^1_u] & \text{si} \quad u \geq t_1 \ . \end{cases}$$

<u>Lemme 6.2.</u> - <u>Il existe une modification de $X^1(u)$ telle que, pour tout u</u> ,

<u>$X^1(u) = \tilde{X}^1(u)$ p.s.</u>

Ce lemme découle de 6.1) et de la possibilité, sous les hypothèses faites, d'intervertir $\int_{t_1}^{t}$ et $E[\,.\,|\mathcal{A}^1_t]$. Si $\mathcal{A}_{[u,t]}$ désigne pour tout $u < t$, la tribu engendrée par $w(s) - w(u)$, $q([u,s]\times A)$, quels que soient s élément de $[u,t]$ et A élément de $\mathcal{B}_o$, on utilise ensuite le fait que $\mathcal{A}^1_t = \mathcal{A}^1_u \vee \mathcal{A}_{[u,t]}$, et que $\mathcal{A}_{[u,t]}$ est conditionnellement indépendante de $\mathcal{A}_u$ par rapport à $\mathcal{A}^1_u$; à l'aide de ces propriétés on en déduit que, pour tout $u \geq t_1$,

$$E[Y^o(u)\,|\,\mathcal{A}^1_t\,] = E[Y^o(u)\,|\,\mathcal{A}^1_u] \ .$$

L'utilisation du lemme 6.2 et de l'inégalité de Jensen conduit à l'inégalité : $\forall t \geq t_1$

$$E[F(t,X^o(t),Y^o(t))\,|\,\mathcal{A}^1_t\,] \geq E[F(t,X^1(t),Y^1(t))] \qquad \text{p.s.}$$

D'où $\forall t \geq t_1$

$$E[F(t,X^o(t),Y^o(t)] = E[F(t,X^1(t),Y^1(t))]$$

et finalement $J_{Y^o} = J_{Y^1}$.

On peut énoncer la propriété

<u>Propriété 6.1.</u> - <u>Sous les hypothèses faites</u>, $J_{Y^o} = J_{Y^1}$. <u>Si, de plus la solution du problème est unique on a, pour presque tout $t \geq t_1$, $Y^o(t) = Y^1(t)$ p.s.</u>

Dans ces conditions on peut interpréter $X^o(t_1)$ comme un résumé exhaustif de l'évolution du système jusqu'au temps t_1 , ce qui donne un sens à l'affirmation du début de ce paragraphe. On obtient ainsi une extension au cas d'évolution continue par rapport au temps d'un résultat connu pour les systèmes à évolution discrète.

A titre de conclusion on peut dire que, bien que l'on puisse douter de l'utilité de théorèmes de maximum pour les problèmes de contrôle stochastique, de tels théorèmes permettent toutefois de préciser de façon non négligeable la nature des stratégies optimales.

BIBLIOGRAPHIE

[1] F. BRODEAU Problèmes de commande optimale stochastique à trajectoires discontinues. Publ. Math. Univ. Bordeaux I 1973-1974 Fasc. 2

[2] F. BRODEAU Problèmes de commande optimale stochastique à trajectoires discontinues. C.R. Acad. Sc. Paris t.278 (1974).

[3] A.V. SKOROHOD Studies in the theory of random processes. Addison-Wesley 1965.

[4] I.I. GIHMAN et A.V. SKOROHOD Stochastic differential equations. Springer Verlag 1972.

[5] H.J. KUSHNER Necessary conditions for continuous parameter stochastic optimization problems. S.I.A.M. Journal Control Vol. 10 N° 3 Aug. 1972.

[6] L.W. NEUSTADT An abstract variational theory with applications to a broad class of optimization problems. S.I.A.M. Journal Control Vol. 4 1966.

[7] M. CANON, C. CULLUM, E. POLAK Theory of Optimal Control and mathematical programming. Mac-Graw-Hill 1970.

[8] J.L. DOOB Stochastic Processes. J. Wiley 1952.

[9] R.M. BLUMENTHAL et R.K. GETOOR - Markov Processes and Potential Theory. Academic Press 1968.

Théorie du Potentiel et Contrôle des Diffusions Markoviennes

Jean-Michel BISMUT, Ingénieur du Corps de Mines
191, rue d'Alésia, 75014 Paris / France

Introduction

L'objet de ce travail est de donner une présentation rapide des principales méthodes de théorie du potentiel utilisées dans [2], qui nous ont permis d'obtenir des résultats d'existence très généraux dans des problèmes de contrôle optimal des diffusions. Afin d'éviter d'alourdir l'exposé, nous ferons des hypothèses simplificatrices chaque fois que cela sera nécessaire, en indiquant toutefois les généralisations possibles.

Nous utilisons largement les pricipaux résultats de Stroock et Varadhan exposés dans [6] et [7]. A partir de ces résultats, nous avions déjà donné dans notre Thèse [1] des résultats d'existence de contrôles optimaux dans certains cas particuliers. Nous avons depuis étendu ces résultats dans [2] sous des hypothèses plus générales.

Soit en effet U un espace compact métrisable. (b,L) désigne une fonction définie sur $[0,+\infty[\times R^n\times U$ à valeurs dans $R^n\times R$, σ une fonction définie sur $[0,+\infty[\times R^d$ à valeurs dans $R^d\otimes R^d$, β un mouvement brownien d-dimensionnel.

On considère l'équation différentielle stochastique:

$$(1)\begin{cases} dx = b(t,x_t,u(t,x_t))dt + \sigma(t,x_t)\cdot d\beta & t\geq s \\ x_s = x \end{cases}$$

où u est une fonction définie sur $[0,+\infty[\times R^d$ à valeurs dans U.

On fera toutes les hypothèses nécessaires pour que (1) ait un sens.

p étant une constante >0, on veut trouver une fonction u minimisant le critère:

$$(2)\qquad E\int_0^{T_A} e^{-pt}\,L(t,x_t,u(t,x_t))\,dt$$

où A est un ensemble inclus dans $[0,+\infty[\times R^d$ et où T_A est le temps où le processus (t,x_t) atteint la cible A.

On pose:

$$(3)\qquad K(t,x) = \{(b(t,x,u),L(t,x,u)) \mid u\in U\}$$

On montre alors qu'on peut se ramener à chercher (b,L) section de la multiapplication K--nous ne nous étendons pas ici sur les questions de mesurabilité--c'est-à dire tel que:

$$(4)\qquad (b(t,x),L(t,x))\in K(t,x)$$

minimisant

$$(5)\quad E\int_0^{T_A} e^{-\rho t} L(t,x_t)\,dt$$

sur l'ensemble des sections de K.

Deux autres questions se posent alors:

a)Le "contrôle" trouvé est-il meilleur que tout autre contrôle non anticipatif, c'est à dire dépendant à l'instant t non seulement de l'état présent du système x_t mais aussi de toutes les valeurs passées de x?

b)Peut-on choisir un contrôle qui convienne simultanément pour tous les points de départ (s,x)?

Pour répondre à toutes ces questions on utilise une méthode généralisable à tous les problèmes de contrôle de diffusions.

I-Rappels sur les diffusions.

Ω est l'espace des fonctions continues définies sur R^+ à valeurs dans R^n. Pour $(s,t)\in R^+\times R^+$ avec $t\geq s$, M_t^s est la sous-tribu de Ω engendrée par $(x_u \mid s\leq u\leq t)$

a est une application définie sur $R^+\times R^d$ à valeurs dans $R^d\otimes R^d$, telle que:

-a est continue.

-a est à valeurs définies positives.

-a est uniformément bornée en norme par une constante A.

b est une fonction borélienne bornée définie sur $R^+\times R^d$ à valeurs dans R^d.

$P_{(s,x)}$ désigne la mesure définie sur Ω, solution du problème des martingales de Stroock et Varadhan associée au couple (a,0) avec (s,x) comme point de départ (pour cette définition, voir [6])

$Q^b_{(s,x)}$ désigne la mesure définie sur Ω, solution du problème des martingales de Stroock et Varadhan associée au couple (a,b) avec (s,x) comme point de départ.

Alors, par le Théorème 6.2 de [6], sur M_t^s, $Q^b_{(s,x)}$ a une densité par rapport à $P_{(s,x)}$ donnée par la formule:

$$(6)\quad \frac{dQ^b_{(s,x)}}{dP_{(s,x)}}\bigg|_{M_t^s} = \exp\left\{\int_s^t \langle b(u,x_u), a^{-1}(u,x_u)\,dx_u\rangle - \frac{1}{2}\int_s^t \langle b(u,x_u), a^{-1}(u,x_u)\,b(u,x_u)\rangle\,du\right.$$

On vérifie alors que $Q^b_{(s,x)}$ ne dépend que de la classe d'équivalence de b pour la mesure de Lebesgue sur $R^+\times R^d$:cela résulte de (6) et de la continuité absolue des potentiels associés au processus de Feller P défini par les mesures $P_{(s,x)}$.

Alors on peut trouver un mouvement brownien unique β^b défini sur $(\Omega, M^s_t, Q^b_{(s,x)})$ tel que si σ désigne la racine carrée positive de a, on a:

(7) $dx = b(t,x_t)dt + \sigma(t,x_t).d\beta^b$

Cela résulte de [6] Corollaire (3.2). On se reportera à [2] pour un résultat similaire quand b est progressivement mesurable borné.

On a alors le résultat suivant:

Proposition 1: Soit $\{L_n\}_{n\in N}$ une suite d'éléments de $L_\infty(R^+ \times R^d)$ convergeant pour la topologie $\sigma(L_\infty, L_1)$ vers $L \in L_\infty(R^+ \times R^d)$ Alors la suite de variables aléatoires

$$\int_0^t L_n(\sigma, x_\sigma)\, d\sigma$$

converge vers $\int_0^T L(\sigma, x_\sigma)d\sigma$ en probabilité pour la mesure $P_{(s,x)}$

Preuve: C'est la proposition IV-4 ([2])

Cette proposition nous permet dans [1] et [2] de montrer que quand b_n converge vers b pour une topologie convenable, $Z_T^{b_n}$ densité de $Q^{b_n}_{(s,x)}$ par rapport à $P_{(s,x)}$ sur M^s_t converge faiblement vers Z_T^b, densité de $Q^b_{(s,x)}$ par rapport à $P_{(s,x)}$. On a en effet:

Théorème 1: Si la suite b_n d'éléments de $L_\infty(R^+ \times R^d)$ converge pour la topologie $\sigma(L_\infty, L_1)$ vers b, alors $Z_T^{b_n}$ converge faiblement vers Z_T^b dans l'ensemble des variables aléatoires intégrables par rapport à la mesure $P_{(s,x)}$

Preuve: Cela résulte du Théorème I-1 de [1] ou du Théorème IV-3 de [2].

Pour $C = (b, L)$, on désigne par $E^b_{(s,x)}$ l'opérateur espérance associée à la mesure $Q^b_{(s,x)}$ et on note V_C la fonction:

$$(8)\quad V_C(s,x) = e^{ps} E^b_{(s,x)} \int_s^{+\infty} e^{-p\sigma} L(\sigma, x_\sigma)\, d\sigma$$

Théorème 2 : L'application $C \to V_C(s,x)$ est continue pour la topologie $\sigma(L_\infty, L_1)$

Preuve : p étant une constante strictement positive, la fonction $C \to V_C(s,x)$ est limite uniforme sur les bornés de $L_\infty(R^+ \times R^d)$ de $C \to e^{ps} E^b_{(s,x)} \int_s^T e^{-p\sigma} L(\sigma, x_\sigma)d\sigma$ quand $T \to +\infty$.

Or

$$(9)\quad E^b_{(s,x)} \int_0^T e^{-p\sigma} L(\sigma, x_\sigma)d\sigma = E^0_{(s,x)} Z_T^b \int_0^T e^{-p\sigma} L(\sigma, x_\sigma)d\sigma$$

Le Théorème est alors démontré par application de la Proposition 1 et du Théorème 1.

Remarquons également que les topologies fines relatives aux différents processus Q^b sont nécessairement identiques.

Les fonctions V_c sont donc nécessairement finement continues--nous ne préciserons plus par rapport à quel processus--puisque ce sont des potentiels.

Si A est un ensemble borélien de $R^+ \times R^d$ on pose:

$$T_A = \inf\{t > s, (t, x_t) \in A\} \text{ et } V'_c(s,x) = e^{ps} E^b_{(s,x)} \int_0^{T_A} e^{-p\sigma} L(\sigma, x_\sigma) d\sigma$$

<u>Théorème 3</u> : L'application $C \to V'_c(s,x)$ est continue pour la topologie $\sigma(L_\infty, L_1)$. De plus , pour tout $C \in L_\infty(R^+ \times R^d)$ la fonction V'_c est finement continue.

<u>Preuve</u> : On vérifie immédiatement l'identité suivante:

$$(10) \quad V'_c(s,x) = V_c(s,x) - e^{ps} E^b_{(s,x)} e^{-pT_A} V_c(T_A, x_{T_A})$$

Or si $T \to \infty$ la suite de fonctions

$$(11) \quad c \to E^b_{(s,x)} 1_{T_A \leq T} e^{-pT_A} V_c(T_A, x_{T_A})$$

converge uniformément vers la fonction

$$E^b_{(s,x)} e^{-pT_A} V_c(T_A, x_{T_A})$$

Or la fonction (11) est M^0_T mesurable. Donc:

$$E^b_{(s,x)} 1_{T_A \leq T} e^{-pT_A} V_c(T_A, x_{T_A}) = E^0_{(s,x)} Z^b_T 1_{T_A \leq T} e^{-pT_A} V_c(T_A, x_{T_A})$$

On déduit alors du Théorème 1 et du Théorème 2 que la fonction (11)

est continue.

On en déduit bien que $C \to E^b_{(s,x)} e^{-pT_A} V_c(T_A, x_{T_A})$ est continue et donc que $C \to V'_c(s,x)$ est aussi continue.

De plus la Proposition (2.8) de II-[3] montre que la fonction

$$e^{ps} E^b_{(s,x)} e^{-pT_A} V_c(T_A, x_{T_A})$$

est p-excessive pour Q^b , donc finement continue.

On en déduit bien que V'_c est finement continue.

<u>II-Un Problème de contrôle des diffusions:le cas convexe.</u>

K désigne maintenant une multiapplication borélienne définie sur $R^+ \times R^d$ à valeurs dans $R^d \times R^+$, non vides, compactes et uniformément bornées.

$\mathcal{U}$ désigne l'ensemble des classes d'équivalence des sections boréliennes de K définies sur $R^+ \, R^d$ valeurs dans $R^d \times R^+$ pour la mesure de Lebesgue de $R^+ \times R^d$.

Si μ est une mesure de probabilité sur $R^+ \times R^d$, P_μ et Q^b_μ sont les mesures correspondant aux processus P et Q^b pour la mesure d'entrée μ

$\tilde{\mathcal{U}}_\mu$ désigne alors l'ensemble des classes d'équivalence des sections progressivement mesurables de $K(t, x_t)$ définies sur (Ω, P_μ)

à valeurs dans $R^d \times R^T$, pour la mesure $dt \otimes d\ell_\mu$ définie sur $R^+ \times \Omega$

Pour tout $C = (b,L)$ dans $\mathcal{L}$ (resp. dans $\tilde{\mathcal{L}}_\mu$) on pose:

$$(12) \quad V_C'(s,x) = e^{\rho s} E^b_{(s,x)} \int_s^{T_A} e^{-\rho\sigma} L(\sigma, x_\sigma)\, d\sigma$$

$$(\text{resp. } (12')) \quad V_C'(\mu) = E^b_\mu e^{\rho s} \int_s^{T_A} e^{-\rho s} L(\sigma,\omega)\, d\sigma$$

Définition 1 : On appelle problème de contrôle Q_μ (resp. $\bar{Q}_\mu$) la recherche de $C \in \mathcal{L}$ (resp. $\tilde{\mathcal{L}}_\mu$) minimisant la fonctionnelle:

$$C \longrightarrow \int V_C'(s,x)\, d\mu(s,x)$$

$$(\text{resp. } C \longrightarrow V_C'(\mu))$$

sur $\mathcal{L}$ (resp. $\tilde{\mathcal{L}}_\mu$)

Définition 2 : On appelle problème de contrôle Q la recherche de $C \in \mathcal{L}$ solution de tous les problèmes Q_μ.

On a alors le Théorème fondamental:

Théorème 4 : Le problème Q a une solution qui est aussi solution de tous les problèmes $\bar{Q}_\mu$.

Pour démontrer ce Théorème, nous allons faire provisoirement l'hypothèse que K est à valeurs convexes.

Pour toute mesure de probabilité μ sur $R^+ \times R^d$, on pose:

$$(13) \quad I_\mu(C) = \int e^{-\rho s} V_C'(s,x)\, d\mu(s,x)$$

Proposition 2 : La fonctionnelle $C \to I_\mu(C)$ possède un minimum sur $\mathcal{L}$.

Preuve : En effet, K étant à valeurs convexes, $\mathcal{L}$ est compact pour la topologie $\sigma(L_\infty, L_1)$. Par le Théorème 3, $C \to I_\mu(C)$ est nécessairement continu.

On va alors représenter le processus $1_{t<T_A} V_C'(t,x_t)$ quand $C \in \mathcal{L}$

Proposition 3 : Pour tout $C = (b,L) \in \mathcal{L}$, le processus

$$(14) \quad 1_{t<T_A} V_C'(t,x_t) - 1_{s<T_A} V_C'(s,x_s) - \int_{s\wedge T_A}^{t\wedge T_A} (L - \rho V_C')(\sigma, x_\sigma)\, d\sigma$$

peut s'écrire sous la forme: $M_{t\wedge T_A} - M_{s\wedge T_A}$

M étant une martingale de carré intégrable fonctionnelle additive pour le processus Q^b.

Preuve: Quand A est réduit à l'ensemble vide, ce résultat est simple. Quand A est un ensemble borélien quelconque, on se réfèrera à [3] T IV-2

Pour trouver des conditions nécessaires et suffisantes pour que $c \in \mathcal{U}$ minimise la fonctionnelle $c \to I_\nu(c)$, nous devons chercher à représenter la martingale de carré intégrable fonctionnelle additive M. Lorsque $\sigma = \sqrt{a}$ est uniformément lipchitzienne, on montre sans difficulté que M peut s'écrire sous la forme d'une intégrale stochastique par rapport à β^b. Dans le cas où on admet que a est seulement continu, la représentation de M est plus difficile. Dans [2], nous avons traité le cas général, qui exige des manipulations très lourdes d'espaces de martingales de carré intégrable fonctionnelles additives orthogonales au sens des martingales à β^b (pour cette définition, voir [4] p 127). L'introduction de ces espaces est également nécessaire pour traiter le cas des diffusions se réfléchissant sur un bord inélastique, cas que nous avons traité dans [2]. Enfin l'espace des martingales orthogonales à β^b apparaît naturellement dans les problèmes plus généraux de contrôle stochastique que nous avons traités dans [1].

Pour simplifier les démonstrations nous allons supposer que σ est uniformément lipchitzienne.

<u>Proposition 4</u> : Toute martingale de carré intégrable fonctionnelle additive pour Q^b M s'écrit:

$$(15) \qquad \int_s^t H(\sigma, x_\sigma)\, d\beta^b_\sigma$$

H étant une fonction borélienne définie sur $R^+ \times R^d$ à valeurs dans R^d telle que pour tout (s,t) $(t \geq s)$ on a :

$$(16) \qquad E^b_{(s,x)} \int_s^t |H(\sigma, x_\sigma)|^2 d\sigma < +\infty$$

<u>Preuve</u>: Supposons tout d'abord que b=0 . Dans ce cas, l'équation différentielle stochastique :

$$(17) \qquad \begin{cases} dx = \sigma(t,x) \cdot d\beta^0 \\ x_s = x \end{cases}$$

a une solution au sens de Ito. Il est alors immédiat de vérifier que x_t et β^0_t engendrent les mêmes tribus. Connaissant le résultat classique de Ito sur la représentation des martingales de carré intégrable pour le mouvement brownien (voir par exemple [4] p 135), on en déduit immédiatement le résultat dans ce cas particulier.

Dans le cas où b est quelconque, on vérifie immédiatement que, pour que M soit une martingale pour $Q^b_{(s,x)}$, il faut et il suffit que si Z^s_t désigne l'expression (6), $M_t Z^s_t$ soit une martingale pour la mesure $P_{(s,x)}$. Un calcul immédiat effectué dans [1] donne alors la représentation (15) .

Si l'on juxtapose les résultats de la Proposition 3 et de la la proposition 4, on peut représenter la martingale decarré intégrable fonctionnelle additive M sous la forme:

$$(18)\quad \int_0^t H_c(\sigma, x_\sigma)\cdot d\beta_\sigma^b$$

On démontre alors une condition nécessaire et suffisante pour que $c \in \mathcal{U}$ minimise I_ν

Proposition 5: Pour que $c \in \mathcal{U}$ minimise I_ν , il faut et il suffit que P_ν p.s., pour presque tout $t < T_A$, on ait:

$$(19)\quad L(t,x_t) + \langle H_c(t,x_t), \sigma^{-1}(t,x_t)\, b(t,x_t)\rangle = \min_{(b',L')\in K(t,x_t)} L' + \langle H_c(t,x_t), \sigma^{-1}(t,x_t)\, b'\rangle$$

Preuve: Par la corollaire 4.3 de [5] , on peut trouver une section borélienne c_1 de K telle que pour tout (t,x) dans $R^+ \times R^d$:

$$(20)\quad L_1(t,x) + \langle H_c(t,x), \sigma^{-1}(t,x)\, b_1(t,x)\rangle = \varphi_c(t,x)$$

$$\text{avec}\quad \varphi_c(t,x) = \min_{(b',L')\in K(t,x)} L' + \langle H_c(t,x), \sigma^{-1}(t,x)\, b'\rangle$$

En effet K est une multiapplication borélienne à valeurs compactes et convexes, et σ^{-1} est une application continue.

De (14) . on tire:

$$(20)\quad 1_{t<T_A} e^{-\rho t} V_c'(t,x_t) - 1_{s<T_A} e^{-\rho s} V_c'(s,x_s) = \int_{s\wedge T_A}^{t\wedge T_A} e^{-\rho u} \langle H_c(u,x_u), d\beta_u^b\rangle - \int_{s\wedge T_A}^{t\wedge T_A} e^{-\rho u} L(u,x_u)\,du$$

De plus, de (7) on tire:

$$(21)\quad \beta_t^b - \beta_s^b = \beta_t^{b_1} - \beta_s^{b_1} + \int_s^t (b_1 - b)(u,x_u)\,du$$

(ces calculs ne sont pas complètement rigoureux. Le lecteur se référera à [2] pour une démonstration plus complète)

On en déduit:

$$(22)\quad 1_{t<T_A} e^{-\rho t} V_c'(t,x_t) - 1_{s<T_A} e^{-\rho s} V_c'(s,x_s) = \int_{s\wedge T_A}^{t\wedge T_A} e^{-\rho u} \langle H_c(u,x_u), d\beta_u^{b_1}\rangle$$

$$- \int_s^t e^{-\rho u} L_1(u,x_u)\,du - \int_{s\wedge T_A}^{t\wedge T_A} e^{-\rho u}\left(L + \langle H_c, \sigma^{-1} b\rangle - \varphi_c\right)(u,x_u)\,du.$$

Soit R_n le temps d'arrêt défini par:

$$(23)\quad R_n = \inf\left\{ t \geq s \,;\ \int_s^t e^{-\rho u} |H_c(u,x_u)|^2 du = n \right\}$$

Nécessairement:

$$(24)\quad E_\nu^{b_1} \int_{s\wedge T_A}^{R_n\wedge T_A} \langle H_c(u,x_u), d\beta_u^{b_1}\rangle = 0$$

Donc:

$$(25)\quad 0 \leq E_\nu^{b_1} e^{-\rho R_n} V_c'(R_n, x_{R_n}) + E_\nu^{b_1} \int_s^{R_n\wedge T_A} e^{-\rho u}\left[L + \langle H_c, \sigma^{-1} b\rangle - \varphi_c\right)(u,x_u)\,du \leq I_\nu(c) - E_\nu^{b_1}\int_s^{R_n\wedge T_A} e^{-\rho u} L_1(u,x_u)\,du$$

Par la Théorème de Fatou on en déduit:

$$(26)\quad I_\nu(c) - I_\nu(c_1) \geq 0$$

Si la relation (19) n'était pas vérifiée, de (25) on tirerait:

$$(27)\quad I_\mu(c) - I_\mu(c_1) > 0$$

C'est impossible si c est optimal pour I_μ .

Inversement suppsosons la relation vérifiée. Soit $c' \in \mathcal{L}$.

On a alors nécessairement:

$$(28)\quad 1_{t<T_A} e^{-pt} V_c'(t,x_t) \geq 1_{s<T_A} e^{-ps} V_c'(s,x_s) + \int_{s\wedge T_A}^{t\wedge T_A} e^{-pu} \langle V_t(u,x_u), db_u^{b'} \rangle - \int_{s\wedge T_A}^{t\wedge T_A} e^{-pu} L'(u,x_u)\,du$$

On en déduit

$$(29)\quad E_\mu^{b'}(e^{-pR_n} V_c'(R_n, x_{R_n})) \geq \int e^{-ps} V_c'(s,x)\,d\mu(s,x) - E_\mu^{b'} \int_s^{R_n\wedge T_A} e^{-pu} L'(u,x_u)\,du$$

Mais la suite de variables aléatoires $e^{-pR_n} V_c'(R_n, x_{R_n})$ est uniformément bornée et tend vers 0 quand $n \to +\infty$.

Donc, en passant à la limite dans (29) , on a:

$$(30)\quad \int e^{-ps} \left(V_c'(s,x) - V_{c'}'(s,x)\right) d\mu(s,x) \leq 0$$

ou encore:

$$(31)\quad I_\mu(c) \leq I_\mu(c')$$

La Proposition 5 est fondamentale. En effet elle nous permet de montrer le résultat suivant:

<u>Proposition 6</u>: Si c minimise I_μ sur $\mathcal{L}$, alors pour tout $\tilde{c} \in \tilde{\mathcal{L}}_\mu$ on a:

$$(32)\quad I_\mu(c) \leq E_\mu^{\tilde{b}}\left(\int_0^{T_A} e^{-pu} \tilde{L}(u,w)\,du\right)$$

<u>Preuve</u>: On utilise les conditions nécessaires et suffisantes établies dans la Proposition 5. Pour une démonstration complète, on se réfèrera à [2] Théorème IV-6 .

On introduit maintenant la fonction coût q qui va jouer un rôle fondamental.

On pose en effet:

$$(33)\quad q = \inf_{c\in\mathcal{L}} V_c'(s,x)$$

On a alors le résultat suivant:

<u>Proposition 7</u> : q est une fonction universellement mesurable et finement semi-continue supérieurement (s.c.s.).

<u>Preuve</u>: $\mathcal{L}$ est compact et métrisable. Il est donc séparable. Soit $\{c_n\}_{n\in N}$ une suite dénombrable et dense dans $\mathcal{L}$. Alors, grâce au Théorème 3, on peut écrire:

$$(34)\quad q = \inf_{n\in N} V_{c_n}'(s,x)$$

q est donc bien universellement mesurable, puisque c'est un inf de fonctions universellement mesurables.

De plus, q étant un inf de fonctions finement continues est nécessairement finement s.c.s.

D'après le Théorème 3 $\mathcal{U}$ étant compact on a nécessairement:

$$(35)\quad q(s,x)=\min_{c\in\mathcal{U}} V_c'(s,x)$$

μ désigne maintenant une mesure de probabilité mutuellement absolument continue avec la mesure de Lebesgue sur $R^+\times R^d$

c désigne un élément de $\mathcal{U}$ minimisant I_μ .

Proposition 8 : (36) $q=V_c'$ μ p.p.

Preuve: On pose:

$$(37)\quad \Gamma(s,x)=\{c\,;\ q(s,x)=V_c'(s,x)\}$$

est une multiapplication à valeurs compactes non vides, dont le graphe est $\mathcal{B}_u(R^+\times R^d)\otimes\mathcal{B}(\mathcal{U})$ mesurable($\mathcal{B}_u(R^+\times R^d)$ désigne la tribu des ensembles universellement mesurables dans $R^+\times R^d$).

En appliquant le Théorème 2 de [5] , on peut trouver une section C μ mesurable de Γ .

A cette section on associe le processus progressivement mesurable $\tilde{c}_{(s,x_s)}(s,x_s)$

On a nécessairement:

$$(38)\quad E_\mu^b\int_0^{T_A} e^{-pt}\,\mathcal{I}(t,\omega)\,dt=\int e^{-ps}\,q(s,x)\,d\mu(s,x)$$

De la Proposition 6, on tire:

$$(39)\quad E_\mu^b\int_0^{T_A} e^{-pt}\,\tilde{\mathcal{I}}(t,\omega)\,dt\geq\int e^{-ps}\,V_c'(s,x)\,d\mu(s,x)$$

Or par (34) , on a:

$$(40)\quad q\leq V_c'$$

On en déduit:

$$q=V_c'\quad \mu\ \text{p.p}$$

On a alors le résultat très important suivant:

Théorème 5 (41) $q=V_c'$

Preuve: On sait que :

$$(42)\quad q=V_c'\quad \mu\ \text{p.p.}$$

Or q est finement s.c.s.. De plus, on sait que les potentiels par rapport aux processus Q^b sont absolument continus par rapport à la mesure μ , grâce au Théorème 8.1 de [6] . Sachant par [3].(4.11) qu'un ensemble complémentaire d'un négligeable est finement dense, on déduit de (42)

$$(43)\quad q\geq V_c'$$

En comparant (34) et (43) , on en déduit:

$$q=V_c'$$

En utilisant la proposition 5, on construit alors dans [2] une fonction borélienne H telle que si c est tel que:

$$(44)\quad L(t,x)+\langle H(t,x),\sigma^{-1}(t,x)\,b(t,x)\rangle=\varphi(t,x)\quad \text{p.p}$$

avec

$$(45)\quad \varphi(t,x) = \min_{(b',L')\in K(t,x)} L' + \langle H(t,x), \sigma^{-1}(t,x)\, b'\rangle$$

alors:

$$\varphi = V_c'$$

De plus un tel choix est possible.

Le Théorème 4 est alors démontré. En effet, si $\varphi = V_c'$, alors nécessairement, pour toute mesure de probabilité μ.

$$(46)\quad I_\mu(c) \leq I_\mu(c')$$

On applique alors laProposition 6, qui montre que si $\tilde{c} \in \tilde{\mathcal{Q}}_{\mu'}$, alors:

$$(47)\quad I_{\mu'}(c) \leq E^{\tilde{b}}_{\mu'} \int_0^{T_A} e^{-\rho t}\, \tilde{L}(t,x_t)\, dt$$

III-Un problème de contrôle des diffusions: le cas général

On se place maintenant dans le cas où K n'est pas nécessairement à valeurs convexes. Nous allons démontrer le Théorème 4.

Preuve du Théorème 4: Soit $\hat{K}(t,x)$ l'enveloppe fermée convexe de $K(t,x)$ La multiapplication $\hat{K}$ est borélienne, par le corollaire 3.3 de [5]

On peut alors résoudre le problème de contrôle $\hat{Q}$ associé à la multiapplication $\hat{K}$. Soit alors H la fonction borélienne construite à la fin de la partie II associée au problème de contrôle $\tilde{Q}$, et soit φ la fonction définie en (45) associée à ce problème.

Soit Δ la multiapplication définie par:

$$(48)\quad (t,x) \xrightarrow{\Delta} \{(b,L)\in K(t,x)\,;\ L + \langle H(t,x), \sigma^{-1}(t,x)\, b\rangle = \varphi(t,x)\}$$

Δ est à valeurs non vides.En effet, $K(t,x)$ et $\hat{K}(t,x)$ ont les mêmes points extrémaux. De plus on vérifie que son graphe est borélien. Par le Théorème 2 de [5] soit $(\tilde{b},\tilde{L})$ une section Lebesgue-mesurable de Δ .

En modifiant $(\tilde{b},\tilde{L})$ sur un ensemble Lebesgue négligeable, on peut trouver une section borélienne (b,L) de K telle que:

$$(b(t,x), L(t,x)) \in \Delta(t,x) \quad p.p.$$

Par le résultat énoncé à la fin de la partie II, $c=(b,L)$ est solution de tous les problèmes $\hat{Q}_\mu$. Or comme $K(t,x) \subset \hat{K}(t,x)\, c$ est a fortiori solution de tous les problèmes Q_μ . On montrera de même que c est solution de tous les problèmes $\overline{Q}_\mu$.

Le Théorème 4 est donc bien démontré dans le cas général.

Si on reprend les grandes étapes de la démonstration, on voit que la méthode utilisée revient à démontrer les points suivants:

1° L'application $c \to V'_c(s,x)$ est continue.

2° L'application $(s,x) \to V'_c(s,x)$ est finement continue.

3° Si K est à valeurs convexes, on introduit une mesure de probabilité μ mutuellement absolument continue avec la mesure de Lebesgue sur $R^+ \times R^d$. Alors:

a)On trouve des conditions nécessaires et suffisantes pour que c minimise I_μ.

b)On en déduit que c est "meilleur" que tout contrôle non anticipatif.

c)On montre que:

$$q = V'_c \quad p.p$$

d)On déduit de c):

$$q = V'_c$$

e)On trouve une condition suffisante, qui peut être vérifiée, pour que

$$q = V'_c$$

4° Grâce à la condition 3° c), on généralise le résultat trouvé en 3°. On peut ainsi choisir un contrôle , qui correspond à l'énoncé du Théorème 4.

Ce schéma de démonstration est tout à fait général et est systématiquement utilisé dans [2] pour des problèmes de contrôle plus difficiles. Nous ne nous sommes par ailleurs pas étendu ici sur des points purement techniques, qui sont traités à fond dans [2] Remarquons enfin que nous avons supposé ici que σ est lipchitzienne. Comme nous l'avons dit, cette hypothèse n'est pas nécessaire, mais sa suppression nous oblige à utiliser dans des espaces de martingales très lourds.

IV-Extensions

Nous donnons dans [2] diverses extensions des résultats précédents. Ils s'appliquent tout d'abord aux processus à mort contrôlée.

a)Contrôle de processus à mort contrôlée

On reprend l'équation (1). ν désigne une variable aléatoire indépendante de x telle que:

$$(48) \quad P(\nu > t) = e^{-t}$$

m est une fonction réelle définie sur $R^+ \times R^d \times U$,telle que $m \geq p > 0$ sur laquelle on fait des hypothèses convenables de continuité et de mesurabilité.

D est le temps d'arrêt défini par:

$$D = \inf\{t \geq s;\ \int_0^t m(\sigma, x_\sigma, u_\sigma)\, d\sigma = V\}$$

On veut minimiser:

$$E\int_s^{D \wedge T_A} L(\sigma, x_\sigma, u_\sigma)\, d\sigma$$

On se ramène alors à la minimisation de:

$$E^b_{(s,x)} \int_s^{T_A} \exp\{-\int_0^t m(\sigma, x_\sigma, u_\sigma)\, d\sigma\}\, L(\sigma, x_\sigma, u_\sigma)\, d\sigma$$

Le problème est résolu dans [2] avec une technique tout à fait comparable à la technique précédente. Le problème de contrôle a bien une solution sous des hypothèses standard.

b) Extension des résultats aux diffusions réfléchies sur un bord totalement inélastique

On utilise la définition des diffusions avec bord donnée dans par Stroock et Varadhan.

La totalité des résultats précédents est étendue dans [2] par ces méthodes .

IV Conclusion

Les méthodes de Théorie du Potentiel semblent donc tout à fait adaptées au traitement des problèmes de contrôle des diffusions. Nous résolvons ainsi dans [2] les problèmes de contrôle de diffusions les plus généraux.

Il est en particulier remarquable que la quasi-totalité des problèmes de contrôle de type classique que l'on peut se poser sur les diffusions markoviennes ont des solutions, sans qu'aucune hypothèse de convexité ne soit nécessaire. Du point de vue de l'analyse convexe, les problèmes de contrôle de diffusion représentent une classe où les méthodes de relaxation donnent toujours un résultat positif.

[1] Bismut J.-M.: Analyse convexe et Probabilités. Thèse. Faculté des Sciences de Paris. Juin 1973.

[2] Bismut J.-M.: Théorie probabiliste du Contrôle des Diffusions. Mémoire des Trans. of Am. Math. Soc. A paraître.

[3] Blumenthal R.M. and Getoor R.K.: Markov Processes and Potential Theory. Acad. Press, New-York and Boston, 1968.

[4] Meyer P.-A.: Intégrales Stochastiques,I,II,III.Séminaire de Probabilités n°1,pp.72,141.Springer Verlag,Lecture Notes in Mathematics,n°39.

[5] Rockafellar R.T.: Measurable Dependence on Convex Sets and Functions on Parameters. J. of Math. Anal. and Appl.Vol.28,1969,pp4-25.

[6] Stroock D.W.and Varadhan S.R.S.:Diffusion Processes with continuous coefficients,Comm. Pure and Appl. Math.,1969,Vol. XXII,PP.345-400, PP. 479-530.

[7] Stroock D.W. and Varadhan S.R.S.:Diffusion Processes with Boundary Conditions. Comm. Pure and Appl. Math.,1971, Vol. XXIV,pp. 147-225.

CONTROLE STATIONNAIRE ASYMPTOTIQUE

Jean-Michel LASRY, C.N.R.S., Equipe de Recherche associée 249,

Mathématiques de la décision, Université de Paris IX Dauphine 75016

Considérons le problème de contrôle optimal

$(P_{s,\varepsilon})$ Minimiser $E\left[\int_0^{+\infty} e^{-st} f(\xi(t),h(\xi(t))dt\right.$ pour $h \in C^1(\mathbb{R}^n, \mathbb{R}^n)$

avec $\xi(0) = x_o$, $d\xi = h(\xi)dt + \varepsilon dw$; où dw est un bruit blanc, E est l'espérance mathématique et où $f \geqslant 0$, $s > 0$, $\varepsilon > 0$ sont donnés (cf. § 1)

C'est un problème stochastique stationnaire. La programmation dynamique permet de le résoudre. Soit $u_{s,\varepsilon}(x_o) = \text{Min } E[\ldots]$ la fonction de Bellman. Elle vérifie l'équation elliptique

$$\frac{\varepsilon^2}{2}\Delta u_{s,\varepsilon} - su_{s,\varepsilon} + g(x, \nabla u_{s,\varepsilon}) = 0$$

La fonction $u_{s,\varepsilon}$ est régulière : la donnée de $u_{s,\varepsilon}$ détermine le contrôle optimal $h_{s,\varepsilon}$. Le minimum est le même lorsqu'on remplace les controles en boucle fermée par des contrôles non anticipatifs. Nous allons étudier le comportement asymptotique de $(P_{s,\varepsilon})$ lorsque $s \searrow 0$ (§ 4). On obtient notamment que le contrôle $h_{s,\varepsilon}$ converge quand $s \searrow 0$ vers un contrôle $h_{0,\varepsilon}$ qui est optimal pour le problème

$(P_{0,\varepsilon})$ Minimiser $\left[\liminf_{T \to \infty} E\frac{1}{T}\int_0^T f(\xi(t),h(\xi(t)))dt\right]$ pour $h \in C^1(\mathbb{R}^n, \mathbb{R}^n)$

Pour ce problème la fonction de Bellman est remplacée par un couple $(\lambda_\varepsilon, v_\varepsilon)$ où λ_ε est une constante et où v_ε est une fonction déterminée à une constante près. L'équation de Bellman est remplacée par l'équation

$$\frac{\varepsilon^2}{2}\Delta v_\varepsilon - \lambda_\varepsilon + g(x, \nabla v_\varepsilon) = 0$$

qui détermine le couple $(\lambda_\varepsilon, v_\varepsilon)$. Lorsque $s \searrow 0$, on a $su_{s,\varepsilon} \to \lambda_\varepsilon$ et $\nabla u_{s,\varepsilon} \to \nabla v_\varepsilon$. La base de la démonstration est l'estimation obtenue au paragraphe 3.

Enfin, au paragraphe 5 nous étudions le cas déterministe "$\varepsilon = 0$" avec $s > 0$ d'abord, puis $s \searrow 0$. On étudie aussi le passage à la limite de $P_{s,\varepsilon}$ à $P_{s,0}$ lorsque $\varepsilon \searrow 0$. Ces résultats sont obtenus modulo une hypothèse de périodicité qui revient à dire que l'on travaille dans les tores plats $\mathbb{R}^n/\mathbb{Z}^n$. Nous conjecturons des résultats du même type dans les variétés Riemanniennes compactes et dans $\mathbb{R}^n$ avec certaines hypothèses de coercivité sur f.

Certaines hypothèses et démonstrations $(H_3$, 5.2, 5.3) sont inspirées de l'artichle de Fleming cité. Je remercie A. Bensoussan pour son aide.

§ 1. Notations, hypothèses

a) Sur la fonction f nous ferons les hypothèses suivantes :

(H_1) la fonction f est de classe C^2 sur $\mathbb{R}^n \times \mathbb{R}^n$, à valeurs positives ($\in \mathbb{R}^+$) Elle est périodique en ce sens que : $f(x+m,y) = f(x,y)$ pour tous x et y dans $\mathbb{R}^n$ et tout m dans Z^n

(H_2) la fonction $y \to f(x,y)$ est convexe. De plus,

$$\sum_{i,j \leq n} \left[\frac{\partial^2}{\partial y_i \partial y_j} f(x,y)\right] z_i z_j > 0$$

pour tous x,y,z dans $\mathbb{R}^n$, et $f(x,y)/\|y\| \to \infty$ lorsque $\|y\| \to \infty$.

(H_3) Il existe une constante $c \in \mathbb{R}^+$ telle que

$$\left\|\frac{\partial}{\partial x} f(x,y)\right\| + \left\|\frac{\partial}{\partial y} f(x,y)\right\| \leq c(1+f(x,y)) \quad \text{pour tous } x,y \text{ dans } \mathbb{R}^n.$$

b) Définition des fonctions ϕ et g.

Soit $p \in \mathbb{R}^n$. D'après l'hypothèse H_2, la fonction $y \to p.y + f(x,y)$ atteint son minimum sur $\mathbb{R}^n$ en un point unique noté $\phi(x,p)$. Ce point est aussi caractérisé par l'équation $p + \frac{\partial}{\partial y} f(x,y) = 0$. Comme $\det \frac{\partial^2}{\partial y^2} f(x,y) \neq 0$, on en déduit, d'après le théorème des fonctions implicites que ϕ est de classe C^1.

Posons

$$g(x,p) = \operatorname{Inf}\{p.y + f(x,y) \mid y \in \mathbb{R}^n\} = p.\phi(x,p) + f(x,\phi(x,p))$$

La fonction composée g est de classe C^1, et $g(x,.)$ est concave. On a les formules de réciprocité habituelles en analyse convexe :

$$f(x,y) = \operatorname{Sup}\{g(x,p) - y.p \mid p \in \mathbb{R}^n\}$$

$$\frac{\partial}{\partial p} g(x,p) = \phi(x,p)$$

c) L'hypothèse H_3 interviendra à travers le :

lemme Pour tous x_o, y_o, x_1, y_1 dans $\mathbb{R}^n$ tels que $\|x_o - x_1\| \leq r$ et $\|y_o - y_1\| \leq r$ on a

$$|f(x_1,y_1) - f(x_o,y_o)| \leq ce^{2rc} \left[\|x_1 - x_o\| + \|y_1 - y_o\|\right] \times [1 + f(x_o,y_o)]$$

démonstration. Posons $\phi(t) = f(x_o + t(x_1 - x_o), y_o + t(y_1 - y_o))$. Compte tenu de H_3, il vient :

$$|\phi'(t)| \leq |(x_1 - x_o).f'_x + (y_1 - y_o).f'_y| \leq K(1 + \phi(t))$$

avec $K = c[\|x_1 - x_o\| + \|y_1 - y_o\|]$. On en déduit que

$$e^{-K}(1 + \phi(0)) \leq 1 + \phi(1) \leq e^K (1 + \phi(0)).$$

Comme $0 \leq K \leq 2rc$, on a $e^{-K} \geq 1 - K$ et $e^K \leq 1 + Ke^{2rc}$. D'où le lemme.

d) Fonctions "périodiques", espaces $C^{\ell,\alpha}(T_n)$.

Dans la suite nous dirons (pour être brefs) qu'une fonction h définie sur $\mathbb{R}^n$ est périodique si elle est périodique par rapport Z^n, c'est à dire si $h(x+m) = h(x)$ pour tous x dans $\mathbb{R}^n$ et tout m dans Z^n. On identifie alors h avec la fonction que l'on obtient par passage au quotient $T_n = \mathbb{R}^n/Z^n$.

Pour $\ell \in \mathbb{N}$ et $\alpha \in [0,1]$ on note $C^{\ell,\alpha}(T_n)$ l'ensemble des fonctions h de $\mathbb{R}^n$ dans $\mathbb{R}$, périodiques, et dont les dérivées jusqu'à l'ordre ℓ sont Holderiennes d'ordre α. L'espace $C^{\ell,\alpha}(T_n)$ est un Banach pour la norme

$$\|h\|_{\ell,\alpha} = \sum_{D \in \mathbb{D}_1} \|Dh\|_\infty + \sum_{D \in \mathbb{D}_2} \|Dh\|_\alpha$$

où $\mathbb{D}_2$ (resp. $\mathbb{D}_1$) est l'ensemble des dérivées d'ordre ℓ (resp. $\leqslant \ell$) et où

$$\|Dh\|_\infty = \text{Sup}\{|Dh(x)| \ ; x \in \mathbb{R}^n\}$$

$$\|Dh\|_\alpha = \text{Sup}\{\left|\frac{Dh(x) - Dh(y)}{|x-y|^\alpha}\right| \ ; \ x \text{ et } y \text{ dans } \mathbb{R}^n, \ x \neq y\}$$

Avec l'identification ci-dessus on pourrait aussi écrire par exemple

$$\|Dh\|_\infty = \text{Sup}\{|Dh(x)| \ ; \ x \in T_n = \mathbb{R}^n/Z^n\}$$

De même $\int_{T_n} h$ désignera l'intégrale de h sur $[0,1]^n$. De même $L^p(T_n)$ désignera l'espace des fonctions h périodiques telles que $\int_{T_n} |h|^p < +\infty$.

e) Estimations à priori.

Soit Ω et Ω' deux ouverts réguliers de $\mathbb{R}^n$ tels que $[-1,+1]^n \supset \Omega \supset \overline{\Omega'} \supset \Omega' \supset [-\frac{1}{2}, +\frac{1}{2}]^n$. On peut appliquer à (Ω, Ω') l'inégalité de Shauder (Ladyzenskaja (1) ch. III § 1, 1.12) et on obtient (avec les notations de Ladyzenskaja et du paragraphe (d) ci-dessus) pour tout $u \in C^{2,\alpha}(T_n)$:

$$\|u\|_{2,\alpha} \leqslant \|u\|_{2,\alpha,\Omega'} \leqslant c_1(\alpha)[\,\|\Delta u\|_{0,\alpha,\Omega} + \max_\Omega |u|] \quad \text{d'où}$$

(1) $$\|u\|_{2,\alpha} \leqslant c_1(\alpha)[\,\|\Delta u\|_{0,\alpha} + \|u\|_\infty]$$

Soit $u \in L^\infty(\mathbb{R}^n)$, périodique, telle que $\Delta u \in C^{0,\alpha}(T_n)$ (au sens des distributions). Soit $\theta_m \in \mathcal{D}(\mathbb{R}^n)$ une unité approchée, et soit $u_m = u * \theta_m$. Comme $u_m \in C^\infty(T_n)$, on a

$$\|u_m\|_{2,\alpha} \leqslant c_1(\alpha)[\,\|\Delta u_m\|_{0,\alpha} + \|u_m\|_\infty] \leqslant c_1(\alpha)[\,\|\Delta u\|_{0,\alpha} + \|u\|_\infty]$$

Comme $u_m \to u$ pour la topologie $\sigma(L^\infty, L^2)$ (par exemple) et que les bornés de $C^{2,\alpha}(T_n)$ sont compacts pour cette topologie on a $u \in C^{2,\alpha}(T_n)$ et l'inégalité (1) est encore vérifiée.

De la même façon, on démontre à partir de l'inégalité III.11.8. de Ladyzenskaja (1) et du théorème de Sobolev (Ladyzenskaja (1)) ch II § 2, théorème 2.1) que pour tout $u \in L^\infty(\mathbb{R}^n)$, périodique, tel que $\Delta u \in L^\infty$ on a :

(2) $\|u\|_{1,\alpha} \leq c_2(\alpha)[\|\Delta u\|_\infty + \|u\|_\infty]$

Soit $h \in L^2(T_n)$. On voit facilement par la transformation de Fourier que, pour tout $\varepsilon > 0$ et tout $s > 0$, il existe un unique $u \in L^2(T_n)$ tel que

$$\frac{\varepsilon^2}{2}\Delta u - su = h \text{ au sens des distributions.}$$

Si $h \in L^\infty$, on a, d'après le principe du maximum $\|su\|_\infty \leq \|h\|_\infty$. On en déduit à l'aide de (1) et (2) que $u \in C^{2,\alpha}(T_n)$ si $h \in C^{0,\alpha}(T_n)$ et que $u \in C^{1,\alpha}(T_n)$ si $h \in L^\infty$, et que l'on a les inégalités

(3) $\|u\|_\infty \leq \frac{1}{s}\|h\|_\infty$

(4) $\|u\|_{1,\alpha} \leq c_3(\alpha,\varepsilon,s)\|\frac{\varepsilon^2}{2}\Delta u - su\|_\infty$

(5) $\|u\|_{2,\alpha} \leq c_4(\alpha,\varepsilon,s)\|\frac{\varepsilon^2}{2}\Delta u - su\|_{0,\alpha}$

avec $c_3(\alpha,\varepsilon,s) = c_2(\alpha)[\frac{4}{\varepsilon^2} + \frac{1}{s}]$ et $c_4(\alpha,\varepsilon,s) = c_1(\alpha)[\frac{2}{\varepsilon^2}(1 + c_3(\alpha,\varepsilon,s)) + \frac{1}{s}]$

f) <u>Contrôles non anticipatifs; contrôles en boucle fermée</u>.

On se donne pour toute la suite un Brownien standard w. On note E l'espérance mathématique

On note A l'ensemble des contrôles non anticipatifs c'est à dire des fonctions aléatoires $a : \mathbb{R}^+ \to \mathbb{R}^n$ adaptées à w et telles que $E\int_0^T \|a(t)\|dt < +\infty$.

Soit $x_o \in \mathbb{R}^n$ et soit $h : \mathbb{R}^n \to \mathbb{R}^n$ une fonction lipschitzienne périodique et soit ξ la diffusion définie par

$$d\xi = h(\xi)dt + \varepsilon dw \quad \text{et} \quad \xi(0) = x_o$$

Soit $a \in A$ défini par $a(t) = h(\xi(t))$. On dira que a est le contrôle en boucle fermée associé à h (sous-entendu : pour x_o et ε donnés).

§ 2. Contrôle stochastique stationnaire ordinaire

Soit $\varepsilon > 0$ et $s > 0$. Soit $x_o \in \mathbb{R}^n$. On pose

(2.1) $$u_{s,\varepsilon}(x_o) = \inf_{a\in A} E\int_0^{+\infty} e^{-st} f(\xi(t);\, a(t))dt$$

où ξ est défini par $\xi(0) = x_o$, $d\xi = a(t)dt + \varepsilon dw$.

<u>Proposition 2.1</u>. <u>La fonction</u> $u_{s,\varepsilon}$ <u>appartient à</u> $C^{2,\alpha}(T_n)$. <u>C'est l'unique solution de l'équation</u>

(2.2) $$\frac{\varepsilon^2}{2}\Delta u_{s,\varepsilon} - su_{s,\varepsilon} + g(x,\nabla u_{s,\varepsilon}) = 0$$

<u>Pour tout</u> $T \in]0, +\infty]$, <u>pour tout contrôle</u> $a \in A$, <u>pour tout</u> $x_o \in \mathbb{R}^n$ <u>on a</u>

(2.3) $$u_{s,\varepsilon}(x_o) \leq E\left[\int_0^T e^{-st} f(\xi(t),a(t))dt + e^{-sT} u_{s,\varepsilon}(\xi(T))\right]$$

où ξ est défini par $\xi(0) = x_o$ et $d\xi = a(t)dt + \varepsilon dw$, avec égalité si et seulement si $a = a_o$ où a_o est le contrôle en boucle fermée associé à la fonction $x \to \phi(x, \nabla u_{s,\varepsilon}(x))$. Autrement dit a_o est l'unique solution du problème d'optimisation (2.1).

La proposition résulte des lemmes 1 et 4 ci-dessous. Dans toute la suite α est un réel quelconque de l'intervalle $]0,1[$.

lemme 1 Soit $v \in C^{2,\alpha}(T_n)$ une fonction vérifiant l'équation :

$$\frac{\varepsilon^2}{2} \Delta v - sv + g(x,\nabla v) = 0.$$

Alors pour tout contrôle $a \in A$, tout $T \in]0, +\infty]$ tout $x_o \in \mathbb{R}^n$ on a

$$v(x_o) \leqslant E \left[\int_0^T e^{-st} f(\xi(t),a(t))dt + e^{-st} v(\xi(T))\right]$$

où ξ est défini par $\xi(0) = x_o$, $d\xi = a(t)dt + \varepsilon dw$. On a égalité si et seulement si a est le contrôle en boucle fermée associé à la fonction $x \to \phi(x,\nabla v(x))$.

Posons $A(t) = e^{-st} v(\xi(t))$. Alors on a $E(A(0)) = v(x_o)$ et

$$E(A(T)) - E(A(0)) = E \int_0^T e^{-st}[-sv(\xi(t)) + a(t).\nabla v(\xi(t)) + \frac{\varepsilon^2}{2}\Delta v(\xi(t))]\, dt.$$

$$= E \int_0^T e^{-st} [a(t).\nabla v(\xi(t)) - g(\xi(t),\nabla v(\xi(t)))]\, dt.$$

D'où, d'après la dualité entre f et g (§ 1.b)

$$E(A(T)) - E(A(0)) \geqslant E \int_0^T e^{-st} f(\xi(t),a(t))dt$$

avec égalité si et seulement si on a presque surement

$$a(t) = \phi(\xi(t),\nabla v(\xi(t))) \text{ pour presque tout } t \in [0,T]$$

D'où le lemme pour $T < +\infty$. On obtient ensuite le cas $T = +\infty$ par passage à la limite.

lemme 2 Soit $\alpha \in]0,1[$. Il existe une constante $M(\alpha)$ telle que pour tout $\tau \in [0,1]$ et pour tout $v \in C^2(T_n)$ vérifiant l'équation

$$\frac{\varepsilon^2}{2} \Delta v - sv + \tau\, g(x, \nabla v) = 0$$

on a $\qquad \|v\|_{1,\alpha} \leqslant M(\alpha)$

En procédant comme pour le lemme 1 on montre que

$$v(x) = \inf_{a\in A} E \int_0^{+\infty} e^{-st} \tau f(\xi(t), \frac{1}{\tau} a(t))dt$$

où ξ est défini par : $\xi(0) = x$ et $d\xi = a(t)dt + \varepsilon dw$.
Comme f est positive, on a $v \geqslant 0$. Soit $M_1 = \text{Sup}\{f(x,0), x \in \mathbb{R}^n\}$. En prenant $a \equiv 0$ on voit que $sv \leqslant M_1$.

Soit x_o et x_1 deux points de $\mathbb{R}^n$ tels que $\|x_o-x_1\| \leqslant 1$. Soit a le contrôle optimal quand le point de départ est x_o(on sait que ce contrôle existe et est en boucle fermée; cf. lemme 1). Soit ξ et η définis par

$$\xi(0) = x_o \qquad d\xi = a(t)dt + \varepsilon\, dw$$

$$\eta(0) = x_1 \qquad d\eta = a(t)dt + \varepsilon\, dw$$

Autrement dit $\eta = \xi + (x_1 - x_o)$ presque surement. Il vient :

$$v(x_o) = E \int_0^{+\infty} \tau f(\xi(t), \frac{1}{\tau}\, a(t))\, e^{-st}\, dt$$

$$v(x_1) \leqslant E \int_0^{+\infty} \tau f(\eta(t), \frac{1}{\tau}\, a(t)) e^{-st} dt.$$

Grâce au lemme 1 § 1 on a

$$v(x_1) - v(x_o) \leqslant E \int_0^{+\infty} \|\xi(t) - \eta(t)\|\; c(1 + f(\xi(t), \frac{1}{\tau}\, a(t))) e^{-st} dt$$

$$\leqslant \|x_o - x_1\|\; c(\frac{\tau}{s} + v(x_o))$$

$$\leqslant \|x_o - x_1\|\; c(1 + M_1)s^{-1} = M_2 \|x_o - x_1\|$$

Comme v est périodique et comme cette inégalité est vraie pour tous x_o, x_1 tels que $\|x_o - x_1\| \leqslant 1$, on a démontré que $\|\nabla v\|_\infty \leqslant M_2$. La constante

$$M_3 = \operatorname{Sup}\{|g(x,p)| \;;\; \|p\| \leqslant M_2\}$$

ne dépend pas de τ. On a $\|v\|_\infty \leqslant M_1 s^{-1}$ et $\|\Delta v\|_\infty \leqslant \frac{2}{\varepsilon^2}(M_1 + M_3)$. D'où l'on déduit (cf. § 1.e) que $\|v\|_{1,\alpha} \leqslant M_4$ où M_4 ne dépend pas de τ . D'où le lemme avec $M(\alpha) = M_4$.

Pour tout $\tau \in [0,1]$, soit Z_τ l'application qui associe à $v \in C^{1,\alpha}(T_n)$ la solution unique u de l'équation

$$\frac{\varepsilon^2}{2}\, \Delta u - su + \tau g(x, \nabla v) = 0.$$

<u>lemme 3</u> <u>L'application</u> Z_τ <u>est continue de</u> $C^{1,\alpha}(T_n)$ <u>dans lui-même et envoie les bornés sur les compacts. L'application</u> $\tau \to Z_\tau$ <u>est continue de</u> $[0,1]$ <u>dans</u> $C^{1,\alpha}(T_n)$ <u>uniformément lorsque</u> v <u>reste dans un borné.</u>

Soit $R_1 > 0$ et supposons que $\|v\|_{1,\alpha} \leqslant R_1$. En particulier on a $\|\nabla v\|_\infty \leqslant R_1$. Soit $R_2 = \operatorname{Sup}\{\|g'_x(x,p)\| + \|g'_p(x,p)\| \;;\; \|p\| \leqslant R_1\}$.

La fonction $x \to g(x, \nabla v(x))$ est de classe $C^{0,\alpha}$. Plus précisément

$$\|g(.,\nabla v(.))\|_{0,\alpha} \leqslant R_2\,(1 + \|\nabla v\|_{0,\alpha}) \leqslant R_2(1 + R_1)$$

On en déduit (en utilisant les inégalités du § 1.e) que pour tout $u = Z_\tau v$ on a, pour une constante c_4 qui ne dépend que de ε de α et de s

$$\|u\|_{2,\alpha} \leqslant c_4 \|\frac{\varepsilon^2}{2}\, \Delta u - su\|_{0,\alpha} \leqslant c_4 \tau \|g(.,\nabla v(.))\|_\alpha \leqslant c_4 R_1 (1 + R_2)$$

Il en résulte que l'application Z_τ envoie les bornés de $C^{1,\alpha}(T_n)$ dans les bornés de $C^{2,\alpha}(T_n)$ donc dans les compacts de $C^{1,\alpha}(T_n)$. D'autrepart, pour $u' = Z_{\tau'} v$ et $u'' = Z_{\tau'} v$ on a

$$\|u' - u''\|_{1,\alpha} \leq \|u' - u''\|_{2,\alpha} \leq c_4 \left\| \frac{\varepsilon^2}{2} \Delta(u' - u'') - s(u' - u'')\right\|_{0,\alpha}$$
$$\leq c_4 \|(\tau' - \tau'')g(.,\nabla v(.))\|_{0,\alpha}$$
$$\leq |\tau' - \tau''| \; c_4 R_1(1 + R_2)$$

L'application $\tau \to Z_\tau v$ est donc lipschitzienne uniformément pour v dans un borné.

Enfin, montrons la continuité de Z_τ (pour τ fixé). Soit v_m et v tels que $\|v_m - v\|_{1,\alpha} \to 0$. Soit $u_m = Z_\tau v_m$. Comme $\{v_m\}$ est borné, $\{u_m\}$ est compact. Quitte à extraire une sous-suite on a $u_m \to u$ dans $C^{1,\alpha}(T_n)$. A la limite on a l'équation

$$\frac{\varepsilon^2}{2} \Delta u - su + g(x, \nabla v) = 0 \quad \text{au sens des distributions.}$$

Donc $u = Z_\tau v$ et le lemme est démontré.

<u>lemme 4</u>. <u>Il existe une fonction</u> $u \in C^{2,\alpha}(T_n)$ <u>telle que</u>

$$\frac{\varepsilon^2}{2} \Delta u - su + g(x, \nabla u) = 0.$$

Soit u un point fixe de Z_τ dans $C^{1,\alpha}(T_n)$. Comme $Z_\tau u \in C^{2,\alpha}(T_n)$ et comme $u = Z_\tau u$ on a $u \in C^{2,\alpha}(T_n)$, et u vérifie l'équation

$$\frac{\varepsilon^2}{2} \Delta u - su + \tau\, g(x,\nabla u) = 0.$$

Il en résulte que $\|u\|_{1,\alpha} \leq M(\alpha)$ d'après le lemme 2. Il en résulte aussi que si $\tau = 0$, alors $u = 0$: Z_o n'admet qu'un point fixe.

De ceci et du lemme 3, il résulte que l'on peut appliquer le théorème de Leray-Shauder (Friedman (1), ch , théorème) à la famille Z_τ , $0 \leq \tau \leq 1$. L'application Z_1 admet donc un point fixe u_1 et le lemme est démontré en prenant $u = u_1$.

§ 3. Estimations de $\|\nabla u_{s,\varepsilon}\|_{1,\alpha}$ indépendante de s et ε

<u>Proposition 3.1</u>. <u>Il existe une constante</u> k_1 <u>indépendante de</u> $s > 0$ <u>et de</u> $\varepsilon > 0$ <u>telle que</u> $\|\nabla u_{s,\varepsilon}\|_\infty \leq k_1$.

Nous avons démontré dans le courant de la démonstration du lemme 2 § 2 que $\|\nabla u_{s,\varepsilon}\| \leq M_1 s^{-1}$ où $M_1 = \text{Sup}\{f(x,0);\ x \in \mathbb{R}^n\}$. Il suffit donc de démontrer la proposition 3.1 pour $0 < s \leq 0,1$.

Nous supposons donc que l'on a $0 < s \leq 0,1$ de telle sorte que l'on a pour $0 \leq t \leq 1$ les inégalités

(3.2) $\qquad 0,5 \leq 1\text{-}st \leq e^{-st} \leq 1\text{-}0,5\, st \leq 1.$

Comme s et ε sont fixés pendant la démonstration, il n'y a pas d'ambiguité à écrire u au lieu de $u_{s,\varepsilon}$.

Soit x_o et x_1 deux points de $\mathbb{R}^n$ tels que $\|x_o - x_1\| \leq 1$.

D'après la proposition 2.1 il existe $a_o \in A_o$ tel que

$$(3.3) \qquad u(x_o) = E\left[\int_0^1 e^{-st} f(\xi(t), a_o(t))dt + e^{-s} u(\xi(1))\right]$$

où ξ est défini par $\xi(0) = x_o$ et $d\xi = a(t)dt + \varepsilon dw$.

Soit η défini par

$$\eta(t) = \xi(t) + (x_1 - x_o)(1-t) \quad \text{pour} \quad 0 \leq t \leq 1.$$

On a : $\eta(0) = x_1$ et $d\eta = a(t)dt + \varepsilon dw$ où $a \in A$ est défini par $a(t) = a_o(t) + x_o - x_1$ pour $0 \leq t \leq 1$ (et $a(t)$ quelconque pour $t \geq 1$). D'après la proposition 2.1. on a

$$u(x_1) \leq E\left[\int_0^1 f(\eta(t), a(t))e^{-st}\, dt + e^{-s} u(\eta(1))\right]$$

Comme $\eta(1) = \xi(1)$ on a

$$u(x_1) - u(x_o) \leq E\int_0^1 (f(\eta,a) - f(\xi,a_o))e^{-st}dt.$$

Or $\|\eta - \xi\| \leq \|x_1 - x_o\| \leq 1$ et $\|a - a_o\| \leq \|x_1 - x_o\| \leq 1$.

On a donc d'après le lemme 1 & 1.c :

$$(3.4) \qquad u(x_1) - u(x_o) \leq \|x_1 - x_o\| \int_0^1 c_1(1 + f(\eta,a))dt$$

où $c_1 = ce^{2c}$ est indépendant de s et de ε.

D'autrepart, d'après (3.2) et (3.3) on a

$$(3.5) \qquad E\int_0^1 0{,}5 \times f(\eta,a)dt \leq u(x_o) - e^{-s} \min u$$

Nous avons vu (cf. § 2) que $0 \leq su \leq M_1$. Donc $|\min u - e^{-s} \min u| \leq s \min u \leq M_1$. Il résulte alors de (3.4) et (3.5) l'inégalité

$$(3.6) \qquad u(x_1) - u(x_o) \leq \|x_1 - x_o\|\, c_1(1 + 2M_1 + 2(u(x_o) - \min u)).$$

Cette inégalité est vraie pour tous x_o et x_1 tels que $\|x_o - x_1\| \leq 1$. Soit $x_2 \in \mathbb{R}^n$. Comme u est périodique, il existe $x_3 \in \mathbb{R}^n$ tel que $\|x_3 - x_2\| \leq 1$ et $u(x_3) = \min u$. En appliquant (3.6) au couple x_2, x_3 on trouve

$$(3.7) \qquad u(x_2) - \min u = u(x_2) - u(x_3) \leq \|x_2 - x_3\|\, c_1\, (1 + 2M_1)$$

Cette inégalité est vraie pour tout $x_2 \in \mathbb{R}^n$. En la reportant dans (3.6) avec $x_2 = x_o$ on trouve

$$(3.8) \qquad u(x_1) - u(x_o) \leq k_1 \|x_1 - x_o\|$$

où $k_1 = c_1\, (1 + 2M_1 + 2c_1\, (1 + 2M_1))$ est une constante indépendante de s et de ε . La proposition est démontrée.

<u>Corollaire 3.1</u>. <u>Il existe une constante</u> k_2 <u>indépendante de</u> $s > 0$ <u>et de</u> $\varepsilon > 0$ <u>telle que</u> $\|\nabla u_{s,\varepsilon}\|_{0,\alpha} \leq k_2 \times c_2(\alpha) \times \varepsilon^{-2}$ <u>où</u> $c_2(\varepsilon)$ <u>est une constante qui ne dépend</u>

que de $\alpha \in]0,1[$ (cf. § 1.e).

Soit $\bar{u}_{s,\varepsilon} = u_{s,\varepsilon} - u_{s,\varepsilon}(0)$. Comme $u_{s,\varepsilon}$ est périodique, et lipschitzienne de rapport k_1, on a

$\|\bar{u}\|_\infty \leqslant nk_1$. Soit $k_3 = \mathrm{Sup}\{|g(x,p)| \; ; \|p\| \leqslant k\}$.

De l'équation $\frac{\varepsilon^2}{2} \Delta\bar{u}_{s,\varepsilon} = su_{s,\varepsilon} + g(x, \nabla u_{s,\varepsilon})$ on déduit l'inégalité :

$\|\Delta\bar{u}_{s,\varepsilon}\|_\infty \leqslant (M_1 + k_3)2\varepsilon^{-2}$. D'où $\|\bar{u}\|_{1,\alpha} \leqslant k_2\, \varepsilon^{-2} c_2(\alpha)$ d'après l'inégalité 2 § 1.e. Le corollaire résulte alors de l'égalité $\nabla\bar{u} = \nabla u$.

§ 4. Contrôle stochastique stationnaire asymptotique

Proposition 4.1. Soit $\varepsilon > 0$. Il existe un unique couple λ_ε, v_ε tel que $\lambda_\varepsilon \in \mathbb{R}$ $v_\varepsilon \in C^{2,\alpha}(T_n)$ pour tout $\alpha \in]0,1[$ et tel que quand $s \searrow 0$ on a

(4.1) $\qquad su_{s,\varepsilon} \to \lambda_\varepsilon$ uniformément

$\qquad \nabla u_{s,\varepsilon} \to \nabla v_\varepsilon$ pour la norme dans $C^{1,\alpha}(T_n, \mathbb{R}^n)$

(4.2) $\qquad$ Pour tout $x_0 \in \mathbb{R}^n$, tout $T \in \mathbb{R}$, tout $a \in A$ et pour ξ défini par $d\xi = a(t)dt + \varepsilon dw$, $\xi(0) = x_0$ on a l'inégalité

$$v_\varepsilon(x_0) \leqslant E\left[\int_0^T (f(\xi(t),a(t)) - \lambda_\varepsilon)dt + v_\varepsilon(\xi(T))\right]$$

avec égalité si et seulement si a est le contrôle a_0 en boucle fermée associé à la fonction $x \to \phi(x, \nabla v_\varepsilon(x))$.

(4.3) $\qquad \frac{\varepsilon^2}{2} \Delta v_\varepsilon - \lambda_\varepsilon + g(x, \nabla v_\varepsilon) = 0.$

Ces trois propriétés sont équivalentes : autrement dit chacune d'entre elles caractérise le couple $(\lambda_\varepsilon, v_\varepsilon)$ dans $\mathbb{R} \times C^{2,\alpha}(T_n)$.

La propriété (4.2) montre en particulier que a_0 est une solution du problème de contrôle optimal asymptotique

$$(\mathbb{P}) \qquad \underset{a \in A}{\text{Minimiser}} \left[\liminf_{T \to \infty} E\frac{1}{T} \int_0^T f(\xi(t),a(t))dt\right]$$

(4.4) et que $\lambda_\varepsilon = \underset{a \in A}{\mathrm{Min}}$ [lim.inf.....] . De plus a_0 est l'unique solution optimale de $\mathbb{P}$ parmi les contrôles en boucle fermée.

On obtient (4.4) en divisant les deux membres de l'inégalité (4.2) par T et en passant à la limite. Le reste de la proposition résulte des lemmes suivants.

lemme 1 Soit $\alpha \in]0,1[$. Soit s_m une suite qui tend vers zéro. Il existe une sous-suite s_{m_k}, un réel λ et une fonction $v \in C^{2,\alpha}(T_n)$ tels que (on pose

$v_k = u_{s_{m_k},\varepsilon}$) :

i) $s_{m_k} v_k \to \lambda$ uniformément

ii) $\|\nabla v_k - \nabla v\|_{1,\alpha} \to 0$

iii) $\frac{\varepsilon^2}{2} \Delta v - \lambda + g(x, \nabla v) = 0$

<u>démonstration</u>. Soit β tel que $\alpha < \beta < 1$. D'après le corollaire 3.1 la suite ∇u_{s_m} est bornée dans $C^{1,\beta}$, donc compacte dans $C^{1,\alpha}$.
D'autrepart, on a les inégalités

$$0 \leqslant su_{s,\varepsilon}(x) \leqslant M_1 \quad \text{et} \quad |u_{s,\varepsilon}(x) - u_{s,\varepsilon}(y)| \leqslant k_1|x - y|$$

pour tous x et y dans $\mathbb{R}^n$, où M et k_1 sont des constantes indépendantes de s (cf. § 2 et proposition 3.1). On en déduit (i) et (ii). Puis à l'aide de ces convergences, on démontre iii en passant à la limite dans l'équation (2.2).

<u>lemme 2</u> <u>Soit</u> v <u>une fonction de</u> $C^{2,\alpha}(T_n)$ <u>et</u> $\lambda \in \mathbb{R}$ <u>vérifiant</u> $\frac{\varepsilon^2}{2}\nabla v - \lambda + g(x,\nabla v)=0$
<u>Soit</u> $x_o \in \mathbb{R}^n$, $a \in A$, <u>et soit</u> ξ <u>défini par</u>

(4.5) $\xi(0) = x_o$, $d\xi = a(t)dt + \varepsilon dw$

<u>Alors on a</u>

(4.6) $$v(x_o) \leqslant E[\int_0^T [f(\xi(t), a(t)) - \lambda]\, dt + v(\xi(T))]$$

<u>avec égalité si et seulement si</u> a <u>est le controle en boucle fermée associé à</u> $x \to \phi(x, \nabla v(x))$

Même démonstration que le lemme 1 § 2.

<u>lemme 3</u> <u>Soit</u> λ_1, λ_2 <u>dans</u> $\mathbb{R}$, v_1 <u>et</u> v_2 <u>dans</u> $C^{2,\alpha}(T_n)$. <u>Supposons que pour tout</u> $T > 0$, <u>pour tout</u> $x_o \in \mathbb{R}^n$, <u>pour tout</u> $a \in A$, <u>pour</u> ξ <u>défini par</u> $\xi(0) = x_o$, <u>et</u> $d\xi = a(t)dt + \varepsilon dw$ <u>et pour</u> $i = 1, 2$ <u>on a</u>

(4.7) $$v_i(x_o) \leqslant E[\int_0^T (f(\xi(t),a(t)) - \lambda_i)dt + v_i(\xi(T))]$$

<u>avec égalité lorsque</u> $a = a_i$ <u>où</u> a_i <u>est le controle en boucle fermée associé à la fonction</u> $x \to \phi(x, \nabla v_i(x))$. <u>Alors</u> $\lambda_1 = \lambda_2$ <u>et</u> $v_1 = v_2$.

En prenant $a = a_j$ $(j = 1,2)$ dans (4.7) on a :

$$v_i(x_o) \leqslant E[\int_0^T (f(\xi_j(t),a_j(t)) - \lambda_i)dt + v_i(\xi(T))]$$

avec égalité si $i = j$. En divisant par T, et en passant à la limite $(T \to \infty)$ on trouve $\lambda_1 = \lambda_2$. On pose $\lambda_1 = \lambda_2 = \lambda$ pour la suite de la démonstration.

En ajoutant une constante à v_2 on peut supposer que $v_1(0) = v_2(0)$. Si $v_1 - v_2$ n'est pas constant il existe un point x_o tel que

$$\gamma = v_1(x_o) - v_2(x_o) \neq v_2(0) - v_2(0) = 0.$$

Supposons $\gamma > 0$. Ajoutons $\gamma/2$ à v_2. On a alors

$$v_1(x_o) - v_2(x_o) > \gamma/2 > 0 \quad >- \gamma/2 > v_1(0) - v_2(0).$$

Il existe donc deux ouverts O_1 et O_2 et $\eta > 0$ (par exemple $\eta \geq \gamma/4$) tel que

$$(4.9) \qquad \begin{aligned} v_1(x) &\geq \eta + v_2(x) \quad \text{pour } x \in O_1 \\ v_2(x) &\geq \eta + v_1(x) \quad \text{pour } x \in O_2 \end{aligned}$$

(Si γ est négatif il en est de même). Comme v_1 et v_2 sont périodiques, on peut supposer que O_1 et O_2 sont des ouverts de $T_n = \mathbb{R}^n/Z^n$.

On définit par récurrence ξ et i_m (aléatoire) :

a) $\xi(0) = x_o$, $i_o = 1$

b) si $i_m = 1$, alors $i_{m+1} = 1$ (resp. 2) si $\xi(m) \notin O_1$ (resp. si $\xi(m) \in O_1$).

c) si $i_m = 2$, alors $i_{m+1} = 2$ (resp. 1) si $\xi(m) \notin O_2$ (resp. si $\xi(m) \in O_2$).

d) pour $m \leq t \leq (m+1)$, ξ vérifie l'équation

$$d\xi = \phi(\xi, \nabla v_{i_m}(\xi))dt + \varepsilon dw$$

D'après (d) et l'hypothèse de l'énoncé, on a

$$(4.10) \qquad v_{i_m}(\xi(m)) = \underset{\text{sachant } \xi(m)}{\text{Espérance}} \left[\int_m^{m+1} (f(\xi, \phi(\xi, \nabla v_{i_m}(\xi))) - \lambda)dt + v_{i_m}(\xi(m+1)) \right]$$

D'autrepart pour tout $x \in T_n$ et tout $m \in \mathbb{N}$ on a

$$v_{i_{m+1}}(x) \geq v_{i_m}(x) + \eta\, 1_{O_{i_m}}(x)$$

où 1_O est la fonction caractéristique de O. En utilisant le fait que $\|\phi(.,\nabla v_i(.))\|_\infty < +\infty$ on montre qu'il existe $\theta > 0$ tel que pour $i = 1,2$ et $m \in \mathbb{N}$ on a

$$E(1_{O_i}(\xi(m))) \geq \theta$$

Par conséquent, en partant de (4.10) il vient

$$\begin{aligned} E(v_{i_m}(\xi(m))) &= E\left[\int_m^{m+1} [cf(\xi, \phi(\nabla, v_{i_m}(\xi))) - \lambda\, dt + v_{i_m}(\xi(m+1))\right] \\ &\geq E\left[\int_m^{m+1} [f(\ldots) - \lambda]\, dt + v_{i_{m+1}}(\xi(m+1)) + \eta\, 1_{O_{i_m}}(\xi(m+1))\right] \\ &\geq E\int_m^{m+1} [f(\ldots) - \lambda]\, dt\,] + E[v_{i_{m+1}}(\xi(m+1))] + \theta\eta \end{aligned}$$

En sommant de $m = 1$ à p ces inégalités, il vient :

$$E\int_0^p (f(\xi(t), a(t)) - \lambda)dt \leq 2(\|v_1\|_\infty + \|v_2\|_\infty) - p\theta\eta$$

où $a \in A$ est défini par $a(t) = \phi(\xi(t), \nabla v_{i_m}(\xi(t)))$ pour $(m-1) \leq t \leq m$.
Mais d'après (4.7) on a

$$E\int_0^p (f(\xi(t), a(t)) - \lambda)dt \geq -2\|v_1\|_\infty$$

D'où une contradiction pour p assez grand puisque $\Theta > 0$ et $\eta > 0$.

§ 5 Controle déterministe stationnaire ordinaire et asymptotique

On note Q l'ensemble des fonctions continues ξ de $\mathbb{R}^+$ dans $\mathbb{R}^n$ dont la dérivée $\dot{\xi}$ est localement sommable. On pose pour $s > 0$ et $x_o \in \mathbb{R}^n$

(5.1) $$u_s(x_o) = \inf_{\xi \in Q, \xi(0) = x_o} \int_0^{+\infty} e^{-st} f(\xi(t), \dot{\xi}(t))dt$$

On remarque tout d'abord que u_s est une fonction périodique (cf. § 1, hypothèse H_1)

Comme $0 \leq f(x,0) \leq M_1$ pour tout $x \in \mathbb{R}^n$, on a

(5.2) $$0 \leq s.u_s(x) \leq M_1$$

Montrons que u_s est lipschitzienne. Soit $x \in \mathbb{R}^n$ tel que $\|x\| \leq 1$ et soit $\alpha > 0$. Il existe $\xi \in Q$ tel que $\xi(0) = x_o$ et

$$\alpha + u_s(x_o) \geq \int_0^{+\infty} e^{-st} f(\xi(t), \dot{\xi}(t))dt$$

Soit $\eta = \xi + x$. On a $\eta \in Q$ et $\eta(0) = x_o + x$; d'où ;

$$u_s(x_o+x) \leq \int_0^{+\infty} e^{-st} f(\xi(t)+x, \dot{\xi}(t))dt$$

En appliquant le lemme 1 § 1 on trouve

$$u_s(x_o+x) - u_s(x_o) \leq \alpha + \int_0^{+\infty} c_1\|x\|(1+f(\xi, \dot{\xi}))e^{-st}dt \leq \alpha + c_2\|x\|$$

avec $c_2 = c_1(M+1)^{s-1}$. Comme $\alpha > 0$ est arbitraire, comme x est un point quelconque de normé ≤ 1, on en déduit que u_s est lipschitzien (de rapport c_2).

<u>Proposition 5.1</u> <u>La fonction u_s est lipschitzienne donc presque partout différentiable. En tout point où ∇u_s existe on a l'équation d'Hamilton-Jacobi-Bellman</u> :

(5.3) $$-su_s + g(x,\nabla u_s) = 0$$

<u>On a</u> : $0 \leq su_s \leq M_1 = \operatorname{Sup}\{f(x,0);\ x \in \mathbb{R}^n\}$ <u>donc</u> $0 \leq g(x,\nabla u_s) \leq M_1$. <u>Ceci implique en particulier que</u> $\|\nabla u_s\| \leq k_3$ <u>où</u> k_3 <u>est une constante indépendante de</u> s.

Le seul point qui reste à démontrer est l'équation (5.3). Avant de le démontrer, nous allons étudier l'existence et la régularité des trajectoires optimales.

Le principe d'optimalité de Bellman s'écrit ici :

(5.4) $$u_s(x_o) = \operatorname*{Inf}_{\xi \in Q;\ (0) = x_o} \int_0^T e^{-st} f(\xi(t), \dot{\xi}(t))dt + e^{-st} u_s(\xi(T))$$

Comme u_s est lipschitzien, le minimum est atteint dans (5.4) donc aussi dans (5.1) en mettant bout à bout des solutions optimales de (5.4) (On peut aussi le démontrer directement).

Soit ξ_o une solution optimale de (5.1). La restriction de ξ_o à l'intervalle $[0,T]$ est une solution optimale de (5.4). En appliquant le principe de

Pontryagin amélioré à l'usage des fonctions lipschitziennes par Clark (1) dans (5.4) puis en faisant tendre T vers $+\infty$ on prouve que si p est défini par

(5.5) $$p(t_o) = \int_{t_o}^{+\infty} e^{-st} f'_x(\xi_o(t), \dot{\xi}_o(t))dt$$

alors on a

(5.6) $$\dot{\xi}_o(t) = \phi(\xi_o(t), p(t))e^{st}$$

c'est-à-dire que $\dot{\xi}_o(t)$ est le point où la fonction $y \to e^{-st}f(\xi_o(t),y) + y.p(t)$ est minimum.

En utilisant alternativement (5.5) et (5.6) on trouve successivement que :

(i) $p(t)$ est continu et $\|p(t)\| \leq c_3 e^{-st}$ (grâce à H_1 § 1)

(ii) $\dot{\xi}_o(t)$ est continu et $\|\dot{\xi}_o(t)\| \leq c_4 e^{st}$ où la constante c_4 ne dépend que de c_3.

(iii) $\dot{p}(t)$ est continu

(iv) $\dot{\xi}_o$ est dérivable et $\ddot{\xi}_o$ est continu.

La constante c_3 (donc aussi c_4) dépend seulement de c_1, de M_1 et de s.

Revenons à la démonstration de (5.3). Supposons que u_s soit différentiable au point x_o. Comme ξ_o est une trajectoire optimale on a

$$u_s(x_o) = \int_0^T e^{-st} f(\xi_o, \dot{\xi}_o)dt + e^{-sT}u_s(\xi_o(T))$$

On fait tendre T vers 0 et on utilise la régularité de ξ_o pour passer à la limite. On trouve

(5.7) $$0 = f(x_o, \dot{\xi}_o(0)) + \nabla u_s(x_o).\ \dot{\xi}_o(0) - su_s(x_o)$$

Soit ξ la trajectoire définie par $\xi(t) = x_o + tx$. En reportant ξ dans (5.4) on trouve

$$u_s(x_o) \leq \int_0^T f(x_o + tx, x)dt + e^{-sT}u_s(x_o + Tx)$$

$$0 \leq T[f(x_o,x) + \nabla u_s(x_o).x - su_s(x_o) + \varepsilon(T)]$$

avec $\varepsilon(T) \to 0$ quand $T \to 0$. D'où, compte tenu de (5.7) :

$$\operatorname*{Inf}_{x\in \mathbb{R}^n} [\, f(x_o,x) + x.\nabla u_s(x_o)]- su_s(x_o) = 0$$

C'est bien l'équation (5.3) cherchée. Comme le mimimum est atteint pour $x = \dot{\xi}_o$ on voit en outre (d'après (5.5)) que

(5.8) $$p(t) = \nabla u_s(x_o)$$

Sans l'hypothèse de différentiabilité de u_s au point x_o on peut écrire

$$f(x_o, \dot{\xi}(0)) = \lim_{T \searrow 0} \frac{1}{T} \int_0^T f(\xi_o(t), \dot{\xi}_o(t))e^{-st}dt$$

$$= \lim_{T \searrow 0} \frac{u_s(x_o) - u_s(\ (T))e^{-sT}}{T} \leqslant k_3\|\dot{\xi}_o(0)\| + su_s(x_o)$$

$$\leqslant k_3\|\dot{\xi}_o(0)\| + M_1$$

On en déduit, d'après l'hypothèse H_1 que $\|\dot{\xi}(0)\| \leqslant c_7$ ou c_7 est une constante qui ne dépend que de k_3 et M_1 donc qui ne dépend pas de s. Comme $t \to \xi(t + t_o)$ est aussi une trajectoire optimale

(5.9) on a $\|\xi(t_o)\| \leqslant c_7$ pour tout $t_o \geqslant 0$.

Cette majoration sera utile pour le passage à la limite quand $s \searrow 0$. Auparavant nous allons étudier le rapport entre les fonctions $u_{s,\varepsilon}$ et u_s.

<u>Proposition 5.2</u> . <u>Soit</u> $s > 0$ <u>fixé. Lorsque</u> $\varepsilon \to 0$ <u>la fonction</u> $u_{s,\varepsilon}$ <u>converge uniformément vers</u> u_s.

<u>Démonstration</u> :

Soit $x_o \in \mathbb{R}^n$ et soit ξ_o une solution optimale de (5.1). Soit $a \in A$ le contrôle défini par $a(\ t) = \dot{\xi}_o(t)$ et soit ξ_ε défini par $\xi_\varepsilon(0) = x_o$ et $d\xi_\varepsilon = a(t)dt + \varepsilon dw$,c'est à dire $\xi_\varepsilon(t) = \xi_o(t) + \varepsilon(w(t) - w(0))$

Soit, d'autrepart $h_\varepsilon(x) = \phi(x, \nabla u_{s,\varepsilon}(x))$ et η_ε défini par $d\eta_\varepsilon = h_\varepsilon(\eta_\varepsilon)dt + \varepsilon dw, \eta_\varepsilon(0) = x_o$. Soit enfin θ_ε défini par $\theta_\varepsilon(0) = x_o$ et $d\theta_\varepsilon = h_\varepsilon(\eta_\varepsilon)dt$. Presque surement $\theta_\varepsilon \in C^1(\mathbb{R}^+, \mathbb{R}^n)$ et $\eta_\varepsilon(t) = \theta_\varepsilon(t) + \varepsilon(w(t) - w(0))$

En tenant compte de (5.1) et de la proposition 2.4, il vient :

$$u_s(x_o) = \int_0^{+\infty} e^{-st} f(\xi_o(t), \dot{\xi}_o(t))dt$$

$$u_{s,\varepsilon}(x_o) = E\int_0^{+\infty} e^{-st} f(\eta_\varepsilon(t), h_\varepsilon(\eta_\varepsilon(t)))dt$$

$$u_s(x_o) \leqslant E\int_0^{+\infty} e^{-st} f(\theta_\varepsilon(t), \dot{\theta}_\varepsilon(t))dt$$

$$u_{s,\varepsilon}(x_o) \leqslant E\int_0^{+\infty} e^{-st} f(\xi_\varepsilon(t), \dot{\xi}_o(t))dt$$

On en déduit que

$$\alpha_\varepsilon \leqslant u_{s,\varepsilon}(x_o) - u_s(x_o) \leqslant \beta_\varepsilon \quad \text{avec}$$

$$\alpha_\varepsilon = E\int_0^{+\infty} e^{-st} f(\eta_\varepsilon, h_\varepsilon(\eta_\varepsilon))dt - E\int_0^{+\infty} e^{-st} f(\theta_\varepsilon, \dot{\theta}_\varepsilon)dt$$

$$\beta_\varepsilon = E\left[\int_0^{+\infty} e^{-st} f(\xi_\varepsilon, \dot{\xi}_o)dt\right] - \int_0^{+\infty} e^{-st} f(\xi_o, \dot{\xi}_o)dt\ ]$$

$$|\beta_\varepsilon| + |\alpha_\varepsilon| \leqslant 2k\ E\int_0^{+\infty} e^{-st}[\,\mathrm{Inf}(1, \varepsilon\|w(t) - w(0)\|)\,]dt$$

où $k = \mathrm{Sup}\{|f'_x(x,y)| ; x \in \mathbb{R}^n$ et $|y| \leqslant K\}$ (et où $K = \mathrm{Sup}\{|\phi(x,p)| ; x \in \mathbb{R}^n$ et $|p| \leqslant k_1\}$, et où k_1 provient de la proposition 3.1). Quand $\varepsilon \to 0$ le second membre tend vers 0 (indépendemment de x_o, ce qu'il fallait démontrer).

Remarque On peut déduire de ce passage à la limite la majoration $\|\nabla u_s\| \leqslant k_1$ indépendante de s, grâce à la même majoration sur $\|\nabla u_{s,\varepsilon}\|$ (cf. proposition 3.1). On retrouve indirectement la conclusion de la proposition 5.1.

Proposition 5.3. Il existe une constante k_2 telle que

$$(5.11) \qquad [u_s(x_o - x) - 2u_s(x_o) + u_s(x_o + x)]\|x\|^{-2} \leqslant k_2$$

pour tous x_o et x dans $\mathbb{R}^n$, et tout $s > 0$.

Il suffit de prouver que pour tout $\varepsilon > 0$ et tout $x \in \mathbb{R}^n$ ($\|x\| = 1$) et tout $x_o \leqslant \mathbb{R}^n$ on a $v \leqslant k_2$ où

$$v = \frac{\partial^2 u_{s,\varepsilon}(x_o)}{\partial x^2} = \text{dérivée seconde parallèlement à } x.$$

Rappelons que $u_{s,\varepsilon}$ vérifie l'équation

$$\frac{\varepsilon^2}{2}\Delta u_{s,\varepsilon} - su_{s,\varepsilon} + g(x, \nabla u_{s,\varepsilon}) = 0$$

Comme g est C^2, la fonction $u_{s,\varepsilon}$ est $C^{3,\alpha}$. En dérivant deux fois par rapport à x_i on obtient

$$\frac{\varepsilon^2}{2}\Delta v - sv + \frac{\partial g(x,\nabla u)}{\partial p} \cdot \nabla v + \psi = 0$$

où

$$\psi = \frac{\partial^2 g}{\partial x^2} + 2 \sum_{j \leqslant n} \frac{\partial^2 g}{\partial x \partial p_j} \frac{\partial^2 u}{\partial x \partial x_j} + \sum_{k,j \leqslant n} \frac{\partial^2 g}{\partial p_k \partial p_j} \frac{\partial^2 u}{\partial x \partial x_j} \frac{\partial^2 u}{\partial x \partial x_k}$$

Soit $q = (q_1,\ldots,q_n)$ avec $q_k = \dfrac{\partial^2 u}{\partial x \partial x_k}$. On a $\psi \leqslant a - b\|q\|^2$ où a et b sont des constantes qui ne dépendent que de la stricte concavité de $g(x,p)$ pour $x \in \mathbb{R}^n$ et $|p| \leqslant k_1$ (la constante de la proposition 3.1)

Donc a et b sont indépendants de x_o, s et ε. Comme $\|q\|^2 \geqslant v^2$ il vient

$$0 \leqslant \frac{\varepsilon^2}{2}\Delta v - sv + \frac{\partial g}{\partial p}.\nabla v + a - b.v^2$$

D'où l'on déduit par le principe du maximum, que $v(x_o) \leqslant a^{1/2}b^{-1/2}$.

Proposition 5.4. Lorsque $s \to 0$ la fonction su_s converge uniformément vers la constante λ suivante

$$(5.12) \qquad \lambda = \inf_{\xi \in Q} \left[\liminf_{T \to \infty} \frac{1}{2}\int_0^T f(\xi(t), \dot{\xi}(t))dt\right]$$

Soit s_m une suite telle que $s_m \to 0$. Il existe une sous-suite $r_k = s_{m_k}$ et une fonction lipschitzienne v telles que (notons $v_k = u_{s_{m_k}}$)

$\nabla v_k \to \nabla v$ <u>presque partout</u>

(5.13) $\quad - \lambda + g(x,\nabla v) = 0$ <u>presque partout</u>

(5.14) $\quad v(x_o) = \underset{\xi \in Q;\ (0)=x_o}{\operatorname{Inf}} \int_0^T [\, f(\xi,\dot{\xi}) - \lambda\,]\, dt + v(\xi(T))]$

<u>Soit ξ_m une suite de trajectoires issues de x_o, où ξ_m est optimale pour le problème 5.1 avec $s = s_m$. Alors il existe une sous-suite $\eta_k = \xi_{m_k}$ et $\eta \in Q$ telles que $\eta_k \to \eta$ et $\eta_k \to \eta$ uniformément sur tout intervalle $[0,T]^k$. La trajectoire η est optimale pour le problème asymptotique, c'est à dire</u>

(5.15) $\quad \lambda = \lim \frac{1}{T}\int_0^T f(\eta,\dot{\eta})dt.$

On déduit de l'inégalité (5.11) que pour tout $x \in \mathbb{R}^n$ avec $\|x\| \leqslant 1$ la dérivée au sens des distributions $\frac{\partial^2 u}{\partial x^2}$ est $\leqslant k_2$. C'est donc une mesure. Comme u est périodique on a $\langle\frac{\partial^2 u}{\partial x^2}, 1\rangle_{\mathcal{D}',\mathcal{D}} = 0$. On en déduit que $\frac{\partial^2 u}{\partial x^2}$ est une mesure bornée; plus précisément

$$\left\|\frac{\partial^2 u}{\partial x^2}\right\|_1 \leqslant \int_{[0,1]^n} k_2 = k_2.$$

En prenant $x = e_i$ et $x = (e_i \neq e_j)\,\frac{1}{2}$ (où $e_1,\ldots,e_n$ est la base canonique de $\mathbb{R}^n$) on trouve :

$$\left\|\frac{\partial^2 u}{\partial x_i^2}\right\|_1 \leqslant k_2 \quad \text{et} \quad \left\|\frac{1}{2}\,\frac{\partial^2 u}{\partial x_i^2} \pm \frac{\partial^2 u}{\partial x_i \partial x_j} + \frac{1}{2}\,\frac{\partial^2 u}{\partial x_j^2}\right\|_1 \leqslant k_2$$

D'où l'on déduit

$$\left\|\frac{\partial^2 u}{\partial x_i \partial x_j}\right\|_1 \leqslant k_2.$$

Comme k_2 ne dépend pas de s, il en résulte que lorsque s varie ∇u_s reste dans un compact de L^1 pour la topologie forte.

D'autrepart on a les estimations $0 \leqslant su_s \leqslant M$ et $\|\nabla u_s\| \leqslant k_3$ (cf. proposition 5.1).

Soit s_m une suite telle que $\lim s_m = 0$. D'après ce qui précède, il existe u une sous-suite $r_k = s_{m_k}$, une constante λ et une fonction lipschitzienne v telles que (en notant $v_k = u_{s_{m_k}}$)

$r_k v_k \to \lambda$ uniformément

$\nabla v_k \to \nabla v$ presque partout

$v_k(x) - v_k(0) \to v(x)$ uniformément.

On obtient l'équation (5.13) à partir de l'équation (5.3) par passage à la limite.

Soit $x_o \in \mathbb{R}^n$ et soit $\xi \in Q$ avec $\xi(0) = x_o$. D'après (5.4) on a

$$v_k(x_o) \leqslant \int_0^T e^{-r_k t} f(\xi(t),\dot\xi(t))dt + e^{-r_k T} v_k(\xi(T))$$

En retranchant $v_k(0)$ des deux membres et en passant à la limite, on obtient :

(5.16) $$v(x_o) \leqslant \int_0^T f(\xi,\dot\xi)dt + v(\xi(T)) - \lambda T$$

Soit ξ_k une trajectoire issue de x_o optimale pour $s = r_k$, on a

$$v_k(x_o) = \int_0^T r^{-r_k t} f(\xi_k,\dot\xi_k)dt + e^{-r_k T} v_k(\xi_k(T))$$

D'après (5.9) on a $\|\dot\xi_k\| \leqslant c_7$. Quitte à extraire une sous-suite (encore notée r_k) il existe $\xi_\infty \in Q$ tel que $\xi_k \to \xi_\infty$ (resp. $\dot\xi_k \to \dot\xi_\infty$) uniformément (resp. faiblement) sur tout intervalle $[0,T]$. On en déduit par passage à la limite que

(5.17) $$v(x_o) \geqslant \int_0^T f(\xi_\infty,\dot\xi_\infty)dt + v(\xi_\infty(T)) - \lambda T$$

De (5.16) et (5.17) on déduit (5.14). En divisant les deux membres de (5.14) par T et en passant à la limite $(T \to \infty)$ on obtient (5.12).

On a montré que pour toute suite s_m qui tend vers zéro , il existe une sous-suite s_{m_k} telle que $s_{m_k} u_{s_{m_k}}$ converge vers la constante λ définie par (5.12) On en déduit que $su_s \to \lambda$ quand $s \to 0$.

Au paragraphe précédent, nous avons défini pour $\varepsilon > 0$, les constantes λ_ε et les fonctions v_ε. On a la

<u>Proposition 5.5.</u> <u>Quand</u> $\varepsilon \to 0$, $\lambda_\varepsilon \to \lambda$. <u>De toute suite</u> ε_k <u>telle que</u> $\varepsilon_k \to 0$, <u>on peut extraire une sous-suite notée encore</u> ε_k <u>telle que</u> $\nabla v_{\varepsilon_k} \to \nabla v$ <u>presque partout où</u> v <u>est une fonction qui vérifie (5.13).</u>

On a vu (cf. démonstration de la proposition 5.3) que $\frac{\partial^2}{\partial x^2} u_{s,\varepsilon} \leqslant k_2$ pour toute direction $x(\|x\| = 1)$. D'où l'on déduit comme dans la démonstration 5.4 que $\left\|\frac{\partial^2}{\partial x_i \partial x_j} u_{s,\varepsilon}\right\|_1 \leqslant k_2$. Comme $\frac{\partial^2}{\partial x_i \partial x_j} u_{s,\varepsilon} \longrightarrow \frac{\partial^2}{\partial x_i \partial x_j} v_\varepsilon$ on a $\left\|\frac{\partial^2}{\partial x_i \partial x_j} v_\varepsilon\right\| \leqslant k_2$.
Il en résulte que ∇v_ε reste dans un compact de L^1 quand $s \to 0$. D'autrepart $0 \leqslant \lambda_\varepsilon \leqslant M$ et $|\nabla v_\varepsilon| \leqslant k_1$. Soit $\varepsilon_k \to 0$. Il existe une sous-suite (encore notée ε_k) telle que $\lambda_\varepsilon \to \mu$ et $\nabla v_\varepsilon \to \nabla v$. Par passage à la limite on obtient $-\mu + g(x,\nabla u) = 0$ à partir de (4.3).

Il suffit de prouver maintenant que $\mu = \lambda$ où λ est la constante définie par 5.12. Pour cela nous allons procéder comme dans la démonstration de la proposition 5.2.

Soit v une fonction qui vérifie (5.13). Soit $x_o \in \mathbb{R}^n$. Soit ξ_o une trajectoire optimale asymptotique issue de x_o et soit a_ε le contrôle en boucle fermée associé à $h_\varepsilon=\phi(\cdot,\nabla v_\varepsilon(\cdot))$. Soit η_ε , θ_ε et ξ_ε défini comme dans la démonstration de la proposition 5.2. On a

(a) $$v(x_o) = \int_0^T f(\xi_o, \dot{\xi}_o) - \lambda]\, dt + v(\xi_o(T))$$

(b) $$v_\varepsilon(x_o) = E\left[\int_0^T [f(\eta_\varepsilon, h_\varepsilon(\eta_\varepsilon)) - \lambda_\varepsilon]\, dt + v_\varepsilon(\eta_\varepsilon(T))\right]$$

(c) $$v(x_o) \leqslant E\int_0^T [f(\theta_\varepsilon, \dot{\theta}_\varepsilon) - \lambda]\, dt + v(\theta_\varepsilon(T))]$$

(d) $$v_\varepsilon(x_o) \leqslant E\int_0^T [f(\xi_\varepsilon, \dot{\xi}_o) - \lambda_\varepsilon]\, dt + v_\varepsilon(\xi_\varepsilon(T))$$

Par différence entre (b) et (c) on obtient

$$-\lambda_\varepsilon + \lambda \leqslant k_5 E\int_0^T \inf(1, \varepsilon|w(t) - w(0)|)dt + \frac{k_6}{T}$$

où $k_6 = (\max v_\varepsilon - \min v_\varepsilon) + (\max v - \min v) \leqslant 2k_1 n$ (puisque v_ε et v donc périodiques et lipschitziennes de rapport k_1) et où $k_5 = \operatorname{Sup} \dfrac{|f(x,z) - f(y,z)|}{|x-y|}$

pour $\|x-y\| \leqslant 1$ et $\|z\|$ $\operatorname{Sup}\{|\phi(x,p)| \; ; \; |p| \leqslant k_1\}$.

Soit $\alpha > 0$. On choisit $T = \dfrac{k_6}{\alpha}$. Pour ε assez petit on a

$$\int_0^T \inf(1, \varepsilon|w(t) - w(0)|)dt < \alpha, \text{ d'où } \lambda - \lambda_\varepsilon \leqslant 2\alpha.$$

A l'aide de (a) et (d) on obtiendra l'inégalité inverse. Donc $\lambda_\varepsilon \to \lambda$.

Bibliographie

CLARK , Thèse , à paraitre.

W.H. FLEMING , The Cauchy problem for a non linear first order partial differential equation , J. of differential equations 5 , 1969 , p. 515 à 530.

A. FRIEDMAN , Partial differential equations of parabolic type , Englewood Cliffs , N. J. Prentice Hall , Inc , 1963.

O.A. LADYZENSKAJA et N.N. URALCEVA , Equations elliptiques linéaires et quasi-linéaires Dunod , Paris , 1967.

ON THE EQUIVALENCE OF MULTISTAGE RECOURSE MODELS IN STOCHASTIC OPTIMIZATION

R.T. ROCKAFELLAR*

The following abstract model covers many stochastic optimization problems requiring a sequence of decisions. In each of N stages, an element x_k is chosen from a space X_k. Ultimately a cost $f(\xi_1,\ldots,\xi_N, x_1,\ldots,x_N)$ must be paid, where the elements ξ_k, belonging to spaces Ξ_k, represent exterior factors beyond the control of the decision maker. The cost may be $+\infty$, as a representation of the fact that certain combinations of $\xi = (\xi_1,\ldots,\xi_N)$ and $x = (x_1,\ldots,x_N)$ are impossible or forbidden. It is assumed that only $(\xi_1,\ldots,\xi_k)$ is known with certainty at the time when x_k must be chosen. The only other information available about the exterior factors is that the occurrences of ξ are governed by a known probability distribution. The problem is to determine "decision rules" such that the overall expected cost is minimized.

To make more precise, let us suppose that each Ξ_k is a Hausdorff topological space, and σ is a regular Borel probability measure on $\Xi_1 \times\ldots\times \Xi_N$. The support of σ will be denoted by Ξ. A function

$$x : \Xi \to X_1 \times\ldots\times X_N$$

is <u>nonanticipative</u> if it is of the form

$$x(\xi) = (x_1(\xi_1)\ ,\ x_2(\xi_1,\xi_2),\ldots,\ x_N(\xi_1,\ldots,\xi_N)). \tag{1}$$

The problem may be formulated as that of minimizing the functional

$$F(x) = \int_\Xi f(\xi,\ x(\xi))\ \sigma\ (d\xi) \tag{2}$$

over some class of nonanticipative functions x.

Of course, the integral (2) may not be well defined, unless further conditions are imposed. But these can be of a very general nature. Assume that the spaces X_k are topological, and that f is Borel measurable. If $x(\xi)$ is measurable in ξ, then the mapping $\xi \to (\xi,\ x(\xi))$ is Borel measurable, and hence $f(\xi,\ x(\xi))$ is measurable in ξ. The following convention is then adopted : if either the positive or negative part of the function $\xi \to f(\xi,\ x(\xi))$ is summable, the integral $F(x)$ is assigned its classical

* Supported in part by grant AF-AFOSR-72-2269.

value (possibly $+\infty$ or $-\infty$), while otherwise $F(x)$ is taken to be $+\infty$.

Observe that under this convention the inequality $F(x) < +\infty$ implies

$$x(\xi) \in D(\xi) \text{ for almost every } \xi \in \Xi, \tag{3}$$

where the set

$$D(\xi) = \{x \mid f(\xi,x) < +\infty\} \tag{4}$$

is the <u>implicit</u> <u>feasible</u> <u>region</u> of the decision space $X_1 \times\ldots\times X_N$. Thus in minimizing F over a class of measurable functions one is, in effect, minimizing subject to the constraint (3).

This note is concerned with two fundamental questions that arise in justifying and analyzing the model. First, to what extent is (3) essentially equivalent to the stronger constraint

$$x(\xi) \in D(\xi) \quad \text{for every } \xi \in \Xi , \tag{5}$$

which in some contexts appears more natural ? In other words, what conditions are needed to insure that a measurable, nonanticipative function x satisfying (3) can be converted to one satisfying (5) by alteration on a set of measure zero ? Secondly, when is it true that the infimum in the problem can be approached by functions x which are actually continuous ? The "nonanticipative" property renders these questions quite difficult for $N > 1$, and no one has previously provided any answers.

Our purpose is to describe some results in this direction in the case where

$$X_1 \times\ldots\times X_N = R^{n_1} \times\ldots\times R^{n_N} = R^n \tag{6}$$

and certain convexity, semicontinuity and summability conditions are satisfied by the cost function

$$f : \Xi \times R^n \to R \cup\{+\infty\}.$$

Most of the proofs will appear elsewhere [1].

The context is delimited by the following basic assumptions.

(A1) The support Ξ of the probability measure σ is compact.

(A2) For each $\xi \in \Xi$, $f(\xi,x)$ is convex and lower-semicontinuous as a function of $x \in R^n$.

(A3) For each $\xi \in \Xi$, the set $D(\xi)$ has a nonempty interior.

(A4) The multifunction $\xi \to \mathrm{c}\ell\, D(\xi)$ is continuous from Ξ to R^n (i.e. lower-semicontinuous with closed graph).

(A5) For each $x \in R^n$, $f(\xi,x)$ is measurable as a function of $\xi \in \Xi$.
(A6) Whenever $U \subset \Xi$ is open (relative to Ξ), $V \subset R^n$ is open, and f is finite on $U \times V$, one has

$$\int_U | f(\xi,x) | \sigma(d\xi) < +\infty \quad \text{for each } x \in V.$$

Example 1.

Let $X \subset R^n$ be a closed convex set, and for $i = 0,1,\ldots,m$ let f_i be a real valued (finite) function on $\Xi \times R^n$ such that $f_i(\xi,x)$ is convex in x. Let

$$f(\xi,x) = \begin{cases} f_o(\xi,x) & \text{if } x \in X \text{ and } f_i(\xi,x) \leq 0,\ i = 1,\ldots,m, \\ +\infty & \text{otherwise.} \end{cases}$$

Then f satisfies (A2), and we have

$$D(\xi) = \{x \in X \mid f_i(\xi,x) \leq 0 \ , \ i = 1,\ldots,m\} \ .$$

Suppose that X has a nonempty interior and $f_i(\xi,x)$ is continuous in ξ for $i = 1,..,m$. If for each $\xi \in \Xi$ the set

$$\{x \in X \mid f_i(\xi,x) < 0 \ , \quad i = 1,\ldots,m\}$$

is nonempty, it follows by routine convexity arguments that (A3) and (A4) hold. If furthermore $f_o(\xi,x)$ is a summable function of $\xi \in \Xi$ for each $x \in X$, then (A6) and (A7) are obviously satisfied as well.

LEMMA. Assumptions (A2), (A3) and (A5) imply in particular that f is Borel measurable on $\Xi \times R^n$, so that F(x) is well-defined in the above sense whenever $x(\xi)$ is measurable in ξ.

Proof.

These assumptions imply that f is a normal convex integrand in the sense of [2,Lemma 2]. On the other hand, every such integrand is Borel measurable [3, Theorem 5].

A function $x : \Xi \to R^n$ will be called essentially nonanticipative if it can be made into a measurable nonanticipative function by altering its values on a set of measure zero. Let $\mathcal{N}_\infty$ denote the set of all such functions which are essentially bounded. If we like, we can identify $\mathcal{N}_\infty$ with a certain closed linear subspace of the Banach space $\mathcal{L}_n^\infty = \mathcal{L}^\infty(\Xi,\sigma ; R^n)$ consisting of all essentially bounded functions $x : \Xi \to R^n$. In fact, $\mathcal{N}_\infty$ is then closed not only with respect to the norm topology, but also the weak topology induced on $\mathcal{L}^\infty_n$ by the natural pairing with $\mathcal{L}^1_n = \mathcal{L}^1(\Xi,\sigma\,;\,R^n)$.

The following preliminary result is derived in [1] from the theory of convex integral functionals.

PROPOSITION.

Under assumptions (A1)-(A6), F *is a convex functional from* $\mathcal{L}^\infty_n$ *to* $R \cup \{+\infty\}$ *which is lower-semicontinuous, not only with respect to the norm topology, but also the weak topology induced on* $\mathcal{L}^\infty_n$ *by* $\mathcal{L}^1_n$. *Furthermore*, F *is (norm) continuous on*

$$\mathcal{W} = \{x\ \mathcal{L}^\infty_n \mid \exists\ \varepsilon > 0 \text{ with } x(\xi) + \varepsilon B \subset D(\xi) \text{ a.e.}\}$$

(*where* B *is the unit ball of* R^n), *and this set* $\mathcal{W}$ *is the nonempty (norm) interior of* $\{x \in \mathcal{L}^\infty_n \mid F(x) < +\infty\}$.

COROLLARY.

Suppose there is a compact set $X \subset R^n$ *such that* $D(\xi) \subset X$ *for all* $\xi \in \Xi$. *Then the infimum of* F(x) *over all* $x \in \mathcal{N}_\infty$ *is attained*.

The corollary, which provides an *existence theorem* for solutions to the stochastic optimization problem, is obtained from the fact that the set of functions $x \in \mathcal{L}^\infty_n$ satisfying $x(\xi) \in X$ almost everywhere is compact in the weak topology induced by $\mathcal{L}^1_n$. The hypothesis is satisfied, of course, in the case of Example 1 if the set X introduced there is bounded.

At all events, note that under the hypothesis of the corollary there is no loss of generality in the basic problem when the minimization is restricted to nonanticipative functions x which are essentially bounded, or in other words, when the problem is identified with that of minimizing F over $\mathcal{N}_\infty$. This formulation appears the best suited for obtaining strong results, at least in terms of convex analysis and duality.

The questions raised earlier concern the "almost everywhere" aspects of this formulation of the problem, as well as the relationship between minimizing over $\mathcal{N}_\infty$ and minimizing over $\mathcal{N}_\mathcal{C}$, the subspace of $\mathcal{N}_\infty$ consisting of the *continuous* nonanticipative functions. Our main result is the following (see [1]).

THEOREM.

Suppose in addition to (A1)-(A6) *that the measure* σ *is* "*laminary*", *as defined below. Then every* $x \in \mathcal{N}_\infty$ *with* $F(x) < +\infty$ *can be converted, by alteration on a set of measure zero, to a measurable, (truly) bounded, (truly) nonanticipative function*

<u>satisfying</u> $x(\xi) \in c\ell\, D(\xi)$ <u>for every</u> $\xi \in \Xi$.

<u>If furthermore</u> $\mathcal{W} \cap \mathcal{N}_\infty \neq \emptyset$, <u>then</u> $\mathcal{W} \cap \mathcal{N}_e \neq \emptyset$ <u>and</u>

$$\inf \{F(x) \mid x \in \mathcal{N}_\infty\} = \inf \{F(x) \mid x \in \mathcal{N}_e\} .$$

To define what is meant by "laminary", we need some notation. For any set $S \subset \Xi$ and index k, $1 \leq k < N$, let

$$(7) \qquad \Lambda_k^S(\xi_1,\ldots,\xi_k) = \{(\xi_{k+1},\ldots,\xi_N) \mid (\xi_1,\ldots,\xi_k\, ,\, \xi_{k+1},\ldots,\xi_N) \in S\} ,$$

$$(8) \qquad S^k = \{(\xi_1,\ldots,\xi_k) \mid \Lambda_k^S(\xi_1,\ldots,\xi_k) \neq \emptyset\} .$$

We say that the measure σ is <u>laminary</u> if it satisfies :

(i) The multifunction Λ_k^Ξ is lower-semicontinuous relative to Ξ^k, and

(ii) Whenever S is a Borel subset of Ξ with $\sigma(S) = \sigma(\Xi)$ such that S^k is a Borel set, then $\Lambda_k^S(\xi_1,\ldots,\xi_k)$ is dense in $\Lambda_k^\Xi(\xi_1,\ldots,\xi_k)$ for almost every $(\xi_1,\ldots,\xi_k)$ in S^k (with respect to the "projection" of σ on Ξ^k).

It is not hard to show, for instance, that σ is laminary if

$$(9) \qquad \sigma(d\xi) = \rho(\xi_1,\ldots,\xi_N)\, \pi_1(d\xi_1) \ldots \pi_N(d\xi_N) ,$$

where π_k is a (nonnegative regular Borel) measure on Ξ_k for $k = 1,\ldots,N$, and the density function ρ is positive on the support of the product measure $\pi_1 \times\ldots\times \pi_N$. (This follows from Fubini's theorem and the fact that in this case the multifunctions Λ_k^Ξ are constant-valued).

Even the first conclusion in the theorem can fail, without the presence of the two properties in the definition of "laminary". This can be demonstrated by counterexamples.

<u>Example 2</u>.

This is a two-stage example where $\Xi_1 = \Xi_2 = R$ and $R^{n_1} = R^{n_2} = R$. Let the interval $[0,1]$ be expressed as the union of two disjoint subsets A and A' of positive measure, such that A is dense and A' is closed, and let

$$T = (A \times [0,2]) \cup (A' \times [0,1]) .$$

Define the Borel measure σ on R^2 by

$$\sigma(S) = \text{mes}\,(S \cap T) \,/\, \text{mes}\, T ,$$

where "mes" denotes Lebesgue measure. Then σ is a probability measure whose support is

$$\Xi = c\ell\, T = [0,1] \times [0,2] .$$

But σ does not satisfy property (ii) in the definition of "laminary" (take $S = T$), even though σ is absolutely continuous with respect to a product measure. Define f on $\Xi \times R$ by

$$f(\xi_1,\xi_2,x_1,x_2) = \begin{cases} x_1 & \text{if } 0 \leq x_2 \leq x_1 - \xi_2, \\ +\infty & \text{otherwise}, \end{cases}$$

so that

$$D(\xi_1,\xi_2) = c\ell\, D(\xi_1,\xi_2) = \{(x_1,x_2) \in R^2 \mid 0 \leq x_2 \leq x_1 - \xi_2\}.$$

It may be verified that assumptions (A1)-(A6) are satisfied and

$$\min \{F(x) \mid x \in \mathcal{N}_\infty\} = F(\bar{x}) < 2 ,$$

where

$$\bar{x}(\xi) = (\bar{x}(\xi_1), \bar{x}_2(\xi_1,\xi_2)) = \begin{cases} (2,0) & \text{if } \xi_1 \in A , \\ (1,0) & \text{if } \xi_1 \in A' . \end{cases}$$

But if the stronger condition (5) is imposed, the minimum is instead $F(\bar{\bar{x}}) = 2$, where

$$\bar{\bar{x}}(\xi) = (\bar{\bar{x}}_1(\xi_1) , \bar{\bar{x}}_2(\xi_1,\xi_2)) \equiv (2,0).$$

Thus the constraint conditions (3) and (5) are <u>not</u> "equivalent" in this case, and in particular the first assertion of the theorem is false for $x = \bar{x}$.

<u>Example 3</u>.

Again we consider a two-stage case with $\Xi_1 = \Xi_2 = R$ and $R^{n_1} = R^{n_2} = R$, but this time it is only property (i) of the definition of "laminary" which is lacking, and still the first assertion of theorem is false. The probability measure is

$$\sigma(S) = \frac{1}{2} \text{ mes } (S \cap T) ,$$

where

$$T = ([0,1] \times [0,1]) \cup ([-1,0] \times [-1,0]) .$$

We have $\Xi = T$, so Λ_1^{Ξ} is not lower-semicontinuous at $\xi_1 = 0$. Let

$$f(\xi_1,\xi_2,x_1,x_2) = \begin{cases} 0 \text{ if } \xi_2 \geq 0 \text{ and } -2+3\xi_2 \leq x_1 \leq 2 , \\ 0 \text{ if } \xi_2 \leq 0 \text{ and } -2 \leq x_1 \leq 2+3\xi_2 , \\ +\infty \text{ otherwise.} \end{cases}$$

Assumptions (A1)-(A6) are satisfied, and

$$\min \{F(x) \mid x \in \mathcal{N}_\infty\} = F(\bar{x}) = 0$$

for the function

$$\bar{x}(\xi) = (\bar{x}_1(\xi_1), \bar{x}_2(\xi_1,\xi_2)) = \begin{cases} (3/2,0) \text{ if } \xi_1 > 0 , \\ (0,0) \text{ if } \xi_1 = 0 , \\ (-3/2,0) \text{ if } \xi_1 < 0. \end{cases}$$

In fact $\bar{x}(\xi) + \frac{1}{2} B \subset D(\xi)$ almost everywhere. But it is easy to see there does not exist any nonanticipative function x whatsoever which satisfies $x(\xi) \in D(\xi)$ for all $\xi \in \Xi$.

The proof of the first assertion of the theorem makes use of convexity mainly just as a matter of convenience in terms of the formulation. However, for the rest of the theorem, concerning approximation of the infimum via continuous recourse functions x, convexity seems to be essential. The basic tool is a theorem of E. MICHAEL [4] on the existence of continuous selections, and convexity is already an important hypothesis in this result, as is well understood.

Of course, the proof is not effected by means of a single continuous selection, but by a certain sequence of N selections, each from a multifunction dependant on the preceding selections. This is the great complication caused by the requirement of nonanticipativity. The multifunctions must be constructed in such a way that Michael's theorem is applicable at each step, and here convexity seems to play a crucial role over and over again.

The argument establishing the first part of the theorem is similarly complicated, but a sequence of measurable, rather than continuous, selections is involved.

These results are motivated especially by applications to Example 1. A theory of Lagrange multipliers and duality for this case, based heavily on them, is outlined in [5]. The multipliers are certain measures, not necessarily absolutely continuous with respect to the underlying probability measure σ. This theory provides an alternative to the approach of R. WETS and the author in [6], where the multipliers in general can take the form of elements of the dual of an $\mathcal{L}^\infty$ space.

REFERENCES

[1] R.T. ROCKAFELLAR and R.J.B. WETS,
"Continuous versus measurable recourse in N-stage stochastic programming",
J. Math. Analysis Appl., to appear.

[2] R.T. ROCKAFELLAR,
"Integrals which are convex functionals",
Pacific J. Math. 24 (1968), 525-539.

[3] R.T. ROCKAFELLAR,
"Measurable dependance of convex sets and functions on parameters",
J. Math. Analysis Appl. 28 (1969), 4-25.

[4] E. MICHAEL,
"Continuous selections, I",
Ann. of Math. 63 (1956), 361-382.

[5] R.T. ROCKAFELLAR,
"Lagrange multipliers for an N-stage model in stochastic programming",
Colloque d'Analyse Convexe (St-Pierre-de-Chartreuse, 1974),
J.P. AUBIN (editor), Springer-Verlag, to appear.

[6] R.T. ROCKAFELLAR and R.J.B. WETS,
"Stochastic convex programming : basic duality" and "Stochastic convex programming : extended duality and singular multipliers",
Pacific J. Math., to appear.

THE INTRINSIC MODEL FOR DISCRETE STOCHASTIC CONTROL: SOME OPEN PROBLEMS

by

H. S. Witsenhausen
Bell Laboratories
Murray Hill, New Jersey

1. INTRODUCTION

An important difference between deterministic and stochastic control is that in the stochastic case much depends on what data are available for each control decision. In the simplest case all control decisions are made from one station where also all data are gathered. The station remembers all its observations and can base each decision on all data gathered up to the time at which the decision must be made. However not all systems are completely centralized in this fashion nor is it always true that unlimited memory is available to store data. There may be several control stations communicating only by signaling through the system itself or through noisy channels which should be considered part of the system. Likewise the storage devices for data are also parts of the system to be controlled. In this view each control selection is a separate event, the selection being based on data provided out of the system specifically for that event. Whether this data comes from sensors, communication channels or from memory devices is only a matter of specific detail.

The data available for a certain decision may be insufficient to determine what the control values chosen at earlier decisions were. Worse yet, the data may be insufficient to determine which decisions have already been made and which are in the future and could possibly have their data dependent upon the decision under consideration. This is because for any agent (device) which is to implement a decision, the time (and place) of that decision may depend upon the random inputs to the system and on the values decided upon by other agents.

What goes on in the system is determined by the realization of the noise variables and the realization of all control variables. The cost whose expectation is to be minimized is some function of this set of variables. For each decision, the data available for that decision are functions of the same set of variables, determining a partition or σ-field in the corresponding space. The cost function and the σ-fields constitute a complete specification of the problem for control purposes.

The theory of this general formulation is at a very early stage of development. In this paper an appropriate mathematical model is described and some of the questions that arise are brought into focus.

2. THE INTRINSIC MODEL

The information structure analysis proposed in [9] for n person games applies, for n = 1, to the stochastic control problem with finite number of actions. In that case, the probabilities of actions of nature being known, no randomization of the policy need be considered. It becomes therefore more convenient to recognize the special role of nature's action as the sole source of uncertainty and consider the set A of n agents which form the controlling organization, excluding nature. In the formulation used in [9] nature would then appear as an $(n+1)^{th}$ agent which is the starting agent.

Formally, let $(\Omega, \mathcal{B}, P)$ be the probability space for actions ω of nature and $(U_\alpha, \mathfrak{F}_\alpha)$ the measurable space in which agent $\alpha \in A$ chooses his action u_α. Assume that $\mathcal{B}$ and the $\mathfrak{F}_\alpha$ contain all singletons.

For $B \subset A$ let H_B be the product space $\Omega \times \prod_{\alpha \in B} U_\alpha$ with the product σ-field $\mathcal{B} \times \prod_{\alpha \in B} \mathfrak{F}_\alpha = \mathfrak{F}_B(B)$. For $C \subset B$ let $\mathfrak{F}_B(C)$ be the subfield of $\mathfrak{F}_B(B)$ which is the inverse image of $\mathfrak{F}_C(C)$ under the canonical projection of H_B onto H_C. Note that for $A \supset B \supset C \supset D$ one has $\mathfrak{F}_B(C) \supset \mathfrak{F}_B(D)$, $\mathfrak{F}_B(C) \cap \mathfrak{F}_B(D) = \mathfrak{F}_B(C \cap D)$, $\mathfrak{F}_B(C) \vee \mathfrak{F}_B(D) = \mathfrak{F}_B(C \cup D)$.

Also, $H_{\emptyset} = \Omega$, and $\mathcal{F}_B(\emptyset)$ is the cylindrical extension of $\mathcal{B}$ to H_B. Let $U_B = \prod_{\alpha \in B} U_\alpha$ with the product σ-field. The abbreviations H for H_A, U for U_A, $\mathcal{F}(B)$ for $\mathcal{F}_A(B)$ will be used.

Thus for $A \supset B \supset C$ one has $H_B = H_C \times U_{B-C}$ and likewise for the σ-fields. In particular, $H = \Omega \times U$.

The information available to agent α is characterized by a subfield $\mathcal{J}_\alpha$ of $\mathcal{F}(A)$. The possible control laws for agent α are the functions γ_α: $H \to U_\alpha$ which are measurable from $\mathcal{J}_\alpha$ to $\mathcal{F}_\alpha$, they form the set Γ_α. The design of control for subset $B \subset A$ can thus be chosen from the set $\Gamma_B = \prod_{\alpha \in B} \Gamma_\alpha$ and the whole design is chosen in $\Gamma = \Gamma_A$.

Rephrasing the definitions given in [9], the system $(A, \Omega, \mathcal{B}, (U_\alpha, \mathcal{F}_\alpha, \mathcal{J}_\alpha)_{\alpha \in A})$ is said to be <u>solvable</u> if for all $\omega \in \Omega$ and $\gamma \in \Gamma$ the simultaneous equations

$$u_\alpha = \gamma_\alpha(h) \equiv \gamma_\alpha(\omega, u) \tag{1}$$

have one and only one solution $u \in U$. Then $u = \Sigma_\gamma(\omega)$ defines the <u>solution map</u>. (A weaker definition would be to require that for each $\gamma \in \Gamma$ and $\omega \in \Omega_\gamma$, (1) has a unique solution and $P(\Omega_\gamma) = 1$. However we are interested in properties valid for all P.)

The <u>causality</u> condition of [9] takes the following form: there exists (at least one) function φ from H into the set S of total orderings of A, with the property that for $1 \leq k \leq n$ and any ordered set $(\alpha_1, \ldots, \alpha_k)$ of distinct elements of A, the set $E \subset H$ on which $\varphi(h)$ begins with $(\alpha_1, \ldots, \alpha_k)$ satisfies(*)

$$\forall F \in \mathcal{J}_{\alpha_k}, \quad E \cap F \in \mathcal{F}(\{\alpha_1, \ldots, \alpha_{k-1}\}).$$

(*) Note that $\mathcal{F}(\emptyset)$ is the cylindrical extension of $\mathcal{B}$ to H in contrast to [9].

It was shown in [9] that causality implies (recursive) solvability with a measurable solution map. The converse is false, but solvability implies at least the absence of self-information, i.e. $\mathcal{J}_\alpha \in \mathfrak{F}(A-\alpha)$ for all α.

If for some ordering $(\alpha_1,\ldots,\alpha_n)$ of A one has $\mathcal{J}_{\alpha_k} \in \mathfrak{F}(\{\alpha_1,\ldots,\alpha_{k-1}\})$ for $1 \leq k \leq n$, then the system is called sequential and satisfies the causality condition with a constant function φ. [10]

The system is classical if it is sequential and in addition $\mathcal{J}_{\alpha_1} \in \mathfrak{F}(\emptyset)$, $\mathcal{J}_{\alpha_{k-1}} \subset \mathcal{J}_{\alpha_k}$ for $k = 2,\ldots,n$. A (static) team is a system with $\mathcal{J}_\alpha \in \mathfrak{F}(\emptyset)$ for all α, i.e., all information depends only on ω. [5]

3. THE CAUSALITY PROBLEM

For a causal system the set Φ of all functions φ from H to total orders on A for which the causality condition holds may be a large one. For instance, when the system is a static team Φ contains in particular all $\mathfrak{F}(\emptyset)$-measurable functions.

The problem is to find a simple characterization of Φ for instance in terms of a function from H to partial orders on A. The advantage of such a characterization is that for a given system Φ is unique.

For each h in H, as φ runs through Φ one obtains a set of total orders $\varphi(h)$ on A. Let $\psi(h)$ be the strongest partial order on A compatible with all these total orders. Is ψ a characterization of Φ, that is, can Φ be recovered from ψ?

4. SUBSYSTEMS

Heuristically, a set of agents form a subsystem if the data received by the members of the set depends only on the action of nature and the actions of the members of the set but not on the actions of nonmembers.

Thus a subset B of A is a subsystem if for all $\beta \in B$, one has $\mathcal{J}_\beta \subset \mathfrak{F}(B)$.

Any subfield $\mathcal{G}$ of $\mathcal{F}(B)$ is the cylindrical extension to H of a subfield $\mathcal{G}_B$ of $\mathcal{F}_B(B)$ on H_B. Thus if B is a subsystem then $(B, \Omega, \mathcal{B}, (U_\beta, \mathcal{F}_\beta, \mathcal{J}_{\beta B})_{\beta \in B})$ is a system in its own right, carried by the space H_B, with $\mathcal{J}_{\beta B}$ the projection of $\mathcal{J}_\beta$ upon H_B.

Lemma 1. The subsystems are just the closed sets of a topology τ on A.

Proof: Since A is finite this reduces to the assertion that the union and the intersection of two subsystems are again subsystems. For B, C subsystems and $\alpha \in B \cup C$ one has either $\mathcal{J}_\alpha \subset \mathcal{F}(B)$ or $\mathcal{J}_\alpha \subset \mathcal{F}(C)$ or both, hence $\mathcal{J}_\alpha \subset \mathcal{F}(B) \vee \mathcal{F}(C) = \mathcal{F}(B \cup C)$ and $B \cup C$ is a subsystem. For $\alpha \in B \cap C$ one has $\mathcal{J}_\alpha \in \mathcal{F}(B) \cap \mathcal{F}(C) = \mathcal{F}(B \cap C)$ so that $B \cap C$ is a subsystem. □

In general even for a causal system there may be no proper subsystems, that is $\tau = \{\emptyset, A\}$. If A is connected under τ the complement of a proper subsystem is never a subsystem. Otherwise A is the union of its connected components which are dynamically decoupled subsystems: the information of members of each component subsystem is independent of the actions of the agents in the other components. A static coupling remains through the common dependence upon ω. [It may happen that the members of component C_i have information fields contained in $\mathcal{B}_i \times \prod_{\alpha \in C_i} \mathcal{F}_\alpha \times \prod_{\alpha \notin C_i} \{\emptyset, U_\alpha\}$ and that the subfields $\mathcal{B}_i$ of $\mathcal{B}$ are independent under distribution P. Then the information decoupling is complete but coupling through the criterion of optimization remains. However, that is a P-dependent property].

If (A, τ) is a Hausdorff space then all subsets are closed and A decomposes into its singletons.i.e., $\mathcal{J}_\alpha \in \mathcal{F}(\{\alpha\})$. If there is no self-information $\mathcal{J}_\alpha \subset \mathcal{F}(A - \{\alpha\})$ as well, hence $\mathcal{J}_\alpha \subset \mathcal{F}(\emptyset)$ which is just the case of (static) team theory. This may be stated as

Lemma 2. The following are equivalent

(i) the system is a (static) team.

(ii) the system has no self-information and τ is Hausdorff.

The τ-closure $\bar{B}$ of a subset B of A is the smallest subsystem containing B. Most important are the closures of singletons. Define a binary relation $\leftarrow$ on A by $\alpha \leftarrow \beta$ iff $\overline{\{\alpha\}} \subset \overline{\{\beta\}}$. Clearly $\leftarrow$ is reflexive and transitive hence a quasiorder. However in general it is not antisymmetric because one may have $\overline{\{\alpha\}} = \overline{\{\beta\}}$ for $\alpha \neq \beta$. For instance, when $\tau = \{\emptyset, A\}$ then $\overline{\{\alpha\}} = A$ for all α, so that $\alpha \leftarrow \beta$ for all α, β.

Recall that a T_o-space is a topological space in which of two distinct points at least one has a neighborhood not containing the other.

Theorem 1. For a system without self-information the following are equivalent

(i) (A, τ) is a T_o space

(ii) the relation $\leftarrow$ is a partial order

(iii) the system is sequential.

Proof: (i) $\Rightarrow$ (ii): Suppose $\alpha \leftarrow \beta$ and $\beta \leftarrow \alpha$, if one had $\alpha \neq \beta$ then by the T_o property one of these points, say α, has a closure $\overline{\{\alpha\}}$ not containing β. A fortiori $\overline{\{\alpha\}}$ does not contain $\overline{\{\beta\}}$ contradicting $\beta \leftarrow \alpha$. Hence $\alpha = \beta$ and relation $\leftarrow$ is antisymmetric. An antisymmetric quasiorder is a partial order.

(ii) $\Rightarrow$ (iii) Any partial order is compatible with at least one total order. That is, there exists an ordering $(\alpha_1, \ldots, \alpha_n)$ of A such that $\alpha_i \leftarrow \alpha_j$ implies $i \leq j$. Then for $1 \leq k \leq n$, one has $\overline{\{\alpha_k\}} = \{\beta | \beta \in \overline{\{\alpha_k\}}\} = \{\beta | \overline{\{\beta\}} \subset \overline{\{\alpha_k\}}\} = \{\beta | \beta \leftarrow \alpha_k\} \subset \{\alpha_i | i \leq k\}$. Since $\overline{\{\alpha_k\}}$ is a subsystem $\mathcal{J}_{\alpha_k} \subset \mathfrak{F}(\overline{\{\alpha_k\}}) \subset \mathfrak{F}(\{\alpha_i | i \leq k\})$. Since there is no self information $\mathcal{J}_{\alpha_k} \subset \mathfrak{F}(A - \{\alpha_k\})$. Thus

$$\mathcal{J}_{\alpha_k} \subset \mathfrak{F}((A - \{\alpha_k\}) \cap \{\alpha_i | i \leq k\}) = \mathfrak{F}(\{\alpha_1, \ldots, \alpha_{k-1}\})$$

which shows that the system is sequential.

(iii) $\Rightarrow$ (i) Suppose the system is sequential under the ordering $(\alpha_1, \ldots, \alpha_n)$. Then for each i $\mathcal{J}_{\alpha_i} \subset \mathfrak{F}(\alpha_1, \ldots, \alpha_{i-1})$ so that

$\{\alpha_1,\ldots,\alpha_k\}$ is closed for all k. Given two distinct elements α_i, α_j in A with say $i < j$, the set $\{\alpha_{i+1},\ldots,\alpha_n\}$ is an open set containing α_j but not α_i, verifying the T_o separation property. □

Sequential systems have many properties.

Lemma 3: For a sequential system

(i) the set of all (total) orderings under which the system is sequential is precisely the set of all those compatible with partial order $\leftarrow$.

(ii) If $(\alpha_1,\ldots,\alpha_n)$ is any compatible ordering then $\overline{\{\alpha_k\}} \subset \{\alpha_1,\ldots,\alpha_k\} = \overline{\{\alpha_1,\ldots,\alpha_k\}}$

(iii) For all α, $\overline{\{\alpha\}} - \{\alpha\}$ is closed

Proof:

(i) If a total order is compatible with $\leftarrow$ the argument used in the proof of Theorem 1 shows that the system is sequential under that order. If an ordering $(\alpha_1,\ldots,\alpha_n)$ is incompatible with $\leftarrow$ there are indices i, j with $i < j$, $\alpha_j \leftarrow \alpha_i$, then the set $\{\alpha_1,\ldots,\alpha_i\}$ does not contain α_j while its closure must. Hence this set is not closed as it would have to be if the system were sequential under that order.

(ii) By the sequential property $\{\alpha_1,\ldots,\alpha_k\}$ is closed, hence contains the closure of each of its elements.

(iii) $B \equiv \overline{\{\alpha\}} - \{\alpha\} = \{\beta \mid \beta \leftarrow \alpha\} - \{\alpha\}$ thus if $\gamma \in B$ then $\gamma \leftarrow \alpha$ and $\gamma \neq \alpha$. By transitivity $\delta \leftarrow \gamma$ implies $\delta \leftarrow \alpha$, and $\delta \neq \alpha$ (for $\delta = \alpha$ one would have $\alpha \leftarrow \gamma$, $\gamma \leftarrow \alpha \Rightarrow \alpha = \gamma \Rightarrow \alpha \in B$ a contradiction). Thus $B \supset \{\delta \mid \delta \leftarrow \gamma\} = \overline{\{\gamma\}}$ which shows that $B = \bigcup_{\gamma \in B} \overline{\{\gamma\}}$ and is closed. □

Note that Lemma 3 (i) characterizes only the constant functions φ for which the causality condition holds. In general, there may be in addition many functions with this property which are not constant.

5. TAXONOMY AND INHERITANCE OF PROPERTIES

For sequential systems, the partial order relation $\leftarrow$ permits a classification of certain special classes of information structures.

A sequential system is quasiclassical when $\alpha \leftarrow \beta$ implies $\mathcal{J}_\alpha \subset \mathcal{J}_\beta$. It is strictly quasiclassical if $\alpha \leftarrow \beta$, $\alpha \neq \beta$ imply $\mathcal{J}_\alpha \vee [\mathcal{F}_\alpha] \subset \mathcal{J}_\beta$, where $[\mathcal{F}_\alpha]$ denotes the cylindrical extension of $\mathcal{F}_\alpha$ to H.

The classical systems now appear as a special case of the quasiclassical ones: they are characterized by the existence of a total order $(\alpha_1,\ldots,\alpha_n)$ compatible with partial order $\leftarrow$, such that $\mathcal{J}_{\alpha_{k+1}} \supset \mathcal{J}_{\alpha_k}$ for $k = 1,\ldots,n-1$ and $\mathcal{J}_{\alpha_1} \subset \mathcal{F}(\emptyset)$.

A strictly classical system is a classical system in which $\mathcal{J}_{\alpha_{k+1}} \supset [\mathcal{F}_{\alpha_k}]$ for $k = 1,\ldots,n-1$.

Given a classical, respectively quasiclassical, system, its strict expansion is the strictly classical [strictly quasiclassical] system obtained by replacing $\mathcal{J}_{\alpha_k}$ by $\mathcal{J}_{\alpha_k} \vee (\bigvee_{i<k} [\mathcal{F}_{\alpha_i}])$, $k = 1,\ldots,n-1$; respectively $\mathcal{J}_\alpha$ by $\mathcal{J}_\alpha \vee (\vee\{[\mathcal{F}_\beta] \mid \beta \in \overline{\{\alpha\}} - \{\alpha\}\})$, for all α.

By successive substitutions of the γ_α any solution map achievable in the strict expansion of a classical or quasiclassical system is achievable in the system itself, and conversely.

The linear systems considered by Ho and Chu [4] are examples of strictly quasiclassical systems. It follows from the above remark that the solution to their problem is still linear if their system is quasiclassical but not strictly so.

Finally a monic system is the simplest type: A has only one element α and $\mathcal{J}_\alpha \subset \mathcal{F}(\emptyset)$. Several properties of an information structure have been defined. (i) monic, (ii) team, (iii) strictly classical, (iv) classical, (v) strictly quasiclassical, (vi) quasiclassical, (vii) sequential, (viii) causal, (ix) solvable, (x) without self-information.

Among properties, the implication (i) $\Rightarrow$ (ii) $\Rightarrow$ (v),

(i) $\Rightarrow$ (iii) $\Rightarrow$ (iv) $\Rightarrow$ (vi) and (iii) $\Rightarrow$ (v) $\Rightarrow$ (vi) $\Rightarrow$ (vii) $\Rightarrow$ (viii) $\Rightarrow$ (ix) $\Rightarrow$ (x) hold and none of these implications is reversible.

These facts follow readily from the definitions and the results in [9].

The properties of a system imply similar properties for its subsystems.

Theorem 2. Any of the properties (i) through (x) that holds for a system holds for all its subsystems.

We prove only the inheritance of causality, the most difficult case, as follows: Assume the system causal with function φ, and let B be a subsystem. Choose $\nu \in U_{A-B}$ and define the injection $f: H_B \to H$ by $f(h_B) = (h_B, \nu)$. For $C \subset A$, the collection of sets E in H for which $f^{-1}(E) \in \mathfrak{F}_B(B \cap C)$ forms a σ-field. Clearly the cylindrical extension to H of any rectangle in the product field $\mathfrak{F}_C(C)$ belongs to this collection. Since these cylinders generate $\mathfrak{F}(C)$ it follows that $E \in \mathfrak{F}(C)$ implies $f^{-1}(E) \in \mathfrak{F}_B(B \cap C)$.

For a sequence s of distinct elements of A define e(s) to be the sequence of those elements of s belonging to B, in the order in which they occur in s. Using this extraction function e, define the function ψ from H_B into orderings of B by $\psi = e \circ \varphi \circ f$. It is claimed that subsystem B satisfies the causality condition with function ψ. This requires that for $(\beta_1, \ldots, \beta_k)$ a sequence of distinct elements of B, and E the set in H_B where $\psi(h_B)$ begins with $(\beta_1, \ldots, \beta_k)$ and $F_B = f^{-1}(F)$, $F \in \mathfrak{J}(\beta_k)$ (which may be considered a subfield of $\mathfrak{F}_B(B)$ by the assumption that B is a subsystem) one has $E \cap F_B \subset \mathfrak{F}_B(\{\beta_1, \ldots, \beta_{k-1}\})$. Consider the sequences s_i of distinct elements of s which end with β_k and satisfy $e(s_i) = \{\beta_1, \ldots, \beta_k\}$. Let E_i be the set in H on which $\varphi(h)$ begins with s_i. Then $E = \bigcup_i f^{-1}(E_i)$, a finite union. Let s_i' be the sequence obtained from s_i by deleting its last element β_k, noting that $e(s_i') = (\beta_1, \ldots, \beta_{k-1})$. By causality of the whole system

$$E_i \cap F \in \mathfrak{F}(s_i') \subset \mathfrak{F}((A-B) \cup \{\beta_1,\ldots,\beta_{k-1}\}).$$

Then $f^{-1}(E_i \cap F) \subset \mathfrak{F}_B([(A-B) \cup \{\beta_1,\ldots,\beta_{k-1}\}] \cap B)$

or

$$f^{-1}(E_i) \cap f^{-1}(F) \subset \mathfrak{F}_B(\beta_1,\ldots,\beta_{k-1}).$$

Taking the union over i yields

$$\bigcup_i f^{-1}(E_i) \cap f^{-1}(F) = E \cap F_B \subset \mathfrak{F}_B(\beta_1,\ldots,\beta_{k-1})$$

as claimed.□

6. OPTIMIZATION AND POLICY INDEPENDENCE

Let K be a nonnegative, $\mathfrak{F}(A)$-measurable function on H, defining for each realization of (ω,u) the corresponding cost $K(h) = K(\omega,u)$.

For a causal system, every choice of design $\gamma \in \Gamma$ defines a $\mathfrak{B}$-measurable solution map Σ_γ. Let $J(\gamma) = E\{K(\omega,\Sigma_\gamma(\omega))\}$ under P. The optimization problem is to determine the infimum J^* of J over Γ and a design γ^* achieving J^* exactly or at least within ε.

In the strictly classical case, the dynamic programming algorithm of Striebel [7] is a possible approach to optimization. It relies on the following fact. If the future control laws have been fixed, then the cost, given the information available at the current step and the decision taken at the current step, has a version which is valid for every choice of the past and present control laws. This was stated without proof in [7] and recently proven for the case where $(\Omega,\mathfrak{B})$ and the $(U_\alpha,\mathfrak{F}_\alpha)$ are standard Borel spaces by Aumasson [1].

For general sequential systems a dynamic programming algorithm in terms of unconditional distributions is theoretically available [10], but often computationally forbidding.

For causal, not necessarily sequential, systems there may be subsystems in which one agent knows the data and action of the other members of the subsystem. To exploit such a situation it would be useful to establish the following conjecture.

Conjecture: For a causal system, with $(\Omega, \mathcal{B})$ and the $(U_\alpha, \mathcal{F}_\alpha)$ standard Borel spaces, let $\mathcal{G}$ be a σ-field contained in $\mathcal{F}(A)$ and containing, for all α, $\mathcal{J}_\alpha$ and $[\mathcal{F}_\alpha]$. Then there exists a real function F on H such that for all $\gamma \in \Gamma$, F is a version of the conditional expectation of K given $\mathcal{G}$ under the probability distribution P_γ that P induces on H via the mapping $\omega \to (\omega, \Sigma_\gamma(\omega))$.

It will be shown elsewhere that for countable $(U_\alpha, \mathcal{F}_\alpha)$ and arbitrary $(\Omega, \mathcal{B})$ a $\mathcal{G}$-measurable function with the above property always exists. For countable $(\Omega, \mathcal{B})$ and arbitrary $(U_\alpha, \mathcal{F}_\alpha)$ such an F also exists but may only be measurable on the intersection of the P_γ completions of $\mathcal{G}$. The conjecture stated above remains open.

7. LINEAR SYSTEMS

The intrinsic model of a finite-dimensional linear system is obtained by observing that the equations of the system make every state variable and every observation variable a known linear function of the noise and control variables.

Hence in the intrinsic model, $(\Omega, \mathcal{B})$ and the $(U_\alpha, \mathcal{F}_\alpha)$ are finite-dimensional linear spaces with their Borel sets and therefore $(H, \mathcal{F}(A))$ is a finite-dimensional linear space with its Borel sets. The probability space is a product of, possibly independent, factors corresponding to the initial state and the dynamic and observation noise vectors. Because of the linear structure of $(\Omega, \mathcal{B})$ it makes sense to talk about gaussian or nongaussian P.

If the cost is originally given as a function of the state and control variables and possibly ω, then substituting for the state its known <u>linear</u> expression in terms of ω and the u_α, one obtains a function K on H which is quadratic when the given function is quadratic.

For each α, an array of available output vectors and controls is given. As the output vectors are fixed linear functions of ω and the u_α, the data available to agent α consists of a number of linear functions on H. Let S_α be the subspace of H on which all these functions vanish and let q_α be the quotient mapping of H onto H/S_α. Then the information field $\mathcal{J}_\alpha$ is the inverse image in H under q_α of the σ-field of Borel sets of the linear quotient space H/S_α. Thus the atoms of $\mathcal{J}_\alpha$ are the translates of S_α. In the classical case the subspaces S_α form a chain under inclusion.

In summary, the intrinsic model is defined by the linear spaces $(\Omega,\mathcal{B})$, $(U_\alpha,\mathcal{F}_\alpha)$ with product $(H,\mathcal{F}(A))$; the distribution P, possibly Gaussian, on $(\Omega,\mathcal{B})$, the measurable function K on $(H,\mathcal{F}(A))$, possibly quadratic, and the subspaces S_α of H.

A control law γ_α, being $\mathcal{J}_\alpha$ measurable, can be considered as a Borel function on the linear space H/S_α and this gives meaning to the statement of linearity or nonlinearity of γ_α.

Constraints on the values u_α can be specified as subsets of the spaces U_α.

The coefficient matrices of the original system are redundant because the corresponding information, as far as it is relevant, is contained in the specification of K and the S_α.

For the case of gaussian P and quadratic K, a solution with linear γ has been obtained in the classical case and also in the quasiclassical situations investigated by Radner [5], Ho and Chu [4] and Sandell and Athans [6].(*)

For sequential LQG systems that are not quasiclassical, it is known that J* may not be a minimum (Bismut, [3]) and that when it is a minimum it may not be attainable by linear γ [8].

A major open problem is the development of bounds and approximation methods for such problems.

(*) The example of the next section is a LQG problem not belonging to any of these cases with linear optimum γ.

8. INFORMATION THEORETIC BOUNDS

The transmission of data over noisy channels under power constraints can be formulated as a nonclassical stochastic control problem to minimize distortion, the control laws being the coder and decoder functions. However, the distortion is measured by comparing source variables with decoded variables at corresponding times, either the same time or with fixed delay. Then the positive results of rate distortion theory [2] are not valid, that is the rate distortion function is usually not achievable even within ε because coding techniques leading to long delays are heavily penalized. However the converse bounds following from the data processing theorem are valid as bounds, though they will usually be too loose.

In very special cases these bounds may turn out to be sharp, as in the following LQG problem. For positive k, minimize over all pairs of Borel functions f,g the expectation of $k^2f(x)^2 + (x-g(v+f(x)))^2$ where x and v are independent scalar zero-mean gaussian random variables. Considering x as the source and v as additive channel noise, a bound follows from the rate distortion function of a gaussian source with square law distortion and the capacity of the gaussian channel with given average input power.

The readily determined best linear choice of f,g meets this bound and is therefore optimal, although the problem is not even quasiclassical. In other problems of this type, while the usual rate distortion bound may be far from achievable (with the delay restriction), the generalized data processing theorem of Ziv and Zakai [11] yields other bounds which, while they do not extend to the block coding situation, may be much closer to the optimum achievable when delay is ruled out.

To what extent this approach can be applied to more general situations is an open problem.

REFERENCES

1. Aumasson, C., Processus Aléatoires à Commande, Publication no. 141, O.N.E.R.A., Chatillon, 1972.

2. Berger, T., Rate Distortion Theory, Prentice Hall, 1971.

3. Bismut, J. M., An example of interaction between information and control, IEEE Trans. AC-18, pp. 63-64, 1973.

4. Ho, Y. C. and Chu, K. C., Team decision theory and information structures in optimal control problems - Part I, IEEE Trans. AC-17, pp. 15-22, 1972.

5. Radner, R., Team Decision Problems, Am. Math. Stat., vol. 33, pp. 857-881, 1962.

6. Sandell, N. R., Jr. and Athans, M., Solution of some nonclassical LQG stochastic control problems, IEEE Trans. AC-19, pp. 108-116, 1974.

7. Striebel, C., Sufficient Statistics in the Optimum Control of Stochastic Systems, Jl. Math. Anal. Appl., vol. 12, pp. 576-592, 1965.

8. Witsenhausen, H. S., A counterexample in stochastic optimum control, SIAM Jl. Contr., vol. 6, pp. 131-147, 1968.

9. ________________, On Information Structures, Feedback and Causality, SIAM Jl. Control, vol. 9, pp. 149-160, 1971.

10. ________________, A Standard Form for Sequential Stochastic Control, Math. Syst. Th., vol. 7, pp. 5-11, 1973.

11. Ziv, J. and Zakai, M., On Functionals satisfying a Data-Processing Theorem, IEEE Trans. IT-19, pp. 275-283, 1973.

FINITE DIFFERENCE METHODS FOR THE WEAK SOLUTIONS OF THE KOLMOGOROV EQUATIONS FOR THE DENSITY OF DIFFUSIONS AND CONDITIONAL DIFFUSIONS*

H.J. Kushner

Divisions of Applied Mathematics and Engineering
Brown University
Providence, Rhode Island 02912

Abstract

The problem of obtaining numerical solutions to the Fokker-Plank equation for the density of a diffusion, and for conditional density (with "white noise" corrupted observations) is treated. We assume the diffusion coefficients to be bounded and uniformly continuous, and that the diffusion has a unique solution in the sense of multivariate distributions. These equations usually have only weak sense solutions. But we show that, if the finite difference approximations are chosen in a natural way, then the finite difference solutions to the formal adjoints yield a sequence of approximations which converges weakly to the weak sense solution to the Fokker-Plank equation or to the "conditional" equation as the difference intervals go to zero.

The approximations are natural for this type of problem. They define the transition functions of a sequence of Markov chains which, in turn, converge weakly to the diffusion, as the difference intervals go to zero. They seem to have more physical significance than the usual approximations. Our method relies heavily on results on the weak convergence of measures on abstract spaces. Some more details appear in [9].

*This research was supported in part by the Office of Naval Research NONR N00014-67-A-0191-0018, in part by the Air Force Office of Scientific Research AF-AFOSR 71-2078B, and in part by the National Science Foundation GK 40493X.

Introduction. Let $w(\cdot)$ denote a standard r-dimensional Wiener process, R^+ denote the real interval $[0,\infty)$, and suppose that $f(\cdot,\cdot)$ and $\sigma(\cdot,\cdot)$ are uniformly continuous and bounded R^r and $(r\times r)$ matrix-valued functions on $R^r \times R^+$, respectively. Let f_i denote the ith component of f, and define the matrix $A = \{a_{ij}, i,j=1,\ldots,r\} = \sigma\sigma'$. Suppose that the Itô stochastic differential equation

$$dx(t) = f(x,t)dt + \sigma(x,t)dw(t) \tag{1}$$

has a unique solution, in the sense of multivariate distributions, on $[s,\infty)$ for each $s \geq 0$ and initial value $x(s) = x\varepsilon R^r$.

Let $p_s(s)$ denote the weak sense density of $p_s(\cdot)$, and suppose that $x(t)$, $t \geq s$, has a weak sense density, which we denote by $p(\cdot,t)$, or by $p(u,s;\cdot,t)$ when $p_s(\cdot) = \delta(x-u)$. Then $p(u,s;\cdot,\cdot)$ and $p(\cdot,\cdot)$ satisfy the Kolmogorov forward (Fokker-Plank) equation (2) with the appropriate initial conditions:

$$\frac{\partial p}{\partial t} = \mathcal{L}^*p = \frac{1}{2}\sum_{i,j}\frac{\partial}{\partial x_i}(a_{ij}(x,t)\frac{\partial}{\partial x_j}p) + \sum_i \frac{\partial}{\partial x_i}(f_i(x,t)p). \tag{2}$$

Suppose one tries to solve (2) by the finite difference method. If A and f are not sufficiently smooth or if A is not uniformly elliptic, then, even if p exists in the strong sense (which it may not), it may not be smooth enough for (2) to have a strong sense meaning, and the finite difference solution may not have a meaning. We will study a finite difference solution to the formal adjoint of (2). That solution will immediately give us an approximation to either $p(\cdot,t)$ or $p(u,s,\cdot,t)$; the approximations converge weakly* to these quantities as the difference intervals go to zero. The proof requires only the conditions of the first paragraph and the conditions (A1)-(A3) below. Note that we include the completely degenerate case $A(\cdot,\cdot) = 0$. The sequence of approximations which we obtain by the finite difference method are also transition probabilities for a sequence of Markov chains which, suitably interpolated, converge weakly (as the difference intervals

*Thus, expectations of continuous bounded functions which are computed with these approximations converge to the true expectations as the difference intervals go to zero.

Part I. The Approximate Solution to (2)

1. Problem formulation. For numerical purposes, something akin to a finite state space is required. In particular, we will assume the following. Let G be a bounded open set with compact closure $\overline{G} = G + \partial G$, and fix s, T (s < T) in R^+.

Let C_T^r denote the space of continuous R^r-valued functions on [s,T], T > 0, with the sup norm topology. We need the following assumptions for the values s, T of interest. $u(\cdot)$ will always denote a generic element of C_T^r. Define the function $(C_T^r \to R^+)$ $\tau_s(u(\cdot)) = \min\{t: t \geq s, u(t) \notin G\}$. If $u(t) \in G$ for all $s \leq t < \infty$, then $\tau_s(u(\cdot)) = \infty$. We simply write $\tau_s(x(\cdot))$ as τ_s, if $x(\cdot)$ solves (1). Approximations to the weak sense densities will be obtained for the process $x(\cdot)$ defined by $\tilde{x}(t) = x(t \cap \tau_s)$.

If the initial vector $x(s)$ *has a density* $p_s(\cdot)$, *then replace the conditioning in* (A1), (A2) *by the condition that* $x(s)$ *has density* $p_s(\cdot)$.

(A1) $P_{x,s}\{\tau_s = T\} = 0$, $x \in G$, *where* $P_{x,s}$ *denotes the probability, given that* $x(s) = x$.

(A2) *The function* $T \cap \tau_s(u(\cdot))$: $C_T^r \to [s,T]$, *is* continuous on* C_T^r w.p. 1** *relative to the measure* μ_C *induced on* C_T^r *by* $x(t)$, $t \in [s,T]$, *where* $x(s) = x$.

Let $k(\cdot)$ denote a real-valued bounded continuous function on R^r. Let I_A denote the indicator function of a set A. The function $V_k(\cdot,\cdot)$ defined by $V_k(x,t) = E_{x,t}k(x(T))I_{\{\tau_t > T\}} + E_{x,t}k(x(T \cap \tau_t))\, I_{\{\tau_t \leq T\}}$, $T \geq t$, $x \in \overline{G}$, formally satisfies

$$\frac{\partial V_k(x,t)}{\partial t} + \mathcal{L} V_k(x,t) = 0, \quad x \in G, \quad t \leq T, \tag{5}$$

$$V_k(x,t) = k(x), \quad x \in G,\ t = T \text{ or } x \in \partial G,\ t \leq T.$$

---

*$T \cap \tau \equiv \min[T,\tau]$.

**Continuity of $\tau_s(u(\cdot)) \cap T$ is equivalent to (a) $\tau_s(u(\cdot)) \neq T$, (b) $u(\cdot)$ is not tangent to ∂G at the point of contact, if any; namely, at the point of contact $u(\cdot)$ actually crosses the boundary. This must hold only at almost all trajectories of (1). This condition is not very stringent. See [4], p. 89-93, for a further discussion. Note that (a), (b) imply that $I_{\{\tau_s(u(\cdot)) > T\}}$ is continuous w.p. 1, relative to μ_C.

go to zero) to the measure induced by (1). Thus the approximations have a natural physical interpretation, unlike direct finite difference solutions to (2), and it seems reasonable that our approach yields a more stable and more suitable solution. The equation (2) or its formal adjoint are thus used only for numerical purposes. No statements are made concerning smoothness of p. The methods of proof are purely probabilistic, and do not rely on classical methods of numerical analysis.

The standard problem of nonlinear filtering will also be treated by the same method. For some integer v, let $g(\cdot,\cdot)$ be a bounded* continuous R^v-valued function on $R^r \times R^+$, and let $z(\cdot)$ denote a standard R^v-valued Wiener process, independent of $w(\cdot)$. Define the Itô process $y(\cdot)$ by

$$dy(t) = g(x(t),t)dt + dz(t). \tag{3}$$

Let $\mathcal{B}_s^t$ denote the minimal σ-algebra generated by $\{y_r, s \le r \le t\}$. Then, <u>purely formally</u>, the conditional density $p(x,t,\omega) \equiv p(x \le x(t) < x+dx \mid \mathcal{B}_s^t)$ $(s \le t)$ satisfies (4) with initial condition $p(x,s,\omega) = p_s(x)$ given (Kushner [8])

$$dp(x,t,\ ) = \mathcal{L}^* p(x,t,\omega)dt + p(x,t,\omega)(dy(t)-E_s^t g(x(t),t)dt)'(g(x,t) - E_s^t g(x(t),t)) \tag{4}$$

where, for any integrable function F, we use

$$E_s^t F = E[F \mid \mathcal{B}_s^t].$$

Our method will yield an interesting weak sense** approximation to $p(x,t,\omega)$, which converges under the same conditions given for the first problem, and the above continuity and boundedness conditions on g.

The basic techniques involve results in weak convergence of measures and their application to the finite difference solutions to degenerate elliptic and parabolic equations, and are based on the ideas in Kushner [1], Kushner and Yu [2].

*Boundedness of f, σ and g is assumed, since, for the numerical method, a finite state space is used. But, if we wish to use finite difference approximations on an infinite difference grid, the boundedness can be weakened.

**The sequence of expectations computed with the approximations converges to the true conditional expectation as the difference interval goes to zero.

$$\mathcal{L} = \frac{1}{2} \sum_{i,j} a_{ij}(x,t) \frac{\partial^2}{\partial x_i \partial x_j} + \sum_i f_i(x,t) \frac{\partial}{\partial x_i} .$$

Let h and Δ denote the spatial and temporal finite difference intervals, respectively. Define R_h^r as the grid $\{x: x=(n_1h,\ldots,n_rh)\}$, where each n_i takes values 0, ±1, ±2,..., and define $\bar{G}_h = R_h^r \cap \bar{G}$ and $G_h = R_h^r \cap G$.

The procedure involves getting a representation for the finite difference solution to the "backward"equation (5), via a natural choice of finite difference approximations. Certain parameters in the representation will immediately yield the desired approximate weak sense solution to (2) and also allow us to calculate an approximation to $V_k(x,t)$ for any k.

2. <u>The finite difference approximations</u>. Let e_i denote the unit vector in the ith coordinate direction of R^r. The finite difference approximations (6)-(9) will be used. The reasons will be given below. See also [1], [2], [9].

$$\partial V_k(x,t)/\partial t \rightarrow [V_k(x,t+\Delta)-V_k(x,t)]/\Delta \tag{6}$$

$$\begin{aligned} \partial V_k(x,t)/\partial x_i &\rightarrow [V_k(x+e_ih,t+\Delta)-V_k(x,t+\Delta)]/h, \text{ if } f_i(x) \geq 0 \\ &\rightarrow [V_k(x,t+\Delta)-V_k(x-e_ih,t+\Delta)]/h, \text{ if } f_i(x) < 0 \end{aligned} \tag{7}$$

$$\partial^2 V_k(x,t)/\partial x_i^2 \rightarrow [V_k(x+e_ih,t+\Delta)-2V_k(x,t+\Delta)+V_k(x-e_ih,t+\Delta)]/h^2 \tag{8}$$

$$\begin{aligned} \partial^2 V_k(x,t)/\partial x_i \partial x_j &\rightarrow \pm[2V_k(x,t+\Delta)+V_k(x+e_ih\pm e_jh,t+\Delta)+V_k(x-e_ih\mp e_jh,t+\Delta)]/2h^2 \\ &\quad \mp [V_k(x+e_ih,t+\Delta)+V_k(t-e_ih,t+\Delta)+V_k(x+e_jh,t+\Delta) \\ &\quad + V_k(x-e_jh,t+\Delta)]/2h^2, \end{aligned} \tag{9}$$

$i \neq j$, where the upper signs are used if $a_{ij}(x,t) \geq 0$, the lower if $a_{ij}(x,t) < 0$.

Define

$$p_t^{h,\Delta}(x,x) \equiv 1 - (\Delta/h^2)[\sum_i h|f_i(x,t)| + 2 \sum a_{ii}(x,t) - \sum_{\substack{i,j \\ i\neq j}} |a_{ij}(x,t)|] .$$

For any real-valued $g(\cdot)$, define $F^+(x) \equiv \max[g(x),0]$, $F^-(x) \equiv \max[-g(x),0]$, and define $p_t^{h,\Delta}(\cdot,\cdot)$ by

$$p_t^{h,\Delta}(x,x\pm e_i h) \equiv (\Delta/h^2)[a_{ii}(x,t) - \sum_{\substack{j\neq i\\ j}} |a_{ij}(x,t)|/2 + hf_i^{\pm}(x,t)],$$

$$p_t^{h,\Delta}(x,x+e_i h-e_j h) \equiv p_t^{h,\Delta}(x,x-e_i h+e_j h) \equiv (\Delta/h^2)a_{ij}^-(x,t)/2, \qquad i \neq j,$$

$$p_t^{h,\Delta}(x,x+e_i h+e_j h) \equiv p_t^{h,\Delta}(x,x-e_i h-e_j h) \equiv (\Delta/h^2)a_{ij}^+(x,t)/2, \qquad i \neq j.$$

$p_t^{h,\Delta}(x,y) = 0$ for all other pairs (x,y). We also need condition (A3).

(A3) Choose Δ, h so that $p_t^{h,\Delta}(x,x) \in [0,1]$ and $p_t^{h,\Delta}(x,x\pm e_i h) \geq 0$ and let

$$a_{ii}(x,t) - \sum_{\substack{j\neq i\\ j}} |a_{ij}(x,t)|/2 + h|f_i(x,t)| \geq 0.$$

3. The finite difference solution to (5). Denote the solution to the finite difference equation by $V_k^{h,\Delta}(x,t)$. Choose Δ so that $T = N\Delta$ for some integer N. Substituting (6)-(9) into (5) and rearranging yields

$$V_k^{h,\Delta}(x,t) = \sum_{y\in R_h^r} p_t^{h,\Delta}(x,y)V_k^{h,\Delta}(y,t+\Delta), \qquad x\in G_h \tag{10}$$

with

$$V_k^{h,\Delta}(x,t) = k(x), \quad x\in G_h,\ t = T \text{ or } x\in R_h^r - G_h,\ t = n\Delta,\ n = N,N-1,\dots . \tag{11}$$

Since the $p_t^{h,\Delta}(x,y)$ are non-negative and they sum (over y) to unity, for each x, they can (and will) be considered to be the transition probability for a non-stationary Markov chain on the state space R_h^r for each fixed Δ, h. Let $\{\xi_n^{h,\Delta}\}$ denote the random variables of the chain. It will turn out that these $p_t^{h,\Delta}$ will provide the desired approximations to both the forward and backward equations.

Define the continuous parameter process $\xi^{h,\Delta}(\cdot)$ by the interpolation $\xi^{h,\Delta}(t) = \xi_i^{h,\Delta}$, $t\in[i\Delta,i\Delta+\Delta)$.

4. The process $\xi^{h,\Delta}(\cdot)$. Fix s and T and define D_T^r to be the space of R^r-

valued functions on [s,T] which are right continuous, left continuous at T, and have left-hand limits. Let D^r_t have the Skorokhod topology [3, pp. 111-115]. Then the paths of the processes $x(\cdot)$ and $\xi^{h,\Delta}(\cdot)$ on [s,T] are elements of D^r_T w. p. 1 (we can ignore any jumps at T). The convergence $u_n(\cdot) \to u(\cdot)$ in the Skorokhod topology implies uniform convergence if $u(\cdot)$ is continuous. Let $F(\cdot)$ denote any bounded real-valued function on D^r_T which is continuous with probability one with respect to $\mu_D(\cdot)$, the measure that $x(\cdot)$ induces on D^r_T. Let $\mu^{h,\Delta}(\cdot)$ denote the measure that $\xi^{h,\Delta}(\cdot)$ induces on D^r_T. Then it can be shown that $\mu^{h,\Delta}(\cdot)$ converges weakly* to $\mu_D(\cdot)$ in the sense that for any x which is in all the grids R^r_h, and for $s \leq T$,

$$E_{x,s}F(\xi^{h,\Delta}(\cdot)) \to E_{x,s}F(x(\cdot)), \qquad \text{as } h,\Delta \to 0. \tag{12}$$

Equation (12) follows from a result in Kushner [1], showing weak convergence of a sequence of Markov chains to a diffusion. It is not hard to verify that the conditions of [1] hold.

Under conditions (A1)-(A2), the real-valued function on D^r_T with values $\tau_s(u(\cdot)) \cap T$ is continuous w.p. 1 relative to $\mu_D(\cdot)$. Thus, $k(u(\tau_s(u(\cdot))) \cap T)I_{\{\tau_s(u(\cdot))\leq T\}}$ and $k(u(T))I_{\{\tau_s(u(\cdot))>T\}}$ are continuous w.p. 1 on D^r_T relative to $\mu_D(\cdot)$. Define $\tau^{h,\Delta}_s \equiv \tau_s(\xi^{h,\Delta}(\cdot))$. Then (13) and (14) hold as $\Delta,h \to 0$. (Since $x(\cdot)$ is continuous w.p. 1, the C^r_T and μ_C in (A2) can be replaced by D^r_T and μ_D, respectively.)

$$E_{x,s}k(\xi^{h,\Delta}(\tau^{h,\Delta}_s \cap T)) \to E_{x,s}k(x(\tau_s \cap T)), \tag{13}$$

$$E_{x,s}k(\xi^{h,\Delta}(T))I_{\{\tau^{h,\Delta}_s>T\}} \to E_{x,s}k(x(t))I_{\{\tau_s>T\}}\,. \tag{14}$$

(13) and (14) imply that the transition probability for the Markov chain $\{\xi^{h,\Delta}_n\}$ gives the required approximation to the solutions of both the forward and backward Kolmogorov equations.

---

*We may say that the sequence $\xi^{h,\Delta}(\cdot)$ converges weakly to $x(\cdot)$.

5. Approximations to the solutions of the forward and backward Kolmogorov equations. Let G_1^h denote the set of states y on R_h^r which communicate in one step with some state in G_h, for some i: $s \leq i\Delta \leq T$. We want to write a computationally convenient expression for the left sides of (13), (14). Thus define the transition probability $\tilde{p}_t^{h,\Delta}(\cdot,\cdot)$ on G_1^h by

$$\begin{aligned}\tilde{p}_t^{h,\Delta}(x,y) &= p_t^{h,\Delta}(x,y), \quad && x\varepsilon G_h,\ y\varepsilon G_1^h,\ t = i\Delta \ \varepsilon\ [s,T]\\ &= \delta_{xy}, && x\varepsilon G_1^h - G_h.\end{aligned}$$

Let $P_i^{h,\Delta}$ denote the transition matrix $\{\tilde{p}_i^{h,\Delta}(x,y),\ x,y\varepsilon G_1^h\}$. $P_i^{h,\Delta}$ is, of course, the transition function for the Markov chain which is stopped on first entrance into $R_h^r - G_h$. Let n, N be integers for which $n\Delta = s$, $N\Delta = T$, and define the matrix $C_{n,N}^{h,\Delta} = \{c_{n,N}^{h,\Delta}(x,y),\ x,y\varepsilon G_1^h\}$ by

$$C_{n,N}^{h,\Delta} = P_n^{h,\Delta},\ldots,P_N^{h,\Delta}\ . \tag{15}$$

Then, e.g., for all $x\varepsilon G_h$,

$$E_{x,s}k(\xi^{h,\Delta}(T))I_{\{\tau_s^{h,\Delta}>T\}} = \sum_{y\varepsilon G_h} c_{n,N}^{h,\Delta}(x,y)k(y) \tag{16}$$

The matrix $C_{n,N}^{h,\Delta}$ provides the desired approximation (convergent as $h,\Delta \to 0$) to either the forward or backward equation. The approximation (15) can be iterated to the left (backward) or to the right (forward).

Observe that, even though (5) may be purely formal, the use of the finite difference approximation still yields the desired quantity in the limit, namely (for each x for which (A1)-(A2) holds) $E_{x,s}k(x(t))I_{\{\tau_s>T\}} + E_{x,s}k(x(\tau_s \cap T))I_{\{\tau_s \leq T\}}$. Our approach has given us an automatic and computationally convenient method for finding an approximating chain, and the approximating transition function.

Suppose that the random variable x(s) is concentrated on G and has the weak sense density $p_s(\cdot)$. For each small h, let $p_s^h(\cdot)$ denote a weak sense density of a random variable with support on G_h, such that $p_s^h(\cdot)$ converges weakly to $p(\cdot)$ as $h \to 0$. Then to get the desired approximation to the formal equation, we simply use the forward iteration

$$C^{h,\Delta}_{n,N+1} = C^{h,\Delta}_{n,N} P^{h,\Delta}_{N+1}, \tag{17}$$

where $C^{h,\Delta}_{n,m}$ are row vectors, and $C^{h,\Delta}_{n,m}$ is simply the vector of components of the initial condition $p^h_s(\cdot)$, arranged in the correct order.

The computation in (17) is of the same order as for any other of the standard finite difference methods of solving (2).

Part II. The Nonlinear Filtering Problem

1. Preliminaries. By a suitable choice of the probability space, weak convergence of $\mu^{h,\Delta}(\cdot)$ to $\mu_D(\cdot)$ implies uniform convergence of $\xi^{h,\Delta}(\cdot)$ to $x(\cdot)$ on $[s,T]$, w.p. 1. There is a metric, which yields the Skorokhod topology, with respect to which D^r_T is complete and separable [3, pp. 114-115]. This, together with the fact that $\{\mu^{h,\Delta}(\cdot)\}$ converges weakly to $\mu_D(\cdot)$, implies that we can use a result of Skorokhod [5, p. 281] which says that there is a probability space $(\hat{\Omega},\hat{\mathcal{B}},\hat{P})$ and separable random functions $\hat{\xi}^{h,\Delta}(\cdot),\hat{x}(\cdot),t \in [s,T]$, defined on that space, so that for each Borel set $B \in D^r_T$,

$$\hat{P}\{\hat{\xi}^{h,\Delta}(\cdot)\in B\} = P\{\xi^{h,\Delta}(\cdot))\in B\},\ \hat{P}\{\hat{x}(\cdot)\in B\} = P\{x(\cdot)\in B\},$$

and $\hat{\xi}^{h,\Delta}(\cdot)$ converges to $\hat{x}(\cdot)$ in the topology of D^r_T w.p. 1. The chain $\hat{\xi}^{h,\Delta}_n$ has the same probability law as $\xi^{h,\Delta}_n$, and it can be shown that $\hat{x}(\cdot)$ satisfies (1) for some Wiener process $\hat{w}(\cdot)$. The convergence is uniform on $[s,T]$, since $x(\cdot)$ is continuous. The underlying probability space is irrelevant here; we can suppose that there is w.p. 1 convergence for all initial conditions $x(s)$ for which (A1)-(A2) hold. The affix ^ will be omitted. Since $\tau_s(u(\cdot)) \cap T$ is also continuous on D^r_T w.p. 1, with respect to $\mu_D(\cdot)$, the $\xi^{h,\Delta}(t \cap \tau^{h,\Delta}_s)$ converge to $x(t \cap \tau_s)$ uniformly on $[s,T]$ w.p. 1, as $h,\Delta \to 0$, if (A1)-(A2) hold for $x(s)$, the initial condition.

Let $\bar{x}(\cdot)$ and $\bar{\xi}^{h,\Delta}(\cdot)$ denote the stopped (at τ_s and $\tau^{h,\Delta}_s$, respectively) processes. To complete the formulation of the filtering problem we need to augment the probability space by adding a continuous (stopped) process $\tilde{x}(\cdot)$, which is independent of $\bar{x}(\cdot)$, but which has the same multivariate distributions. Thus,

$$\bar{x}(t) = x + \int_s^{t\cap\tau_s} f(\bar{x}(\rho),\rho)d\rho + \int_s^{t\cap\tau_s} \sigma(\bar{x}(\rho),\rho)d\bar{w}(\rho),$$

$$\tilde{x}(t) = x + \int_s^{t\cap\tilde{\tau}_s} f(\tilde{x}(\rho),\rho)d\rho + \int_s^{t\cap\tilde{\tau}_s} \sigma(\tilde{x}(\rho),\rho)d\tilde{w}(\rho),$$

where $\tilde{\tau}_s \equiv \min\{t: \tilde{x}(t)\varepsilon\partial G\}$, and $\tilde{w}(\cdot)$ and $w(\cdot)$ are independent. Define $\tilde{y}(\cdot)$ analogously to $y(\cdot)$ by

$$d\tilde{y}(t) = g(\tilde{x}(t),t)dt + dz(t),$$

where $z(\cdot)$ is a standard R^r-valued Wiener process, independent of $\bar{w}(\cdot)$ and $\tilde{w}(\cdot)$ and $g(\cdot,\cdot)$: $R^r \times R^+ \to R^v$ is bounded and continuous. Define

$$R(s,T) = \exp \{\int_s^T g'(\bar{x}(\rho),\rho)d\tilde{y}(\rho) - \frac{1}{2}\int_s^T |g(\bar{x}(\rho),\rho)|^2 \, d\rho \, . \tag{18}$$

Let $k(\cdot)$ denote, henceforth, a bounded continuous function: $R^r \to R$. Define the random function $V_k(\cdot,\cdot,\cdot)$ with values $(T \geq s)$

$$V_k(x,s,\omega) = E_{x,s}\{R(s,T)k(\bar{x}(T))|\tilde{\mathcal{B}}_s^T\} ,$$

where $\tilde{\mathcal{B}}_s^T$ is the minimum σ-algebra which measures $\tilde{y}(\rho)$, $s \leq \rho \leq T$. Then, it is known (Kushner [6], Zakai [7]) that

$$E_{x,s}\{k(\tilde{x}(T))|\tilde{\mathcal{B}}_s^T\} = \frac{V_k(x,s,\omega)}{V_1(x,s,\omega)} \, . \tag{19a}$$

If $x(s)$ is a random variable whose weak sense density $p_s(\cdot)$ is supported on G, then

$$E\{k(\tilde{x}(T))|\tilde{\mathcal{B}}_s^T\} = \int V_k(x,s,\omega)p_s(x)dx/\int V_1(x,s,\omega)p_s(x)dx. \tag{19b}$$

We will develop an approximation to (19b) which will yield the desired weak sense solution to (4). In particular, the results of Part I lead to a natural method of approximation $V_k(x,s,\omega)$, which converges as $h,\Delta \to 0$. It is important to keep in mind that the $\bar{x}(\cdot)$ and $\tilde{y}(.)$ in (18) are independent.

Formally, $V_k(x,s,\omega)$ satisfies

$$0 = dV_k(x,t,\omega) + \mathcal{L}V_k(x,t,\omega)dt + g'(x,t)dy(t)V_k(x,t,\omega), \quad x\varepsilon G,\ s \le t \le T$$

$$V_k(x,T,\omega) = k(x), \quad V_k(x,t,\omega) = k(x) \text{ for } x\varepsilon\partial G. \tag{20}$$

Equation (20) will be treated in almost the same way that we treated (5), and the finite difference approximations, to be denoted by $V_k^{h,\Delta}(x,t,\omega)$, will converge to $V_k(x,t,\omega)$ as $h,\Delta \to 0$.

Define $\delta\tilde{y}_n \equiv \tilde{y}(n\Delta+\Delta)-\tilde{y}(n\Delta)$. Using (6)-(9), and rearranging, we get

$$V_k^{h,\Delta}(x,n\Delta,\omega) = \sum_{y\varepsilon R_h^r} p_{n\Delta}^{h,\Delta}(x,y)\, V_k^{h,\Delta}(y,n\Delta+\Delta,\omega)$$

$$+ \delta\tilde{y}_n' g(x,n\Delta)V_k^{h,\Delta}(x,n\Delta+\Delta,\omega), \qquad x\varepsilon G_h \tag{21}$$

A somewhat more convenient difference equation can be obtained by adding the small term

$$\delta\tilde{y}_n' g(x,n\Delta)[\sum_{y\varepsilon R_h^r} p_{n\Delta}^{h,\Delta}(x,y)V_k^{h,\Delta}(y,n\Delta+\Delta,\omega) - V_k^{h,\Delta}(x,n\Delta+\Delta,\omega)]$$

to the right-hand side of (21) and rearranging to get the "approximation"

$$V_k^{h,\Delta}(x,n\Delta,\omega) = (1+\delta\tilde{y}_n' g(x,n\Delta)) \sum_{y\varepsilon R_h^r} p_{n\Delta}^{h,\Delta}(x,y)V_k^{h,\Delta}(y,n\Delta+\Delta), \qquad x\varepsilon G_h, \tag{22}$$

with boundary conditions

$$V_k^{h,\Delta}(x,n\Delta,\omega) = k(x)\ , \quad (n\Delta = N\Delta = T,\ x\varepsilon G_1^h \text{ or } n < N,\ x\varepsilon G_1^h-G_h). \tag{23}$$

The term is merely $\delta\tilde{y}_n'(x,n\Delta)$ times the average one-step change in $V_k^{h,\Delta}(x,n\Delta,\omega)$.

As in Part I, the $p_t^{h,\Delta}(x,y)$ are one-step transition probabilities for a Markov chain, whose variables we will denote by $\{\xi_n^{h,\Delta}\}$. We can suppose that the chains are all independent of the $\tilde{y}(\cdot)$ process. By Part I, the interpolations $\xi^{h,\Delta}(\cdot)$ converge to $x(\cdot)$ w.p. 1, uniformly on any interval $[s,T]$, and for each s, T, the stopped processes $\bar{\xi}^{h,\Delta}(\cdot)$ converge similarly to the stopped process $\bar{x}(\cdot)$.

2. <u>The solution to (22), (23).</u> Fix s, T and let $n\Delta = s$, $N\Delta = T$, and define

$$R^{h,\Delta}(n,N) = \prod_{i=n}^{N-1} (1+\delta\tilde{y}_i' g(\bar{\xi}_i^{h,\Delta}, i\Delta)).$$

Then the unique solution to (22), (23) can be written as

$$V_k^{h,\Delta}(x,n\Delta,\omega) = E_{x,n\Delta}\{R^{h,\Delta}(n,N)k(\bar{\xi}_N^{h,\Delta})|\tilde{\mathcal{B}}_{n\Delta}^{N\Delta}\}. \quad (24)$$

In (24), the expectation is only over the $\{\tilde{\xi}_i^{h,\Delta}\}$. The $\{\delta\tilde{y}_i\}$ are "fixed" due to the conditioning. Because of the boundary condition (23), the $\tilde{p}_{n\Delta}^{h,\Delta}(x,y)$ actually can replace the $p_{n\Delta}^{h,\Delta}(x,y)$ in (22).

It is not hard to show that

$$R^{h,\Delta}(n,N) \to R(t,T) \quad (25)$$

in probability as $\Delta,h \to 0$.

Let $k(\cdot)$ be a bounded and continuous function on R^r. The second moments of the $R^{h,\Delta}(n,N)$ are uniformly bounded (in h,Δ). Then (25) and the w.p. 1 convergence of $\bar{\xi}^{h,\Delta}(\cdot)$ to $\bar{x}(\cdot)$ yields

$$E_{x,s}|R^{h,\Delta}(n,m)k(\bar{\xi}^{h,\Delta}(T)) - R(s,T)k(\bar{x}(T))| \to 0 \quad \text{as } h,\Delta \to 0 . \quad (26)$$

By (26),

$$E_{x,s}[R^{h,\Delta}(n,m)k(\bar{\xi}^{h,\Delta}(T)) - R(x,T)k(\bar{x}(T))|\tilde{\mathcal{B}}_s^T] \xrightarrow{P} 0 \quad (27)$$

as $h,\Delta \to 0$. Equation (27) suggests that (24), the finite difference solution to (22), is indeed the desired approximation to $V_k(x,s,\omega)$, from which the approximate conditional expectations may be obtained from the formula $V_k^{h,\Delta}(x,s,\omega)/V_1^{h,\Delta}(x,s,\omega)$. If x(s) has a weak sense density $p_s(\cdot)$ for which (A1)-(A2) hold, then (27) holds if the expectation there reflects that fact.

3. <u>A computational procedure for the conditional density.</u> We only need give the analogue of (15)-(17) to complete the computational procedure. Fix h,Δ in this section. Let M denote the number of points on G_1^h. For arbitrary M vectors V, X

and an arbitrary (M×M) square matrix F, define

$$V\circ F = \begin{Bmatrix} V_1F_{11} & & V_1F_{1M} \\ \vdots & & \\ V_MF_{M1} & \cdots & V_MF_{MM} \end{Bmatrix}, \quad V\circ X = \begin{Bmatrix} V_1X_1 \\ \vdots \\ V_MX_M \end{Bmatrix}.$$

Order the points of G_h^1 as they are ordered in the definition of $P_n^{h,\Delta}$ (defined above (15)) and define K, $\mathcal{J}_n$ and V(n,k) to be the vectors formed by the $\{k(x),\ x\varepsilon G_h^1\}$, $\{1+\delta\tilde{y}_n'g(x),\ x\varepsilon G_h^1\}$, and $\{V_k^{h,\Delta}(n,x,\omega), x\varepsilon G_h^1\}$, respectively. Then, we can write (22), (23) in the form

$$V(n,k) = \mathcal{J}_n \circ P_n^{h,\Delta}\, V(n+1,k), \qquad V(N,k) = K. \tag{28}$$

For each integer n,m, $m \geq n$, define the M×M matrix (with the states ordered as above) $Q^{h,\Delta}(n,m,\omega) = q_{xy}^{h,\Delta}(n,m,\omega),\ x,y\varepsilon G_1^h\}$ by

$$Q^{h,\Delta}(n,n,\omega) = I,$$

the identity, and for $m > n$,

$$\begin{aligned} Q^{h,\Delta}(n,m,\omega) &= \mathcal{J}_n \circ P_n^{h,\Delta}\mathcal{J}_{n+1} \circ P_{n+1}^{h,\Delta} \cdots \mathcal{J}_{m-1} \circ P_{m-1}^{h,\Delta} \\ &= \mathcal{J}_n \circ P_n^{h,\Delta}\, Q^{h,\Delta}(n+1,m,\omega) \\ &= Q^{h,\Delta}(n,m-1,\omega)\mathcal{J}_{m-1} \circ P_{m-1}^{h,\Delta}. \end{aligned} \tag{29}$$

Then $V(n,k) = Q^{h,\Delta}(n,N,\omega)K$ and

$$V_k^{h,\Delta}(x,n,\omega) = \sum_{y\varepsilon G_1^h} q_{xy}^{h,\Delta}(n,N,\omega)k(y), \quad x\varepsilon G_h. \tag{30}$$

For $x\varepsilon G_h$, denote the corresponding row of $Q^{h,\Delta}(n,N,\omega)$ as $Q^{h,\Delta}(n,N,\omega,x)$. By (30), this row is an "unnormalized" weak sense conditional density. The row evolves as

$$Q^{h,\Delta}(n,m,\omega,x) = Q^{h,\Delta}(n,m-1,\omega,x)\mathcal{J}_{m-1} \circ P_{m-1}^{h,\Delta} \tag{31a}$$

a computation of the order of that involved in (17).

If x(s) has a density $p_s(\cdot)$, such that (A1)-(A2) hold, then we proceed as in the development of (17b). Define $p_s^h(\cdot)$ as above (17b) and $\tilde{Q}^{h,\Delta}(n,n,\omega)$ to be the vector of components of $p_s^h(\cdot)$ arranged in the correct order. Then define the row vectors and components $\tilde{Q}^{h,\Delta}(n,m,\omega) \equiv \{\tilde{q}_y^{h,\Delta}(n,m,\omega), y\varepsilon G_1^h\}$ by

$$\tilde{Q}^{h,\Delta}(n,m,\omega) = \tilde{Q}^{h,\Delta}(n,m-1,\omega)\mathcal{S}_{m-1} \circ P_{m-1}^{h,\Delta} \tag{31b}$$

$\tilde{Q}^{h,\Delta}(n,N,\omega)$ is the desired unnormalized conditional density and

$$\frac{\sum_{y\varepsilon G_1^h} \tilde{q}_y^{h,\Delta}(n,m,\omega)k(y)}{\sum_{y\varepsilon G_1^h} \tilde{q}_y^{h,\Delta}(n,m,\omega)} \overset{P}{\rightarrow} E[k(\tilde{x}(T)) | \tilde{\mathcal{B}}_s^T] \tag{32}$$

as $h,\Delta \rightarrow 0$, as desired.

References

1. H.J. Kushner, "On the weak convergence of interpolated Markov chains to a diffusion", Ann. Prob. 2, 1974, pp. 40-50.
2. H.J. Kushner and C.F. Yu, "Approximations, existence, and numerical procedures for optimal stochastic controls", C.D.S. Report 73-1, Brown University, 1973; to appear in J. Math. Anal. and Applic., 1974.
3. H.J. Kushner, "Probability limit theorems and the convergence of finite difference approximations of partial differential equations", J. Math. Anal. and Applic. 32, 1970, pp. 77-103.
4. P. Billingsley, Convergence of Probability Measures, John Wiley, New York, 1968.
5. A.V. Skorokhod, "Limit theorems for stochastic processes", Theory of Probability and its Applications 1, 1956, pp. 261-290 (English translation).
6. H.J. Kushner, "Dynamical equations for nonlinear filtering", J. Diff. Eqs. 3, 1967, pp. 179-190.
7. M. Zakai, "On the optimal filtering of diffusion processes", Z. Wahrscheinlichkeits Theorie verw. Geb. 11, 1969, pp. 230-243.
8. H.J. Kushner, "On the differential equation satisfied by conditional probability densities of Markov processes", SIAM J. Control, Ser. A 2, 1964, pp. 106-119.
9. H.J. Kushner, "Finite difference methods for the weak solutions of the Kolmogorov equations for the density of both diffusion and conditional diffusion processes", C.D.S. Report, Brown University, 1974; submitted to SIAM J. on Applied Math.

ON THE RELATION BETWEEN STOCHASTIC AND DETERMINISTIC OPTIMIZATION

Roger J-B. Wets*

Abstract

It is shown that stochastic optimization problems ressemble deterministic optimization problems up to the nonanticipativity restriction on the choice of the control policy. This condition can be introduced explicitly in the form of a constraint. Doing so, lead to an optimization problem for which it is possible to derive optimality criteria. The convex case is considered here. It is shown that the variables associated with the nonanticipativity restriction form a martingale. These variables can be used to formulate an equivalent problem allowing for pointwise optimization, a result strongly akin to the Maximum principle. Finally two simple examples are used to illustrate the main results.

1. Introduction

The last decade has seen the development of an abstract theory for (deterministic) optimization problems. This general theory is applicable to mathematical programs, control problems as well as the classical family of optimization problems fitting in the framework of the calculus of variations. Consult for example the work of Neustadt [1] and Gamkrelidze-Kharatishvili [2] on necessary conditions, Rockafellar [3] on convex optimization problems. Roughly speaking, the framework for this general theory consists in studying the following problem:

(1.1) Find $x \in C \subset \mathcal{X}$ such that $g(x)$ is minimized when C is a subset of $\mathcal{X}$ a linear topological space and g is a functional on $\mathcal{X}$.

Stochastic optimization problems differ from the standard deterministic optimization problems in a number of ways, but the essential difference is the sequential nature of the problem: A decision $u \in \mathcal{U}$ must be selected *before* the outcome of some random event $\theta \in \Theta$ revealed. (There are wide variations as to the amount of information made available to the decision maker: complete or partial knowledge of the distribution of the random events,. . .). The return is a real valued function g defined on $\mathcal{U} \times \Theta$. The decision criterion is to minimize expected returns. Thus assuming that the probability space $(\Theta, \mathcal{F}, \nu)$, on which the random variable $\underset{\sim}{\theta}$ is defined, is known, it seems natural that an "abstract" formulation of a stochastic optimization problem could be

(1.2) Find $u \in \mathcal{U}$ such that $(u, \underset{\sim}{\theta}) \in C$ almost surely and $E\{g(u, \underset{\sim}{\theta})\}$ is minimized.

This time C is a subset of $\mathcal{U} \times \Theta$ and represents the constraints of the problem and $\mathcal{U}$ is a linear topological space.

Naturally problem (1.2) is a special case of (1.1) and it is not clear that there is any need for a specialized theory for stochastic optimization problems. However we have given to (1.2) so little structure

* Supported in part by the National Science Foundation under grant GP–31551.

that only results with very little bearing on "real" stochastic optimization problems can be derived. As pointed out above, the "time" element is a significant component of stochastic optimization model. (Causal system have also been discussed cfr. [4] and references mentioned therein).

We introduce "time" as follows: we think of $\underset{\sim}{\theta}$ as representing a stochastic process and let $\mathcal{U}$ be a function space on $(\Theta, \mathfrak{F}, \nu)$. Naturally, we may demand that the decision functions u posess a certain number of qualifications. To fix the ideas, let T represent the time-space. The random variable $\underset{\sim}{\theta}$ is a stochastic process on T, thus element of Θ are "paths" $\theta(t)$, which do not need to be specified any further at this time. The decision space $\mathcal{U}$ is a linear space of functions on $(\Theta, \mathfrak{F}, \nu)$

$$\mathcal{U} = F((\Theta, \mathfrak{F}, \nu);\ E(T;B))$$

with values in E(T,B), a linear space of "time"-dependent functions which are B-valued with B an appropriate linear space. The time-space T is a closed (naturally ordered) subset of **R**, specifically T might be an *interval* (continuous time problems), a *discrete set* (stage problems).

The choice of u is now to occur as a function of θ which is itself time dependent.

For purposes of illustration consider the following standard stochastic control problem, a variant of the problem described in [5]. The dynamics of the system are given by the stochastic integral equations:

(1.3) $$x(t) = x^{o} + \int_0^t h(s,x(s),v(s))ds + \int_0^t \sigma(s,x(s))dw(s), \qquad t \in [0,1]$$

with x^{o} the initial conditions. Here w is brownian motion. The objective is to select a *control policy*

(1.4) $$u : T \times \mathcal{X} \rightarrow \mathcal{V}$$

where $\mathcal{X}$ is the state space and $\mathcal{V}$ is the control space. This control policy must be selected so as to minimize

(1.5) $$\mathcal{J}(u) = E\{ \int_0^1 L(t,x(t),v(t))dt\} .$$

We now impose sufficient Lipschitz-continuity on h and σ so that to each continuous control policy correspond a solution to (1.3) which is almost surely unique. This control policy and a corresponding solution to (1.3) determine the value of the objective. The problem can thus be cast in the general framework (1.2) with in this case $\mathcal{U}$ the space of acceptable control policies and g the term appearing between curly brackets in (1.5) with u and x as defined by (1.4) and (1.3) respectively.

With $(\Theta, \mathfrak{F}, \sigma)$ and $\mathcal{U}$ as defined before, problem (1.2) now incorporates "time" but in this form it still ignores the important requirement that decisions must occur before observations. It is essential that we restrict the class of functions $\mathcal{U}$ to the subclass of *nonanticipative functions* . Let D denote the subet of $\mathcal{U}$ consisting of these nonanticipative functions. Restricting the class of admissible functions u to lie in D is innate to stochastic optimization problems and distinguishes them from the standard optimization problems. The purpose of this paper is to investigate the consequence of this restriction. We shall thus study the following problem:

(1.7) Find $u \in D \subset \mathcal{U}$ such that $E\{F(u, \underset{\sim}{\theta})\}$ is minimized where the function

(1.8) $$F : U \times \Theta \to]-\infty, +\infty]$$

is allowed to take on the value $+\infty$ in order to incorporate the constraints in the objective. The relation between F and g and C appearing in (1.2) is as follows

$$F(u,\theta) = \begin{cases} g(u,\theta) \text{ if } (u,\theta) \in C \subset \mathcal{U} \times \Theta \\ +\infty \quad \text{otherwise.} \end{cases}$$

This model covers a large class of stochastic optimization problems. There are however some significant features of certain classes of control problems which are not included in the present formulation, such as: noncomplete observation (this can be partially remedied by replacing the domain of definition of u by one obtained from $(\Theta, \mathfrak{F}, \nu)$ by a transformation representing information transmission); the existence of a control policy defined on the σ-fields generated by the state of the system rather than defined on the σ-field of the underlying probability space; the possible dependence of the stochastic process on the control policy, for example when the control affects the variance or mean of a brownian process ;. . .

The results presented here are part of a study of stochastic optimization problems undertaken in cooperation with R. T. Rockafellar. A more complete exposition will appear in a joint paper. Here we limit ourselves to give one form of the basic theorem and some of its implications.

2. A duality theorem

In order to exploit duality to its fullest, we shall only consider the convex case, i.e. when F is a convex function u. We shall also give F a more specific representation in terms of a "time"-integral. We set

(2.1) $$F(u,\theta) = \int_T f(u(\theta(t)),\theta(t))dt.$$

The following hypothesis will be in force throughout the rest of this paper: the stochastic process $\underset{\sim}{\theta}(t)$ is a measurable stochastic process with values in Ξ; the class of admissible policies

$$\mathcal{U} = \mathcal{L}^{\infty}((\Theta,\mathfrak{F},\nu); \mathcal{L}^{\infty}((T,dt); R^n))$$

is the class of (ν-)essentially bounded measurable functions with $(\Theta,\mathfrak{F},\nu)$ for domain and for range, the R^n-valued (dt-)essentially bounded measurable functions on T; for each $\xi \in \Xi$, the function

$$u \mapsto f(u,\xi)$$

viewed as a function from R^n into $R \cup \{+\infty\}$ is convex, lower semicontinuous, not identically $+\infty$ and the interior of its effective domain $\{u \mid f(u,\xi) < +\infty\}$ is nonempty; for each $u \in R^n$, the function

$$\theta(t) \mapsto f(u,\theta(t))$$

is $\nu \times dt$ measurable. Finally, we shall assume that there exists $\epsilon > 0$ and a function $a(\theta(t)) \in D \subset \mathcal{U}$ such that for each $u \in R^n$ with $\|u\|_2 < \epsilon$, the function $f(u+a(\theta(t)),\theta(t))$ is finite and bounded in θ

and $t \in T$, where D is the class of nonanticipative functions, introduced in §1 and to be given a precise meaning in this context later on. This last assumption will be known as the *strict feasibility condition.* Justification of this terminology comes from the observation that feasibility of the original problem corresponds essentially to finiteness of f and the preceeding condition demands that there a function $a(\theta(t))$ which is "strictly" inside the feasible region.

From the above, (among others) it follows [6, Lemma 2] that f is a *normal convex integrand.* As in [6] a function h on $R^N \times S$, with (S,μ) a measure space, is a convex integrand if for each $s \in S$, $h(\cdot, s)$ is a lower semicontinuous convex function on R^N with values in $]-\infty, +\infty]$ not identically $+\infty$, and if furthermore there exist a sequence of measurable functions

$$z^k : S \to R^N, \; k=1,2,\ldots$$

such that $h(s,z^k(s))$ is measurable in s for each k, while for each fixed s the set of points of the form $z^k(s)$ lying in

$$\mathrm{dom}\; h(s,\cdot) = \{z \in R^N \mid h(s,z) < +\infty\}$$

are dense in the latter set. The study of normal convex integrands was initiated by Rockafellar [6], [7]. They have been used in stochastic optimal control theory by Bismut [8], [9] and in stochastic programming by Rockafellar and Wets [10], [11]. Of immediate interest here is the fact that $f(u(\theta(t)), \theta(t))$ is $\nu \times dt$-measurable for every measurable function u from $T \times \Theta$ to R^n [6, Lemma 5].

The integral functional

$$\text{(2.2)} \qquad I_f(u) = \int_{\Theta \times T} (\nu \times dt) f(u(\theta(t)), \theta(t)) = E\{\int_T f(u(\theta(t)), \theta(t))dt\}$$

is therefore well defined. Its value is real or $-\infty$ for every measurable function u such that $f(u(\theta(t)), \theta(t))$ is majorized by a summable function of $\theta(t)$; when $f(u(\theta(t)), \theta(t))$ is not majorized by an integrable function its value is set at $+\infty$ by convention. Now, from the strict feasibility condition we may conclude [6, Theorem 3] that the functional

$$u \mapsto I_f(u)$$

on $\mathcal{U}$, with values in $]-\infty, +\infty]$, is convex, lower semicontinuous, not identically $+\infty$. Moreover, I_f is continuous in the norm topology of $\mathcal{U}$ at any strictly feasible point.

Let P_t be a *projection* operator mapping a function $\mathcal{V}(s)$ into its "t-first" coordinates. More precisely, consider the function $s \to \mathcal{V}(s)$ from T into B, then the projection P_t is defined by

$$\mathcal{V} \mapsto P_t\mathcal{V} = \{\mathcal{V}(s), \; s \leq t\}.$$

The *tail projection*, denoted by $(I-P_t)$, is defined in a similar way

$$\mathcal{V} \mapsto (I-P_t)\mathcal{V} = \{\mathcal{V}(s), s \geq t\}.$$

A control policy u evaluated for some path $\theta \in \Theta$ determines a function on T into R^n. It is sometimes convenient to write $u_\theta(t)$ for this function on T, indicating more clearly than the notation $u(\theta(t))$ the functional dependence. The class of nonanticipative functions on $\mathcal{U}$ is the subfamily of functions in D, where

(2.3) $D = \{u \in \mathcal{U} \mid P_t u_\theta = P_t u_{\bar\theta}$ a.e., for almost all $(\theta, \bar\theta) \in \Theta$ such that $P_t\theta = P_t\bar\theta$ a.e.$\}$ where a.e. stands for almost everywhere dt.

With these definitions, we have now a concrete form of the original problem (1.8) which we can also write as:

(2.4) Find $u \in D \subset \mathcal{U}$ such that $I_f(u)$ is minimum.

This is an optimization problem whose only explicit constraint is the nonanticipativity restriction (the other constraints are burried in the $+\infty$ value for f). Restriction to D is a nontrivial restriction and obviously

$$\operatorname*{Inf}_{u \in \mathcal{U}} I_f(u) \leq \operatorname*{Inf}_{u \in D \subset \mathcal{U}} I_f(u) .$$

Consequently there is a "price" to be payed for this limitation. The classical Lagrangian theory asserts that this "price" can be expressed in terms of a *price system* associated with the constraints. Let

$$\mathcal{V} = \mathcal{L}^1((\Theta, \mathfrak{F}, \nu); \mathcal{L}^1((T,dt); R^n))$$

be a space put in duality with $\mathcal{U}$ by the pairing

(2.5) $$\langle u,p \rangle = E\{ \int_T u(\theta(s)) \cdot p(\theta(s))ds \}.$$

Moreover let $\mathfrak{F}_t$ $(t \in T)$ be the nested family of σ-fields induced on $\mathfrak{F}$ by the stochastic process $\underset{\sim}{\theta}(t)$ such that for any $s,t \in T$ we have that

$$s < t \text{ implies } \mathfrak{F}_s \subset \mathfrak{F}_t \subset \mathfrak{F} .$$

For conditional expectation we write

$$E\{ \cdot \mid \mathfrak{F}_t \}$$

which stands for conditional expectation given the information field represented by $\mathfrak{F}_t$.

As a corollary to the main result of this paper we obtain the following:

COROLLARY 2. *A control policy $\bar u \in D \subset \mathcal{U}$ is an optimal solution of the stochastic optimization problem*

$$\text{Find inf } E\{ \int_T f(u_\theta(t), \theta(t))dt \} \text{ for } u \in D \subset \mathcal{U}$$

if and only if there exist a stochastic process $\bar p_\theta$ with $\bar p \in \mathcal{V}$ such that the pair $(\bar u, \bar p)$ satisfy the following conditions

(a) *$\bar u$ minimizes the Lagrangian*

(2.6) $$E\{ \int_T [f(u_\theta(t), \theta(t)) - u_\theta(t) \cdot \bar p_\theta(t)]\, dt \}$$

on the space $\mathcal{U}$;

(b) *for every* $t \in T$, *the stochastic process* $\bar{p}_\theta$ *satisfies the martingale property*

(2.7) $$E\{(I-P_t)\bar{p}_\theta \,|\, \mathcal{F}_t\} = 0 \qquad \text{a.e.}$$

Clearly this corollary characterizes the structure of the dual variables to be associated with nonanticipativity. We obtain also for problem (2.4) a optimality criterion strongly akin to the deterministic maximum principle which allows for pointwise optimization, cfr. Corollary 3. First, we establish some of the properties for D – the class of nonanticipative functions – which have direct bearing on the main theorem.

PROPOSITION 1. *The class* D *of nonanticipative functions is a closed linear subspace of* $\mathcal{U}$.

PROOF. It is clear that D is a linear subspace. To show that it is closed, we consider a sequence $\{u_k\}$ in D converging to u_o in $\mathcal{U}$ and establish that u_o is also in D. Convergence of $\{u_k\}$ to u_o implies that for k sufficiently large

$$\|u_k(\theta)-u_o(\theta)\| \leqslant \mathbf{1}\cdot 1/m \qquad \text{a.e.}$$

for every $\theta \in \Theta$ except possibly on a set B_m such that $\nu(B_m) = 0$, where $\mathbf{1}$ is the constant map from (T,dt) into R^n with all of its n components equal to 1. Let $B = \cup B_m$ then $\nu(B) = 0$ and $(u_k(\theta)-u_o(0)) \to \mathbf{0}$ on $\Theta \setminus B$. The convergence of $(u_n(\theta)-u_o(\theta))$ to the $\mathbf{0}$ function is also in the ess. sup norm with respect to dt. Repeating the same argument than above it is easy to see that

$$u_k(\theta(t)) \to u_o(\theta(t))$$

pointwise on $\Theta \times T$ except possibly on a subset A of $\nu \times dt$-measure zero. These properties remain valid for the projections. In particular we have that

$$P_t\, u_n(\theta(s)) \to P_t\, u_o(\theta(s)) \qquad \text{a.e. s.}$$

and for almost all $\theta \in \Theta$. This is obviously requires the function u^o to be nonanticipative.

Depending on the pairing between a space $\mathcal{V}$ and the space $\mathcal{U}$, different duality results can be derived. Here we have selected the pairing given by (2.5) which can also be written as

$$\langle u,p\rangle = \int (\nu \times dt)\, u_\theta(t)\cdot p_\theta t$$

this gives to $\mathcal{U}$ the weak*-topology, induced on $\mathcal{U}$ by $\mathcal{V}$.

COROLLARY. *D is weakly closed in the topology induced on* $\mathcal{U}$ *by* $\mathcal{V}$ *by the apiring (2.5).*

Let $D^\perp$ denote the subset of $\mathcal{V}$ annihilating D, i.e.

(2.8) $$D^\perp = \{p \in \mathcal{V} \,|\, \langle u,p\rangle = 0 \ \text{ for all } \ u \in D\}.$$

PROPOSITION 2. *The subset* $D^\perp$ *consists of all functions p satisfying a martingale property, namely*

(2.9) $$D^\perp = \{p \in \mathcal{V} \,|\, \text{for all } t \in T \quad E\{(I-P_t)\,p_\theta \,|\, \mathcal{F}_t\} = 0 \quad \text{a.e.} \quad s\}.$$

PROOF. It is clear that for all $p \in \mathcal{V}$, $\langle u,p \rangle$ is finite thus, in particular, we can apply Fubini's theorem to write (2.7) as

$$\langle u,p \rangle = \int_T E\{u(\theta(s)) \cdot p(\theta(s))\} ds.$$

Applying the iterated condition expectation formula, which is again Fubini's theorem in a convenient form, the above yields

$$\langle u,p \rangle = \int_T E[E\{u(\theta(s)) \cdot p(\theta(s)) \,|\, \tilde{\mathcal{F}}_t\}] ds. \tag{2.10}$$

From (2.10) it is obvious that if $p \in D^{\perp}$ then for every $u \in D$ $\langle u,p \rangle = 0$. To see that it is also necessary consider the class of simple nonanticipative functions i.e. those functions obtained as countable sums of nonanticipative characteristic functions such that $u(\theta(s)) = 0$ if $s \leq t$. For these classes of functions it is clear that $\langle u,p \rangle = 0$ for all the members of this class only if $p \in D^{\perp}$, from which the proposition follows.

LEMMA: I_f is (weakly) lower semicontinuous at a point $u^o \in D$.

PROOF. This follows from the existence of a point u^o strictly feasible since with this restriction, the functionals I_f and I_{f*} are convex functions conjugate to each other [6, Theorem 3] where

$$I_{f*}(p) = \int_{\Theta} (\nu * dt)\, f^*(p(\theta(t), \theta(t))) \tag{2.11}$$

and

$$f^*(p,\xi) = \operatorname*{Sup}_u [u \cdot p - f(u,\xi)]$$

for every $p \in R^n$ and $\xi \in \Xi$. Thus I_f is lower semicontinuous at x^o at which it is finite.

Henceforth, we shall also assume that I_f is *continuous in the Mackey-topology* at strictly feasible points u^o. This will certainly be the case whenever $f(u,\xi)$ is finite and $(\nu \times dt)$-summable. (When the effective domain of f is constant in u and ξ, the problem can also be reduced to this case)

DUALITY THEOREM. With the above hypotheses

$$\operatorname{Inf}\{I_f(u) \,|\, u \in D\} = \operatorname{Max}\{-I_{f*}(p) \,|\, p \in D^{\perp}\}. \tag{2.12}$$

PROOF. From proposition 2 it follows that D is closed in its Mackey topology. Thus the indicator function

$$\psi_D(u) = 0 \text{ if } u \in D \quad \text{and} \quad \psi_D(u) = +\infty \text{ otherwise}$$

is also convex, lower semicontinuous. Observe that ψ_D finite at u^o. We can thus reformulate problem (2.4) as

$$\text{Find inf } I_f(u) + \psi_D(u), \; u \in \mathcal{U}.$$

We apply Fenchel's duality theorem which asserts that

$$\operatorname{Inf}[I_f(u) + \psi_D(u)] = \operatorname{Max}[-\psi_D^*(p) - I_f^*(p)] \tag{2.13}$$

since there exists a point u^o at which I_f is (weakly) continuous and ψ_D is finite [12, Theorem 1]. The theorem follows from (2.13), (2.12) and the observation that

$$\psi^*_D(p) = 0 \text{ if and only if } p \in D^{\perp}.$$

COROLLARY 1. *A control policy* $\bar{u}$ *is a point where* $I_f(u)$ *achieves a minimum on D if and only if there exists* $\bar{p} \in D^{\perp}$ *such that*

$$-\bar{p} \in \partial I_f(\bar{u}) .$$

Moreover such points p are precisely those at which $-I_{f^*}$ *achieves its maximum on* $D^{\perp}$*, where* ∂h *denotes the subdifferential of a functional h.*

PROOF. A subgradient of a functional h at $x \in E$ is an element $x^* \in E^*$ such that

$$h(z) \geq h(x) + \langle z-x, x^* \rangle \text{ for all } y \in E. \tag{2.14}$$

The sudifferential $\partial h(x)$ of h at x is the set of subgradients of h at x. From the definition of conjugate function and (2.14) it follows that

$$x^* \in \partial h(x) \text{ if and only if } h(x) + h^*(x^*) \leq \langle x,x^* \rangle . \tag{2.15}$$

Thus $\bar{p} \in \partial\psi_D(\bar{u})$ and $-\bar{p} \in \partial I_f(\bar{u})$ – which is equivalent to $0 \in \partial I_f(\bar{u}) + \partial\psi_D(\bar{u})$ – implies

$$I_f(\bar{u}) + I_{f^*}(\bar{p}) \leq \langle \bar{u}, -\bar{p} \rangle$$

and

$$\psi_D(\bar{u}) + \psi_{D^{\perp}}(\bar{p}) \leq \langle \bar{u}, \bar{p} \rangle .$$

But the reverse inequalities are always true, it follows that

$$I_f(\bar{u}) + I_{f^*}(\bar{p}) = \langle \bar{u}, -\bar{p} \rangle = -\psi_D(\bar{u}) - \psi_{D^{\perp}}(\bar{p}) . \tag{2.16}$$

Thus

$$I_f(\bar{u}) + \psi_D(\bar{u}) = -I_{f^*}(\bar{p}) - \psi_{D^{\perp}}(\bar{p}) . \tag{2.17}$$

The corollary now follows directly from the duality theorem.

Corollary 2 announced above is just another version of corollary 1. We give a short proof.

PROOF OF COROLLARY 2. In view of proposition 2, $\bar{p}$ is an element of $D^{\perp}$ if and only condition (b) of Corollary 2 is satisfied. On the other hand $\bar{u}$ minimizes the Lagrangian (2.6) if and only if

$$0 \in \partial\, (I_f(\bar{u}) + \psi_D(\bar{u})).$$

Since I_f is continuous at a point where ψ_D is finite, the above is equivalent to the existence of $\bar{p}$ in $D^{\perp}$ such that $-p \in \partial I_f(\bar{u})$.

COROLLARY 3. (Maximum Principle) *Suppose that for each* $\xi \in \Xi$*, the functions*

$$u \mapsto f(u,\xi)$$

are finite. Moreover suppose that $I_f(u)$ is finite for each $u \in \mathcal{U}$. Then a control policy $\bar{u} \in D \subset \mathcal{U}$ is an optimal solution of the stochastic optimization problem

$$\text{Find } \inf E\{\int_T f(u_\theta(t), \theta(t))dt\} \quad \text{for } u \in D \subset \mathcal{U}$$

if and only if there exists a stochastic process $\bar{p}_\theta$ with $\bar{p} \in \mathcal{V}$ such that the pair $(\bar{u}, \bar{p})$ satisfy the following conditions

(a) for almost all θ and almost everywhere, $\bar{u}(\theta(t))$ minimizes the expression

$$f(u, \theta(t)) - \langle u, \bar{p}(\theta(t))\rangle$$

for $u \in R^n$

(b) for every $t \in T$, the stochastic process $\bar{p}_\theta(s)$ satisfies the martingale property

$$E\{(I - P_t)\bar{p}_\theta \mid \mathfrak{F}_t\} = 0 \quad \text{a.e.}$$

PROOF. In view of Corollary 2 it suffices to show that in this case, we can replace condition (a) of Corollary 2 by condition (a) of Corollary 3. This follows directly from [7, Corollary 1B] which when applied to this case asserts that an element $v \in \mathcal{V}$ is a subgradient of $I_f(\bar{u})$ if and only if for almost all θ

$$v_\theta(t) \in \partial f(\bar{u}_\theta(t), \theta(t)) \quad \text{a.e.} \quad t.$$

(We do not have to worry about "singular" components under the hypothesis of the theorem). Thus $\bar{p} \in \partial I_f(\bar{u})$ if and only if

$$0 \in \partial f(\bar{u}(\theta(t)), \theta(t)) - \bar{p}(\theta(t))$$

for almost all θ and almost everywhere. Clearly this is equivalent to condition (a) above.

The conditions of Corollary 3 will be satisfied when the original problem satisfies a condition comparable to complete controlability for control problems (satisfied for example by systems described by stochastic differential equations modelling a Wiener process) or comparable to the complete recourse condition in stochastic programming. In particular, this conditions is satisfied by the stochastic control problem described in section 1 or more to the point, when the function f is an *Hamiltonian* which already incorporates the usual constraints (by opposition to the nonanticipative constraints) associated with the appropriate multipliers, see §4.

3. A discrete time control problem

For prupose of illustration we consider a very simple control problem. We consider only a discrete time problem (T is a finite set) in order to avoid some of the technical difficulties inherent to continuous time problem. Let us consider the following stochastic control problem: Given

the initial state : x_o,

the dynamics: $x_t = ax_{t-1} + v_t + \eta_t \quad t = 1, \ldots, K,$

the stochastic process : η_t with a finite number of states,

the control constraints: $|v_t| \leqslant 1.$

The goal is to steer the system from x_o to a point as close as possible to the origin, the penalty is expressed by bx_k^2 where $b > 0$. We shall assume here that $b = 1/2$. It is easy to see that the dynamics of the system imply that

$$x_k = a^k x_o + \sum_{s=1}^{k} a^{k-s}(\mathcal{V}_s + \eta_s).$$

The stochastic optimization problem is:

Find a nonanticipative control policy u satisfying $-1 \leq u_\eta(t) \leq 1$ for every η and all $t = 1,\ldots,K$ and minimizing $E_\eta\{\frac{1}{2}[a^k x_o + \sum_{s=1}^{k} a^{k-s}(u_\eta(s) + \eta_s)]^2\}$.

Given a fixed path $\tilde{\eta}$ - a specific realization of the stochastic process - then for the corresponding deterministic control problem a control $\tilde{\mathcal{V}}$ is optimal if there exists $\alpha(\tilde{\eta})$ and $\beta(\tilde{\eta})$ such that

(a') $0 \leq \tilde{\mathcal{V}}_s \leq M$; $(\alpha_s(\tilde{\eta}), \beta_s(\tilde{\eta}) \geq 0$; $\alpha_s(\tilde{\eta})\mathcal{V}_s = 0$, $\beta_s(\tilde{\eta})(M - \tilde{\mathcal{V}}_s) = 0$ for $s = 1,\ldots,K$

and

(b') $\tilde{\mathcal{V}}$ minimizes $\frac{1}{2}[a^k x_o + \Sigma_1^k a^{k-s}(\mathcal{V}_s + \tilde{\eta}_s)]^2 + \Sigma_1^k [\alpha_s(\tilde{\eta})(-\mathcal{V}_s - 1) + \beta_s(\tilde{\eta})(\mathcal{V}_s - 1)]$.

Let us set

$$\psi_s(\tilde{\eta}) = \beta_s(\tilde{\eta}) - \alpha_s(\tilde{\eta})$$

we can rexpress (a') and (b') as follows.

(a) $|\tilde{\mathcal{V}}_s| \leq 1$ and $\psi_s(\tilde{\eta}) = 0$ whenever $|\tilde{\mathcal{V}}_s| < 1$, $\psi_s(\tilde{\eta}) \geq 0$ if $\tilde{\mathcal{V}}_s = 1$, $\psi_s(\tilde{\eta}) \leq 0$ if $\tilde{\mathcal{V}}_s = -1$

(b) $\tilde{\mathcal{V}}$ minimizes $\frac{1}{2}[a^k x_o + \Sigma_1^k a^{k-s}(\mathcal{V}_s + \tilde{\eta}_s)]^2 + \Sigma_1^k \psi_s(\tilde{\eta})\mathcal{V}_s$.

Using the above conditions in conjunction with the "pointwise" conditions obtained in Corollary 3, we have that a control polcy $\bar{u}_\eta$ is optimal if there exists integrable functions ψ and p such that

(i) for almost all η, $|\bar{u}_\eta(s)| \leq 1$, $\psi_\eta(s) = 0$ if $|\bar{u}_\eta(s)| < 1$ with $\psi_\eta(s) \geq 0$ whenver $\bar{u}_\eta(s) = 1$ and $\psi_\eta(s) \leq 0$ whenever $\bar{u}_\eta(s) = -1$.

(ii) $E_\eta\{[a^k x^o + \Sigma_1^k a^{k-s}(\bar{u}_\eta(s) + \eta(s))]a^{k-\ell} + \psi_\eta(\ell) \mid \eta(1),\ldots,\eta(\ell)\} = 0$ for $\ell = 1,\ldots,k$.

4. Stochastic Programming

We consider a simplified version of a problem investigated in [10], [11]. Let $(\Xi, \mathfrak{F}, \sigma)$ be a probability space;

$$(\xi, x_1, x_2) \mapsto f_i(\xi, x_1, x_2) \quad i=0,\ldots,m$$

finite convex functions in (x_1,x_2), measurable and bounded in ξ with $\xi \mapsto f_0(\xi,x_1,x_2)$ summable. The problem consists in finding $x_1 \in R^{n_1}$, $x_2 \in \mathcal{L}^\infty((\Xi, \mathfrak{F}, \sigma); R^{n_2}) = \mathcal{L}^\infty_{n_2}$ satisfying almost surely

(4.1) $$f_i(\xi, x_1, x_2(\xi)) \leq 0 \qquad i=1,\ldots,m$$

and minimizing

$$E\{f_o(\xi,x_1,x_2(\underset{\sim}{\xi}))\}$$

The nonanticipativity condition is hidden in the requirement that x_1 does not depend on ξ. A pointwise condition can be obtained if we introduce the nonanticipativity condition and then via the duality theorem and its corollaries we can derive the appropriate conditions.

We give here a *very rough* treatment. A different approach has been used in [13]. The "Hamiltonian" associated with this stochastic program can be written as

$$H(\xi,x_1,x_2,y) = f_0(\xi,x_1,x_2) + \Sigma_1^m y_i f_i(\xi,x_1,x_2)$$

for any fixed y, the function H is finite and thus we can use Corollary 3. For fixed $\bar{y}(\xi) \in \mathcal{L}^1_m$, the pair $(\bar{x}_1,\bar{x}_2(\xi)) \in R^{n_1} \times \mathcal{L}^\infty_{n_2}$ minimize

$$E\{H(\xi,x_1,x_2(\xi), \bar{y}(\xi))\}$$

if there exist a function $\bar{p} = (\bar{p}_1, \bar{p}_2) \in \mathcal{L}^1_{n_1} \times \mathcal{L}^1_{n_2}$ such that for almost all ξ, $(\bar{x}_1,\bar{x}_2(\xi))$ minimizes

(4.2) $$H(\xi,x_1,x_2(\xi), \bar{y}(\xi)) - x_1 \cdot \bar{p}_1(\xi) - x_2 \cdot \bar{p}_2(\xi)$$

where $\bar{p}$ satisfies the martingale property

(4.3) $$E\{\bar{p}_1(\xi)\} = 0$$

and

(4.4) $$E\{\bar{p}_2(\xi) \mid \xi\} = 0 \quad \Rightarrow \quad \bar{p}_2(\xi) = 0 \text{ for all } \xi .$$

Now for the sake of exposition let us assume that the functions

$$(x_1,x_2) \mapsto f_i(\xi,x_1,x_2)$$

for i=0,. . .,m are differentiable and that the stochastic program is a stochastic program with complete recourse, i.e. for all ξ in Ξ and $x_1 \in R^{n_1}$ there exists $x_2 \in R^{n_2}$ such that the conditions

$$f_i(\xi,x_1,x_2) \leq 0 \qquad i=1,\ldots,m$$

can be satisfied. (This last conditions allows us to avoid introducing singular functions [11] for the multipliers $\bar{y}(\xi)$ which are associated with the constraints (4.1).) The optimality conditions (4.2), (4.3) and (4.4) become:

$$\nabla_{x_1} f_o(\xi,\bar{x}_1,\bar{x}_2(\xi)) + \Sigma_1^m \bar{y}_i(\xi) \nabla_{x_1} f_i(\xi,\bar{x}_1,\bar{x}_2(\xi)) = \bar{p}_1(\xi) \quad \text{a.s.}$$

$$\nabla_{x_2} f_o(\xi,\bar{x}_1,\bar{x}_2(\xi)) + \Sigma_1^m \bar{y}_i(\xi) \nabla_{x_2} f_i(\xi,\bar{x}_1,\bar{x}_2(\xi)) = 0 \quad \text{a.s.}$$

and

$$E\{\bar{p}_1(\xi)\} = 0.$$

Assuming that the multipliers $\bar{y}(\xi)$ are Lagrange (optimal) multipliers associated with pointwise optimization of f_o subject to $f_i \leq 0$. Then the above yield optimality criteria for this stochastic program (compare with the basic Kuhn-Tucker conditions of [13, Corollary C]).

REFERENCES

[1] L. W. Neustadt, "A general Theory of Extremals", *J. Comput. System Sci.*, **3**(1969), 57–92.

[2] R. V. Gamkrelidze et G. L. Kharatishvili, "Conditions Nécessaires du Premier Ordre dans les Problèmes d'Extremum", **Actes, Congrès intern. Math., T.3**(1970) 169–176.

[3] R. T. Rockafellar, **Conjugate Duality and Optimization,** SIAM. Monograph Series (1974).

[4] H. S. Witsenhamsen, "On Information Structures, Feedback and Causality", *SIAM J. Control,* **9** (1971), 149–160.

[5] W. H. Felming, "Optimal Stochastic Control", **Actes, Congrès intern. Math. T.3**(1970), 163–167.

[6] R. T. Rockafellar, "Integral which are convex functionals", *Pacific J. Math.*, **24**(1968), 525–539.

[7] R. T. Rockafellar, "Integrals which are convex functionals II", *Pacific J. Math.,* **39**(1971), 439–469.

[8] J. M. Bismut, "Intégrals Convexes et Probabilités", *J. Math. Anal. Appl.,* **42**(1973), 639–673.

[9] J. M. Bismut, "Conjugate Convex Functions in Optimal Stochastic Control", **44**(1973), 384–404.

[10] R. T. Rockafellar and R. Wets, "Stochastic Convex Programming: Basic Duality", to appear.

[11] R. T. Rockafellar and R. Wets, "Stochastic Convex Programming: Extended Duality and Singular Multipliers", to appear.

[12] R. T. Rockafellar, "Extension of Fenchel's Duality Theorem for Convex Functions", *Duke Math. Journal,* **33**(1966), 81–89.

[13] R. T. Rockafellar and R. Wets, "Stochastic Convex Programming: Kuhn-Tucker Conditions", to appear.

University of Kentucky
Department of Mathematics
Lexington, Ky., U.S.A.

SOLUTION NUMÉRIQUE
DE L'ÉQUATION DIFFÉRENTIELLE DE RICCATI
RENCONTRÉE EN THÉORIE DE LA COMMANDE OPTIMALE
DES SYSTÈMES HÉRÉDITAIRES LINÉAIRES*

M. DELFOUR
Centre de Recherches Mathématiques
Université de Montréal, Montréal, Québec H3C 3J7, Canada

1. INTRODUCTION. Si l'on adopte le point de vue que les systèmes à retard, et de façon plus générale les systèmes héréditaires, sont des cas particuliers de systèmes gouvernés par des équations aux dérivées partielles (ÉDP), le problème de commande optimale de ces systèmes pour une fonction coût linéaire-quadratique n'est rien d'autre qu'un problème de commande optimale d'un système gouverné par une ÉDP. On en vient alors de façon assez naturelle à utiliser certaines techniques rencontrées dans l'étude des ÉDP de type parabolique. La théorie de la commande optimale des systèmes à retard ne constitue cependant pas un cas particulier de celle des systèmes paraboliques, car on ne retrouve pas de bons opérateurs coercifs. Malgré cela la méthode directe de J.L. LIONS reste applicable et on obtient le même genre de résultats que pour les systèmes paraboliques. Il n'y a qu'à adapter pour tenir compte de l'absence de coercivité. On obtient en particulier une équation différentielle de Riccati (EDR) pour un certain opérateur $\Pi(t)$. Le but de cette communication est de présenter une méthode numérique de résolution de cette équation (et du problème de commande optimale). Pour cela on a choisi d'adapter au cas des systèmes héréditaires linéaires la méthode de J.C. NÉDELEC pour la résolution numérique de l'ÉDR pour les systèmes paraboliques linéaires. La méthode de J.C. NÉDELEC n'est pas liée à l'approximation très simple du système héréditaire (méthode d'Euler) que l'on a adoptée et des approximations plus compliquées peuvent être considérées sans difficultés. On pourrait songer par exemple à utiliser la méthode des projections de BANKS-MANITIUS dans le cas autonome.

Il est important de remarquer que la méthode numérique de cette communication ne nécessite aucune hypothèse supplémentaire sur les matrices qui entrent dans la définition du système (mesurables et bornées). Elle est basée sur une approximation directe de l'ÉDR plutôt que sur une discrétisation d'un système couplé d'équations différentielles ordinaires et aux dérivées partielles pour le noyau de l'opérateur Π (cf. ELLER-AGGARWAL-BANKS). La méthode de J.C. NÉDELEC ne nécessite pas un tel système d'équations pour le noyau de Π. On s'apercevra d'ailleurs en étudiant le noyau de

* Ces travaux ont été réalisés au Centre de Recherches Mathématiques dans le cadre d'une subvention FCAC (1974-1975) du Ministère de l'Education de la Province de Québec et dans celui de la subvention A 8730 du Conseil National de Recherches du Canada.

$\Pi(t)$ que ce système d'équations ne peut être obtenu que dans le cas où les matrices $A_i(t)$ (cf. §2 et 3.2) possèdent une dérivée. Ces considérations seront confirmées par les exemples 5 et 6 de la §7. Enfin on remarquera que la méthode peut se généraliser de façon immédiate à des fonctions coût définies sur l'état du système.

Un algorithme qui se rapproche assez de celui présenté ici a été suggéré par SOLIMAN-RAY [1],[2]. On y utilise une semi-discrétisation (variable d'espace) qui transforme l'ÉDP en équation différentielle ordinaire et l'ÉDR pour l'opérateur $\Pi(t)$ en ÉDR matricielle pour laquelle des algorithmes de résolution numérique existent déjà. Les auteurs indiquent cependant qu'ils ne savent ni dans quel sens précis leur algorithme converge, ni comment démontrer qu'il converge. Dans cette communication on précise dans quel sens et l'on prouve que l'algorithme converge en utilisant une discrétisation complète (variables de temps et d'espace) de l'ÉDP.

La programmation du schéma et les essais numériques ont été faits par G. PAYRE[1].

Notations. Soit $\mathbb{R}$ le corps des nombres réels, $\mathbb{R}^n$ l'espace Euclidien de dimension n ($n\geq 1$, un entier), $\mathcal{L}(X,Y)$ l'espace des applications linéaires bornées de X dans Y, deux espaces de Banach sur $\mathbb{R}$. Soit L^* dans $\mathcal{L}(Y,X)$ l'*opérateur adjoint* de L dans $\mathcal{L}(X,Y)$. Lorsque $X=Y$, on écrira simplement $\mathcal{L}(X)$ au lieu de $\mathcal{L}(X,X)$ et I_X sera l'élément identité dans $\mathcal{L}(X)$. Si X et Y sont des Hilberts, L dans $\mathcal{L}(X)$ est dit symétrique (resp. positive ou ≥ 0) si $L^*=L$ (resp. $\forall\, x \in X,\ (Lx,x)_X\geq 0$). Soit F un sous ensemble convexe fermé de $\mathbb{R}^n$ et E un Banach sur $\mathbb{R}$. Soit $L^p(F;E)$ le Banach des classes d'équivalence des applications $F\to E$ mesurables au sens de Lebesgue qui sont p-intégrables ($1\leq p<\infty$) ou essentiellement bornées ($p=\infty$) et soit $\|\ \|_p$ la norme dans $L^p(F;E)$ et $(\ ,\ )_2$ le produit scalaire pour $p=2$. Soit $C(F;E)$ le Banach des applications continues et bornées de F dans E. On pose $I(a,b)=\mathbb{R}\cap[a,b]$ pour $-\infty\leq a<b\leq+\infty$. Lorsque $F=I(a,b)$ on écrira $L^p(a,b;E)$. Soit E un Hilbert et $H^1(a,b;E)$ l'espace de Sobolev des éléments x de $L^2(a,b;E)$ possèdant une dérivée distribution Dx dans $L^2(a,b;E)$. Enfin on définit pour $-\infty<a<b<+\infty$

$$\mathcal{P}(a,b) = \{(t,s)\in[a,b]\times[a,b]:t\geq s\}.$$

2. DESCRIPTION DU SYSTÈME. Soit N, m et n trois entiers positifs. Soit $X=\mathbb{R}^n$ (resp. $U=\mathbb{R}^m$) de produit scalaire $(\cdot,\cdot)$ (resp. $(\cdot,\cdot)_U$) et de norme $|\cdot|$ (resp. $|\cdot|_U$). Soit $T\geq a>0$ deux réels et soit $a\leq b\leq\infty$. On considère le système héréditaire différentiel

$$(2.1)\quad \begin{cases} \dfrac{dx}{dt}(t) = A_0(t)x(t) + \displaystyle\sum_{i=1}^{N} A_i(t)x(t+\theta_i) + \int_{-b}^{0} A_{01}(t,\theta)x(t+\theta)d\theta \\ \qquad\qquad + B(t)v(t) + f(t), \quad \text{dans }]0,T[\,, \\ x(\theta) = h(\theta) \quad \text{dans } I(-b,0). \end{cases}$$

$A_0,A_1,\ldots,A_N$ (resp. B) sont des fonctions matricielles $[0,T]\to\mathcal{L}(X)$ (resp. $\mathcal{L}(U,X)$) qui sont mesurables et bornées. $A_{01}:[0,T]\times I(-b,0)\to\mathcal{L}(X)$ est aussi mesurable et bornée et

---

[1] Coopérant militaire français au Centre de Recherches Mathématiques dans le cadre des échanges France-Québec.

il existe une famille $\{K(t):t\in[0,T[\}$ de sous-ensembles compacts de $I(-b,0)$ tel que (i) $K(t_1)\subset K(t_2)$, $t_1\le t_2$, (ii) $\{\theta\in I(-b,0):A_{01}(t,\theta)\neq 0\}\subset K(t)$. On se donne aussi f dans $L^2(0,T;X)$ et la fonction de commande v dans $L^2(0,T;U)$.

On choisit comme espace des données initiales le produit $X\times L^2(-b,0;X)$ avec

$$(2.2)\qquad ((h,k)) = (h^0,k^0) + \int_{-b}^{0}(h^1(\theta),k^1(\theta))d\theta,\quad \|h\|^2 = ((h,h)).$$

Pour utiliser l'espace produit on doit considérer le problème plus général qui consiste à trouver x dans $H^1(0,T;X)$ qui satisfasse l'équation suivante

$$(2.3)\qquad \begin{cases} \dfrac{dx}{dt}(t) = A_0(t)x(t) + \sum\limits_{i=1}^{N} A_i(t)\begin{Bmatrix} x(t+\theta_i), & t+\theta_i\ge 0 \\ h^1(t+\theta_i), & \text{sinon}\end{Bmatrix} + B(t)v(t) \\ \qquad + \displaystyle\int_{-b}^{0} A_{01}(t,\theta)\begin{Bmatrix} x(t+\theta), & t+\theta\ge 0 \\ h^1(t+\theta), & \text{sinon}\end{Bmatrix}d\theta + f(t),\quad \text{dans }]0,T[, \\ x(0) = h^0. \end{cases}$$

Dans ce qui suit on écrira M^2 au lieu de $X\times L^2(-b,0;X)$. On montre que le problème (2.3) possède une solution unique et que l'application $(h,v)\mapsto x : M^2\times L^2(0,T;U)\to H^1(0,T;X)$ est affine et continue.

Dans le cas où h appartient à $H^1(-b,0;X)$ on peut transformer l'équation (2.1) en une EDP si l'on introduit la fonction de deux variables $y(t,\theta)=x(t+\theta)$ définie dans $[0,T]\times I(-b,0)$:

ÉDP $\qquad \dfrac{\partial}{\partial t}y(t,\theta) = \dfrac{\partial}{\partial\theta}y(t,\theta)$, dans $]0,T[\times]-b,0[$

CONDITION AUX LIMITES

$$\frac{\partial}{\partial t}y(t,0) = A_0(t)y(t,0) + \sum_{i=1}^{N}A_i(t)y(t,\theta_i) + B(t)v(t) + \int_{-b}^{0}A_{01}(t,\theta)y(t,\theta)d\theta + f(t),\quad \text{dans }]0,T[,$$

CONDITION INITIALE $\qquad y(0,\theta) = h(\theta)$, dans $I(-b,0)$.

Cette formulation mène assez naturellement à une équation différentielle opérationnelle. Pour cela on définit l'*état* du système (2.3) comme un élément $\tilde{x}(t)$ de M^2,

$$(2.4)\qquad \tilde{x}(t)^0 = x(t),\quad \tilde{x}(t)^1(\theta) = \begin{Bmatrix} x(t+\theta), & -t<\theta\le 0 \\ h^1(t+\theta), & \theta\in I(-b,-t)\end{Bmatrix},$$

on se donne le sous-espace $V = \{(h(0),h):h\in H^1(-b,0;X)\}$ de M^2, les opérateurs $\tilde{A}_0(t):V\to X$, $\tilde{A}_1(t):V\to L^2(-b,0;X)$, $\tilde{A}(t):V\to M^2$, $\tilde{B}(t):U\to M^2$ et $\tilde{f}(t)$ dans M^2:

$$(2.5)\qquad \begin{cases} \tilde{A}_0(t)h = A_0(t)h(0) + \sum_{i=1}^{N} A_i(t)h(\theta_i) + \int_{-b}^{0} A_{01}(t,\theta)h(\theta)d\theta \\ (\tilde{A}_1(t)h)(\theta) = \frac{dh}{d\theta}(\theta) \end{cases}$$

$$(2.6)\qquad \tilde{A}(t)h = (\tilde{A}_0(t)h,\tilde{A}_1(t)h),\quad \tilde{B}(t)w = (B(t)w,0),\quad \tilde{f}(t) = (f(t),0).$$

Théorème 2.1. Pour tout h dans V et f dans $L^2(0,T;X)$, $\tilde{x}$ est la solution unique dans $W(0,T)=\{z\in L^2(0,T;V):Dz\in L^2(0,T;M^2)\}$ de l'équation

$$(2.7)\qquad \frac{d}{dt} z(t) = \tilde{A}(t)z(t) + \tilde{B}(t)v(t) + \tilde{f}(t),\quad \text{dans}\quad]0,T[,\ z(0) = h,$$

et l'application $(h,v)\mapsto\tilde{x}:V\times L^2(0,T;X)\to W(0,T)$ est affine et continue (V avec la topologie de H^1). Par densité il existe un redressement de cette application en une application affine et continue $M^2\times L^2(0,T;U)\to C(0,T;M^2)$. De plus, il existe un opérateur d'évolution $\tilde{\Phi}:P(0,T)\to\mathcal{L}(M^2)$ tel que (i) $\forall\ h\in M^2$, $(t,s)\mapsto\tilde{\Phi}(t,s)h$ soit continue, que (ii) $\forall\ (r,s)$, $0\le r\le s\le t\le T$, $\tilde{\Phi}(t,r)=\tilde{\Phi}(t,s)\tilde{\Phi}(s,r)$, et que (iii) la solution de (2.7) puisse s'écrire

$$(2.8)\qquad \tilde{x}(t) = \tilde{\Phi}(t,0)h + \int_0^t \tilde{\Phi}(t,r)[\tilde{B}(r)v(r)+\tilde{f}(r)]dr.\qquad \blacksquare$$

On aura aussi besoin de l'*état adjoint*. Soit $(M^2)'$ (resp. V') le dual topologique de M^2 (resp. V). On identifie les éléments de M^2 et $(M^2)'$ et l'on se donne les injections denses et continues $\Lambda:V\to M^2$ et $\Lambda^*:M^2\to V'$ comme dans la théorie des équations différentielles opérationnelles (cf. LIONS-MAGENES et J.L. LIONS [2]).

Théorème 2.2. Soit k dans M^2 et g dans $L^2(0,T;M^2)$. Alors l'équation

$$(2.9)\qquad \frac{dz}{dt}(t) + \tilde{A}(t)^*z(t) + \Lambda^*g(t) = 0,\quad \text{dans}\quad]0,T[,\ z(T) = k,$$

possède une solution unique $p(\cdot;k,g)$ dans l'espace

$$(2.10)\qquad W^*(0,T) = \{z:[0,T]\to M^2:\forall\ h\in M^2(\text{resp. } V), s\mapsto((h,z(s)))\ \text{est continue (resp. dans } H^1(0,T;\mathbb{R}))\}.$$

On montre aussi que Dz appartient à $L^2(0,T;V')$. L'application $(k,g)\mapsto p(\cdot;k,g):M^2\times L^2(0,T;M^2)\to W^*(0,T)$ est linéaire et continue. De plus

$$(2.11)\qquad p(t;k,g) = \tilde{\Phi}(T,t)^*k + \int_t^T \tilde{\Phi}(r,t)^*g(r)dr.\qquad \blacksquare$$

On trouvera les démonstrations des Théorèmes 2.1 et 2.2 dans DELFOUR [2].

3. PROBLÈME DE COMMANDE. Soient l'équation d'état pour (2.3) et la fonction coût

$$(3.1)\qquad \frac{d}{dt}\tilde{x}(t) = \tilde{A}(t)\tilde{x}(t) + \tilde{B}(t)v(t) + \tilde{f}(t),\ \text{dans}\quad]0,T[,\ \tilde{x}(0) = h,$$

$$(3.2)\qquad J(v,h) = (Fx(T)+2\ell,x(T))+\int_0^T [(Q(t)x(t)+2q(t),x(t))+(N(t)v(t)+2n(t),v(t))_U]dt,$$

où $\ell \in X$, $q \in L^2(0,T;X)$, $n \in L^2(0,T;U)$, $F \in \mathcal{L}(X)$, $Q \in L^\infty(0,T;\mathcal{L}(X))$ et $N \in L^\infty(0,T;\mathcal{L}(U))$. On supposera que F, Q(t) et N(t) sont symétriques et positives et qu'il existe une constante $c>0$ tel que $(w,N(t)w)_U \geq c|w|_U^2$ pour tout t et tout w. La fonction coût (3.2) peut aussi s'écrire à partir de l'état du système

$$(3.3) \qquad ((\tilde{F}\tilde{x}(T)+2\tilde{\ell},\tilde{x}(T)))+\int_0^T [((\tilde{Q}(t)\tilde{x}(t)+2\tilde{q}(t),\tilde{x}(t)))+(N(t)v(t)+2n(t),v(t))_U]dt,$$

où $\tilde{Q}(t)$ et $\tilde{F}$ (resp. $\tilde{q}(t)$ et $\tilde{\ell}$) sont des éléments de $\mathcal{L}(M^2)$ (resp. M^2) définis comme suit

$$(3.4) \qquad \tilde{Q}(t)h = (Q(t)h^0,0), \quad \tilde{F}h = (Fh^0,0), \quad \tilde{q}(t) = (q(t),0), \quad \tilde{\ell} = (\ell,0).$$

On peut montrer que pour tout h il existe une seule commande u dans $L^2(0,T;U)$ qui minimise J(v,h) par rapport aux éléments v de $L^2(0,T;U)$. Cette commande, dite optimale, est complètement caractérisée par le système d'optimalité suivant:

$$(3.5) \qquad \frac{d\tilde{x}}{dt}(t) = \tilde{A}(t)\tilde{x}(t) + \tilde{B}(t)u(t) + \tilde{f}(t), \quad \text{dans }]0,T[, \ \tilde{x}(0) = h,$$

$$(3.6) \qquad \frac{dp}{dt}(t) + \tilde{A}(t)^*p(t) + \Lambda^*[\tilde{Q}(t)\tilde{x}(t)+\tilde{q}(t)] = 0, \quad \text{dans }]0,T[, \ p(T) = \tilde{F}x(T) + \tilde{\ell},$$

$$(3.7) \qquad u(t) = -N(t)^{-1}[\tilde{B}(t)^*p(t)+n(t)], \quad \text{p.p. dans } [0,T],$$

(cf. DELFOUR-MITTER [3, Thm. 4.1], DELFOUR [2, Thm. 5.2] et J.L. LIONS [1]).

3.1. <u>"DÉCOUPLAGE" DU SYSTÈME D'OPTIMALITÉ</u>. Pour "découpler" on considère le problème de minimisation (3.1)-(3.2) dans l'intervalle [s,T] pour s dans [0,T[. C'est la méthode directe de J.L. LIONS [1]. On part de (3.1) avec la donnée initiale h au temps s et l'on associe à h et v (dans $L^2(s,T;U)$) la fonction coût

$$(3.8) \quad J_s(v,h)=((\tilde{F}\tilde{x}(T)+2\tilde{\ell},\tilde{x}(T)))+\int_s^T [((\tilde{Q}(t)\tilde{x}(t)+2\tilde{q}(t),\tilde{x}(t)))+(N(t)v(t)+2n(t),v(t))_U]dt.$$

Etant donné h on cherche à minimiser $J_s(v,h)$ par rapport à v dans $L^2(0,T;U)$.

<u>Proposition 3.1</u>. Soient $\tilde{x}$ et p les solutions de (3.5)-(3.6)-(3.7). Il existe une famille d'opérateurs linéaires $\Pi(s):M^2 \to M^2$ et des éléments r(s) de M^2, $0 \leq s \leq T$, tel que

$$(3.9) \qquad p(s) = \Pi(s)\tilde{x}(s) + r(s).$$

$\Pi(s)$ et r(s) sont obtenus de la façon suivante. (i) On résout

$$(3.10) \qquad \frac{d\beta}{dt}(t) = \tilde{A}(t)\beta(t)-\tilde{R}(t)\gamma(t), \quad \text{dans }]s,T[, \ \beta(s)=h, \ \tilde{R}(t)=\tilde{B}(t)N(t)^{-1}\tilde{B}(t)^*$$

$$(3.11) \qquad \frac{d\gamma}{dt}(t) + \tilde{A}(t)^*\gamma(t) + \Lambda^*\tilde{Q}(t)\beta(t) = 0, \quad \text{dans }]s,T[, \ \gamma(T) = \tilde{F}\beta(T),$$

et $\Pi(s)h = \gamma(s)$. (ii) On résout

$$(3.12)\qquad \frac{d\eta}{dt}(t) = \tilde{A}(t)\eta(t) - \tilde{R}(t)\xi(t) + \tilde{f}(t), \text{ dans }]s,T[,\ \eta(s) = 0,$$

$$(3.13)\qquad \frac{d\xi}{dt}(t) + \tilde{A}(t)^*\xi(t) + \Lambda^*[\tilde{Q}(t)\eta(t)+\tilde{q}(t)] = 0, \text{ dans }]s,T[,\ \xi(T) = \tilde{\ell},$$

et $r(s) = \xi(s)$ (cf. DELFOUR-MITTER [3, Lem. 4.5], DELFOUR [2, Thm. 5.2] et J.L. LIONS [1]. ■

3.2. <u>CARACTÉRISATION DE L'OPÉRATEUR Π</u>. D'après la Proposition 3.1 on peut supposer que f, q, n et ℓ sont nuls pour l'étude de Π. En effet dans ce cas r=0.

<u>Proposition 3.2</u>. (i) Soit β (resp. $\bar{\beta}$) la solution de l'équation

$$(3.14)\qquad \frac{d\beta}{dt}(t) = [A(t)-R(t)\Pi(t)\Lambda]\beta(t), \text{ dans }]s,T[,\ \beta(s) = h \text{ (resp. } \bar{h}).$$

Alors pour tout h et $\bar{h}$ dans M^2

$$(3.15)\qquad ((\Pi(s)h,\bar{h})) = ((\tilde{F}\beta(T),\bar{\beta}(T))) + \int_s^T (([\tilde{Q}(t)+\Pi(t)\tilde{R}(t)\Pi(t)]\beta(t),\bar{\beta}(t)))dt.$$

(ii) $\Pi(s)$ est un élément positif et symétrique de $\mathcal{L}(M^2)$ et il existe c>0 tel que $(\forall\ s)(\forall\ h)\|\Pi(s)h\|\le c\|h\|$. Il est équivalent à la matrice d'opérateurs

$$(3.16)\qquad \begin{bmatrix}\Pi_{00}(s) & \Pi_{01}(s)\\ \Pi_{10}(s) & \Pi_{11}(s)\end{bmatrix}\quad \begin{matrix}\Pi_{00}(s)\in\mathcal{L}(X),\ \Pi_{01}(s)\in\mathcal{L}(L^2(-b,0;X),X)\\ \Pi_{10}(s)\in\mathcal{L}(X,L^2(-b,0;X)),\ \Pi_{11}(s)\in\mathcal{L}(L^2(-b,0;X)),\end{matrix}$$

où $\Pi_{00}(s)^*=\Pi_{00}(s)\ge 0$, $\Pi_{01}(s)=\Pi_{10}(s)^*$, $\Pi_{11}(s)^*=\Pi_{11}(s)\ge 0$ et en particulier

$$(3.17)\qquad J_s(u,h) = ((\Pi(s)h,h)) = (\Pi_{00}(s)h^0,h^0) + 2(\Pi_{01}(s)h^1,h^0) + (\Pi_{11}(s)h^1,h^1)_2.$$

Il existe $\Pi_{01}(s,\cdot)$ dans $L^2(-b,0;\mathcal{L}(X))$ et $\Pi_{11}(s,\cdot,\cdot)$ dans $L^2(I(-b,0)\times I(-b,0);\mathcal{L}(X))$ tel que

$$(3.18)\qquad \Pi_{01}(s)h^1 = \int_{-b}^0 \Pi_{01}(s,\alpha)h^1(\alpha)d\alpha \quad\text{et}\quad (\Pi_{11}(s)h^1)(\alpha) = \int_{-b}^0 \Pi_{11}(s,\alpha,\beta)h^1(\beta)d\beta.$$

(iii) Soit $\tilde{\Phi}_\Pi(t,s)$ l'opérateur d'évolution correspondant à $\tilde{A}(t)-\tilde{R}(t)\Pi(t)\Lambda$. On peut maintenant écrire le côté droit de l'équation (3.15) sous la forme

$$(3.19)\qquad ((\tilde{F}\tilde{\Phi}_\Pi(T,s)h,\tilde{\Phi}_\Pi(T,s)\bar{h})) + \int_s^T (([\tilde{Q}(t)+\Pi(t)\tilde{R}(t)\Pi(t)]\tilde{\Phi}_\Pi(t,s)h,\tilde{\Phi}_\Pi(t,s)\bar{h}))dt.$$

(iv) Pour tout h et k dans M^2 (resp. V), l'application $s\mapsto((\Pi(s)h,k))$ est continue (resp. dérivable) dans [0,T]. Π possède une dérivée $D\Pi:[0,T]\to\mathcal{L}(V,V')$ au sens des distributions tel que $\forall\ h\in V$, l'application $s\mapsto D\Pi(s)h$ soit dans $L^\infty(0,T;V')$. On vérifie également que

$$(3.20)\qquad \begin{cases}\dfrac{d\Pi}{dt}(t) + \tilde{A}(t)^*\Pi(t)\Lambda + \Lambda^*\Pi(t)\tilde{A}(t) + \Lambda^*[\tilde{Q}(t)-\Pi(t)\tilde{R}(t)\Pi(t)]\Lambda = 0, \text{ dans }]0,T[,\\ \Pi(T) = \tilde{F}.\end{cases}$$

(v) Enfin le système (3.20) possède une solution unique dans l'espace

$$(3.21)\quad W(0,T;M^2,V,V') = \{K:[0,T]\to\mathcal{L}(M^2):\forall\ h,\bar{h}\ \text{dans}\ M^2\ (\text{resp. } V)$$

$$t\mapsto((K(t)h,\bar{h}))\ \text{est continue (resp. dérivable) dans } [0,T]\}.$$

Démonstration. Cf. DELFOUR [3]. On établit d'abord l'identité (3.15). On utilise alors les propriétés de l'opérateur $\tilde{\Phi}_\Pi$ pour en déduire celles de Π. On vérifie ensuite que Π est une solution de (3.20) et il ne reste plus qu'à prouver l'unicité. Cette méthode ne nécessite donc pas une étude directe de (3.20). ■

Dans la plupart des travaux publiés à date, on ne s'est pas intéressé à l'obtention de (3.20), mais plutôt à celle d'un système compliqué d'équations pour les fonctions matricielles $\Pi_{00}(t)$, $\Pi_{01}(t,\alpha)$ et $\Pi_{11}(t,\alpha,\beta)$. L'espace des fonctions continues y est utilisé comme espace des données initiales et on y résout le problème par des méthodes du type Carathéodory-Hamilton-Jacobi. Invariablement on y suppose (i) que la fonction coût optimale est d'une forme particulière et (ii) que les matrices entrant dans la définition de la fonction coût sont dérivables par rapport à chacune de leurs variables. En général pour le système (2.3) l'hypothèse (i) est correcte. l'hypothèse (ii) n'est cependant pas toujours vérifiée. En fait on ne peut pas toujours obtenir un système d'équations différentielles pour $\Pi_{00}(t)$, $\Pi_{01}(t,\alpha)$ et $\Pi_{11}(t,\alpha,\beta)$.

L'étude des propriétés des matrices $\Pi_{00}(t)$, $\Pi_{01}(t,\alpha)$ et $\Pi_{11}(t,\alpha,\beta)$ se fait à partir de l'identité (3.19) et de la caractérisation de l'opérateur d'évolution $\tilde{\Phi}_\Pi$ en fonction de la matrice $\Phi^0(t)$ des solutions fondamentales du système

$$(3.22)\qquad \frac{dx}{dt}(t) = [\tilde{A}(t)-\tilde{R}(t)\Pi(t)]^0\tilde{x}(t),\quad \text{dans }\ ]0,T[,\ \tilde{x}(s) = h,$$

et des matrices $A_1(t),\dots,A_N(t)$, $A_{01}(t,\theta)$, $R(t)$ et $\Pi_{01}(t,\theta)$ dont on connait certaines propriétés. De façon plus précise on utilise les identités suivantes (cf. DELFOUR-MITTER [2]):

$$(3.23)\qquad [\tilde{\Phi}_\Pi(r,t)h]^0 = \Phi(r,t)h,\ [\tilde{\Phi}_\Pi(r,t)h]^1(\theta) = \begin{cases}\Phi(r+\theta,t)h, & r+\theta\geq t\\ h^1(r+\theta-t), & \text{sinon}\end{cases}$$

$$(3.24)\qquad \Phi(r,t)h = \Phi^0(r,t)h^0 + \Phi^1(r,t)h^1,\ \Phi^1(r,t)h^1 = \int_{-b}^{0}\Phi^1(r,t,\alpha)h^1(\alpha)d\alpha$$

$$(3.25)\qquad \Phi^1(r,t,\alpha) = \sum_{i=1}^{N}\begin{cases}\Phi^0(r,t+\alpha-\theta_i)A_i(t+\alpha-\theta_i), & \theta_i\leq\alpha\leq\theta_i+r-t\\ 0, & \text{sinon}\end{cases}$$

$$+\int_{\max\{-b,\alpha+t-r\}}^{\alpha}\Phi^0(r,t+\alpha-\theta)[A_{01}(t+\alpha-\theta,\theta)-R(t+\alpha-\theta)\Pi_{01}(t+\alpha-\theta,\theta)]d\theta.$$

En faisant $h = (h^0,0)$ et $\bar{h} = (\bar{h}^0,0)$ dans l'expression (3.19) on obtient un expression explicite par la matrice $\Pi_{00}(t)$ et l'on vérifie qu'elle possède une dérivée. Pour obtenir $\Pi_{01}(t,\alpha)$ on fait $h = (0,h^1)$ et $\bar{h} = (\bar{h}^0,0)$ dans (3.19) et l'on obtient une expression de la forme

$$(3.26)\quad \Pi_{01}(t,\alpha)=\sum_{i=1}^{N}\begin{cases}[K_i(t,\alpha)+\Phi^0(T,t)^*F\Phi^0(T,t+\alpha-\theta_i)]A_i(t+\alpha-\theta_i), & \theta_i\le\alpha<\theta_i+T-t\\ 0 & ,\ \text{sinon}\end{cases}+K(t,\alpha),$$

où $K(t,\alpha)$ et $K_i(t,\alpha)$ sont des matrices dérivables en t et α et $K_i(t,\ \theta_i+T-t)=0$ pour $T+\theta_i\le t\le T$. On en déduit que l'application $\alpha\to\Pi_{01}(t,\alpha)$ est la somme d'une application dérivable et d'applications à support dans l'intervalle $[\theta_i,\min(0,\theta_i+T-t)]$ modulées par la matrice $A_i(t)$, $i=1,\dots,N$. Ceci indique que dans le cas général la fonction matricielle $\Pi_{01}(t,\alpha)$ ne sera pas dérivable par rapport à t ou α si l'une des fonctions matricielles $A_i(t)$ n'est pas dérivable. Lorsque les fonctions matricielles $A_i(t)$ sont continues (resp.dérivables), l'application $\alpha\to\Pi_{01}(t,\alpha)$ sera continue (resp. dérivable) par morceaux avec des sauts de hauteur $\Pi_{00}(t)A_i(t)$ aux points $\alpha=\theta_i$ et des sauts de hauteur $-\Phi^0(T,t)^*FA_i(T)$ aux points $\alpha=\theta_i+T-t$ pour $T+\theta_i\le t\le T$.

Dans le cas où les A_i sont dérivables, on peut obtenir un ensemble d'équations différentielles pour $\Pi_{00}(t)$, $\Pi_{01}(t,\alpha)$ et $\Pi_{11}(t,\alpha,\beta)$. Pour cela on utilise la même technique que dans DELFOUR-McCALLA-MITTER pour l'obtention d'équations semblables pour l'opérateur solution de l'équation algébrique de Riccati. Ici on obtient les résultats suivants.

$$(3.27)\quad \begin{cases}\frac{d\Pi}{dt}00\ (t)\ +\Pi_{00}(t)A_{00}(t)+A_{00}(t)^*\Pi_{00}(t)-\Pi_{00}(t)R(t)\Pi_{00}(t)\\ \qquad +Q(t)+\Pi_{01}(t,0)+\Pi_{01}(t,0)^*=0,\ \text{dans}\ \ ]0,T[,\ \Pi_{00}(T)=F.\end{cases}$$

Soit $D_i=\{(t,\theta_i):t\in[0,T]\}\cup\{(t,\theta_i+T-t):t\in[\max(0,T+\theta_i),T]\}$. $D=\cup\{D_i:i=1,\dots,N\}$ est l'ensemble des points de $[0,T]\times I(-b,0)$ où l'application $(t,\alpha)\mapsto\Pi_{01}(t,\alpha)$ présente des discontinuités. Dans le complément de D on a

$$(3.28)\quad \begin{cases}\left(\frac{\partial}{\partial t}-\frac{\partial}{\partial\alpha}\right)\Pi_{01}(t,\alpha)+[A_{00}(t)-R(t)\Pi_{00}(t)]^*\Pi_{01}(t,\alpha)\\ \qquad\qquad +\Pi_{00}(t)A_{01}(t,\alpha)+\Pi_{11}(t,0,\alpha)=0,\\ \Pi_{01}(T,\alpha)=0,\ \Pi_{01}(t,-b)=\begin{cases}\Pi_{00}(t)A_N(t), & b=a\\ 0 & ,\ b>a\end{cases}.\end{cases}$$

On remarquera que lorsque b=a la discontinuité au point $\theta_N=-a$ coincide avec la condition initiale en -a. On peut aussi définir pour $\Pi_{11}(t,\alpha,\beta)$ l'ensemble des points où cette fonction présente des discontinuités et sur le complément de cet ensemble on obtient

$$(3.29)\quad \begin{cases} \left(\frac{\partial}{\partial t}-\frac{\partial}{\partial\alpha}-\frac{\partial}{\partial\beta}\right)\Pi_{11}(t,\alpha,\beta)+\Pi_{01}(t,\alpha)^*A_{01}(t,\beta)+A_{01}(t,\alpha)^*\Pi_{01}(t,\beta) \\ \qquad -\Pi_{01}(t,\alpha)^*R(t)\Pi_{01}(t,\beta)=0 \\ \Pi_{11}(T,\alpha,\beta)=0,\ \Pi_{11}(t,\alpha,-b)=\begin{Bmatrix}\Pi_{01}(t,\alpha)^*A_N(t), & b=a \\ 0 & , b>a\end{Bmatrix} \\ \Pi_{11}(t,\alpha,\beta)=\Pi_{11}(t,\beta,\alpha)^*. \end{cases}$$

Dans le cas où b=a, F=0, N=1 et A_1 dérivable, toutes les discontinuités disparaissent. Ceci est le seul cas dans lequel l'hypothèse (ii) soit vérifiée et l'on obtient bien des résultats semblables à ceux de ALEKAL-BRUNOVSKY-CHYUNG-LEE, KUSHNER-BARNEA et A. MANITIUS (cf. aussi leur bibliographie respective).

3.3. CARACTÉRISATION DE LA FONCTION r. D'après la Proposition 3.1 on peut se ramener au cas h=0 et utiliser le système d'optimalité (3.12)-(3.13) avec s=0, $\eta=\tilde{x}$ et $\xi=p$. On sait aussi par l'identité (3.9) que $r(t)=p(t)-\Pi(t)\tilde{x}(t)$ et on en déduit facilement que r est la solution unique dans $W_*(0,T)$ de l'équation

$$(3.30)\quad \begin{cases} \frac{dr}{dt}(t)+(\tilde{A}(t)-\tilde{R}(t)\Pi(t)\Lambda)^*r(t)+\Lambda^*[\Pi(t)(\tilde{f}(t)-\tilde{B}(t)N(t)^{-1}n(t))+\tilde{q}(t)]=0,\text{dans }]0,T[, \\ r(T)=\tilde{\ell}. \end{cases}$$

4. APPROXIMATION DU SYSTÈME. Dans ce chapitre on construit une approximation de l'espace des données initiales (§4.1) et une approximation du système (2.3) (§4.2). A partir de cela on construit une approximation de l'équation d'évolution de l'état du système (2.3) (§4.3). En raison des hypothèses sur A_{01} dans §2, on peut toujours supposer que b est fini car le système (2.3) est défini sur l'intervalle compact [0,T]. On supposera aussi qu'il existe des entiers positifs non nuls $M,L,L_1,\dots,L_N$ et un pas de discrétisation $\delta>0$ tel que $T=M\delta$, $b=N\delta$, $\theta_i=-N_i\delta$, $i=1,\dots,N$. Dans le cas contraire on pourra utiliser un pas de discrétisation variable et les résultats demeurent valables.

4.1. APPROXIMATION DE L'ESPACE DES DONNÉES INITIALES. On choisit l'approximation interne $H^\delta=X^{L+1}$ de l'espace $H=X\times L^2(-b,0;X)$ et le produit scalaire

$$(4.1)\qquad (\underline{h},\underline{k})_{H^\delta}=\sum_{\ell=-L}^{-1}(h_\ell,k_\ell).$$

On définit les applications

$$(4.2)\quad h=(h^0,h^1)\mapsto r_H(h)=(h^0,h^1_{-1},\dots,h^1_{-L}):H\to H^\delta,\ h^1_\ell=\delta^{-1}\int_{\ell\delta}^{(\ell+1)\delta}h^1(\theta)d\theta,\ -L\le\ell\le-1,$$

$$(4.3)\quad \underline{h}=(h_0,h_{-1},\dots,h_{-L})\mapsto q_H(\underline{h})=(h_0,\sum_{\ell=-L}^{-1}h_\ell\chi_\ell),$$

où χ_ℓ est la fonction caractéristique de l'intervalle $[\ell\delta,(\ell+1)\delta[$. On vérifiera aisément que $\|q_H r_H(h)\| \le \|h\|$.

4.2. APPROXIMATION DE (2.3). On associe à $A_0, A_1, \ldots, A_N$, les familles de matrices

$$(4.4) \qquad A_i^m = \delta^{-1}\int_{m\delta}^{(m+1)\delta} A_i(t)dt, \quad m = 0,\ldots,M-1, \quad i = 0,1,\ldots,N,$$

à A_{01} la famille de matrices

$$(4.5) \quad \begin{cases} A_{01}^{m,0} = \dfrac{1}{\delta^2}\displaystyle\int_{m\delta}^{(m+1)\delta} dt \int_{-(t-m\delta)}^{0} d\theta \quad A_{01}(t,\theta), A_{01}^{m,-L} = \dfrac{1}{\delta^2}\int_{m\delta}^{(m+1)\delta} dt \int_{-b}^{-b+(m+1)\delta-t} d\theta \quad A_{01}(t,\theta), \\ A_{01}^{m,\ell} = \dfrac{1}{\delta^2}\displaystyle\int_{m\delta}^{(m+1)\delta} dt \int_{(m+\ell)\delta-t}^{(m+\ell+1)\delta-t} d\theta \; A_{01}(t,\theta), \end{cases}$$

et à f une famille d'éléments de X

$$(4.6) \qquad f^m = \delta^{-1}\int_{m\delta}^{(m+1)\delta} f(t)dt, \; m = 0,\ldots,M-1.$$

On construit aussi une approximation interne $U^\delta = U^M$ de l'espace $L^2(0,T;U)$ des fonctions de commande et les applications

$$(4.7) \quad v \mapsto r_U(v) = (v_0,\ldots,v_{M-1}) : L^2(0,T;U) \to U^\delta, \; v_m = \delta^{-1}\int_{m\delta}^{(m+1)\delta} v(t)dt, \; m = 0,\ldots,M-1,$$

$$(4.8) \qquad \underline{v} = (v_0,\ldots,v_{M-1}) \mapsto q_U(\underline{v}) = \sum_{m=0}^{M-1} v_m\chi_m : U^\delta \to L^2(0,T;U),$$

où χ_m est la fonction caractéristique de l'intervalle $[m\delta,(m+1)\delta[$. On vérifiera que $\|q_U r_U(v)\|_2 \le \|v\|_2$. Finalement on associe à B la famille de matrices

$$(4.9) \qquad B^m = \delta^{-1}\int_{m\delta}^{(m+1)\delta} B(t)dt, \; m = 0,1,\ldots,M-1.$$

Etant donné $\underline{h}$ dans H^δ, on considère le schéma explicite suivant dans X

$$(4.10) \quad \begin{cases} x_{m+1}-x_m = \delta\left[A_0^m x_m + \displaystyle\sum_{i=1}^N A_i^m \begin{Bmatrix} h_{m-L_i}, & m-L_i<0 \\ x_{m-L_i}, & \text{sinon} \end{Bmatrix} + B^m v_m \right. \\ \qquad\qquad \left. + \displaystyle\sum_{\ell=-L}^{0} \delta\, A_{01}^{m,\ell} \begin{Bmatrix} h_{m+\ell}, & m+\ell<0 \\ x_{m+\ell}, & \text{sinon} \end{Bmatrix} + f^m\right], \; m = 0,\ldots,M-1, \\ x_0 = h_0. \end{cases}$$

Ce schéma possède une solution unique $(x_0,\ldots,x_m)$ à partir de laquelle on construit les fonctions suivantes:

$$x^\delta(t) = \sum_{m=0}^{M-1} x_m \chi_m(t), \quad x^\delta(T) = x_M \tag{4.11}$$

$$y^\delta(t) = \sum_{m=0}^{M-1} \left[x_m+(x_{m+1}-x_m)\left(\frac{t-m\delta}{\delta}\right)\right]\chi_m(t), \quad Dy^\delta(t) = \sum_{m=0}^{M-1} \delta^{-1}(x_{m+1}-x_m)\chi_m(t). \tag{4.12}$$

<u>Proposition 4.1</u>. Soit $(x_0,\ldots,x_m)$ la solution du schéma (4.10) correspondant à la donnée initiale $\underline{h}=r_H(h)$, $h\in H$, et soit x^δ et $y^\delta:[0,T]\to X$ les fonctions construites à partir de celle-ci. (i) (Stabilité.) Lorsque δ tend vers zéro il existe une constante $c>0$ tel que pour tout h, f, v et δ

$$\|x^\delta\|_\infty + |x^\delta(T)| + \|y^\delta\|_\infty + \|Dy^\delta\|_2 \le c[\|h\|+\|f\|_2+\|v\|_2]. \tag{4.13}$$

(ii) (Convergence.) Les approximations x^δ et y^δ convergent vers x dans $L^\infty(0,T;X)$, $x^\delta(T)=y^\delta(T)$ converge vers x(T) dans X et Dy^δ converge vers Dx dans $L^2(0,T;X)$, où x est la solution de (2.3).

<u>Démonstration</u>. Cf. DELFOUR [3]. ■

4.3. <u>APPROXIMATION DE L'ÉQUATION D'ÉVOLUTION DE L'ÉTAT</u>. On aura besoin non seulement d'une approximation de H, mais aussi d'une approximation externe $V^\delta=X^{(L+1)}$ de l'espace V avec comme produit scalaire

$$(\underline{h},\underline{k})_{V^\delta} = \sum_{\ell=-L}^{0} (h_\ell,k_\ell) + \delta^{-2}\sum_{\ell=-L}^{-1} (h_{\ell+1}-h_\ell,k_{\ell+1}-k_\ell). \tag{4.14}$$

Pour cela on introduit l'espace $\mathbb{F} = L^2(-b,0;X)\times L^2(-b,0;X)$ de produit scalaire

$$((f^0,f^1),(g^0,g^1))_{\mathbb{F}} = (f^0,g^0)_2 + (f^1,g^1)_2 \tag{4.15}$$

et les applicationy $h\mapsto\pi(h)=(h,Dh):V\to\mathbb{F}$ (Dh la dérivée distribution de h),

$$\underline{h} = (h_0,h_{-1},\ldots,h_{-L}) \mapsto q_V(\underline{h}) = \left(\sum_{\ell=-L}^{-1} h_\ell\chi_\ell,\ \sum_{\ell=-L}^{-1}\delta^{-1}(h_{\ell+1}-h_\ell)\chi_\ell\right):V^\delta\to\mathbb{F}, \tag{4.16}$$

$$h\mapsto r_V(h)=(h_0,h_{-1},\ldots,h_{-L}):V\to V^\delta,\ h_0=h(0),\ h_\ell=\delta^{-1}\int_{\ell\delta}^{(\ell+1)\delta} h(\theta)d\theta,\ -L\le\ell\le-1. \tag{4.17}$$

On vérifie (cf. J.P. AUBIN) que pour tout h dans V $\|q_V r_V(h)\|_{\mathbb{F}} \le 2\|h\|_V$ et que $q_V r_V(h)$ tend vers $\pi(h)$ dans $\mathbb{F}$ quand δ tend vers zéro.

La construction du schéma d'approximation de l'équation d'état (3.1) se fait à partir du schéma (4.10) (approximation de l'équation (2.3)) en utilisant le fait que les "caractéristiques" de l'équation (3.1) sont des lignes de pente -1 dans le domaine $[0,T]\times I(-b,0)$. Etant donné $\underline{h}$ dans V^δ, on considère le schéma suivant où l'on cherche à déterminer $\{x_{m,n}:0\le m\le M,-L\le n\le 0\}$:

$$
(4.18)\quad \begin{cases} x_{m+1,0}-x_{m,0} = \delta\Big[A_0^m x_{m,0} + \sum_{i=1}^{N} A_i^m x_{m,-L_i} + B^m v_m \\ \qquad\qquad + \delta \sum_{\ell=-L}^{0} A_{01}^{m,\ell} x_{m,\ell} + f^m\Big],\ m = 0,\dots,M-1, \\ x_{0,\ell} = h_\ell,\quad -L \le \ell \le 0, \\ x_{m+1,\ell} = x_{m,\ell+1},\ 0 \le m \le M-1,\ -L \le \ell \le 0. \end{cases}
$$

Pour simplifier l'aspect de ce schéma, on définit les matrices $\tilde{A}^m$ (de dimension $(L+1)n\times(L+1)n$) et $\tilde{B}^m$ (de dimension $(L+1)n\times m$) et les vecteurs $\tilde{x}_m$ et $\tilde{f}^m$ dans H^δ

$$
(4.19)\quad \begin{cases} [\tilde{A}^m]_0 = A_0^m h_0 + \sum_{i=1}^{N} A_i^m h_{-L_i} + \delta \sum_{\ell=-L}^{0} A_{01}^{m,\ell} h_\ell,\ [\tilde{B}^m]_0 = B^m, \\ [\tilde{A}^m]_\ell = \delta^{-1}(h_{\ell+1}-h_\ell),\ [\tilde{B}^m]_\ell = 0,\ -L \le \ell \le -1, \end{cases}
$$

$$
(4.20)\quad \tilde{x}_m = (x_{m,0},x_{m,-1},\dots,x_{m,-L}),\ \tilde{f}^m = (f^m,0,\dots,0).
$$

A l'aide de ces définitions on peut maintenant écrire le schéma (4.18) sous la forme

$$
\tilde{x}_{m+1}-\tilde{x}_m = \delta[\tilde{A}^m\tilde{x}_m+\tilde{B}^m v_m+\tilde{f}^m],\ 0 \le m \le M-1,\ \tilde{x}_0 = \underline{h}.
$$

Par construction, ce schéma possède une solution unique $(\tilde{x}_0,\dots,\tilde{x}_M)$ pour chaque $\underline{h}$ dans V^δ. A partir de celle-ci on définit l'application $\tilde{x}^\delta:[0,T]\to H$

$$
(4.22)\quad \begin{cases} \tilde{x}^\delta(t)^0 = \sum_{m=0}^{M-1} x_{m,0}\chi_m(t) & 0 \le t < T \\ \tilde{x}^\delta(t)^1(\theta) = \sum_{m=0}^{M-1}\sum_{n=-L}^{-1} [x_{m,n}\chi_{m,n}^\ell(t,\theta)+x_{m,n+1}\chi_{m,n}^u(t,\theta)] & -b \le \theta \le 0 \\ \tilde{x}^\delta(T) = q_H(\tilde{x}_M), \end{cases}
$$

où χ_m est la fonction caractéristique de l'intervalle $[m\delta,(m+1)\delta[$, $\chi_{m,n}^\ell$ celle de l'ensemble $\{(t,\theta)\in[m\delta,(m+1)\delta[\times[n\delta,(n+1)\delta[:t+\theta<(m+n+1)\delta\}$ et $\chi_{m,n}^u$ celle de l'ensemble $\{(t,\theta)\in[m\delta,(m+1)\delta[\times[n\delta,(n+1)\delta[:(m+n+1)\delta\le t+\theta\}$.

<u>Proposition 4.2</u>. Soit $(\tilde{x}_0,\dots,\tilde{x}_M)$ la solution du schéma (4.21) correspondant à

$\underline{h}=r_V(h)$ pour h dans V et soit $\tilde{x}^\delta$ la fonction définie en (4.22) à partir de cette solution. Soit $\tilde{x}$ la solution de (3.1). (i) (Stabilité.) Lorsque δ tend vers zéro, il existe une constante $c>0$ tel que pour tout δ, h, v et f

$$(4.23) \qquad \max\{\|q_H(\tilde{x}_m)\|:m=0,\ldots,M\} + \|\tilde{x}^\delta\|_\infty \le c[\|h\|+\|f\|_2+\|v\|_2].$$

(ii) (Convergence.) Quelque soit h, f, v et s dans [0,T] lorsque δ tend vers zéro et $m\delta=s$

$$(4.24) \quad q_H(\tilde{x}_m) \to \tilde{x}(s) \text{ dans } H,\ \tilde{x}^\delta \to x \text{ dans } L^\infty(0,T;H) \text{ et } \tilde{x}^\delta(T) \to \tilde{x}(T) \text{ dans } H.$$

Démonstration. Cf. DELFOUR [3]. ■

5. APPROXIMATION DU PROBLÈME DE COMMANDE. On considère l'approximation (4.10) de l'équation (2.3) pour la donnée initiale $\underline{h}$ dans V^δ et la commande $\underline{v}$ dans U^δ. On définit une fonction coût approchée compatible avec l'approximation (4.10)

$$(5.1) \qquad J_\delta(\underline{v},\underline{h}) = (Fx_M+2\ell,x_M) + \delta\sum_{m=0}^{M-1}[(Q^m x_m+2q^m,x_m)+(N^m v_m+2n^m,v_m)_U],$$

$$(5.2) \qquad Q^m = \frac{1}{\delta}\int_{m\delta}^{(m+1)\delta} Q(t)dt,\ N^m = \frac{1}{\delta}\int_{m\delta}^{(m+1)\delta} N(t)dt,\ m = 0,\ldots,M-1,$$

$$(5.3) \qquad q^m = \frac{1}{\delta}\int_{m\delta}^{(m+1)\delta} q(t)dt,\ n^m = \frac{1}{\delta}\int_{m\delta}^{(m+1)\delta} n(t)dt,\ m = 0,\ldots,M-1.$$

Le problème de commande approché consiste à minimiser la fonctionnelle $J_\delta(\underline{v},\underline{h})$ par rapport à $\underline{v}$ dans U^δ. On peut aussi partir de l'approximation (4.21) de l'équation (3.1) et minimiser par rapport à $\underline{v}$ dans U^δ la fonction coût approchée

$$(5.4) \qquad J_\delta(\underline{v},\underline{h}) = (\tilde{F}\tilde{x}_M+2\tilde{\ell},\tilde{x}_M)_{H^\delta} + \delta\sum_{m=0}^{M-1}[(\tilde{Q}^m\tilde{x}_m+2\tilde{q}^m,\tilde{x}_m)_{H^\delta} + (N^m v_m+2n^m,v_m)_U],$$

avec $\tilde{F}$ et $\tilde{Q}^m$ dans $\mathcal{L}(H^\delta)$ et $\tilde{\ell}$ et $\tilde{q}^m$ dans H^δ

$$(5.5) \qquad \begin{cases} [\tilde{F}]_0 = F,\ [\tilde{Q}^m]_0 = Q^m,\ \tilde{\ell}_0 = \ell,\ [\tilde{q}^m]_0 = q^m \\ [\tilde{F}]_\ell = 0,\ [\tilde{Q}^m]_\ell = 0,\ \tilde{\ell}_\ell = 0,\ [\tilde{q}^m]_\ell = 0,\ \ell = -L,\ldots,-1. \end{cases}$$

Lemma 5.1. Soit $\underline{h}$ dans V^δ, le problème de commande optimale approché (4.21)-(5.4) possède une solution unique $\underline{u}$ dans U^δ. Cette solution est complètement caractérisée par le système d'optimalité:

$$(5.6) \qquad \tilde{x}_{m+1}-\tilde{x}_m = \delta[\tilde{A}^m\tilde{x}_m+\tilde{B}^m u_m+\tilde{f}^m],\ 0 \le m \le M-1,\ \tilde{x}_0 = \underline{h}$$

(5.7) $$p_{m+1}-p_m + \delta[(\tilde{A}^m)^* p_{m+1}+\tilde{Q}^m\tilde{x}_m+\tilde{q}^m] = 0,\ 0 \le m \le M-1,\ p_M = \tilde{F}\tilde{x}_M + \tilde{\ell},$$

[5.8] $$u_m = -(N^m)^{-1}[(\tilde{B}^m)^* p_{m+1}+n^m],\ m = 0,\dots,M-1.$$ ■

Proposition 5.2. Soit h dans M^2, le problème de commande optimale approché de donnée initiale $\underline{h}=r_H(h)$ pour h dans H possède une solution unique $\underline{u}$ dans U^δ. Lorsque δ tend vers zéro $q_U(\underline{u})$ converge vers u dans $L^2(0,T;U)$, où u est la commande optimale du problème de minimisation (3.1)-(3.2). On définit les applications $\tilde{x}^\delta$ et $p^\delta:[0,T]\to H$

(5.9) $$\begin{cases} \tilde{x}^\delta(t) = \sum_{m=0}^{M-1} q_H(\tilde{x}_m)\chi_m(t),\ \tilde{x}^\delta(T) = q_H(\tilde{x}_M) \\ p^\delta(t) = \sum_{m=0}^{M-1} (p_{m,0}, \sum_{n=-\ell}^{-1} \delta^{-1} p_{m,n}\chi_n)\chi_m(t) \end{cases}$$

(5.10) $$p^\delta(T) = (p_{M,0}, \sum_{n=-L}^{-1} \delta^{-1} p_{M,n}\chi_n),$$

où $(\tilde{x}_0,\dots,\tilde{x}_M)$ et $(p_0,\dots,p_M)$ sont les solutions du système d'optimalité (5.6)-(5.7)-(5.8) avec $\underline{h}=r_H(h)$ comme donnée initiale. Quand δ tend vers zéro

(5.11) $$J_\delta(\underline{u},r_H(h)) \to J(u,h)$$

(5.12) $$\tilde{x}^\delta \to \tilde{x}(\cdot;u,h) \text{ dans } L^\infty(0,T;H),\ q_H(\tilde{x}_M) \to \tilde{x}(T;u,h) \text{ dans } H$$

(5.13) $$\begin{cases} \forall\, h \in H, \text{ l'application } t \mapsto ((p^\delta(t),h)) \text{ converge vers} \\ \qquad \text{l'application } t \mapsto ((p(t),h)) \text{ dans } L^\infty(0,T;\mathbb{R}) \\ \qquad p^\delta(T) \to p(T) \text{ dans H faible,} \end{cases}$$

où $\tilde{x}$ et p sont les solutions du système d'équations (3.5)-(3.6)-(3.7). ■

On obtient aussi l'analogue du Théorème 3.2.

Proposition 5.3. Soit $\{\tilde{x}_m\}$ et $\{p_m\}$ la solution du système d'optimalité (5.6)-(5.7)-(5.8). Il existe une famille de matrices $\{\Pi_m^\delta : m=0,\dots,M\}$ dans $\mathcal{L}(H^\delta)$ et une famille d'éléments $\{\rho_m : m=0,\dots,M\}$ dans H^δ tel que

(5.14) $$p_m = \Pi_m^\delta \tilde{x}_m + \rho_m,\ m = 0,\dots,M.$$

Les matrices Π_m^δ et les éléments ρ_m sont obtenus de la manière suivante. (i) On résout

(5.15) $$\beta_{m+1}-\beta_m = \delta[\tilde{A}^m\beta_m-\tilde{R}^m\gamma_{m+1}],\ r \le m \le M-1,\ \beta_r = \underline{h},$$

(5.16) $$\gamma_{m+1}-\gamma_m + \delta[(\tilde{A}^m)^*\gamma_{m+1}+\tilde{Q}^m\beta_m] = 0,\ r \le m \le M-1,\ \gamma_M = \tilde{F}\beta_M,$$

et $\Pi_{r-}^{\delta}h=\gamma_r(\widetilde{R}^m=\widetilde{B}^m(N^m)^{-1}(\widetilde{B}^m)^*)$. (ii) On résout

$$\eta_{m+1}-\eta_m = \delta[\widetilde{A}^m\eta_m-\widetilde{R}^m\xi_{m+1}+\widetilde{f}^m-\widetilde{B}^m(N^m)^{-1}n^m],\ r \le m \le M,\ n_r = 0, \tag{5.17}$$

$$\xi_{m+1}-\xi_m + \delta[(\widetilde{A}^m)^*\xi_{m+1}+\widetilde{Q}^m\eta_m+\widetilde{q}^m] = 0,\ r \le m \le M,\ \xi_M = \widetilde{F}\eta_M+\widetilde{\ell} \tag{5.18}$$

et $\rho_r=\xi_r$. ■

Les démonstrations des résultats de cette section se trouvent dans DELFOUR [3].

6. APPROXIMATION DES ÉQUATIONS POUR Π et r. L'étude de la famille de matrices Π_m^{δ} parallèle celle de l'opérateur Π en §3.2. On peut supposer f, q, n et ℓ nuls. On fixe r, $0 \le r \le M$,et on considère le problème de commande approché

$$\widetilde{x}_{m+1}-\widetilde{x}_m = \delta[\widetilde{A}^m\widetilde{x}_m+\widetilde{B}^m v_m],\ r \le m \le M-1,\ \widetilde{x}_r = h \tag{6.1}$$

$$J_{\delta}^r(\underline{v},\underline{h}) = (\widetilde{F}\widetilde{x}_M,\widetilde{x}_M)_{H^{\delta}} + \delta \sum_{m=r}^{M-1} [(\widetilde{Q}^m\widetilde{x}_m,\widetilde{x}_m)_{H^{\delta}} + (N^m v_m,v_m)_U], \tag{6.2}$$

où l'on cherche à minimiser $J_{\delta}^r(\underline{v},\underline{h})$ par rapport à $\underline{v}$ dans U^{δ}. Comme précédemment ce problème admet une solution unique $\underline{u}$ caractérisée par le système d'optimalité (5.15)-(5.16) avec $\widetilde{x}_m=\beta_m$ et $p_m=\gamma_m$.

Proposition 6.1. On fixe s, $0 \le s \le T$, et $\delta>0$ tel qu'il existe au moins un entier $r\ge1$ tel que $s=r\delta$. (i) Si $\underline{u}$ est la commande optimale du problème (6.1)-(6.2), alors

$$J_{\delta}^r(\underline{u},\underline{h}) = (\Pi_{r-}^{\delta}\underline{h},\underline{h})_{H^{\delta}},\ (\Pi_r^{\delta})^* = \Pi_r^{\delta} \ge 0. \tag{6.3}$$

(ii) Lorsque δ tend vers zéro, il existe une constante $c>0$ tel que

$$\forall\ h \in H,\ (\Pi_r^{\delta}r_H(h),r_H(h))_{H^{\delta}} \le c\|h\|^2. \tag{6.4}$$

(iii) Si $\widetilde{x}_m$ et p_m, $r \le m \le M-1$ sont les solutions du système (5.15)-(5.16), alors

$$p_m = \Pi_m^{\delta}\widetilde{x}_m,\ 0 \le m \le M. \tag{6.5}$$ ■

Proposition 6.2. (Convergence.) On se place sous les hypothèses de la Proposition 6.1 avec les notations de la Proposition 3.2.

On définit $\Pi^{\delta}(s) \in \mathcal{L}(H)$ et $\Pi_{00}^{\delta}(s)$, $\Pi_{01}^{\delta}(s,\theta)$ et $\Pi_{11}^{\delta}(s,\alpha,\theta)$ dans $\mathcal{L}(X)$ comme suit:

$$\begin{cases} \Pi_{00}^{\delta}(s) = (\Pi_r^{\delta})_{00},\ \Pi_{01}^{\delta}(s,\theta) = \sum_{n=-L}^{-1} \delta^{-1}(\Pi_r^{\delta})_{0n}\chi_n(\theta) \\ \Pi_{11}^{\delta}(s,\alpha,\theta) = \sum_{\ell=-L}^{-1}\sum_{n=-L}^{-1} \delta^{-2}(\Pi_r^{\delta})_{\ell n}\chi_{\ell}(\alpha)\chi_n(\theta), \end{cases} \tag{6.6}$$

$$(6.7)\quad \begin{cases} [\Pi^{\delta}(s)]_{00} = \Pi^{\delta}_{00}(s),\ [\Pi^{\delta}(s)]_{01}h^1 = \int_{-b}^{0} \Pi^{\delta}_{01}(s,\theta)h^1(\theta)d\theta \\ ([\Pi^{\delta}(s)]_{11}h^1)(\alpha) = \int_{-b}^{0} \Pi^{\delta}_{11}(s,\alpha,\theta)h^1(\theta)d\theta. \end{cases}$$

Lorsque δ tend vers zéro

(i) $\forall\ h,k \in H,\ ((\Pi^{\delta}(s)h,k)) \to ((\Pi(s)h,k))$,

(ii) $\Pi^{\delta}_{00}(s) \to \Pi_{00}(s)$ dans $\mathcal{L}(X)$,

(iii) $\forall\ h^1 \in L^2(-b,0;X)$, $\Pi^{\delta}_{01}(s)h^1 \to \Pi_{01}(s)h^1$ dans X,

(iv) $\forall\ h^1 \in L^2(-b,0;X)$, $\Pi^{\delta}_{11}(s)h^1 \to \Pi_{11}(s)h^1$ dans $L^2(-b,0;X)$ faible. ■

<u>Proposition 6.3</u>. Lorsque δ est assez petit, la famille de matrices Π^{δ}_m, $0 \le m \le M$, définie par le système (5.15)-(5.16) est solution du schéma suivant

$$(6.8)\quad \begin{cases} \Pi_m = (I+\delta\tilde{A}^m)^*\Pi_{m+1}(I+\delta\tilde{R}^m\Pi_{m+1})^{-1}(I+\delta\tilde{A}^m) + \delta\tilde{Q}^m,\ 0 \le m \le M-1, \\ \Pi_M = \tilde{F},\quad (I,\ \text{l'élément identité dans } \mathcal{L}(H^{\delta})). \end{cases}$$ ■

<u>Remarques</u>. 1) La matrice $Y = \Pi_{m+1}Z_{\delta}^{-1}$ est symétrique et positive si la matrice $Z_{\delta} = I + \delta\tilde{R}^m\Pi_{m+1}$ est inversible.

2) Si la matrice Z_{δ} est inversible pour un $\delta_0>0$, elle l'est pour tout δ tel que $0<\delta\le\delta_0$.

3) La matrice Z_{δ} est inversible si et seulement si la matrice $D=I_X+\delta R^m(\Pi_{m+1})_{00}$ est inversible (I_X, l'élément identité dans $\mathcal{L}(X)$ et $R^m=B^m(N^m)^{-1}(B^m)^*$).

L'étude de la famille d'éléments ρ_m parallèle celle de la fonction r en §3.3.

<u>Proposition 6.4</u>. (i) Lorsque δ tend vers zéro, la famille d'éléments $(\rho_0,\rho_1,\dots,\rho_M)$ de H dont l'existence a été assurée par la Proposition 5.3 est la solution du schéma suivant

$$(6.9)\quad \begin{cases} \rho_m = (I+\delta\tilde{A}^m)^*[(I+\delta\tilde{R}^m\Pi_{m+1})^{-1}]^*\rho_{m+1} \\ \qquad + \delta[\tilde{q}^m+(I+\delta\tilde{A}^m)^*\Pi_{m+1}(I+\delta\tilde{R}^m\Pi_{m+1})^{-1}(\tilde{f}^m-\tilde{B}^m(N^m)^{-1}n^m)],\ 0 \le m \le M-1, \\ \rho_M = \tilde{\ell}. \end{cases}$$

(ii) (Stabilité.) Lorsque δ tend vers zéro, il existe des constantes $c_1>0$ et $c_2>0$ tel que

$$(6.10)\qquad \|\rho^{\delta}(s)\| \le c_1\|h\| + c_2,$$

où l'application $\rho^{\delta}:[0,T]\to H$ est définie comme suit

$$(6.11)\quad \rho^{\delta}(s) = \sum_{m=0}^{M-1} (\rho_{m,0},\ \sum_{\ell=-L}^{-1} \rho_{m,\ell}\chi_{\ell})\chi_m(s),\ 0\le s\le T,\ \rho^{\delta}(T) = (\rho_{M,0},\ \sum_{\ell=-L}^{-1} \rho_{M,\ell}\chi_{\ell}).$$

(iii) (Convergence.) Pour tout k dans H et tout s dans [0,T], on a lorsque δ tend vers zéro

(6.12) $$\rho^{\delta}(s) \to r(s) \text{ dans H faible.} \quad \blacksquare$$

Les démonstrations des Propositions 6.1, 6.2 et 6.3 se trouvent dans DELFOUR [3].

7. EXEMPLES. On se limitera à quelques exemples pour $X = \mathbb{R}$ afin d'illustrer les caractéristiques essentielles des fonctions matricielles $\Pi_{00}(t)$, $\Pi_{01}(t,\alpha)$, $\Pi_{11}(t,\alpha,\beta)$.

Exemple 1. Soit le système à un retard

(7.1) $$\dot{x}(t) = x(t-1) + v(t),\ 0 \le t \le 2,\ x(s) = 1,\ -1 \le s \le 0,$$

en conjonction avec la fonction coût

(7.2) $$J(v,1) = F[x(2)]^2 + \int_0^2 [v(t)]^2 dt.$$

Cet exemple est assez simple pour calculer la solution du problème de commande optimale analytiquement:

(7.3) $$u(t) = -1.05d\begin{cases} 2-t, & 0 \le t \le 1 \\ 1, & 1 < t \le 2 \end{cases},\quad \begin{matrix} J(u,1) = 3.675d,\ d = (1+0.3F^{-1})^{-1} \\ x(2) = 3.5(1-d). \end{matrix}$$

(7.4) $$\Pi_{01}(s,\alpha) = K(s)w(s,\alpha),\ \Pi_{11}(s,\theta,\alpha) = K(s)w(s,\theta)w(s,\alpha),$$

où K(s) est une fonction absolument continue et

(7.5) $$w(s,\alpha) = \begin{cases} \begin{cases} 1, & -s \le \alpha \le 0 \\ 1-(\alpha+s), & -1 \le \alpha < -s \end{cases}, & 0 \le s \le 1, \\ \begin{cases} 0, & 1-s \le \alpha \le 0 \\ 1, & -1 \le \alpha < 1-s \end{cases}, & 1 < s \le 2. \end{cases}$$

L'évolution de l'application $\alpha \mapsto \Pi_{01}(t,\alpha)$ en fonction du temps t se trouve en Figure 1 pour F=1. En Figure 7 on trouve la solution analytique et la solution approchée (δ=.1) pour x et u. En Figure 8, on trouve une étude de l'erreur e dans la fonction coût et du temps de calcul t_c de la famille de matrices $\{\Pi_m^{\delta}\}$ en fonction du pas de discrétisation δ. On remarque que e est proportionnelle à δ et que t_c est proportionnel à δ^{-3}.

Exemple 2. Soit le système à deux retards avec A_0=1 et F=0 dans (7.6) et (7.7):

(7.6) $$\dot{x}(t) = A_0 x(t) + x(t-.5) + x(t-1) + v(t),\ 0 \le t \le 2,\ x(s) = 1,\ -1 \le s \le 0,$$

(7.7) $$J(v,1) = \int_0^2 \{[x(t)]^2 + [v(t)]^2\}dt + F[x(2)]^2.$$

Exemple 3. On considère le système à deux retards avec A_0=0 et F=1 dans (7.6)-(7.7).

Exemple 4. Soit le système à quatre retards avec F=4 dans la fonction coût (7.2):

$$\dot{x}(t) = x(t-.2)+x(t-.4)+x(t-.8)+x(t-1)+v(t),\ 0\le t\le 2,\ x(s) = 1,\ -1\le s\le 0. \tag{7.8}$$

Exemple 5. Soit le système à un retard

$$\dot{x}(t) = A_1(t)x(t-1)+v(t),\ 0\le t\le 2,\ x(s) = 1,\ -1\le s\le 0, \tag{7.9}$$

avec la fonction coût (7.2) pour F=1 et

$$A_1(t) = \begin{cases} 0 & , \ n/10 \le t < (n+1)/10,\ n = 0,2,4,6,8 \\ 1 & , \ n/10 \le t < (n+1)/10,\ n = 1,3,5,7,9 \\ 1-4(t-1) & , \ 1 \le t \le 1.25 \\ 4(t-1.25) & , \ 1.25 < t \le 1.50 \\ 1-4(t-1.5) & , \ 1.50 < t \le 1.75 \\ 4(t-1.75) & , \ 1.75 < t \le 2. \end{cases} \tag{7.10}$$

Exemple 6. On considère le système (7.9) avec la fonction coût (7.7) pour F=0 et

$$A_1(t) = \begin{cases} 1 \ , & n/10 \le t < (n+1)/10,\ \ n \text{ pair}, \\ 0 \ , & n/10 \le t < (n+1)/10,\ \ n \text{ impair}. \end{cases} \tag{7.11}$$

Les quatre premiers exemples illustrent les propriétés de l'application $\alpha\mapsto\Pi_{01}(t,\alpha)$ en fonction de t. On remarquera les discontinuités fixes aux retards et dans le cas $F\neq 0$ la présence de discontinuités additionnelles qui se propagent avec une vitesse 1 de $\alpha=\theta_i$ au temps T vers $\alpha=0$ au temps $T+\theta_i$. Les deux derniers exemples illustrent l'effet "modulateur" des matrices $A_i(t)$ tel que prédit en §3.2.

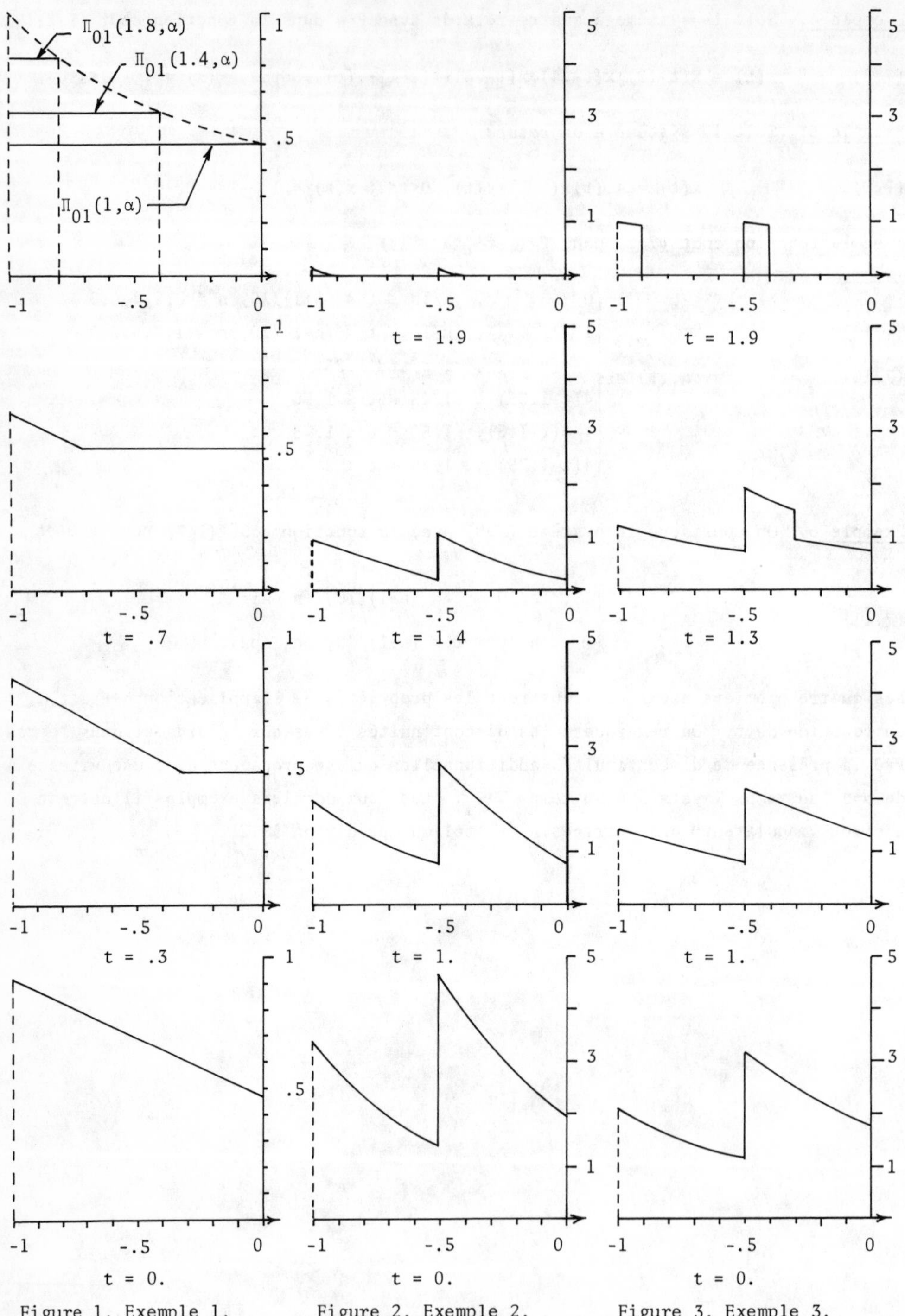

Figure 1. Exemple 1. Figure 2. Exemple 2. Figure 3. Exemple 3.

Schémas de l'application $\alpha \mapsto \Pi_{01}(t,\alpha)$ en fonction du temps t.

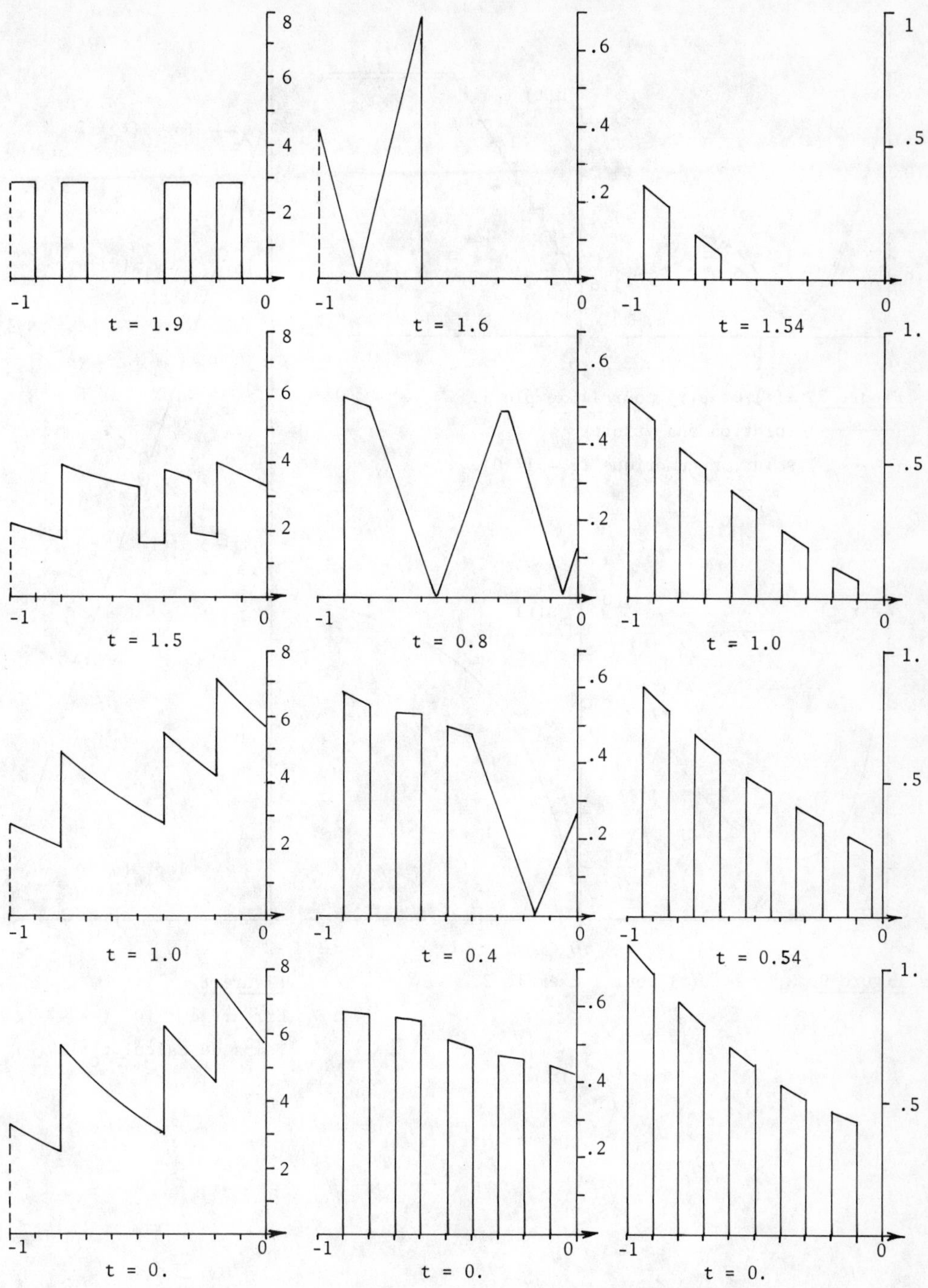

Figure 4. Exemple 4. Figure 5. Exemple 5. Figure 6. Exemple 6.

Schémas de l'application $\alpha \mapsto \Pi_{01}(t,\alpha)$ en fonction du temps t.

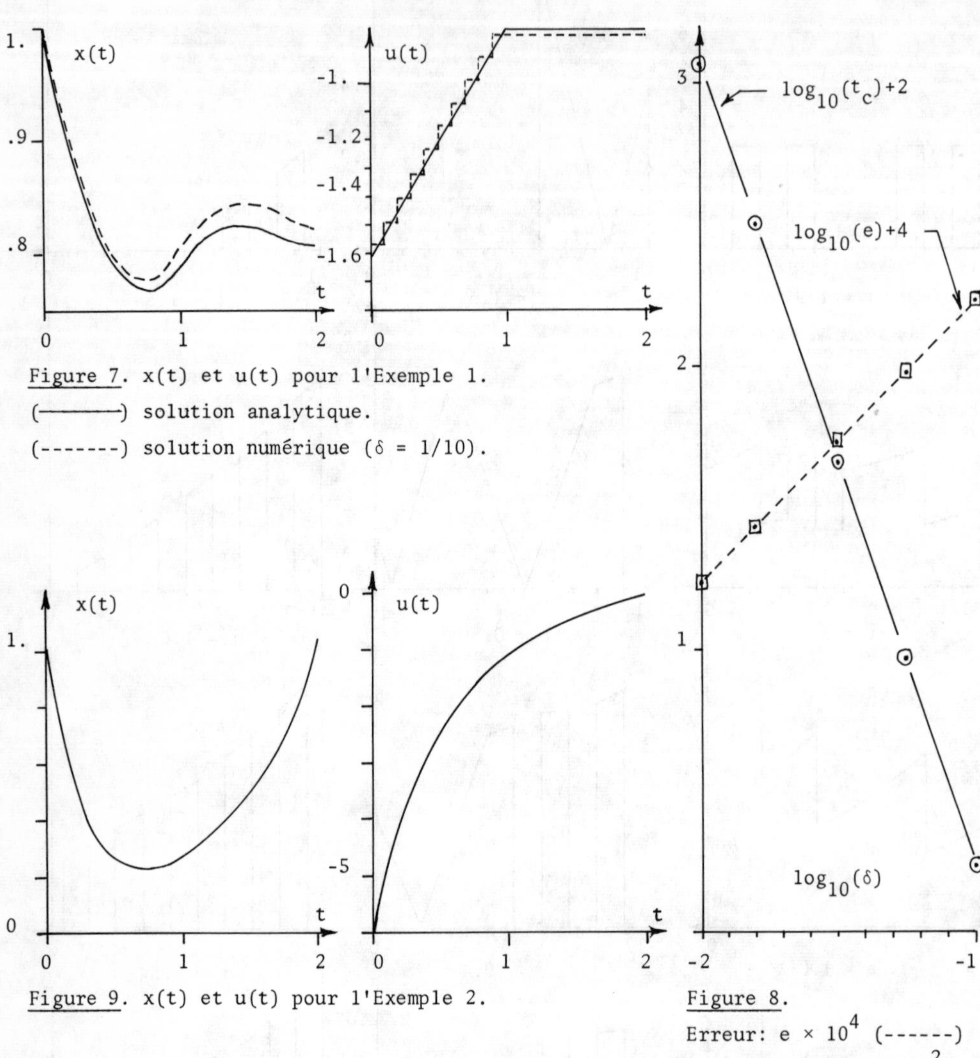

Figure 7. x(t) et u(t) pour l'Exemple 1.
(———) solution analytique.
(-------) solution numérique (δ = 1/10).

Figure 9. x(t) et u(t) pour l'Exemple 2.

Figure 8.
Erreur: $e \times 10^4$ (------)
Temps de calcul: $t_c \times 10^2$ (—)

RÉFÉRENCES.

Y. ALEKAL, P. BRUNOVSKY, D.H. CHYUNG et E.B. LEE, The quadratic problem for systems with time delays, IEEE Trans. on Automatic Control, AC-16 (1971), 673-687.

J.P. AUBIN, Approximation of Elliptic Boundary-Value Problems, Wiley-Interscience, New York 1972.

H.T. BANKS et A. MANITIUS, Projection series for retarded functional differential equations with applications to optimal control problems, à paraître.

A. BENSOUSSAN, M.C. DELFOUR et S.K. MITTER, Notes on infinite dimensional system theory, livre en préparation.

G. DA PRATO, Equations d'évolution dans les algèbres d'opérateurs et applications à des équations quasi-linéaires, J. Math. Pure Appl., 48 (1969), 59-107.

M.C. DELFOUR [1], Linear hereditary differential systems and their control, à paraître dans les "Proceedings of the 14th Biennial Seminar of the Canadian Mathematical Congress on Optimal Control and its Applications", London, Canada, août 1973.
[2], State theory of linear hereditary differential systems, Rapport interne CRM-395, Centre de Recherches Mathématiques, Université de Montréal, Montréal, Canada.
[3], Numerical solution of the optimal control problem for linear hereditary differential systems with a linear-quadratic cost function and approximation of the Riccati differential equation, Rapport interne CRM-408, Centre de Recherches Mathématiques, Université de Montréal, Montréal, Canada.

M.C. DELFOUR et S.K. MITTER [1], Hereditary differential systems with constant delays. I. General case, J. Differential Equations, 12 (1972), 213-235.
[2], Hereditary differential systems with constant delays. II. A class of affine systems and the adjoint problem, J. Differential Equations, à paraître.
[3], Controllability, observability and optimal feedback control of hereditary differential systems, SIAM J. on Control, 10 (1972), 298-328.
[4], Control of affine systems with memory, in "5th Conference on Optimization Techniques", R. Conti et A. Ruberti eds., Springer-Verlag, New York 1973, 292-303.

D.H. ELLER, J.K. AGGARWAL et H.T. BANKS, Optimal control of linear time-delay systems, IEEE Trans. on Automatic Control 14 (1969), 678-687.

H.J. KUSHNER et D.I. BARNEA, On the optimal control of a linear functional-differential equation with quadratic cost, SIAM J. on Control, 8 (1970), 257-272.

J.L. LIONS [1], Contrôle optimal de systèmes gouvernés par des équations aux dérivées partielles, Dunod, Paris, 1968.
[2], Problèmes aux limites dans les équations aux dérivées partielles, Séminaire de Mathématiques Supérieures, Les Presses de l'Université de Montréal, Montréal, 1967.

J.L. LIONS et E. MAGENES, Problèmes aux limites non homogènes et applications, vol. 1,2,3, Dunod, Paris, 1968, 1970.

A. MANITIUS, Optimum control of linear time-lag processes with quadratic performance indices, Proc. 4th IFAC Congress, Warsaw, Poland, 1969.

J.C. NÉDELEC, Schémas d'approximation pour des équations intégro-différentielles de Riccati, Thèse de doctorat d'état, Paris, 1970.

D.W. ROSS et I. FLÜGGE-LOTZ [1], Optimal control of systems described by differential-difference equations, Division of Engineering Mechanics, Stanford University, Technical Report no. 177, 1967.
[2], An optimal control problem for systems with differential-difference equation dynamics, SIAM J. Control, 7 (1969), 609-623.

M.A. SOLIMAN et W.H. RAY [1], Optimal control of multivariable systems with pure time delays, Automatica, 7 (1971), 681-689.
[2], Optimal feedback control for linear-quadratic systems having time delays, Int. J. Control, 15 (1972), 609-627.

REDUCTION OF THE OPERATOR RICCATI EQUATION

Lennart Ljung

Division of Automatic Control
Lund Institute of Technology
S-220 07 Lund, Sweden.

John Casti

Departments of Mathematics, Systems&Industrial Engineering, University of Arizona, Tucson, Arizona 85721 and IIASA, Laxenburg, Austria.

1. INTRODUCTION.

The fundamental role of the Riccati equation for optimal control and optimal filtering of linear systems is well known. For linear distributed parameter systems it has been shown by J. L. Lions [8], and in several other papers, see e.g. [12], [13], [5], that there is an operator Riccati equation (ORE), (or, equivalently, an integro-differential equation of Riccati type for the kernel) associated with various control and filtering problems.

The solution to the ORE is an operator in the state space. In the case of an infinite dimensional state space numerical solution of the ORE may therefore be cumbersome and time consuming, or even impossible. The interesting operator for the control problem maps the state space into the space of the control vector, and it is constructed using the solution to the ORE. In [4] it has been shown that, under certain conditions, it is possible to solve for this operator directly. This leads to less extensive computations for the numerical solution if the space of the control vector is of less dimensionality than the state space. In many practical situations there are only a finite number of control variables and also only a finite number of observations. In those cases, therefore, a substantial decrease in computational effort may be achieved.

While [4] treats general control and observation spaces and distributed control, we will here consider boundary control applied at a finite number of points. Many real-life, distributed-parameter control processes seem to be of this type.

2. THE OPTIMAL CONTROL PROBLEM.

Consider the following formulation, [12]:

Let D be a connected, open domain of a r-dimensional Euclidian space E^r, and let S be the boundary of D. The spatial coordinate vector will be denoted by

$$x = (x_1, \ldots, x_r) \in D$$

and the time will be denoted by $t \in T = (t_o, t_1)$. Consider a time invariant matrix differential operator

$$A_x = \sum_{i=1}^{r} \frac{\partial}{\partial x_i} \left(\sum_{j=1}^{r} A_{ij}(x) \frac{\partial}{\partial x_j} \right) + A_o(x)$$

where $A_{ij}(x)$ and $A_o(x)$ are n×n self-adjoint matrix functions, with $A_{ij} = A_{ji}$.

Introduce the derivative with respect to the co-normal of the surface S relative to the operator A_x

$$\frac{\partial}{\partial n_A} = \sum_{j=1}^{r} \left(\sum_{i=1}^{r} A_{ij}(\xi) \cos(\nu, x_i) \right) \frac{\partial}{\partial x_j}$$

where ν is the exterior normal to the surface at a point $\xi \in S$, and $\cos(\nu, x_i)$ is the i:th direction cosine of ν.

Consider a linear, distributed-parameter system described by

$$\frac{\partial z(t,x)}{\partial t} = A_x z(t,x) \; ; \quad z(t_o, x) = z_o(x) \tag{1a}$$

where $z(t,x)$ is a n-vector state function of $t \in T$ and $x \in D$. The boundary condition is given by

$$F(\xi) z(t,\xi) + \frac{\partial z(t,\xi)}{\partial n_A} = \sum_{i=1}^{m} \delta(\xi - \xi_i) B_i u_i(t) \tag{1b}$$

where $\xi \in S$, $F(\xi)$ is an n×n matrix function and $u_i(t)$ is a ℓ×1 control vector applied at the point ξ_i at the boundary.

Introduce

$u(t) = col(u_1(t), \ldots, u_m(t))$ ($\ell m\times 1$) matrix)

Assume that the system described by (1) is well posed in the sense of Hadamard. A p-dimensional output vector is observed:

$$y(t) = \int_D C(x)z(t,x)dx$$

where $C(x)$ is a $p\times n$ matrix function.

The control problem is to determine $u_i(t)$ so that the cost functional

$$J = \int_{t_o}^{t_1} [y(t)^T Q_1 y(t) + u(t)^T Q_2 u(t)]dt$$

is minimized. Q_1 and Q_2 are positive definite symmetric matrices of dimensions $p\times p$ and $\ell m\times \ell m$ respectively.

In [12] it is shown that the solution can be written

$$u(t) = - \int_D K(t,x)z(t,x)dx$$

where

$$K(t,x) = Q_2^{-1} \tilde{P}(t,x) \qquad (\ell m\times n \text{ matrix})$$

where

$$\tilde{P}(t,x) = col(B_1^T P(t,\xi_1,x), \ldots, B_m^T P(t,\xi_m,x)) \qquad (\ell m\times n \text{ matrix})$$

and $P(t,x,x')$ is the solution to the Riccati equation

$$\frac{\partial P(t,x,x')}{\partial t} = -A_x P(t,x,x') - [A_{x'} P(t,x,x')]^T - \qquad (2a)$$
$$- C(x)^T Q_1 C(x') + \tilde{P}(t,x)^T Q_2^{-1} \tilde{P}(t,x')$$

with the initial and boundary conditions

$$P(t_1,x,x') = 0 \qquad (2b)$$
$$F(\xi)P(t,\xi,x) + \frac{\partial P(t,\xi,x)}{\partial n_A} = 0 \qquad \xi \in S,\ x \in D$$

Addition of distributed control terms leads to straightforward modifications. These are given in [12].

3. THE OPTIMAL FILTERING PROBLEM.

The problem of optimal filtering of linear distributed-parameter systems has been treated in detail and rigorously by Bensoussan [2]. Here we will follow the formulation of [13].

In most practical situations the control equipment cannot be regarded as perfect. To the right hand side of (1b) therefore should be added a noise term

$$\sum_{i=1}^{m} \delta(\xi-\xi_i)H_i\, w_i(t)$$

where $w_i(t)$ is a q×1 vector white gaussian process with covariance

$$E\, w_i(t)\, w_i(t)^T = I, \qquad E\, w_i(t)\, w_j(t)^T = 0 \qquad i \neq j$$

Furthermore, the variable y(t) cannot be observed exactly, but a noise-currupted measurement

$$\tilde{y}(t) = y(t) + e(t)$$

is obtained, where e(t) is a p×1 vector white gaussian process with covariance

$$E\, e(t)\, e(t)^T = R_2$$

An appropriate estimate, $\hat{z}(t,x)$, of the state function, z(t,x), is then obtained from $\tilde{y}(t)$ as follows:

$$\frac{\partial z(t,x)}{\partial t} = A_x\hat{z}(t,x) + \tilde{K}(t,x)[\tilde{y}(t) - \int_D C(x)\hat{z}(t,x)dx] \tag{3}$$

$$\hat{z}(t_o,x) = z_o(x)$$

where

$$\tilde{K}(t,x)^T = R_2^{-1} \int_D C(x')\, P(t,x,x')dx'$$

and

$P(t,x,x')$ is the solution of

$$\frac{\partial P(t,x,x')}{\partial t} = A_x\, P(t,x,x') + [A_{x'}\, P(t,x,x')]^T -$$

$$- \int_D P(t,x,\xi)\, C(\xi)^T d\xi\, R_2^{-1} \int_D C(\xi)\, P(t,\xi,x')d\xi + \tag{4a}$$

$$+ \sum_{i=1}^{m} H_i\, H_i^T\, \delta(x-\xi_i)\, \delta(x'-\xi_i)$$

with initial and boundary conditions

$$P(t_o,x,x') = 0$$

$$F(\xi)\, P(t,x,\xi) + \frac{\partial P(t,x,\xi)}{\partial n_A} = 0 \qquad x \in D, \quad \xi \in S \tag{4b}$$

Clearly, the Riccati equation (4) is quite analogous to (2). It corresponds to an optimal control problem with boundary observations and distributed control applied from a finite number of sources.

4. REDUCTION OF THE RICCATI EQUATION.

The equations (2) and (4) are partial differential equations in three variables: time and two space variables. However, the sought functions $K(t,x)$ and $\tilde{K}(t,x)$ depend only on one space variable. We will derive equations directly for these functions. The derivation will be formal inasmuch as existence and uniqueness of solutions to the discussed equations will not be verified. The idea of the reduction originated from certain problems in radiative transfer, see e.g. [3] and [6] for a corresponding treatment of the finite dimensional problem.

Consider eq. (2). Differentiate it with respect to t:

$$P_{tt} = -A_x\, P_t - [A_{x'}\, P_t]^T + \tilde{P}_t^T\, Q_2^{-1}\, \tilde{P} + \tilde{P}^T\, Q_2^{-1}\, \tilde{P}_t \tag{5a}$$

where

$$P_t = \frac{\partial}{\partial t} P$$

and the arguments have been suppressed.

If $\tilde{P}$ were known, eq. (5a) could be regarded as a linear equation for P_t with initial and boundary conditions

$$P_t(t_1x,x') = -C(x)^T Q_1 C(x')$$

$$F(\xi) P_t(t,\xi,x) + \frac{\partial}{\partial n_A} P_t(t,\xi,x) = 0 \quad \xi \in S, \quad x \in D \tag{5b}$$

Consider the equation

$$\frac{\partial}{\partial t} L(t,x) = -A_x L(t,x) + \tilde{P}(t,x)^T Q_2^{-1} \bar{L}(t) \tag{6a}$$

where

$$\bar{L}(t) = \mathrm{col}(B_1^T L(t,\xi_1),\ldots,B_m^T L(t,\xi_m)) \qquad (\ell m\times p \text{ matrix})$$

with initial and boundary conditions

$$L(t_1,x) = C(x)^T \sqrt{Q_1} \qquad (n\times p \text{ matrix})$$

$$F(\xi) L(t,\xi) + \frac{\partial L(t,\xi)}{\partial n_A} = 0 \; ; \quad \xi \in S \tag{6b}$$

It is straightforward to see that

$$P_t(t,x,x') = -L(t,x) L(t,x')^T$$

with $L(t,x)$ defined by (6) satisfies eq. (5).

Since

$$K(t,x) = Q_2^{-1} \tilde{P}(t,x)$$

and

$$\tilde{P}_t(t,x) = \mathrm{col}(B_1^T L(t,\xi_1) L(t,x)^T,\ldots,B_m^T L(t,\xi_m) L(t,x)^T) =$$

$$= -\bar{L}(t) L(t,x)^T$$

the equations

$$\frac{\partial}{\partial t} K(t,x) = -Q_2^{-1} \bar{L}(t) L(t,x)^T$$

(7a)

$$\frac{\partial}{\partial t} L(t,x) = -A_x L(t,x) + K(t,x)^T \bar{L}(t)$$

with initial and boundary conditions

$$K(t_1,x) = 0$$

$$L(t_1,x) = C(x)^T \sqrt{Q_1} \qquad (7b)$$

$$F(\xi)\, L(t,\xi) + \frac{\partial}{\partial n_A} L(t,\xi) = 0 \qquad \xi \in S$$

determine the sought gain function $K(t,x)$ directly without first determining $P(t,x,x')$

Analogously, eq. (4) can be reduced to

$$\frac{\partial}{\partial t} \tilde{K}(t,x)^T = R_2^{-1}\left[\int_D C(x')\, \tilde{L}(t,x')dx'\right] \tilde{L}(t,x)^T$$

(8a)

$$\frac{\partial}{\partial t} \tilde{L}(t,x) = A_x\, \tilde{L}(t,x) - \tilde{K}(t,x) \int_D C(x')\, \tilde{L}(t,x')dx'$$

with intial and boundary conditions

$$\tilde{K}(t_o,x) = 0 \qquad (n\times p \ \ \text{matrix})$$

$$\tilde{L}(t_o,x) = (H_1\, \delta(x-\xi_1),\ldots,H_m\, \delta(x-\xi_m)) \qquad (n\times qm \ \ \text{matrix}) \qquad (8b)$$

$$F(\xi)\, \tilde{L}(t,\xi) + \frac{\partial}{\partial n_A} \tilde{L}(t,\xi) = 0 \qquad \xi \in S$$

5. NUMERICAL SOLUTION.

Numerical solution of the ORE has been discussed in several papers. The following three methods seem to be predominating:

Finite difference approximations
Galerkin methods
Eigenfunction expansion

The best choice of solution method probably depends on the problem formulation, and it is difficult to give any general rules. We will here briefly discuss how the mentioned methods can be used to solve the reduced Riccati equation (7) and how the computational effort is reduced.

Finite difference approximation has been discussed in e.g. [1]. When it is applied to eq. (2) with, say, $n = 1$ and $r = 1$, a three-dimensional space must be discretized. For stability reasons the time step must be kept quite small, and consequently numerical solution will be time consuming. When the same method is applied to the reduced equation (7), the computation required is of an order of magnitude less, since one space variable has been eliminated.

Galerkin methods have been applied to the ORE in [11], and for more general control problems for parabolic systems in [10]. Then the kernel $P(t,x,x')$ is expanded in some coordinate function system:

$$P(t,x,x') = \sum_{i,j} P_{ij}(t)\, \Psi_i(x)\, \Psi_j(x')$$

The expansion is truncated at some suitable number N, and ordinary differential equations are derived for $P_{ij}(t)$, so that the orthogonal projection of the residual on the subspace spanned by $\Psi_i(x)$, $i=1,\ldots,N$ is zero. The resulting equations have some resemblance with the matrix Riccati equation. The same technique can be applied to the reduced equation (7), which leads to $N(\ell m+p)$ ordinary differential equations instead of $N(N+1)/2$.

The eigenfunction method is discussed in e.g. [12], [13], [5] and [14]. It can be regarded as a special case of Galerkin's method with the coordinate funtions being eigenfunctions of the operator A_x. Consequently, the same reduction of the computational effort is obtained as for this method. In [9] a similar reduction is discussed for the case when the operator A_x is diagonal.

6. A NUMERICAL EXAMPLE.

Consider a heat rod of length Λ (45cm) with diffusion constant κ (1.16 cm^2/sec), conductivity constant μ (3.8 W/cm °C) and cross section area S(1.54 cm^2). The heat flow at the left endpoint is controlled, while the right endpoint is isolated. At either endpoint the heat flow is disturbed by white gaussian noise, $w_1(t)$ and $w_2(t)$, with variances r_1 ($3.4 \cdot 10^{-3}$ W^2). The output variable is the temperature at $\Lambda/4$ from the right endpoint, and the temperature measurements are corrupted by white gaussian noise with variance r_2 ($4\cdot10^{-8}\ {}^oC^2$).

The numerical values given in paranthesis apply for a laboratory process at the Division of Automatic Control, Lund Institute of Technology, see e.g. [7].

Let the temperature at distance x from the left endpoint at time t be denoted by z(t,x). Then

$$\frac{\partial z(\ ,x)}{\partial t} = \kappa \frac{\partial^2 z(t,x)}{\partial x^2}$$

$$z(0,x) = z_o(x) \qquad \text{(known)}$$

$$S\mu \frac{\partial z(t,0)}{\partial x} = u(t) + \sqrt{r_1}\ w_1(t)$$

$$S\mu \frac{\partial z(t,\Lambda)}{\partial x} = \sqrt{r_1}\ w_2(t)$$

$$y(t) = z(t,3\Lambda/4) + e(t)$$

where

$$E\ w_i(t)\ w_j(s) = \delta_{ij}\ \delta(t-s)$$

$$E\ e(t)\ e(s) = r_2\ \delta(t-s)$$

$$E\ w_i(t)\ e(s) = 0$$

Let the objective of the control be to minimize

$$E \int_0^{t_1} [q_1 u^2(t) + q_2 z^2(t,3\Lambda/4)]dt$$

According to Sections 2-4 the solution is given by

$$u(t) = \int_0^{\Lambda} K(t,x)\ \hat{z}(t,x)dx$$

where

$$\frac{\partial}{\partial t}\hat{z}(t,x) = \kappa \frac{\partial^2}{\partial x^2}\hat{z}(t,x) + \tilde{K}(t,x)[y(t) - \hat{z}(t,3\Lambda/4)]$$

$$\hat{z}(0,x) = z_o(x)$$

$$\hat{z}(t,0) = u(t)$$

$$\hat{z}(t,\Lambda) = 0$$

The function K(t,x) is the solution of

$$\frac{\partial}{\partial t} K(t,x) = -1/q_1 \; L(t,0) \; L(t,x)$$

$$\frac{\partial}{\partial t} L(t,x) = -\kappa \frac{\partial}{\partial x^2} L(t,x) + L(t,0) \; K(t,x)$$

$$K(t_1,x) = 0$$

$$L(t_1,x) = \delta(x-3\Lambda/4)\cdot\sqrt{q_2}$$

$$\frac{\partial}{\partial x} L(t,0) = \frac{\partial}{\partial x} L(t,\Lambda) = 0$$

and the function K(t,x) is the solution of

$$\frac{\partial}{\partial t}\tilde{K}(t,x) = r_2^{-1}[\tilde{L}_1(t,3\Lambda/4)\;\tilde{L}_1(t,x) + \tilde{L}_2(t,3\Lambda/4)\;\tilde{L}_2(t,x)]$$

$$\frac{\partial}{\partial t}\tilde{L}_1(t,x) = \kappa \frac{\partial^2}{\partial x^2}\tilde{L}_1(t,x) - \tilde{L}_1(t,3\Lambda/4)\;\tilde{K}(t,x)$$

$$\frac{\partial}{\partial t}\tilde{L}_2(t,x) = \kappa \frac{\partial^2}{\partial x^2}\tilde{L}_2(t,x) - \tilde{L}_2(t,3\Lambda/4)\;\tilde{K}(t,x)$$

$$\tilde{K}(0,x) = 0$$

$$\tilde{L}_1(0,x) = \sqrt{r_1}\;\delta(x)$$

$$\tilde{L}_2(0,x) = \sqrt{r_1}\;\delta(x-\Lambda)$$

$$\frac{\partial}{\partial x}\tilde{L}_i(t,0) = \frac{\partial}{\partial x}\tilde{L}_i(t,\Lambda) = 0 \qquad i = 1,2$$

These equations have been solved using a straightforward difference scheme with 100 grid points in the x coordinate. The numerical values of the constants are those given in paranthesis. Furthermore, $q_1 = 1$ (W^{-2}), $q_2 = 8.5\cdot 10^{-4}$ ($^oC^{-2}$) and $t_1 = 1000$ sec. The solution is shown in fig. 1.

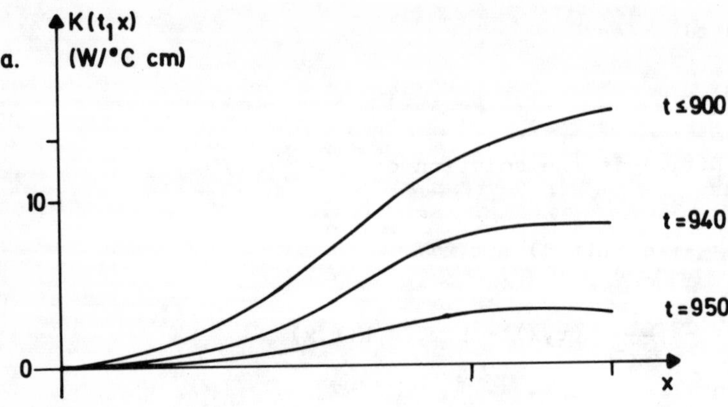

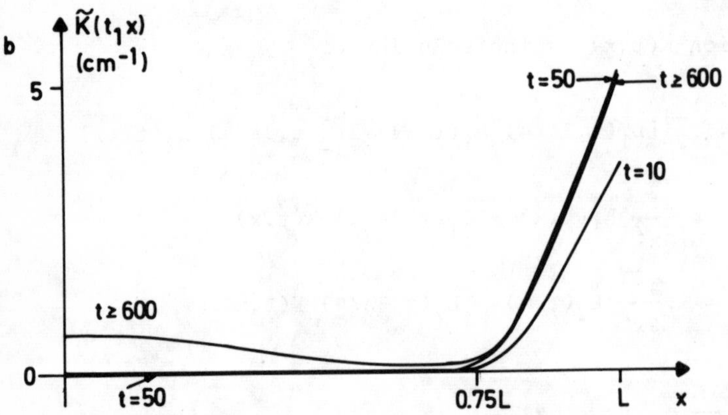

Fig. 1. a) Optimal feedback function K(t,x) for the heat regulation problem for various t. The final time t_1 is 1000 sec, and the numerical values of the constants are given in Section 6.

b) Optimal gain function $\tilde{K}(t,x)$ for the Kalman filter associated with the heat regulation problem stated in Section 6.

7. CONCLUSIONS.

It has been shown that the operator Riccati equation can, under certain, frequently occuring conditions be reduced to equations that contain fewer independent variables.

Therefore, numerical solution of the reduced equation requires less computation effort than solution of the original equation.

A case with boundary control at a finite number of points has been considered. Combining this with the results of [4] for distributed control, the general case with both distributed and boundary control can be treated straightforwardly.

It should be noticed that the reduction has been shown only for the initial condition $P(t_o,x,x') = 0$, corresponding to no terminal cost for the control problem or known initial state for the filtering problem.

There seems to be no possibility to extend the results to general initial conditions, using the same approach.

However, when computing the steady-state solutions, which are frequently used in the control strategies, the initial condition can as well be taken as zero.

REFERENCES.

[1] Alvardo, F., R. Mukundan, An optimization problem in distributed parameter systems, Int. J. Control, 9, No. 6 (1969), 665-677.

[2] Bensoussan, A., Filtrage Optimal des Systèmes Linéaires, Dunod, Paris 1971.

[3] Casti, J., Matrix Riccati equations, dimensionality reduction and generalized X-Y functions, to appear Utilitas Mathematica, 1974.

[4] Casti, J., L. Ljung, Some new analytic and computational results for operator Riccati equations, Report 7344, December 1973, Lund Institute of Technology, Division of Automatic Control. To appear in SIAM J. Control.

[5] Erzberger, H., M. Kim, Optimum boundary control of distributed parameter systems, Information and Control 9, (1966) 265-278.

[6] Kailath, T., Some new algorithms for recursive linear estimation in constant linear systems, IEEE Trans. on information theory IT-19, Nov., (1973).

[7] Leden, B., M. H. Hamza, M. A. Sheirah, Different methods for estimation of thermal diffusivity of a heat diffusion process, Proc. of the 3rd IFAC Symposium on Identification and System Parameter Estimation, the Hague (1973).

[8] Lions, J. L., The optimal control of systems governed by partial differential equations, Springer-Verlag, New York, 1971.

[9] Mayhew, M. J. E., A. J. Pritchard, Reduction of the Riccati equation for distributed parameter systems, Elec. Letters 7, No. 20, (1971), 628-629.

[10] McKnight, R. S., W. E. Bosarge, The Ritz-Galerkin procedure for parabolic control problems, SIAM J. Control 11, No. 3 (1973), 510-524.

[11] Prabhu, S. S., I. McCausland, Optimal control of linear diffusion processes with quadratic error criteria, Automatica 8 (1972), 299-308.

[12] Sakawa, Y., A matrix Green's formula and optimal control of Linear distributed-parameter systems, J, of Optimization Theory and Applications 10, No. 5 (1972), 290-299.

[13] Sakawa, Y., Optimal filtering in linear distributed-parameter systems, Int. J. Control 16, No. 1 (1972), 115-127.

[14] Tzafestas, S. G., J. M. Nightingale, Optimal control of a class of linear stochastic distributed-parameter systems, proc. IEE 115, No. 8 (1968).

ALGORITHME D'IDENTIFICATION RECURSIVE UTILISANT LE CONCEPT DE POSITIVITE

I.D. LANDAU
Maître de Conférences Associé
Laboratoire d'Automatique
Institut National Polytechnique de Grenoble

RESUME.

On présente un algorithme d'identification pour les processus affectés par des perturbations additives. L'identification des paramètres du procédé se fait à l'aide d'un modèle d'estimation de type parallèle dont les paramètres sont obtenus récursivement à l'aide d'un algorithme d'adaptation à gain variable. La structure de ce dernier a été déterminée en utilisant une extension du lemme sur le caractère réel positif pour les systèmes échantillonnés à paramètres variables. On présente des résultats expérimentaux permettant d'évaluer les performances de l'algorithme par rapport à d'autres algorithmes. Dans la partie finale, on indique l'utilisation de cette approche pour l'étude de la convergence de divers algorithmes connus.

ABSTRACT.

An algorithm for the parameter identification of processes affected by additive perturbations is presented. The identification is done using a parallel adjustable model. The parameters of this adjustable model are obtained recursively using an adaptation algorithm with a variable gain. The structure of the adaptation algorithm was obtained by using an extension of the positive real lemma for time-varying discrete systems. Experimental results are shown allowing comparison with other identification methods. In the final part, the use of this approach for the analysis of the convergence of various known algorithms is indicated.

I - INTRODUCTION.

L'identification récursive des paramètres dynamiques caractérisant divers types de procédés, fait intervenir l'utilisation d'un modèle construit à l'aide des paramètres estimés. En fonction de la configuration de ces modèles d'estimation, de leur position relative par rapport aux entrées-sorties du procédé et du mode d'obtention de l'erreur "modèle d'estimation - procédé" on distingue trois types principaux de méthodes d'identification :

- méthode de l'erreur de sortie (output-error method) (Fig. 1.a)
- méthode de l'équation de l'erreur (equation-error method) (Fig. 1.b)
- méthode de l'erreur d'entrée (input-error method) (Fig. 1.c)

Fig. 1.a, b, c.

Ces trois configurations correspondent en fait |1| |2| à des structures de systèmes adaptatifs avec modèle de référence dans lesquels le modèle de référence est constitué par le procédé à identifier et le système ajustable est constitué par les divers types de modèles d'estimation.

La correspondance entre les méthodes d'identification et les configurations de systèmes adaptatifs à modèle de référence (S.A.M.R.) est la suivante :

- méthode de l'erreur de sortie ⟷ S.A.M.R. "parallèle"
- méthode de l'équation de l'erreur ⟷ S.A.M.R. "série-parallèle"
- méthode de l'erreur d'entrée ⟷ S.A.M.R. "série"

Dans le cas des algorithmes d'adaptation des S.A.M.R. nous rencontrons deux situations :

a) Adaptation dans le cas uniquement d'une différence initiale entre les paramètres du modèle et du système ajustable.
b) Adaptation dans le cas des perturbations fréquentes sur les valeurs des paramètres du système ajustable ou du modèle de référence.

Dans le cas de l'identification des procédés pouvant être modélisés par des systèmes linéaires à paramètres invariant, c'est le premier cas qui nous intéresse.

De nombreuses recherches sur l'identification à l'aide des systèmes adaptatifs à modèle de référence utilisant des algorithmes de type (b) ont montré qu'en présence de bruits additifs sur les sorties du procédé à identifier, une bonne précision s'obtient en utilisant des S.A.M.R. de type "parallèle" avec un faible gain alors que l'utilisation des S.A.M.R. de type "série-parallèle" conduit à l'apparition systématique d'un biais sur l'estimation des paramètres même pour des faibles gains d'adaptation |3| |4|. D'autre part ces mêmes recherches ont montré que la vitesse d'adaptation, en absence des perturbations s'accroît quand les gains d'adaptation

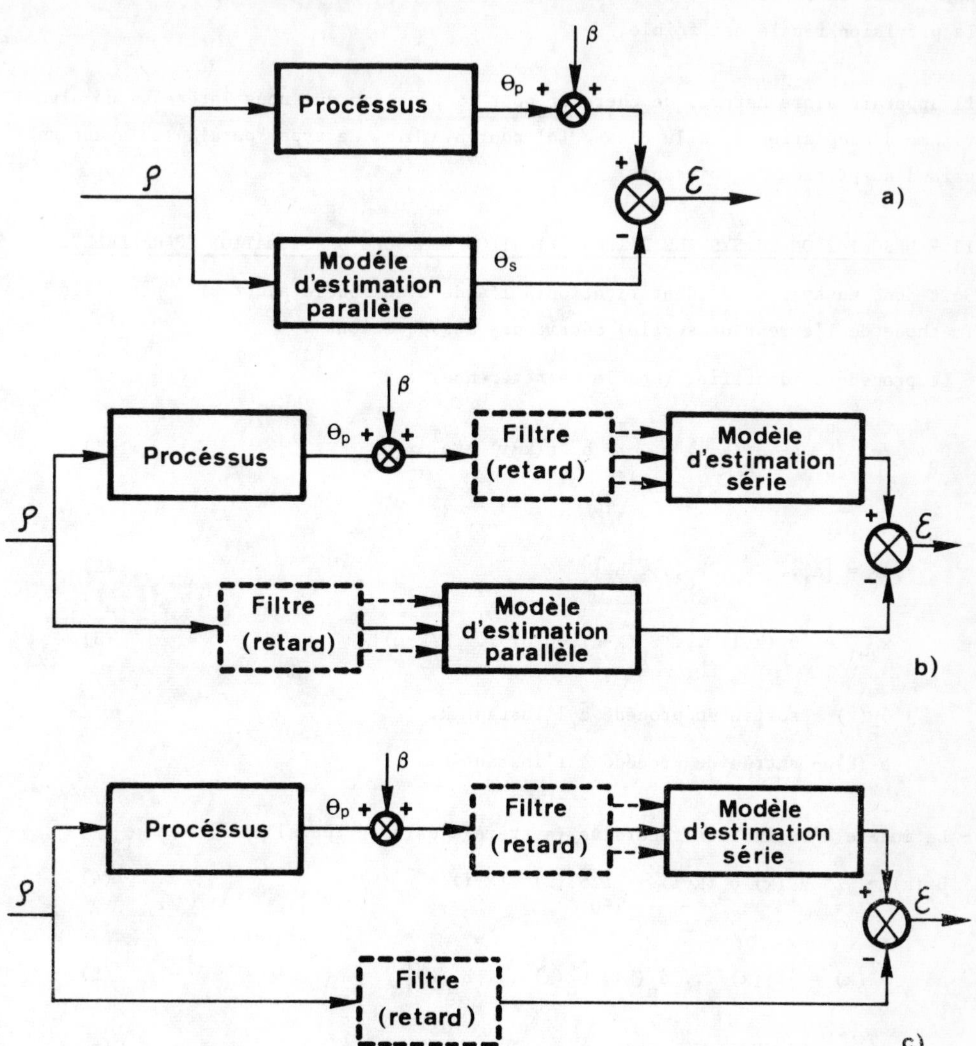

β
Procéssus
Θp
+
+
ρ
+
ε
a)
–
Modéle
d'estimation
parallèle
Θs
β
Θp
+
+
Procéssus
Filtre
(retard)
Modèle
d'estimation
série
ρ
+
ε
–
Filtre
(retard)
Modèle
d'estimation
parallèle
b)
β
Θp
+
+
Procéssus
Filtre
(retard)
Modèle
d'estimation
série
ρ
+
ε
–
Filtre
(retard)
c)

Figure . 1

sont plus grands. En présence de perturbations un grand gain d'adaptation assure une réduction rapide dans la phase initiale, de l'erreur sur les paramètres, mais la précision finale est faible.

Il apparaît alors naturel de chercher pour le problème qui nous intéresse un algorithme d'adaptation dans la classe (a) pour S.A.M.R. de type "parallèle" ayant un gain d'adaptation décroissant.

II - DESCRIPTION DU SYSTEME D'IDENTIFICATION A MODELE D'ESTIMATION "PARALLELE".

Soit donc un système d'identification à l'aide d'un modèle ajustable "parallèle" (méthode de l'erreur de sortie) décrit pas les équations :

- Le procédé à identifier (modèle de référence) :

$$\theta_p(k) = \sum_{i=1}^{n} a_i\, \theta_p(k-i) + \sum_{i=0}^{m} b_i\, \rho(k-i) = p^T x_{k-1} \qquad (1)$$

où

$$p^T = [a_1 \ldots a_n,\ b_o, \ldots b_m] \qquad (2)$$

$$x_{k-1} = [\theta_p(k-1) \ldots \theta_p(k-n),\ \rho(o) \ldots \rho(k-m)] \qquad (3)$$

$\theta_p(k)$ - sortie du procédé à l'instant k.

$\rho\,(k)$ - entrée du procédé à l'instant k.

- Le modèle d'estimation récursive (système ajustable "parallèle")

$$\theta_s(k) = \sum_{i=1}^{n} \hat{a}_i(k)\, \theta_s(k-i) + \sum_{i=0}^{m} \hat{b}_i(k)\, \rho(k-i) = \hat{p}^T(k)\, y_{k-1} \qquad (4)$$

$$\text{où} \quad \hat{p}^T(k) = |\hat{a}_1(k) \ldots \hat{a}_n(k),\ \hat{b}_o(k) \ldots b_m(k)| \qquad (5)$$

$$y^T_{k-1} = |\theta_s(k-1) \ldots \theta_s(k-n),\ \rho(o) \ldots \rho(k-m)| \qquad (6)$$

$\theta_s(k)$ - sortie du modèle d'estimation à l'instant k.

- L'erreur généralisée

$$\varepsilon_k = \theta_p(k) - \theta_s(k) \qquad (7)$$

- L'algorithme d'adaptation

$$\nu_k = \varepsilon_k + \sum_{i=1}^{n} c_i\, \varepsilon_{k-i} = \varepsilon_k + c^T e_{k-1} \qquad (8)$$

$$c^T = [c_1 \;\ldots\; c_n] \qquad (9)$$

$$e^T_{k-1} = [\varepsilon_{k-1} \;\ldots\; \varepsilon_{k-n}] \qquad (10)$$

$$\hat{p}(k) = \hat{p}(k-1) + f(\nu_k) \qquad (11)$$

La relation (8) définit un bloc linéaire de traitement de l'erreur généralisée ε_k dont la sortie est effectivement utilisée pour l'adaptation du vecteur des paramètres estimés $\hat{p}(k)$.

Avec les notations vectorielles introduites et en passant de l'instant k à l'instant k+1 on obtient pour les équations (7) (8) (11) les expressions :

$$\varepsilon_{k+1} = a^T_M\, e_k = [p - \hat{p}(k+1)]\, y_k \qquad (12)$$

$$\nu_{k+1} = \varepsilon_{k+1} + c^T e_k \qquad (13)$$

$$\hat{p}(k+1) = \hat{p}(k) + f\,(\nu_{k+1}) = \hat{p}(k) + \tilde{f}\,(\nu^o_{k+1}) \qquad (14)$$

$$\text{où } a^T_M = [a_1,\, a_2 \;\ldots\; a_n] \qquad (15)$$

Le signal d'erreur ν^o_{k+1} est calculé à partir de l'erreur généralisée "a priori" à l'instant (k+1) notée avec ε^o_{k+1} qui s'obtient avec les paramètres estimés à l'instant k :

$$\varepsilon^o_{k+1} = [p - \hat{p}(k)]\, x_k \qquad (16)$$

$$\nu^o_{k+1} = \varepsilon^o_{k+1} + c^T e_k \qquad (17)$$

Le problème qui se pose est donc de déterminer les paramètres $c_1 \ldots c_n$ intervenant dans le traitement de l'erreur généralisée et la loi d'adaptation $f(\nu^o_{k+1})$ afin que $\lim_{k\to\infty} \varepsilon_k = 0$ quel que soit l'écart initial des paramètres $(\hat{p}(o)-p)$.

En introduisant la variable auxiliaire

$$z_k = [\hat{p}(k) - p] \tag{18}$$

Les équations (12) (13) (14) décrivant le système d'identification à modèle d'estimation parallèle peuvent s'écrire alors :

$$\varepsilon_{k+1} = a_M^T e_k - y_k^T z_{k+1} = a_M^T e_k + \mu_{k+1}^1 \tag{19}$$

$$\nu_{k+1} = \varepsilon_{k+1} + c^T e_k \tag{20}$$

$$\mu_{k+1} = - \mu_{k+1}^1 = y_k^T z_k + y_k^T f(\nu_{k+1}) = y_k^T z_k + y_k^T \tilde{f} (\nu_{k+1}^o) \tag{21}$$

Ces équations correspondent à un système équivalent à contre-réaction constitué d'un bloc linéaire (19) (20) et d'un bloc à paramètres variables dans le temps (21) (Fig. 2).

<u>Fig. 2</u>

Pour que ce système soit asymptotiquement hyperstable (respectivement asymptotiquement stable globalement) il suffit que les deux blocs soient des opérateurs de type positif (cf. théorème d'hyperstabilité de Popov |5|) . En effet si le bloc à paramètres variables dans le temps (eqs (21)) est de type positif, le produit entrée sortie vérifie une inégalité de la forme :

$$\eta(o_1\ k_1) = \sum_{k=0}^{k_1} \mu_{k+1}\ \nu_{k+1} \geq - \gamma_o^2 \tag{22}$$

Dans ces conditions on peut appliquer le théorème d'hyperstabilité (Popov), l'hyperstabilité asymptotique étant alors assurée si la fonction de transfert du bloc linéaire équivalent défini par les équations (19) (20) :

$$G(z) = \frac{1 + \sum_{i=1}^{n} c_i z^{-i}}{1 - \sum_{i=1}^{n} a_i z^{-i}} \tag{23}$$

est une fonction de transfert strictement réelle positive (bloc linéaire de type positif).

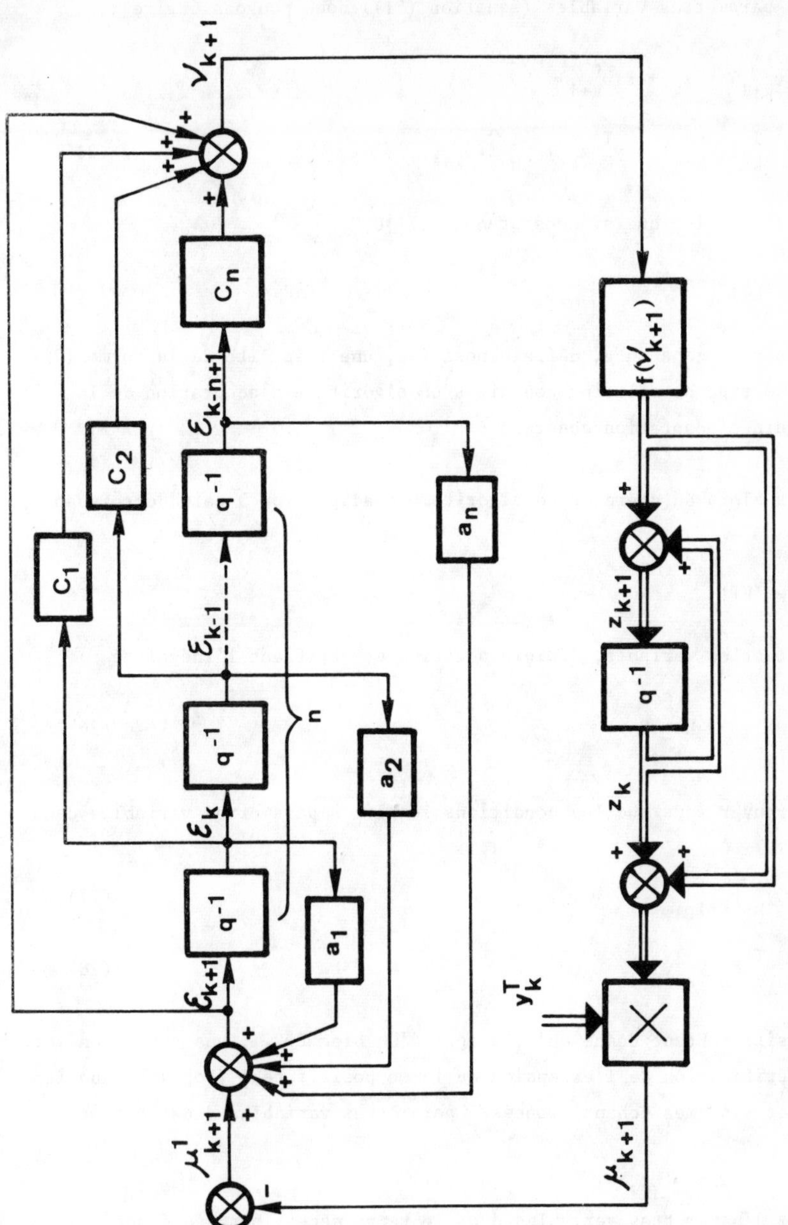

C1
C2
Cn
q-1
q-1
q-1
q-1
a1
a2
an
n
f(νk+1)
νk+1
εk+1
εk
εk-1
εk-n+1
μ1k+1
μk+1
zk+1
zk
yTk

Figure . 2

III - POSITIVITE DU BLOC EQUIVALENT A PARAMETRES VARIABLES DANS LE TEMPS.

Pour le bloc à paramètres variables (équation (21)) nous pouvons écrire :

$$z_{k+1} = z_k + f(\nu_{k+1}) = z_k + \tilde{f}(\nu^o_{k+1}) \quad (24)$$

$$\mu_{k+1} = y_k^T z_k + y_k^T f(\nu_{k+1}) = y_k^T z_k + y_k^T \tilde{f}(\nu^o_{k+1}) \quad (21')$$

Il est facile à vérifier que si nous prenons |7||8|

$$f(\nu_{k+1}) = F\, y_k\, \nu_{k+1}$$

où F est une matrice constante, définie positive, une inégalité de la forme (22) est vérifiée. Ce type de solution conduit à un algorithme d'adaptation de la classe (b) à gain d'adaptation constant.

Nous essaierons alors de chercher un algorithme d'adaptation à gain décroissant de la forme

$$f(\nu_{k+1}) = F_k\, y_k\, \nu_{k+1} \quad (25)$$

où F_k est une matrice variable, définie positive et vérifiant l'inégalité

$$F_k \leq F_{k-1} \qquad k \geq 0 \quad (26)$$

Il faut donc trouver sous quelles conditions le bloc à paramètres variables dans le temps (P.V.T.)

$$z_{k+1} = z_k + F_k\, y_k\, \nu_{k+1} \quad (27)$$

$$\mu_{k+1} = y_k^T z_k + y_k^T F_k\, y_k\, \nu_{k+1} \quad (28)$$

est de type positif. Etant donné qu'il s'agit d'un bloc P.V.T. une condition suffisante est la vérification de l'extension du lemme positif réel (Popov-Yakubovitch-Kalman) pour les systèmes échantillonnés à paramètres variables dans le temps |5| |8| |9|.

Soit un système à paramètres variables dans le temps décrit par les équations :

$$x_{k+1} = A_k\, x_k + B_k\, u_k \quad (29)$$

$$v_k = C_k\, x_k + J_k\, u_k \quad (30)$$

où A_k, B_k, C_k, J_k sont des matrices définies pour $k \geq 0$.

Lemme |5| |8| : Pour que le système échantillonné à paramètres variables dans le temps défini par les équations (29) (30) soit de type positif il suffit qu'il existe une matrice symétrique définie positive P_k, et des matrices arbitraires W_k et L_k vérifiant pour tout $k \geq 0$ le système d'équations

$$A_k^T P_{k+1} A_k - P_k = - L_k L_k^T \tag{31}$$

$$B_k^T P_{k+1} A_k + W_k^T L_k^T = C_k \tag{32}$$

$$W_k^T W_k = J_k + J_k^T - B_k^T P_{k+1} B_k \tag{33}$$

Si W_k est une matrice inversible le système d'équations (31) (32) (33) peut se réduire à une seule équation. En effet de l'équation (32) on obtient :

$$- L_k = (A_k^T P_{k+1} B_k - C_k^T) W_k^{-1} \tag{34}$$

et respectivement en utilisant (3) :

$$L_k L_k^T = (A_k^T P_{k+1} B_k - C_k^T) (W_k^T W_k)^{-1} (B_k^T P_{k+1} A_k - C_k) \tag{35}$$

En introduisant l'équation (35) dans l'équation (31) on obtient :

$$A_k^T P_{k+1} A_k - P_k = - (A_k^T P_{k+1} B_k - C_k^T) (J_k^T + J_k - B_k^T P_{k+1} B_k)^{-1}(B_k^T P_{k+1} A_k - C_k) \tag{36}$$

qui remplace le système d'équations (31) (32) (33).

Si on revient au système considéré (27) (28) et on tient compte que la propriété de positivité est indépendante du signal d'entrée on peut faire le changement de variable.

$$\varepsilon_{k+1} = u_k \; ; \; \mu_{k+1} = v_k \tag{37}$$

et les équations (27) (28) s'écrivent alors :

$$z_{k+1} = z_k + F_k y_k u_k \tag{38}$$

$$v_k = y_k^T z_k + \frac{1}{2} y_k^T F_k y_k u_k + \frac{1}{2} y_k^T F_k y_k u_k \tag{39}$$

Ceci permet la décomposition de ce bloc en deux blocs en parallèle, l'un ayant la sortie

$$v_k^1 = y_k^T z_k - \frac{1}{2} y_k^T F_k y_k u_k \quad (40)$$

et l'autre la sortie

$$v_k^2 = \frac{1}{2} y_k^T F_k y_k u_k \quad (41)$$

Pour que le bloc ayant l'entrée u_k et la sortie v_k^2 vérifie une inégalité de la forme (22) il faut que F_k soit une matrice définie positive (ou non-négative) pour $k \geq 0$.

Pour le bloc ayant l'entrée u_k et la sortie v_k^1 on peut appliquer le lemme positif réel donné plus haut. En faisant la correspondance :

$$x_k \to z_k \; ; \; A_k \to I \; , \; B_k \to F_k y_k \; ; \; C_k \to y_k^T ; \; J_k \to \frac{1}{2} y_k^T F_k y_k$$

dans l'équation (36) on obtient la condition que F_k doit être une matrice définie positive vérifiant pour tout k l'équation :

$$P_{k+1} - P_k = -(P_{k+1} F_k - I) y_k \left| y_k^T (F_k - F_k P_{k+1} F_k) y_k \right|^{-1} . \; y_k^T (F_k P_{k+1} - I) \quad (42)$$

Cette équation où P_k est une matrice définie positive admet comme solution en dehors de $F_k = F > 0$ respectivement $P_k = P_{k+1} = F_k^{-1}$ (algorithme d'adaptation à gain constant), la solution :

$$F_{k+1}^{-1} = F_k^{-1} + y_k y_k^T \quad (43)$$

et respectivement

$$P_k = F_k^{-1} \; ; \; P_{k+1} = F_k^{-1} + y_k y_k^T \quad (44)$$

qui nous intéresse.

En utilisant le lemme d'inversion |8| on obtient de (43) la relation :

$$F_{k+1} = F_k - \frac{F_k y_k y_k^T F_k}{1 + y_k^T F_k y_k} \quad (45)$$

où F_o est une matrice arbitraire définie positive. On observe que F_{k+1} donné par (45) vérifie la condition (26) imposée au départ sur les gains d'adaptation. On obtient alors la loi d'adaptation des paramètres :

$$\hat{p}(k+1) = \hat{p}(k)+F_k\, y_k\, \nu_{k+1} \tag{46}$$

Néanmoins dans la dérivation de l'algorithme d'identification il faut tenir compte que ν_{k+1} n'est pas accessible pour le calcul de $\hat{p}(k+1)$ et que la modification des paramètres du modèle d'estimation à l'instant (k+1) doit se faire à partir de la mesure de l'erreur "modèle d'estimation-procédé" obtenue en utilisant les paramètres estimés à l'instant k (ε^o_{k+1} respectivement ν^o_{k+1}).

Des équations (19) (20) (16) (17) (46) on obtient alors :

$$a_M^T\, e_k - \varepsilon^o_{k+1} = y_k^T\, z_k \tag{47}$$

$$a_M^T\, e - \varepsilon_{k+1} = y_k^T\, z_k + y_k\, Fy_k\, (\varepsilon_{k+1} + c^T\, e_k) \tag{48}$$

et respectivement

$$\varepsilon^o_{k+1} - \varepsilon_{k+1} = y_k^T\, F_k\, y_k\, (\varepsilon_{k+1} + c^T\, e_k) \tag{49}$$

En retranchant dans les deux membres de l'égalité (49) le terme $c^T\, e_k$ et tenant compte de (20) on obtient :

$$\varepsilon^o_{k+1} - \nu_{k+1} = y_k^T\, F_k\, y_k\, \nu_{k+1} - c^T e_k \tag{50}$$

et tenant compte de (17) on obtient finalement la relation :

$$\nu_{k+1} = \frac{\nu^o_{k+1}}{1 + y_k^T\, F_k\, y_k} \tag{51}$$

L'algorithme d'adaptation des paramètres s'écrit alors :

$$\hat{p}(k+1) = p(k) + \frac{F_k\, y_k}{1 + y_k\, F_k\, y_k}\, \nu^o_{k+1} = \hat{p}(k)\; F_{k+1}\, y_k\, \nu^o_{k+1} \tag{52}$$

où F_{k+1} est donné par la relation (45) et ν^o_{k+1} par la relation (17).

IV - POSITIVITE DU BLOC LINEAIRE EQUIVALENT.

L'application du théorème d'hyperstabilité permet de déterminer des conditions suffisantes pour lesquelles la convergence de l'algorithme d'identification est assurée quelle que soit la différence initiale entre les paramètres du procédé et les paramètres du modèle d'estimation. L'application de ce théorème conduit donc à la condition que la fonction de transfert $G(z)$ donnée par (23) soit une fonction de transfert strictement réelle positive. Ceci conduit tout d'abord à la condition que les pôles du procédé à identifier doivent être situés à l'intérieur du cercle unité ($|z| < 1$).

Dans l'hypothèse de la connaissance des paramètres a_i, pour calculer les paramètres c_i, deux principales méthodes peuvent être utilisées :

1) Utilisation des critères de positivité de type fréquentiel ou algébrique développés pour les systèmes continus par l'intermédiaire de la transformation bilinéaire $z = \frac{1 + s}{1 - s}$ |9| |10|.

2) Représentation du bloc linéaire équivalent (19) (20) sous forme d'équation en variables d'état et utilisation directe du lemme positif réel (Popov-Yakubovitch-Kalman) pour systèmes échantillonnés linéaires et invariants |5| |8| |9|.

En introduisant le vecteur d'état :

$$x_k^T = \varepsilon_{k-n}, \varepsilon_{k-n+1} \ldots \varepsilon_{k-1} \tag{53}$$

on obtient pour les équations (19) (20) la représentation :

$$x_{k+1} = A\, x_k + b\, u_k \tag{54}$$

$$\nu_k = (c + a_M)^T x_k + u_k \tag{55}$$

où

$$A = \begin{bmatrix} 0 & 1 & & & 0 \\ & & 1 & & \\ & & & 1 & \\ & & & & 1 \\ a_n & a_{n-1} & \ldots & & a_1 \end{bmatrix} \quad (56) \qquad b = \begin{bmatrix} 0 \\ 0 \\ \vdots \\ 1 \end{bmatrix} \quad (57)$$

$$c^T = [c_n, c_{n-1} \ldots c_1] \quad (58) \qquad a^T = [a_n \,;\, a_{n-1} \ldots a_1] \quad (59)$$

En utilisant le lemme positif réel pour systèmes échantillonnés linéaires et invariant (ce lemme s'obtient en remplaçant dans les équations (31) (32) (33) A_k, B_k, C_k, J_k, P_k, W_k, L_k par des matrices constantes) et tenant compte de la forme particulière de A et b on obtient le système d'équations :

$$A^T PA - P = -1.1^T \tag{60}$$

$$p_{nn} a_M^T + [0, p_{n1} \dots p_{n,\, n-1}] + w.1^T = a_M^T + c^T \tag{61}$$

$$w^2 = 2 - p_{nn} \tag{62}$$

où

$$P = \begin{bmatrix} p_{11} & p_{n-1,1} & p_{n1} \\ \vdots & & \\ p_{n1} \dots p_{n,n-1} & & p_{nn} \end{bmatrix} \tag{63}$$

Les coefficients c_i sont donnés alors par l'expression :

$$c_i = (p_{nn} - 1)\, a_i + 1_{n+1-i} \sqrt{2 - p_{nn}} + p_{n,\, n-i} \qquad i = 1 \cdot\cdot n-1 \tag{64}$$

respectivement

$$c_n = (p_{nn} - 1)\, a_n + 1_1 \sqrt{2 - p_{nn}} \qquad i = n \tag{64'}$$

où p_{nn}, $p_{n,n-i}$ sont les termes de la matrice définie positive P, solution de l'équation de Lyapunov (60) obtenue pour un vecteur 1, tel que $p_{nn} \leq 2$ (des solutions particulières intéressantes s'obtiennent pour $p_{nn} = 1$ respectivement $p_{nn} = 2$).

En réalité les paramètres a_i ne sont pas connus, par conséquent on procède de la manière suivante :

On calcule les coefficients c_i, pour des coefficients $\hat{a}_i(o)$ d'initialisation du modèle ajustable (obtenu éventuellement à l'aide de l'algorithme des moindres carrés). Les coefficients c_i ainsi calculés garantissent (condition suffisante mais non nécessaire) la convergence de l'algorithme d'identification pour l'ensemble des valeurs des a_i vérifiant le lemme positif réel (équations (60) (61) (62)) pour un vecteur c donné (ou autrement dit, pour l'ensemble des valeurs de a_i, pour lesquelles l'équation :

$$A^T PA - P = -\frac{1}{2 - b^T Pb}\left[\tilde{a}_M + \tilde{c} - A^T Pb\right]\left[\tilde{a}_M + \tilde{c} - A^T Pb\right]^T$$

admet comme solution une matrice P définie positive le vecteur c étant donné). Dans le cas ou une instabilité apparaît soit on recalcule les c_i pour une autre valeur du vecteur l soit on refait le calcul en modifiant les valeurs initiales de $\hat{a}_i(o)$.

V - ALGORITHME D'IDENTIFICATION A GAIN DECROISSANT POUR LA METHODE DE L'EQUATION DE L'ERREUR.

Dans le cas de la méthode de l'équation de l'erreur le modèle d'estimation qui est de type série-parallèle est décrit par l'équation :

$$\theta_s(k) = \sum_{i=1}^{n} \hat{a}_i(k)\, \theta_p(k-i) + \sum_{i=0}^{m} \hat{b}_i(k)\, \rho(k-i) = \hat{p}^T(k)\, x_{k-1} \tag{65}$$

En appliquant la même procédure de dérivation de l'algorithme d'identification on obtient un système équivalent à contre réaction décrit par les équations :

$$\varepsilon_{k+1} = \left[p - \hat{p}(k+1)\right]\; x_k \tag{66}$$

$$\nu_{k+1} = \varepsilon_{k+1} + c^T e_k \tag{67}$$

$$\hat{p}(k+1) = \hat{p}(k) + f(\nu_{k+1}) \tag{68}$$

et respectivement en utilisant le changement de variable (18) on obtient :

$$\varepsilon_{k+1} = - z_{k+1}\, x_k = \mu^1_{k+1} \tag{69}$$

$$\nu_{k+1} = \varepsilon_{k+1} + c^T e_k \tag{70}$$

$$\mu_{k+1} = - \mu^1_{k+1} = x_k^T z_k + x_k^T f(\nu_{k+1}) \tag{71}$$

Le bloc linéaire équivalent étant défini par les équations (69) (70) est caractérisé par la fonction de transfert :

$$G(z) = 1 + \sum_{i=1}^{n} c_i\, z^{-i} \tag{72}$$

Pour que $G(z)$ soit strictement réelle positive il suffit donc de prendre $c_i = 0, \forall i$ et par conséquent :

$$\nu_{k+1} = \varepsilon_{k+1} \tag{73}$$

bloc à paramètres variables dans le temps défini par l'équation (71) est similaire à celui défini par l'équation (21) si on remplace y_k par x_k. On obtient alors l'algorithme d'identification

$$\hat{p}(k+1)=\hat{p}(k) + \frac{F_k\, x_k\, \varepsilon^o_{k+1}}{1 + x_k^T F_k\, x_k} \tag{74}$$

$$F_{k+1} = F_k - \frac{F_k\, x_k\, x_k^T F_k}{1 + x_k^T F_k\, x_k} \tag{75}$$

$$\varepsilon^o_{k+1} = \theta_p(k+1) - \hat{p}(k)^T x_k \tag{76}$$

On reconnait tout de suite l'algorithme des moindres carrés récursif utilisé pour l'identification à l'aide de la méthode de l'équation de l'erreur mais qui bien entendu conduit à des estimations biaisées des paramètres en présence de perturbations additives sur la sortie.

En comparant l'algorithme pour modèle d'estimation parallèle (52) (45) (16) (17) et l'algorithme des moindres carrés (74) (75) (76) on constate une certaine ressemblance structurelle. La différence essentielle entre les deux algorithmes provient du fait que dans le cas de l'identification avec modèle parallèle on utilise les sorties du système ajustable à divers instants (y_k) qui ne sont pas directement entachées par le bruit alors que dans le cas du modèle série-parallèle (moindres carrés) on utilise divers échantillons successifs de la sortie bruitée du procédé à identifier. C'est de cette différence que proviennent les très bonnes performances obtenues avec l'algorithme (52) (45) (17) en présence de bruit.

VI - RESULTATS EXPERIMENTAUX.

Pour illustrer les performances qui peuvent être obtenues en présence de bruit sur les mesures avec l'algorithme développé pour l'estimation à l'aide d'un modèle parallèle ,nous avons considéré un exemple en simulation et l'identification d'un système réel constitué par le transfert débit de papier (m^3/h) et humidité du papier (%H_2O) d'une machine à papier expérimentale (Centre Technique du Papier).

Identification d'un système du 4ème ordre

Le système que nous avons considéré est décrit par l'équation aux différences

$$\theta_p(k) = \sum_{i=1}^{4} a_i \, \theta_p(k-i) + b_o \, \rho(k)$$

où : $a_1 = 1,3$; $a_2 = 0,22$; $a_3 = 0,832$; $a_4 = 0,269$; $b_0 = 1$

Ce système est caractérisé par une réponse oscillante faiblement amortie. Le système a été excité par une séquence binaire pseudo aléatoire (+1, -1) et la mesure de la sortie a été entachée d'un bruit de mesure (bruit gaussien centré de variance 0,81).

Le vecteur c^T intervenant dans le correcteur linéaire (voir équation (17) a été $c^T = |0,65, -0,4, -0,3, -0,15|$

Pour apprécier globalement la qualité de l'identification nous avons calculé à chaque pas le carré de la distance de structure définie par :

$$D^2(k) = (p - \hat{p}(k))^T (p - \hat{p}(k)) = \sum_{i=1}^{4} (a_i - \hat{a}_i(k))^2 + (b_o - \hat{b}_o(k))^2$$

L'évolution du carré de la distance de structure est illustrée figure 3 où pour comparaison ont été représentés aussi les résultats obtenus avec la méthode de l'équation de l'erreur utilisant l'algorithme des moindres carrés.

Fig. 3

Dans la figure 4 nous avons représenté l'évolution de la distance de structure pour les méthodes suivantes* :

- Gain d'adaptation variable et modèle d'estimation parallèle (LANDAU)
- Moindres carrés
- Moindres carrés généralisés
- Variable instrumentale (YOUNG)
- Variable instrumentale à paramètres retardés.

Nous mentionnons que pour l'exemple considéré la méthode de la variable instrumentale à observations retardées ne converge pas, certainement à cause de la réponse peu amortie du système

Fig. 4

* Ces résultats ont été obtenus en utilisant le programme conversationnel développé par M. BETHOUX (Alsthom) |12|.

Figure 3

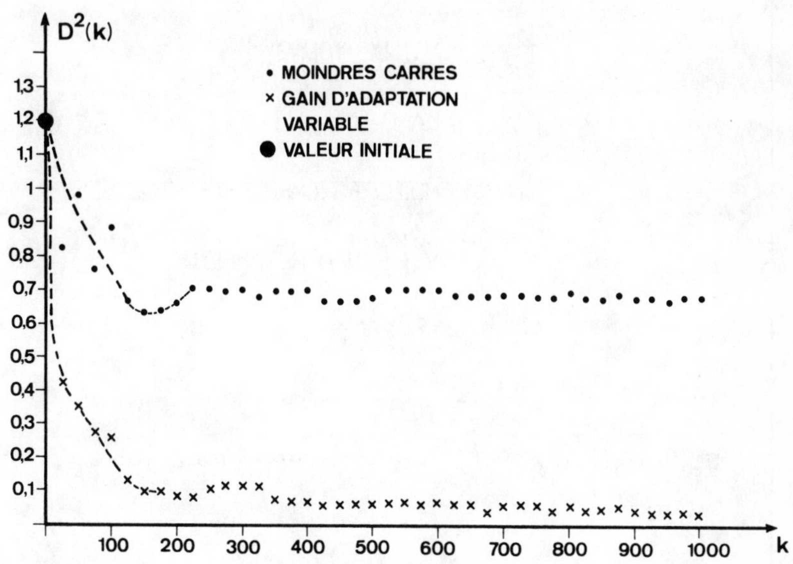

$D^2(0) = 1,2$

$D^2(k)$ \ k	100	200	500	1000	2000	3500
Moindres carrés	0,886	0,659	0,681	0,672	0,705	0,680
Modèle-parallèle gain variable	0,267	0,093	0,061	0,037	0,015	0,003

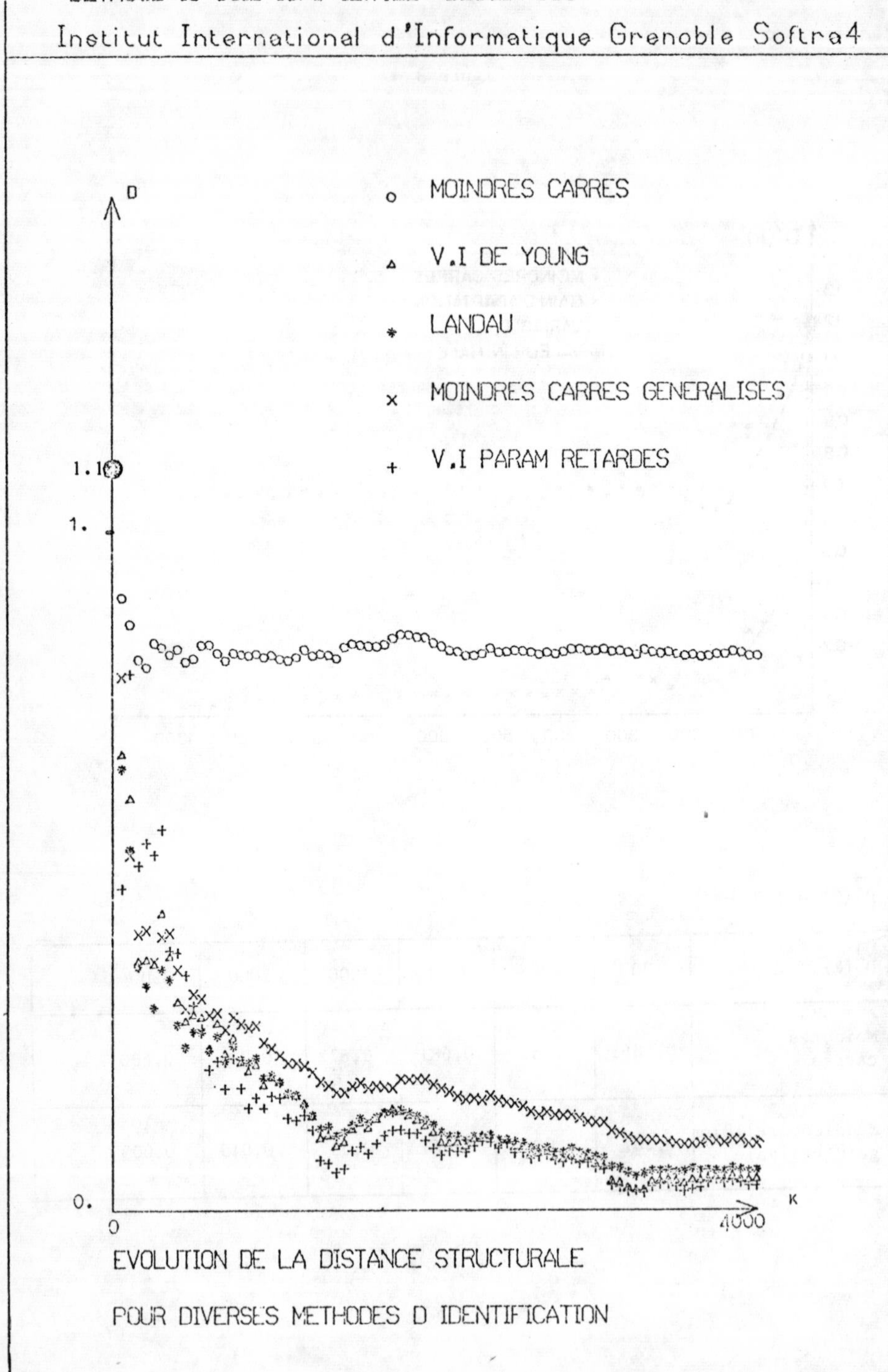

BETHOU2 10 JUIL 1973 12h.22 DESSIN 001
Institut International d'Informatique Grenoble Softra4
D
MOINDRES CARRES
V.I DE YOUNG
LANDAU
MOINDRES CARRES GENERALISES
V.I PARAM RETARDES
1.10
1.
0.
0
4000
K
EVOLUTION DE LA DISTANCE STRUCTURALE
POUR DIVERSES METHODES D IDENTIFICATION

Figure 4

L'analyse de ces résultats montre que en ce qui concerne le biais sur l'estimation des paramètres l'algorithme proposé est nettement supérieur à l'algorithme des moindres carrés et à celui des moindres carrés généralisés, ses performances étant comparables à celles de la méthode de la variable instrumentale de YOUNG, ou de la variable instrumentale à paramètres retardés.

Néanmoins par rapport aux méthodes de variable instrumentale elle présente l'avantage de pouvoir assurer sous certaines conditions (liées au correcteur linéaire) la convergence de l'algorithme quelle que soit la valeur initiale des paramètres du modèle d'estimation et la nature de la réponse du système (pour aucune méthode de variable instrumentale on ne peut pas garantir la convergence des paramètres).

Identification du transfert "débit papier-humidité papier" d'une machine à papier (C.T.P. - Grenoble)

Parmi les différents transferts existant dans une machine à papier (voir référence |11|), le transfert "débit papier-humidité papier" a été choisi pour vérifier l'algorithme d'identification proposé.

L'entrée est constituée par un signal de valeur moyenne nulle (la vanne de commande du débit a été excitée par une séquence binaire pseudoaléatoire) superposé sur une valeur moyenne correspondant à un point de fonctionnement (0,01 m^3/h), la valeur moyenne de la sortie étant de 5,44 % H_2O. Les mêmes méthodes d'identification considérées dans l'exemple précédent ont été utilisées.

Les meilleurs résultats ont été obtenus avec une modélisation du procédé par une fonction de transfert de la forme :

$$H(z) = z^{-r} \frac{b_o + b_1 z^{-1}}{1 - a_1 z^{-1} - a_2 z^{-2}}$$

où le retard r est égal à 13 périodes d'échantillonnage (période d'échantillonnage : 5 secondes).

Pour l'appréciation de la qualité de l'identification on a utilisé le critère :

$$J = \frac{1}{N} \sum_{k=1}^{N} \left[\theta_p(k) - \hat{\theta}_s(k) \right]^2$$

où $\hat{\theta}_s(k) = \sum_{i=1}^{n} \hat{a}_i \hat{\theta}_s(k-i) + \sum_{i=0}^{m} \hat{b}_i \rho(k-i)$

$\hat{a}_i$, $\hat{b}_i$ étant les paramètres du modèle du procédé obtenus à l'aide des divers algorithmes d'identification.

Les résultats obtenus avec différentes méthodes sur la base de 245 mesures sont donnés dans le tableau I.

La méthode des "moindres carrés" a été initialisée avec $a_1=a_2=b_o=b_1=0$. Toutes les autres méthodes ont été initialisées avec les valeurs données par la méthode des moindres carrés. La matrice initiale des gains d'adaptation a été dans tous les cas F_o = diag [10,10,10,10].

	Moindres carrés	Gain variable Modèle d'estimation parallèle (LANDAU)	Variable instrumentale de YOUNG	Variable instrumentale à paramètres retardés(retard=1)	Moindres carrés généralisés (CLARKE)
a_1	1,598	1,768	1,770	1,772	1,679
a_2	-0,639	-0,800	-0,804	-0,807	-0,717
b_o	0,162	0,161	0,165	0,165	0,161
b_1	0,035	0,025	0,026	0,026	0,030
Critère (J)	0,105	0,081	0,084	0,084	0,089

TABLEAU 1

Par rapport au critère considéré l'algorithme d'identification à gain d'adaptation variable et modèle d'estimation parallèle donne le meilleur résultat (J=0,081) ; viennent ensuite les méthodes de variable instrumentale avec modèle auxiliaire (J=0,084) et enfin l'algorithme des moindres carrés généralisés (J=0,089). Les résultats obtenus sur cet exemple réel confirme les résultats obtenus précédemment en simulation.

VII - EXTENSION DE LA METHODE DE SYNTHESE D'ALGORITHMES D'IDENTIFICATION PROPOSEE.

La particularité de la méthode de synthèse que nous avons utilisée réside dans la possibilité de déterminer des algorithmes d'identification pour lesquels on obtient d'emblée les conditions de convergence vis-à-vis des valeurs d'initialisation des paramètres des divers modèles d'estimation. Elle peut être utilisée par conséquent pour l'analyse de la convergence de divers algorithmes connus. En utilisant cette approche M.G. BETHOUX |13| a démontré la convergence de la méthode des moindres carrés généralisés. On peut aussi expliquer à l'aide de cette approche pourquoi les méthodes de "variable instrumentale" peuvent diverger. En effet dans les méthodes de "variable instrumentale" l'algorithme des moindres carrés (74) (75) (76) devient :

$$\hat{p}(k+1) = \hat{p}(k) + \frac{F_k \tilde{y}_k}{1 + \tilde{y}_k^T F_k x_k} \varepsilon^o_{k+1} \tag{77}$$

$$F_{k+1} = F_k - \frac{F_k \tilde{y}_k x_k^T F_k}{1 + \tilde{y}_k^T F_k x_k} \tag{78}$$

$$\varepsilon^o_{k+1} = \theta_p(k+1) - \hat{p}(k)^T x_k \tag{79}$$

où $\tilde{y}_k$ est la variable instrumentale. Dans le cas de la variable instrumentale à observations retardées :

$$\tilde{y}_k = x_{k-r} \tag{80}$$

et dans le cadre de la variable instrumentale de YOUNG $\tilde{y}_k$ s'obtient à l'aide d'un modèle auxiliaire de type parallèle régi par l'équation :

$$\theta_{vi}(k+1) = \hat{p}(k)\, \tilde{y}_k \tag{81}$$

$$\tilde{y}_k = \left[\theta_{vi}(k) \ldots \theta_{vi}(k-n+1), \rho(k+1) \ldots \rho(k-m+1)\right] \tag{82}$$

Les méthodes de "variable instrumentale" conduisent à un schéma équivalent à contre-réaction dont la partie linéaire est un gain constant et la partie à paramètres variables dans le temps est décrite par les équations :

$$z_{k+1} = z_k + F_k \tilde{y}_k u_k \tag{83}$$

$$v_k = x_k^T z_k + \frac{1}{2} y_k F_k x_k u_k + \frac{1}{2} \tilde{y}_k F_k x_k u_k \tag{84}$$

Il est facile de voir que les conditions de positivité sont violées si

$$\tilde{y}_k \not\approx x_k \tag{85}$$

ce qui peut entraîner l'apparition d'instabilités dans le système. Pour la méthode de la variable instrumentale à observations retardées ceci conduit à la conclusion que la convergence de l'algorithme peut être certainement assurée si :

$$\tilde{y}_k = x_{k-r} \approx x_k \tag{86}$$

(dans le cadre de l'exemple considéré cette condition n'était pas respectée ni même poru r = 1).

VIII - CONCLUSIONS.

L'algorithme d'identification présenté qui a donné de très bons résultats pour l'identification en présence de bruit a été obtenu à partir des configurations utilisées dans les systèmes adaptatifs avec modèle de référence et pour lesquels la synthèse à partir des concepts de stabilité s'est avérée déjà dans le passé extrêmement utile. La solution concrète pour le problème posé a été obtenue à partir de l'utilisation de l'extension du lemme positif réel. Il est devenu possible à l'aide de cette approche d'interpréter divers algorithmes connus du point de vue de la configuration du système d'identification mais surtout d'analyser la convergence de ces algorithmes vis-à-vis des valeurs initiales des estimations.

Si on essaie de donner maintenant une interprétation de l'algorithme développé dans l'optique de la théorie de l'estimation,plus familière aux spécialistes de l'identification, cet algorithme apparaît comme un type spécial de méthode de variable instrumentale pour lequel des conditions de convergence de l'algorithme,indépendantes des estimations initiales des paramètres,sont données.

*

* *

L'auteur tient à remercier MM. BETHOUX, LEBEAU et MULLER pour leur concours apporté à la vérification de l'algorithme d'identification proposé.

BIBLIOGRAPHIE

|1| ASTRÖM K.J., EYKOFF P. - "System Identification. A Survey" Automatica, Vol. 7, n°2, pp. 123-162, Mars 1971.

|2| LANDAU I.D. - "A Survey of Model Reference Adaptive Techniques (Theory and Applications)", Automatica, Vol. 10, n°4, Juillet 1974.
(Proc. of 3rd IFAC Symposium on "Sensitivity, Adaptivity and Optimality", 18-21 Juin 1973, Ischia, pp. 15-42.

|3| HIRSCH J.J., PELTIE P. - "Real Time Identification Using Adaptive Discrete Model", Proc. 3rd IFAC Symposium on "Sensitivity, Adaptivity and Optimality" 18-21 Juin 1973, Ischia, pp.290-297.

|4| SINNER Ed. - "Etude d'un régulateur adaptatif à variables d'états pour la commande des procédés industriels", Thèse de Docteur-Ingénieur, N° CNRS A.O. 8631, I.N.P. Grenoble, Juillet 1973.

|5| POPOV V.M. "Hyperstabilité des systèmes automatiques". Dunod, Paris 1973.

|6| LANDAU I.D. "Algorithme pour l'identification à l'aide d'un modèle ajustable parallèle", Compte-rendu Acad. Sci. Paris, 16 Juillet 1973, t. 277, Série A, pp. 197-200.

|7| LANDAU I.D. - "Design of Discrete Model Reference Adaptive Systems Using the Positivity Concept", Proc. of 3rd IFAC Symposium on "Sensitivity Adaptivity and Optimality", 18-21 Juin 1973, Ischia, pp. 307-314.

|8| LANDAU I.D. - "Sur une méthode de synthèse des systèmes adaptatifs avec modèle utilisés pour la commande et l'identification d'une classe de procédés physiques" Thèse de Docteur Es Sciences Physiques, N° CNRS A.O. 8495, U.S.M.G. Grenoble, Juin 1973.

|9| FAURRE P. - "Réalisations markoviennes de processus stationnaires", Thèse Es Sciences Mathématiques - Université Paris VI - Décembre 1972, (Rap. de Rech. I.R.I.A. n°13, Mars 1973).

|10| KARMARKAR J. - "Application of Positive Real Functions in Hyperstable Discrete Model Reference Adaptive System Design" Proc. 5th Internat. Hawaï Conf. on System Sciences, Honolulu, Janvier 1972, pp. 382-384.

|11| GENTIL S., SANDRAZ J.P., FOULARD C. - "Different Methods for Dynamic Identification of an Experimental Paper Machine", Proc. IFAC Symposium on System Identification", Juin 1973, La Haye, pp. 473-483.

|12| BETHOUX G. - "Etude comparative de plusieurs méthodes d'identification" Note Interne, Direction des Recherches ALSTHOM, Grenoble, Juillet 1973.

|13| BETHOUX G. - "Identification en temps réel des procédés multidimensionnels par la méthode des moindres carrés généralisés dans l'optique des systèmes adaptatifs discrets hyperstables", Note Interne, Direction des Recherches ALSTHOM, Grenoble, Août 1973.

PROBLEMES DE CONTROLE DES COEFFICIENTS DANS DES EQUATIONS AUX DERIVEES PARTIELLES

PAR Luc TARTAR
Maitre de Conférence
Université Paris IX

I - INTRODUCTION

1°) Certains problèmes de recherche d'un domaine optimal peuvent se mettre sous la forme suivante :

- Ω étant un ouvert borné de R^N (N=2 ou 3 en général) on cherche une partie Ω_1 de Ω telle que

(1.) si $a(x)$ vaut α sur Ω_1 et β sur $\Omega-\Omega_1$,($0 < \alpha < \beta$) et $u(x)$ est la solution du problème de Dirichlet.

(2.)
$$-\sum_i \frac{\partial}{\partial X_i}\left(a(x)\frac{\partial u}{\partial X_i}\right) = f \text{ dans } \Omega$$

$$u = 0 \text{ sur le bord } \partial\Omega \text{ de } \Omega.$$

la fonction

(3.)
$$J(a) = \int_\Omega F(x,u(x))dx$$

atteigne son minimum.

2°) Si on impose à Ω_1 d'avoir certaines propriétés de régularité, on peut obtenir l'existence d'une solution optimale D. CHENAIS [1].

Si on n'impose aucune condition alors il peut arriver (cela dépend de la fonction coût J) qu'il n'y ait pas de solution optimale. F.MURAT [2].

C'est à ce phénomène que nous allons nous intéresser.

II - LE PROBLEME LIMITE

1°) Considérons une suite minimisante $a_n(x)$ $n \in N$ du problème précédent; c'est à dire telle que $J(a_n)$ décroisse vers la borne inférieure de $J(a)$.

Si f est une fonction de $L^2(\Omega)$ (c'est à dire $\int_\Omega |f|^2 dx < +\infty$) alors, d'après des propriétés classiques du problème de Dirichlet, les solutions u_n demeurent dans un borné de $H_o^1(\Omega) = \{v \in L^2(\Omega) : \frac{\partial v}{\partial x_i} \in L^2(\Omega)\ i = 1,\dots,N \quad v|_{\partial\Omega} = 0\}$.

Ceci permet d'affirmer qu'une sous-suite extraite de la suite u_n convergera faiblement dans $H_o^1(\Omega)$ vers une fonction u de $H_o^1(\Omega)$.

2°) Malheureusement u n'est pas, en général, solution d'un problème du type 2 avec a ne prenant que les valeurs α ou β (ni même avec a prenant des valeurs dans l'intervalle $[\alpha,\beta]$).

Heureusement on a quelques renseignements concernant cette solution u. S. SPAGNOLO [3] .F.MURAT-L.TARTAR [4].

3°) Il existe une sous suite a_m extraite de la suite a_n et des fonctions $a_{ij}(x)$ $i,j = 1,\dots N$ (indépendantes de f) vérifiant

(4.) $\quad a_{ij}(x) = a_{ji}(x)$ presque partout dans Ω.

(5.) $\quad \alpha \sum_i \xi_i^2 \leqslant \sum_{i,j} a_{ij}(x)\, \xi_i\, \xi_j \leqslant \beta \sum_i \xi_i^2$ presque partout dans Ω et $\forall \xi_i \in R^N$.

telles que la solution u_m converge (dans $H_o^1(\Omega)$ faible) vers la solution u du problème de Dirichlet

(6.)
$$-\sum_{i,j} \frac{\partial}{\partial x_i} a_{ij}(x) \frac{\partial u}{\partial x_j} = f \text{ dans } \Omega$$
$$u = 0 \text{ sur } \partial\Omega$$

La détermination des coefficients $a_{ij}(x)$ du problème limite 6. à partir de la suite a_n étant un problème local que l'on ne sait pas résoudre explicitement (sauf le cas N=1) dans le cas général.

4°) On peut préciser ces résultats de la manière suivante :

En plus de 4 , 5 la matrice $a_{ij}(x)$ a la propriété supplémentaire

(7.) Si $\alpha \leqslant \lambda_1(x) \leqslant \lambda_2(x) \dots \leqslant \lambda_N(x) \leqslant \beta$ désignent les valeurs propres (réelles) de la matrice $a_{ij}(x)$ on a

$$\frac{\alpha\beta}{\alpha+\beta-\lambda_N(x)} \leqslant \lambda_1(x) \quad \text{presque partout dans } \Omega.$$

quand aux convergences des différents termes on a

$$\frac{\partial u_m}{\partial x_i} \rightharpoonup \frac{\partial u}{\partial x_i} \quad \text{dans } L^2(\Omega) \text{ faible} \quad i = 1 \ldots N$$

(8.) $$a_m \frac{\partial u_m}{\partial x_i} \rightharpoonup \sum_j a_{ij} \frac{\partial u}{\partial x_j} \quad \text{dans } L^2(\Omega) \text{ faible} \quad i = 1 \ldots N$$

$$a_m \sum_i \left(\frac{\partial u_m}{\partial x_i} \right)^2 \rightharpoonup \sum_i a_{ij} \frac{\partial u}{\partial x_i} \frac{\partial u}{\partial x_j} \quad \text{vaguement dans les mesures sur } \Omega.$$

5°) Le problème inverse de savoir quelles sont les matrices $a_{ij}(x)$ qui peuvent être obtenues de la manière précédente n'est pas complètement résolu, sauf pour $N = 1$ ou 2.

- Pour $N = 1$: Si $\frac{1}{a_m}(x) \rightharpoonup \frac{1}{a}(x)$ dans $L^\infty(\Omega)$ faible étoile alors c'est $a(x)$ qui apparait comme coefficient du problème limite.

Donc toute fonction vérifiant $a(x) \in [\alpha,\beta]$ satisfait à la question.

- Pour $N = 2$. Toute matrice $a_{ij}(x)$ satisfaisant 4 et 7 (et donc 5) peut **être** obtenue comme coefficients d'un problème limite.

- Pour $N > 2$ on ne sait pas si toutes les matrices vérifiant 4 et 7 sont obtenues. On peut cependant exhiber quelques matrices admissibles comme par exemple :

Les matrices a_{ij} dont les valeurs propres vérifient

(9.) $$\alpha \leqslant \lambda_1(x) \leqslant \lambda_2 = \lambda_3 = \ldots = \lambda_N(x) \leqslant \beta$$

$$\frac{\alpha\beta}{\alpha+\beta-\lambda_N(x)} = \lambda_1(x)$$

et

(10) Les matrices $a_{ij}(x) = a(x)\delta_{ij}$ avec $a(x) \in [\alpha,\beta]$

III- INTERPRETATION PHYSIQUE

1°) On suppose qu'on a à notre disposition 2 matériaux diélectriques dont les perméabilités diélectiques sont α et β.

- On veut remplir le domaine Ω par un corps, construit à l'aide des 2 matériaux ci-dessus de manière à minimiser une certaine fonction du potentiel électrostatique $u(x)$ en présence d'une distribution de charge $\frac{1}{4\pi} f(x)$.

2°) Le potentiel électrostatique u est donné par l'équation 2 (le bord de Ω étant à la masse) où a(x) vaut α ou β suivant qu'on est dans une zone de l'un ou l'autre matériau.

. $E_i = -\dfrac{\partial u}{\partial x_i}$ désigne la $i^{\text{ème}}$ composante du champ électrique.

. $D_i = a\,E_i$ la $i^{\text{ème}}$ composante du champ d'induction électrique

. $\dfrac{1}{4\pi}\, a \sum_i \left(\dfrac{\partial u}{\partial X_i}\right)^2 = \dfrac{E.D}{4\pi}$ désignant l'énergie locale du champ.

3°) Le cas où il n'y a pas de solution optimale correspond au cas où, pour minimiser la fonction J, on a intérêt à disséminer à l'extrême les matériaux de manière à "s'approcher" d'un corps hétérogène anisotrope (en général) qui physiquement ressemblerait plutôt à un alliage formé à partir des 2 matériaux de départ.

Du point de vue microscopique on aura toujours un corps formé des 2 matériaux, mais du point de vue macroscopique on aura obtenu un corps ayant des propriétés très différentes.

4°) Ce corps limite, en général anisotrope, aura un tenseur de perméabilité diélectique a_{ij} (non nécessairement de la forme $a\delta_{ij}$ qui correspond à un corps isotrope).

On conçoit aisément que la détermination de a_{ij} au voisinage d'un point ne dépend que de la manière de disposer les différents morceaux des 2 matériaux au voisinage de ce point.

Les convergences mathématiques données par 8 expriment que, macroscopiquement, les valeurs du champ électrique, du champ d'induction et de l'energie convergent vers les valeurs correspondant au corps hétérogène limite.

Notons pour finir que les matrices vérifiant 9 correspondent à des corps feuilletés dans la direction du vecteur propre correspondant à $\lambda_1(x)$, c'est à dire qui, localement sont formés de tranches de matériau α ou β dont le plan est perpendiculaire au vecteur propre (les proportions respectives des deux matériaux déterminant la valeur de λ_1).

IV - CONDITIONS NECESSAIRES D'OPTIMALITE

1°) On suppose qu'il existe une solution optimale $a_o(x)$, ne prenant que les valeurs α ou β , correspondant à $u_o(x)$.

On suppose donc

(11.) $$J(a) \geq J(a_o) \quad \forall a \text{ vérifiant } 1.$$

On suppose $F(x,u)$ dérivable en u et on introduit $w_o(x)$ solution de

(12.) $$- \Sigma \frac{\partial}{\partial x_i} a_o(x) \frac{\partial w_o}{\partial x_i} = \frac{\partial F}{\partial u}(x,u_o) \quad \text{dans } \Omega$$

$$w_o = 0 \quad \text{sur } \partial\Omega$$

2°) Si a_{ij} est admissible (c'est-à-dire obtenu comme limite) on peut définir $J(a_{ij}) = \int_\Omega F(x,u)\,dx$ où u est la solution de 6 .

Alors (si F est régulier, c'est-à-dire $u \to \int_\Omega F(x,u)\,dx$ continue sur $L^2(\Omega)$) on a

(13.) $$J(a_{ij}) \geq J(a_o) = J(a_o \delta_{ij})$$

3°) L'application $a_{ij} \to u$ est dérivable ainsi que l'application $u \to J(a_{ij})$. Donc si a_o est optimale on en déduit que les dérivées de J dans les directions d'accroissement admissible sont ≥ 0 .

Un calcul simple montre que si Δa_{ij} est une direction d'accroissement admissible (c'est-à-dire $\exists\, t_n > 0\, t_n \to 0$ a^n_{ij} admissible tels que $\frac{a^n_{ij} - a_o \delta_{ij}}{t_n} \to \Delta a_{ij}$)

On a

(14.) $$\Sigma_{ij} \int_\Omega \Delta a_{ij} \frac{\partial u_o}{\partial x_j} \frac{\partial w_o}{\partial x_i}\,dx \leq 0$$

4°) Grâce à 9 on connait certaines directions d'accroissement admissible :

Là où $a_o(x) = \alpha$, Δa_{ij} peut être toute matrice ayant une valeur propre égale à $\frac{\alpha}{\beta}$, les autres étant égales à + 1 .

Là où $a_o(x) = \beta$, Δa_{ij} aura une valeur propre égale à $-\frac{\beta}{\alpha}$ les autres étant -1 .

. Dans le cas où $a_o(x) = \alpha$ et e est un vecteur propre unitaire correspondant à la valeur propre $\frac{\alpha}{\beta}$, la condition 14 devient

$$(\text{grad } u_o , \text{grad } w_o) + (\frac{\alpha}{\beta} - 1)\,(\text{grad } u_o , e)\,(\text{grad } w_o , e) \leq 0$$

On choisit e tel que cette quantité soit maximum (si les vecteurs grad $u_o(x)$ et grad $w_o(x)$ ne sont pas nuls on trouve la bissectrice extérieure de l'angle qu'il forment) et on trouve la condition

(15.) là où $a_o(x) = \alpha$ ou bien $|\text{grad } u_o(x)|\ |\text{grad } w_o(x)| = 0$
ou bien l'angle θ de ces vecteurs vérifie

$$\cos\theta \leqslant -\frac{\beta-\alpha}{\beta+\alpha}$$

De même

(16.) là où $a_o(x) = \beta$ ou bien $|\text{grad } u_o(x)|\ |\text{grad } w_o(x)| = 0$
ou bien l'angle θ de ces vecteurs vérifie

$$\cos\theta \geqslant \frac{\beta-\alpha}{\beta+\alpha}$$

5°) Dans le cas $N = 2$ on connait l'ensemble des a_{ij} où l'on trouve obligatoirement la solution du problème de minimisation. On peut donc donner des conditions nécessaires pour que $(a^o_{ij}(x)u_o)$ soit optimal.

Notons w_o la solution de

(12.) bis

$$-\sum_{ij} \frac{\partial}{\partial x_i} a_{ij} \frac{\partial w_o}{\partial x_j} = \frac{\partial F}{\partial u}(x,u_o)$$

$$w_o = 0 \quad \text{sur} \quad \partial\Omega$$

On trouve alors la même condition d'optimalité 14 .

. En plus des cas où $a^o_{ij}(x) = \alpha\delta_{ij}$, qui donne la condition 15 et $a^o_{ij}(x) = \beta\delta_{ij}$ qui donne la condition 16 on doit examiner le cas où $\alpha < \frac{\alpha\beta}{\alpha+\beta-\lambda^o_2} = \lambda^o_1 < \lambda^o_2 < \beta$ et le cas où $\frac{\alpha\beta}{\alpha+\beta-\lambda^o_2} < \lambda^o_1$

. Dans le deuxième cas toutes les directions d'accroissement sont admissibles et on obtient

(17.)

$$\text{là où} \quad \alpha < \frac{\alpha\beta}{\alpha+\beta-\lambda^o_2(x)} < \lambda^o_1(x) < \lambda^o_2(x) < \beta \quad \text{on a}$$

$$|\text{grad } u_o(x)|\ |\text{grad } w_o(x)| = 0$$

. Dans le premier cas, soit (e_1 , e_2) une base orthonormée de vecteurs propres correspondant aux valeurs propres λ^o_1 , λ^o_2 .

On vérifie que les matrices $\pm \begin{pmatrix} 0 & 1 \\ 1 & 0 \end{pmatrix}$ sont des directions d'accroissement admissible de même que la matrice

$$\begin{pmatrix} \frac{\alpha\beta}{(\alpha+\beta-\lambda^o_2)^2} & 0 \\ 0 & 1 \end{pmatrix}$$

En portant ces trois matrices dans la condition 14 on trouve finalement là où $\lambda < \frac{\alpha\beta}{\alpha+\beta-\lambda_2^o} = \lambda_1^o < \lambda_2^o < \beta$ on a :

ou bien $|\text{grad } u_o(x)|\ |\text{grad } w_o(x)| = 0$

(18.) ou bien la bissectrice intérieure de l'angle formé par les vecteurs grad u_o et grad w_o est vecteur propre pour λ_2^o et l'angle ϑ de ces vecteurs vérifie $\cos \vartheta = \frac{(\lambda_1^o)^2 - \alpha\beta}{(\lambda_1^o)^2 + \alpha\beta}$.

B I B L I O G R A P H I E

[1] D. CHENAIS CRAS. t. 276 - pp. 547-550.

[2] F. MURAT CRAS. t. 273 - pp. 708-711.
CRAS. t. 274 - pp. 395-398.

[3] S. SPAGNOLO Ann. Scu. Norm. Sup. Pisa 21 (1967) 657-699.
" 22 (1968) 571-597.
E. DE GIOGI - SPAGNOLO " 23 (1969) 657-673.

[4] F. MURAT - L. TARTAR A paraître.

ETUDE DE LA METHODE DE BOUCLE OUVERTE ADAPTEE POUR LE CONTROLE DE SYSTEMES DISTRIBUES

J.P. YVON
IRIA - LABORIA
78 - ROCQUENCOURT (FRANCE)

RESUME.

Le principe de la méthode dite de la "boucle ouverte adaptée" consiste à recalculer un contrôle en boucle ouverte en tenant compte des mesures faites sur le système réel à intervalles de temps fixes. L'objet de cet article est d'étudier cette méthode appliquée au contrôle d'équations aux dérivées partielles paraboliques, et, plus précisément, d'étudier le comportement du contrôle obtenu lorsque le pas d'échantillonnage tend vers zéro. Pour les démonstrations des résultats donnés ci-dessous, on pourra se reporter à A.BAMBERGER-C.SAGUEZ-J.P. YVON [1].

I. INTRODUCTION.

Nous allons commencer par exposer le problème de contrôle linéaire-quadratique en utilisant la formulation générale donnée par LIONS [1].

Soient V et H deux espaces de Hilbert, V dense dans H avec injection continue. En identifiant H à son dual, on a :

(1.1) $$V \subset H \subset V'$$ chaque espace étant dense dans le suivant avec injection continue.

Dans toute la suite, on notera :

$\|\cdot\|$ $((.,.))$ la norme et le produit scalaire dans V
$|\cdot|$ $(.,.)$ la norme et le produit scalaire dans H
$\|\ \|_*$ la norme de V'
$(.,.)$ la dualité entre V' et V .

Par ailleurs, on donne un opérateur $A \in \mathcal{L}(V;V')$ vérifiant :

(1.2) $$(A\varphi,\varphi) \geqslant \alpha\|\varphi\|^2 \qquad \forall\ \varphi \in V$$

On considère de plus deux espaces de Hilbert E et F et :

(1.3) $B \in \mathcal{L}(E;V')$

(1.4) $C \in \mathcal{L}(V;F)$

(1.5) $N \in \mathcal{L}(E;E)$ vérifiant $(Ne,e)_E \geqslant \nu \| e\|_E^2 \quad \forall e \in E \; (\nu>0)$

Soit $T>0$ donné. L'état du système est donné par :

$$(1.6) \quad \begin{cases} \dfrac{dy}{dt} + Ay = f + Bv & \text{sur }]0,T[\\ y(o) = y_o \end{cases}$$

où

$$(1.7) \quad \begin{cases} v \text{ donné dans } L^2(0,T;E) \\ f \text{ donné dans } L^2(0,T;V') \\ y_o \text{ donné dans } H . \end{cases}$$

Rappelons le :

THEOREME 1.1 (LIONS [1]) : Sous les hypothèses (1.1) ... (1.7), il existe une unique solution $y(v)$ à (1.6) vérifiant :

$$(1.8) \quad \begin{cases} y \in L^2(0,T;V) \cap L^\infty(0,T;H) \\ \dfrac{dy}{dt} \in L^2(0,T;V') \end{cases}$$ ∎

Nous introduisons maintenant la fonction coût :

$$(1.9) \quad J(v) = \int_0^T |C\, y(t,v)|_F^2 dt + \int_0^T (Nv(t),v(t))_E dt.$$

On a alors le :

THEOREME 1.2 : Soit E_{ad} un convexe fermé de E, alors il existe un unique contrôle u tel que :

$$(1.10) \quad \begin{cases} u(t) \in L^2(0,T;E_{ad}) \\ J(u) \leqslant J(v) \qquad \forall v \leqslant L^2(0,T,E_{ad}) \end{cases}$$

Tel que le problème est formulé (situation complètement déterministe), il existe un grand nombre d'algorithmes de résolution utilisant, soit la programmation mathématique (cf. par exemple : LEROY [1] , YVON [1] etc..), soit la résolution des équations de Riccati [(1)] (solution feedback donc, cf. NEDELEC [1], TARTAR [1] , LIONS [1] , etc..).

Le problème que nous nous posons est de savoir si, ne connaissant pas f à l'avance, on peut utiliser un algorithme de type boucle ouverte et une prédiction de f pour calculer un contrôle qui, évidemment ne sera plus optimal. Une réponse (partielle) à cette question est l'algorithme de boucle ouverte adaptée que nous allons étudier maintenant.

2. LE PRINCIPE GENERAL DE LA METHODE.

Dans toute la suite T est fixé et N est un entier positif destiné à tendre vers l'infini.

Posons :

$$\tau = \frac{T}{N} \tag{2.1}$$

et considérons un découpage de $[0,T]$ en N intervalles :

$$[0,\tau[\,, [\tau,2\tau[\,, \quad \ldots \;, [(N-1)\tau, T] \quad .$$

La formulation générale de la méthode consiste à construire le contrôle successivement sur chaque intervalle $[k\tau, (k+1)\tau[$ en tenant compte des observations sur le système jusqu'à l'instant $t = k\tau$.

Pour simplifier, nous supposerons que "f" peut être prédit comme ayant la valeur zéro (ce qui est toujours possible, à une translation près).

Pour simplifier, on peut schématiser la situation de la façon suivante :

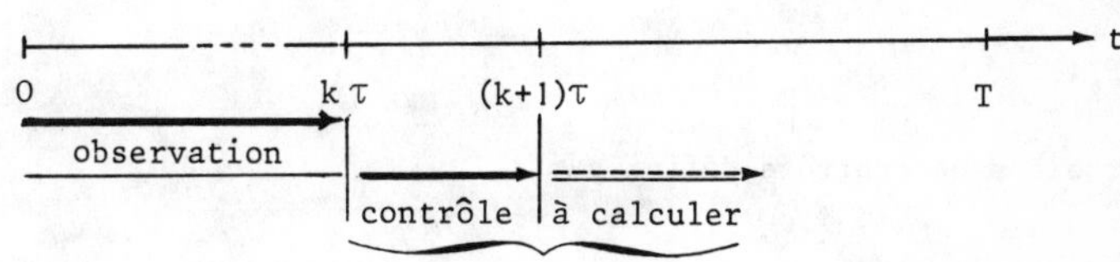

La formulation précise de la méthode est la suivante :

On considère le Problème de contrôle initial mais avec $f = 0$.

---

(1) Dans le cas $E_{ad} \equiv E$ essentiellement.

PROBLEME P_o.

C'est le problème de contrôle défini par :

(2.2)
$$\begin{cases} \dfrac{d\varphi}{dt} + A\varphi = Bv & \text{sur }]0,T[\\ \varphi(o) = \varphi_o \end{cases}$$

(2.3)
$$J_o(v) = \int_0^T | C\,\varphi(v)|_F^2 \, dt + \int_0^T (Nv,v)_E dt$$

(2.4)
$$\underset{v\in L^2(0,T;E_{ad})}{\text{Min}} \quad J_o(v)$$

Ce problème, d'après les théorèmes 1.1 et 1.2, admet une solution unique u_o^N à laquelle correspond l'état du modèle (2.2) :

$$\hat{y}_o^N = \varphi(u_o^N) \quad \text{solution de (2.2).}$$ ∎

Supposons donc que nous ayons construit un contrôle $u^N(t)$ jusqu'au temps $t = k\tau$. Nous allons prolonger $u^N(t)$ sur $[k\tau\ ,\ (k+1)\tau[$ en introduisant un nouveau problème de contrôle. On aura ainsi construit par récurrence le contrôle sur $(0,T)$.

Le système réel a été commandé pendant la période $[o,k\tau[$ par le contrôle $u^N(t)$, donc l'<u>état réel du système</u> $y^N(t)$ est solution de :

(2.5)
$$\begin{cases} \dfrac{dy^N}{dt} + Ay^N = f + B\,u^N & \text{sur }]o,\ k\tau\ [\\ y^N(o) = y_o\ . \end{cases}$$

Supposons alors que l'on puisse observer $y^N(k\tau)$, alors nous pouvons introduire le

PROBLEME P_k.

C'est le problème de contrôle défini par :

(2.6)
$$\begin{cases} \dfrac{d\varphi}{dt} + A\,\varphi = Bv & \text{sur }]\,k\tau,\ T\ [\\ \varphi(k\tau) = y^N(k\tau). \end{cases}$$

(2.7)
$$J_k(v) = \int_{k\tau}^T |c\,\varphi(v)|_F^2 \, dt + \int_{k\tau}^T (Nv,v)_E dt$$

(2.8) $$\underset{v \in L^2(k\tau,T;E_{ad})}{\text{Min}} J_k(v)$$

Le Problème P_k admet une solution unique u_k^N à laquelle correspond l'état

(2.9) $$\hat{y}_k^N = \varphi(u_k^N) \quad \text{sur} \quad]k\tau, T[$$

On pose donc

$$u^N(t) = u_k^N(t) \quad \text{pour} \quad t \in [kT,(k+1)\tau[\qquad \blacksquare$$

En résumé, on a défini sur $[0,T[$ un contrôle $u^N(t)$ calculé par morceaux en résolvant les problèmes $P_o, P_1, \ldots, P_{N-1}$; on a donc :

(2.10) $$u^N(t) = u_k^N(t) \quad \text{si} \quad t \in [k\tau, (k+1)\tau[$$

ceci pour $k = 0,1,\ldots,N-1$.

Il lui correspond l'état du système réel y^N solution de :

(2.11) $$\begin{cases} \dfrac{dy^N}{dt} + Ay^N = f + Bu^N \\ y^N(o) = y_o. \end{cases}$$

Enfin, on dispose de la solution du modèle de simulation, soit $\hat{y}^N(t)$ donné par :

(2.12) $$\hat{y}^N(t) = \hat{y}_k^N(t) \quad \text{si} \quad t \in [k\tau, (k+1)\tau[$$

et $\hat{y}_k^N$ défini par (2.9). En clair, $\hat{y}^N(t)$ satisfait par morceaux aux équations différentielles :

(2.13) $$\begin{cases} \dfrac{d\hat{y}^N}{dt}(t) + A\hat{y}^N(t) = Bu^N(t) \quad \text{sur} \quad]k\tau, (k+1)\tau[\\ \hat{y}^N(k\tau) = y^N(k\tau) \end{cases}$$

pour $k = 0,1,\ldots,N-1$. $\blacksquare$

3. CONVERGENCE DE L'ALGORITHME.

De façon à pouvoir donner un résultat général de convergence lorsque $N \to \infty$, nous allons nous placer dans le cas où

(3.1) $$B \in \mathcal{L}(E;H) \quad ,$$

c'est-à-dire lorsque le <u>contrôle est distribué</u>.

Dans le cas général (1.3), en particulier lorsque le contrôle est frontière, il faut étudier le problème directement.

Le point fondamental du raisonnement est d'obtenir une estimation sur u^N dans $L^\infty(0,T;E)$. Ce résultat est fourni par la :

PROPOSITION 3.1.

Considérons le problème de contrôle (hypothèses et notations du N° 1) :

$$\left\{\begin{aligned} &\frac{d\varphi}{dt} + A\varphi = Bv \qquad \text{sur }]0,T[\qquad B \in \mathcal{L}(E;H). \\ &\varphi(o) = h \end{aligned}\right. \tag{3.2}$$

$$J^h(v) = \int_0^T \|c\,\varphi(v;h)\|_F^2 dt + \int_0^T (Nu,v)_E dt \tag{3.3}$$

dont le contrôle optimal u_h est solution de :

$$J^h(u_h) \leqslant J^h(v) \qquad \forall\, v \in \mathcal{U}_{ad}.$$

Alors on a les estimations suivantes :

$$J^h(u_h) \leqslant C_o[1 + |h|^2] \tag{3.4}$$

et

$$\|u_h(t)\|_E^2 \leqslant C_1[1 + |h|^2] \tag{3.5}$$

où C_o et C_1 ne dépendent que de T. ∎

Ce résultat est utilisé de façon plus ou moins directe dans la théorie du découplage telle que celle exposée par LIONS [1].

On a alors la :

PROPOSITION 3.2.

Sous les hypothèses des Numéros 1 et 2, l'hypothèse (3.1), on a :

$$\left\{\begin{aligned} &y^N \quad \text{borné dans} \quad L^\infty(0,T;H) \cap L^2(0,T;V) \\ &u^N \quad \text{borné dans} \quad L^\infty(0,T;E) \end{aligned}\right. \tag{3.6}$$

lorsque $N \to \infty$.

On déduit alors de la Proposition précédente le :

THEOREME 3.1. Sous les hypothèses de la proposition précédente, on peut, lorsque $N \to \infty$, extraire des sous-suites telles que :

(3.7) $$u^{N'} \to \bar{u} \quad \text{dans} \quad L^{\infty}(0,T;E) \text{ faible } * .$$

(3.8) $$y^{N'} \to y(\bar{u}) \quad \text{dans} \quad L^{\infty}(0,T;H) \text{ faible } * \text{ et } L^{2}(0,T;V) \text{ faible.}$$

(3.9) $$\hat{y}^{N'} \to y(\bar{u}) \quad \text{dans} \quad L^{\infty}(0,T;E) \text{ faible } * .$$

lorsque $N' \to \infty$.

De plus dans le cas où $E_{ad} = E$ (pas de contraintes sur le contrôle) on a le résultat plus précis suivant :

THEOREME 3.2. Dans le cas sans contrainte sur le contrôle, le contrôle $\bar{u}$ donné par (3.7) est unique et caractérisé par :

(3.10) $$\bar{u}(t) = -N^{-1} \Lambda_E^{-1} B^* P \bar{y}(t)$$

où P est l'opérateur de Riccati solution de :

(3.11) $$\begin{cases} -\dfrac{dP}{dt} + AP + PA^* + PD_1P = D_2 \\ P(T) = 0 \end{cases}$$

avec

Λ_E isomorphisme canonique de E sur E'

Λ_F isomorphisme canonique de F sur F'

$D_1 = BN^{-1} \Lambda_E^{-1} B^*$

$D_2 = C^* \Lambda_F C$. ■

Remarque 3.1.

Il faut bien remarquer que $\bar{u}$ n'est pas en général le contrôle optimal u du problème (1.8)(1.9)(1.10). De façon à bien mettre en évidence cette différence, écrivons les lois de feedback obtenues dans les deux cas.

Le contrôle optimal u est donné par :

(3.12) $$u(t) = -N^{-1}\Lambda_E^{-1} B^* [Py(t) + r(t)]$$

où P est également solution de (3.11) et r est solution de :

(3.13) $$\begin{cases} -\dfrac{dr}{dt} + Ar + PD_1 r = Pf. \\ r(T) = 0 \end{cases}$$

Si nous réunissons les résultats obtenus concernant les deux trajectoires y (optimale) et $\bar{y}$ (adaptée), on a :

(3.14) $$\begin{cases} \dfrac{dy}{dt} + Ay + D_1 Py = f\ D_1 r. \\ y(o) = y_o \end{cases}$$ où r est solution de (3.13)

et

(3.15) $$\begin{cases} \dfrac{d\bar{y}}{dt} + A\bar{y} + D_1 P\bar{y} = f \\ \bar{y}(o) = y_o \end{cases}$$

cette dernière équation provenant de (3.10) et du passage à la limite sur (2.11).

Donc, si l'on pose : $z = \bar{y} - y$, on a :

$$\begin{cases} \dfrac{dz}{dt} + Az + D_1 Pz = D_1 r \\ z(o) = 0. \end{cases}$$

ce qui permet d'obtenir une "estimation de l'erreur" sur l'état optimal. ∎

Remarque 3.2. :
Dans le cas avec contrainte, on peut caractériser également le contrôle $\bar{u}$ obtenu par une loi de feedback. Pour s avec $0 \leqslant s < T$, on introduit le problème de contrôle sur $]s,T[$ ainsi défini :

(3.16) $$\begin{cases} \dfrac{d\varphi}{dt} + A\varphi = Bv \quad \text{sur} \quad]s,T[\\ \varphi(s) = h \qquad h \in H , \end{cases}$$

équation dont la solution est notée $\varphi(v;s,h)$.

(3.17) $$J_h^s(v) = \int_s^T \|C\varphi(v;s,h)\|_F^2\, dt + \int_s^T (Nv,v)_E\, dt.$$

Le problème (3.16)(3.17) admet un contrôle optimal $u(t;s,h) \in \mathcal{U}_{ad}$ solution de

$$J_s^h(u(.;s,h)) \leqslant J_s^h(v) \qquad \forall\, v \in \mathcal{U}_{ad}.$$

Le contrôle $\bar{u}(t)$ donné par le Théorème 3.1 consiste à prendre comme loi de feedback :

$$\bar{u}(t) = u(t;t,\bar{y}(t)). \qquad \blacksquare$$

4. VARIANTES POSSIBLES DE LA METHODE.

L'inconvénient fondamental de la méthode qui vient d'être exposée réside dans le fait qu'elle n'est pas implémentable "en ligne". En effet, on n'est pas capable de résoudre en temps réel les Problèmes P_k (n° 2). Cette difficulté peut être surmontée de plusieurs façons.

Le principe général des variantes possibles peut être illustré par le schéma ci-dessous :

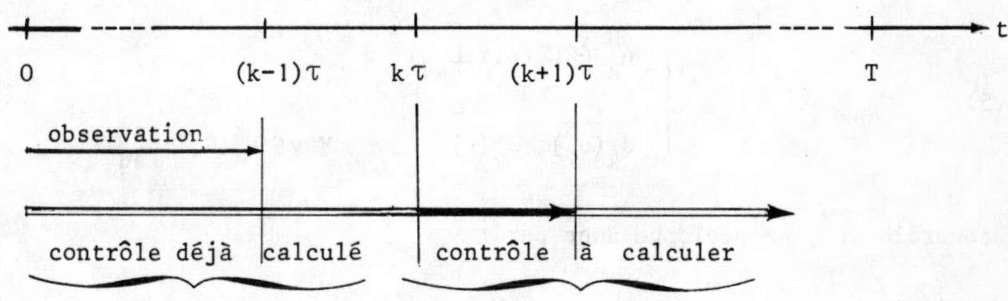

On suppose toujours que le contrôle peut être calculé par récurrence sur $k=0,1,2,\ldots$ $\ldots,(N-1)$, mais cette fois-ci, on utilise les observations sur $[0,(k-1)\tau[$ pour calculer le contrôle à venir sur l'intervalle $[k\tau,(k+1)\tau[$. Donc, si l'on choisit τ assez grand, on pourra effectuer en temps réel le calcul.

De manière précise, les hypothèses sont les suivantes :

- on dispose d'un contrôle $u^N(t)$ sur $[0, k\tau[\ (k \geqslant 1)$
- on connaît la valeur de l'état du système au temps $t = (k-1)\tau$:

$$\text{(4.1)} \qquad \begin{cases} \dfrac{dy^N}{dt} + Ay^N = f + Bu^N \quad \text{sur}^{(1)}\ ]0,(k-1)\tau[\\ y^N(o) = y_o. \end{cases}$$

- ce qui fournit :

$$\text{(4.2)} \qquad y^N((k-1)\tau)$$

Pour calculer la valeur ultérieure du contrôle u^N pour $t \geqslant k\tau$, on introduit le :

---

(1) Evidemment si $k=1$ l'équation (4.1) se réduit à $y^N((k-1)\tau)=y^N(o) = y_o$!

PROBLEME E_k .

C'est un problème de contrôle sur $[k\tau, T]$ dont l'état est donné par :

$$(4.3)\qquad \begin{cases} \dfrac{d\varphi}{dt} + A\varphi = Bv & \text{sur }]k\tau, T[\\ \varphi(k\tau) = \varphi_k . \end{cases}$$

et la fonction économique par :

$$(4.4)\qquad J_k(v) = \int_{k\tau}^{T} \| C\varphi(t;v)\|_F^2 \, dt + \int_{k\tau}^{T} (Nv,v)_E dt$$

où φ_k est un élément de H à choisir.

Ce problème de contrôle possède une solution u_k^N telle que :

$$(4.5)\qquad \begin{cases} u_k^N \in L^2(0,T;E_{ad}) \\ J_k(u_k^N) \leq J_k(v) & \forall\, v \in L^2(0,T;E_{ad}). \end{cases}$$

Le contrôle u^N se prolonge donc par :

$$(4.6)\qquad u^N(t) = u_k^N(t) \qquad \text{pour } t \in [k\tau, (k+1)\tau[\qquad \blacksquare$$

Remarque 4.1.
Pour pouvoir appliquer la méthode précédente, il faut naturellement avant le départ du processus, donner une valeur du contrôle sur $[0, \tau[$. Ceci peut être fait hors ligne en résolvant par exemple le Problème P_o du N° 2. Toute autre valeur arbitraire de la commande convient. ■

Remarque 4.2.
La fonction φ_k figurant dans (4.3) représente une estimation de l'état réel du système à l'instant $t = k$. C'est sur le choix de φ_k que se différencient les algorithmes possibles. ■

Choix de la fonction φ_k.

Première possibilité : on prend

$$(4.7)\qquad \varphi_k = y^N((k-1)\tau).$$

Il ne s'agit là d'une estimation grossière car on estime que l'état du système est resté le même sur l'intervalle $[(k-1)\tau, k\tau]$.

Seconde possibilité : on prend :

(4.8) $$\varphi_k = \eta_k^N(k\tau)$$

où η_k^N est une approximation de y^N sur $](k-1)\tau,k\tau[$,solution de :

(4.9) $$\begin{cases} \dfrac{d\eta_k^N}{dt} + A\,\eta_k^N = Bu^N & \text{sur }](k-1)\tau,k\tau[\\ \eta_k^N((k-1)\tau) = y^N((k-1)\tau) \end{cases}$$ ■

Comme au Numéro 2, on dispose ainsi de trois fonctions caractérisant l'algorithme : y^N, u^N et $\hat{y}^N$, cette dernière représentant l'état optimal du modèle et étant solution par morceaux des équations :

(4.10) $$\begin{cases} \dfrac{d\hat{y}^N}{dt} + A\,\hat{y}^N = Bu^N & \text{sur }]k\tau,(k+1)\tau[\\ \hat{y}^N(k\tau) = \varphi_k . \end{cases}$$

où φ_k est donné par (4.7) ou bien par (4.8).
Les résultats du N° 3 restent valables.

THEOREME 4.1. Sous l'hypothèse (3.1) et avec les notations précédentes, on a :

$$u^N \to \bar{u} \quad \text{dans } L^\infty(0,T;E) \text{ faible } *$$
$$y^N \to \bar{y} = y(\bar{u}) \quad \text{dans } L^\infty(0,T;H) \text{ faible } * \text{ et } L^2(0,T;V) \text{ faible}$$
$$\hat{y}^N \to \bar{y} \quad \text{dans } L^\infty(0,T;E) \text{ faible } *$$

lorsque $N \to \infty$. Le contrôle $\bar{u}$ et l'état correspondant $\bar{y}$ étant toujours uniques caractérisés au N° 3 (Théorème 3.2 et Remarque 3.2). ■

5. CONTROLE FRONTIERE.

La présentation générale de l'algorithme est la même (cf. Numéros 2 et 4). Cependant, les résultats du Numéro 3 et le Théorème 4.1 ne sont plus valables. Considérons alors un exemple précis où $B \in \mathcal{L}(E;V')$: contrôle frontière.

Soit Ω un ouvert de $\mathbb{R}^n$ de frontière régulière Γ, ; on note :

$$Q = \Omega \times]0,T[\quad , \quad \Sigma = \Gamma \times]0,T[.$$

On pose : $H = L^2(\Omega)$ $V = H^1(\Omega)$. Soit A un opérateur différentiel elliptique, l'état du système est donné par :

$$
(5.1) \quad \begin{cases} \dfrac{\partial y}{\partial t} + Ay = f & \text{dans} \quad Q \\[2mm] \dfrac{\partial y}{\partial \nu_A} = v & \text{sur} \quad \Sigma \ \dfrac{\partial}{\partial \nu_A} \text{ dérivée conormale associée à A.} \\[2mm] y(x,o) = y_o \end{cases}
$$

où f et y_o sont donnés dans $L^2(Q)$ et $L^2(\Omega)$ respectivement.
Ici $E = L^2(\Gamma)$ de sorte que $v \in L^2(0,T;L^2(\Gamma)) \equiv L^2(\Sigma)$.
Supposons que l'observation soit distribuée :

$$
(5.2) \qquad J(v) = \int_Q |y(x,t;v)|^2 \, dudt + \nu \int_\Sigma |v(\sigma)|^2 \, d\sigma
$$

alors on peut appliquer au problème

$$
(5.3) \qquad \operatorname*{Inf}_{v \in L^2(0,T;E_{ad})} J(v)
$$

les algorithmes des Numéros 2 et 4. Mutatis mutandis, on dispose de $u^N \in L^2(\Sigma)$, $y^N \in L^2(0,T;V) \cap L^\infty(0,T;H)$ et $\hat{y}^N$ qui est solution par morceaux de :

$$
(5.4) \quad \begin{cases} \dfrac{\partial \hat{y}^N}{\partial t} + A\hat{y}^N = 0 & \text{sur} \quad \Omega \times]k\tau, (k+1)\tau[\,, \\[2mm] \dfrac{\partial \hat{y}^N}{\partial \nu_A} = u^N & \text{sur} \quad \Gamma \times]k\tau, (k+1)\tau[\,, \\[2mm] \hat{y}^N(k\tau) = y^N(k\tau) & \text{dans} \quad \Omega \,, \end{cases}
$$

lorsque l'on prend l'algorithme exposé au Numéro 2.
Bien que les méthodes de démonstration soient différentes, les résultats des Numéros 3 et 4 demeurent vrais ; de manière précise, on a le :

<u>THEOREME</u> 5.1. <u>Sous les hypothèses précédentes, on a</u> :

$$
\begin{array}{llll}
u^N \to \bar{u} & \underline{\text{dans}} & L^\infty(0,T;L^2(\Gamma)) & \text{faible } * \\
y^N \to \bar{y} = y(\bar{u}) & \underline{\text{dans}} & L^\infty(0,T;H) \cap L^2(0,T;V) & \text{faible} \\
\hat{y}^N \to \bar{y} & \underline{\text{dans}} & L^\infty(0,T;H) & \text{faible } *
\end{array}
$$

<u>lorsque</u> $N \to \infty$. <u>Le contrôle</u> $\bar{u}$ <u>étant unique et pouvant être caractérisé de manière analogue au Théorème 3.1.</u> ∎

6. REMARQUES FINALES.

6.1. Problèmes non linéaires.

Lorsque l'équation d'état est non linéaire, on peut appliquer encore le principe des méthodes qui viennent d'être exposées. Une étude générale de la convergence n'est cependant pas possible car on ne peut pas aborder systématiquement les équations non linéaires par une théorie globale. Cependant des résultats analogues aux Théorèmes 3.1 et 4.1 peuvent être obtenus sur des exemples particuliers (cf. A. BAMBERGER-C. SAGUEZ-J.P. YVON [1]).

6.2. Problèmes stochastiques.

Ce type de méthode a été développé surtout pour le cas stochastique. Ici, nous nous sommes placés dans une situation déterministe, mais il est évident que la fonction f intervenant dans l'équation d'état du système devrait être considérée comme un bruit. Naturellement, les solutions "open-loop" que nous avons considérées jusqu'ici ne conviendront plus. On est amené à prendre une solution feedback fonction de l'observation. En utilisant les méthodes stochastiques développées par A. BENSOUSSAN [1] [2] , on peut étudier le problème avec f de type bruit blanc et développer le même type d'algorithme : cf. J.P. YVON [2] .

REFERENCES

A. BAMBERGER-C. SAGUEZ-J.P.YVON [1] Quelques algorithmes de contrôle utilisables en temps réel. A paraître.

A. BENSOUSSAN [1] Filtrage optimal des systèmes linéaires. DUNOD Paris (1971).

[2] Contrôle optimal stochastique de systèmes paraboliques.Rendiconti di Matematica (1-2) Vol. 2, Série VI (1969).

D. LEROY [1] Thèse de 3ème cycle, Paris (1972).

J.L. LIONS [1] Contrôle optimal des systèmes distribués. DUNOD, Paris (1968).

[2] Quelques méthodes de résolution de Problèmes non linéaires. DUNOD, Paris (1974).

J.C. NEDELEC [1] Thèse d'Etat, Paris (1971).

L. TARTAR [1] Note C.R. Ac. Sc.

J.P. YVON [1] Thèse d'Etat, Paris (1973).

[2] Une méthode de contrôle stochastique de systèmes distribués. A paraître.

-- § --

ESTIMATION DES PERMEABILITES RELATIVES ET DE LA PRESSION CAPILLAIRE DANS UN ECOULEMENT DIPHASIQUE

G. CHAVENT
Université de PARIS IX
IRIA-LABORIA

P. LEMONNIER
IRIA-LABORIA - IFP

INTRODUCTION

Etant donnés deux fluides non miscibles (eau et huile par exemple), on s'intéresse au déplacement de l'un des fluides par l'autre au sein d'un milieu poreux. Les applications pratiques de ce phénomène sont nombreuses dans l'industrie pétrolière (injection d'eau dans les gisements pour récupérer l'huile).

Par suite de l'introduction des concepts de perméabilités relatives et de pressions capillaires, la séparation entre les zones occupées par chacun des fluides n'est pas brutale ; on passe progressivement d'une teneur maxima en eau à une teneur minima : c'est un problème de frontière libre "floue".

Les perméabilités relatives et les pressions capillaires sont des fonctions de la saturation difficiles à déterminer par une expérimentation directe. Il était donc intéressant de pouvoir déterminer ces fonctions indirectement à partir de mesures de pressions et de débits faites au cours d'une expérience de déplacement d'un fluide par l'autre.

Des recherches ont été entreprises à l'IRIA-LABORIA sur ce thème, pour le compte de l'Institut Français de Pétrole, dont nous présentons ici les premiers résultats. Les équations utilisées ici sont celles provenant directement de l'analyse physique du phénomène, telles qu'elles sont utilisées par les ingénieurs pétroliers [1][2]. En l'absence d'une formalisation plus adaptée du problème, nous avons choisi ces équations, pour lesquelles un programme de résolution numérique existait déjà à l'IFP [3]. D'autre part, ces équations, dégénérées, assez complexes, font intervenir un grand nombre de variables dépendantes, et il n'a pas été possible d'en faire une étude mathématique rigoureuse (pas de théorème d'existence ni d'unicité).

La méthode utilisée pour l'ajustement des perméabilités relatives et de la pression capillaire consiste à utiliser (nécéssairement formellement !) la théorie du contrôle pour minimiser un critère d'erreur (cf. [4] à [7]), les non-linéarités étant déterminées "point par point" à l'aide de la technique proposée dans [8].

Bien qu'entièrement formels, les résultats obtenus ici nous paraissent intéressants pour deux raisons :

- ils montrent la possibilité de l'utilisation, même formelle, de la théorie du contrôle dans des problèmes concrets, mais aussi ses limites (l'ajustement est possible en dimension 1 d'espace, mais difficilement réalisable en dimension 2).
- cette étude a permis aux auteurs de mieux cerner ces équations, et dans un autre article, nous donnerons une nouvelle formulation du même problème, susceptible d'un traitement mathématique rigoureux et permettant d'envisager une résolution numérique en dimension 2 ou 3 : les résultats du présent article serviront alors de point de comparaison.

2. NOTATIONS

1 = indice de l'eau

2 = indice de l'huile

y_1, y_2 = pressions dans l'eau et dans l'huile

S_1, S_2 = saturations en eau et huile

$$o \leqslant S_{i_{min}} \leqslant S_i \leqslant S_{i_{max}} \leqslant 1 \qquad i = 1.2$$

$\Phi(y_2)$ = porosité, dépend de la pression huile

$\rho_1(y_1)$, $\rho_2(y_2)$ = masses volumiques de chaque fluide

$B_1(y_1)$, $B_2(y_2)$ = facteurs de volume de chaque fluide

$$B_i(y_i) = \frac{\rho_i(P_{ref})}{\rho_i(y_i)}$$

où P_{ref} est la pression de référence à laquelle tous les volumes sont calculés.

$\mu_1(y_1)$, $\mu_2(y_2)$ = viscosité de chaque fluide

K(x) = perméabilité intrinsèque du milieu

$K_1(S_1)$, $K_2(S_1)$ = perméabilités relatives de chaque fluide

K(S_1)
$K_2(S_1)$
$K_1(S_1)$
0 S_{min} S_{max} 1 S_1

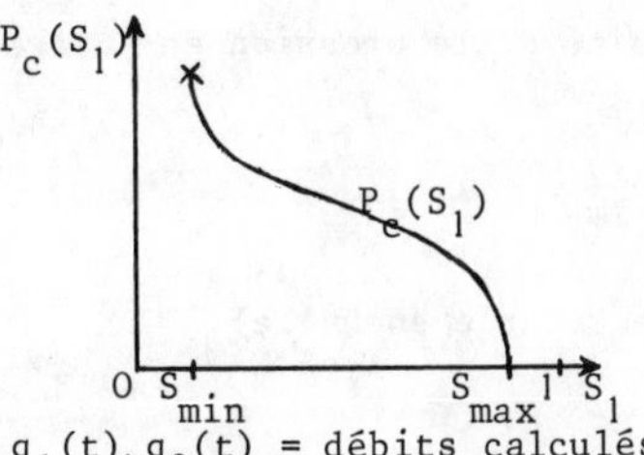

$P_c(S_1)$ = pression capillaire (fonction de la saturation en eau, par exemple)

$q_1(t)$, $q_2(t)$ = débits calculés d'eau et d'huile sortant en x = 1

$z_1(t), z_2(t)$ = débits mesurés d'eau et d'huile sortant en $x = 1$

$p_1(x_1 t)$, $p_2(x,t)$ = état adjoint correspondant à $y_1(x_1 t)$ et $y_2(x_2 t)$

3. LES EQUATIONS D'ETAT

Considérons un domaine unidimensionnel $\Omega =]o,1[$, correspondant à une "carotte" horizontale, initialement imbibée d'huile ($S_1(x,o) = S_{1min}$), dans laquelle on injecte en x=o de l'eau, et dont on récupère de l'huile (avec éventuellement de l'eau) en x=1. Les équations utilisées sont les suivantes :

- dans Ω :

(1) $$\frac{\partial}{\partial t}\left[\Phi(y_2)\frac{S_1}{B_1(y_1)}\right] - \frac{\partial}{\partial x}\left[\frac{K(x)K_1(S_1)}{\mu_1(y_1)B_1(y_1)}\frac{\partial y_1}{\partial x}\right] = o$$

(2) $$\frac{\partial}{\partial t}\left[\Phi(y_2)\frac{S_2}{B_2(y_2)}\right] - \frac{\partial}{\partial x}\left[\frac{K(x)K_2(S_1)}{\mu_2(y_2)B_2(y_2)}\frac{\partial y_2}{\partial x}\right] = o$$

(3) $S_1 + S_2 = 1$

(4) $y_2 - y_1 = P_c(S_1)$

- en x=o : alimentation en eau à la pression $\varphi_o(t)$:

(5) $S_1(o,t) = S_{1max}$ (alimentation en eau)

(6) $y_1(o,t) = \varphi_o(t) - P_c(S_{1max})$ (pression capillaire)

(7) $y_2(o,t) = \varphi_o(t)$ (continuité de la pression en huile)

- en x=1 : débit dans un fluide à la pression $\varphi_1(t)$:

(8) $S_1(1,t) = S_{1max}$ (effet de bord)

(9) $y_1(1,t) = \varphi_1(t) - P_c(S_{1max})$ (pression capillaire)

(10) $y_2(1,t) = \varphi_1(t)$ (continuité de la pression en huile)

- en t=o : condition initiale :

(11) $S_1(x,o) = S_{1min}$ (carotte saturée en huile)

(12) $y_1(x,o) = y_o(x)$

(13) $y_2(x,o) = y_o(x) + P_c(S_{1min})$

Ces équations ont permis une bonne simulation numérique des écoulements polyphasiques (cf. [3]) et représentent un progrès certain par rapport aux modèles antérieurs. D'un point de vue strictement mathématique, ces équations possèdent cependant un certain nombre de faiblesses, qui sont probablement à l'origine d'un certain nombre de difficultés rencontrées lors de l'application de la théorie du contrôle pour l'ajustement des paramètres.
Ces points faibles sont plus précisément :

- On constate que dans l'équation (1)(resp (2)), le terme de diffusion disparait lorsque $S_1 \equiv S_{1min}$(resp $S_2 \equiv S_{2min}$) c'est à dire sur des domaines à priori inconnus ; ceci correspond au fait que, à la saturation en eau (resp. en huile) minimum, la pression y_1 de l'eau (resp. y_2 de l'huile) n'a pas de sens physique.

- Les équations ne sont pas symétriques en les fluides 1 et 2 : la porosité Φ dépendant de la pression, cela oblige à la choisir dépendant arbitrairement d'une des deux pressions (en général la pression y_2 de l'huile).

- Les conditions aux limites du type "pression imposée" ne se formulent pas de façon symétrique en chacun des fluides : à l'extérieur de la carotte, en x=o ou x=1, ne règne qu'une seule pression φ_o ou φ_1 : comment la relier aux <u>deux pressions</u> y_1 et y_2 dans la carotte ? Le choix de la continuité de la pression huile détruit la symétrie en y_1 et y_2.

- Enfin, dans le cas de fluides incompressibles avec une porosité constante, les premiers termes des équations (1) et (2) ne sont plus indépendants, ce qui nécessite de toujours introduire pour la résolution numérique une compressibilité arbitrairement faible.

4. L'OBSERVATION, LA FONCTION COUT

On observe les débits $q_1(t)$ et $q_2(t)$ d'eau et d'huile sortant en x=1 :

$$(14) \quad \begin{cases} q_1(t) = - \dfrac{K(x)K_1(S_1)}{\mu_1(y_1)B_1(y_1)} \left.\dfrac{\partial y_1}{\partial x}\right|_{x=1} \\ q_2(t) = - \dfrac{K(x)K_2(S_2)}{\mu_2(y_2)B_2(y_2)} \left.\dfrac{\partial y_2}{\partial x}\right|_{x=1} \end{cases}$$

<u>Remarque</u> : la condition en x=1 sur les saturations étant $S=S_{1max}$, c'est à dire $K_2=o$, la sortie d'huile ($q_2 \neq o$) en x=1 nécessite que la pression y_2 vérifie $\left.\dfrac{\partial y_2}{\partial x}\right|_{x=1} = -\infty$. ■

La fonction coût sera alors :

$$J(K_1, K_2, P_c) = \int_o^T (q_1(t) - z_1(t))^2 dt + \int_o^T (q_2(t) - z_2(t))^2 dt \tag{15}$$

La minimisation numérique de cette fonction coût sur un ensemble de paramètres admissibles permettra d'ajuster le modèle.

<u>Remarque</u> : au lieu des débits eau et huile, les pétroliers utilisent souvent :

- le débit total $Q(t) = q_1(t) + q_2(t)$
- le rapport eau/huile $WOR(t) = q_1(t)/q_2(t)$

auxquels on peut naturellement associer une fonctionnelle du même type que (15) :

$$J(K_1 K_2, P_c) = \int_o^T \left(\frac{q_1(t)}{q_2(t)} - WOR(t)\right)^2 dt + \int_o^T (q_1(t) + q_2(t) - Q(t))^2 dt \tag{16}$$ ■

5. ETAT ADJOINT, GRADIENT

Les calculs permettant d'obtenir le gradient de J à partir des équations (1) à (13) sont fastidieux. Nous ne donnerons ici que le résultat. Dans les équations ci-dessous, les saturations S_1 et S_2 sont considérées comme fonctions des pressions y_1 et y_2 :

$$\left\{\begin{array}{l} S_1 = P_c^{-1}(y_2 - y_1) \\ S_2 = 1 - P_c^{-1}(y_2 - y_1) \end{array}\right. \tag{17}$$

c'est à dire qu'on a considéré dans le système direct y_1 et y_2 comme variables dépendantes principales, dont p_1 et p_2 sont les variables adjointes.

<u>Le système adjoint</u> (correspondant à la fonction coût (15)) est le suivant :

- <u>Dans Ω</u> :

$$-\Phi \frac{\partial}{\partial y_1}\left(\frac{S_1}{B_1}\right)\frac{\partial p_1}{\partial t} - \frac{\Phi}{B_2}\frac{\partial S_2}{\partial y_1}\frac{\partial p_2}{\partial t} - \frac{K_1}{\mu_1 B_1}\frac{\partial}{\partial x}\left(K\frac{\partial p_1}{\partial x}\right) + \left(\frac{1}{\mu_2 B_2}\frac{\partial K_2}{\partial y_1}\frac{\partial p_2}{\partial x} - \frac{1}{\mu_1 B_1}\frac{\partial K_1}{\partial y_2}\frac{\partial p_1}{\partial x}\right)K\frac{\partial y_2}{\partial x} = o \tag{18}$$

$$-\frac{1}{B_1}\frac{\partial}{\partial y_2}(\Phi S_1)\frac{\partial p_1}{\partial t} - \frac{\partial}{\partial y_2}\left(\Phi\frac{S_2}{B_2}\right)\frac{\partial p_2}{\partial t} - \frac{K_2}{\mu_2 B_2}\frac{\partial}{\partial x}\left(K\frac{\partial p_2}{\partial x}\right) + \left(\frac{1}{\mu_1 B_1}\frac{\partial K_1}{\partial y_2}\frac{\partial p_1}{\partial x} - \frac{1}{\mu_2 B_2}\frac{\partial K_2}{\partial y_1}\frac{\partial p_2}{\partial x}\right)K\frac{\partial y_1}{\partial x} = o \tag{19}$$

- <u>en x=o</u> :

$$K_1 p_1 = K_2 p_2 = o \tag{20}$$

- <u>en x=1</u> :

$$K_1 p_1 = -2K_1(q_1(t) - z_1(t)) \tag{21}$$

$$K_2 p_2 = -2K_2(q_2(t) - z_2(t)) \tag{22}$$

- <u>en t = T</u> :

(23) $p_1(x_1,T) = p_2(x_2,T) = o$

On remarque dans les conditions aux limites la présence d'un facteur K_1 (resp K_2) dans les deux membres, due au fait que le terme de diffusion disparaissant lorsque K_1=o, il n'y a alors pas lieu d'imposer de conditions aux limites.

Le gradient par rapport aux paramètres se déduit de :

$$(24) \qquad \frac{\partial J}{\partial K_1}\delta K_1 = \int_Q \delta K_1(S)\,\frac{K}{\mu_1 B_1}\,\frac{\partial y_1}{\partial x}\frac{\partial p_1}{\partial x}\,dx\,dt$$

$$(25) \qquad \frac{\partial J}{\partial K_2}\delta K_2 = \int_Q \delta K_2(S)\,\frac{K}{\mu_2 B_2}\,\frac{\partial y_2}{\partial x}\frac{\partial p_2}{\partial x}\,dx\,dt$$

$$(26) \qquad \frac{\partial J}{\partial P_c}\delta P_c = -\int_Q \delta P_c(S)\frac{\Phi}{P'_c}\left(\frac{1}{B_1}\frac{\partial p_1}{\partial t} - \frac{1}{B_2}\frac{\partial p_2}{\partial t}\right)dx\,dt - \left[\int_\Omega \delta P_c(S)\frac{\Phi}{P'_c}\left(\frac{p_1}{B_1} - \frac{p_2}{B_2}\right)dx\right]\Big|_{t=o}$$

à l'aide d'un changement de variables (cf. [8]) ramenant les intégrales en x,t apparaissant au 2ème membre de (24) (25) (26) à des intégrales en S.

6. DISCRETISATION

- La discrétisation en espace des équations d'état (1)-(13) et des équations adjointes (18)-(23) a été faite par différences finies. Nous avons utilisé pour la discrétisation en temps un schéma totalement implicite.
- Le système direct est non linéaire. Il a été résolu par l'IFP (cf. [3]) en prenant comme inconnues principales la pression en huile y_2 et la saturation en eau S_1 et en linéarisant. La linéarisation utilisée permet d'employer de grands pas de temps en fin de simulation.
- Le système adjoint est linéaire. Sa résolution numérique présente des difficultés qui résident dans son mauvais conditionnement et dans le choix d'une discrétisation satisfaisante des dérivées premières qui interviennent dans les équations (18)(19). Les considérations physiques qui avaient prévalu dans la discrétisation du système direct ne peuvent être d'aucune aide car il est difficile de donner une signification physique aux variables adjointes.
- La discrétisation du gradient de la fonction coût dépend de la forme sous laquelle on recherche les perméabilités relatives.

Dans le cas d'une recherche sous forme <u>paramétrique</u> quelconque, par exemple :

$$(27) \qquad K_i(S_1) = \sum_{j=1}^{k} \beta_j^i\, W_j(S_1) \qquad\qquad i = 1,2.$$

où les W_j sont des fonctions connues de S_1 et où les β_j sont des paramètres scalaires à déterminer, le gradient par rapport aux β_j s'exprime facilement :

(28) $$\frac{\partial J}{\partial \beta^i_j} = \int_Q W_j(S_1) \frac{K}{\mu_i B_i} \frac{\partial y_i}{\partial x} \frac{\partial p_i}{\partial x} \, dx \, dt \qquad j = 1,k \quad \text{et} \quad i = 1,2.$$

Lorsque les perméabilités relatives sont recherchées sous forme libre, c'est à dire déterminées aux N points de discrétisation S_i de l'intervalle de variation de la saturation en eau S, le calcul numérique du gradient par rapport au $K_1(S_i)$ et $K_2(S_i)$, i=1,N, revêt certaines difficultés.

Nous allons indiquer la manière de procéder, dont on pourra trouver un exposé détaillé dans [8].

Les expressions (24)(25) s'écrivent en toute généralité :

(29) $$\frac{\partial J}{\partial K} . \delta K = \int_Q \delta K(S) \varphi(x,t) dx \, dt$$

On peut faire apparaître une "dérivée partielle fonctionnelle" de J par rapport à K. Pour cela posons pour tout $\zeta \in \mathbb{R}$:

(30) $$Q_\zeta = \{(x,t) \in Q \mid S(x,t) \geqslant \zeta\}$$

(31) $$\gamma(\zeta) = \int_{Q_\zeta} \varphi(x,t) dx \, dt$$

On a alors le résultat suivant (cf [8]) :

La fonction γ est bornée et continue sur $\mathbb{R}$. Elle définit en particulier une distribution sur $\mathbb{R}$.

(32) $$J'(K).\delta K = \langle \frac{\partial \gamma}{\partial \zeta}, \delta K \rangle$$

Divisons R en intervalles de longueur h et posons $S_i = ih$, $i \in \mathbb{Z}$. La formulation (32) est alors équivalente à :

(33) $$J'(K).\delta K = \lim_{h \to 0} \sum_{i=-\infty}^{+\infty} \delta K(S_i)(\gamma(S_{i+1}) - \gamma(S_i))$$

L'analogue discret de (33) permet alors de calculer numériquement le gradient de J(K). On voit que cela nécessite d'intégrer numériquement des fonctions entre les lignes de niveau de la saturation en eau, S(x,t). Nous nous sommes limités au cas de la dimension 1 en espace. Pour la dimension 2 nous avons identifié les perméabilités sous forme paramètrique.

Le système direct est plus ou moins sensible pour certaines valeurs de la saturation en eau à de petites variations des perméabilités relatives. On constate d'autre part dans (24)(25) que les points où le gradient de la pression ou celui de l'état adjoint a de faibles valeurs interviendront peu dans le calcul du gradient de la fonctionnelle J. Cela entraîne dans le cas d'une recherche sous forme libre un ajustement ponctuel non uniforme des perméabilités. L'ajustement se fait plus ou moins rapidement suivant les valeurs des points de discrétisation de l'intervalle de variation de la saturation. On obtient alors des courbes discontinues en dents de scie. Les discontinuités se résorbent difficilement au cours des itérations de l'algorithme d'identification et elles peuvent entraîner un arrêt de convergence. C'est pourquoi nous avons introduit des contraintes de régularité en imposant des bornes à la dérivée seconde des perméabilités relatives, de la forme :

$$(34) \qquad C_1 \leqslant K''(S) \leqslant C_2$$

Si on sait physiquement que la perméabilité est convexe, cette condition s'introduit facilement en prenant $C_1 = o$.
Le problème d'estimation des perméabilités relatives revient alors à un problème de minimisation avec contraintes linéaires qui a été résolu par la méthode de Franck et Wolfe.

7. RESULTATS NUMERIQUES

Dimension 1

Les données numériques correspondent à un des exemples traités dans [3] restreint à la dimension 1. Pour tester l'algorithme d'identification, nous avons fait une simulation avec des valeurs de perméabilités relatives données. Puis nous avons pris les débits en eau et huile obtenus au cours de la simulation comme observation. Nous avons alors cherché à retrouver les valeurs précédentes des perméabilités relatives appelées perméabilités exactes par opposition aux perméabilités identifiées (donc trouvées par l'algorithme d'identification). Les valeurs extrêmes des perméabilités relatives pouvant être mesurées physiquement sont supposées connues.

Les essais numériques présentés ont été effectués avec une discrétisation de 30 points en espace et 70 points en temps.

La figure 1 présente schématiquement l'expérience ainsi que des profils de la saturation en eau à divers instants en temps.

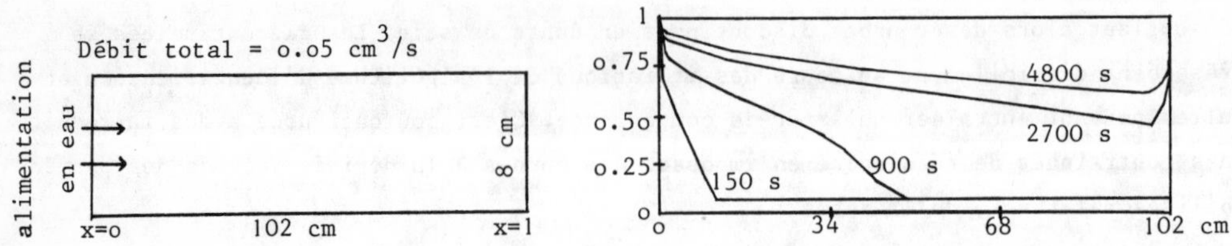

Figure 1 : Schéma de l'expérience et profils de saturation.

Recherche sous forme paramétrique

Les perméabilités relatives sont recherchées sous forme d'un polynôme du second degré. Le fait d'augmenter le degré des polynômes n'a pas amélioré les résultats. Les courbes de perméabilité et de WOR obtenues après 3 itérations de gradient sont présentées sur la figure 2. Le temps de calcul sur CDC 7600 est de 41 secondes. La fonctionnelle J a décru de 3.40 à 0.001.

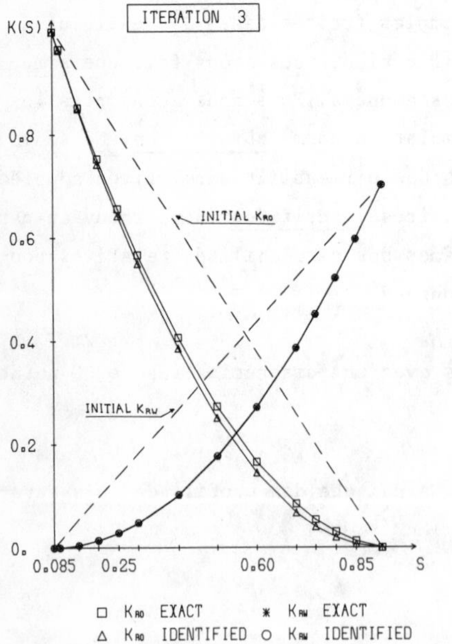

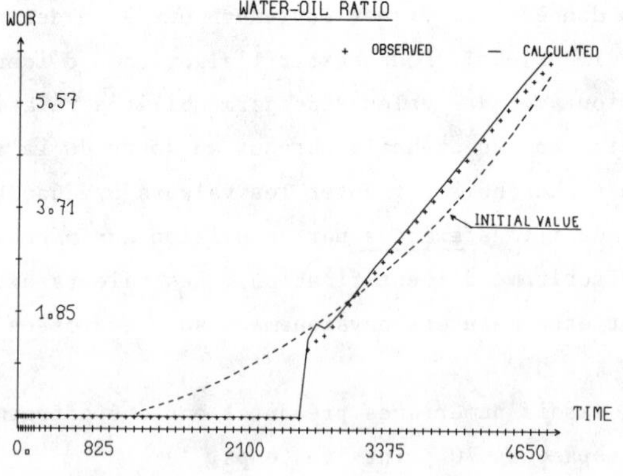

Figure 2 : Recherche sous forme paramétrique

Recherche sous forme libre sans contraintes

Les essais qui suivent ont été réalisés en recherchant les perméabilités relatives sous forme libre, c'est à dire en déterminant leurs valeurs aux points de discrétisation de l'intervalle de variation de la saturation. Les courbes de perméabilités relatives et de WOR obtenues après 6 itérations de gradient sont tracées sur la figure 3. Les discontinuités qui apparaissent dans les courbes de perméabilité sont dues à la plus ou moins grande sensibilité des perméabilités relatives suivant les valeurs de la saturation. Ces discontinuités se résorbent difficilement et provoquent un arrêt de la convergence dans un minimum local.

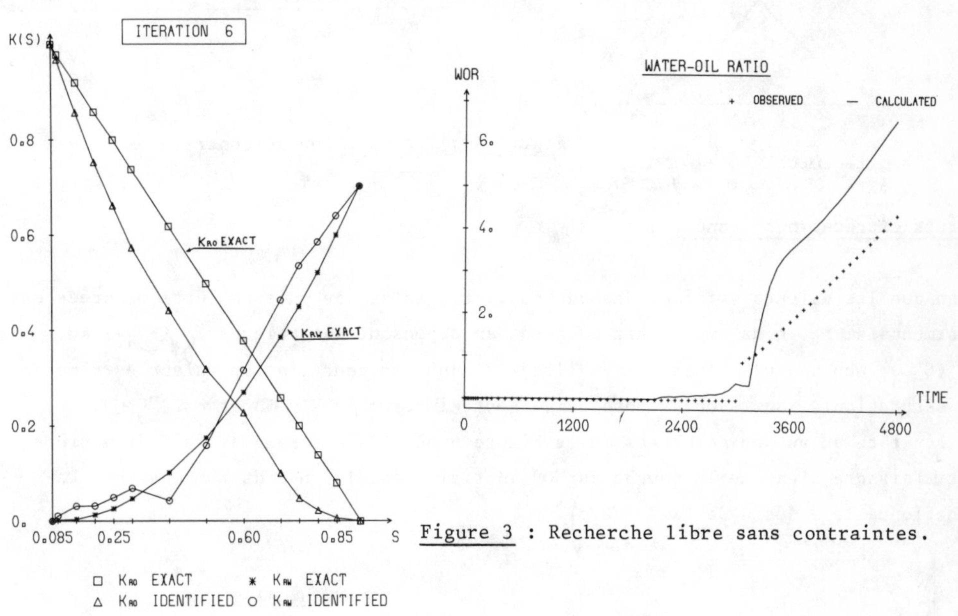

Figure 3 : Recherche libre sans contraintes.

Introduction des contraintes

Pour remédier aux difficultés rencontrées précédemment, nous avons introduit des contraintes de régularité sur la dérivée seconde des perméabilités de la forme :

$$|K''_W| \leqslant 3 \quad \text{et} \quad |K''_O| \leqslant 3.5.$$

On utilise alors la méthode de Franck et Wolfe pour minimiser la fonction coût. Les résultats obtenus après 11 itérations sont présentés sur la figure 4 . L'introduction de contraintes permet d'obtenir un ajustement parfait. Ces contraintes d'autre part s'introduisent naturellement et leur grandeur peut-être précisée par le physicien. La fonctionnelle a décru de 3.40 à $5.5.10^{-4}$. Le temps de calcul est de 1 minute 30 secondes sur CDC 7600.

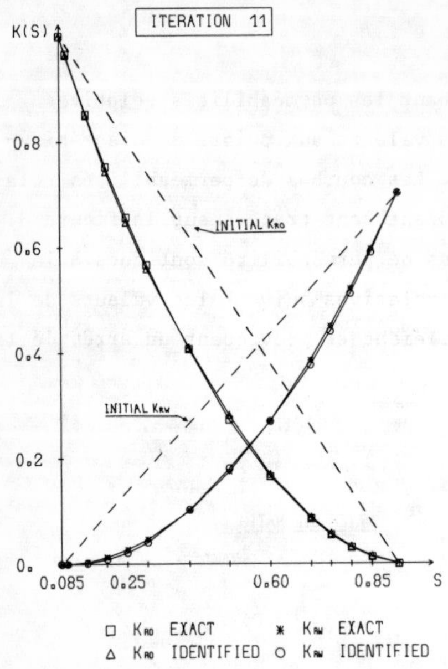

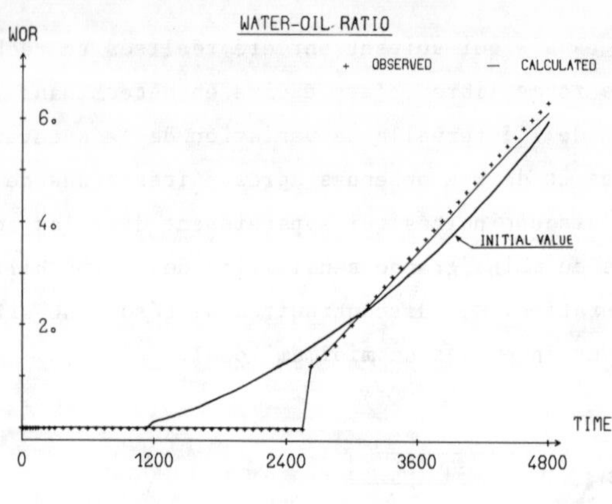

Figure 4 : Introduction de contraintes.

Points extrêmes non connus.

Bien que les valeurs extrêmes des perméabilités relatives puissent être mesurées expérimentalement, nous avons fait un essai en supposant les valeurs $K_{RW}(S_{max})$ et $K_{R_O}(S_{min})$ non connues. (Les perméabilités s'annulent pour l'autre valeur extrême de la saturation). Nous avons imposé comme contraintes : $o \leqslant K''_{RW} \leqslant 3$ et $o \leqslant K''_{R_O} \leqslant 3$.
On constate au vu des résultats de la Figure 5 que l'ajustement se fait de manière satisfaisante bien que la courbe du WOR initial soit éloignée du WOR observé. La fonctionnelle a décru de 12.7 à o.o7.

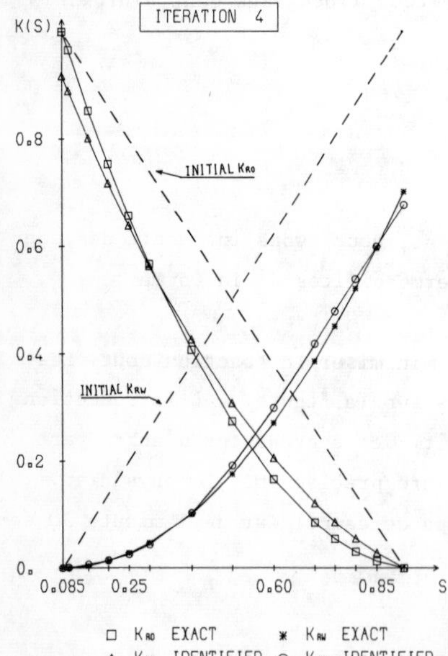

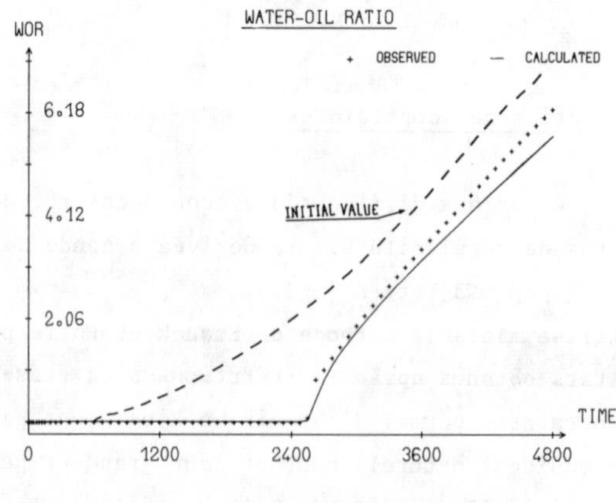

Figure 5 : Points extrêmes non connus.

Ajustement sur des données expérimentales

Les différents essais numériques précédents ont permis de tester la méthode d'ajustement proposée. Les résultats étant satisfaisants, nous avons traité un exemple avec des données réelles provenant d'expériences de laboratoire sur des modèles pétrophysiques.

La figure 6 présente les courbes de perméabilités relatives obtenues et les courbes de WOR observé et calculé après 4 itérations de l'algorithme. La fonctionnelle a décru de 4.10^{-3} à 5.10^{-5}. L'ajustement est convenable compte tenu de la précision relative des mesures expérimentales faites sur le modèle pétrophysique considéré.

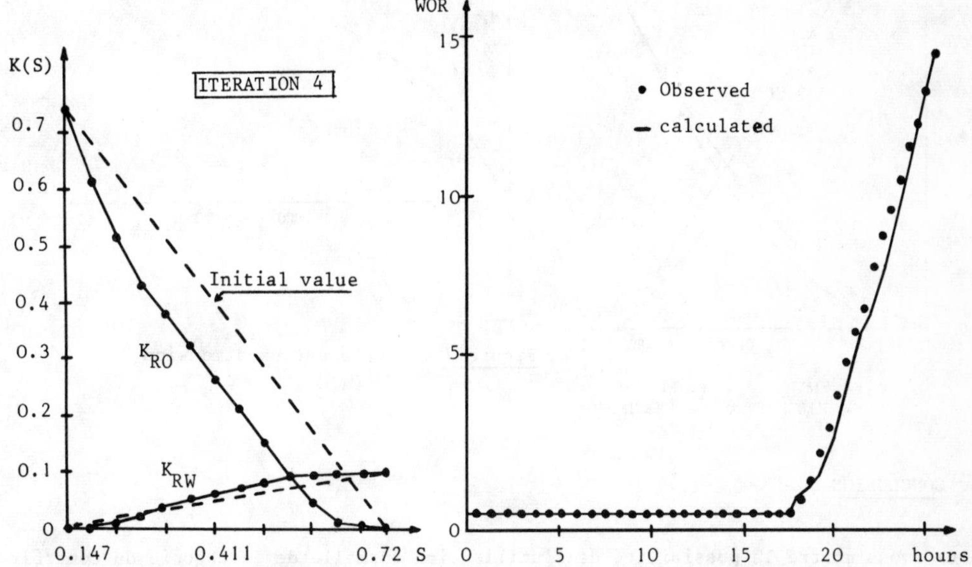

Figure 6 : Ajustement avec des données expérimentales.

Ajustement de la pression capillaire.

Nous avons constaté numériquement que l'influence de la pression capillaire était négligeable dans les exemples précédents. Aussi nous n'avons pas actuellement de résultats numériques relatifs à l'ajustement de la pression capillaire. Il semble qu'il sera nécessaire, pour pouvoir identifier les deux perméabilités relatives et la pression capillaire d'avoir les données expérimentales correspondant à deux expériences réalisées à des vitesses d'écoulement différentes.

Dimension 2

Comme nous l'avons dit, nous avons procédé en dimension 2 uniquement à une recherche sous forme paramétrique avec des polynômes d'ordre 2. Nous présentons les résultats numériques à titre indicatif. Il semble difficile de poursuivre les essais du fait des temps de calcul nécessaires pour intégrer les équations d'états, de la complexité et de la grande instabilité numérique des équations adjointes.

Les résultats sont présentés sur la figure 7. La courbe de WOR exacte est proche de celle obtenue après 14 itérations bien que les perméabilités relatives ne soient pas parfaitement ajustées. Le temps de calcul est de 4 minutes sur CDC 7600.

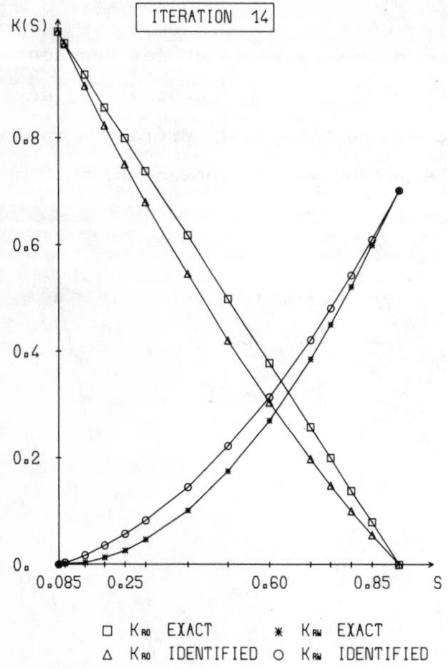

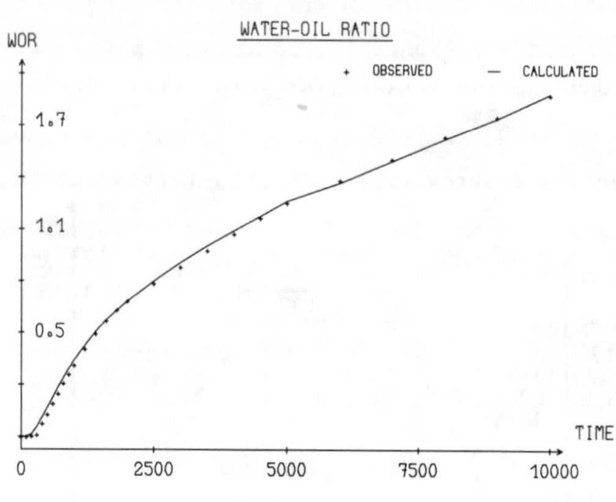

Figure 7 : Ajustement en dimension 2

8. CONCLUSION

Nous avons montré la possibilité de l'utilisation formelle de la théorie du contrôle pour l'ajustement de fonctions d'une variable dépendante apparaissant dans un système de deux équations paraboliques couplées dégénérées. Les résultats obtenus sont satisfaisants en dimension 1 d'espace, et très partiels en dimension 2. Les difficultés rencontrées proviennent sans doute des lacunes constatées dans la formalisation mathématique du problème.

BIBLIOGRAPHIE

[1] MARLE C. : Cours de production. Tome IV. Ed. Technip 1965.

[2] COLLINS R.E. : "Flow of fluids through porous materials". Reinhold Publishing Corporation, New York 1961.

[3] SONIER F., BESSET Ph., OMBRET R. : "A numerical model of multiphase flow around a well". Soc. Pet. Eng. J. (Déc. 1973).

[4] LIONS J.L. : "Contrôle optimal de systèmes gouvernés par des équations aux dérivées partielles". Dunod 1968 (English translation by S.K. MITTER, Grundlehren, Springer 170, 1971).

[5] CHAVENT G. : "Analyse fonctionnelle et identification de coefficients répartis dans les équations aux dérivées partielles". Thèse, Paris 1971.

[6] CHAVENT G. : "Identification of distributed parameters". Proc. of the 3rd IFAC Symposium on Identification. The Hague, 1973.

[7] CHAVENT G., DUPUY M., LEMONNIER P. : "History matching by use of optimal control theory". SPE 4627 presented at the 48th Annual Fall Meeting of the SPE of AIME. Las Vegas 1973.

[8] CHAVENT G., LEMONNIER P. : "Identification de la non-linéarité d'une équation parabolique". Rapport Laboria N° 45, à paraître dans "Applied Math. and Opt. 1974.

UNE METHODE D'OPTIMISATION DE FORME DE DOMAINE

Application à l'écoulement stationnaire à travers une digue poreuse

Ph. MORICE

IRIA-LABORIA

Nous présentons une méthode d'optimisation d'un domaine plan qui consiste essentiellement à contrôler une transformation quasi-conforme devant mettre en correspondance biunivoque un domaine fixe avec le domaine variable à optimiser. Nous exposerons la méthode sur un problème de filtration dans une digue poreuse décrit en §2. Nous donnerons en §3 les détails de la transformation qui permet d'aboutir à un problème de contrôle dans le domaine fixe (§4) et indiquerons en §5 et 6 les grandes lignes de la résolution de ce problème en terminant par des résultats numériques.

2. LE PROBLEME MODELE

Soit D ouvert de $\mathbb{R}^2$ représentant la section (constante) d'une digue séparant deux bassins de niveaux y_1 et y_2, et Ω_v la partie mouillée de cette section.

Nous supposons que $\Gamma=\partial\Omega_v$ est composée de quatre arcs de classe C^1 : $\Gamma_1, \Gamma_2, \Gamma_3, \Gamma_4$ pouvant être décrits par les fonctions :

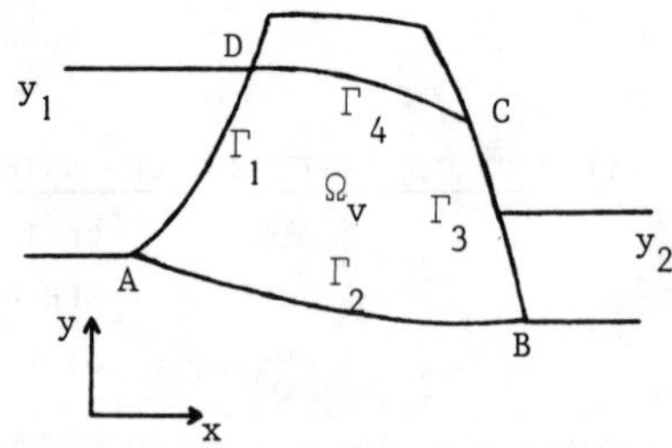

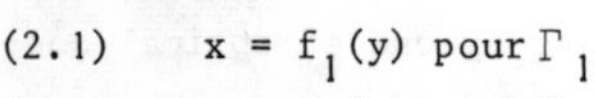

(2.1) $\quad x = f_1(y)$ pour Γ_1

(2.2) $\quad y = f_2(x)$ pour Γ_2

(2.3) $\quad x = f_3(y)$ pour Γ_3

(2.4) $\quad y = \varphi(x)$ pour Γ_4

Les trois fonctions f_1, f_2, f_3 sont données et φ est une fonction représentant la surface libre à déterminer.

Pour ce faire, on dispose de l'équation que doit vérifier z(x,y), potentiel des vitesses de l'écoulement stationnaire :

(2.5) $$\frac{\partial^2 z}{\partial x^2} + \frac{\partial^2 z}{\partial y^2} = o \quad \text{dans } \Omega_v$$

(2.6) $$z = y_1 \quad \text{sur } \Gamma_1$$

(2.7) $$\frac{\partial z}{\partial n} = o \quad \text{sur } \Gamma_2$$

(2.8) $z = \max(y_2,y)$ sur Γ_3

(2.9) $\frac{\partial z}{\partial n} = o$ sur Γ_4

et aussi

(2.10) $z = y$ sur Γ_4

La condition (2.9) traduit la fait que Γ_4 est une ligne de courant et (2.10) le fait que sur la surface libre la pression hydrostatique est constante.
Il s'agit de déterminer la fonction φ régulière telle que z solution de (2.5)-(2.9) vérifie également (2.10).
L'existence et l'unicité d'une telle fonction est démontrée dans [1] avec, il est vrai, certaines restrictions sur la forme de D. Ces auteurs se ramènent à une inéquation quasi-variationnelle ou à une suite d'inéquations variationnelles pour obtenir ces résultats théoriques et des méthodes numériques qui nous ont servi de point de comparaison.

3. LA TRANSFORMATION DE DOMAINE.

Nous allons considérer un changement de variables qui applique Ω_v sur un carré $K = [o,1]\times[o,1]$ et ceci pour toute fonction φ "raisonnable".

Soient $\xi(x,y)$ et $\eta(x,y)$ deux fonctions définies sur Ω_v, nous noterons :

(3.1) $X = (x,y)$ et $\Theta = (\xi,\eta)$

(3.2) $\hat{F}_v$ l'application : $X \to \Theta$ de Ω_v sur K

(3.3) F_v l'application : $\Theta \to X$ de K sur Ω_v

(3.4) $\hat{T} = D(\hat{F}_v(X)) = \begin{bmatrix} \xi'_x & \xi'_y \\ \eta'_x & \eta'_y \end{bmatrix}$; $T = D(F_v(\Theta)) = \begin{bmatrix} x'_\xi & x'_\eta \\ y'_\xi & y'_\eta \end{bmatrix}$

(3.5) $J = \det(T)$; $I = \det(\hat{T})$

En toute généralité, pour $o < |J| < +\infty$, on a les relations :

(3.6) $\hat{T} = T^{-1}$; $I = J^{-1}$ et

(3.7) $\xi'_x = y'_\eta I$; $\xi'_y = -x'_\eta I$; $\eta'_x = -y'_\xi I$; $\eta'_y = x'_\xi I$.

Faute de pouvoir choisir - comme dans BEGIS-GLOWINSKI [2]pour le cas des f_i constantes - explicitement ξ et η en fonction de φ, nous choisirons de définir implicitement ces fonctions comme solutions uniques des deux équations aux dérivées partielles suivantes :

(3.8)
$$\begin{cases} -\Delta\xi = o \text{ dans } \Omega_v \\ \xi = o \text{ sur } \Gamma_1 \; ; \; \dfrac{\partial\xi}{\partial n} = o \text{ sur } \Gamma_2 \; ; \; \xi = 1 \text{ sur } \Gamma_3 \; ; \; \dfrac{\partial\xi}{\partial n} = o \text{ sur } \Gamma_4 \end{cases}$$

(3.9)
$$\begin{cases} -\Delta\eta = o \text{ dans } \Omega_v \\ \dfrac{\partial\eta}{\partial n} = o \text{ sur } \Gamma_1 \; ; \; \eta = o \text{ sur } \Gamma_2 \; ; \; \dfrac{\partial\eta}{\partial n} = o \text{ sur } \Gamma_3 \; ; \; \eta = 1 \text{ sur } \Gamma_4 \end{cases}$$

Il est clair grâce au choix des conditions aux limites que $\eta(x,y)$ est proportionnelle à la fonction conjuguée de $\xi(x,y)$ et on a les relations d'association du type Cauchy :

(3.10) $$\frac{\partial\xi}{\partial x} = \mu\frac{\partial\eta}{\partial y}$$

(3.11) $$\frac{\partial\xi}{\partial y} = -\mu\frac{\partial\eta}{\partial x}$$

avec

(3.12) $$\mu = \int_{\widehat{P_2P_4}} \left(-\frac{\partial\xi}{\partial y}\,dx + \frac{\partial\xi}{\partial x}\,dy\right) \; \forall P_2 \in \Gamma_2,\; P_4 \in \Gamma_4$$

On peut montrer à l'aide du principe du maximum que l'application $\hat{F}_v(X)$ est biunivoque inversible sur Ω_v. Mais nous désirons travailler dans le domaine K et surtout préserver les relations suivantes issues de (3.7), (3.10), (3.11) :

(3.13) $$y'_\eta = \mu x'_\xi$$

(3.14) $$y'_\xi = -\frac{1}{\mu}x'_\eta$$

Ces relations d'association se traduisent par

(3.15) $$\mu\frac{\partial^2 x}{\partial\xi^2} + \frac{1}{\mu}\frac{\partial^2 x}{\partial\eta^2} = o \quad \text{dans } \Omega_v$$

(3.16) $$\mu\frac{\partial^2 y}{\partial\xi^2} + \frac{1}{\mu}\frac{\partial^2 y}{\partial\eta^2} = o \quad \text{dans } \Omega_v$$

Pour atteindre ce but, il est nécessaire de lier les conditions aux limites à vérifier par x et y sur $\gamma = \partial K$.

Nous avons choisi le couplage suivant :

(3.17) $$x|_{\gamma_1} = v_1(\eta);\frac{\partial x}{\partial n}\Big|_{\gamma_2} = \mu v'_2(\xi) \; ; \; x|_{\gamma_3} = v_3(\eta);\frac{\partial x}{\partial n}\Big|_{\gamma_4} = -\mu v'_4(\xi)$$

(3.18) $$\frac{\partial y}{\partial n}\Big|_{\gamma_1} = \frac{1}{\mu}v'_1(\eta);y|_{\gamma_2} = v_2(\xi);\frac{\partial y}{\partial n}\Big|_{\gamma_3} = -\frac{1}{\mu}v'_3(\eta) \; ; \; y|_{\gamma_4} = v_4(\xi)$$

Ces conditions aux limites associées à (3.15), (3.16) permettent de définir $x(\xi,\eta)$ et $y(\xi,\eta)$ de façon unique donc aussi $F_v[\Theta]$.

Si on avait résolu (3.8), (3.9) ces fonctions v_i seraient connues. Ce n'est pas le cas et nous prendrons ces quatre fonctions inconnues pour contrôle.

Les conditions (3.17) et (3.18) traduisent la condition (3.14) et elle seule sur γ, tout comme les conditions aux limites de (3.8) et (3.9) traduisaient (3.10) sur Γ.

Il reste à utiliser (3.13) pour calculer μ grâce à une intégration par parties :

$$(3.19)\qquad o = \int_K (y'_\eta - \mu x'_\xi)\,d\Theta = \int_o^1 [v_4(\xi) - v_2(\xi)]\,d\xi - \mu\int_o^1 [v_3(\eta) - v_1(\eta)]\,d\eta$$

d'où

$$(3.20)\qquad \mu = \frac{\int_o^1 (v_4 - v_2)\,d\xi}{\int_o^1 (v_3 - v_1)\,d\eta}$$

On vérifie sans peine que grâce à (3.13), (3.14) la fonction $z(\xi,\eta)$ correspondant à (2.5)-(2.9) vérifie :

$$(3.21)\qquad \mu\frac{\partial^2 z}{\partial \xi^2} + \frac{1}{\mu}\frac{\partial^2 z}{\partial \eta^2} = o \text{ dans } K$$

$$(3.22)\qquad z\big|_{\gamma_1} = y_1 \;;\; \frac{\partial z}{\partial n}\Big|_{\gamma_2} = o \;;\; z\big|_{\gamma_3} = \max(y_2, y\big|_{\gamma_3}) \;;\; \frac{\partial z}{\partial n} = o \text{ sur } \gamma_4$$

Enfin, notons que dans Ω_v le débit de la digue est : $d = -\int_C \frac{\partial z}{\partial n}\,ds = \text{cste}$ pour toute courbe C joignant un point de Γ_2 à un point de Γ_4 et en particulier sur C telle que $\xi(x,y) = \xi_o$.

Ce qui fournit dans K :

$$d = -\int_o^1 \mu\frac{\partial z}{\partial \xi}(\xi_o,\eta)\,d\eta = -\mu\int_K \frac{\partial z}{\partial \xi}\,d\Theta$$

ou encore

$$(3.23)\qquad d = \mu\Big[y_1 - \int_{\gamma_3} z(1,\eta)\,d\eta\Big]$$

Tout ceci reste assez formel ; on pourra consulter [8] pour plus de précisions.

<u>Remarque 3.1</u>

Au lieu de (3.17) et (3.18) on pourrait choisir le contrôle et les conditions aux limites qui assurent la satisfaction de (3.13) au lieu de (3.14) qui servirait alors à calculer μ. Des essais sont en cours pour vérifier l'avantage éventuel d'un tel choix.

Remarque 3.2

Si les relations (3.13) et (3.14) ne sont pas respectées sur ∂K (par exemple en imposant des conditions de Dirichlet sur ∂K indépendemment pour x et y), on ne peut voir ces relations vérifiées dans K et on obtient alors à la place de (3.15) et (3.16) un système de deux équations quasi-linéaires couplées en x,y. C'est ce que fait WINSLOW [6] et à sa suite CHU [3] pour transformer en polygone à maillage triangulaire équilatéral un domaine de forme quelconque fixe. On pourra aussi consulter GODUNOV, PROKOPOV [5] dont les idées sont peut-être plus proches des nôtres, en ce sens que (sur le problème discret), ces auteurs contrôlent la transformation (mais sans utiliser les méthodes du contrôle optimal) par les conditions aux limites plutôt que par les coefficients de l'opérateur définissant x et y. Toutefois, eux aussi s'en tiennent au cas du domaine fixe. Nous souhaitons, dans un prochain travail, étudier ces possibilités dans le cadre de l'optimisation de domaine.

4. LE PROBLEME DE CONTROLE

Nous pouvons maintenant formuler un problème de contrôle dans le carré K.
L'état sera le triplet $\{x(u), y(u), z(u)\}$ où x(u), y(u), z(u) sont les solutions uniques des trois équations suivantes :

$$(4.1)\quad \left| \begin{array}{l} A(\mu)x = \mu^2 \dfrac{\partial^2 x}{\partial \xi^2} + \dfrac{\partial^2 x}{\partial \eta^2} = o \\ x|_{\gamma_1} = u_1(\eta)\ ;\ \dfrac{\partial x}{\partial n}\Big|_{\gamma_2} = \mu u_2(\xi)\ ;\ x|_{\gamma_3} = u_3(\eta)\ ;\ \dfrac{\partial x}{\partial n} = -\mu u_4'(\xi) \end{array} \right.$$

$$(4.2)\quad \left| \begin{array}{l} A(\mu)y = o \\ \dfrac{\partial y}{\partial n}\Big|_{\gamma_1} = \dfrac{1}{\mu} u_1'(\eta)\ ;\ y|_{\gamma_2} = u_2(\xi)\ ;\ \dfrac{\partial y}{\partial n}\Big|_{\gamma_3} = -\dfrac{1}{\mu} u_3'(\eta)\ ;\ y|_{\gamma_4} = u_4(\xi) \end{array} \right.$$

$$(4.3)\quad \left| \begin{array}{l} A(\mu)z = o \\ z|_{\gamma_1} = y_1\ ;\ \dfrac{\partial z}{\partial n}\Big|_{\gamma_2} = o\ ;\ z|_{\gamma_3} = \max(y_2, y(u)|_{\gamma_3})\ ;\ \dfrac{\partial z}{\partial n}\Big|_{\gamma_4} = o \end{array} \right.$$

Le seul couplage entre ces trois équations provient de la condition en z sur γ_3. Rappelons que l'on choisit :

$$(4.4)\quad \mu = \frac{\int_{\gamma_4} u_4 d\xi - \int_{\gamma_2} u_2 d\xi}{\int_{\gamma_3} u_3 d\eta - \int_{\gamma_1} u_1 d\eta}$$

Si le <u>contrôle</u> $u = (u_1, u_2, u_3, u_4)$ est pris dans un convexe fermé et borné $\mathcal{U}_{ad}$ de $H^1(\gamma_1)\times H^1(\gamma_2)\times H^1(\gamma_3)\times H^1(\gamma_4)$, on peut remplacer les équations (4.1) à (4.3) par une formulation variationnelle de sorte que x(u), y(u), z(u) appartiennent à $H^1(K)$.

Ainsi en introduisant :

$$V_1(K) = \{\varphi \in H^1(K) \; ; \varphi|_{\gamma_1 \cup \gamma_3} = o\} \; ; \; V_2(K) = \{\varphi \in H^1(K) \; ; \varphi|_{\gamma_2 \cup \gamma_4} = o\}$$

et $\bar{x}(\xi, \eta)$ (resp. $\bar{y}$, $\bar{z}$) un relèvement dans $H^1(K)$ des conditions de Dirichlet à vérifier par $\bar{x}$ (resp $\bar{y}$, $\bar{z}$) sur $\gamma_1 \cup \gamma_3$ (resp. $\gamma_2 \cup \gamma_4$, $\gamma_1 \cup \gamma_3$) on peut remplacer (4.1), (4.2), (4.3) par :

Trouver x(u), y(u), $z(u) \in H^1(K)$ tels que

$$(4.5) \qquad x(u) - \bar{x} \in V_1, \; y(u) - \bar{y} \in V_2, \; z(u) - \bar{z} \in V_1$$

et que

$$(4.6) \qquad \int_K \left(\mu^2 \frac{\partial x}{\partial \xi}\frac{\partial \psi}{\partial \xi} + \frac{\partial x}{\partial \eta}\frac{\partial \psi}{\partial \eta}\right) d\,\Theta = \mu \int_{\gamma_2} u_2' \,\psi\, d\xi - \mu \int_{\gamma_4} u_4' \psi\, d\,\xi \qquad \forall \psi \in V_1$$

$$(4.7) \qquad \int_K \left(\mu^2 \frac{\partial y}{\partial \xi}\frac{\partial \varphi}{\partial \xi} + \frac{\partial y}{\partial \eta}\frac{\partial \varphi}{\partial \eta}\right) d\,\Theta = \mu \int_{\gamma_1} u_1' \,\varphi\, d\eta - \mu \int_{\gamma_3} u_3' \varphi\, d\,\eta \qquad \forall \varphi \in V_2$$

$$(4.8) \qquad \int_K \left(\mu^2 \frac{\partial z}{\partial \xi}\frac{\partial \psi}{\partial \xi} + \frac{\partial z}{\partial \eta}\frac{\partial \psi}{\partial \eta}\right) d\,\Theta = o \qquad \forall \psi \in V_1$$

On imposera sur u_1, u_2, u_3 les seules contraintes :

$$(4.9) \qquad \left| \begin{array}{l} u_1(o) = x_A, \; u_1(1) = x_D \\ u_2(o) = y_A, \; u_2(1) = y_B \\ u_3(o) = x_B, \; u_3(1) = f_3(u_4(1)) \end{array} \right.$$

Pour u_4 on impose

$$(4.10) \qquad \left| \begin{array}{l} u_4(o) = y_D = y_1 \\ \\ u_4'(\xi) \leq o \quad \text{avec égalité possible seulement en C.} \end{array} \right.$$

Les conditions (4.9), (4.10) définissent $\underline{\mathcal{U}_{ad}}$.

On a choisi assez naturellement le <u>critère</u> suivant :

$$(4.11) \qquad J(u) = \int_{\gamma_1} [u_1 - f_1(y|_{\gamma_1})]^2 d\eta + \int_{\gamma_2} [u_2 - f_2(x|_{\gamma_2})]^2 d\xi + \int_{\gamma_3} [u_3 - f_3(y|_{\gamma_3})]^2 d\eta + \int_{\gamma_4} (u_4 - z|_{\gamma_4})^2 d\xi$$

qui s'annule si les images des γ_i sont les Γ_i données (i=1,2,3) et si, sur Γ_4, z=y. Grâce à la simplicité de la formule (4.4) on peut montrer l'existence d'un contrôle optimal.

Par contre, l'unicité semble hors d'atteinte et d'ailleurs nous verrons dans le problème discrétisé qu'il y a non unicité et comment on y remédie pratiquement.

Enfin nous avons besoin pour minimiser J(u) de déterminer son gradient. Nous nous contenterons ici de donner son expression.

Pour cela, on doit calculer l'état adjoint constitué du triplet (p(u), q(u), r(u)) défini par les équations suivantes :

$$(4.12)\quad \left|\begin{array}{l} A(\mu)p = o \quad \text{dans } K \\ p|_{\gamma_1} = o \; ; \; \frac{\partial p}{\partial n}|_{\gamma_2} = (f_2(x) - u_2)f_2'(x) \; ; \quad p|_{\gamma_3} = o \quad \frac{\partial p}{\partial n}|_{\gamma_4} = o \end{array}\right.$$

$$(4.13)\quad \left|\begin{array}{l} A(\mu)q = o \quad \text{dans } K \\ \frac{\partial q}{\partial n}|_{\gamma_1} = \frac{1}{\mu^2}(f_1(y)-u_1)f_1'(y) \; ; \; q|_{\gamma_2} = o \; ; \; \frac{\partial q}{\partial n}|_{\gamma_3} = \frac{1}{\mu^2}(f_3(y)-u_3)f_3'(y) - \frac{\partial r}{\partial n}s(y) ; q|_{\gamma_4} = o \end{array}\right.$$

$$(4.14)\quad \left|\begin{array}{l} A(\mu)r = o \quad \text{dans } K \\ r|_{\gamma_1} = o \; ; \; \frac{\partial r}{\partial n}|_{\gamma_2} = o \; ; \; r|_{\gamma_3} = o \; ; \; \frac{\partial r}{\partial n}|_{\gamma_4} = (z-u_4) \end{array}\right.$$

avec dans (4.13) $s(y)|_{\gamma_3} = \begin{cases} o \text{ si } y(1,\eta) \leq y_2 \\ 1 \text{ si } y(1,\eta) > y_2 \end{cases}$

Notant

$$(4.15)\quad g|_{\gamma} = \begin{cases} u_1 - f_1(y) \text{ sur } \gamma_1 \\ u_2 - f_2(x) \text{ sur } \gamma_2 \\ u_3 - f_3(y) \text{ sur } \gamma_3 \\ u_4 - z \quad \text{ sur } \gamma_4 \end{cases}$$

$$(4.16)\quad C = \frac{\left[-2\mu\int_K \left(\frac{\partial x}{\partial \xi}\frac{\partial p}{\partial \xi} + \frac{\partial y}{\partial \xi}\frac{\partial q}{\partial \xi} + \frac{\partial z}{\partial \xi}\frac{\partial r}{\partial \xi}\right)d\Theta + \int_{\gamma_1} u_1' q d\eta + \int_{\gamma_2} u_2' p d\xi - \int_{\gamma_3} u_3' q d\eta - \int_{\gamma_4} u_4' q d\xi\right]}{\int_o^1 (u_3 - u_1) d\eta}$$

$$(4.17)\quad e = \left(\mu\frac{\partial p}{\partial \xi} - \frac{\partial q}{\partial \eta} + C\right)$$

On obtient

$$(4.18)\quad \frac{1}{2}J'(u).v = \int_{\gamma_1} (g+\mu e)\, v_1 d\eta + \int_{\gamma_2} (g-e)v_2 d\xi + \int_{\gamma_3} (g-\mu e)v_3 d\eta + \int_{\gamma_4} (g+e)v_4 d\xi$$

La condition nécessaire d'optimalité pour $u \in \mathcal{U}_{ad}$ est que

(4.19) $J'(u).(v-u) \geqslant o \qquad \forall v \in \mathcal{U}_{ad}$

5. LE PROBLEME DE CONTROLE DISCRETISE

Sur le carré $K : [o,1]x[o,1]$ on prend un maillage carré de pas $h = \frac{1}{2^k} = \frac{1}{N}$ pour discrétiser en différences finies le contrôle u et les équations d'état.
Le réseau n'est pas "décalé" et l'on a pris l'approximation classique de la condition de Neuman dans ce cas : par exemple pour γ_1 et y, (3.16) et (3.18) donnent :

(5.1) $\mu^2(2y_{i,j}-2y_{i,j+1})+2y_{i,j}-y_{i+1,j}-y_{i-1,j}=\mu([u_1]_{i+1}-[u_1]_{i-1})$

γ_1 K
i+1,j
i,j i,j+1
i-1,j

Pour résoudre les équations d'états (et les équations adjointes) nous avons utilisé une méthode directe particulièrement rapide valable pour les opérateurs à coefficient constant : la variante de BUNEMAN de la "réduction cyclique par blocs" décrite par BUZBEE, GOLUB, NIELSON [3].
Pour le calcul du gradient J'(u), nous avons utilisé la formule des trapèzes dans la discrétisation des intégrales sur le bord avec approximations centrées aux noeuds, et, pour l'intégrale dans K de termes du type $\frac{\partial x}{\partial \xi}\frac{\partial p}{\partial \xi}$, une approximation au centre des mailles.
La méthode d'optimisation choisie est celle de Frank et Wolfe (voir par exemple CEA [4]) qui comprend la résolution d'un programme linéaire pour déterminer une direction de descente compatible avec les contraintes. Pour cette étape, nous avons utilisé non un programme de simplexe mais la méthode du gradient réduit programmée par LEMARECHAL et YVON.
Les premiers résultats numériques ont montré que, pour le problème ainsi formulé, il n'y avait pas unicité de la solution et que les différents contrôles "optimaux" obtenus suivant les valeurs initiales de u différaient essentiellement par la position du point C sur Γ_3. Nous avions le choix pour résoudre cette "indétermination" entre deux possibilités :
1) prendre $\mathcal{U}_{ad}$ plus "petit" en introduisant des contraintes supplémentaires (mais non-linéaires).
2) ajouter au critère un terme qui puisse augmenter la convexité de J(u).
Nous avons retenu la deuxième solution en utilisant la connaissance du comportement analytique de $x(\xi,\eta)$ et $y(\xi,\eta)$ en C. Plus précisément, la courbe Γ_4 devant se raccorder tangentiellement à Γ_3 en C, la fonction $\eta \to y(1,\eta)$ doit se comporter au voisinage de $\eta = 1$ comme $y(1,1) - \alpha(1-\eta)^2$; ce qui peut se traduire en discret par :

(5.2) $3y_{N+1,N+1} - 4y_{N,N+1} + y_{N-1,N+1} = g(y_h) = o$

Nous avons donc ajouté au critère le terme $[g(y_h)]^2$ qui a été suffisant pour l'obtention d'un contrôle optimal discret unique.

Le caractère assez artificiel de cette approche devrait pouvoir être évité dans la variante évoquée à la remarque 3.1.

6. RESULTATS NUMERIQUES

Pour chaque forme de digue étudiée on a procédé de la façon suivante : un premier essai est effectué avec $h = \frac{1}{8}$ et une solution initiale raisonnable mais arbitraire. La solution obtenue est alors interpolée pour servir de solution initiale à l'essai avec $h = \frac{1}{16}$.

Dans tous les cas le temps de calcul observé correspondait sur la 10070 CII à 36 à 38 itérations à la minute pour $h = \frac{1}{8}$ et 10 ou 11 itérations à la minute pour $h = \frac{1}{16}$.

Le premier programme écrit l'a été pour une digue rectangulaire (problème modèle de BAIOCCHI et alt [1]) avec y_1=3.22 ; y_2=o.84 ; $a = x_B - x_A = 1.62$.

Pour $h = \frac{1}{16}$, le critère est passé de $.35.10^{-3}$ à $.44.10^{-6}$ en 26 itérations donnant $y_c = 2.019$ et débit = 2.982.

Nous donnons sur la figure 1a) l'image de la grille de discrétisation et les lignes équipotentielles de z dans ce cas et également pour a = 1 et a = 2.5. (Figures 1b) et 2).

Le programme général issu de la méthode décrite dans cet article a ensuite été utilisé d'abord pour une forme prismatique testée par COMINCIOLI [6] Les résultats sont donnés figure 3.

Un autre essai avec une forme arbitraire composée d'arcs de paraboles est donné figure 4.

7. CONCLUSION

Nous avons exposé une méthode assez générale dans son principe pour l'optimisation de domaine plan sur un cas concret. Si son application à des formes de domaine très compliquées semble délicate - les difficultés provenant surtout du nombre et de la forme des "coins" ainsi que de la multiconnexité - dans de nombreux cas des aménagements doivent être possibles. Il reste que, sur le problème envisagé, cette méthode se compare plutôt favorablement du point de vue de son efficacité numérique avec des méthodes antérieures.

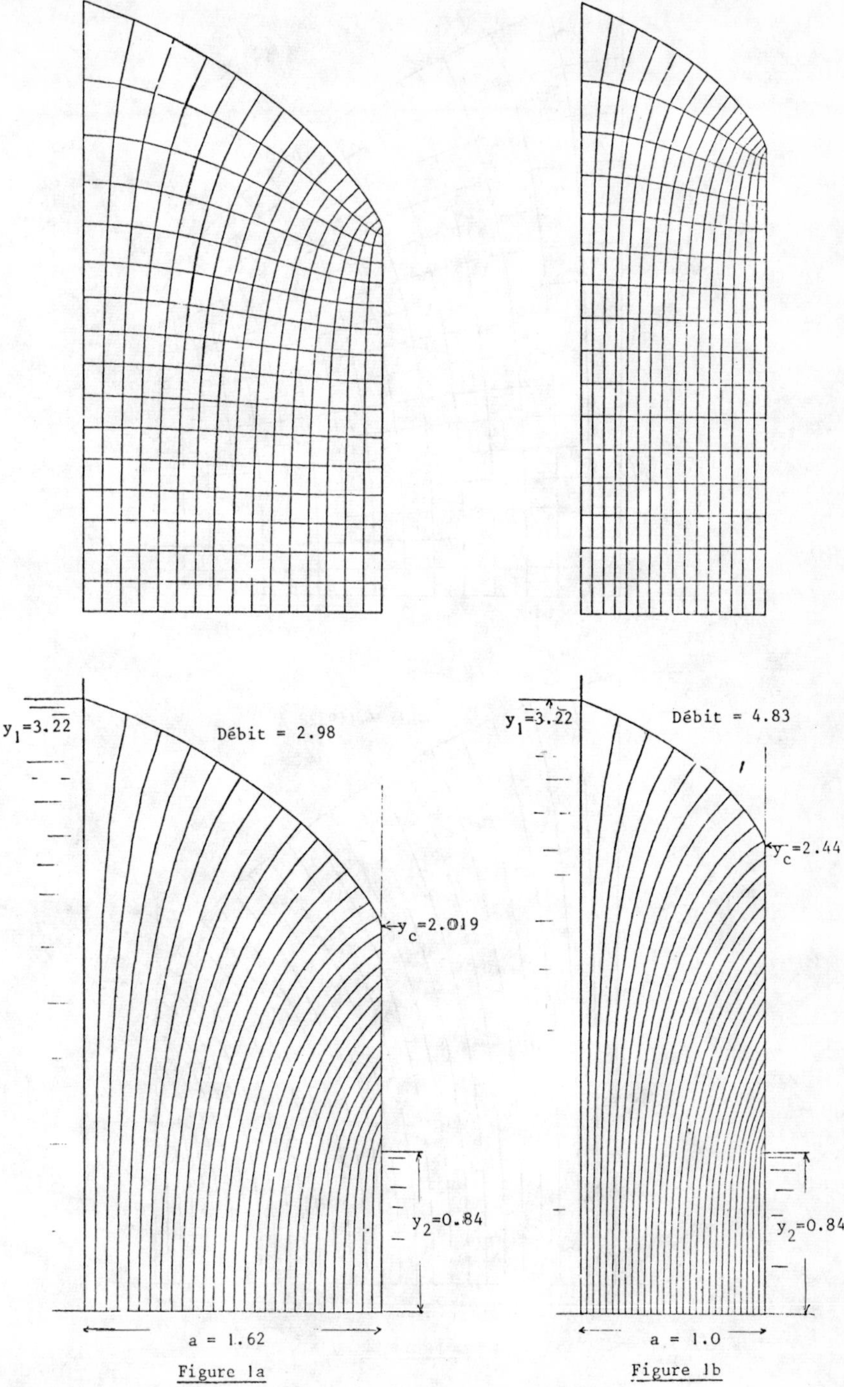

Figure 1a

Figure 1b

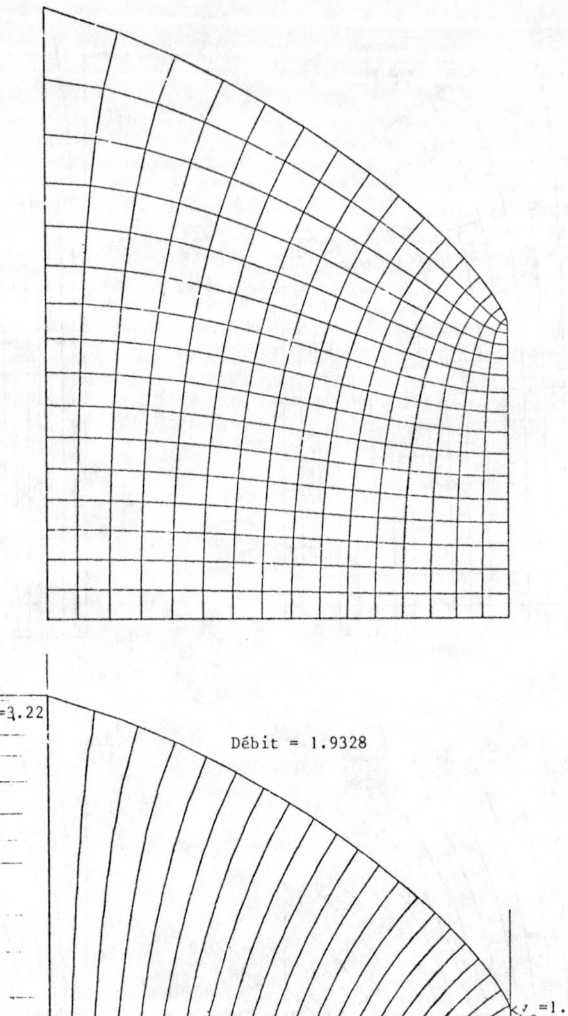

$y_1 = 3.22$

Débit = 1.9328

$y_c = 1.55$

$y_2 = 0.84$

$a = 2.5$

Figure 2

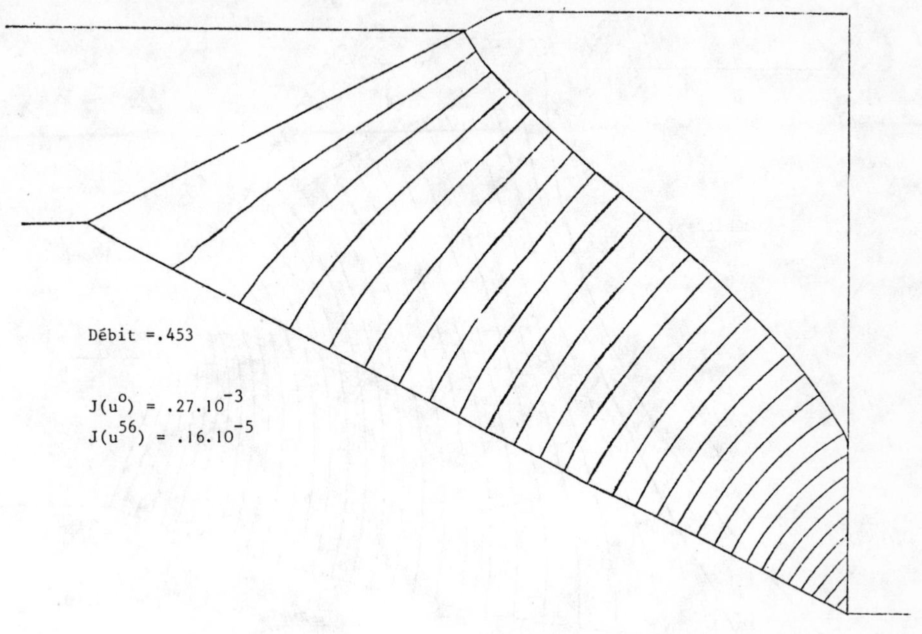

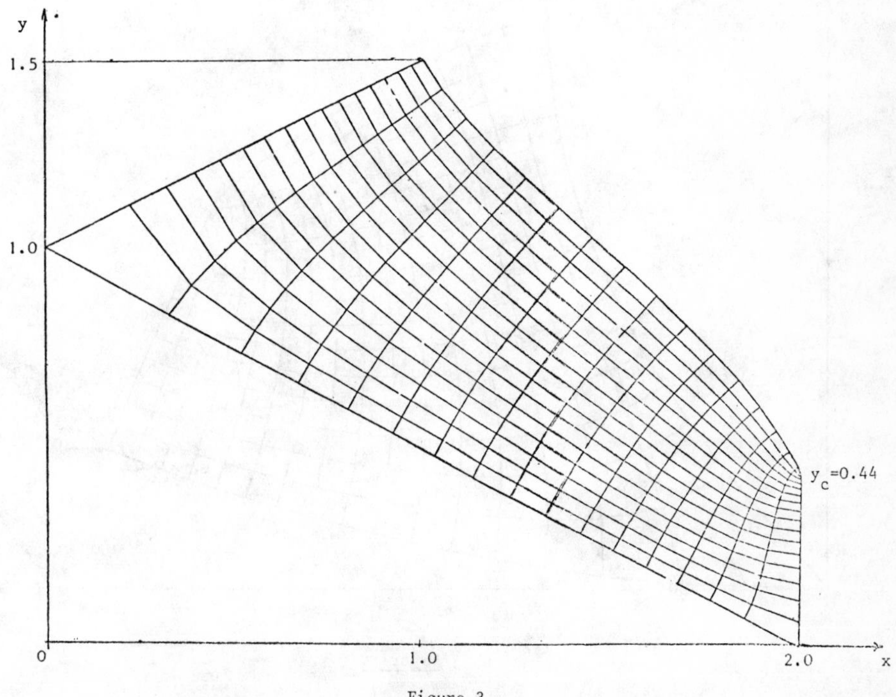

Figure 3

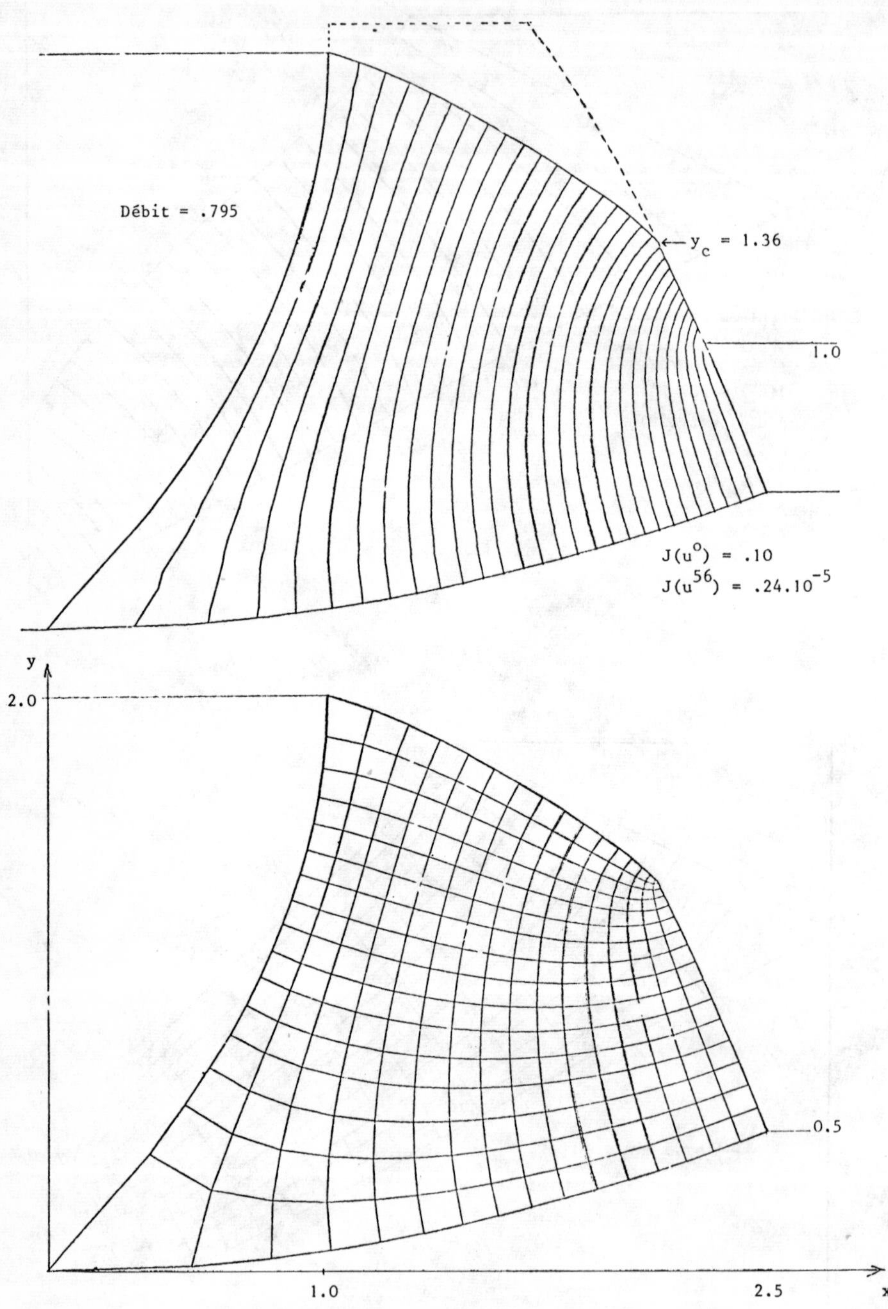

Figure 4

BIBLIOGRAPHIE

[1] C. BAIOCCHI, V. COMINCIOLI, L. GUERRI, G. VOLPI : "Free Boundary Problems in the theory of fluid flow through porous media : a numerical approach".
Calcolo Vol. 10-1 (1973).

[2] D. BEGIS, R. GLOWINSKI : Application de la méthode des éléments finis à la résolution d'un problème de domaine optimal.
International Symposium on Computing Methods in Engineering. IRIA Décembre 1973.

[3] B.L. BUZBEE, G.H. GOLUB, C.W. NIELSON : On direct methods for solving Poisson's equations.
SIAM J. Num. Anal. Vol. 7 N°4 (Déc. 1970).

[4] J. CEA : Optimisation.
Dunod, Paris (1971).

[5] W.H. CHU : "Development of a general finite difference approximation for a general domain, Part 1 : Machine transformation".
J. Computational Phys. 8, (1971) p. 392-408.

[6] V. COMINCIOLI : "On some oblique derivative problems arising in the fluid flow of porous media. A theoretical and numerical approach".
"Applied Mathematics and Optimization" journal.pp.1-33

[7] S.K. GODUNOV, G.P. PROKOPOV : "The calculation of conformal mappings and the construction of difference networks".
USSR Computational Math. Math. Phys. 7 (1967),5.

[8] Ph. MORICE : Thèse (à paraître).

[9] A.M. WINSLOW : Numerical solution of the quasilinear Poisson equation in a non uniform triangular mesh".
J. Computational Phys. 2. (1966) pp. 149-172.

FILTERING FOR SYSTEMS EXCITED BY POISSON WHITE NOISE

Huibert Kwakernaak

Department of Applied Mathematics

Twente University of Technology

Enschede, The Netherlands

Summary

The application of martingale theory to filtering problems for linear systems, excited by Poisson white noise but with Gaussian observation noise, is described. Stochastic differential equations for the conditional density function and moments are derived, and two approximate methods for solving these equations are developed. Numerical results are presented. Preliminary results are given for the smoothing problem, and for filtering problems for distributed parameter systems excited by Poisson white noise.

1. Introduction and motivation

The work described in this paper has been motivated by the following environmental problem. A river is occasionally polluted by deposits of an undesired chemical. The total number of deposits in a section of the river of infinitesimal length dr (where r is the distance coordinate along the river) behaves according to a Poisson process with rate parameter $\lambda(r)dr$, where $\lambda(.)$ is a given function. The numbers of deposits in non-overlapping sections are independent processes. The amounts of chemical that are deposited each time are independent stochastic variables with given distributions. The chemical is dispersed in the river by diffusion and transport. Along the river measuring stations are installed, which continuously record the observed local concentrations of chemical. The problem is to reconstruct the times, locations and amounts of deposits from the measured data.

As the dispersion of the pollutant in the river is best described by a diffusion equation, the problem as sketched can be represented as a filtering problem for a distributed parameter system. What distinguishes this problem from other filtering problems for distributed parameter systems described in the literature (see e.g. Benssoussan, 1971) is the Poisson nature of the disturbing noise. As filtering problems for finite-dimensional systems excited by Poisson white noise appear to have received little or no attention, it seems fitting to study such problems first before tackling the distributed-parameter version.

The following problem will be considered. The stochastic process X_t, $t \geq t_o$, is the solution of the stochastic differential equation

$$dX_t = \alpha X_t dt + h dP_t, \qquad t \geq t_o, \tag{1}$$

where α and h are given constants, and P_t, $t \geq t_o$, is a Poisson process with constant parameter λ(with $\lambda \geq 0$). X_{t_o} is a given stochastic variable, independent of the process P. The <u>observed</u> process Y_t, $t \geq t_o$, is defined by

$$\begin{aligned} dY_t &= \gamma X_t dt + dW_t, \qquad t \geq t_o \\ Y_{t_o} &= 0, \end{aligned} \tag{2}$$

where γ is another constant, and where W_t, $t \geq t_o$, is Brownian motion with constant parameter μ(with $\mu > 0$), independent of X_{t_o} and the process P.

Martingale theory will be used to solve the problem of reconstructing the process X from the observed process Y. Recent publications of Wong (1973) and others (Fujisaka, Kallianpur, and Kunita, 1972; Clark, 1973) have made it abundantly clear that martingales are a very effective and natural tool in the solution of stochastic filtering problems. Recent extensions of martingale theory with respect to discontinuous martingales (Doléans-Dade and Meyer, 1970) complete the framework in which the present paper is set. In the account to follow, familiarity with the papers of Wong

(1973) and Doléans-Dade and Meyer (1970) is assumed, although the level of rigor maintained by the latter two authors will by no means be approached.

In Section 2 of the paper, martingale theory will be used to derive the stochastic differential equation satisfied by $p_{t|t}(x)$, which is the conditional density function of X_t, given $\mathcal{Y}_t$. Here $\mathcal{Y}_t$ is the sigma field generated by Y_s, $t_o \leq s \leq t$. This stochastic partial differential equation, which will be referred to as the conditional Fokker-Planck equation, is quite similar in form to that obtained in case the Poisson process in (1) is replaced with Brownian motion, except for the form of the forward operator. This difference makes it impossible to find an explicit solution to the conditional Fokker-Planck equation, such as may be determined in the case of Brownian motion. Therefore, in Section 3 two methods are developed for obtaining approximate solutions. Some properties of the approximate solution methods are analyzed, and simulation results are presented.

In Section 4, the extension of the results of Section 2 to multi-dimensional systems is described, while also some results are given for the distributed-parameter filtering problem that constitutes the motivation for the work of this paper. Finally, in Section 5 the smoothing problem is discussed, and equations are given satisfied by $p_{t|s}(x)$, the conditional density function of X_t given $\mathcal{Y}_s$, with $s \geq t$.

2. The conditional Fokker-Planck equation

In this section, the stochastic partial differential equation satisfied by $p_{t|t}(x)$, the conditional density function of X_t given $\mathcal{Y}_t$, will be derived, using martingale theory. Furthermore, stochastic differential equations will be obtained for the central moments of this probability density function.

First the following theorem is stated, which may be considered as fundamental in filtering theory for problems with additive, Gaussian white noise. This theorem may be found in Van Schuppen (1973). It is similar to a theorem of Fujisaka,

Kallianpur and Kunita (1972) (see also Clark, 1973), except that a more general case is considered.

Let Q_t, $t \geq t_o$, be the vector-valued semi-martingale defined by

$$dQ_t = R_t dt + dM_t, \qquad t \geq t_o, \tag{3}$$

where M is a vector-valued martingale with respect to a growing family of sigma fields $\mathcal{F}_t$, $t \geq t_o$, and where R is a process that is adapted to $\mathcal{F}$. Let furthermore the vector-valued process Y_t, $t \geq t_o$, be the semi-martingale defined by

$$dY_t = H_t dt + dW_t, \qquad t \geq t_o \tag{4}$$

where H is another vector-valued process that is adapted to $\mathcal{F}$. W is vector-valued Brownian motion and a martingale with respect to $\mathcal{F}$, such that $E(dW_t dW'_t) = V(t)dt$. The prime denotes the transpose, and $V(t) > 0$ for $t \geq t_o$. Define $\mathcal{Y}_t$, $t \geq t_o$, as the growing family of sigma fields generated by the process Y, and let

$$\hat{Q}_t \overset{df}{=} E(Q_t|\mathcal{Y}_t), \quad \hat{R}_t \overset{df}{=} E(R_t|\mathcal{Y}_t), \quad \hat{H}_t \overset{df}{=} E(H_t|\mathcal{Y}_t), \quad \widehat{Q_t H'_t} \overset{df}{=} E(Q_t H'_t|\mathcal{Y}_t). \tag{5}$$

<u>Fundamental filtering theorem for white, Gaussian observation noise</u>: The process $\hat{Q}_t$, $t \geq t_o$, satisfies

$$d\hat{Q}_t = \hat{R}_t dt + [\widehat{Q_t H'_t} - \hat{Q}_t \hat{H}'_t + \hat{C}_t]V^{-1}(t)[dY_t - \hat{H}_t dt], \qquad t \geq t_o. \quad \blacksquare \tag{6}$$

Here $\hat{C}_t \overset{df}{=} E(C_t|\mathcal{Y}_t)$, while $C_t \overset{df}{=} \frac{d}{dt} \langle M^c, W\rangle_t$. M^c is the continuous part of the martingale M. If M_1 and M_2 are two vector-valued continuous martingales or semi-martingales, $\langle M_1, M_2\rangle$ is a matrix stochastic process, the (i,j)-th element of which is given by $\langle M_{1i}, M_{2j}\rangle \overset{df}{=} \frac{1}{2}(\langle M_{1i}+M_{2j}, M_{1i}+M_{2j}\rangle - \langle M_{1i}, M_{1i}\rangle - \langle M_{2j}, M_{2j}\rangle)$. In the notation of Wong (1973), $\langle N,N\rangle$ is the increasing process associated with the continuous martingale or semi-martingale N.

<u>Outline of proof</u> (Fujisaka, Kallianpur, and Kunita, 1972; Clark, 1973; Van Schuppen, 1973; Wong, 1973): The proof consists of the following steps.

(a) First it is shown that the innovations process I, defined by $dI_t = dY_t - \hat{H}_t dt$, $I_{t_o} = 0$, is a martingale with respect to Y, and, moreover, is a Brownian motion with $d\langle I,I\rangle_t = V(t)dt$. (b) The second step is to prove that the process M^*, defined by $dM_t^* = d\hat{Q}_t - \hat{R}_t dt$, $M_{t_o}^* = 0$, is also a martingale with respect to Y. (c) Next, a representation theorem is invoked, which says that if M^* is a martingale with respect to Y, while Y is generated by the semi-martingale

$$\int_{t_o}^{t} \hat{H}_s ds + I_t, \tag{7}$$

with I Brownian motion, M^* may be written as

$$M_t^* = M_{t_o}^* + \int_{t_o}^{t} L_s dI_s, \tag{8}$$

where L is a matrix-valued process that is adapted to Y. Rewriting (8) in differential form it follows that $d\hat{Q}_t = \hat{R}_t dt + L_t dI_t$. (d) It finally is established that L may be expressed as in (6) by proving and using the equality $E[d(Q_t Y_t')|Y_t] = E[d(\hat{Q}_t Y_t')|Y_t]$. ■

In step (d) of the proof, use is made of the differentiation rule of Doléans-Dade and Meyer (1970) for discontinuous semi-martingales. Since this rule will be used repeatedly in the sequel, it is restated here. Let X_t be an n-dimensional vector-valued, discontinuous semi-martingale, and let ϕ be a twice differentiable function defined on R^n. Then the process $Q_t \overset{df}{=} \phi(X_t)$ is also a semi-martingale, such that

$$dQ_t = \phi_x'(X_{t-})dX_t + \tfrac{1}{2}tr[\phi_{xx}(X_{t-})d\langle X^c, X^c\rangle_t] + d\sum_{t_o \le s \le t}[\phi(X_s) - (X_{s-}) - \phi_x'(X_{s-})(X_s - X_{s-})]. \tag{9}$$

Here ϕ_x is the gradient of the function ϕ, ϕ_{xx} its Jacobian, while X^c is the continuous part of X, and the summation is carried out over those values of s where X jumps.

After these preliminaries, the filtering problem described in Section 1 can be dealt with. Let the process X and Y be defined as in (1) and (2), and let

$$Q_t \overset{df}{=} e^{iuX_t}, \tag{10}$$

where $i = \sqrt{-1}$, and where u is real. Application of the differentiation rule yields

$$dQ_t = iu\, e^{iuX_t}(\alpha X_t dt + h dP_t) + d \sum_{t_o \leq s \leq t} [e^{iu(X_{s-}+h)} - e^{iuX_{s-}} - iue^{iuX_{s-}} h], \tag{11}$$

where one uses $\langle X^c, X^c\rangle = h^2 \langle P^c, P^c\rangle = 0$, and $X_s - X_{s-} = hP - hP_{s-} = h$ if a jump occurs at s. Since for any function ψ one has

$$d \sum_{t_o \leq s \leq t} \psi(X_{s-}) = d \int_{t_o}^{t} \psi(X_s) dP_s = \psi(X_t) dP_t, \tag{12}$$

(11) may be rewritten as

$$dQ_t = iu\alpha X_t e^{iuX_t} dt + e^{iuX_t}(e^{iuh}-1) dP_t. \tag{13}$$

The semi-martingale nature of Q may be emphasized by rearranging the right-hand side of (13) as

$$dQ_t = [iu\alpha X_t e^{iuX_t} + \lambda e^{iuX_t}(e^{iuh}-1)]dt + e^{iuX_t}(e^{iuh}-1)(dP_t - \lambda dt). \tag{14}$$

Comparison with (3) shows that the adapted process R and the martingale N may be identified as follows:

$$\begin{aligned} R_t &= iu\alpha X_t e^{iuX_t} + \lambda e^{iuX_t}(e^{iuh}-1), \\ dN_t &= e^{iuX_t}(e^{iuh}-1)(dP_t - \lambda dt). \end{aligned} \tag{15}$$

Moreover, comparison of the observation equation $dY_t = \gamma X_t dt + dW_t$ with (4) leads to the identification $H_t = \gamma X_t$. Direct application of the basic filtering equation (6) gives

$$\begin{aligned} d(\widehat{e^{iuX_t}}) &= iu\alpha \widehat{X_t e^{iuX_t}} dt + \lambda \widehat{e^{iuX_t}}(e^{iuh}-1)dt \\ &\quad + \frac{\gamma}{\mu}(\widehat{X_t e^{iuX_t}} - \hat{X}_t \widehat{e^{iuX_t}})(dY_t - \gamma \hat{X}_t dt), \end{aligned} \tag{16}$$

where, as before, providing a stochastic variable with a caret signifies taking its conditional expectation given $\mathcal{Y}_t$. Now, since

$$\widehat{e^{iuX_t}} = \int_{-\infty}^{\infty} e^{iux} p_{t|t}(x)dx,$$

$$\widehat{X_t e^{iuX_t}} = \int_{-\infty}^{\infty} e^{iux} x p_{t|t}(x)dx, \tag{17}$$

inverse Fourier transformation of (16) yields the following stochastic partial differential-difference equation satisfied by $p_{t|t}$:

$$dp_{t|t}(x) = Lp_{t|t}(x)dt + \frac{\gamma}{\mu}(x-\hat{X}_t)p_{t|t}(x)(dY_t-\gamma\hat{X}_t dt), \qquad t \geq t_o, \tag{18}$$

where L is the linear operator defined by

$$Lp(x) \stackrel{df}{=} -\frac{\partial}{\partial x}[\alpha x p(x)] + \lambda[p(x-h)-p(x)], \qquad -\infty < x < \infty \tag{19}$$

and where

$$\hat{X}_t \stackrel{df}{=} \int_{-\infty}^{\infty} x p_{t|t}(x)dx. \tag{20}$$

The conditional Fokker-Planck equation is the same as when the system is excited by Brownian motion, except for the form of the operator L. In the Brownian motion case L is given by

$$Lp(x) = -\frac{\partial}{\partial x}[\alpha x p(x)] + \tfrac{1}{2}\lambda\frac{\partial^2 p(x)}{\partial x^2}, \qquad -\infty < x < \infty, \tag{21}$$

where λ is the parameter of the Brownian motion V in the system equation $dX_t = \alpha X_t dt + dV_t$.

It is of interest to derive the stochastic differential equation satisfied by the conditional expectation $\hat{X}_t$. This equation may be obtained by multiplying the conditional Fokker-Planck equation (18) by x and integrating over $(-\infty, \infty)$, or by applying the basic filtering theorem directly to $Q_t \stackrel{df}{=} X_t$. In either case it follows that

$$d\hat{X}_t = (\alpha\hat{X}_t+\lambda h)dt + \frac{\gamma}{\mu}P_{2t}(dY_t-\gamma\hat{X}_t dt), \qquad t \geq t_o, \tag{22}$$

where P_{2t} is the second conditional central moment. The n-th conditional central moment is defined by

$$P_{nt} \stackrel{df}{=} E[(X_t-\hat{X}_t)^n|Y_t] = \int_{-\infty}^{\infty} (x-\hat{X}_t)^n p_{t|t}(x)dx, \quad n = 0,1,\ldots\ldots . \tag{23}$$

It is noted that (22) has precisely the form of the Kalman filter, except that P_{2t} is realization-dependent.

It is possible to derive stochastic differential equations for all the conditional central moments by either applying the differentiation rule to $(x-\hat{X}_t)^n p_{t|t}(x)$ and then integrating over $(-\infty, \infty)$, or by applying the basic filtering theorem directly to $Q_t \stackrel{df}{=} (X_t-\hat{X}_t)^n$. It follows that

$$dP_{nt} = \alpha n P_{nt}dt + \sum_{k=0}^{n-2} \binom{n}{k} P_{kt} h^{n-k} dt + \frac{\gamma^2}{2\mu} n(n-1)P_{n-2,t}P_{2t}^2 dt$$

$$- \frac{\gamma^2}{\mu} nP_{nt}P_{2t}dt + \frac{\gamma}{\mu}(P_{n+1,t} - nP_{n-1,t}P_{2t})(dY_t - \gamma\hat{X}_t dt), \qquad n = 1,2,\ldots\ldots, \tag{24}$$

where $P_{0t} = 1$ and $P_{1t} = 0$, $t \geq t_o$. In particular, it follows for n = 2 and n = 3,

$$dP_{2t} = 2\alpha P_{2t}dt + \lambda h^2 dt - \frac{\gamma^2}{\mu} P_{2t}^2 dt + \frac{\gamma}{\mu} P_{3t}(dY_t - \gamma\hat{X}_t dt), \tag{25}$$

$$dP_{3t} = 3\alpha P_{3t}dt + \lambda h^3 dt - \frac{\gamma^2}{\mu} 3P_{3t}P_{2t}dt + \frac{\gamma}{\mu}(P_{4t} - 3P_{2t}^2)(dY_t - \gamma\hat{X}_t dt). \tag{26}$$

The differential equation for P_{nt} involves besides lower-order moments also $P_{n+1,t}$, which means that one is faced with an infinite set of simultaneous stochastic differential equations. Contrary to the Gaussian case, one cannot hope that the equations reduce to a finite set because all odd moments are zero.

Obtaining exact solutions to the filtering problem is out of the question. Numerical experience has shown that naive truncation of the equations leads to unstable filters and, generally, gives very poor results. In the next section, two approaches to obtaining approximate solutions are developed.

3. Two approximate solution methods for the filtering problem

In this section two methods for obtaining approximate solutions to the filtering problem will briefly be described, as well as some simulation results. The first method is a type of Ritz-Galerkin method for approximately solving the conditional Fokker-Planck equation, while the second method is based on truncation of the stochastic differential equations satisfied by the cumulants of the conditional density function. In passing it is remarked that both methods are also applicable in the case of linear filtering problems with Gaussian white noise but with non-Gaussian initial statistics, a problem which has recently been discussed in the literature (Lainiotis, 1971; Lo, 1972).

In the first method, an approximate solution $\tilde{p}_{t|t}(x)$ of the conditional Fokker-Planck equation (18) is sought of the form

$$\tilde{p}_{t|t}(x) = \sum_{i=0}^{N} S_{it} h_{it}(x) \phi_t(x), \qquad t \geq t_o, \qquad -\infty < x < \infty. \tag{27}$$

Here the coefficients S_{it}, $t \geq t_o$, $i = 0,1,\ldots,N$, with N to be determined, are semi-martingales to be suitably chosen. The functions $h_{it}(x)$, $i = 0,1,\ldots,N$, are given by

$$h_{it}(x) = \frac{1}{i!} He_i\left[\frac{x-\tilde{X}_t}{p^{\frac{1}{2}}(t)}\right], \tag{28}$$

with $He_i(x)$, $i = 0,1,\ldots$, Hermite polynomials, orthogonal with respect to the weight function $\exp(-\frac{1}{2}x^2)$ (see Abramowitz and Stegun, 1965, for the properties of Hermite polynomials). The function $\phi_t(x)$ is a Gaussian probability density function with mean $\tilde{X}_t$ and variance p(t), while, finally, $\tilde{X}$ and p are the solutions of the differential equations

$$d\tilde{X}_t = (\alpha\tilde{X}_t+\lambda h)dt + \frac{\gamma}{\mu}p(t)(dY_t-\gamma\tilde{X}_t dt),$$

$$\dot{p}(t)= 2\alpha p(t) + \lambda h^2 - \frac{\gamma^2}{\mu}p^2(t), \tag{29}$$

both for $t \geq t_o$. The form of (27) indicates that an approximate solution is sought by modifying the solution derived from the Kalman filter, which is the best linear filter for the problem at hand. Equations (29) describe the Kalman filter.

The processes S_{it}, $i = 0,1,\ldots,N$, will be chosen in such a way that at each instant t,

$$d\tilde{p}_{t|t}(x) - L\tilde{p}_{t|t}(x)dt - \frac{\gamma}{\mu}(x-\hat{X}_t)\tilde{p}_{t|t}(x)(dY_t-\gamma\hat{X}_t dt), \tag{30}$$

where now

$$\hat{X}_t = \int_{-\infty}^{\infty} x\tilde{p}_{t|t}(x)dx, \tag{31}$$

is minimized in some suitable sense. Let

$$dS_{it} = U_{it}dt + dS^m_{it}, \qquad i = 0,1,\ldots,N, \tag{32}$$

where U is adapted to Y, and where S^m_i is a Y-martingale. Substitution of (27) into (30) shows that the martingale part in (30) can be canceled completely by choosing

$$dS^m_{it} = \frac{\gamma}{\mu}[(\tilde{X}_t-\hat{X}_t)S_{it}+p^{\frac{1}{2}}(t)S_{i+1,t}](dY_t-\gamma\hat{X}_t dt), \qquad i = 0,1,\ldots,N-1,$$

$$dS^m_{Nt} = \frac{\gamma}{\mu}[(\tilde{X}_t-\hat{X}_t)S_{Nt}(dY_t-\gamma\hat{X}_t dt). \tag{33}$$

Using various properties of Hermite polynomials, a rather lengthy calculation shows that with this choice of the martingale parts of the S_i, (30) is given by

$$\begin{aligned} &d\tilde{p}_{t|t}(x) - L\tilde{p}_{t|t}(x)dt - \frac{\gamma}{\mu}(x-\hat{X}_t)\tilde{p}_{t|t}(x)(dY_t-\gamma\hat{X}_t dt) \\ &= \sum_{i=0}^{N}[U_{it}+S_{it}(\alpha-\lambda i+\frac{\gamma^2}{\mu}pi)]h_{it}(x)\phi_t(x)dt + \frac{\lambda h}{p^{\frac{1}{2}}}\sum_{i=0}^{N} S_{it}(i+1)h_{i+1,t}(x)\phi_t(x)dt \\ &\quad + \frac{\lambda h^2}{2p}\sum_{i=0}^{N} S_{it}(i+2)(i+1)h_{i+2,t}(x)\phi_t(x)dt - \lambda\tilde{p}_{t|t}(x-h)dt. \end{aligned} \tag{34}$$

Define

$$d\tilde{p}_{t|t}(x)-L\tilde{p}_{t|t}(x)dt-\frac{\gamma}{\mu}(x-\hat{X}_t)\tilde{p}_{t|t}(x)(dY_t-\gamma\hat{X}_t dt) \stackrel{df}{=} e_t(x)dt. \tag{35}$$

Then the Ritz-Galerkin approach (see e.g. Collatz, 1966) to obtaining an approximate solution of the stochastic partial differential-difference equation consists of choosing at each fixed t the unknown quantities U_{it}, $i = 0,1,\ldots,N$, such that an error criterion of the form

$$\tfrac{1}{2}\int_{-\infty}^{\infty} e_t^2(x)w_t(x)dx \tag{36}$$

is minimized, where w_t is a suitable weighting function. A convenient choice turns out to be

$$w_t(x) = \frac{1}{\phi_t(x)}, \qquad -\infty < x < \infty. \tag{37}$$

Differentiation of (36) with respect to U_{it}, $i = 0,1,\ldots,N$, setting these derivatives equal to zero, and further use of the properties of Hermite polynomials yields the following explicit expression for the U_i:

$$U_{jt} = j[\alpha-\frac{\gamma^2}{\mu}p(t)]S_{jt} + \lambda\sum_{i=0}^{j-3}\binom{j}{i}\left[\frac{h}{p^{\frac{1}{2}}(t)}\right]^{j-i} S_{it}, \qquad j = 0,1,\ldots,N. \tag{38}$$

Here the convention is adopted that a summation cancels if the lower limit exceeds the upper limit. The derivation may be completed by obtaining expressions for the first three moments of the approximate density function (27). It is found that

$$\int_{-\infty}^{\infty} \tilde{p}_{t|t}(x)dx = S_{0t},$$

$$\hat{X}_t = \int_{-\infty}^{\infty} x\tilde{p}_{t|t}(x)dx = S_{0t}\tilde{X}_t + S_{1t}p^{\frac{1}{2}}(t),$$

$$M_{2t} \stackrel{df}{=} \int_{-\infty}^{\infty}(x-\hat{X}_t)^2\tilde{p}_{t|t}(x)dx = (S_{0t}+S_{2t})p(t) + 2S_{1t}(\tilde{X}_t-\hat{X}_t)p^{\frac{1}{2}}(t)$$
$$+ S_{0t}(\tilde{X}_t-\hat{X}_t)^2. \tag{39}$$

Moreover, it is easily established that (32), (33, (38) possess the solution $S_{0t} = 1$, $t \geq t_o$, which may be used to simplify some of the relations. The complete approximate filtering algorithm may now be summarized as follows:

$$\hat{X}_t = \tilde{X}_t + p^{\frac{1}{2}}(t)S_{1t}, \tag{40}$$

$$M_{2t} = p(t) - (\tilde{X}_t - \hat{X}_t)^2 + p(t)S_{2t}, \tag{41}$$

$$dS_{it} = i[\alpha - \frac{\gamma^2}{\mu}p(t)]S_{it}dt + \lambda \sum_{k=0}^{i-3} \binom{i}{k} [\frac{h}{p^{\frac{1}{2}}(t)}]^{i-k} S_{kt}dt$$

$$+ \frac{\gamma}{\mu}[(\tilde{X}_t - \hat{X}_t)S_{it} + p^{\frac{1}{2}}(t)S_{i+1,t}](dY_t - \gamma\hat{X}_t dt), \quad i = 0,1,\ldots,N, \tag{42}$$

$$d\tilde{X}_t = (\alpha\tilde{X}_t + \lambda h)dt + \frac{\gamma}{\mu}p(t)(dY_t - \gamma\tilde{X}_t dt), \tag{43}$$

$$\dot{p}(t) = 2\alpha p(t) + \lambda h^2 - \frac{\gamma^2}{\mu}p^2(t). \tag{44}$$

Here, by convention, $S_{N+1,t} = S_{N+2,t} = 0$ for $t \geq t_o$, while the first of the differential equations for the S_i may be replaced with $S_{0t} = 1$, $t \geq t_o$. It may be shown that the approximate algorithm possesses the following properties:

(a) The approximate conditional expectation $\hat{X}_t$ and the approximate conditional moments M_{nt}, where

$$M_{nt} \overset{df}{=} \int_{-\infty}^{\infty} (x - \hat{X}_t)^n \tilde{p}_{t|t}(x)dx, \tag{45}$$

satisfy the filtering equation (22) and the exact moment equations (24) for $n = 2,3,\ldots,N$.

(b) For $N = 0$, $\hat{X}_t = \tilde{X}_t$ for $t \geq t_o$, while for $N = 1$ and $N = 2$,

$$\lim_{t\to\infty} (\hat{X}_t - \tilde{X}_t) = 0 \tag{46}$$

with probability one, at least for initial deviations that are sufficiently small. This means that only for $N \geq 3$ essential improvement over the Kalman filter may be expected.

Numerical experiences with the approximate filter are reported at the end of this section.

The second approximate method of solving the filtering problem is based on representing the conditional characteristic function

$$\psi_t(u) \overset{df}{=} \widehat{e^{iuX_t}} \tag{47}$$

as

$$\psi_t(u) = \exp\Big[\sum_{k=1}^{\infty} C_{kt}\frac{(iu)^k}{k!}\Big] , \tag{48}$$

with u real. The numbers C_{kt}, $k = 1,2,\ldots,N$, are called the cumulants of the conditional density function $p_{t|t}(x)$ (Cramér, 1961). In terms of the characteristic function ψ_t the stochastic differential equation (16) may be rewritten in the form

$$d\psi_t(u) = \alpha u\psi_t'(u)dt + \lambda\psi_t(u)(e^{iuh}-1)dt - \frac{\gamma}{\mu}i[\psi_t'(u)-\psi_t'(0)\psi_t(u)][dY_t+\gamma i\psi_t'(0)dt], \tag{49}$$

where $\psi_t'(u)$ denotes the partial derivative of $\psi_t(u)$ with respect to u. Substitution of (48) into (49) and expansion of e^{iuh} in powers of u shows that the cumulants C_{kt}, $k = 1,2,\ldots$, obey the stochastic differential equations

$$dC_{kt} = \alpha kC_{kt}dt + \lambda h^k dt - \frac{\gamma^2}{2\mu}\sum_{j=1}^{k-1}\binom{k}{j}C_{j+1,t}C_{k-j+1,t}dt + \frac{\gamma}{\mu}C_{k+1,t}(dY_t-\gamma C_{1,t}dt),$$

$$k = 1,2,\ldots, \qquad t \geq t_o. \tag{50}$$

The first and second conditional moments are respectively given by

$$\hat{X}_t = C_{1t},\ P_{2t} = C_{2t}, \qquad t \geq t_o . \tag{51}$$

As in the case of the conditional moments, (50) forms an infinite set of coupled stochastic differential equations. An approximate solution may be obtained by setting $C_{kt} = 0$ for $k = N+1, N+2,\ldots$, and integrating the resulting N stochastic differential equations for C_{kt}, $k = 1,2,\ldots,N$. This method of truncating has been proposed in a slightly different context by Zeman (1971). The resulting approximate algorithm has the following properties:

(a) The approximate conditional expectation and the approximate conditional moments

M_{nt} satisfy the filtering equation (22) and the exact moment equations (24) for $n = 2,3,\ldots,N-1$.

(b) For $N = 2$ the filter reduces to the Kalman filter (29). Thus, only for $N \geq 3$ improvement over the Kalman filter may be expected.

The performance of the two approximate filtering algorithms has been studied by simulation for the data $\alpha = -4$, $h = 1$, $\lambda = 1$, $\gamma = 1$, and $\mu = 0.02$. The process was simulated over the interval [0,2]. The filtering equations (42-44) and (50) were integrated by straightforward application of Eulers integration rule, with a time step $dt = 0.005$. In each case, the initial conditions were chosen corresponding to the steady-state solution of the (unconditional) Fokker-Planck equation $dp = Lpdt$. The "Ritz-Galerkin" approximate filter was simulated for $N = 0,3,4,\ldots,10$, and the "cumulant" approximate filter for $N = 2,3,\ldots,10$. The simulations showed that, at least for the present data, the more sophisticated approximate filtering algorithms give only a small improvement over the Kalman filter. In both cases, the best results were obtained for N equal to about 7. For greater values of N, the time step $dt = 0.005$ gets too large for the value of N, which results in instabilities in the computed solutions. These instabilities are probably similar to those experienced by Clements and Anderson (1973). In Fig. 1, the trajectory of X_t, $0 \leq t \leq 2$, is compared to the filtered reconstruction obtained by the Kalman filter. Fig. 2 gives the corresponding result for the Ritz-Galerkin approximate filter for N=7. In Fig. 3 the behavior of the approximate conditional variance M_{2t} for the Ritz-Galerkin approximation with N=7 is compared to the behavior of the Kalman filter variance. Finally, in the Table the root mean square reconstruction errors over the intervals [0,0.5], [0.5,1], [1,1.5], and [1.5,2] as well as over the entire interval [0,2] are given for the Kalman

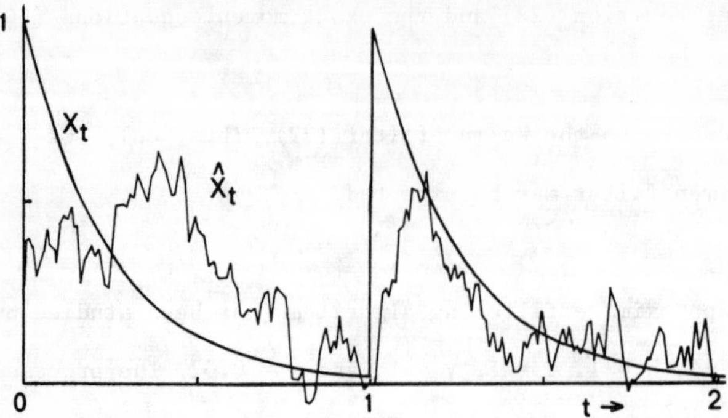

Fig. 1: A realization of X_t and its Kalman filter estimate $\hat{X}_t$.

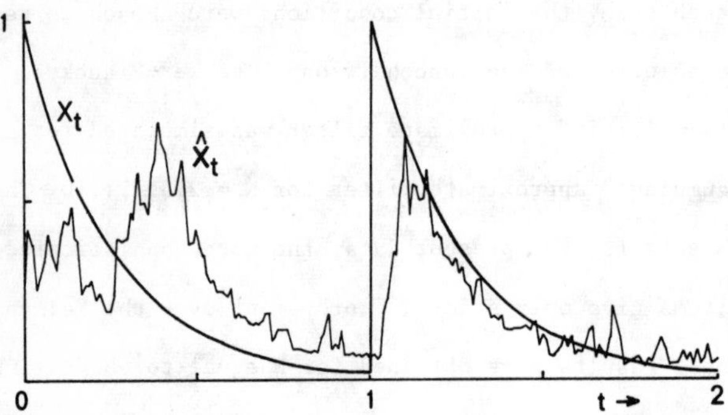

Fig. 2: A realization of X_t and its Ritz-Galerkin estimate $\hat{X}_t$ (N=7).

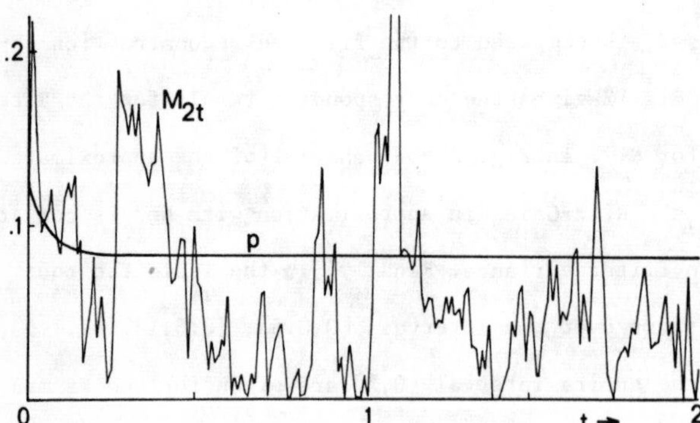

Fig. 3: Kalman variance p(t) and the approximate conditional second central moment M_{2t} for the Ritz-Galerkin estimate of Fig. 2.

RMS RECONSTRUCTION ERRORS FOR VARIOUS FILTERS OVER VARIOUS PERIODS

Filter	Rms reconstruction error over the interval				
	[0,0.5]	[0.5,1]	[1,1.5]	[1.5,2]	[0,2]
Kalman	0.322	0.156	0.307	0.079	0.239
Ritz-Galerkin, N=7	0.334	0.109	0.305	0.044	0.233
Cumulants, N=7	0.333	0.123	0.316	0.049	0.239

filter and the Ritz-Galerkin and cumulant approximate filters (both for N=7).

It is seen that the improvement over the Kalman filter is marginal or nil. The Ritz-Galerkin filter is better than the cumulant filter. In both cases, improvement is obtained during the times when X_t is small. In the case of the cumulant filter, the improvement is offset by a deterioration during the periods when X_t is large. The numbers as given may be compared to the square root of the steady-state variance of the Kalman filter, which is $\sqrt{0.08248} \simeq 0.287$.

4. Extensions

In this section, the results obtained in Section 2 for scalar linear systems with Poisson white noise input will first be extended to multi-dimensional linear systems with Poisson white noise inputs, and next to a particular distributed parameter system with a distributed Poisson white noise input.

Let X_t, $t \geq t_o$, be an n-dimensional vector stochastic process, given by

$$dX_t = A(t)X_t dt + dP_t, \qquad t \geq t_o. \tag{52}$$

Here A(.) is a time-varying matrix, the elements of which are continuous functions of t, while P_t, $t \geq t_o$, is a vector stochastic process, the elements of which are independent, non-homogeneous compound Poisson processes. The i-th component P_{it} of P_t has the intensity function λ_{it}, $t \geq t_o$, while its jumps have the probability distribution of the stochastic variable H_i, which possesses the distribution function $G_i(.)$. The observed process Y_t, $t \geq t_o$, is an m-dimensional vector stochastic process, given by

$$dY_t = C(t)X_t dt + dW_t, \qquad t \geq t_o, \tag{53}$$

where $C(.)$ is another time-varying matrix with continuous elements and where W_t, $t \geq t_o$, is vector Brownian motion with matrix intensity function $V(t)$, $t \geq t_o$. The processes P and W are independent.

To find the appropriate Fokker-Planck equation the basic filtering theorem is applied to the semi-martingale

$$Q_t = e^{iu'X_t}, \qquad t \geq t_o, \tag{54}$$

where $u = \mathrm{col}(u_1, u_2, \ldots, u_n)$ has real elements, and where the prime denotes the transpose. It is easily found by applying the differentiation rule that Q_t may be represented by

$$dQ_t = R_t dt + dM_t, \qquad t \geq t_o, \tag{55}$$

where R is the adapted process

$$R_t \overset{df}{=} iu'A(t)X_t e^{iu'X_t} + \sum_{j=1}^{n} \lambda_{jt} E(e^{iu_j H_j} - 1)e^{iu'X_t}, \qquad t \geq t_o \tag{56}$$

while M is the martingale defined by

$$M_t \overset{df}{=} \sum_{t_o \leq s \leq t} (e^{iu'X_s} - e^{iu'X_{s-}}) - \int_{t_o}^{t} \sum_{j=1}^{n} \lambda_{js} E(e^{iu_j H_j} - 1)e^{iu'X_s} ds, \qquad t \geq t_o. \tag{57}$$

Application of the basic filtering theorem to the semi-martingale Q yields

$$d\hat{Q}_t = \hat{R}_t dt + (\widehat{Q_t X_t'} - \hat{Q}_t \hat{X}_t')C'(t)V^{-1}(t)[dY_t - C(t)\hat{X}_t dt], \tag{58}$$

or

$$d(\widehat{e^{iu'X_t}}) = iu'A(t)\widehat{X_t e^{iu'X_t}}dt + \sum_{j=1}^{n} \lambda_{jt} E(e^{iu_j H_j}-1)\widehat{e^{iu'X_t}}dt$$

$$+ (\widehat{X_t' e^{iu'X_t}} - \hat{X}_t' \widehat{e^{iu'X_t}})C'(t)V^{-1}(t)[dY_t - C(t)\hat{X}_t dt]. \tag{59}$$

In terms of the conditional density function $p_{t|t}(x)$ of X_t, given $\mathcal{Y}_t$, one thus has

$$dp_{t|t}(x) = -\frac{\partial'}{\partial x}[A(t)xp_{t|t}(x)]dt - \sum_{j=1}^{n} \lambda_{jt} p_{t|t}(x)dt$$

$$+ \sum_{j=1}^{n} \lambda_{jt} \int_{-\infty}^{\infty} p_{t|t}(x_1,..,x_{j-1},x_j-h,x_{j+1},..,x_n)dG_j(h)dt$$

$$+ p_{t|t}(x)(x-\hat{X}_t)'C'(t)V^{-1}(t)[dY_t - C(t)\hat{X}_t dt], \qquad t \geq t_o. \tag{60}$$

Here

$$\frac{\partial'}{\partial x}q(x) \stackrel{df}{=} \sum_{j=1}^{n} \frac{\partial q_j(x)}{\partial x_j}, \tag{61}$$

if $q(x) = \mathrm{col}[q_1(x), q_2(x), \ldots, q_n(x)]$. Eq. (60) is the conditional Fokker-Planck equation for the multi-dimensional case.

As another demonstration of the power of the fundamental filtering theorem its application to the distributed filtering problem described in Section 1 is considered. The time evolution of the stochastic field $X_t(r)$, $0 \leq r \leq L$, is described by the partial differential equation

$$\frac{\partial X_t(r)}{\partial t} = D\frac{\partial^2 X_t(r)}{\partial r^2} - V\frac{\partial X_t(r)}{\partial r} + Z_t(r), \qquad 0 \leq r \leq L, \qquad t \geq 0, \tag{62}$$

Here $X_t(r)$ is the concentration of the chemical at time t and location r along the river, D is the dispersion coefficient, and V the water velocity, while $Z_t(r)$ is the rate of increase of the concentration at location r and time t as a result of the deposits of chemical into the water. Let v be a suitable, but otherwise arbitrary real function on [0,L], and consider the inner product

$$(v,X_t) \stackrel{df}{=} \int_0^L v(r)X_t(r)dr, \qquad t \geq t_o. \tag{63}$$

It follows from (62) that

$$d(v,X_t) = (v,KX_t)dt + dP_t(v), \qquad t \geq t_o. \tag{64}$$

K is the partial differential operator represented by the first two terms on the right hand side of (62), while $P_t(v)$ is the process

$$P_t(v) \stackrel{df}{=} \int_0^L v(r)dr \int_0^t Z_\theta(r)d\theta, \qquad t \geq 0. \tag{65}$$

As a result of the assumptions on the polluting process as described in Section 1, P is a jump process with stationary, independent increments, with characteristic function

$$E(e^{iuP_t}) = \exp[t \int_0^L \lambda(r)\{\phi_r[uv(r)]-1\}dr], \qquad t \geq 0. \tag{66}$$

Here $\phi_r(z) \stackrel{df}{=} E(e^{izH_r})$ is the characteristic function of the stochastic variable H_r that describes the amount of chemical per unit cross sectional area deposited at location r. The result (66) may intuitively be derived by approximating

$$P_t(v) = \sum_{j=1}^{n} v(r_j)P_{jt}, \tag{67}$$

where P_{jt} describes the total amount of chemical deposited in a river section of length d_jr, positioned at the location r_j. P_{jt} is a compound Poisson process with rate parameter $\lambda(r_j)d_jr$ and jumps distributed as H_{r_j}.

It follows from the form of (66) that $P_t(v)$ is a compound Poisson process with rate parameter $\bar{\lambda} \stackrel{df}{=} \int_0^L \lambda(r)dr$, while its jumps have the characteristic function

$$\frac{1}{\bar{\lambda}} \int_0^L \lambda(r)\phi_r[uv(r)]dr. \tag{68}$$

The mean value function of P is

$$E(P_t) = t \int_0^L \lambda(r)E(H_r)v(r)dr = (v,\lambda m)t, \qquad t \geq 0, \tag{69}$$

where $m(r) \overset{df}{=} E(H_r)$. As a result, the process $P_t(v)-(v,\lambda m)t$, $t \geq 0$, is a martingale.

Assuming that the observation equation is given by

$$dY_t = (c,X_t)dt + dW_t, \tag{70}$$

where Y is an m-dimensional process, c(.) an m-dimensional vector function on [0,L], and W an m-dimensional Brownian motion, independent of Z, with constant intensity matrix V, application of the basic filtering theorem to $Q_t \overset{df}{=} e^{i(v,X_t)}$ leads to the filtering equation

$$d[\widehat{e^{i(v,X_t)}}] = i\widehat{(v,KX_t)e^{i(v,X_t)}}dt + \widehat{e^{i(v,X_t)}}\int_0^L \lambda(r)\{\phi_r[v(r)]-1\}dr\, dt$$
$$+ [\widehat{(c,X_t)e^{i(v,X_t)}} - (c,\hat{X}_t)\widehat{e^{i(v,X_t)}}]'V^{-1}[dY_t-(c,\hat{X}_t)dt]. \tag{71}$$

The analogy of this equation to (59) is obvious. It does of course not permit inverse Fourier transformation, as in the finite-dimensional case. The problem is how to go on from this point to obtain practical solutions. By expanding the exponentials and ϕ_r in (71) with respect to v, the moment equations can be generated. The first of these is

$$d\hat{X}_t(r) = \widehat{KX_t(r)}dt + \lambda(r)m(r)dt + \widehat{(c,X_t-\hat{X}_t)[X_t(r)-\hat{X}_t(r)]}'V^{-1}[dY_t-(c,\hat{X}_t)dt]. \tag{72}$$

The results as given are also valid in case K is a nonlinear operator.

5. The smoothing problem

In this section the smoothing problem for the scalar problem outlined in Section 1 will be considered. It is easy to obtain results for the fixed point smoothing problem, that is, the problem of determining the conditional density function $p_{s|t}(x)$ of X_s given $\mathcal{Y}_t$, with s fixed, and $t \geq s$. Define $Z_t \overset{df}{=} X_s$, $t \geq s$. Then X_t and Z_t simultaneously satisfy for $t \geq s$

$$\begin{aligned} dX_t &= \alpha X_t dt + h dP_t, \\ dZ_t &= 0. \end{aligned} \tag{73}$$

The observation equation for the joint process $\mathrm{col}(X_t, Z_t)$ still is

$$dY_t = \alpha X_t dt + dW_t, \qquad t \geq s. \tag{74}$$

Application of the results obtained in the preceding section for vector-valued processes yields that the simultaneous conditional density function $q_{t|t}(x,z)$ of X_t and Z_t, given $\mathcal{Y}_t$, satisfies the conditional Fokker-Planck equation

$$\begin{aligned} dq_{t|t}(x,z) = &- \frac{\partial}{\partial x}[\alpha x q_{t|t}(x,z)]dt + \lambda[q_{t|t}(x-h,z) - q_{t|t}(x,z)]dt \\ &+ \frac{\gamma}{\mu}(x-\hat{X}_t)q_{t|t}(x,z)(dY_t - \gamma \hat{X}_t dt), \qquad t \geq s, \end{aligned} \tag{75}$$

where, as before,

$$\hat{X}_t = \int_{-\infty}^{\infty} x p_{t|t}(x)dx = \int_{-\infty}^{\infty}\int_{-\infty}^{\infty} x q_{t|t}(x,z)dxdz. \tag{76}$$

The initial condition for (75) is

$$q_{s|s}(x,z) = p_{s|s}(x)\delta(z-x). \tag{77}$$

It is attempted to find a solution to (75) and (77) of the form

$$q_{t|t}(x,z) = r(t,x;s,z)p_{t|t}(x), \tag{78}$$

where r is to be determined. Substitution of (78) into (75) and (77) shows that these equations are satisfied if r obeys the relations

$$\frac{\partial r(t,x;s,z)}{\partial t} = \frac{1}{p_{t|t}(x)} L_x[p_{t|t}(x)r(t,x;s,z)] - \frac{r(t,x;s,z)}{p_{t|t}(x)} L_x p_{t|t}(x),$$

$$t \geqslant s, \qquad (79)$$

$$r(s,x;s,z) = \delta(x-z). \qquad (80)$$

Here L_x is the operator defined by (19), with the subscript x added to indicate that it operates on the argument x. The above equations show first of all that r is a differentiable process. They moreover show that $r(t,x;s,z)$, is nothing but the Green's function for the initial value problem defined by (79) and (80). It follows that as a function of s and z, the function r satisfies the adjoint equation

$$\frac{\partial r(t,x;s,z)}{\partial s} = \frac{r(t,x;s,z)}{p_{s|s}(z)} L_z p_{s|s}(z) - p_{s|s}(z) L_z^*[\frac{r(t,x;s,z)}{p_{s|s}(z)}], \qquad s \leqslant t, \qquad (81)$$

where L^* is the adjoint of L, i.e.,

$$L^* p(x) \stackrel{df}{=} \alpha x \frac{\partial p(x)}{\partial x} + \lambda[p(x+h)-p(x)]. \qquad (82)$$

The transition to the adjoint equation is made because it is desired to find the solution to the fixed interval smoothing problem (where t is fixed and s variable, with $s \leqslant t$), which is more interesting. One has for the conditional density function $p_{s|t}(z)$ of X_s, given Y_t, with $s \leqslant t$,

$$d_s p_{s|t}(z) = d_s \int_{-\infty}^{\infty} q_{t|t}(x,z)dx = d_s \int_{-\infty}^{\infty} r(t,x;s,z)p_{t|t}(x)dx$$

$$= \int_{-\infty}^{\infty} p_{t|t}(x) d_s r(t,x;s,z)dx. \qquad (83)$$

With the aid of (81) it easily follows that

$$\frac{\partial p_{s|t}(z)}{\partial s} = \frac{p_{s|t}(z)}{p_{s|s}(z)} L_z p_{s|s}(z) - p_{s|s}(z) L_z^*[\frac{p_{s|t}(z)}{p_{s|s}(z)}], \qquad s \leqslant t. \qquad (84)$$

The corresponding terminal condition follows by specifying $p_{t|t}(z)$.

An alternative presentation of the solution of the smoothing problem is to define

$$k_{s|t}(z) \overset{\mathrm{df}}{=} \frac{p_{s|t}(z)}{p_{s|s}(z)}, \qquad t_o \leq s \leq t. \tag{85}$$

Application of the differentiation rule yields with the use of (84) and (18)

$$d_s k_{s|t}(z) = -L_z^* k_{s|t}(z)ds - \frac{\gamma}{\mu}(z-\hat{X}_s)k_{s|t}(z)(dY_t-\gamma\hat{X}_s ds)+\frac{\gamma^2}{\mu}(z-\hat{X}_s)^2 k_{s|t}(z)ds, \quad s \leq t,$$

$$k_{t|t}(z) = 1. \tag{86}$$

Except for the form of the operator L, the expressions derived in this section are similar to results obtained for Gaussian white noise by Liptser and Shiryaev (1968), Loh and Eyman (1970), Lo (1970), Leondes, Peller and Stear (1970), and Lee (1971). Most of these authors use Bucy's representation theorem (Bucy and Joseph, 1968) as a starting point for their derivations. As in the case of the filtering problem, approximate numerical methods have to be developed to obtain practical solutions.

6. Conclusions

In this paper it has been found that martingale theory and martingale calculus constitute excellent tools for the solution of filtering problems for linear systems excited by Poisson white noise. Various questions have been considered: filtering problems for scalar, multi-dimensional and distributed linear systems, as well as the smoothing problem for the scalar case. Conditional Fokker-Planck equations, moment and cumulant evolution equations are easily determined. The optimal filters are all infinite-dimensional because no closed finite set of moment or cumulant equations may be obtained. Two approximate methods of solving the filtering equations have been developed, but their performance in a numerical simulation study proved to be disappointing in terms of improvement over the Kalman filter. It would be interesting to find theoretical bounds on the possible improvement over the Kalman filter. Approximate methods for the numerical solution of the smoothing problem still have to be developed.

References

M. Abramowitz and I.A. Stegun, eds. (1965), Handbook of Mathematical Functions. Dover, New York.

A. Benssoussan (1971), Filtrage Optimal des Systèmes Lineaires. Dunod, Paris.

R.S. Bucy, P.D. Joseph (1968), Filtering for Stochastic Processes, with Applications to Guidance. Interscience, New York.

J.M.C. Clark (1973), Two recent results in nonlinear filtering theory. In: Recent Mathematical Developments in Control, edited by D.J. Bell. Academic Press, London.

D.J. Clements, B.D.O. Anderson (1973), Well-behaved Itô equations with simulations that always misbehave. IEEE Trans. Automatic Control 18, 6, 676-677.

L. Collatz (1966), The numerical treatment of differential equations. Springer, Berlin.

H. Cramér (1961), Mathematical Methods of Statistics. Ninth printing. Princeton University Press, Princeton.

C. Doléans-Dade, P. Meyer (1970), Intégrales stochastiques par rapport aux martingales locales. In: Seminaire de probabilités IV, Lecture Notes in Mathematics, vol. 124. Springer, Berlin.

M. Fujisaka, G. Kallianpur, H. Kunita (1972), Stochastic differential equations for the nonlinear filtering problem. Osaka J. Math., 9, 1, 19-40.

D.G. Lainiotis (1971), Optimal non-linear estimation. Int. J. Control 14, 6, 1137-1148.

G.M. Lee (1971), Nonlinear interpolation. IEEE Trans. Information Theory 17, 1, 45-49.

C.T. Leondes, J.B. Peller, E.B. Stear (1970), Nonlinear smoothing theory. IEEE Trans. Sys. Sc. Cyb. 6, 1, 63-71.

R. Sh. Liptser, A.N. Shiryaev (1968), Non-linear interpolation of components of Markov diffusion processes (direct equations, effective formulas). Th. of Prob. and its Appl. XIII, 4, 564-583.

J.T. Lo (1970), On optimal nonlinear estimation - Part I: Continuous observation. Proc. of the 8th Annual Allerton Conf. on Circuit and Systems Theory, Urbana, Ill.

J.T. Lo (1972), Finite-dimensional sensor orbits and optimal nonlinear filtering. IEEE Trans. Information Theory 18, 5, 583-588.

N.K. Loh, E.D. Eyman (1970), Nonlinear smoothing for stochastic processes. Proc. Fourth Asilomar Conference on Circuits and Systems, edited by S.R. Parker, Pacific Grove, Calif., 639-643.

J.H. van Schuppen (1973), Estimation theory for continuous time processes, a martingale approach. Electronics Research Laboratory Memorandum No. ERL-M405, College of Engineering, University of California, Berkeley, Calif.

E. Wong (1973), Recent progress in stochastic processes - a survey. IEEE Trans. Information Theory 19, 3, 262-275.

J.L. Zeman (1971), Approximate Analysis of Stochastic Processes in Mechanics. Udine Courses and Lectures No. 95, Springer, Berlin.

A Minimum Principle for Controlled Jump Processes

by

Raymond Rishel
University of Kentucky

In Queueing Theory and many other fields problems of control arise for stochastic processes with piecewise constant paths. In this paper the validity of optimality conditions analagous to the Pontryagin Maximum Principle for deterministic control problems is investigated for this type of stochastic process. A minimum principle which involves the conditional jump rate, the conditional state jump distribution, system performance rate, and the conditional expectation of the remaining performance is obtained. The conditional expectation of the remaining performance plays the role of the adjoint variables. This conditional expectation satisfies a type of integral equation and an infinite system of ordinary differential equations.

Whether these conditions will lead to computational results for specific systems is still an open question. It appears that obtaining the conditional expectation of the remaining performance may be a complicated matter. An example is given showing the implication of the maximal principle for the case of a controlled Poisson source with unknown input rate.

An approach directed toward control of Jump processes using martingales and stochastic integrals to define the processes involved has been initiated by Boel, Varaiya and Wong in [4] and [5]. The approach of this paper is quite different in that the processes are defined through a simple correspondence between a jump process and the corresponding discrete process consisting successive states and times between jumps to these states of the jump process.

Part I Jump Processes

For jump Markov processes the correspondence between the process itself and an associated discrete process consisting of the successive states of the process $x(t)$ and times between jumps to these states has been used as an effective tool [1] p. 334, [2] p. 354 in both the construction of the processes and their analysis. Many of the properties of this correspondence carry over to general jump processes. The objective of part 1 will to be to define this correspondence and then to show that the analytical properties necessary for the optimal control discussion of part II may be deduced from it. A jump process is defined to be a stochastic process with piecewise constant, right continuous paths, so that the times between successive jumps of the process are finite with probability one.

Jump Processes and Corresponding Discrete Processes

To exhibit the correspondence, let $x(t)$ be a jump process with values in E^k. Let $\tau_1, \tau_2, \cdots$ denote the times of the successive jumps of the sample functions of $x(t)$. Define the times between jumps s_i by

$$s_1 = \tau_1 \text{ and } s_n = \tau_n - \tau_{n-1} \quad \text{if } n > 1$$

Define the successive states by

$$x_o = x(0) \text{ and } x_n = x(\tau_n) \text{ if } n \geqslant 1$$

It can be seen from the right continuity of the sample functions of x(t) that

(1) $$x_o, s_1, x_1, \cdots, s_n, x_n, \cdots$$

is a sequence of random elements. Hence this sequence is a discrete stochastic process corresponding to x(t).

The probability measure induced on this sequence is completely determined [3] p. 364 when the conditional distribution functions

$$P\{s_{n+1} < a \mid x_o, s_1, x_1, \cdots, s_m, x_n\}$$

and

$$P\{x_{n+1} \in A \mid x_o, s_1, x_1, \cdots, s_n, x_n, s_{n+1}\}$$

are specified.

Conversely suppose there is a discrete sequence

$$x_o, s_1, x_1, \cdots, s_n, x_n \cdots$$

of random elements so that x_i has values in E^k and s_i has values in $R^+ = [0,\infty)$. Define x(t) by

(2) $$x(t) = \begin{cases} x_o & \text{if } 0 \leq t < s_1 \\ x_n & \text{if } \sum_{i=1}^{n} s_i \leq t < \sum_{i=1}^{n+1} s_i . \end{cases}$$

This defines a stochastic process x(t) on the random interval $0 \leq t < \sum_{n=1}^{\infty} s_n$.

Next a sufficient condition that (2) define a process x(t) on $0 \leq t < \infty$ with probability one will be given. In the remainder of the paper the notations

$$X_n = (x_o, s_1, x_1, \cdots, s_n, x_n)$$

and

$$\widetilde{X}_{n+1} = (x_o, s_1, x_1, \cdots, s_n, x_n, s_{n+1})$$

will be used to condense the writing of certain formulas. Consider the case in which the conditional distribution of s_{n+1} has a density; that is

$$P\{s_{n+1} < a \mid X_n\} = \int_o^a f(s \mid X_n) ds.$$

Denote $P\{x_{n+1} \in A \mid \widetilde{X}_{n+1}\}$ by $P\{x_{n+1} \in A \mid \widetilde{X}_{n+1}\} = \Pi(A \mid \widetilde{X}_{n+1})$.

If there is a constant K so that

(3) $$f(a \mid X_n) < K$$

it follows that

$$P\{\sum_{i=1}^{n} s_i \leq t\} = \int_o^t \int_{E^k} \cdots \int_o^{t-\tau_{n-1}} \int_{E^k} f(s_1 \mid X_o)\, \Pi(dx_1 \mid \widetilde{X}_1) \cdots f(s_n \mid X_{n-1})\, \Pi(dx_n \mid \widetilde{X}_n) ds_n \cdots ds_1$$
$$\leq K^n \int_o^t \int_o^{t-s_1} \cdots \int_o^{t-\sum_{i=1}^{n-1} s_i} ds_n \cdots ds_1 = \frac{K^n t^n}{n!} .$$

Hence in this case

$$P\{\sum_{i=1}^{\infty} s_i \leq t\} = \lim_{n\to\infty} P\{\sum_{i=1}^{n} s_n \leq t\} = 0 \quad .$$

This implies that (3) is a sufficient condition for (2) to define a jump process on the interval $[0,\infty)$.

Jump Rates

The conditional probability $P\{a \leq s_n < a + \Delta a \mid X_{n-1}, s_n > a\}$ is given by

(4) $$P\{a \leq s_n < a + \Delta a \mid X_{n-1}, s_n > a\} = \frac{\int_a^{a+\Delta a} f(s \mid X_{n-1}) ds}{\int_a^{\infty} f(s \mid X_{n-1}) ds} \; .$$

Define the conditional density $q(a \mid X_{n-1})$ by

$$q(a \mid X_{n-1}) = \frac{f(a \mid X_{n-1})}{\int_a^{\infty} f(s \mid X_{n-1}) ds}$$

The relation (4) implies

$$P\{a \leq s_n < a + \Delta a \mid X_{n-1}, s_n > a\} = q(a \mid X_{n-1})\, \Delta a + o(\Delta a)$$

Thus under the correspondence (2) between (1) and $x(t)$; $q(a|X_{n-1})$ is the conditional rate at which jumps of $x(t)$ occur, at time $t = a + \sum_{i=1}^{n-1} s_i$, given $\{X_{n-1}, s_n > a\}$.

An elementary computation shows

(5) $$e^{-\int_s^t q(a \mid X_{n-1}) da} = \frac{\int_t^{\infty} f(a \mid X_{n-1}) da}{\int_s^{\infty} f(a \mid X_{n-1}) da}$$

and that f and q are related by

(6) $$f(a \mid X_{n-1}) = q(a \mid X_{n-1})\, e^{-\int_0^a q(s \mid X_{n-1}) ds}$$

Functionals on a Jump Process and Their Representations

Let $x(t)$ be a jump process and σ_t the σ–field generated by the random elements

$$\{x(\tau) \; ; 0 \leq \tau \leq t\} .$$

A random variable $h(t)$ will be called a **functional on the past of** $x(\cdot)$ **up to time** t, if $h(t)$ is σ_t measurable. Generally for brevity the phrase "on the past of x up to time t" will be omitted and $h(t)$ will be called a **functional on** $x(\cdot)$.

Let σ_n denote the σ–field generated by

$$X_n = \{x_0, s_1, x_1, \cdots, s_n, x_n\} \; .$$

Define n(t) by

(7) $$n(t) = \sup\{n: \sum_{i=1}^{n} s_i \leq t\}$$

It can be seen that

(8) $$\sigma_t \cap \{n(t) = n\} \subset \sigma_n \cap \{n(t) = n\} .$$

To see this notice that $\sigma_t \cap \{n(t) = n\}$ is a σ–field on the set $\{n(t) = n\}$ and it is generated by sets of the form

(9) $$\{x_j \in A, \sum_{i=1}^{j} s_i \leq \tau \leq \sum_{i=1}^{j+1} s_i\}$$

where $0 \leq \tau \leq t$ and $j = 1, \cdots, n$. Each of these sets is the intersection of a σ_n measurable set with $\{n(t) = n\}$.

From (8) it follows that the restriction to $\{n(t) = n\}$ of a σ_t measurable function is the restriction to $\{n(t) = n\}$ of a σ_n measurable function. A sketch of the argument to show this is the following. The relationship (8) implies, if χ defined **on** $\{n(t) = n\}$ is the characteristic function of a set in $\{n(t) = n\} \cap \sigma_t$, that χ is the characteristic function of a set in $\{n(t) = n\} \cap \sigma_n$ and hence the restriction of the characteristic function of a σ_n measurable set to $\{n(t) = n\}$. Since this is true the usual uniform approximation agrument by simple functions implies the more general statement.

A σ_n measurable function [1] p.395 can be written as a Borel function of the random elements

$$X_n = \{x_o, s_1, x_1, \cdots, s_n, x_1\}$$

Thus there are Borel functions $h(t,X_n)$ defined on $E^{k(n+1)} \times R^{+n}$ so that

(10) $$h(t) = \sum_{n=0}^{\infty} \chi_{\{n(t) = n\}} h(t, x_o, s_1, x_1, \cdots s_n, x_n).$$

In (10) and in the rest of the paper χ_A denotes the characteristic function of the set A. An equivalent notation for (10) is

(11) $$h(t) = h(t, x_o, s_1, x_1, \cdots s_{n(t)}, x_{n(t)}).$$

The common abuse of using X_n to denote both a vector of variables of $E^{k(n+1)} \times R^{+n}$ and a sequence of random elements with values in $E^{k(n+1)} \times R^{+n}$ has been used above and will be followed in the rest of the paper.

A sequence of functions $h(t,X_n)$ so that (10) or (11) holds will be called a **representation** of h(t). Define a norm of a representation by

$$|| h(t) || = \sup_{X_n} \frac{|h(t,X_n)|}{1 + || x_n ||^2}$$

A Formula for the Conditional Expectation of a Functional

For a functional h(t) on x(·) consider

$$E\{h(t) \mid \sigma_s\} \tag{12}$$

Formula (12) defines a functional on x(·) in the variable s.

Using (5) (6) and (10) we see, if $s < t$, and the integrals involved are finite, that (12) has the representation

$$E\{[h(t)]\,(s,X_j) = h(t,X_j)\, e^{-\int_{s-\tau_j}^{t-\tau_j} q(a_1 \mid X_j)da_1} \tag{13}$$

$$+ \int_{s-\tau_j}^{t-\tau_j} \int_{E^k} h(t,X_{j+1})\, q(s_{j+1} \mid X_j)\, e^{-\int_{s-\tau_j}^{s_{j+1}} q(a_1 \mid X_j)da_1}\, \Pi(dx_{j+1} \mid \widetilde{X}_{j+1})\, e^{-\int_o^{t-\tau_{j+1}} q(a_2 \mid X_{j+1})da_2} ds_{j+1}$$

$$+ \int_{s-\tau_j}^{t-\tau_j} \int_{E^k} \int_o^{t-\tau_{j+1}} h(t,X_{j+2})\, q(s_{j+1} \mid X_j)\, e^{-\int_{s-\tau_j}^{s_{j+1}} q(a_1 \mid X_j)da_1}\, \Pi(dx_{j+1} \mid \widetilde{X}_{j+1})$$

$$q(s_{j+2} \mid X_{j+1})\, e^{-\int_o^{s_{j+2}} q(a_2 \mid X_{j+1})da_2}\, \Pi(dx_{j+2} \mid \widetilde{X}_{j+2})\, e^{-\int_o^{t-\tau_{j+2}} q(a_3 \mid X_{j+2})da_3}\, ds_{j+2}\, ds_{j+1} + \cdots .$$

In (13) and the rest of the paper s_{j+1}, x_{j+1}, and X_{j+1} are related by

$$X_{j+1} = (x_o, s_1, x_1, \cdots, s_{j+1}, x_{j+1})$$

and a similar convention applies to s_{j+2}, x_{j+2}, and X_{j+2} and corresponding terms with successively higher indicies.

Theorem 1. If there are constants K and M so that for every j

$$q(a \mid X_j) < K \tag{14}$$

and

$$\int (1 + ||x_{j+1}||^2)\, \Pi(dx_{j+1} \mid \widetilde{X}_{j+1}) \leq M(1 + ||x_j||^2), \tag{15}$$

then

$$|| E[h(t)](s) || \leq e^{MK(t-s)} || h(t) || \tag{16}$$

Proof: Applying (14) and (15) in formula (13) gives

$$| E[h(t)]\,(s,X_j) | \leq || h(t) ||\, (1 + || x_j ||)^2)$$

$$[1 + \int_{s-\tau_j}^{t-\tau_j} KM\, ds_{j+1} + \int_{s-\tau_j}^{t-\tau_j} \int_o^{t-\tau_{j+1}} (KM)^2 ds_{j+2}\, ds_{j+1} + \int_{s-\tau_j}^{t-\tau_j} \int^{t-\tau_{j+1}} \int_o^{t-\tau_{j+2}} (KM)^3\, ds_{j+3}\, ds_{j+2}\, ds_{j+1} + \cdots]$$

$$\leq || h(t) ||\, (1 + || x_j ||^2)\, [1 + KM(t-s) + \cdots + \frac{[KM(t-s)]^n}{n!} + \cdots .$$

Theorem 2: If $\|h(t)\| < \infty$ and the hypothesis of Theorem 1 hold then

$$(17)\quad E[h(t)]\,(s,X_j) = h(t,X_j) + \int_{s-\tau_j}^{t-\tau_j} \int_{E^k} E[h(t)]\,(\tau_{j+1},X_{j+1})\,q(s_{j+1}\mid X_j)\,e^{-\int_{s-\tau_j}^{s_{j+1}} q(a\mid X_j)da}\,\Pi(dx_{j+1}\mid \widetilde{X}_{j+1})ds_{j+1}$$

$$-\int_{s-\tau_j}^{t-\tau_j} h(t,X_j)\,q(a_2\mid X_j)\,e^{-\int_{s-\tau_j}^{a_2} q(a_1\mid X_j)da_1}\,da_2$$

Proof: The hypotheses imply formula (13) holds and an interchange of summation and integration is valid. Hence

$$(18)\qquad E[h(t)]\,(s,X_j) = h(t,X_j)\,e^{-\int_{s-\tau_j}^{t-\tau_j} q(a\mid X_j)da}$$

$$+\int_{s-\tau_j}^{t-\tau_j}\int_{E^n} E[h(t)]\,(\tau_{j+1},X_{j+1})\,q(s_{j+1}\mid X_j)\,e^{-\int_{s-\tau_j}^{s_{j+1}} q(a\mid X_j)da}\,\Pi(dx_{j+1}\mid \widetilde{X}_{j+1})ds_{j+1}.$$

Applying formula (19)

$$(19)\qquad e^{-\int_{s-\tau_j}^{t-\tau_j} q(a\mid X_j)da} = 1-\int_{s-\tau_j}^{t-\tau_j} q(a_2\mid X_j)\,e^{-\int_{s-\tau_j}^{a_2} q(a_1\mid X_j)da_1}\,da_2$$

in (18) gives (17).

Recall that

$$(\tau_j,X_j) = \Big(\sum_{i=1}^{j} s_i,\, x_o, s_1, x_1 \cdots s_j, x_j\Big).$$

Corollary 1:

If $$\lim_{t\downarrow s} \|h(t)-h(s)\| = 0$$

then $$\lim_{t\downarrow s} \|E[h(t)]\,(s,X_j) - h(s,X_j)\| = 0$$

and $$E[h(t)]\,(\tau_j,X_j)$$

is continuous in norm as a function of s_j.

Proof. Apply (16) in (17).

For a function $g(x)$ defined on E^k define $\|g\| = \sup_x \dfrac{|g(x)|}{1+\|x\|^2}$ and for a measure M on E^k define $\|M\|$ by

$$\|M\| = \sup_{\|g\|\leq 1} \left|\int g(x)\,M(dx)\right|.$$

Theorem 3: If $q(a \mid X_j)$ is right continuous in norm as a function of a and the measure

$$\Pi(dx_{j+1} \mid \tilde{X}_{j+1}) = \Pi(dx_{j+1} \mid X_j, s_j)$$

satisfies

$$\lim_{s_j' \downarrow s_j} \| \, \Pi(dx_{j+1} \mid X_j, s_j) - \Pi(dx_{j+1}) \mid X_j, s_j' \,) \, \| = 0$$

and the representation of the functional $h(t_j)$ has a partial derivative $h_t(t, X_j)$ in the sense that

$$\lim_{\Delta t \downarrow 0} \left\| \frac{h(t+\Delta t, X_j) - h(t, X_j)}{\Delta t} - h_t(t, x_j) \right\| = 0$$

then

$$(20) \quad \frac{E[h(t+\Delta t] \ (t, X_j) - h(t, X_j)}{\Delta t} - \int_{E^k} h(t, X_{j+1}) \, q(t-\tau_j \mid X_j) \, \Pi(dx_{j+1} \mid \tilde{X}_j) + h(t, X_j) \, q(t-\tau_j \mid X_j - h_t(t, X_j)$$

converges to zero in norm. In (20)

$$X_{j+1} = (x_o, s_1, x_1, \cdots, s_j, x_j, t-\tau_j, x_{j+1})$$

and

$$\tilde{X}_{j+1} = (x_o, s_1, x_1, \cdots, s_j, x_j, t-\tau_j)$$

Proof: Using (17), (20) can be written

$$\frac{1}{\Delta t} \, [h(t+\Delta t, X_j) - h(t, X_j)] - h_t(t, X_j) +$$

$$(21) \quad \frac{1}{\Delta t} \int_{t-\tau_j}^{t+\Delta t-\tau_j} \int_{E^k} E[h(t+\Delta t)] \, (\tau_{j+1}, X_j, s_{j+1}, x_{j+1}) \, q(s_{j+1} \mid X_j) \, e^{-\int_{t-\tau_j}^{s_j} q(a \mid x_j,) da} \, \Pi(dx_{j+1} \mid X_j, s_{j+1}) ds_{j+1}$$

$$-\frac{1}{\Delta t} \int_{t-\tau_j}^{t+\Delta t-\tau_j} h(t+\Delta t, X_j) \, q(a_2 \mid X_j) \, e^{-\int_{t-\tau_j}^{a_2} q(a_1 \mid X_j) \, da_1} \, da_2$$

$$-\frac{1}{\Delta t} \int_{t-\tau_j}^{t-\Delta t-\tau_j} \int_{E^k} h(t, X_j, t-\tau_j, x_{j+1}) \, q(t-\tau_j \mid X_j) \, \Pi(dx_j \mid X_j, t-\tau_j) ds_{j+1} + \frac{1}{\Delta t} \int_{t-\tau_j}^{t-\Delta t+\tau} h(t, X_j) \, q(t-\tau_j \mid X_j) da_2$$

In (21) in the second term X_{j+1} and $\tilde{X}_{j+1}$ are written out as

$$X_{j+1} = (X_j, s_{j+1}, x_j) \text{ and } \tilde{X}_j = (X_j, s_{j+1})$$

and in the fourth term X_{j+1} and $\tilde{X}_{j+1}$ are written out as

$$X_{j+1} = (X_j, t-\tau_j, x_j) \text{ and } \tilde{X}_{j+1} = (X_j, t-\tau_j).$$

This is done to emphasize that in the first term these quantities involve the variable of integration s_{j+1} while in the third they involve $t-\tau_j$ which is independent of the variable of integration s_{j+1}.

It can be shown that as Δt approaches zero the integrands of (21) with respect to s_{j+1} and with respect to a_2 each converge in norm to zero. The argument uses the boundness in norm of the terms involved and is essentially the argument for proving the continuity of a triple product. Using this shows (21) converges in norm to zero.

Integrals of Functionals

Let h(t) be a measurable stochastic process so that for each t, h(t) is a functional on the past of x(·) up to time t. Suppose

$$\sup_t \| h(t) \| < \infty \tag{22}$$

consider the functional

$$\phi(t) = E\{\textstyle\int_t^T h(s)\, ds \mid \sigma_t\} \tag{23}$$

Theorem 4: The functional $\phi(t)$ has finite norm satisfying

$$\| \phi(t) \| \leqslant \frac{1}{MK}\, [e^{MK(T-t)} - 1] \sup_s \| h(s) \| \tag{24}$$

and $\phi(t)$ has the representation

$$\phi(t,X_j) = \int_t^T h(s,X_j)ds \tag{25}$$

$$+ \int_{t-\tau_j}^{T-\tau_j} q(s_{j+1} \mid X_j)\, e^{-\int_{t-\tau_j}^{s_{j+1}} q(a \mid X_j)da} \Big[\int_{E^k} (\tau_j+s_{j+1}, X_{j+1})\, \Pi(dx_{j+1} \mid \widetilde{X}_{j+1}) - \int_{\tau_j+s_{j+1}}^T h(s,X_j)ds\Big]\, ds_{j+1}$$

Proof: Interchanging integration and conditional expectations gives

$$\phi(t) = \int_t^T E\{h(s) \mid \sigma_t\}\, ds$$

On the set $n(t) = j$, $E\{h(s) \mid \sigma_t\}$ agrees with the function of X_j given by

$$E[h(s)](t,X_j)\,.$$

Hence

$$\phi(t,X_j) = \int_t^T E[h(s)]\ (t,X_j)ds \tag{26}$$

and using Theorem 1

$$\begin{aligned} \| \phi(t) \| &\leqslant \int_t^T \| E[h(s)]\ (t) \|\, ds \\ &\leqslant \int_t^T \sup_{t \leqslant s \leqslant T} \| h(s) \|\, e^{KM(s-t)} ds \\ &= \sup_{t \leqslant s \leqslant T} \| h(s) \|\ \frac{1}{KM}\, [e^{KM(T-t)} - 1] \end{aligned}$$

Substituting (17) in (26) and interchanging orders if integration in the last two integrals gives (25).

Differentiating (25) with respect to t gives the following corollary.

Corollary 2: The elements $\phi(t,X_j)$ of the representation of $\phi(t)$ satisfy the infinite system of ordinary differential equations

$$\phi(t,X_j) = -h(t,X_j) - q(t-\tau_j \mid X_j) \int_{E^k} \phi(t,X_{j+1})\, \Pi(dx_{j+1} \mid \widetilde{X}_{j+1}) + q(t-\tau_j, X_j)\, \phi(t,X_j) \tag{27}$$

where $$X_{j+1} = (x_o,s_1,x_1,\cdots,s_j,x_j,t-\tau_j,x_{j+1})$$

and $$\tilde{X}_{j+1} = (x_o,x_1,x_1,\cdots,s_j,x_j,t-\tau).$$

Part II

Controlled Jump Processes

Let $x(t)$ be a jump process. Suppose $x(t)$ consists of two component processes $x(t) = (y(t), z(t))$. Consider the component $y(t)$ as the component which can be observed and the component $z(t)$ as being unobserable. For each t, let A_t denote a given subset of $[0,t]$. We shall define the concept of $x(t)$ being a process corresponding to a feedback control $u(t)$ based on measurements of $y(\cdot)$ at times A_t.

Write $E^k = E^{k_1} \times E^{k_2}$. Let $\mathcal{Y}_\infty$ be the space of right continuous step functions defined on $[0,\infty)$ with values in E^{k_1}. Let B_∞ denote the σ–field of cylinder sets of $\mathcal{Y}_\infty$. Let $\sigma(A_t)$ be the σ–field of B_∞ generated by the sets $\{y(s) \in B: s \in A_t\}$ where B denotes a borel set of E^{k_1}. Let U be a Borel subset of E^m. A feedback control will be defined by a measurable function $u(t,y(\cdot))$ defined on $[0,\infty) \times \mathcal{Y}_\infty$ with values in U, such that for each fixed t $u(t,y(\cdot))$ is $\sigma(A_t)$ measurable as a function of $y(\cdot)$.

Since $u(t,y(\cdot))$ is $\sigma(A_t)$ measurable, $u(t,y(\cdot))$ depends only on the restriction of $y(\cdot)$ to $[0,t]$. Let $(y_o,s_1,y_1,\cdots,s_n,y_n)$ denote a sequence of successive values in E^{k_1} and times. Let $S(Y_n)$ denote the mapping which takes the sequence

$$(y_o,s_1,y_1,\cdots,s_n,y_n)$$

into the function $y(\cdot)$ defined by

$$y(s) = \begin{cases} y_o \text{ if } 0 \leqslant s \leqslant s_1 \\ y_j \text{ if } \sum_{i=1}^{j} s_i \leqslant s < \sum_{i=1}^{j+1} s_i \\ y^n \text{ if } \sum_{i=1}^{n} s_i \leqslant s \end{cases}$$

If $u(t,y(\cdot))$ is a measurable function defined on $[0,t) \times \mathcal{Y}_\infty$ with values in U such that $u(t,y(\cdot))$ is $\sigma(A_t)$ measurable for each t, the sequence of functions

$$u(t,Y_n) = u(t,S(Y_n))$$

will be defined to be a representation of a feedback control.

If for a jump process $x(t)$, there are functions $q(a,u \mid X_n)$ and $\Pi(A, u \mid \tilde{X}_{n+1})$ and a representation of a feedback control $u(t,Y_n)$ such that the associated discrete process

$$x_0,s_1,x_1,\cdots,s_n,x_n\cdots$$

corresponding to $x(t)$ has conditional jump rates $q(a, u(\tau_n+a,Y_n) \mid X_n)$ and conditional state jump distribution

$$\Pi(A,\ u(\tau_{n+1}, Y_n) \mid X_{n+1}),$$

the process will be said to be a controlled process corresponding to the feedback control $u(t) = u(t, y(\cdot))$.

A collection of controlled jump processes $x(t)$ corresponding to a given class of controls,such that in the correspondence the conditional jump rate function

$$q(t,u \mid X_n)$$

and conditional state distribution

$$\Pi(A,u \mid \tilde{X}_{n+1})$$

are the same for each process,will be called a **controlled system of jump processes.**

Construction of a Controlled Process

Suppose functions $q(a,u \mid X_n)$ and measures $\Pi(A,u \mid \tilde{X}_{n+1})$ such that $q(a,u \mid X_n) < K$ for some constant K, depending on an m—dimensional vector of control parameters u are given. Let $u(t,Y_n)$ be a representation of a feedback control. A controlled process $x(t)$ corresponding this control can be constructed in the following manner.

Form the conditional jump rate and contiional state distribution functions

$$q(a,u(a+\tau_n, Y_n) \mid X_n) \quad \text{and} \quad \Pi(A,u(\tau_{n+1}, Y_n) \mid \tilde{X}_{n+1})$$

From these conditional expectations construct the probability measure for a discrete process

$$x_o, s, x, \cdots, s_n, x_n \cdots$$

From this process construct $x(t)$ by (2).

Technical Conditions Assumed and Statement of the Control Problem

Consider a system whose processes correspond to a given class of controls and are constructed in the manner described in the previous paragraph in terms of given conditional jump rate and conditional state distribution functions

$$q(a,u \mid X_n) \text{ and } \Pi(A,u \mid \tilde{X}_{n+1}).$$

The following assumptions will be made for this system.

1) The information pattern of the system is locally increasing. That is for each t there is an $\eta > 0$ so that

$$A_t \subset A_{t+\epsilon} \quad \text{if } 0 \leqslant \epsilon \leqslant \eta$$

An example of a locally increasing information pattern for which the sets A_t are not increasing is given by $A_t = \{[t]\}$ where

$$[t] = \text{greatest integer} \leqslant t.$$

2) The class of controls is closed under concatenations. To define this concept, for each control $u(t)$ of the class, time τ, value v of the control set U, set $B \subset \sigma-\{y(s): s \in A_\tau\}$ and $\epsilon > 0$ define a control $u_\epsilon(t)$ by specifying the representation $u_\epsilon(t,Y_n)$ of $u_\epsilon(t)$ by

(28) $$u_\epsilon(t,Y_n) = v\chi_{[\tau,\tau+\epsilon)\times B}(t,S(Y_n)) + [1-\chi_{[\tau,\tau+\epsilon)\times B}(t,S(Y_n))]u(t,Y_n)$$

The class of controls is closed under concatenations if for some $\eta > 0$, $u_\epsilon(t)$ is in the class for $0 \leqslant \epsilon \leqslant \eta$.

Notice that if η is the positive number in the definition of locally increasing for the time τ and $0 \leqslant \epsilon \leqslant \eta$ that

$$\chi_{[\tau,\tau+\epsilon]\times B}(t,y(\cdot))$$

is $\sigma(A_t)$ measurable in $y(\cdot)$ for each t. Hence (28) has the appropriate form for a representation of a feedback control in this case.

3) For each control of the system there are constants K and M (the constants may differ for different controls) such that

$$q(a, u(\tau_n+a, Y_n) \mid X_n) < K$$

and

$$\int_{E^k}(1+\|x_{n+1}\|^2)\,\Pi(dx_{n+1}, u(\tau_{n+1}Y_n) \mid \widetilde{X}_{n+1}) \leqslant M(1+\|x_n\|^2)$$

4) For each control of the system the conditional jump distribution

$$q(a, u(\tau_n+a, Y_n) \mid X_n)$$

is right continuous in a in the norm topology for functions and conditional state jump distribution

$$\Pi(A, u(\tau_n+s_n, Y_n) \mid X_n, s_n)$$

is right continuous in s_n in the norm topology for measures that is

$$\lim_{s_n' \downarrow s_n} \sup_{\|g\| \leqslant 1} \int g(x_{n+1})[\Pi(dx_{n+1}, u(\tau_n+s_n', Y_n) \mid X_n, s_n') - \Pi(dx_{n+1}, u(\tau_n+A_n, Y_n) \mid X_n, s_n)] = 0.$$

Let T denote a fixed terminal time. Let $f(t,x,u)$ be a Borel measurable function on E^{k+m+1} and $g(x)$ a Borel measurable function on E^k. The performance index $J[u]$ of a control $u(t)$ whose corresponding jump process is $x(t)$ is defined by

$$J[u] = E\{g(x(t)) + \int_0^T f(t,x(t),u(t))dt\}$$

Assume for each control there is some constant R so that

5) $$\|f(t)\| = \sup_{X_n} \frac{|f(t,x_n,u(t,Y_n))|}{(1+\|x_n\|^2)} < R$$

$$||g|| = \sup_{x \in E^k} \frac{|g(x)|}{1+||x||^2} < R$$

and that $f(t,x_n,v)$ for each v in U and $f(t,x_n,u(t,Y_n))$ for each control u(t) are right continuous in the norm topology as functions of t.

Consider the optimal control problem of choosing the control in the class of controls of the system for which the performance index is a minimum. Such a control will be called optimal.

The Minimum Principle

For a control u(t) with corresponding trajectory x(t) let

$$J[u](t) = E\{g(x(T)) + \int_t^T f(s,x(s),u(s))ds \mid \sigma_t\} \tag{29}$$

Let $J[u](t,X_n)$ denote the functions of the representation of J[u](t). For a value v of the control set U let H(t,v) denote the functional on x(t) which has the representation

$$H(t,v,X_n) = f(t,x_n,v) + \int_{E^k} J[u](t,X_{n+1})q(T-\tau_n,v \mid X_n)\,\Pi(dx_{n+1},v \mid \tilde{X}_{n+1}) - J[u](t,X_n)q(t-\tau_n,v \mid X_n) \tag{30}$$

where in (30)

$$X_n = (x_0,s,\cdots,s_n,x_n)$$

$$X_{n+1} = (x_0,s_1,x_1,\cdots,s_n,x_n,t-\tau_n,x_{n+1}) = (X_n,t-\tau_n,x_{n+1})$$

and

$$\tilde{X}_{n+1} = (x_0,s_1,x_1,\cdots,s_n,x_n,t-\tau_n) = (X_n,t-\tau_n)$$

Since the class of controls of a system are assumed to be closed under concatenations the mapping

$$J(u_\epsilon)$$

which takes a nonegative number ϵ into the performance index corresponding to a control u_ϵ is defined on some interval $0 \leqslant \epsilon \leqslant \eta$. The next theorem evaluates the right derivative of this mapping at $\epsilon = 0$.

Theorem 5: At $\epsilon = 0$ the right derivative of $J(u_\epsilon)$ exists and has the value

$$\frac{d}{d\epsilon} J(u_\epsilon)\Big|_{\epsilon=0} = E\{\chi_B[H(\tau,v) - H(\tau,u(\tau))]\} \tag{31}$$

Proof: Let

$$I[u] = g(x(t)) + \int_0^T f(s,x(s),u(s))ds$$

Then

$$J[u_\epsilon] - J[u] = E\{I[u_\epsilon] - I[u]\} = E\{E\{I[u_\epsilon] - I[u] \mid \sigma_t\}\}$$

Let $x^{\epsilon}(t)$ denote the jump process corresponding to $u_{\epsilon}(t)$. Since $u_{\epsilon}(t) = u(t)$ if $0 \leq t < \tau$

$$I[u_{\epsilon}] = \chi_B[\int_0^{\tau} f(s,x(s),u(s))ds + \int_{\tau}^{\tau+t} f(s,x^{\epsilon}(s),v)ds + \int_{\tau+\epsilon}^{T} f(s),x^{\epsilon}(s),u^{\epsilon}(s)ds + g(x^{\epsilon}(t))]$$

$$+(1-\chi_B)[\int_0^T f(s,x(s),u(s))ds + g(x(t))]$$

Hence

$$J[u_{\epsilon}] - J[u] = E\{E\{\chi_B[\int_{\tau}^{\tau+\epsilon} f(s,x^{\epsilon}(s),v)ds - \int_{\tau}^{\tau+\epsilon} f(s,x(s),u(s))ds]\,\sigma_{\tau}\}\} \tag{32}$$

$$+ E\{E\{J[u_{\epsilon}](\tau+\epsilon)\,|\,\sigma_{\tau}\}\} - E\{E\{J[u]\ (\tau+\epsilon)\,|\,\sigma_{\tau}\}\}$$

Since the controls $u(t)$ and $u_{\epsilon}(t)$ have the same representation if $t \geq \tau + t$, it follows from definition (29) of $J[u]\ (\tau+\epsilon)$ and (13) that

$$J[u_{\epsilon}]\ (\tau+\epsilon) \text{ and } J[u]\ (\tau+\epsilon)$$

have the same representation. Hence

$$J[u_{\epsilon}]\ (\tau+\epsilon) = J[u]\ (\tau+\epsilon, x_0^{\epsilon}, s_1^{\epsilon}, x_1^{\epsilon}, \cdots s_{n(\tau+\epsilon)}^{\epsilon}, x_{n(\tau+\epsilon)}) \tag{33}$$

It follows from (33) and (20) that

$$\frac{1}{\epsilon}[E\{J[u_{\epsilon}]\ (\tau+\epsilon)\,|\,(\tau,X_j)\} - J[u]\ (\tau,X_j)] + f(\tau,x_j,u(\tau,X_j)) \tag{34}$$

$$-\int_{E^k} J[u]\,(\tau,X_{j+1})q(\tau-\tau_j,v\,|\,X_j)\,\Pi(ds_{j+1},v\,|\,\tilde{X}_{j+1}) + J[u]\,(\tau,X_j)q(\tau-\tau,v\,|\,X_j)$$

converges to zero in norm. It follows from (20) that

$$\frac{1}{\epsilon}[EJ[u]\,(\tau+\epsilon)\,|\,(\tau,X_j) - J[u]\ (\tau,X_j)] + f(\tau,x_j,u(\tau,X_j)) \tag{35}$$

$$-\int_{E^k} J[u]\ (\tau,X_{j+1})\,q(\tau-\tau_j,u(\tau,Y_j)\,|\,X_j)\,\Pi(dx_{j+1},u(\tau,Y_j)\,|\,\tilde{X}_{j+1}) + J[u]\,(\tau,X_j)q(\tau-\tau_j,u(\tau,Y_j)\,|\,X_j)$$

converges in norm to zero. Using (25) and the continuity in norm of $f(t,x_j,v)$ and $f(t,x_j,u(t,Y_j))$ it can be shown that each of

$$\frac{1}{\epsilon}E\{\chi_B\int_{\tau}^{\tau+\epsilon} f(s,x^{\epsilon}(s),v)ds\,|\,(\tau,X_j)\} - \chi_B f(\tau,x_j,v) \tag{36}$$

and

$$\frac{1}{\epsilon}E\{\chi_B\int_{\tau}^{\tau+\epsilon} f(s,x(s),u(s))ds\,|\,(\tau,X_j)\} - \chi_B f(\tau,x_j,u(\tau,Y_j)) \tag{37}$$

converge in norm to zero.

Since (34)–(37) converge in norm to zero, from Theorem 1 it follows that the expected value of each of (34)–(37) converges to zero. Applying this with (32) then gives (31).

Theorem 6: (Minimum principle) Necessary conditions that u(t) can be an optimal control in the class of controls of a controlled system of jump processes are:

For each $t \in [0,T]$ and each $v \in U$

(38) $$\min_{v \in U} E\{H(t,v) \mid \sigma-\{y(s): s_\epsilon A_t\}\} = E\{H(t,u(t)) \mid \sigma-\{y(s): s_\epsilon A_t\}\}$$

where

(39) $$H(t,v,X_j) = f(t,x_j,v)+\int_{E^k} J[u](t,X_{j+1})q(t-\tau_j,v \mid X_j)\Pi(dx_{j+1},v \mid \tilde{X}_{j+1}) - J[u](t,X_j)q(t-\tau_j,v,X_j)$$

and $J[u](t,X_j)$ is a solution of the infinite system of differential equations

(40) $$\dot{J}[u](t,X_j) = -H(t, u(t,Y_j),X_j)$$

with boundary condition

(41) $$J[u](T,X_j) = g(x_j)$$

In (38)–(40), $X_j, X_{j+1}, \tilde{X}_{j+1}, t$, and x_j are related by

$$X_{j+1} = (x_0,s_1,x_1,\cdots,s_j,x_j,t-\tau_j,x_{j+1}) = (X_j,t-\tau_j,x_{j+1})$$

and

$$\tilde{X}_{j+1} = (x_1,s_1,x_1,\cdots,s_j,x_j,t-\tau_j) = (X_j,t-\tau_j).$$

Proof: If u(t) is an optimal control and $u_\epsilon(t)$ is defined by (28).

(42) $$J[u_\epsilon] - J[u] \geqslant 0.$$

By Theorem 5, $\frac{d}{d\epsilon} J_{u\epsilon}\Big|_\epsilon = 0$ exists. By (42) it must be nonnegative. Hence from (31)

(43) $$E\{\chi_B [H(\tau,v) - H(\tau,u(\tau))]\} \geqslant 0$$

Now since χ_B is $\sigma\{y(s) : s\epsilon A_\tau\}$measurable

(44) $$E\{\chi_B[H(\tau,v) - H(\tau,u(\tau))]\} =$$

$$E\{\chi_B E\{H(\tau,v) - H(\tau,u(\tau)) \mid \sigma-\{y(s):s\epsilon A_\tau\}\}$$

Since (43) is true for any $\sigma-\{y(s):s\epsilon A_\tau\}$ measurable set B it must follow from (43) and (44) that

$$E\{H(\tau,v) - H(\tau,u(\tau)) \mid \sigma-\{y(s):s\epsilon A_\tau\} \geqslant 0$$

which gives (38).

Statements (40) and (41) follow from Theorem 4 and Corollary 2.

Controlled Poisson Source

The paper will be concluded by giving a very simple example and showing the implications of the maximal principle in this case. Consider a system which has as an input process a Poisson process m(t) of arrivals. Consider controling the input process to form an output process $k(t)$. Let the control $u(t)$ take values in $0 \leqslant u \leqslant 1$. Suppose that if the control is $u(t)$ and there is an arrival at time t, the probability that the arrival is allowed to join the output process in $u(t)$. Suppose it is desired to keep the rate of entries into the output process $k(t)$ near a level r and to achieve this the quantity

(45) $$E\{\int_0^T (k(t) - rt)^2 dt\}$$

has been selected as a performance index. Suppose the rate parameter λ of the Poisson process considered as an unknown random variable with a known probability density $f(\lambda)$. Suppose neither λ nor the output process $k(t)$ can be measured but that it is possible to measure the input process $m(t)$. Suppose measurments made at times A_t of $m(t)$ are available to the controler at time t. Two possible choices for A_t are $A_t = [0,t]$ and $A_t = \{[t]\}$. The first corresponds to recording each arrival and the second corresponds to recording the cumulative number which have arrived at each unit of time.

To begin to describe the quantities involved in this problem let

$$X_j = (m_o,k_o,s_1,m_1,k_1,\cdots,s_j,m_j,k_j)$$

As before let $\tau_1 = s_1$ and $\tau_j = \sum_{i=1}^{j} s_i$ if $j = 1$. The conditional density of λ given interarrival times $s_1,\cdots s_j$, $f(\lambda \mid s_1 \cdots s_j)$ is given by $f(\lambda_1 \mid s_1,\cdots s_j) = \dfrac{\lambda^j e^{-\lambda \sum_{i=1}^{j} s_i} f(\lambda)}{\int \lambda^j e^{-\lambda \sum_{i=1}^{j} s_i} f(\lambda) d\lambda}$

Hence

(46) $$q(a,u \mid X_j) = \frac{\int_0^\infty \lambda^{j+1} e^{-\lambda(a + \sum_{i=1}^{j} s_i)} f(\lambda) d\lambda}{\int_0^\infty \int_0^\infty \lambda^{j+1} e^{-\lambda(s+\sum_{i=1}^{j} s_i)} f(\lambda) d\lambda\, ds}$$

Notice that $q(a,u \mid X_j)$ is independent of u. Denote it merely by $q(a \mid X_j)$. From the definition of $u(t)$

(47) $$\Pi(x_{j+1} = (m_j + 1, k_j + 1), u \mid \tilde{X}_{j+1}) = u$$

$$\Pi(x_{j+1} = (m_j+1,k_j), u \mid \tilde{X}_{j+1}) = 1 - u$$

$$\Pi(x_{j+1} \neq (m_j + k_j+1) \text{ nor } (m_{j+1},k_j), u \mid \tilde{X}_{j+1}) = 0$$

Let $X_j^1 = (X_j,t-\tau_j,m_j+1,k_j+1) = (m_0,k_0,s_1,m_1,\cdots,k_j,t-\tau_j,m_j+1,k_j+1)$

and

$$X_j^0 = (X_j,t-\tau_j,m_j+1,k_j) = (m_0,k_0,s_1,m_1,k_1,\cdots,s_j,m_j,k_j,t-\tau_j,m_j+1k_j)$$

From (45)–(47) and (39)

(48) $$H(t,v,X_j) = (m_j - rt)^2 + J[u](t,X_j^1)q(t-\tau_j \mid X_j)v + J[u](t,X_j^0)q(t-\tau_j \mid X_j)(1-v) - J[u](t,X_j)q(t-\tau_j \mid X_j)$$

Notice from (46) that

$$q(t-\tau_j \mid X_j) = \frac{\int_0^\infty \lambda^{j+1} e^{-\lambda t} f(\lambda) d\lambda}{\int_0^\infty \int_0^\infty \lambda^{j+1} e^{-\lambda s} f(\lambda) ds}$$

and thus is a function only of t. Let

$$\sigma_{Mt} = \sigma - \{m(s): s \epsilon A_t\}$$

denote the σ field generated by the measurments. Since $q(t-\tau_j \mid X_j)$ is a function of t, (48) and (38) imply that an optimal control u(t) must satisfy

$$u(t) = \begin{cases} 1 & \text{if } E\{J[u](t, X^1_{n(t)}) \mid \sigma_{Mt}\} < E\{J[u](t, X^0_{n(t)}) \mid \sigma_{Mt}\} \\ 0 & \text{if } E\{J[u](t, X_{n(t)}) \mid \sigma_{Mt}\} > E\{J[u](t, X^0_{n(t)} \mid \sigma_{Mt}\} \end{cases}$$

References

1. Breiman, L. "Probability Theory, "Addison Wesley, Reading Massachusetts (1968).

2. Gikman, I.I. and Skorokhod, A.V. "Introduction to the Theory of Random Processes, " W.B. Saunders Co., Philadelphia (1969).

3. Loeve, M. "Probability Theory" 3rd Ed. Van Nostrand New York (1963).

4. Boel, R. Varaiya, P. and Wong E., "Martingales on Jump Process I: Representation Results, " University of California, Electronics Research Laboratory, Memorandum No. ERL–M407 (1973).

5. Boel, R. Varaiya, P. and Wong E. "Martingales on Jump Process II: Applications, "University of California, Electronics Research Laboratory, Memorandum No. ERL–M409 (1973).

FILTERING AND CONTROL OF JUMP PROCESSES

Pravin Varaiya

Department of Electrical Engineering and Computer Sciences
and the Electronics Research Laboratory
University of California, Berkeley, California 94720

1. Summary

This paper presents results in the filtering and control of jump processes. The paper is divided into three parts. The first part consists of a precise definition of jump processes and a characterization of the set of all martingales generated by a jump process. Essentially, a jump process is any piecewise-constant process with values in an arbitrary state space such that all the jumps are totally inaccessible and such that every sample path has only a countable number of discontinuities. The generated martingales are described by exhibiting a set of "basis" martingales and showing that all the other martingales are stochastic integrals of the basis martingales.

In the second part of the paper we use this "representation" theorem to obtain the form of the optimum (least-squares) filter. The argument is similar to the innovations approach which has been used for Wiener processes.

The final part of the paper presents Dynamic Programming conditions for the optimal control of jump processes when the control affects the rates at which jumps occur.

The first two parts of the paper are summaries of work reported in [1] and [2]. The last part is a "progress" report on current research being conducted jointly with René Boel.

2. Jump processes and their martingales

2.1 Definition of a jump process

Throughout Ω is a fixed sample space and the time interval is $R_+ = [0,\infty)$. Let $\mathcal{F}_t$, $t \in R_+$, be an increasing family of σ-fields of Ω and let $\mathcal{F} = \bigvee_t \mathcal{F}_t$. Let $\mathcal{P}$ be a probability measure on $(\Omega, \mathcal{F})$.

Let $(Z, \mathcal{Z})$ be a measurable space. A <u>process</u> (with values in Z) is a family $(x_t, \mathcal{F}_t, \mathcal{P})$ such that for each t x_t is a $\mathcal{F}_t$ measurable function from Ω into Z.[1] The family (x_t) defines a different process if either $(\mathcal{F}_t)$ or $\mathcal{P}$ is different. In particular, if $(x_t, \mathcal{F}_t, \mathcal{P})$ is a process so is $(x_t, \mathcal{F}_t^x, \mathcal{P})$ where $\mathcal{F}_t^x$ is the σ-field generated by the sets $\{x_s \in B\}$, $s \leq t$, $B \in \mathcal{Z}$. When no confusion results, we write $(x_t, \mathcal{F}_t, \mathcal{P})$ as $(x_t, \mathcal{F}_t)$ or $(x_t, \mathcal{P})$ or (x_t). The process $(x_t, \mathcal{F}_t, \mathcal{P})$ is a <u>fundamental jump process</u> if

(i) $(Z, \mathcal{Z})$ is a Blackwell space,

(ii) for each $\omega \in \Omega$, the sample path $t \mapsto x_t(\omega)$ is piece-wise constant, constant to the right, and has only a finite number of discontinuities in every finite interval,

(iii) the jump times of the process, T_n, n=0,1,..., are all totally inaccessible stopping times of the family $(\mathcal{F}_t)$; here $T_0 \equiv 0$, and

$$T_{n+1}(\omega) = \begin{cases} \inf\{t \mid t > T_n(\omega),\ x_t(\omega) \neq x_{T_n(\omega)}(\omega) \\ \infty \text{ if the set above is empty.}^2 \end{cases}$$

---

[1] Processes are real-valued unless stated otherwise.

[2] Assumption (i) guarantees that the T_n are indeed stopping times.

Research supported in part by the National Science Foundation Grant GK-10656X3

2.2 Martingales of a jump process and stochastic integrals.

For each $B \in \mathcal{Z}$ let

$$P(B,t) = \sum_{s \leq t} I_{\{x_{s-} \neq x_s\}} I_{\{x_s \in B\}}$$

be the number of jumps of x which occur before t and which end in the set B. (P(B, t)) is a counting process.[3] Evidently P(B,t) is $\mathcal{F}_t^x$ measurable. Associated with (P(B,t)) are two unique, continuous processes $(\tilde{P}(B,t)) \in \mathcal{A}^+(\mathcal{F}_t, \mathcal{P})$ and $(\tilde{P}^x(B,t)) \in \mathcal{A}^+(\mathcal{F}_t^x, \mathcal{P})$ such that

$$(Q(B,t)) = (P(B,t) - \tilde{P}(B,t)) \in \mathcal{M}(\mathcal{F}_t, \mathcal{P}), \tag{2.1}$$

$$(Q^x(B,t)) = (P(B,t) - \tilde{P}^x(B,t)) \in \mathcal{M}(\mathcal{F}_t^x, \mathcal{P}) \tag{2.2}$$

Here $\mathcal{A}^+(\mathcal{F}_t, \mathcal{P})$, respectively $\mathcal{A}^+(\mathcal{F}_t^x, \mathcal{P})$, is the set of all <u>increasing</u>, <u>locally integrable</u>, <u>$\mathcal{F}_t$-predictable</u>, respectively <u>$\mathcal{F}_t^x$-predictable</u> processes, and $\mathcal{M}(\mathcal{F}_t, \mathcal{P})$, respectively $\mathcal{M}(\mathcal{F}_t^x, \mathcal{P})$, is the set of all <u>local</u>, <u>$\mathcal{F}_t$-adapted</u>, respectively $\mathcal{F}_t^x$-adapted, martingales.

Let $L^1(\tilde{P})$ be the set of all functions $f(z,\omega,t)$ which are measurable with respect to $\mathcal{Z} \times \mathcal{F} \times \mathcal{B}$ such that $(f(z,\cdot,t))$ is a $\mathcal{F}_t$-predictable process for each z , and such that there exists a sequence of stopping times $S_k \uparrow \infty$ with

$$E \int_Z \int_{R_+} |f(z,t)| I_{\{t \leq S_k\}} \tilde{P}(dz,dt) < \infty \tag{2.3}$$

The family $L^1(\tilde{P}^x)$ is defined in the same manner replacing $(\mathcal{F}_t)$ by $(\mathcal{F}_t^x)$ and $\tilde{P}$ by g $\times \tilde{P}^x$. The integral above is to be interpreted as a Lebesgue-Stieltjes integral. It turns out that (2.3) holds if and only if

$$E \int_Z \int_{R_+} |f(z,t)| I_{\{t \leq S_k\}} P(dz,dt) < \infty \tag{2.4}$$

Hence for each $f \in L^1(\tilde{P})$ $(f \in L^1(\tilde{P}^x))$, we can define the <u>stochastic integral</u> of f with respect to $Q(Q^x)$,

$$(f \circ Q)_t = \int_Z \int_0^t f(z,s) Q(dz,ds) = \int_Z \int_0^t f(z,s) P(dz,ds - \int_Z \int_0^t f(z,s) \tilde{P}(dz,ds)$$

$$((f \circ Q^x)_t = \int_Z \int_0^t f(z,s) Q^x(dz,ds)).$$

It is easy to see that $((f \circ Q)_t) \in \mathcal{M}(\mathcal{F}_t, \mathcal{P})$ and $((f \circ Q^x)_t) \in \mathcal{M}(\mathcal{F}_t^x, \mathcal{P})$.

2.3 Martingale representation theorem

<u>Theorem 2.1</u> $(m_t) \in \mathcal{M}(\mathcal{F}_t^x, \mathcal{P}) \Leftrightarrow$ there exists $f \in L^1(\tilde{P}^x)$ such that

$$m_t - m_0 = (f \circ Q^x)_t \quad \text{a.s. for all } t$$

The theorem asserts that the martingales $(Q^x(B,t))$, $B \in \mathcal{Z}$, form a "basis" for

[3] A counting process is a fundamental jump process with integer values, which starts at 0 and has jumps of size +1.

the space of all local martingales generated by the process $(x_t, \mathcal{F}_t^x, \mathcal{P})$.

2.4 An example

Let $(x_t, \mathcal{F}_t, \mathcal{P})$ be a fundamental jump process with values in $(Z, \mathcal{Z})$ and suppose that from each $z \in Z$ the process can make at most n transitions where n is a fixed finite number. Thus the transitions can be represented in the state-transition diagram of Figure 1 where they are labeled $\sigma_1, \ldots, \sigma_n$. Define the counting processes $p_i(t), i=1,\ldots,n,$

$p_i(t)$ = number of transitions of type σ_i made by the process x_t in time $[0,t]$

Then there exist unique, continuous processes $(\tilde{p}_i(t)) \in \mathcal{A}^+(\mathcal{F}_t, \mathcal{P})$ and $(\tilde{p}_i^x(t)) \in \mathcal{A}^+(\mathcal{F}_t^x, \mathcal{P})$ such that

$$(q_i(t)) = (p_i(t) - \tilde{p}_i(t)) \in \mathcal{M}(\mathcal{F}_t, \mathcal{P}),$$

$$(q_i^x(t)) = (p_i(t) - \tilde{p}_i^x(t)) \in \mathcal{M}(\mathcal{F}_t^x, \mathcal{P})$$

and, most importantly, $(m_t) \in \mathcal{M}(\mathcal{F}_t^x, \mathcal{P})$ if and only if there exist $(f_i(t)) \in L^1(\tilde{p}_i^x)$ such that

$$m_t - m_0 = (foq^x)_t = \sum_{i=1}^n \int_0^t f_i(s) q_i^x(ds)$$

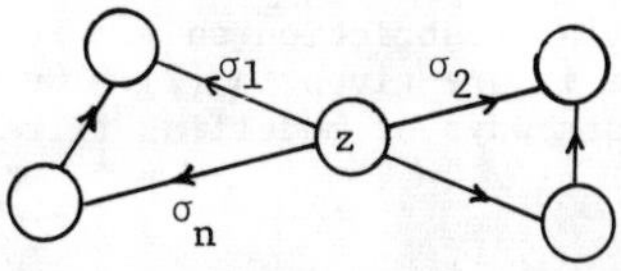

Figure 1. State-transition diagram for example

This example covers many jump processes of practical interest.

2.5 The likelihood ratio and transformations of probability

Let $(x_t, \mathcal{F}_t, \mathcal{P})$ be a fundamental jump process with values in $(Z, \mathcal{Z})$ and let $\mathcal{P}_1$ be another probability measure on $(\Omega, \mathcal{F})$ such that $\mathcal{P}_1 \sim \mathcal{P}$. Define the (conditional) likelihood ratio

$$L_t = E\left(\frac{d\mathcal{P}_1}{d\mathcal{P}} \middle| \mathcal{F}_t^x\right)$$

Evidently $(L_t) \in \mathcal{M}(\mathcal{F}_t^x, \mathcal{P})$. The representation theorem together with a result due to Doléans-Dade [3] permits the following assertion.

<u>Theorem 2.2</u> There exists $\phi(z,\omega,t) \in L^1(\tilde{P}^x)$ such that $1 + \phi \geqq 0$ and

$$L_t = \prod_{\substack{s<t \\ x_{s-} \neq x_s}} [1 + \phi(x_s, s)] \exp\left[-\int_Z \int_0^t \phi(z,s)\, \tilde{P}^x(dz,ds)\right], \quad t \in R_+ \tag{2.5}$$

Furthermore, the increasing processes $(\tilde{P}_1^x(B,t))$ associated with the fundamental jump process $(x_t, \mathcal{F}^x_t, \mathcal{P}_1)$ are given by

$$\tilde{P}^x_1(B,t) = \int_B \int_0^t [1 + \phi(z,s)]\, \tilde{P}^x(dz,ds) \tag{2.6}$$

The next result is in a way a converse to the above. It is an application in the context of jump processes of a technique of transformations of processes developed by Girsanov [4] for the case of Brownian movement.

Theorem 2.3 Let $(x_t, \mathcal{F}^x_t, \mathcal{P})$ be a fundamental jump process with values in $(Z, \mathcal{Z})$. Let $\phi \in L^1(\tilde{P}^x)$ with $1 + \phi \geq 0$ and define (L_t) by (2.5). Let $L = \lim_{t\to\infty} L_t$. Then $L \geq 0$ a.s. and $E(L) \leq 1$. Suppose that

$$E(L) = 1 \tag{2.7}$$

and define the probability measure $\mathcal{P}_1$ on $(\Omega, \mathcal{F}^x)$ by

$$\mathcal{P}_1(A) = \int_A L(\omega)\, \mathcal{P}(d\omega)$$

Then $(x_t, \mathcal{F}^x_t, \mathcal{P}_1)$ is a jump process with associated increasing processes given by (2.6).

The condition (2.7) is a non-trivial restriction on ϕ. Several sufficient and verifiable conditions which guarantee it are given in [2]. One of the interesting uses of this result is that it suggests ways of modelling filtering and control problems as we see below.

3. Filtering for jump processes

We suppose that there is a "state" process (y_t) and statistically related to it is an "observed" jump process (x_t). We want to obtain the least-squares estimate of some function of y_t, say $g(y_t)$, given x_s, $s \leq t$. The statistical relationship is specified through two assumptions the second one being stated later.

Assumption 3.1 $(\Omega, \mathcal{F}_t)$, $t \in R_+$, is a family of spaces and $\mathcal{P}$, $\mathcal{P}_1$ are two probability measures on $(\Omega, \mathcal{F})$. x_t and y_t are measurable functions on $(\Omega, \mathcal{F}_t)$ with values in $(Z, \mathcal{Z})$ and $(Y, \mathcal{Y})$ respectively.

(i) Z is a Borel subset of R^n for some n and $\mathcal{Z}$ is the Borel field. Y is a locally compact Hausdorff space, $\mathcal{Y}$ is the Borel field. Also $\mathcal{F}_t = \mathcal{F}^x_t \times \mathcal{F}^y_t$.

(ii) Under measure $\mathcal{P}$

(a) $(x_t, \mathcal{F}_t, \mathcal{P})$ is a fundamental jump process with independent increments i.e., $x_t - x_s$ is independent of $\mathcal{F}_s$ (under $\mathcal{P}$) for $s \leq t$,

(b) $(y_t, \mathcal{F}_t, \mathcal{P})$ is a Markov process whose sample paths are right-continuous and have left-hand limits, and the jump times of y are totally inaccessible,

(c) the processes (x_t) and (y_t) are independent i.e., $\mathcal{F}^x$ and $\mathcal{F}^y$ are independent.

(iii) $\mathcal{P}_1 \ll \mathcal{P}$, there exists $f(z,\omega,t) \in L^1(\tilde{P})$ with a representation

$$f(z,\omega,t) = \phi(z, y_{t-}(\omega), \omega, t),$$

such that $\phi(\cdot, y, \cdot, \cdot)$ is $\mathcal{F}^x_t$-predictable for each $y \in Y$, and there also exist $\mathcal{F}^x_t$-predictable processes $(\mu(B,t))$ for $B \in \mathcal{Z}$ such that

$E(|f(z,t)|) + E_1(|f(z,t)|) < \infty$ for all z, t and

$$L_t = E(\frac{d\mathcal{P}_1}{d\mathcal{P}} | \mathcal{F}_t) = \prod_{\substack{s<t \\ x_{s-}\neq x_s}} [1 + \phi(x_s, y_{s-}, s)] \exp[- \int_Z\int_0^t \phi(z, y_{s-}, s)\, \mu(dz,ds)]$$

If $\tilde{P}$ and $\tilde{P}^x$ are the processes associated with $(x_t, \mathcal{P})$ and $\tilde{P}_1$, $\tilde{P}_1^x$ the processes associated with $(x_t, \mathcal{P}_1)$ then it is easy to show that

$$\tilde{P}(B,t) = \tilde{P}^x(B,t) = \mu(B,t),$$

whereas,

$$\tilde{P}_1(B,t) = \int_B\int_0^t (1 + f(z,s))\mu(dz,ds), \quad \tilde{P}_1^x(B,t) = \int_B\int_0^t (1+\hat{f}(z,s))\, \mu(dz,ds),$$

where

$$\hat{f}(z,t) = E_1(f(z,t)|\mathcal{F}_t^x).$$

Now let $\mathcal{G}$ be the family of all bounded, measurable, real-valued functions g on Y. We seek to obtain a (recursive) expression for $E_1(g(y_t)|\mathcal{F}_t^x)$. Now

$$E_1(g(y_t)|\mathcal{F}_t^x) = \frac{E(L_t g(y_t)|\mathcal{F}_t^x)}{E(L_t|\mathcal{F}_t^x)} \tag{3.1}$$

the denominator of which does not depend on g. We will therefore determine the numerator instead. So define the family of operators π_t, $t \in R_+$, by

$$\pi_t(g) = E(L_t g(y_t)|\mathcal{F}_t^x).$$

<u>Lemma 3.1</u> Under Assumption 3.1, for each $g \in \mathcal{G}$, the process $(\pi_t(g))$ satisfies

$$\pi_t(g) = Eg(y_t) + \int_Z\int_0^t \pi_s(\phi(z,\cdot,s)\, H_{t,s}(g))\, Q(dz,ds), \tag{3.2}$$

where

$$H_{t,s}(g) = E_1(g(y_t)|y_s) = E(g(y_t)|y_s), \tag{3.3}$$

$$Q(B,t) = P(B,t) - \tilde{P}(B,t) \tag{3.4}$$

<u>Assumption 5.2</u> The operators $H_{t,s}$ in (3.3) have the following properties:

(i) $\lim_{s\uparrow t} H_{t,s} = I$, the identity operator on $\mathcal{G}$;

(ii) there exist operators A_t, $t \geq 0$ on $\mathcal{G}$ such that

$$\lim_{\varepsilon\downarrow 0} \frac{1}{\varepsilon}(H_{t+\varepsilon,s} - H_{t,s})(g) = H_{t,s}A_t(g)$$

With this assumption (3.2) can be strengthened as follows.

Theorem 3.1 Under Assumptions 3.1, 3.2, for each $g \in \mathcal{G}$.

$$\pi_t(g) = \pi_0(g) + \int_0^t \pi_s(A_s g)\, ds + \int_Z \int_0^t \pi_s(\phi(z,\cdot,s)g)\, Q(dz,ds) \tag{3.5}$$

4. Optimal control of jump processes

4.1 A model of controlled jump processes

Let $(x_t, \mathcal{F}_t^x, \mathcal{P})$ be a fundamental jump process with values in $(Z,\mathcal{Z})$. x_t is the "state" of the system to be controlled. Let $(\tilde{P}^x(B,t))$ be the associated increasing processes. To illustrate the meaning of $(\tilde{P}^x(B,t))$ suppose that its sample paths are absolutely continuous and let $\gamma(B,t) = \frac{d\tilde{P}^x}{dt}(B,t)$. Then $\gamma(B,t)\, dt$ is the probability that the process (x_s) makes a jump in the interval $[t,t+dt]$ with $x_{t+dt} \in B$, conditioned on the past, $\mathcal{F}_t^x$. Thus $(\gamma(B,t))$ tells us the rates per unit time at which the jumps of (x_t) occur. Recall that since the jump times are totally inaccessible by definition we cannot tell from past events exactly when a jump will occur in the future.

We are going to study the process (x_t) when it is being continuously controlled by a control or input process (u_t). It is assumed that the controlling action can only affect (x_t) by changing its "rates." Thus the sample paths of the controlled process are the same as the uncontrolled one[4]. Let $\mathcal{P}^u$ be the law of the process when control u is used. Assume further that $\mathcal{P}_u << \mathcal{P}$. Then by the results mentioned earlier, the increasing processes $(\tilde{P}_u^x(B,t), \mathcal{P}_u)$ must be of the form

$$\tilde{P}_u^x(B,t) = \int_B \int_0^t [1 + \phi^u(z,s)]\, \tilde{P}^x(dz,ds) \tag{4.1}$$

This should motivate why we assume a function Φ such that the effects of any control process (u_t) on the state process is given via (4.1) where

$$\phi^u(z,\omega,s) = \Phi(z,\omega,s,u_s(\omega)) \tag{4.2}$$

Let $(Q_u^x(B,t) = (P(B,t) - \tilde{P}_u^x(B,t))$ be the basis martingales associated with $(x_t, \mathcal{P}_u)$.

The next modelling decision concerns the formulation of the information available to the controller upon which the control u_t is based at each time t. Let $\mathcal{F}_t^y \subset \mathcal{F}_t^x$ be the information available at time t. We want $\mathcal{F}_t^y$ to be generated by a jump process (y_t). For example, if x_t is a vector then y_t could be some of the components of x_t. However, for our purpose here we will consider the simpler case of "complete" information i.e., we assume

$$y_t = x_t \tag{4.3}$$

We can now summarize precisely. Let U be a locally compact separable Hausdorff space.

Definition 4.1 An (admissible) control process is any process $(u_t, \mathcal{F}_t^x, \mathcal{P})$ such that $u_t(\omega) \in U$ for all ω,t, and such that

$$E\, L^u = 1$$

---

[4] This implies that we cannot consider controls which are "impulsive" i.e., which cause an instantaneous jump in the (y_t) process as in many inventory problems. It turns out however that by a proper choice of the (y_t) process we can formulate these inventory models as "rate" controlled processes.

where $L^u = \lim_{t\uparrow\infty} L_t^u$ and

$$L_t^u = \prod_{\substack{s\leq t \\ x_{s-}\neq x_s}} [1 + \phi_u(x_s,s)] \exp[-\int_Z\int_0^t \phi_u(z,s)\ \tilde{P}^x(dz,ds)].$$

Let $\mathcal{U}$ be the set of all control processes. For each $(u_t) \in \mathcal{U}$, the controlled process is the process $(x_t, \mathcal{F}_t^x, \mathcal{P}_u)$ where

$$\mathcal{P}_u(A) = \int_A L^u(\omega)\ \mathcal{P}(d\omega), \quad A \in \mathcal{F}^x.$$

The final modelling step is to formulate the "cost" associated with each control process (u_t). We make the cost vary with the rates of jumps as

$$c^u(z,s)\ \tilde{P}_u^x(dz,ds) = C(z,\omega,s,u_s)\ \tilde{P}_u^x(dz,ds) \tag{4.4}$$

It is assumed that C is non-negative and bounded. Since we are optimizing over an infinite time interval it is essential to introduce a "discounting" factor. For each $s \in R_+$ let $(r_t^s, \mathcal{F}_t^x, \mathcal{P})$, $t \in R_+$, be a process with $0 \leq r_t^s < 1$ for $t < s$, $r_t^s r_\tau^t = r_\tau^s$ for all s, t, τ, and $r_s^s \equiv 1$.

Definition 4.2 The cost associated with $(u_t) \in \mathcal{U}$ is

$$J(u) = E_u \left[\int_0^\infty\int_Z r_0^s\ c^u(z,s)\ \tilde{P}_u^x(dz,ds)\right] \tag{4.5}$$

The optimal control problem is to "find" $u^* \in \mathcal{U}$ such that

$$J^* = J(u^*) \leq J(u) \quad \text{for all } u \in \mathcal{U}$$

4.2 Principle of optimality

It is easy to see that if u, v are in $\mathcal{U}$ and $t \in R_+$, then the process $w \in \mathcal{U}$ where

$$w_s = \begin{cases} u_s, & s \leq t \\ v_s, & s > t \end{cases}$$

Next define the random variables

$$\psi(u,v,t) = E_w\left[\int_t^\infty\int_Z r_0^s\ c^v(z,s)\ \tilde{P}_v^x(dz,ds)\,|\,\mathcal{F}_t^x\right]$$

Thus $\psi(u,v,t)$ is the value at time 0 of the future cost of v when u has been adopted up to t. Note that $\psi(u,v,t)$ does not depend on u since

$$\psi(u,v,t) = \frac{E[L_t^u (L_t^v)^{-1} L^v \int_t^\infty \int_Z r_0^s \, c^v(z,s) \, \tilde{P}_v^x(dz,ds) | \mathcal{F}_t^x]}{E[L_t^u (L_t^v)^{-1} L^v | \mathcal{F}_t^x]}$$

$$= \frac{E[(L_t^v)^{-1} L^v \int_t^\infty \int_Z r_0^s \, c^v(z,s) \, \tilde{P}_v^x(dz,ds) | \mathcal{F}_t^x]}{E[(L_t^v)^{-1} L_v | \mathcal{F}_t^x]}$$

Define the infimium of ψ as v varies over $\mathcal{U}$,

$$W_t = \bigwedge_{v \in \mathcal{U}} \psi(u,v,t)$$

The process (W_t) is the value at time 0 of the minimum cost beyond t.

Theorem 4.1 (principle of optimality) For all $u \in \mathcal{U}$, $t \in R_+$, $h \geq 0$

$$W_t \leq E_u[\int_t^{t+h} \int_Z r_0^s c^u(z,s) \, \tilde{P}_u^x(dz,ds) | \mathcal{F}_t^x] + E_u[W_{t+h} | \mathcal{F}_t^x] \tag{4.6}$$

Furthermore, equality holds in (4.6) if and only if u is optimal.

4.3 The Hamilton-Jacobi conditions

We now sketch a proof for deriving the Hamilton-Jacobi equations for the optimal control. Let $u \in \mathcal{U}$ be arbitrary. Using Meyer's results [4] on the decomposition of supermartingales we can show that there exist processes $(\Lambda^u W_t) \in \mathcal{A}(\mathcal{F}_t^x, \mathcal{P}_u)$[5] and $(m_t^u) \in \mathcal{M}(\mathcal{F}_t^x, \mathcal{P}_u)$ such that

$$W_t = J^* + \Lambda^u W_t + m_t^u \tag{4.7}$$

By the martingale representation theorem there exists a process f^u such that

$$m_t^u = \int_0^t \int_Z f^u(z,s) \, Q^u(dz,ds)$$

$$= \int_0^t \int_Z f^u(z,s)[P(dz,ds) - (1+\phi^u(z,s)) \, \tilde{P}^x(dz,ds)] \tag{4.8}$$

From (4.6), (4.7)

$$W_t - E_u(W_{t+h} | \mathcal{F}_t^x) = \Lambda^u W_t - E_u(\Lambda^u W_{t+h} | \mathcal{F}_t^x) \leq E_u[\int_t^{t+h} \int_Z r_0^s c^u(z,s) \tilde{P}_u^x(dz,ds) | \mathcal{F}_t^x]$$

---

[5] $\mathcal{A}(\mathcal{F}_t^x, \mathcal{P}_u) = \{(a_t^1 - a_t^2) | (a_t^i) \in \mathcal{A}^+(\mathcal{F}_t^x, \mathcal{P}_u), \ i=1,2\}$

It follows, again from Meyer's results, that for all u, $t \in R_+$, $h \geq 0$

$$\Lambda^u W_{t+h} - \Lambda^u W_t + \int_t^{t+h}\int_Z r_0^s c^u(z,s)\ \tilde{P}_u^x(dz,ds) \geq 0 \qquad (4.9)$$

Now, u = u* is optimal if and only if we have equality in (4.9). Thus

$$\Lambda^{u*} W_{t+h} - \Lambda^{u*} W_t + \int_t^{t+h}\int_Z r_0^s c^{u*}(z,s)\ \tilde{P}_{u*}^x(dz,ds) = 0 \qquad (4.10)$$

and

$$W_t = J^* + \Lambda^{u*} W_t + m_t^{u*}$$

$$= J^* + \Lambda^{u*} W_t + \int_0^t\int_Z f^{u*}(z,s)\ [P(dz,ds) - (1+\phi^{u*}(z,s))\ \tilde{P}^x(dz,ds)] \qquad (4.11)$$

From (4.7), (4.8), (4.11) and Meyer's result, we can conclude that

$$f^u(z,s) = f^{u*}(z,s) = \nabla W(z,s), \text{ say}$$

and,

$$\Lambda^u W_t - \int_0^t\int_Z \nabla W(z,s)\ (1 + \phi^u(z,s))\ \tilde{P}^x(dz,ds)$$

$$= \Lambda^{u*} W_t - \int_0^t\int_Z \nabla W(z,s)\ (1 + \phi^{u*}(z,s))\ \tilde{P}^x(dz,ds)$$

$= \Lambda\ W_t$, say.

Hence from (4.9), (4.10) we obtain the "Hamilton-Jacobi" conditions

$$\Lambda\ W_{t+h} - \Lambda\ W_t + \int_t^{t+h}\int_Z [\nabla W(z,s) + r_0^s c^u(z,s)]\ (1 + \phi^u(z,s))\ \tilde{P}^x(dz,ds) \geq 0$$

$$= \Lambda\ W_{t+h} - \Lambda\ W_t + \int_t^{t+h}\int_Z [\nabla W(z,s) + r_0^s c^{u*}(z,s)]\ (1 + \phi^{u*}(z,s)\ \tilde{P}^x(dz,ds)$$

References

1. R. Boel, P. Varaiya, E. Wong, Martingales on Jump Processes I: Representation Results, Memo No. ERL-M407, Electronics Res. Lab., Univ. of California, Berkeley, Sept. 6, 1973.
2. R. Boel, P. Varaiya, E. Wong, Martingales on Jump Processes II: Applications, Memo No. ERL-M409, Electronics Res. Lab., Univ. of California, Berkeley, Dec. 12. 1973.
3. C. Doléans-Dade, Quelques applications de la formule de changement de variables pour les martingales, Z. Wahr. v. Greb. 16, 181-194, 1970.
4. P. A. Meyer, Probabilités et potentiel, Paris: Hermann, 1966
5. I. V. Girsanov, On transforming a class of stochastic processes by absolutely continuous substitution of measures, Th. Prob. Appl. 5, 285-301, 1960.

THE MARTINGALE THEORY OF POINT PROCESSES OVER THE REAL HALF LINE ADMITTING AN INTENSITY

P. BREMAUD

0. - INTRODUCTION

In [2] we have shown how the frame of martingale theory and stochastic integration could be systematically used in the theory of point processes, and, more generally, of jump processes. This new point of view has proven to be particularly adapted to problems of a dynamical nature such as filtering and optimal stopping for instance. It established a formal bridge between the theory of Wiener processes with a drift and point processes with an intensity through the martingale point of view. The methods used in the first case in the problems of detection and filtering can be applied word for word to solve the corresponding problems in the theory of jump processes : a striking example of this situation is the extension of the Fujisaki, Kallianpur and Kunita [11] martingale method of filtering (see [7] [18] and [19]). The martingale representation theorem plays a central role in the martingale point of view for jump processes. In this article we obtain it through the theorem of transformation of probabilities of [2] and recent results of J. Jacod [14] . The method seems to be a "natural" one and will find other applications.

1. - MARTINGALES AND STOCHASTIC STIELTJES INTEGRATION

Let (Ω,F) be a measurable space and $\underline{X} = (X_t,\ t\varepsilon R^+)$ a real-valued process on it. We denote by $\sigma(\underline{X},t)$ the σ-field generated by the family of random variables $(X_s,\ s\varepsilon[o,t])$, and by $\sigma(\underline{X})$ the increasing family of σ-fields $(\sigma(\underline{X},t),t\varepsilon R^+)$. By definition $\sigma(\underline{X})$ is called the <u>internal history</u> of $\underline{X}$. More generally, a <u>history</u> of $\underline{X}$ is an increasing family $\underline{F} = (F_t,t\varepsilon R^+)$ of sub σ-fields of F such that for all $t\varepsilon R^+$, $F_t \supseteq \sigma(\underline{X},t)$.

$\underline{F}$ being a history of $\underline{X}$, $\underline{X}$ is said to be $\underline{F}$-<u>predictable</u> if the mapping $(t,\omega) \to X_t(\omega)$ of $R^+ x\Omega$ into R is $P(\underline{F})/B(R)$ measurable, where B(R) is the borelian σ-field of R and $P(\underline{F})$ is the σ-field on $R^+x\Omega$ generated by the sets of the form $]t,\infty)xA,t\varepsilon R^+,A\varepsilon F_t$.

$\underline{F}$ being a history of $\underline{X}$ and P a probability on (Ω,F), $\underline{X}$ is called a$(P,\underline{F})$-submartingale if and only if, for all $0 \leqslant s \leqslant t$:

(1) a) $E\{|X_t|\} < \infty$

(2) b) $E\{X_t/F_s\} \leqslant X_s$, Pa.s.

If in condition a) the symbol $\leqslant$ is replaced by $=$, $\underline{X}$ is called a $(P,\underline{F})$ <u>martingale</u>.

$\underline{F}$ being an increasing family of sub-σ-fields of F, the random variable T taking its values in $\bar{R}^+ = R^+ \cup \{\infty\}$ is called an $\underline{F}$-stopping time, if for all $t\varepsilon R^+$:

$$\{T \leqslant t\} \varepsilon F_t \tag{3}$$

T being an $\underline{F}$-stopping time, the σ-field F_T is defined by

$$F_T = \{A\varepsilon F \ / \ A\cap\{T \leqslant t\} \varepsilon F_t,\ \forall t\varepsilon R^+\} \tag{4}$$

A stochastic process $\underline{X}$ defined on the probability space (Ω,F,P) is called a $(P,\underline{F})$ local martingale if there exists a family $(T_n, n\in N^+)$ of $\underline{F}$ stopping times increasing P a.s. to infinity and such that, for all $n\in N^+$, the process $X^{Tn} = (X_{t\wedge Tn}, t\in R^+)$ is a $(P,\underline{F})$ martingale.

The following lemma, linking the non negative local martingales and the submartingales is well known :

Lemma 1

If $\underline{X}$ is a non negative $(P,\underline{F})$ local martingale, then it is a $(P,\underline{F})$ submartingale

Proof :

The proof follows trivially from application of Fatou's lemma to the equation :

$$(5) \qquad E\{X_{t\wedge T_n} / F_s\} = X_{s\wedge T_n} , \quad \forall o \leqslant s \leqslant t , \quad \forall n \in \mathbb{N}^+$$

where the T_n's are the "localizing times" in the definition of local martingales. ■

Next, we state a result which is fundamental to this paper : it links the stochastic Stieltjes Lebesgue integrals of predictable processes with respect to martingales of bounded variation and the martingales. Before stating this theorem, we wish to attract the attention of the reader to the fact that the symbol $\int$ refers exclusively in this paper to Stieltjes-Lebesgue integrals.

Proposition 2 (C. Doléans - Dade and P.A Meyer [10])

Let $\underline{M} = (M_t, t\in R^+)$ be a $(P,\underline{F})$ martingales with paths P a.s. of bounded variation and $\underline{C} = (C_t, t\in R^+)$ a $\underline{F}$-predictable process such that, for all $t\in R^+$:

$$(6) \qquad E\left\{\int_o^t |C_s| \; |dM_s|\right\} < \infty$$

Then the process $\underline{CoM}$ defined for all $t\in R^+$, by $(CoM)_t = \int_o^t C_s \, dMs$, is a $(P,\underline{F})$ martingale.

Proof : We refer to [10] (Proposition 2, p. 89-90) ■

2. - POINT PROCESSES AND MARTINGALES

Definition 3

Let (Ω,F) be a measurable space and $\underline{N} = (N_t,\ t\varepsilon R^+)$ a counting process on (Ω,F), that is to say a process with right-continuous paths taking its values in $\mathbb{N}^+$, such that $N_o = o$ and $\Delta N_t = N_t - N_{t-} = o$ or 1, for all $t\varepsilon R^+$. A probality P on (Ω,F) being given, the pair $(\underline{N},P)$ is called a stochastic point process (S.P.P.).

Sometimes, we shall abuse notations and refer to $\underline{N}$ as a point process. In the following, we will restrict our attention to SPP's that are non-explosive, that is to say : such that $N_t < \infty$,$\forall t\varepsilon R^+$, P a.s.

A trivial application of P.A. Meyer's decomposition theorem [16] (or of C. Dellacherie's predictable projection theorem [8]) yields the following :

Theorem 4

Let $(\underline{N},P)$ be a (non-explosive) SPP on (Ω,F) and $\underline{F}$ a history of $\underline{N}$. Then there exists, one and only one $\underline{F}$ predictable increasing process $\underline{A} = (A_t,\ t\varepsilon R^+)$ such that $Ao = 0$ and verifying :

$$N_t - A_t = M_t\ ,\ \forall t\varepsilon R^+ \tag{7}$$

where $\underline{M}$ is a (P,F) local martingale.
Moreover the localizing times for $\underline{M}$ can be chosen to be :

$$T_n = \text{"}\inf \{t/N_t = n\} \text{ or } \infty\text{"} \tag{8}$$

Proof :

Consider the bounded sub-martingale $\underline{M}^{(n)} = (N_{t\wedge Tn},\ t\varepsilon R^+)$. By P.A. Meyer's decomposition theorem [15] there exists one and only one increasing $\underline{F}$ predictable process $\underline{A}^{(n)} = (A_t^{(n)}, t\varepsilon R^+)$ such that $(N_{t\wedge Tn} - A_t^{(n)},\ t\varepsilon R^+)$ is a $(P,\underline{F})$ martingale.

Now by Proposition 2, $(N_{t\wedge T_n} - A^{(n+m)}_{t\wedge T_n}, t\varepsilon R^+)$ is a $(P,\underline{F})$ martingale since :

(9) $$N_{t\wedge T_n} - A^{(n+m)}_{t\wedge T_n} = \int_o^t I\{s \leqslant T_n\} \; d(N_{S\wedge T_{n+m}} - A_s^{(n+m)})$$

By the unicity of P.A. Meyer's decomposition :

(10) $$A_t^{(n+m)} = A_t^{(n)} \quad \text{on} \quad t \leqslant T_n$$

We can therefore define a increasing $\underline{F}$ predictable $\underline{A}$ by :

(11) $$A_t = A_t^{(n)} \quad \text{on} \quad t \leqslant T_n$$

This process is $\underline{F}$-predictable and satifies. That it is unique is obvious from the uniqueness of the $\underline{A}^{(n)}$'s. ■
J. Jacod has shown in [14]that one can choose a version of $\underline{A}$ that has no jump strictly greater than 1 and gave in the case where $\underline{F} = (\underline{N})$ the explicit form of A_t on $N_t = n$:

(12) $$A_t = \sum_{i=o}^{n-1} \int \frac{G^i(dx)}{G^i([x,\infty])} \; I\{x \leqslant S_{i+1}\} + \int \frac{G^n(dx)}{G^i([x,\infty])} \; I\{x \leqslant t-T_n\}$$

where $G^n(.)$ is a regular version of the conditional probability law of $\{S_{n+1} \overset{d}{=} T_{n+1} - T_n \;\epsilon.\}$ with respect to $(\underline{N},T_n)$. This expression of $\underline{A}$ is due in a special case to F Papangelou [17].

Also, J. Jacod has given in [14] a fundamental theorem of existence and uniqueness which complements Theorem 4. In order to state this result a definition is needed :

Definition 5 :

The basic measurable space (Ω,F) of point processes on the half line R^+ is defined by :

a) $\omega \in \Omega \Longleftrightarrow \omega = (N_t,\ t\in R^+)$

where $(N_t,\ t\in R^+)$ is a counting function on R^+.

b) F is the smallest σ-field on Ω that makes the mappings $\omega \to N_t$ measurable for all $t\in R^+$.

Theorem 6 (J. Jacod [14])

Let (Ω,F) be the basic measurable space of point processes on R^+ and F_t the σ-field generated by the random variables $\omega \to N_s$, $s \in [o,t]$.
Let $\underline{A}$ be a right continuous $\underline{F}$-predictable increasing process such that $A_o = o$ and $\Delta A_t = A_t - A_{t-} \leqslant 1$.

There exists one and only one probability on (Ω,F) such that $(N_t - A_t,\ t\in R^+)$ is a $(P,\underline{F})$ local martingale.

Proof : We refer to [14] (Theorem 4 pp. 13-18) ■

The following is an immediate consequence of equation (12) :

Corollary 7

Let (Ω,F) be a measurable space, $\underline{N}$ a counting process on it, P and P' two probalities on (Ω,F) such that $(N_t - A_t,\ t\in R^+)$ is a $(P,\underline{F})$ local martingale and $(N_t - A'_t\ ,\ t\in R^+)$ is a $(P',\underline{F})$ local martingale where E is a history of $\underline{N}$, and $\underline{A}$ and $\underline{A}'$ are two increasing $\underline{F}$-predictable processes.

Suppose that $P' << P$. Then $\underline{A}' << \underline{A}$ in the sense that :

$$(13) \qquad A'_t = \int_o^t \mu_s \, d\, A_s \ , \ \forall t\in R^+$$

for some non negative process μ adapted to $\underline{F}$. Moreover μ can be chosen to b $\underline{F}$-predictable.

Proof :

If $P' << P$, then, necessarily the conditional probability under P' of S_{n+1} given $\sigma(\underline{N},T_n)$ and $S_{n+1}>x$ is absolutely continuous with respect to the same conditional probability under P so that :

$$(14) \qquad \int_o^t \frac{G'^n\ (dx)}{G'^n([x,\infty])} = \int_o^t \lambda_x^{(n)} \frac{G^n\ (dx)}{G'^n\ ([x,\infty])}$$

for some non negative $\underline{\lambda}^{(n)}$ depending on $\sigma(\underline{N},T_n)$. Therefore by (12) $A'_t = \int_o^t \mu_t \, d\, A_t$ for some $\underline{\mu}$ that can be chosen $\sigma(\underline{N})$ predictable by the dual predictable projection theorem of [8]. ■

Definition 8 :

a - Let $(\underline{N},P)$ be a SPP on (Ω,F) , $\underline{F}$ a history of $\underline{N}$ and $\underline{A}$ the (unique) increasing $\underline{F}$-predictable process such that $A_o = o$ and $(N_t - A_t,\ t\epsilon R^+)$ is a $(P,\underline{F})$ local martingale. The pair $(\underline{A},\underline{F})$ is called a description of $(\underline{N},P)$. If $(N_t - A_t,\ t\epsilon R^+)$ is a martingale, and not merely a local martingale, we say that $(\underline{A},\underline{F})$ is a simple description of $(\underline{N},P)$.

b - If $(\underline{A},\underline{F})$ is a description of $(\underline{N},P)$ where $\underline{F}$ is $\equiv \sigma(\underline{N})$ we say that it is an internal description of $(\underline{N},P)$. We will also speak of a simple internal description.

c - If $\underline{A}$ is such that $A_t = \int_o^t \lambda_s\, d_s$ for all $t\epsilon R^+$, we say that $(\underline{N},P)$ admits the $\underline{F}$-intensity $\lambda = (\lambda_t\ ,\ t\epsilon R^+)$. We can also define in an obvious manner a simple $\underline{F}$-intensity.

Remark 1 :

If $\underline{\lambda}$ is an $\underline{F}$-intensity for $(\underline{N},P)$ (simple or not), one can check that, for all $o \leqslant s \leqslant t$:

$$(15) \qquad E\left\{ N_t - N_s \,/\, F_s \right\} = E\left\{ \int_s^t \lambda_u \, du \,/\, F_s \right\}$$

Remark 2 :

The $\underline{F}$-intensity $\underline{\lambda}$ of a SPP can always be chosen to be $\underline{F}$-predictable since, if it is not, we can replace it by a version $\tilde{\lambda}$ which is $\underline{F}$-predictable and such that :

$$E\left\{\int_S^T \lambda_s ds | F_S\right\} = E\int_S^T \tilde{\lambda}_s ds | F_S, \text{ P a.s.}$$

for all $\underline{F}$ stopping times $S \leq T$ (cf Dellacherie [8])

Remark 3 :

When (Ω,F) is the basic measurable space of point processes and $F = \sigma(\underline{N})$, a necessary and sufficient condition for a right continuous increasing process $\underline{C}$ to be $\sigma(\underline{N})$ predictable is that C_t depends measurably on $T_1, T_2, .., T_n$, t on $(T_n \leq t)$, i.e. that there exists a sequence $(f_n, n\varepsilon N^+)$ of borelian functions

(16) $\quad f_n : R_{n+1} \longrightarrow R \quad , \quad n \in N^+$

such that

(17) $\quad C_t = f_n (T_1,..,T_n,t)$ on $T_n \leqslant t$ for all $n \epsilon N^+$, $t \epsilon R^+$.

(see [14] , P. 16).

Remark 4 :

Jacod's result as stated in [14] is more general than theorem 6. In particular, it includes the doubly stochastic point processes. However in the present work, the restricted form is enough for our purpose.

Also we have stated the existence and uniqueness theorem in the special setting of the basic measurable space of point processes. As it is clear from a reading of the proof of theorem 4 of [14] , the proof of the uniqueness is valid in the setting of $(\Omega, \sigma(\underline{N},\infty))$ where (Ω,F) is a general measurable space on which is defined a counting process $\underline{N}$. This remark has its interest in the proof of the separation theorem (corollary 13).

3. - ABSOLUTELY CONTINUOUS CHANGE OF MEASURE

In this section, we produce the other basic result of the theory : it relates the Radon-Nikodym derivative $\frac{dP'}{dP}$ of an absolutely continuous change of probability measure $P \rightarrow P'$, to the change of intensity $\underline{\lambda} \rightarrow \underline{\lambda}'$ that it provokes in a given point process $\underline{N}$. As such, it is a twin to the I. Girsanov's theorem [9] .

Theorem 9 : (Theorem of change of Law)

Let $(\underline{N},P)$ be a SPP with the $\underline{F}$-intensity $\underline{\lambda}$ and let $\underline{\mu}$ be a non negative $\underline{F}$-predictable process such that $\int_0^t \mu_s \lambda_s \, ds \quad < \quad , \forall t \in R^+$, P a.s.

Define, for all $t \in R^+$:

(18) $$L_t = \left(\prod_{s \leqslant t} \mu_s \, I \left\{ N_s \neq N_{s-} \right\} \right) \exp \left(- \int_0^t (\mu_s - 1) \lambda_s \, ds \right)$$

Then, the process $\underline{L}$ is a non negative $(P,\underline{F})$ local martingale.
If in addition :

(19) $$E \left\{ L_1 \right\} = 1$$

then $(L_{t \wedge 1} \; , t \in R^+)$ is a non negative $(P,\underline{F})$ martingale. Under condition (19) one can define P' by :

(20) $$\frac{dP'}{dP} = L_1$$

Then $(N_{t \wedge 1} - \int_0^{t \wedge 1} \mu_s \lambda_s \, ds, \; t \in R^+)$ is a $(P',\underline{F})$ local martingale.

Proof :

We need a lemma, which is an almost obvious corollary of Proposition 2.

Lemma 10 :

a) Let $(\underline{N},P)$ be a SPP with the simple F-intensity λ
Let $\underline{C}$ be an $\underline{F}$ predictable process such that for all $t \in R^+$:

(21) $$E \left\{ \int_0^t \mid C_s \mid \lambda_s \, ds \right\} <$$

Then $\left(\sum_{s\leq t} C_s\ I\{N_s \neq N_{s-}\} - \int_0^t C_s\ \lambda_s\ ds,\ t\in R^+\right)$ is a $(P,\underline{F})$ martingale.

b) If $(\underline{N},P)$ is a SPP with the (not necessarily simple) $\underline{F}$-intensity $\underline{\lambda}$ and $\underline{C}$ is an $\underline{F}$-predictable process such that, for all $t\in R^+$:

(22) $$\int_0^t |C_s|\ \lambda_s\ ds < \infty,\ P \text{ a.s.}$$

Then $\left(\sum_{s\leq t} C_s\ I\{N_s \neq N_{s-}\} - \int_0^t C_s\ \lambda_s\ ds,\ t\ R^+\right)$ is a $(P,\underline{F})$ local martingale.

Proof :

a) We need only show that :

(23) $$E\left\{\int_0^t |C_s|\ \lambda_s\ ds\right\} < \infty \Rightarrow E\left\{\int_0^t |C_s|\ |dM_s|\right\} < \infty$$

where $M_t = N_t - \int_0^t \lambda_s\ ds$

We consider $\underline{C}^{(n)}$ defined by $C_t^{(n)} = |C_t|\ I\{|C_t| \leq n\}$ for all $t\in R^+$, and we apply Proposition 2 to obtain, after remarking that $|dM_t| = dN_t + \lambda_t\ dt$:

(24) $$E\left\{\sum_{s\leq t} |C_s|\ I\{|C_t| \leq n\}\quad I\{N_s \neq N_{s-}\}\right\} = E\left\{\int_0^t |C_s|\ I\{|C_s| \leq n\}\ ds\right\}$$

Letting n go to infinity, we get :

(25) $$E\left\{\int_0^t |C_s|\ \lambda_s\ ds\right\} = \frac{1}{2}\ E\left\{\int_0^t |C_s|\ |dM_s|\right.$$

which yields the desired result.

b) Let $(T_n,\ n\in N^+)$ be a sequence of $\underline{F}$ stopping times increasing P a.s. to ∞ and such that $\left(N_{t\wedge Tn} - \int_0^{t\wedge Tn} \lambda_s ds,\ t\in R^+\right)$ is a $(P,\underline{F})$ martingale, and $E\int_0^{t\wedge Tn} |C_s|\lambda_s\ ds \leq n$, for all $n\in N^+$. Such a sequence exists and, obviously, is the "localizing" sequence we are looking for. ∎

Proof of Theorem 9 :

We rewrite L_t as :

(26) $$L_t = 1 + \int_0^t L_{s-} \; (\mu_s - 1)\, d\, M_s$$

where $M_t = N_t - \int_0^t \lambda_s \, ds$. Defining :

$$T_n = \text{"}\inf\{t \mid N_t = n \text{ or } L_{t-} \geqslant n \text{ or } \int_0^t (\mu_s - 1)\lambda_s \, ds = n\} \text{ or } \infty\text{"}$$

We see, by application of Lemma10 that $(L_{t \wedge T n},\ t \in R^+)$ is a $(P, \underline{F})$ martingale. Also, clearly, $T_n \uparrow \infty$ P a.s. Therefore $\underline{L}$ is a $(P,\underline{F})$ local martingale. By Lemma 1, it is a $(P,\underline{F})$ sub martingale. Also, clearly, if $E\{L_1\} = 1 \; (= E\{L_o\})$, $\underline{L}$ is a $(P,\underline{F})$ martingale. To prove the last assertion of Theorem 4, it is sufficient to prove that

$$(N_{t\wedge 1} - \int_0^{t\wedge 1} \mu_s \lambda_s \, ds)\, L_{t\wedge 1},\ t \in R^+) \text{ is a } (P,\underline{F}) \text{ local martingale.}$$

To do this we define $M'_t = N_t - \int_0^t \mu_s \, \lambda_s \, ds$ and we integrate by parts to obtain :

(27) $$M'_t \, L_t = \int_0^t M'_s \, dL_s + \int_0^t L_{s-} \, dM'_s$$

Rewriting $M'_s = M_s - \int_0^t (\mu_s - 1)\lambda_s \, ds$

$$dL_s = L_{s-} \; (\mu_s - 1) \; dM_s$$

and noting that :

$$M'_s = M'_{s-} + (M'_s - M'_{s-}) = M'_{s-} + \Delta N_s$$

we obtain after a rearrangement of the terms :

(28) $$M'_t \, L_t = \int_0^t \Big[L_{s-} \, \mu_s + M_{s-} \, L_{s-} \, (\mu_s - 1) \Big] \, d\, M_s$$

We define the stopping times :

$$S_n = \text{"inf } \left\{ t \;\middle|\; L_{s-} \geq n \text{ or } M_{s-} \geq n \text{ or } \int_0^t (\mu_s + 1)\lambda_s \, ds = n \text{ or } N_t = n \right\} \text{ or } \infty$$

and apply the Lemma 10 to prove that $(M'_{t\wedge Sn} \; L_{t\wedge Tn} \; , \; t\in R^+)$ is a $(P,\underline{F})$ martingale. We remark that $S_n \uparrow \infty$ P a.s., thus proving the final result. ∎

Remark 5 :

The "formula" :

$$(29) \quad E\left\{\frac{dP'}{dP} / F_t\right\} = \left(\prod_{s\leqslant t} I\{N_s + N_{s-}\}\mu_s\right) \exp\left(-\int_0^t (\mu_s - 1)\lambda_s \, ds\right), \; t \leqslant 1$$

is well known in the case where $\underline{\mu}$ and $\underline{\lambda}$ are deterministic processes (see M. Brown [5] , for instance).

Remark 6 :

The two ways of writing L_t, namely :

$$(30) \quad L_t = \left(\prod_{s\leqslant t} I\{N_s + N_{s-}\} \mu_s\right) \exp\left(-\int_0^t (\mu_s - 1)\lambda_s \, ds\right)$$

and $L_t = 1 + \int_0^t L_{s-} (\mu_s - 1) \, dM_s$,

constitute a trivial form of C. Doléans-Dade exponential formula [7] .

Remark 7 :

The condition $E\{L_1\} = 1$ restricts the scope of theorem 10.
Some simple sufficient conditions have been found. See [1] and [19] for instance.
A special case of importance where the condition holds is when $\lambda_t \equiv 1$ and $\underline{\mu}$ is bounded.
To show this, we first remark that if $\lambda_t = 1$ then (N,P) is "Poisson".

More precisely :

Theorem 11 : (S. Watanabé [16])

If $(\underline{N},P)$ is a SPP such that $(N_t - t, t \in R^+)$ is a $(P,\underline{F})$ local martingale, then, for all $o \leq s \leq t$

a) F_s is independent of $N_t - N_s$

b) $N_t - N_s$ is a Poisson random variable with parameter t-s

Proof : See [16] or, for a more elementary proof [2] and [3]. ■

Next, from :

$$(31) \qquad L_t \leq A^{X_t} \exp t$$

where A is an upper bound of μ_t, we see, using the fact that X_t is a Poisson r.v. with parameter t, that :

$$(32) \qquad E\left\{\int_o^t \left|L_{s-} \ (\mu_s - 1)\right| ds\right\} \leq E\left\{A^{X_t+1}\right\} \exp t <$$

Thus, by Lemma 10 and (26), $\underline{L}$ is a $(P,\underline{F})$ martingale.

Some other sufficient conditions can be found. We will not, for the time being, investigate this problem, only attracting the attention of the reader that the problem is still open. The same problem arises in I. Girsanov Theorem and has not, so far been solved. In the more general terms of C. Doléans-Dade exponentiation formula, it would be interesting to find necessary and sufficient conditions on a local martingale $\underline{M}$ so that $\underline{L}$ defined by $L_t = 1 + \int_o^t L_{s-} \, dM_s$ where $\int$ denotes stochastic integration be a non negative martingale.

Remark 8 :

We will insist on the following point : in formula (18) the version of $\underline{\mu}$ to be chosen must be predictable, otherwise the conclusion of the theorem does not hold. For instance if we start with a Poisson process ($\lambda_t \equiv 1$) and if we chose $\underline{\mu}$ such that :

$$(33) \qquad \mu_t = \begin{cases} 2 & \text{if } t < T_1 = \inf\{t/\ N_t = 1\} \\ 1 & \text{otherwise} \end{cases}$$

then the corresponding $\underline{L}$ is given by :

$$(34) \qquad L_t = \begin{cases} e^{-t} & \text{if } \ t \leqslant T_1 \\ e^{-t_1} & \text{if } \ t > T_1 \end{cases}$$

and one can check that it is not a martingale. But if instead of a right continuous version, we had chosen the left continuous (therefore predictable) version.

$$(33') \qquad \mu_t = \begin{cases} 2 & \text{if } \ t \leqslant T_1 \\ 1 & \text{otherwise} \end{cases}$$

then :

$$(34') \qquad L_t = \begin{cases} e^{-t} & \text{if } \ t \leqslant T_1 \\ 2e^{-t_1} & \text{if } \ t > T_1 \end{cases}$$

Such an $\underline{L}$ is indeed a martingale.
The above remark is very important in applications such as in theory of filtering of signals modulating a point process, where the use of a "bad" version of $\underline{\mu}$ yields wrong results.

Remark 9

We wish to show that the form of L_t is expressible in terms of the G^n's of (12). The formulas obtained then look more familiar. We will consider the case where $\lambda_t \equiv \lambda t$ (therefore by theorem 11, $(\underline{N},P)$ is Poissonian with intensity λ). Let G'^n be the conditional law of S_{n+1} under P' and g'^n the density of G'^n.

Then it is a trivial exercise to discover that :

$$(35) \qquad L_t = \prod_{i \leqslant n} \frac{g^i(S_i)}{\exp(-\lambda S_i)} \cdot \frac{1 - G^n(t - T_n)}{\exp(-\lambda(t - T_n))} \quad \text{on } (N_t = n)$$

The general form would be, if G^n and g^n are the conditional distributions and densities of S_{n+1} under P :

$$(35') \qquad L_t = \prod_{i \leqslant n} \frac{g'^i(S_i)}{g^i(S_i)} \quad \frac{1 - G'^n(t - T_n)}{1 - G^n(t - T_n)} \quad \text{on } (N_t = n)$$

4. - THE RADON NIKODYM DERIVATIVE CONVERSE THEOREM

The result of this section is the converse of theorem 9.

Theorem 12 :

Let P' and P be two probabilities on (Ω, F) and $\underline{N}$ a counting process on it. Suppose that :

$$P' \ll P \quad \text{on } \sigma(\underline{N}, 1)$$

and that $(\underline{N}, P)$ has the $\sigma(\underline{N})$ intensity $\underline{\lambda}$

Then :

a) there exists a $\sigma(\underline{N})$ predictable process $\underline{\mu}$ such that : $(\underline{N}, P')$ has the $\sigma(\underline{N})$ intensity $\underline{\lambda}'$ where $\lambda'_t = \mu_t \lambda_t$, $\forall t \in [o,1]$

b) $E\left\{ \frac{dP'}{dP} \middle| \sigma(\underline{N}, 1) \right\} = \prod_{s \leqslant t} \mu_s \exp\left(-\int_o^t (\mu_s - 1)\lambda_s \, ds\right)$

Remark 10

By theorem 10, 2) implies that :

$$(36) \qquad E\left\{\frac{dP'}{dP}\,\middle|\,\sigma(\underline{N},t)\right\} = \prod_{s\leqslant t}\mu_s \exp\left(-\int_o^t (\mu_s-1)\lambda_s\,ds\right) \text{ on } t\in[o,1]$$

Proof :

1) follows from corollary 7

2) Define $L_t = \prod_{s\leqslant t}\mu_s \exp\left(-\int_o^t (\mu_s-1)\lambda_s\,ds\right)$. By theorem 9 , $\underline{L}$ is a $(P,\sigma(\underline{N}))$ local martingale. Let $(T_n, n\in N^+)$ be the localizing times. Define P'^n by $E\left\{\frac{dP'^n}{dP}\,\sigma(\underline{N},1)\right\} = L_{1\wedge T_n}$. Then by theorem 9 :

$$(37) \qquad N_{t\wedge T_n} = \int_o^{t\wedge T_n} \mu_s\lambda_s\,ds = (P'^n,\ \sigma(\underline{N}))\text{ martingale}$$

Therefore by the uniqueness theorem 6, P' and P'^n agree on $\sigma(\underline{N},T_n)$.

Therefore

$$(38) \qquad E\left\{\frac{dP'}{dP}\,\middle|\,\sigma(\underline{N},\ T_n)\right\} = L_{1\wedge T_n}$$

Letting $n\uparrow\infty$ yields the announced result. ∎

A corollary of this result is known as the theorem of separation of detection and filtering :

Corollary 13 :

Let (Ω,F) be a measurable space on which is given a SPP $(\underline{N},P')$ such that :

$$(39) \qquad N_t - \int_o^t \lambda_s\,ds = (P',\underline{F})\text{ local martingale}$$

where $\underline{F}$ is an history of $\underline{N}$.

Suppose that the restriction of P' to $(\Omega,\sigma(\underline{N},1))$ is absolutely continuous to a probability P on $(\Omega,\sigma(\underline{N},1))$ such that $N_t - t = (P,\sigma(\underline{N}))$ local martingale on [o,1].

Then :

$$(40)\qquad E\left\{\frac{dP'}{dP}\,\middle|\,\sigma(\underline{N},t)\right\} = \prod_{s\leqslant t}\hat{\lambda}_s \quad \exp\left(-\int_0^t (\hat{\lambda}_s - 1)\,ds\right)$$

where $(\hat{\lambda}_t,\ t\in R^+)$ is a $\sigma(\underline{N})$-predictable version of $(E'\{\lambda_t \mid \sigma(\underline{N},t),\ t\in R^+)$.

Proof :

The proof is based on the following lemma :

Lemma 14 :

Let $(\underline{N},P)$ be a SPP with the $\underline{F}$-intensity $\underline{\lambda}^F$. Let $\underline{G}$ be a history of $\underline{N}$ included in $\underline{F}$, i.e such that :

$$(41)\qquad G \subseteq F_t\ ,\ \forall t\in R^+$$

Let $\lambda_t^G = E\{\lambda_t \mid G_t\}$ for all $t\in R^+$. Then $\underline{\lambda}^G$ is a G-intensity of $(\underline{N},P)$.

Remark 11 : This result is interesting in the case where $\underline{G} = \sigma(\underline{N})$.

Proof :

Let T_n = "inf $\{t \mid N_t = n\}$ or ∞". Then :

$$(42)\qquad E\left\{N_{t\wedge T_n} - N_{s\wedge T_n} \,\middle|\, F_s\right\} = E\left\{\int_s^t \lambda_u^F\, I\{u\leqslant T_n\}\,du \,\middle|\, F_s\right\}$$

by $F_s \supseteq G_s$, Fubini's theorem and the fact that T_n is a $\sigma(\underline{W})$ stopping time :

$$(43)\qquad E\left\{N_{t\wedge T_n} - N_{s\wedge T_n} \,\middle|\, G_s\right\} = E\left\{\int_s^t \lambda_u^G\, I\{u\leqslant T_n\}\,du \,\middle|\, G_s\right\} \blacksquare$$

Proof of the corollary :

It suffices to apply Lemma 14 and Theorem 12. ■

5. - THE REPRESENTATION THEOREM AND THE PROJECTION PRINCIPLE

The next corollary gives the necessary form of the $(P,\sigma(\underline{N}))$ martingales, when $(\underline{N},P)$ is a S.P.P. with an intensity.

Corollary 15 (MHA. Davis [7], J. Van Schuppen [19])

Let $(\underline{N},P)$ be a S.P.P. with the $\sigma(\underline{N})$-intensity $\underline{\lambda}$ and $\underline{M}$ a $(P,\sigma(\underline{N}))$ martingale on $[0,1]$. Then there exists a $\sigma(\underline{N})$-predictable process $\underline{\Phi}$ such that :

$$M_t = E\{M_o\} + \int_o^t \phi_s \; (dN_s - \lambda_s \, ds), \; \forall t \in [o,1] \tag{44}$$

Proof :

For any $t \in [o,1]$

$$M_t = \{E \, M_1 / \sigma(\underline{N},t)\} = E\{M_1^+ / \sigma(\underline{N},t)\} + E\{M_1^- / \sigma(\underline{N},t)\} \tag{45}$$

where $M_1^+ = \begin{cases} M_1 \text{ if } M_1 > o \\ o \text{ othewise} \end{cases}$

and $M_1^- = M_1 - M_1^+$

one can identify $\dfrac{M_1^+}{E\{M_1^+\}}$ or $\dfrac{M_1^-}{E\{M_1^-\}}$ to a

Radon-Nikodym derivative $\dfrac{dP'}{dP}$ and apply Theorem 12. ∎

Remark 12

From the proof of corollary 15 and Theorem 12, we see that if $\underline{M}$ is a positive martingale, then :

$$M_t = E\{M_o\} + \int_o^t M_{s-} \; (\mu_s - 1) \; (dN_s - \lambda_s \, ds) \tag{46}$$

where $\underline{\mu}$ is $\sigma(\underline{N})$-predictable. This form is more specialized than (44) and yields some technical results that may be of interest : for instance if $\underline{M}$ is a bounded martingale the corresponding $\underline{\Phi}$ in (44) is bounded from below.

An important consequence of corollary 15 is the :

Projection principle

Let $(\underline{N},P)$ be a SPP with the $\sigma(\underline{N})$-intensity $\underline{\lambda}$ and X be a P-integrable r.v. Let $\hat{X}_{t_o}$ be a P-integrable $\sigma(\underline{N},t_o)$ measurable random variable. Then $\hat{X}_{t_o} = E\{X \mid \sigma(\underline{N},t_o)\}$ if and only if :

(47) $$E\left\{X - \hat{X}_{t_o}\right\} = o$$

(48) $$E\left\{(X - \hat{X}_{t_o}) \int_o^{t_o} \Phi_s \, (dN_s - \lambda_s \, ds)\right\} = o$$

for all $\sigma(\underline{N})$-predictable $\underline{\Phi}$ such that the process $(\int_o^t \Phi_s \, (dN_s - \lambda_s \, ds), \, t \in R^+)$ is a bounded $(P,\sigma(\underline{N}))$ martingale on $[o, t_o]$.

Proof :

In order for $\hat{X}_{t_o}$ to be $= E\{X / \sigma(\underline{N},t_o)\}$ it is necessary and sufficient that :

(49) $$E\left\{(X - \hat{X}_{t_o}) \, Y\right\} = o$$

for any bounded random variable Y which is $\sigma(\underline{N},t_o)$ measurable.
Now, the rest of the proof follows from the fact that, for any $\sigma(\underline{N},to)$-measurable integrable random variable Y :

(50) $$Y = M_{t_o}$$

where $\underline{M} = (E\{Y / \sigma(\underline{N},t)\}, \, t \in R^+)$ is a $(P,\sigma(\underline{N}))$ martingale. ■

Remark 13 :

This principe has been implicitly used by E. Wong [22] et al. [1, 7, 19] in filtering theory in order to calculate the gain of non linear filters.

6. - CASE OF THE MULTIVARIATE POINT PROCESSES - EXAMPLE OF THE QUEUING PROCESSES

A k-variate point process on (Ω,F) is a pair $(\underline{N},P)$ where P is a probability measure and $\underline{N} = (N_t^{(1)}, \dots, N_t^{(k)}, \, t \in R^+)$, each of the $\underline{N}^{(i)}$ being a counting process. The results of J. Jacod [14] do hold in such a case, and the rephrasing of all the results of the previous paragraphs does not present any difficulty. We wish only insist on the 2-variate point process associated to a stochastic queuing process. More presisely :

a stochastic queuing process on (Ω,F) is a pair $(\underline{Q},P)$ where P is a probability measure on (Ω,F) and $\underline{Q} = (Q_t, t\epsilon R^+)$ is a right continuous process taking its values on $\mathbb{N}^+$, such that $Q_o = o$, $Q_t - Q_t = o, +1$ or -1 for all $t\epsilon R^+$. Note that $Q_t \geqslant o$ for all $t\epsilon R^+$ since $\underline{Q}$ takes its values in $\mathbb{N}^+$.

Define

(51) $\quad N_t^{(1)}$ = number of upward jumps of $\underline{Q}$ in $[o,t]$

(52) $\quad N_t^{(2)}$ = number of downward jumps of $\underline{Q}$ in $[o,t]$

We can then associate to $(\underline{Q},P)$ the 2-variate SPP $(\underline{N},P)$ defined by $\underline{N} = (N_t^{(1)}, N_t^{(2)}, t\epsilon R^+)$.

$\underline{N}^{(1)}$ is called the <u>input</u> process and $\underline{N}^{(2)}$ the <u>output</u> process. To $\underline{N}$ and to each history $\underline{F}$ of $\underline{Q}$ we can in turn associate an $\underline{F}$ predictable increasing process $\underline{A} = (A_t^{(1)}, A_t^{(2)}, t\epsilon R^+)$ such that

$$(N_t^{(1)} - A_t^{(1)}, t\epsilon R^+) \text{ and } (N_t^{(2)} - A_t^{(2)}, t\epsilon R^+) \text{ are } (P,\sigma(\underline{Q})) \text{ local}$$

martingales. Il should be noted that $A_t^{(2)}$ can be rewritten as :

(53) $$A_t^{(2)} = \int_o^t I\left\{Q_s \neq o\right\} dA_s^{(2)}$$

Indeed, if we let V_n be the nth time that $\underline{Q}$ hits 0 and V_n the first time after V_n at which $\underline{Q}$ jumps again, then by Doob's stopping theorem :

(54) $$O = E\left\{N_{V_n}^{(2)} - N_{U_n}^{(2)}\right\} = E\left\{A_{V_n}^{(2)} - A_{U_n}^{(2)}\right\}$$

therefore $A_{V_n}^{(2)} = A_{V_n}^{(2)}$ since $\underline{A}^{(2)}$ is increasing.

If $A_t^{(1)} = \int_o^t \lambda^{(1)} ds$ and $A_t^{(2)} = \int_o^t \lambda_s^{(2)} I\left\{Q_s \neq o\right\} ds$ for some non negative processes $\underline{\lambda}^{(1)}$ and $\underline{\lambda}^{(2)}$, we say that $(\underline{Q},P)$ has the <u>F-parameters</u> $\underline{\lambda}^{(1)}$ and $\underline{\lambda}^{(2)}$.

If $\lambda_t^{(1)} = \lambda^{(1)}$ and $\lambda_t^{(2)} = \lambda^{(2)}$, for all $t \in R^+$, where $\lambda^{(1)}$ and $\lambda^{(2)}$ are non negative constants we say that ($\underline{Q}$,P) is a $\underline{F}$-M/M/1 queue with parameters $\lambda^{(1)}$ and $\lambda^{(2)}$.

The analogous of theorem 9 is a very important result of practical interest. It reads in a particular case :

Let ($\underline{Q}$,Po) be a S Q P on (Ω,F) which is a $\underline{F}$-M/M+1 queue with parameters $\lambda^{(1)}$ and $\lambda^{(2)}$. Let $\underline{\lambda}^{(1)}$ and $\underline{\lambda}^{(2)}$ be two $\underline{F}$-predictable non negative processes such that $E\left\{L_1\right\} = 1$ where :

$$L_1 = \prod_{s \leqslant 1} \frac{\lambda_s^{(1)}}{\lambda^{(1)}} I\left\{Q_s = Q_{s-} + 1\right\} \prod_{s \leqslant 1} \frac{\lambda_s^{(2)}}{\lambda^{(2)}} I\left\{Q_s = Q_s - 1\right\} \times \exp\left(-\int_0^1 (\lambda_s^{(1)} - \lambda^{(1)}, ds\right) \ldots$$

$$\ldots \exp\left(-\int_0^1 (\lambda_s^{(2)} - \lambda^{(2)})\, I\left\{Q_s \neq 0\right\} ds\right)$$

If we define P' on (Ω,F) by $\frac{dP'}{dP} = L_1$, then ($\underline{Q}$,P) has the $\underline{F}$-parameters $\lambda^{(1)}$ and $\lambda^{(2)}$ on [0,1].

CONCLUSION

We have presented a new approach to the martingale theory of point processes on the real half line admitting an intensity. In this approach, the martingale representation (corollary 15) is obtained as a corollary of the Radon Nikodym derivative converse theorem (theorem 19). The latter is in turn obtained from the two basic results of the theory : the existence and uniqueness theorem of J. Jacod [14] and the constructive Radon Nikodym theorem of [2,3] of the Girsanov-Cameron-Martin type [12,6]. This new approach differs notably from the approach of [2] where the martingale point of view was introduced in the problems involving jump processes, and of the subsequent literature [1, 7, 18, 19] It gives additional results and will presumably find other applications in the general theory of jump processes.

REFERENCES

[1] R. BOEL ; P. VARAIYA ; E. WONG
Martingales on Jump Processes
I : Representation Results and II : Applications ; Memos ERL M407 and M409
Department of Electrical Engineering
University of California at Berkeley (1973)

[2] P. BREMAUD
A martingale approach to point processes
Ph. D. Thesis, University of California at Berkeley
Department of Electrical Engineering and Computer Science (1972)
Memo ERL - M - 345

[3] " "
Notes sur le point de vue des martingales dans les processus ponctuels et de sauts
Cahiers des mathématiques de la Décision ; centre d'études et de Recherche des mathématiques de la Décision : Université de Paris IX Dauphine (mai 1974)

[4] " "
Filtrage de l'état d'une file d'attente par rapport à sa sortie
Mai 1974 (à paraître)

[5] M. BROWN
Discrimination of Point processes
Ann. Math. Statist. 42 (1971)

[6] H. CAMERON ; W. MARTIN
Transformation of Wiener integrals under translations
Ann. Math. 45 (1944)

[7] M.H.A DAVIS
Detection of Signals with Point process observation. Publication 73/8 (Research Report), Department of Computing and Control, Imperial College, London (1973)

[8] C. DELLACHERIE
Capacités et processus stochastiques, Springer, Berlin (1972)

[9] C. DOLEANS-DADE
Quelques applications de la formule de changement de variables pour les semi martingales, Z. für Wahrscheinlischkeitsheorie 16, 3, (1970)

[10] C. DOLEANS-DADE ; P.A. MEYER
Integrales stochastiques par rapport aux martingales locales, Séminaire de Probabilité IV, Université de Strasbourg, Lecture notes in Maths 124, Springer-Verlag, Berlin (1970)

[11] M. FUJISAKI ; G. KALLIANPUR ; H. KUNITA
Stochastic differential equations for the non linear filtering problem
Osaka Math. 9 (1) (1972)

[12] I. GIRSANOV
On transforming a certain class of stochastic processes by absolutely continuous substitution of measures, Theory of Prob. and Appl. (Russian) 5,3 (1960)

[13] GRIGELIONIS
On Non Linear Filtering Theory and the absolute Continuity of Probability measures corresponding to Stochastic Processes ; in Procedings of the 2nd Japan USSR Symposium on Probability Theory, Lect. Notes in Maths, Springer Verlag (1973)

[14] J. JACOD
On the stochastic intensity of a random point process over the half-line, Tech. Report 51, Series 2, Department of Stat..., Princeton University (1973)

[15] H. KUNITA ; S. WATANABE
On square integrable martingales
Nagoya Math. J. 30 (1967)

[16] P. A. MEYER
Probabilités et Potentiel
Hermann, Paris, 1966

[17] F. PAPANGELOU
Integrability of expected increments of point processes and a related change of scale, Trans. Am. Math. Soc. 165 (1972)

[18] A. SEGALL
A martingale approach to modeling, estimation and detection of jump processes.
Ph. D. Thesis, Stanford U. Tech. Rep. 7050-21, Center for Systems Research, Stanford U. (1973)

[19] J. Van SCHUPPEN
Estimation theory for continuous time processes, a martingale approach,
Ph. D. Thesis, Dpet of El. Eng. University of California at Berkeley (1973)
Memo ERL-M405

[20] J. VAN SCHUPPEN ; E. WONG
Translation of local martingales under a change of probability Law, Memo ERL M-Department of Electrical Engineering and Computer Science, University of California at Berkeley (1973)

[21] S. WATANABE
On discontinuous additive functionals and Levy measures of a Markov Process, Jap. J. Math 34 (1964)

[22] E. WONG
Recent progress in sotchastic processes - A survey IEEE Trans. Inf. Theory IT-19, 3 (1973)

RESPONSE TIME OF A FIXED-HEAD DISK TO VARIABLE-LENGTH TRANSFERS

E. GELENBE*, J. LENFANT+, D. POTIER

IRIA - LABORIA

I - INTRODUCTION

Even though the trend in modern computer systems is to replace rotating secondary memory devices by core or semiconductor memories, drums and more particularly fixed head disks remain in wide usage while assuring important input-output functions in newer systems. The analysis of their performance remains of interest also because newer "rotating" memory devices such as magnetic bubble memories retain some of their information accessing and transfer characteristics.

The purpose of this paper is to analyze the response time of fixed-head (as opposed to moveable arm) rotating secondary memory device when the length of the records to be transferred obeys some arbitrary distribution function and when the records are transferred in FCFS order. The method of analysis of such devices, and the results obtained, will depend to a large extent on the scheduling algorithm used a well as on the nature of the transfers. Thus Coffman [1] has examined the behaviour of a drum whose circumference is divided into equal sized sectors each containing a page, and for which page transfers are requested singly (as opposed to batched requests). In his model, the drum (or fixed-head disk) is scheduled using the Eschenbach scheme [2] with one queue per sector. In [3], an algorithm for optimally scheduling variable length transfers for a drum has been given and it has been shown that it is a solution of a special case of the travelling salesman problem. A comparison of the algorithm in [3] with the classical SLTF (shortest-latency-time-first) algorithm has been presented by STONE and FULLER [5].

In this paper, in addition to our assumption of arbitrarily distributed length of records to be transferred, we also suppose that the starting address on the drum is at a fixed angular position on the circumference for <u>all</u> records. This is a realistic assumption in many cases. Many computer operating systems adopt this policy because of the great ease and simplicity it implies. The records to be transferred in such a system might be segments, file records or groups of pages belonging to the same program or system module for which "pre-paging" is used .

* Université de Paris XIII

+ Université de Rennes.

In the sequel, we shall obtain the long run probability distribution for the number of requests waiting for transfer in the system we have described and use it to compute the expected response time of the disk with the assumptions we have made regarding its operation; the arrival process of requests for transfers is assumed to be Poisson and the length of each transfer obeys an arbitrary distribution with finite first and second moments.

Assuming that the starting addresses of records are drawn from a uniform distribution around the circumference of the disk, Fuller [11] obtains the mean response time with the shortest-latency-time-first and other "optimal" schedules for given probability distribution functions of the size of records to be transferred using simulation experiments.

II - THE MODEL

The system we wish to analyze is shown in Figure 1. A fixed starting address at a given angular position, is established on the surface of the fixed-head disk (or drum, henceforth called the disk) for all records. For the cylindrical drum this address would correspond to a line, along the axis of rotation, on its surface. Concentric circles on the disk correspond to tracks. Even though the physical length of the tracks diminishes as the center of the disk is approached, they all contain the same number of words. A record beginning on one track may span several of them.

The lengths of records to be transferred are independent and identically distributed random variables with an arbitrary distribution function $F(x)$ having finite first and second moments. We shall define the probability r_n that a record is transferred in n rotations of the disk, beginning from its starting address :

$$(1) \qquad \begin{aligned} r_n &= \text{Prob}\,[(n-1)\,T < x \leqslant nT] \\ &= F(Tn) - F(T(n-1)),\ n=1,2,\ldots \end{aligned}$$

where T is the time necessary for one complete disk rotation. We have implicitly assumed that the length x of a record is given in units of time, and that it is equal to the time necessary for the disk heads to move form its starting address to its ending address.

We assume that the interarrival times of transfer requests are independent and identically distributed random variables with distribution function

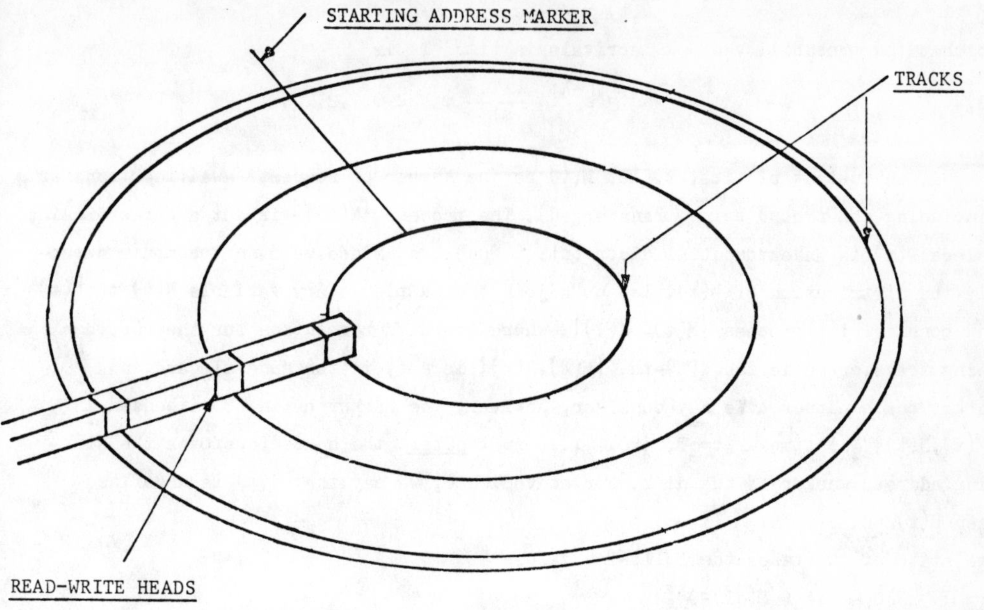

a) FIXED-HEAD DISK

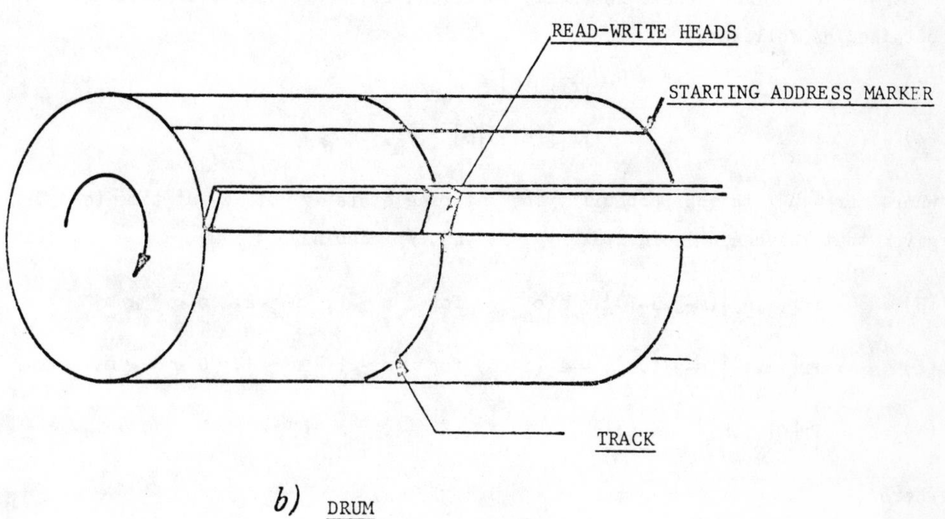

b) DRUM

Figure 1

$$G(t) = 1 - e^{-\lambda t}$$

so that the probability of k arrivals in time T is

(2) $$a_k = e^{-\lambda t} \frac{(\lambda T)^k}{k!}, \quad k=0,1,2,\ldots$$

At an instant of time, t, let $M(t)$ be the number of requests awaiting transfer (including the record being transferred). The process $\{M(t)\}$ is not a Markov chain, unless $F(x)$ is an exponential distribution function. Since we cannot compute directly the distribution of $M(t)$, let us adjoin the supplementary variable $N(t)$ to $M(t)$ and consider the process $\{M(t), N(t)\}$, where the disk revolution for the record being transferred is the $N(t)$-th. $\{M(t),N(t)\}$ is not, in the general case, a Markov chain either . We may consider, however, the Markov chain C imbedded in $\{M(t),N(t)\}$ at time $t=qT$, $q=0,1,\ldots$, just <u>after</u> the heads pass over the starting address marker of the disk. For convenience, we say that $N(t)$ is undefined when $M(t)=0$.

The state of C takes the following values :

0 if $M(qT)=0$

(m,n) if $M(qT) = m>0$ and the disk revolution starting at qT is the n-th for the record being transferred,

for $q=0,1,2,\ldots$ and $n=1,2,\ldots$

That C is a Markov chain is easily verified. Its transition probabilities are obtained as follows :

(3) $$\Pr[0|0] = a_o$$

(4) $$\Pr[(m,1)|0] = a_m, \quad m \geqslant 1$$

where $\Pr[v|u]$ is the probability of entering state v of C at time $(q+1)T$ given that the chain is in state u at qT. We also have :

(5) $$\Pr[(m,n)|(m-j,n-1)] = a_j c_{n-1} \quad \text{for} \quad m \geqslant 1, \; n \geqslant 2, \; 0 \leqslant j \leqslant m$$

(6) $$\Pr[(m,1)|(m-j+1,n)] = a_j(1-c_n) \text{ for } m \geqslant 1, \; n \geqslant 1, \; 0 \leqslant j \leqslant m.$$

(7) $$\Pr[0|(1,n)] = a_o(1-c_n)$$

where

(8) $$c_n = \Pr[x > Tn \mid x > T(n-1)] = \frac{1-F(Tn)}{1-F(T(n-1))}.$$

Notice that we also have

$$(9) \qquad c_n = \frac{\sum_{j=n+1}^{\infty} r_j}{\sum_{j=n}^{\infty} r_j} .$$

The transition probabilities are zero for all cases not covered by equations (3) to (7).

If i, j are states of C $(i,j \in \{0\} \cup \{(m,n)\}$ where $m \geq 1,\ n \geq 1)$, let $\Pi(i,j,q)$ be the q-step transition probability of C from i to j. From [4] we know that

$$\Pi_j = \lim_{q \to \infty} \Pi(i,j,q)$$

exists and is independent of i if

(a) C is aperiodic and irreducible and

(b) There exist real numbers $\Pi_i > 0$ satisfying the equations

$$\sum_{\text{all } i} \Pi_i = 1$$

$$\Pi_j = \sum_{\text{all } i} \Pi_i \Pi(i,j,1), \text{ for each } j .$$

That C is aperiodic and irreducible is easily verified : it suffices to show, using (3) to (7), that there exists an integer $k > 0$ such that

$$\Pi(i,j,q) > 0$$

for all $q \geq k$ [4], for all i,j. If (a) and (b) are satisfied, we say that C is ergodic.

<u>Theorem</u> 1

Denote Π_i by $\gamma(m,n)$ if $i=(m,n)$ and by $\gamma(0)$ if $i=0$. If $E\{n\}\ \lambda\ T < 1$, where

$$E\{n\} = \sum_{n=1}^{\infty} n r_n$$

then C is ergodic and

$$\gamma(0) = 1 - \lambda T\ E\{n\}.$$

Furthermore, $P_n(z)$ the generating function

$$P_n(z) = \sum_{m=1}^{\infty} \gamma(m,n)\ z^m$$

is given by

$$P_n(z) = [1-F((n-1)T)]\ P_1(z)\ [A(z)]^{n-1}$$

for $n \geq 1$, where

$$A(z) = \sum_{k=0}^{\infty} a_k z^k = e^{-\lambda T(1-z)}$$

a_k being given by (2). Also

$$P_1(z) = \gamma(0)z \left[\frac{A(z)-1}{z-R(A(z))}\right]$$

where

$$R(z) = \sum_{n=1}^{\infty} r_n z^n.$$

Theorem 1, as well as the next result we present, will be proved in later sections.

The result we give above yields the joint probability distribution, at the stationary state, for the number of records awaiting transfer and for the index of the current revolution for the record being transferred, at instants of time right after the disk heads pass over the starting address. We would also like to have information regarding the number of records awaiting transfer at _any_ instant of time. For this purpose, we proceed as follows.

Let $p_m(u)$ be the stationary probability of finding m records waiting to be transferred at instants of time $qT + uT$, $q=0,1,2,\ldots$, where $0 < u < 1$. Let $P(z,u)$ be the generating function

$$P(z,u) = \sum_{m=0}^{\infty} p_m(u) z^m. \tag{10}$$

Consider the function $P(z)$ defined as the mean of $P(z,u)$ over the interval $[0,1]$

$$P(z) = \int_0^1 P(z,u)du.$$

The expected number of requests waiting for transfer (including the one being transferred) at an arbitrary instant of time is then $E\{M\}$, given by

$$E\{M\} = \lim_{z \to 1} \frac{d}{dz} P(z).$$

Theorem 2

If $\lambda T E\{n\} < 1$, then

$$E\{M\} = \lambda T \left[E\{\ell\} + \frac{1}{2}\right] + \frac{(\lambda T)^2}{2} \frac{E\{n^2\}}{[1-\lambda T E\{n\}]}$$

where, x being the length of a transfer, ℓ is given by

$$\ell = \frac{x}{T}$$

and

$$E\{n^2\} = \sum_{n=1}^{\infty} n^2 r_n .$$

Remark It is known that certain results on the queuing analysis of input-output devices can be obtained as corollaries of a theorem of Skinner [7]. Theorem 1 of this paper cannot be obtained from Skinner's results, but Theorem 2 can be deduced as a special case from [7].

III - PROOF OF THEOREM 1

We omit the proof that C is aperiodic and irreducible. We shall only show that there exist numbers $\Pi_i > 0$ satisfying the equations

$$\sum_{\text{all } i} \Pi_i = 1$$

and

$$\Pi_j = \sum_{\text{all } i} \Pi_i \, \Pi(i,j,1)$$

for each j under the conditions stated in Theorem 1. Adopting the notations of Theorem 1, we have

$$\gamma(0) = a_o \gamma(0) + \sum_{n=1}^{\infty} a_o (1-c_n) \gamma(1,n) \tag{11}$$

from (3) and (7). (4) and (6) yield for $m \geq 1$

$$\gamma(m,1) = a_m \gamma(0) + \sum_{j=0}^{m} \sum_{n=1}^{\infty} a_j (1-c_n) \gamma(m-j+1,n). \tag{12}$$

Finally, from (5) we have

$$\gamma(m,n) = \sum_{j=0}^{m-1} a_j \, c_{n-1} \gamma(m-j,n-1) \tag{13}$$

for $m \geq 1$, $n > 1$.

We now use these equations to compute the generating functions $P_n(z)$:

$$P_n(z) = \sum_{m=1}^{\infty} \gamma(m,n) z^m .$$

From (11) and (12) we have

$$z[\gamma(0) + P_1(z)] = a_o \, \gamma(0) z + \sum_{n=1}^{\infty} a_o (1-c_n) \gamma(1,n) + \sum_{m=1}^{\infty} z^{m+1} [a_m \gamma(0) + \tag{14}$$

$$+ \sum_{j=0}^{m} \sum_{n=1}^{\infty} a_j (1-c_n) \gamma(m-j+1,n)] = z \, \gamma(0) A(z) + \sum_{n=1}^{\infty} (1-c_n) A(z) P_n(z).$$

Using (13), we obtain for $n > 1$

$$P_n(z) = \sum_{m=1}^{\infty} \sum_{j=0}^{m-1} a_j c_{n-1} \gamma(m-j,n-1) z^m$$

yielding

$$P_n(z) = c_{n-1} A(z) P_{n-1}(z), \quad n=2,3,\ldots$$

which when solved gives us for $n > 1$

$$(16) \qquad P_n(z) = \left[\prod_{j=1}^{n-1} c_j\right] P_1(z) [A(z)]^{n-1}$$

Notice that by (8) we have

$$\prod_{j=1}^{n-1} c_j = 1-F(T(n-1))$$

so that $P_n(z)$ is of the form

$$(17) \qquad P_n(z) = [1-F(T(n-1))] P_1(z) [A(z)]^{n-1}$$

which was to be shown.

From (14) we obtain

$$P_1(z) = A(z) [\gamma(0) + \frac{1}{z} \sum_{n=1}^{\infty} (1-c_n) P_n(z)] - \gamma(0)$$

and using (17)

$$\sum_{n=1}^{\infty} (1-c_n) P_n(z) = P_1(z) \sum_{n=1}^{\infty} [1-F(T(n-1))] (1-c_n) [A(z)]^{n-1}.$$

But (1) and (8) yield

$$[1-F(T(n-1))] (1-c_n) = r_n$$

and

$$R(A(z)) = \sum_{n=1}^{\infty} r_n [A(z)]^n.$$

Therefore $P_1(z)$ is of the form

$$(18) \qquad P_1(z) = \gamma(0) A(z) + \frac{1}{z} P_1(z) R(A(z)) - \gamma(0)$$

$$= z\, \gamma(0) \left[\frac{A(z)-1}{z-R(A(z))}\right]$$

which was to be shown. To obtain $\gamma(0)$ we use the equality

$$\gamma(0) + \lim_{z \to 1} \sum_{n=1}^{\infty} P_n(z) = 1$$

which means that the sum of stationary probabilities over all states is 1. Using (17) we have

$$\lim_{z\to 1} \sum_{n=1}^{\infty} P_n(z) = P_1(1) \sum_{n=1}^{\infty} [1-F(T(n-1))]$$

$$= P_1(1) \sum_{n=1}^{\infty} [1-\sum_{j=1}^{n-1} r_j]$$

$$= P_1(1) \sum_{n=1}^{\infty} \sum_{j=n}^{\infty} r_j$$

$$= P_1(1) \sum_{n=1}^{\infty} nr_n.$$

Therefore

$$\gamma(0) = 1-E\{n\}\, P_1(1).$$

We compute $P_1(1)$ from (18) applying l'Hôpital's rule to obtain

$$P_1(1) = \gamma(0) \frac{\lambda T}{1-\lambda T E\{n\}}.$$

Finally

$$\gamma(0) = 1 - \lambda T\, E\{n\}.$$

To insure ergodicity we must have $\gamma(0) > 0$, therefore the condition

$$\lambda\, T\, E\{n\} < 1$$

must be satisfied.

IV - PROOF OF THEOREM 2

Consider the generating function $P(z,u)$ defined in (10). $p_m(u)$ is given by the following equation, in which $a_j(u)$ is the probability of j arrivals in time uT :

$$(19)\quad p_m(u)=\gamma(0)a_m(u) + \sum_{n=1}^{\infty}\sum_{k=1}^{m} (1-g_n(u))\gamma(k,n)a_{m-k}(u) + \sum_{n=1}^{\infty}\sum_{k=1}^{n} g_n(u)\gamma(k,n)a_{m-k+1}(u)$$

where $g_n(u)$, the probability that a transfer is no longer than $T(n-1+u)$ given that it is longer than $T(n-1)$, is given by

$$(20)\qquad g_n(u) = \frac{F(T(n-1+u))-F(T(n-1))}{1-F(T(n-1))}$$

Let $A_u(z)$ be the generating function

$$(21) \qquad A_u(z) = \sum_{j=0}^{\infty} a_j(u)\, z^j = e^{\lambda Tu(z-1)} .$$

Then $P(z,u)$ is obtained from (19) and (21) as

$$(22) \qquad P(z,u) = \gamma(0)A_u(z) + \sum_{n=1}^{\infty} (1-g_n(u))A_u(z)P_n(z) + \sum_{n=1}^{\infty} g_n(u)A_u(z)\,\frac{P_n(u)}{z}$$

and

$$(23) \qquad E\{M\} = \lim_{z\to 1} \int_0^1 \frac{\partial}{\partial z}\, P(z,u)du.$$

Notice that

$$\lim_{z\to 1} \frac{d}{dz}\, A_u(z) = \lambda Tu$$

and that

$$\sum_{n=1}^{\infty} P_n(1) = 1 - \gamma(0).$$

Using these relations, we obtain after some algebra that

$$E\{M\} = \frac{\lambda T}{2} + \sum_{n=1}^{\infty} \frac{d}{dz}\, p_n(z)\Big|_{z=1} - \int_0^1 du \sum_{n=1}^{\infty} g_n(u)P_n(1).$$

Notice from Theorem 1 and (21) that

$$P_n(1)g_n(u) = [F(T(n-1+u)) - F(T(n-1))]\; P_1(1)$$

and

$$\int_0^1 du\; P_n(1)g_n = P_1(1) \int_{n-1}^{n} [F(yT)-F((n-1)T)]\; dy.$$

We also have

$$\sum_{n=1}^{\infty} \int_0^1 du\; P_n(1)g_n(u) = P_1(1)\; [\; - \sum_{n=1}^{\infty} \int_{n-1}^{n} (1-F(yT))dy + \sum_{n=1}^{\infty} (1-F((n-1)T))]$$

$$= \lambda T\; [\; - \int_0^{\infty} (1-F(yT))dy + \sum_{n=1}^{\infty} n\; G(n)]$$

where we have used $P_1(1) = \lambda T$ and where

$$G(n) = F(nT) - F((n-1)T)$$

is the probability that a transfer will be complete during its n-th rotation, and we have used the fact that

$$1-F((n-1)T) = \sum_{m=n}^{\infty} G(m).$$

Notice that

$$\int_0^\infty [1-F(yT)]dy = \int_0^\infty \frac{du}{T} [1-F(u)]$$

$$= \frac{u}{T} [1-F(u)] \Big|_0^\infty$$

$$+ \int_0^\infty \frac{u}{T} \, d\, F(u)$$

is obtained by a change of variables and an integration by parts. This yields

$$\int_0^\infty [1-F(yT)] \, dy = E \{\ell\}$$

where x being the length of a transfer, ℓ is given by $\ell = \frac{x}{T}$.

We now have

$$E\{M\} = \frac{\lambda T}{2} + \lambda T \, E\{\ell\} - \lambda TE \{n\} + \sum_{n=1}^{\infty} \frac{d}{dz} P_n(z)\Big|_{z=1}.$$

The remaining algebra to complete the proof of Theorem 2 is straightforward, and we omit presenting it here.

V - NUMERICAL EXAMPLES AND CONCLUSIONS

In this paper, we have analyzed the behaviour of a fixed-head disk or drum used for storing variable length records. The initial address of each record corresponds to a fixed angular position on the disk surface (or drum circumference). This placement method has considerable advantages over "optimal" scheduling methods because it simplifies the addressing problems of the secondary memory device.

It is also known that SLTF or optimal schedules for variable length records cannot be implemented efficiently unless special hardware features, which are often non-existent on commercially available equipment, are installed. The major disadvantage of the policy we have analyzed seems to be the wastage in disk space which may be incurred, although it avoids the complicated placement algorithms one would use if an attempt were made to minimize wasted space on disk.

The policy we propose has to be compared with other placement policies or scheduling methods with respect to three points :

(a) the response time

(b) the wastage (or internal fragmentation) of disk space

(c) the complexity of the addressing and scheduling policy.

It is clear that with respect to (c) it is difficult to find policies more economical than ours.

For (a) we have compared on Figures 2 and 3 our results with FULLER'S [11] simulations of various "optimal" scheduling strategies. In [11] the initial addresses of the records are assumed to be uniformly distributed along a disk track. On the same figures the performance of a first-come-first served scheduling algorithm with the same assumptions regarding the distribution of records lengths and the distribution of the initial address of records is shown. For all of the cases considered we see that our policy yields mean response times which are sligtly better than with first-come-first-served but considerably worse than "optimal" policies with the same distribution function of record lengths.

In [11] results for large record lengths are not given : in this case one might expect that optimal policies will yield response times comparable to that of first-come-first-served scheduling.

Suppose that records are transferred on a FCFS basis from the disk, and that they are located at random on the disk surface. The appropriate model in this case is the M/G/1 queue [4] with a service time which is the sum of two independent random variables : an access time uniformly distributed between 0 and T, and a transfer time x with distribution function $F(x)$. By a straightforward application of the formula of Pollactek and Khintchine we obtain the disk response time as :

$$(24) \qquad W_R = T\left(E\{\ell\} + \frac{1}{2} + \lambda T \frac{E\{\ell\} + \frac{1}{3} + E\{\ell^2\}}{2(1-\lambda T(\frac{1}{2} + E\{\ell\}))}\right).$$

Using Little's formula and Theorem 2 we obtain the expected response time for the algorithm we have analyzed :

$$(25) \qquad W_F = T\left[E\{\ell\} + \frac{1}{2} + \lambda T \frac{E\{n^2\}}{2(1-\lambda T\, E\{N\})}\right].$$

Let us compare W_R and W_F for exponentially distributed record lengths. Normalizing to T=1 we set $F(x)=1-e^{-\mu x}$, so that $E\{\ell\}=\mu^{-1}$, $E\{\ell^2\} = 2\mu^{-2}$, and we derive

$$(26) \qquad E\{n\} = (1-e^{-\mu})^{-1}, \quad E\{n^2\} = (1+e^{-\mu})(1-e^{-\mu})^{-2}.$$

For large average record lengths, that is $\mu \ll 1$, it is easily seen that

$$E\{n\} \cong \mu^{-1} + 1/2, \quad E\{n^2\} \cong 2\mu^{-2} + \mu^{-1} - 1$$

correct to $O(\mu)$ so that $W_F \cong W_R$. The approximation is very good even for moderate values of average record length (μ^{-1} of the order of 4 or 5).

As far as point (b) is concerned first note that for small average record sizes with random placement, the "2/3 rule" of KNUTH [9], [10] will come into effect so that on the average one third of disk space will be wasted due to the phenomenon of external fragmentation. An exact analysis of this is unavailable for primary storage since it is a very difficult problem and we do not attempt to solve here the still more complex case for disk space. For small average record size (smaller than T) our policy will waste an amount of space which is large compared to the utilised area. However for large average record lengths the wasted space will be approximately one half sector per record : one of the authors has shown elsewhere [12] that this result is exact for probability density functions of record lengths which have rational Laplace transforms (or equivalently which can be expressed as convex combinations of Erlang densities). Note that these density functions can approximate the j first moments of a given density where j is finite but arbitrarily large : these densities are sometimes called "almost general".

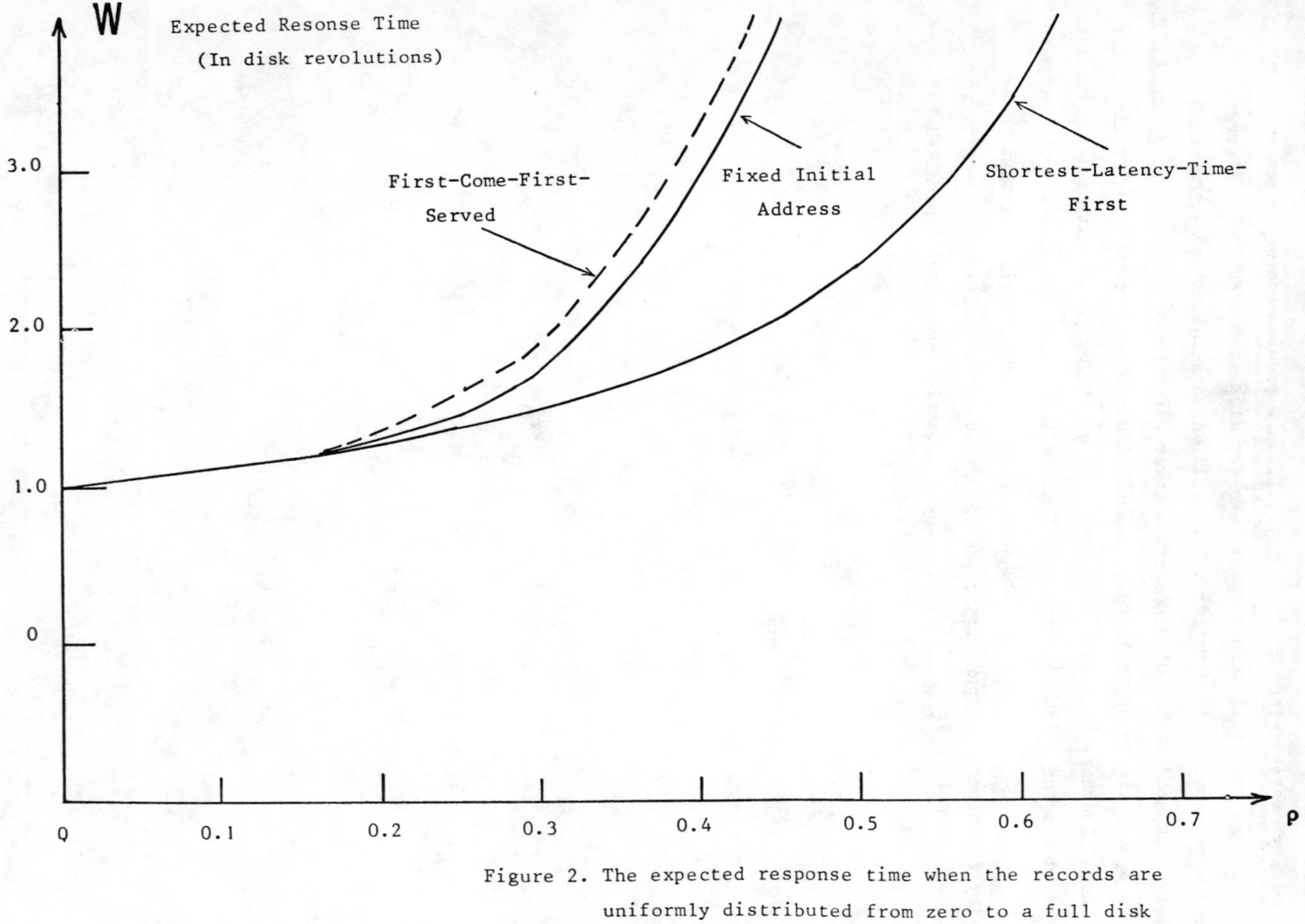

Figure 2. The expected response time when the records are uniformly distributed from zero to a full disk revolution.

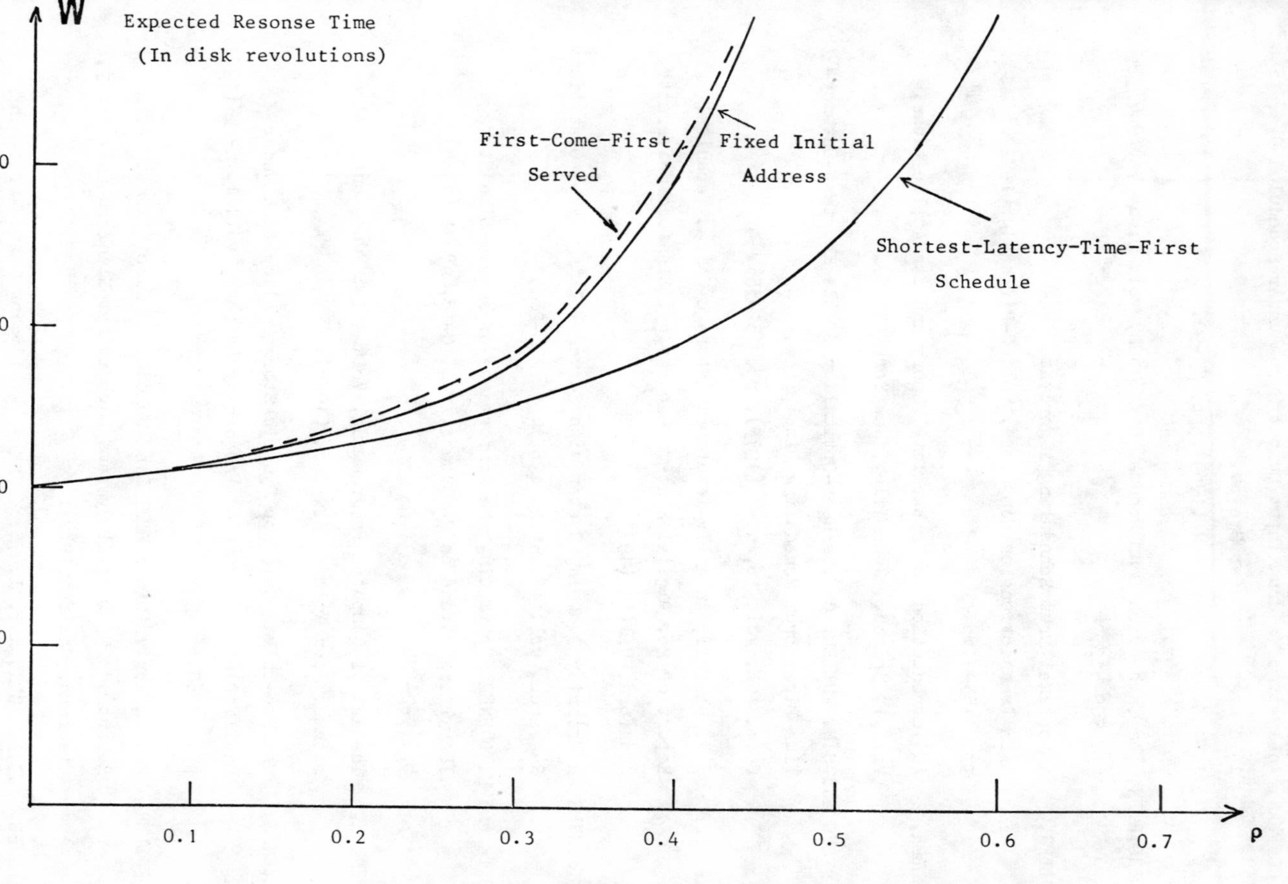

Figure 3. The expected waiting time when all the records are 1/2 the disk's circumference in length

REFERENCES

[1] E.G. COFFMAN, Analysis of a drum Input/Output Queue under Scheduled Operation in a Paged Computer System,
J. ACM, vol. 16, n° 1 (1969), pp. 73-90.

[2] A. WEINGARTEN, The Eschenbach Drum Scheme, Comm. ACM, vol. 9, n° 7 (1966), pp. 509-512.

[3] S. FULLER, An optimal Drum Scheduling Algorithm,
IEEE Transactions on Computers, vol. C-21, n° 11 (1972)
pp. 1153-1165.

[4] W. FELLER, An Introduction to Probability Theory and its Applications, vol. 1, 3rd ed., John Wiley, New York (1968).

[5] H.S. STONE and S.H. FULLER, On the Near-Optimality of the Shortest-Latency-Time-First Drum Scheduling Discipline,
Comm. ACM, vol. 16, n° 6 (1973), pp. 352-353.

[6] S.H. FULLER and F. BASKETT, An Analysis of Drum Storage Units. Technical Report n° 26, Digital Systems Laboratory, Stanford University, Stanford, Calif. (1972).

[7] C.E. SKINNER, A Priority Queueing System with Server-Walking Type. Operations Research, vol. 15, n° 2 (1967) pp. 278-285.

[8] J. ABATE and H. DUBNER, Optimizing the Performance of a Drum-Like Storage. IEEE Transactions on Computers, vol. C-18, n° 11 (1969)
pp. 992-996.

[9] D.E. KNUTH, The Art of Computer Programming, Volume I : Fundamental Algorithms. John Willey, New York (1969).

[10] E. GELENBE, The Two-Thirds Rule for Dynamic Storage Allocation Under Equilibrium. Information Processing Letters, vol. 1, n° 2 (1971) pp. 59-60.

[11] S.H. FULLER, Random Arrivals and MTPT Disk Scheduling Disciplines. Technical Report n° 29, Digital Systems Laboratory, Stanford University, Stanford, Calif. (1972).

[12] E. GELENBE, J.C.A. BOEKHORST, J.L.W. KESSELS, Minimizing Wasted Space in Partitioned Segmentation. Communications of the ACM, vol. 16, n° 6 (1973), pp. 343-349.

STOPPING TIME PROBLEMS AND THE SHAPE OF THE DOMAIN OF CONTINUATION

By

Avner Friedman
Northwestern University
Evanston, Illinois, U.S.A.

Section 1. Stopping time problems and parabolic variational inequalities

Let $x(t)$ be an n-dimensional diffusion process defined as a solution of a system of stochastic differential equations

$$(1.1) \qquad dx(t) = \sigma(x(t),t)dw(t) + b(x(t),t)dt$$

with the n-vector $b(x,t)$ and the $n \times n$ matrix $\sigma(x,t)$ continuous in $(x,t) \in R^n x(-\infty,\infty)$ and uniformly Lipschitz continuous in x. Consider the *reward* function

$$(1.2) \qquad J_{x,s}(\tau) = E_{x,s}\left[\int_s^\tau f(x(t),t)\, dt + \varphi(x(\tau),\tau)\right]$$

with say, f, φ, φ_x, φ_{xx}, φ_t continuous and bounded. Here τ varies over all the stopping times subject to some restrictions. One usually restricts τ by $\tau \leq \tau^*$ where τ^* is either the exit time t_Ω from a given domain $\Omega \subset R^n$, or $\tau^* = t_\Omega \wedge T$ where T is a finite positive number. For definiteness we shall henceforth take $\tau^* = t_\Omega \wedge T$. Let

$$(1.3) \qquad V(x,s) = \sup_{\tau \leq \tau^*} J_{x,s}(\tau).$$

This work is partially supported by National Science Foundation Grant GP-35347 X.

The problem of finding the optimal time τ which maximizes $J_{x,s}(\tau)$ is called a stopping time problem. It is known ([1],[3],[4]) that, if $\sigma\sigma^*$ is uniformly positive definite, the function V is the unique solution of the parabolic variational inequality

$$
(1.4)\quad
\begin{aligned}
& u \geq \varphi \text{ a.e. in } \Omega \times (-\infty, T), \\
& (-u_t - Lu)(v-u) \geq f(v-u) \quad \text{a.e. in } \Omega \times (-\infty,T) \text{ for} \\
& \qquad \text{any } v \geq \varphi, \\
& u = \varphi \text{ if } x \in \partial\Omega, \quad t < T, \\
& u(x,T) = \varphi(x,T) \text{ if } x \in \Omega, \\
& u,\ u_x,\ u_{xx},\ u_t \text{ belong to } L^\infty(\alpha,T;\ \tilde{L}^p(\Omega)) \\
& \qquad \text{for any } -\infty < \alpha < T,\ p > 1
\end{aligned}
$$

where $\tilde{L}^p$ is the space L^p with respect to any measure $\exp(-\lambda|x|)dx$, $\lambda > 0$; here L is the elliptic operator given by

$$
Lu = \frac{1}{2} \Sigma a_{ij} \frac{\partial^2 u}{\partial x_i \partial x_j} + \Sigma b_i \frac{\partial u}{\partial x_i} \qquad ((a_{ij}) = \sigma\sigma^*).
$$

Further, an optimal τ is the hitting time of the set $\{V = \varphi\}$; this set is called the stopping set. The set $\{V > \varphi\}$ is called the domain of continuation. Let Γ_0 = boundary of the set $\{V>\varphi\} \cap \{t<T\}$. The closure Γ of Γ_0 is called the free boundary associated with the stopping time problem (1.1), (1.2).

One can also take other reward functions, such as

$$
(1.5)\quad J_{x,s}(\tau) = E_{x,s}\left\{ \left[\int_s^\tau f(x(t),t)\, dt + \varphi(x(\tau),\tau) \right] \chi_{\tau<T} + h(x(T))\chi_{\tau=T} \right\}.
$$

The corresponding variational inequality for V changes only in that the condition $u(x,T) = 0$ in (1.4) is replaced by the condition

$$u(x,T) = h(x) \quad \text{if} \quad x \in \Omega.$$

Reward functions involving two stopping times σ and τ (where τ is chosen so as to maximize the reward and σ is chosen so as to minimize it) have been considered in [3].

The problem we wish to consider here is the shape of the domain of continuation and the smoothness of the free boundary Γ. For simplicity we shall henceforth take $b \equiv 0$, $\sigma \equiv I$ (the identity matrix), i.e., $x(t)$ is n-dimensional Brownian motion.

Section 2. A class of problems with time decreasing free boundary

We consider the reward function (1.2) with

$$\Omega = R^n - G_0 \tag{2.1}$$

where G_0 is a smooth bounded domain. Let D be another smooth bounded domain such that $\overline{G}_0 \subset D$ and let $G = D - \overline{G}_0$. We shall assume:

$$f(x,t) > 0 \quad \text{if} \quad x \in \overline{G} - \partial D, \quad t < T, \tag{2.2}$$

$$f(x,t) \leq 0 \quad \text{if} \quad x \in R^n - D, \quad t < T, \tag{2.3}$$

$$f(x,t) \leq -k_0 < 0 \quad \text{if} \quad |x| \text{ is sufficiently large}, \quad t < T, \tag{2.4}$$

where k_0 is a constant. Further,

$$f_t(x,t) \leq 0 \quad \text{if} \quad R^n - G_0, \quad t < T \tag{2.5}$$

and

$$\varphi(x,T) = 0 \quad \text{if} \quad x \in R^n - G_0,$$

$$\varphi(x,t) = 0 \quad \text{if} \quad x \in R^n - D, \quad t < T,$$

(2.6)
$$\varphi_t(x,t) \leq 0 \quad \text{if} \quad x \in \partial G_0, \quad t < T,$$

$$\varphi_t + \frac{1}{2}\Delta\varphi + f < 0 \quad \text{if} \quad x \in G, \quad t < T.$$

Theorem 2.1. *If* (2.1)-(2.6) *hold then the family of sets* $\Omega(t) = \{x \in R^n - G_0;\ V(x,t) > \varphi(x,t)\}$ *is monotone decreasing in* t, *and the sets* $\{\Omega(s)\}$ *are uniformly bounded for* s *in bounded intervals* $\alpha \leq s \leq T$.

Denote by (ρ,θ) the polar coordinates in R^n and assume that $0 \in G_0$ and

$$f(x,t) = 0 \quad \text{if} \quad x \in \partial D, \tag{2.7}$$

$$f(x,T) = -\varphi_t(x,T) \quad \text{if} \quad x \in \partial G_0, \tag{2.8}$$

$$(\rho^2 f(x,t))_\rho < 0 \quad \text{if} \quad x \in R^n - G_0,\ t < T. \tag{2.9}$$

Notice that (2.9) for $x \in \overline{G}$ and (2.7) imply that D is a star domain. For simplicity we shall further restrict D, requiring that

$$D = \{x;\ |x| < R\}. \tag{2.10}$$

Denote by A the Laplace operator restricted to ∂D, and assume

$$\varphi_t(x,T) - \varphi_t(x,t) - \frac{1}{2R^2}(A\varphi)(x,t) > 0 \quad \text{if} \quad x \in \partial D, \quad t < T. \tag{2.11}$$

Theorem 2.2. *If* (2.1)-(2.11) *hold then the free boundary can be*

<u>written in the form</u>

$$\rho = \rho^*(\theta,t)$$

<u>where</u> $\rho^*(\theta,t)$ <u>is continuous in</u> (θ,t) <u>and Lipschitz continuous in</u> θ.

Recall, by Theorem 2.1, that $\rho^*(\theta,t)$ is monotone decreasing in t. Notice that Theorem 2.2 implies, in particular, that $\Omega_0(t) = \bar{G}_0 \cup \{x \in R^n - G_0;\ V(x,t) > \varphi(x,t)\}$ is a star domain.

Theorems 2.1, 2.2 are due to Friedman and Kinderlehrer [6].

When

$$f(x,t) = \begin{cases} h(x) & \text{if } x \in G, \\ -k_0 & \text{if } x \in R^n - G, \end{cases}$$

u is a solution of the parabolic variational inequality if and only if $\partial u/\partial t$ is the solution of a one-phase Stefan problem; see Duvaut [2] and also [6].

Section 3. A class of problems with piecewise monotone boundary

In this section we consider the reward function (1.5) in the special case where $\Omega = R^n$, $n = 1$. We assume

(3.1) $\qquad h(x) \geq 0 \;\text{ if }\; x_1 < x < x_2,$

$\qquad\qquad h(x) = 0 \;\text{ if }\; x < x_1 \;\text{ or if }\; x > x_2,$

(3.2) $\qquad f - (\varphi_t - \frac{1}{2}\varphi_{xx}) = -1$

and $h''(x)$ is a bounded function if $x_1 < x < x_2$. The condition (3.2) ensures that the domain of continuation is a bounded set.

<u>Theorem 3.1.</u> <u>If</u> $h'(x)$ <u>changes sign a finite number of times, say</u> 2N-1, <u>then the free boundary intersects every line</u> $t = \lambda$ <u>only at a finite number of points,</u> $s_1(\lambda) < s_2(\lambda) < \dots < s_\ell(\lambda)$, <u>where</u> $\ell \leq 2N$. <u>If, in particular,</u> $N = 1$ <u>then the boundary of the domain of continuation consists of two curves</u> $s_1(t), s_2(t)$.

There are example where $h'(x)$ changes sign $2N-1$ times and $\Gamma \cap \{t = \lambda\}$ consists, in fact, of $2N$ points.

Assume next that

$$(3.3) \qquad h'(x) > 0 \text{ if } x_1 < x < x_0,$$

$$h'(x) < 0 \text{ if } x_0 < x < x_2,$$

$$(3.4) \qquad 1 - \frac{1}{2} h''(\alpha_i) = 0 \text{ if } x_1 < \alpha_1 < \alpha_2 < \dots < \alpha_k < x_2$$

$$1 - \frac{1}{2} h''(x) \neq 0 \text{ if } x_1 < x < x_2, \ x \neq \alpha_i.$$

<u>Theorem 3.2.</u> <u>If</u> (3.1)-(3.4) <u>hold then</u> $s_1(t), s_2(t)$ <u>are continuous and piecewise strictly monotone. The total number of times that the direction of monotonicity changes is</u> $\leq k+2$. <u>Finally,</u> $s_i(t)$ $(i = 1,2)$ <u>is continuously differentiable in every</u> t-<u>interval where</u> $(-1)^i s_i(t)$ <u>is monotone increasing.</u>

Assume next that

$$(3.5) \qquad h'''(x) \text{ is continuous if } x_1 \leq x \leq x_2$$

and either

$$(3.6) \qquad \frac{1}{2} h''(x) > -1 \text{ if } x_1 < x < x_2$$

or

$$h(-x) = h(x) \quad \text{for all} \quad x,$$

(3.7)

$$h''(x) > 0 \quad \text{if} \quad x_1 < x < -\beta \quad \text{or if} \quad \beta < x < x_2,$$

$$h''(x) < 0 \quad \text{if} \quad -\beta < x < \beta.$$

Theorem 3.3. *Let* (3.1)-(3.5) *hold and assume that either* (3.6) *or* (3.7) *holds. Then* $s_1(t)$ *and* $s_2(t)$ *are continuously differentiable for all* $t_0 < t < T$, *where* $s_1(t_0) = s_2(t_0)$. *If further* $h''(x_1) = h''(x_2) = 2$ *then* $s_1(t)$ *and* $s_2(t)$ *are continuously differentiable also at* $t = T$.

Theorem 3.3 under the assumption (3.7) was proved by van Moerbeke [8], [9], by reducing the stopping time problem to a Stefan type problem for $\partial V/\partial t$, and then solving a nonlinear integro-differential equation of Volterra type for $s_2(t)$. $\partial V/\partial t$ can be interpreted as the temperature of the water; this temperature is positive in some regions and negative (i.e., super cooled water) in other regions.

Recently Friedman [5] has proved Theorems 3.1, 3.2 by methods which involve variational inequalities. Theorem 3.3 (with either (3.6) or (3.7)) is also proved in [5]; the proof is based on the approach of van Moerbeke [9], but it is much simpler. In particular, the probabilistic arguments developed in [9] have been eliminated.

Jensen [7] has recently considered a finite difference approximation of the variational problems considered above. He is able to approximate the piecewise monotone free boundary (established in Theorem 3.2) by piecewise monotone polygonal curves.

REFERENCES

[1] Bensoussan, A., and J. L. Lions, Problèmes de temps d'arrêt optimal et inéquations variationelles paraboliques, Applicable Analysis, to appear.

[2] Duvaut, G., Résolution d'un problème de Stefan (Fusion d'un bloc de glace à zero degre), C. R. Acad. Sc. Paris, 276 (1973), 1461 - 1463.

[3] Friedman, Stochastic games and variational inequalities, Archive Rat. Mech. Analys., 51 (1973), 321 - 346.

[4] Friedman, A., Regularity theorems for variational inequalities in unbounded domains and applications to stopping time problems, Archive Rat. Mech. Analys., 52 (1973), 134 - 160.

[5] Friedman, A., Parabolic variational inequalities in one space dimension and the smoothness of the free boundary, to appear.

[6] Friedman, A., and D. Kinderlehrer, A one phase Stefan problem, to appear.

[7] Jensen, R., Finite difference approximation to the free boundary of a parabolic variational inequality, to appear.

[8] Van Moerbeke, P., Optimal stopping and free boundary problems, Archive Rat. Mech. Analys., to appear.

[9] Van Moerbeke, P., An optimal stopping problem for linear reward, Acta Math., 132 (1974), 1 - 41.

PROBLEMES DE TEMPS D'ARRET OPTIMAUX ET DE PERTURBATIONS SINGULIERES DANS LES INEQUATIONS VARIATIONNELLES

A. BENSOUSSAN

Université de Paris IX et LABORIA

J.L. LIONS

Collège de France et LABORIA

INTRODUCTION

Soit A_ε un opérateur elliptique de la forme

$$A_\varepsilon = \frac{\varepsilon^2}{2} A_o + B$$

où A_o est un opérateur elliptique du 2ème ordre (par exemple) et B un opérateur du 1er ordre.

L'étude du comportement de la solution d'un problème aux limites du type Dirichlet attaché à A_ε lorsque $\varepsilon \to 0$ est classique (cf. en particulier l'exposé de VISIK-LIUSTERNIK [1] qui a été suivi de nombreux travaux).

Lorsque, au lieu d'un problème aux limites du type Dirichlet, on considère des conditions aux limites unilatérales, l'étude du problème correspondant est beaucoup moins avancée ; des résultats partiels sont donnés dans LIONS [2]. Motivés par des problèmes de temps d'arrêt optimal, nous considérons ici des problèmes unilatéraux attachés à A_ε mais avec des conditions d'inégalité à l'intérieur du domaine et non seulement sur le bord.

Plus précisément soit $\mathcal{O}$ un ouvert de R^n ; on considère sous des conditions convenables, la solution u_ε de

$$A_\varepsilon u_\varepsilon - f \leq 0 \quad u_\varepsilon \leq 0$$

$$(A_\varepsilon u_\varepsilon - f)u_\varepsilon = 0 \qquad \text{dans } \mathcal{O}$$

$$u_\varepsilon = 0 \quad \text{sur } \partial\mathcal{O}.$$

Notre objet est l'étude de u_ε lorsque $\varepsilon \to 0$.

Nous utiliserons à la fois des méthodes probabilistes et des méthodes d'analyse fonctionnelle. Elles ont des conditions d'applicabilité assez différentes.

D'autres résultats de perturbations singulières dans les inéquations variationnelles ou les inéquations quasi variationnelles issues des problèmes de temps optimaux ou de contrôle impulsionnel sont donnés dans BENSOUSSAN-LIONS, C.R.A. Sc. Paris, Juillet 1974.

Le plan de ce travail est le suivant :

I - POSITION DU PROBLEME. ENONCE DES RESULTATS.

1.1 - Notations, hypothèses.

1.2 - Le problème.

II - DEMONSTRATION DES RESULTATS ET VERIFICATION DES HYPOTHESES.

2.1 - Démonstration du théorème 1.1.

2.2 - Vérification des hypothèses.

III - ETUDE DIRECTE DE LA CONVERGENCE.

3.1 - Notations. Enoncé du résultat.

3.2 - Démonstration du théorème 3.1.

I - POSITION DU PROBLEME. ENONCE DES RESULTATS.

1.1 - Notations. Hypothèses.

Soit $\mathcal{O}$ un ouvert de R . On se donne une application $g : R^n \to R^n$ telle que

(1.1) $g \in C^1(R^n)$ bornée et à gradient borné dans R^n, $\|\frac{\partial g}{\partial x}\| \leq C_o$.

Seules les valeurs de $g(x)$ pour $x \in \bar{\mathcal{O}}$ interviendront par la suite.

On considère la solution $y_x(t)$ de l'équation différentielle ordinaire

$$(1.2) \qquad \begin{cases} \frac{dy}{dt} = g(y), & t > 0, \\ y(0) = x \in \bar{\mathcal{O}} . \end{cases}$$

Celle-ci est définie de manière unique grâce à (1.1).

Soit Σ le complémentaire de $\mathcal{O}$ dans R^n et $\overset{o}{\Sigma}$ l'intérieur de Σ.

On pose pour $x \in \mathcal{O}$

$$(1.3) \qquad \theta(x) = \text{Inf } \{t > 0 | y_x(t) \in \Sigma\}$$

$$(1.4) \qquad \Lambda(x) = \text{Inf } \{t > 0 | y_x(t) \in \overset{o}{\Sigma}\} .$$

On n'exclut pas la possibilité $\theta(x) = +\infty$ (sauf mention explicite du contraire). Si $\theta(x) = +\infty$, cela signifie que $y_x(t) \in \mathcal{O}\ \forall t$. Dans ce cas on a aussi $\Lambda(x) = +\infty$. Si $\theta(x) < +\infty$, la continuité de la trajectoire $y_x(t)$ implique $y_x(\theta(x)) \in \partial \mathcal{O}$.

Il est clair que $\Lambda(x) \geq \theta(x)$. On fera l'hypothèse

(1.5) $$\theta(x) = \Lambda(x) \quad \forall\, x \in \mathcal{O} .$$

On utilisera également dans certains cas les hypothèses

(1.6) $$\theta(x') \to \theta(x) \quad \text{lorsque} \quad x' \to x,\ x,\ x' \in \mathcal{O} \quad (\text{resp } \bar{\mathcal{O}})$$

(1.7) $\theta(x) < +\infty \quad \forall x \in \mathcal{O}$ et $\theta(x)$ est Lipschitzienne sur les compacts de $\mathcal{O}$.

On donnera au nº 2.2 des conditions suffisantes sur l'ouvert $\mathcal{O}$ et sur g pour que (1.5)(1.6)(1.7) soient satisfaites.

On considère aussi une fonction $f \in C^1(R^n)$ vérifiant

(1.8) $$\left|\frac{\partial f}{\partial x}\right| \leq C_1 .$$

1.2 - Le problème

Des problèmes de temps d'arrêt optimaux (cf. BENSOUSSAN-LIONS [1] et le nº2 ci-après) conduisent à la recherche, pour $\varepsilon > 0$ fixé, de u_ε solution des inéquations suivantes :

(1.9) $$\begin{cases} -\frac{\varepsilon^2}{2}\Delta u_\varepsilon - \sum g_i \frac{\partial u_\varepsilon}{\partial x_i} + \alpha\, u_\varepsilon - f \leq 0,\ u_\varepsilon \leq 0,\ u_\varepsilon\left[-\frac{\varepsilon^2}{2}\Delta u_\varepsilon - \sum g_i \frac{\partial u_\varepsilon}{\partial x_i} + \alpha u_\varepsilon - f\right] = 0 \\ \text{dans } \mathcal{O}, \\ u_\varepsilon = 0 \quad \text{sur } \Gamma . \end{cases}$$

<u>Pour simplifier l'exposé</u> (cf. Remarque 1.1 ci-après), <u>on supposera que</u>

(1.10) $$\alpha + \frac{1}{2}\sum_{i=1}^{n} \frac{\partial g_i}{\partial x_i} \geq c > 0 \quad \text{et} \quad \alpha \geq 2\ c_o .$$

Si l'on pose, pour $u, v \in H_o^1(\mathcal{O})$

(1.11) $$\Pi_\varepsilon(u,v) = \frac{\varepsilon^2}{2}\int_{\mathcal{O}} \text{grad } u \text{ grad } v\, dx + \alpha\int_{\mathcal{O}} u\, v\, dx - \int_{\mathcal{O}} (g.\text{grad } u) v\, dx.$$

On voit sans peine que (1.9) équivaut à l'Inéquation Variationnelle (I.V.).

(1.12) $$\begin{cases} \Pi_\varepsilon(u_\varepsilon, v-u_\varepsilon) \geq (f, v-u_\varepsilon) \quad \forall\, v \in H_o^1(\mathcal{O}),\ v \leq 0 \ \text{ p.p. sur } \mathcal{O}, \\ u_\varepsilon \leq 0 \ \text{ p.p. sur } \mathcal{O},\ u_\varepsilon \in H_o^1(\mathcal{O}). \end{cases}$$

Mais

$$\Pi_\varepsilon(v,v) = \frac{\varepsilon^2}{2}\int_{\mathcal{O}} |\text{grad } v|^2 dx + \int_{\mathcal{O}} (\alpha + \tfrac{1}{2}\sum \frac{\partial g_i}{\partial x_i}) v^2 dx$$

et donc d'après (1.10),

(1.13) $$\Pi_\varepsilon(v,v) \geq \min(\frac{\varepsilon^2}{2}, c)\ \|v\|^2_{H_0^1(\mathcal{O})} .$$

Donc d'après LIONS-STAMPACCHIA [1], <u>l'I.V.</u> (1.2) <u>admet une solution unique</u>. On supposera dans la suite que les conditions sur f, g et $\mathcal{O}$ sont telles que

(1.14) $\qquad u_\varepsilon$ est continue sur $\bar{\mathcal{O}}$ et $u_\varepsilon \in H^2(\mathcal{O})$.

D'après BREZIS-STAMPACCHIA [1], (1.14) a lieu si $\mathcal{O}$ est un ouvert régulier. Par le même type de démonstration, on vérifie que (1.14) a encore lieu si $\mathcal{O}$ est par exemple un cube.

Le problème consiste à étudier la limite de u_ε lorsque $\varepsilon \to 0$.

On pose

(1.15) $$\Gamma^- = \{x \in \Gamma \mid \Lambda(x) = \theta(x) = 0\}.$$

On va démontrer les résultats suivants

<u>Théorème</u> 1.1

<u>Sous les hypothèses</u> (1.1), (1.5), (1.8), (1.9) <u>et</u> (1.14) <u>on a</u>

(1.16) $$u_\varepsilon(x) \underset{\varepsilon \to 0}{\to} u(x) \qquad \forall x \in \mathcal{O} \cup \Gamma^- .$$

<u>Si on fait de plus l'hypothèse</u> (1.6) <u>alors</u>

(1.17) $$u(x) \text{ est continue sur } \mathcal{O} \text{ (resp. } \bar{\mathcal{O}}).$$

<u>Si on fait de plus, l'hypothèse</u> (1.17) <u>alors</u>

(1.18) $$u(x) \text{ est lipschitzienne sur tout compact } \subset \mathcal{O} .$$

<u>Elle est p.p. différentiable dans</u> $\mathcal{O}$ <u>et elle est solution du problème</u>

(1.19) $$\begin{cases} \sum_i g_i \frac{\partial u}{\partial x_i} \in L^\infty(\mathcal{O}), \ u \in C(\bar{\mathcal{O}}) \\ -\sum_i g_i \frac{\partial u}{\partial x_i} + \alpha u \le f \text{ p.p. dans } \mathcal{O} \\ u \le 0 \quad \text{dans } \bar{\mathcal{O}} \\ u[-\sum_i g_i \frac{\partial u}{\partial x_i} + \alpha u - f] = 0 \quad \text{p.p. dans } \mathcal{O} \qquad u|_{\Gamma^-} = 0. \end{cases}$$

<u>Si</u> $\mathcal{O}$ <u>est régulier, la solution de</u> (1.19) <u>est unique.</u> ∎

<u>Remarque</u> 1.1

Effectuons le changement de fonction inconnue

(1.20) $$u_\varepsilon(x) = e^{-\rho(x)} w_\varepsilon(x),$$

où ρ est à <u>déterminer.</u>

Alors(1.9) devient

(1.21) $$\begin{cases} -\frac{\varepsilon^2}{2} \Delta w_\varepsilon - \sum(g_i - \frac{\varepsilon^2}{2} \frac{\partial \rho}{\partial x_i}) \frac{\partial w_\varepsilon}{\partial x_i} + \alpha_\varepsilon w_\varepsilon - e^\rho f \le 0 , \\ \alpha_\varepsilon = \alpha + \frac{\varepsilon^2}{2} \Delta\rho + \sum g_i \frac{\partial \rho}{\partial x_i} - \frac{\varepsilon^2}{2} (\text{grad } \rho)^2, \\ w_\varepsilon \le 0, \\ w_\varepsilon[-\frac{\varepsilon^2}{2} \Delta w_\varepsilon - \sum (g_i - \frac{\varepsilon^2}{2} \frac{\partial \rho}{\partial x_i}) \frac{\partial w_\varepsilon}{\partial x_i} + \alpha_\varepsilon w_\varepsilon - e^\rho f] = 0, \quad w_\varepsilon = 0 \text{ sur } \Gamma . \end{cases}$$

Donc α est remplacé par α_ε , g_i par $g_i - \frac{\varepsilon^2}{2}\frac{\partial\rho}{\partial x_i}$ et f par $e^\rho f$. Alors $\alpha+\frac{1}{2}\sum\frac{\partial g_i}{\partial x_i}$ devient

(1.22) $$\alpha + \frac{1}{2}\sum\frac{\partial g_i}{\partial x_i} + \sum g_i \frac{\partial\rho}{\partial x_i} - \frac{\varepsilon^2}{2}(\text{grap } \rho)^2 + \frac{\varepsilon^2}{4}\Delta\rho.$$

Si donc l'on choisit ρ régulière de manière que

(1.23) $$\sum g_i \frac{\partial\rho}{\partial x_i} + \alpha + \frac{1}{2}\sum\frac{\partial g_i}{\partial x_i} = c > 0 ,$$

alors l'expression (1.22) vaut

$$c - \frac{\varepsilon^2}{2}(\text{grad } \rho)^2 + \frac{\varepsilon^2}{4}\Delta\rho \geq \frac{c}{2} > 0$$

pour ε assez petit.

Par conséquent, on est ramené à un problème

(1.24) $$\begin{cases} -\frac{\varepsilon^2}{2}\Delta w_\varepsilon - \sum g_{i\varepsilon}\frac{\partial w_\varepsilon}{\partial x_i} + \alpha_\varepsilon w_\varepsilon - \tilde{f} \leq 0, \; w_\varepsilon \leq 0 \\ w_\varepsilon \left[-\frac{\varepsilon^2}{2}\Delta w_\varepsilon - \sum g_{i\varepsilon}\frac{\partial w_\varepsilon}{\partial x_i} + \alpha_\varepsilon w_\varepsilon - \tilde{f}\right] = 0 \text{ dans } \mathcal{O}, \\ w_\varepsilon = 0 \quad \text{sur } \Gamma, \end{cases}$$

où $$\alpha_\varepsilon + \frac{1}{2}\sum\frac{\partial g_{i\varepsilon}}{\partial x_i} \geq c > 0.$$

Les démonstrations ci-après s'étendent à cette situation, avec quelques difficultés techniques supplémentaires dues à la dépendance de $g_{i\varepsilon}$ en ε .

II - DEMONSTRATION DES RESULTATS ET VERIFICATIONS DES HYPOTHESES

2.1 - Démonstration du Théorème 1.1

La méthode suivie consiste à utiliser l'interprétation en terme de contrôle optimal stochastique de u_ε. Soit $(\Omega,\mathcal{A},P)$ un espace de probabilité et $w(t)$, $t \geq 0$ un processus de Wiener (standard) à valeurs dans R^n. On note $y_x^\varepsilon(t)$ la solution de l'équation de Ito :

(2.1) $$\begin{cases} dy^\varepsilon = g(y^\varepsilon)dt + \varepsilon\, dw(t), \; t > 0, \\ y_x^\varepsilon(0) = x \in \bar{\mathcal{O}} . \end{cases}$$

On désigne par $\mathcal{F}_t$ la famille croissante de σ-algèbres engendrée par $w(s)$, $0 \leq s \leq t$. Tous les temps d'arrêt considérés par la suite sont relatifs à la famille $\mathcal{F}_t$. On pose

(2.2) $$\theta^\varepsilon(x) = \text{Inf}\{t > 0 \mid y_x^\varepsilon(t) \in \Sigma\} .$$

Grâce aux propriétés de régularité (1.14) et en raisonnant comme dans BENSOUSSAN-

LIONS [1] (cf. aussi FRIEDMAN [1]) on vérifie facilement[(1)] que l'on a

$$(2.3)\qquad u_\varepsilon(x) = \inf_{0 \le \tau \le \theta^\varepsilon(x)} E\int_0^\tau e^{-\alpha t} f(y_x^\varepsilon(t))dt,\ x \in \bar{\mathcal{O}}$$

où τ est un temps d'arrêt. En outre, il existe un temps d'arrêt optimal défini par

$$(2.4)\qquad s^\varepsilon(x) = \operatorname{Inf}\ \{t \ge 0 |\ u_\varepsilon(y_x^\varepsilon(t)) = 0\ \}\ .$$

On <u>définit</u> alors $u(x)$ par la formule

$$(2.5)\qquad u(x) = \inf_{0 \le s \le \theta(x)} \int_0^s e^{-\alpha t} f(y_x(t))dt.$$

Nous allons vérifier que la fonction $u(x)$ satisfait aux propriétés énoncées dans le théorème. Naturellement dans (2.5) s est déterministe. Il est utile de remarquer que l'on peut aussi écrire

$$(2.6)\qquad u(x) = \inf_{\tau \ge 0} E\int_0^{\tau\wedge\theta(x)} e^{-\alpha t} f(y_x(t))dt$$

où τ est un temps d'arrêt quelconque (non forcément déterministe). Posons

$$z(t) = y_x^\varepsilon(t) - y_x(t) = \int_0^t [g(y_x^\varepsilon) - g(y_x)]ds + \varepsilon w(t).$$

En appliquant la formule de Ito à $e^{-\alpha t}|z(t)|^2$ et tenant compte de (1.1) on obtient

$$E\, e^{-\alpha t}|z(t)|^2 + (\alpha - 2C_0)E\int_0^t e^{-\alpha s}|z(s)|^2 ds \le \frac{\varepsilon^2 n}{\alpha}\ .$$

Grâce à (1.10) on en déduit l'éstimation

$$(2.7)\qquad E\int_0^\infty e^{-\alpha t}|y_x^\varepsilon(t) - y_x(t)|^2 dt \le \frac{n}{\alpha(\alpha - 2C_0)}\ \varepsilon^2\ .$$

On a aussi

$$e^{-\alpha t} z(t) = \int_0^t e^{-\alpha s}(g(y_x^\varepsilon(s)) - g(y_x(s)) - \alpha z)ds + \varepsilon\int_0^t e^{-\alpha s} dw(s)$$

d'où

$$\sup_{0 \le t \le T} e^{-2\alpha t}|z(t)|^2 \le \frac{4}{\alpha}\int_0^T e^{-\alpha s}(C_0^2 + \alpha^2)|z(s)|^2 ds + 2\varepsilon^2 \sup_{0 \le t \le T} \left|\int_0^t e^{-\alpha s} dw(s)\right|^2\ .$$

En utilisant (2.7) et $E \sup_{0 \le t \le T} \left|\int_0^t e^{-\alpha s} dw(s)\right|^2 \le 4\int_0^T e^{-2\alpha s}\, ds \le \frac{2}{\alpha}$ on obtient

$$E \sup_{0 \le t \le T} e^{-2\alpha t}|z(t)|^2 \le C\ \varepsilon^2$$

où C est une constante indépendante de ε et T. En faisant tendre $T \to \infty$, on obtient

(1) Dans ces articles on se place dans R^n, mais les transformations ne présentent aucune difficulté..

(2.8) $$E \sup_{0 \le t < \infty} e^{-2\alpha t} |y_x^\varepsilon(t) - y_x(t)|^2 \le C \; \varepsilon^2 .$$

On peut extraire une sous-suite encore notée $y_x^\varepsilon(t)$ telle que $\forall\, \eta \;\exists\, \bar\varepsilon(\eta,\omega)$, $\omega \notin \Omega_o$ $(P(\Omega_o) = 0)$

(2.9) $$\varepsilon \le \bar\varepsilon\,(\eta,\omega) \Rightarrow e^{-\alpha t}|y_x^\varepsilon(t) - y_x(t)| \le \eta \quad \forall\, t \ge 0 .$$

On va vérifier que pour cette sous-suite on a

(2.10) $$\theta^\varepsilon(x) \underset{\varepsilon \to 0}{\to} \theta(x),\ \omega \notin \Omega_o \quad (\text{pour } x \in \mathcal{O} \cup \Gamma^-).$$

Grâce à (1.5) on a, si $\theta(x) < \infty$ et $x \in \mathcal{O} \cup \Gamma^-$

$$\theta(x) = \Lambda(x) = \text{Inf } \{t>0 \,|\, y_x(t) \in \overset{o}{\Sigma}\}$$

$$= \text{Inf } \{t > \theta(x) \,|\, y_x(t) \in \overset{o}{\Sigma}\}$$

$$= \theta(x) + \text{Inf } \{s>0 \,|\, y_{y_x(\theta(x))}(s) \in \overset{o}{\Sigma}\}$$

$$= \theta(x) + \Lambda(y_x(\theta(x)) = \theta(x) + \theta(y_x(\theta(x)))$$

d'où $\theta(y_x(\theta(x))) = 0$, ce qui prouve que $y_x(\theta(x)) \in \Gamma^-$.
Considérons tout d'abord le cas $\theta(x) = +\infty$, donc $x \in \mathcal{O}$ et $y_x(t) \in \mathcal{O}$ $\forall\, t \ge 0$. Soit T arbitraire. L'ensemble

$$K_T = \{y_x(t),\ 0 \le t \le T\}$$

est un compact contenu dans $\mathcal{O}$. Donc $\exists\, d_T > 0$ tel que $\xi \in K_T$ et $|\bar\xi - \xi| < d_T \Rightarrow \bar\xi \in \mathcal{O}$. Si dans (2.9) on prend $\eta = d_T\, e^{-\alpha T}$ et $\varepsilon_T = \bar\varepsilon(\eta,\omega)$ on voit que

$$\varepsilon \le \varepsilon_T(\omega) \Rightarrow |y_x^\varepsilon(t) - y_x(t)| \le d_T \quad \forall\, 0 \le t \le T$$

donc $y_x^\varepsilon(t) \in \mathcal{O}$, $\forall\, t \in [0,T]$, c'est-à-dire $\theta_x^\varepsilon(\omega) > T$. On a ainsi démontré que $\theta_x^\varepsilon(\omega) \to +\infty$, i.e. (2.10) pour $\theta(x) = +\infty$. Supposons maintenant que $0 < \theta(x) < +\infty$. Soit $0 < \delta < \theta(x)$ et $K_\delta = \{y_x(t),\ 0 \le t \le \theta(x) - \delta\}$ qui est un compact $\subset \mathcal{O}$. Donc $\exists d_\delta$ tel que $\xi \in K_\delta$, $|\xi - \bar\xi| < d_\delta \Rightarrow \bar\xi \in \mathcal{O}$. En prenant dans (2.9) $\eta = d_\delta\, e^{-\theta(x)}$ et posant $\varepsilon_\delta^1 = \bar\varepsilon(\eta,\omega)$ on a $\varepsilon \le \varepsilon_\delta^1 \Rightarrow |y_x^\varepsilon(t) - y_x(t)| \le d_\delta$ pour $t \in [0,\theta(x)]$; donc pour $t \in [0,\theta(x)-\delta]$ on obtient $y_x^\varepsilon(t) \in \mathcal{O}$ dès que $\varepsilon \le \varepsilon_\delta^1$. Par conséquent on a

(2.11) $$\varepsilon \le \varepsilon_\delta^1 \Rightarrow \theta^\varepsilon(x) \ge \theta(x) - \delta .$$

Par ailleurs, comme $\theta(x) = \Lambda(x)$, il existe t_δ tel que $t_\delta \in [\theta(x),\theta(x) + \delta]$ et

$y_x(t_\delta) \in \overset{\circ}{\Sigma}$. Comme $y_x^\varepsilon(t_\delta) \to y_x(t_\delta)$, $\exists\ \varepsilon_\delta^2 > 0$ tel que $\varepsilon \leq \varepsilon_\delta^2 \Rightarrow y_x^\varepsilon(t_\delta) \in \overset{\circ}{\Sigma}$.

Mais alors

$$t_\delta \geq \Lambda^\varepsilon(x) \geq \theta^\varepsilon(x)\ ^{(1)} ;$$

donc

$$\varepsilon \leq \varepsilon_\delta^2 \Rightarrow \theta^\varepsilon(x) \leq \theta(x) + \delta .$$

Prenant $\varepsilon_\delta = \min(\varepsilon_\delta^1, \varepsilon_\delta^2)$ on voit que $\varepsilon \leq \varepsilon_\delta \Rightarrow |\theta^\varepsilon(x) - \theta(x)| \leq \delta$.

D'où (2.10). Si enfin $\theta(x) = 0$, on montre que $\varepsilon \leq \varepsilon_\delta^2 \Rightarrow 0 \leq \theta^\varepsilon(x) \leq \delta$, d'où aussi (2.10). Si τ est un temps d'arrêt on a

$$(2.12)\ \begin{cases} \left|E\displaystyle\int_0^\tau e^{-\alpha t} f(y_x(t))dt - E\int_0^{\tau\Lambda\theta^\varepsilon(x)} e^{-\alpha t} f(y_x^\varepsilon(t))dt\right| \leq \\ \leq \left| E \displaystyle\int^{(\tau\Lambda\theta(x))\Lambda(\tau\Lambda\theta^\varepsilon(x))} e^{-\alpha t}[f(y_x(t)) - f(y_x^\varepsilon(t))]dt \right| + \\ + \left| E \displaystyle\int_{(\tau\Lambda\theta(x))\Lambda(\tau\Lambda\theta^\varepsilon(x))}^{\tau\Lambda\theta(x)} e^{-\alpha t} f(y_x(t))dt \right| + \left|E \displaystyle\int_{(\tau\Lambda\theta(x))\Lambda(\tau\Lambda\theta^\varepsilon(x))}^{\tau\Lambda\theta^\varepsilon(x)} e^{-\alpha t} f(y_x^\varepsilon(t))dt\right| \\ = I_1(\tau) + I_2(\tau) + I_3(\tau) . \end{cases}$$

Si $M = \underset{\xi \in \bar{\mathcal{O}}}{\operatorname{Max}} |f(\xi)|$, on a

$$I_2(\tau) \leq \frac{M}{\alpha}\ E[e^{-\alpha(\tau\Lambda\theta(x))\Lambda(\tau\Lambda\theta^\varepsilon(x))} - e^{-\alpha\tau\Lambda\theta(x)}]$$

$$\leq \frac{M}{\alpha}\ E|e^{-\alpha\theta^\varepsilon(x)} - e^{-\alpha\theta(x)}| \underset{\varepsilon\to 0}{\to} 0 \quad \text{(uniformément en } \tau).$$

De même

$$I_3(\tau) \underset{\varepsilon\to 0}{\to} 0 \quad \text{(uniformément en } \tau).$$

Enfin, d'après (2.7)

$$I_1(\tau) \leq C_1\ E\int_0^\infty e^{-\alpha t}|y_x(t) - y_x^\varepsilon(t)|dt$$

$$\leq \frac{C_1}{\alpha}\ E\int_0^\infty e^{-\alpha t}|y_x(t) - y_x^\varepsilon(t)|^2\ dt \to 0 \ \text{ uniformément en } \tau .$$

Il en résulte que $u_\varepsilon(x) \to u(x)$. Comme la limite ne dépend pas de la suite extraite,

---

(1) $\Lambda^\varepsilon(x)$ est défini comme $\Lambda(x)$ avec $y_x^\varepsilon(t)$ remplaçant $y_x(t)$

on a bien démontré (1.16). Montrons (1.17). Un calcul similaire à celui fait pour obtenir (2.7) et (2.8) montre que

$$(2.13)\qquad \int_0^\infty e^{-\alpha t}|y_x(t) - y_{x'}(t)|^2 dt \le |x-x'|^2$$

$$(2.14)\qquad e^{-\alpha t}|y_x(t) - y_{x'}(t)|^2 \le |x-x'|^2 \quad \forall\, x,x' .$$

Or

$$\left|\int_0^{s\wedge\theta(x)} e^{-\alpha t} f(y_x(t)) - \int_0^{s\wedge\theta(x')} e^{-\alpha t} f(y_{x'}(t))dt\right| \le$$

$$\le \left|\int_0^{(s\wedge\theta(x))\wedge(s\wedge\theta(x'))} e^{-\alpha t}[f(y_x(t)) - f(y_{x'}(t))]dt\right| +$$

$$+ \left|\int_{(s\wedge\theta(x))\wedge(s\wedge\theta(x'))}^{s\wedge\theta(x)} e^{-\alpha t} f(y_x(t))dt\right| + \left|\int_{(s\wedge\theta(x))\wedge(s\wedge\theta(x'))}^{s\wedge\theta(x')} e^{-\alpha t} f(y_{x'}(t))dt\right|$$

$$= I_1(s) + I_2(s) + I_3(s).$$

On vérifie comme pour (2.12) que

$$(2.15)\qquad \begin{cases} I_2(s),\ I_3(s) \le \frac{M}{\alpha}\left|e^{-\alpha\theta(x')} - e^{-\alpha\theta(x)}\right| \\ I_1(s) \le C_1 \displaystyle\int_0^\infty e^{-\alpha t}|y_x(t) - y_{x'}(t)|dt \le \frac{C_1}{\alpha}|x-x'|^2 . \end{cases}$$

Grâce à (1.6), on voit que $u(x') \to u(x)$, lorsque $x' \to x$. On a bien démontré (1.17). Il résulte de (2.15) et (1.7) que u est lipschitzienne sur les compacts de $\mathcal{O}$ et donc par le théorème de Rademacher, elle est p.p. différentiable dans $\mathcal{O}$. Montrons que $u(x)$ est solution de (1.19). Soit $s(x)$ un point où le inf dans (2.5) est atteint (il existe car $\theta(x) < +\infty$). On considère un point $x \in \mathcal{O}$ où $u(x)$ et $\theta(x)$ sont différentiables. Naturellement $\theta(x) > 0$ et $s(x) \le \theta(x)$. Nous allons calculer $\frac{\partial u}{\partial x}(x)$. Supposons tout d'abord que $s(x) = \theta(x)$. Soit $h \in R^n$, $h \to 0$, $x+h \in \bar{\mathcal{O}}$. On a

$$(2.18)\qquad u(x+h) - u(x) \le \int_0^{\theta(x+h)} e^{-\alpha t} f(y_{x+h}(t))dt - \int_0^{\theta(x)} e^{-\alpha t} f(y_x(t))dt .$$

Grâce à (1.7), (1.8) et puisque u est différentiable en x, on déduit de (2.18) que

$$\frac{\partial u}{\partial x}\, h + 0_1(h) \le \int_0^{\theta(x)} e^{-\alpha t}\, \frac{\partial f}{\partial x}(y_x(t)).\frac{\partial y_x}{\partial x}(t)h\, dt + e^{-\alpha\theta(x)} f(y_x(\theta(x)))\theta'(x).h + 0_2(h)$$

où $0_1(h)$, $0_2(h) \to 0$ avec h. Il en résulte, puisque h est <u>quelconque</u> $\to 0$ que si $s(x) = \theta(x)$, on a :

$$(2.19)\qquad \frac{\partial u}{\partial x} = \int_0^{\theta(x)} e^{-\alpha t}\, \frac{\partial y_x^*}{\partial x}(t)\, \frac{\partial f}{\partial x}(y_x(t))dt + e^{-\alpha\theta(x)} f(y_x(\theta(x)))\theta'(x)$$

où $\frac{\partial y_x^*}{\partial x}$ désigne la matière adjointe de $\frac{\partial y_x}{\partial x}$.

Si maintenant $s(x) < \theta(x)$, alors pour h assez petit on a $s(x) \leq \theta(x+h)$ donc

$$u(x+h) - u(x) \leq \int_0^{s(x)} e^{-\alpha t} f(y_{x+h}(t))dt - \int_0^{s(x)} e^{-\alpha t} f(y_x(t))dt$$

d'où l'on déduit comme plus haut, que

$$(2.20) \qquad \frac{\partial u}{\partial x} = \int_0^{s(x)} e^{-\alpha t} \frac{\partial y_x^*}{\partial x}(t) \frac{\partial f}{\partial x}(y_x(t))dt, \text{ si } s(x) < \theta(x) .$$

Si l'on pose

$$z_j(t) = \frac{\partial y_x(t)}{\partial x_j}$$

alors z_j est solution de

$$\frac{dz_j}{dt} = g'(y_x(t))z_j$$
$$z_j(0) = e_j$$

et

$$\left[\frac{\partial y_x^*(t)}{\partial x} \frac{\partial f}{\partial x}(y_x(t))\right]_j = z_j(t). \frac{\partial f}{\partial x}(y_x(t)).$$

Un calcul facile montre que, grâce à (1.1) et (1.8), on a

$$(2.21) \quad \left(\int_0^{\theta(x)} e^{-\alpha t} \left|\left[\frac{\partial y_x^*(t)}{\partial x} \frac{\partial f}{\partial x}(y_x(t))\right]_j\right| dt\right)^2 \leq \frac{1}{\alpha} \int_0^{\infty} e^{-\alpha t} |z(t)|^2 \left|\frac{\partial f}{\partial x}(y_x(t))\right|^2 dt \leq C.$$

On obtient donc, compte tenu de (2.19), (2.20), (2.21)

$$(2.22) \qquad \left|\sum_i g_i \frac{\partial u}{\partial x_i}\right| \leq C\left[1+\left|\sum_i g_i \frac{\partial \theta}{\partial x_i}\right|\right] .$$

Or pour $\delta \leq \theta(x)$

$$(2.23) \qquad \theta(x) = \delta + \theta(y_x(\delta)).$$

Donc puisque θ est différentiable, on obtient

$$\delta \sum_i g_i(x) \frac{\partial \theta}{\partial x_i} + \delta + O(\delta) = 0.$$

Divisant par δ et faisant tendre δ vers 0 on obtient $\sum_i g_i(x) \frac{\partial \theta}{\partial x_i} = -1$ de sorte que (2.22) implique bien

$$(2.24) \qquad \sum_i g_i \frac{\partial u}{\partial x_i} \in L^{\infty}(\mathcal{O}) .$$

Soit $x \in \mathcal{O}$ tel que $u(x)$ soit différentiable. Comme $\theta(x) > 0$, on peut prendre

$0 < \delta \leq \theta(x)$. Alors

$$(2.25)\quad u(x) \leq \operatorname*{Inf}_{s\geq\delta} \int_0^{s\wedge\theta(x)} e^{-\alpha t} f(y_x(t))dt = \int_0^{\delta} e^{-\alpha t} f(y_x(t))dt + $$

$$+ \operatorname*{Inf}_{s\geq\delta} \int_{\delta}^{s\wedge\theta(x)} e^{-\alpha t} f(y_x(t))dt.$$

Mais

$$\operatorname*{Inf}_{s\geq\delta} \int_{\delta}^{s\wedge\theta(x)} e^{-\alpha t} f(y_x(t))dt = \operatorname*{Inf}_{s'\geq 0} \int_{\delta}^{(s'+\delta)\wedge(\delta+\theta(y_x(\delta)))} e^{-\alpha t} f(y_x(t))dt$$

$$= e^{-\alpha\delta} \operatorname*{Inf}_{s'\geq 0} \int_0^{s'\wedge\theta(y_x(\delta))} e^{-\alpha t} f(y_{y_x(\delta)}(t))dt$$

$$= e^{-\alpha\delta}\, u(y_x(\delta)).$$

On déduit aisément de (2.25) et du fait que u est différentiable en x, que lorsque $\delta \to 0$

$$(2.26)\qquad -\sum_i g_i \frac{\partial u}{\partial x_i} + \alpha u \leq f .$$

Il est évident que $u(x) \leq 0$. Soit x un point de $\mathcal{O}$ tel que u soit différentiable en x et $u(x) < 0$. On a

$$u(x) = \int_0^{s(x)} e^{-\alpha t} f(y_x(t))dt$$

et $s(x) > 0$. On peut donc prendre $0 < \delta \leq s(x)$. On a

$$u(x) = \int_0^{\delta} e^{-\alpha t} f(y_x(t))dt + \int_{\delta}^{s(x)} e^{-\alpha t} f(y_x(t))dt$$

$$= \int_0^{\delta} e^{-\alpha t} f(y_x(t))dt + e^{-\alpha\delta} \int_0^{s(x)-\delta} e^{-\alpha t} f(y_{y_x(\delta)}(t))dt \geq \int_0^{\delta} e^{-\alpha t} f(y_x(t))dt +$$

$$+ e^{-\alpha\delta} u(y_x(\delta)).$$

Comparant avec (2.25) on a en fait

$$u(x) = \int_0^{\delta} e^{-\alpha t} f(y_x(t))dt + e^{-\alpha\delta} u(y_x(\delta)) \qquad 0 \leq \delta \leq s(x).$$

Raisonnant comme pour (2.26) on obtient

$$-\sum_i g_i \frac{\partial u}{\partial x_i} + \alpha u = f.$$

On a bien montré que u est solution de (1.19). Reste à montrer <u>l'unicité</u>. Soient u_1, u_2 deux solutions. On déduit aisément de (1.19) écrit pour u_1, u_2 et par différence que

$$(2.27) \qquad [-\sum_i g_i \frac{-\partial}{\partial x_i}(u_1-u_2) + \alpha(u_1-u_2)]\,(u_1-u_2) \le 0$$

Soit en posant $w = u_1-u_2$

$$-\tfrac{1}{2}\sum_i g_i \frac{\partial w^2}{\partial x_i} + \alpha w^2 \le 0 \quad \text{p.p. dans } \mathcal{O}$$

On intègre sur $\mathcal{O}$ et on utilise la formule de Green, ce qui est loisible puisque $\mathcal{O}$ est régulier. On obtient

$$(2.28) \qquad \int_{\mathcal{O}} w^2(\alpha+\tfrac{1}{2}\sum_i \frac{-\partial}{\partial x_i} g_i)dx - \tfrac{1}{2}\sum_i [g_i w^2 n_i]_\Gamma \le 0$$

où n=normale orientée vers l'extérieur de $\mathcal{O}$.

On démontre facilement que Γ^- défini par (1.15) vérifie :

$$\{x\in\Gamma \mid \sum_i g_i n_i > 0\} \subset \Gamma^- .$$

Comme $w=0$ sur Γ^-, il en résulte que $\sum_i [g_i w^2 n_i]_\Gamma \le 0$ donc (2.28) entraîne

$$\int_{\mathcal{O}} w^2(\alpha + \tfrac{1}{2}\sum_i \frac{-\partial g_i}{\partial x_i})dx \le 0.$$

D'après (1.9) il en résulte que $w= 0$.

2.2 - Vérification des hypothèses

On va donner des exemples d'ouverts où (1.5), (1.6), (1.7) sont satisfaites

2.2.1. Le cube

On prend $\mathcal{O} =]0,1[^n$; l'hypothèse (1.14) a lieu. Pous simplifier un peu, on prend $n=2$ et $g(x) =\binom{-1}{0}$. Si $x = (x_1,x_2)$, il est clair que

$$(2.29) \qquad \theta(x) = x_1 \quad \text{pour } x \in 0$$

de sorte que (1.5), (1.6), (1.7) sont évidentes.

L'unicité de la solution de (1.19) est également vraie, bien que $\mathcal{O}$ ne soit pas régulier, comme on le voit directement en appliquant le même principe de démonstration que dans le cas d'un ouvert. Nous donnerons au §3 une méthode directe de démonstration d'une variante du théorème 1.1, dans le cas du cube, sans utiliser l'interprétation probabiliste de u_ε.

2.2.2. Ouvert défini par une fonction régulière

On suppose que $\mathcal{O} = \{x|\Phi(x) < 0\}$ où $\Phi(x) \in C^2$. On suppose de plus que

$$(2.30) \qquad \begin{array}{l} \forall\, z \in \partial\mathcal{O} \text{ tel que } \frac{\partial\Phi}{\partial x}(x).g(z) = 0;\ \exists\, \rho = \rho(z) \\ \text{tel que } |\xi-z| \le \rho \Rightarrow \frac{\partial^2\Phi}{\partial x^2}(\xi)g(\xi).g(\xi) + \frac{\partial\Phi}{\partial x}(\xi).\frac{\partial g}{\partial x}(\xi)g(\xi) > 0 \end{array}$$

Il est clair que (2.30) sera vérifiée si par exemple Φ est strictement convexe et g constant. On vérifie aisément que

$$(2.31) \qquad \Gamma^- = \{x|\Phi(x) = 0, \frac{\partial\Phi}{\partial x} \cdot g \geq 0\} .$$

<u>Lemme</u> 2.1

<u>Sous l'hypothèse</u> (2.30), (1.5) <u>est vérifiée.</u>

<u>Démonstration</u>

Posons $z=y_x(\theta(x))$, $\theta=\theta(x)$, $\Lambda=\Lambda(x)$. On sait (cf. démonstration théorème 1.1) que $z\in\Gamma^-$. De plus $\Lambda \geq \theta$. Supposons $\Lambda > \theta$, donc $t \in [\theta,\Lambda] \Rightarrow y_x(t)=y(t) \in \bar{\mathcal{O}}$. Considérons d'abord le cas $\frac{\partial\Phi}{\partial x}(z).g(z) > 0$. Pour $\Lambda - \theta > \delta > 0$, on a

$$\Phi(y(\theta+\delta)) = \delta \frac{\partial\Phi}{\partial x} (y(s)). g(y(s)), \; s \in [\theta,\theta + \delta] .$$

Pour δ assez petit on a donc $\Phi(y(\theta+\delta)) > 0$, donc $y(\theta+\delta) \in \overset{o}{\Sigma}$, ce qui contredit le fait que $\theta+\delta < \Lambda$. Supposons maintenant que $\frac{\partial\Phi}{\partial x}(z).g(z) = 0$. On choisit $\delta < \Lambda - \theta$ assez petit de façon que $|y(s)-z| \leq \rho$ pour $s \in[\theta,\theta+\delta]$. Comme alors

$$\Phi(y(\theta+\delta)) = \frac{\delta^2}{2}[\frac{\partial^2\Phi}{\partial x^2} (y(s))g(y(s)).g(y(s))+ \frac{\partial\Phi}{\partial x}(y(s)).\frac{\partial g}{\partial x}(y(s))g(y(s))]$$

avec $s \in [\theta,\theta+\delta]$ il résulte de (2.30) que l'on a encore $y(\theta+\delta) \in \overset{o}{\Sigma}$, d'où la même contradiction que ci-dessus. ∎

<u>Lemme</u> 2.2

<u>Sous l'hypothèse</u> (2.30), (1.6) <u>est vérifiée.</u>

<u>Démonstration</u>

On opère suivant les techniques du Lemme 2.1 et du théorème 1.1, où (2.14) joue un rôle analogue à (2.8). Les détails sont un peu fastidieux, car il faut envisager de nombreux cas, mais ne présentent pas de difficulté fondamentale.

<u>Lemme</u> 2.3

<u>On suppose outre</u> (2.30), <u>que</u> $\theta(x) < +\infty \quad \forall x \in \mathcal{O}$ et

$$(2.32) \qquad \frac{\partial\Phi}{\partial x} (y_x(\theta(x))).g(y_x(\theta(x))) \neq 0 \qquad \forall x \in \mathcal{O} ;$$

<u>alors</u> (1.7) <u>est vérifiée.</u>

<u>Démonstration</u>

Soit K un compact $\subset \mathcal{O}$ et $x_o \in K$. On pose $\theta_o = \theta_{x_o}$ et $\psi(x,\theta) = \Phi(y_x(\theta))$.

On vérifie aisément que grâce à (2.32) les conditions d'application du théorème des fonctions implicites sont remplies pour ψ en x_o, θ_o. Il en résulte que $\theta(x)$ est différentiable en x_o et

$$(2.33) \qquad \theta'(x_o) = - \frac{\sum_i \frac{\partial\Phi}{\partial x_i} (y_{x_o}(\theta_o)) \frac{\partial y_{x_{oi}}(\theta_o)}{\partial x}}{\sum_i \frac{\partial\Phi}{\partial x_i} (y_{x_o}(\theta_o)) g_i (y_{x_o}(\theta_o))} .$$

Comme K et x_o sont arbitraires, θ est différentiable en tout point de $\mathcal{O}$ et la formule (2.33) montre que $\theta'(x)$ est continue.

III - ETUDE DIRECTE DE LA CONVERGENCE

3.1 - Notations - Enoncé du résultat

On se place dans le cadre du n°2.2. On écrit $\{x_1, x_2\} = \{x, y\}$. On va considérer une hypothèse pour f plus faible que (1.8) soit

$$(3.1) \qquad f \in L^2(\mathcal{O}), \; \frac{\partial f}{\partial y} \in L^2(\mathcal{O}).$$

On note $D = \frac{\partial}{\partial x}$. Il est clair que $\Gamma^- = \{x=0, \; 0 \leq y \leq 1\}$.

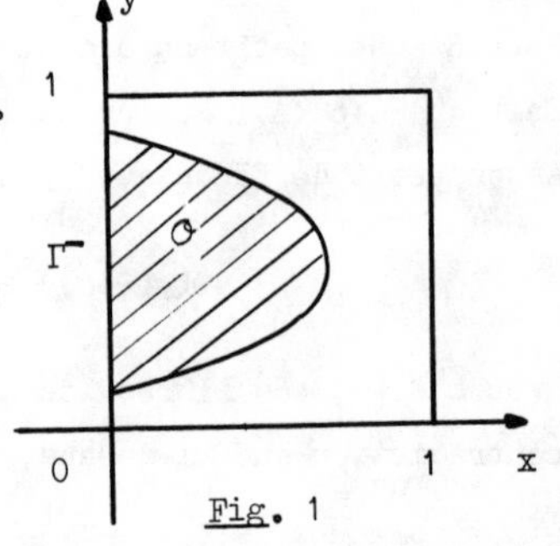

Fig. 1

Théorème 3.1

Hypothèses du n° 2.2.1. et 3.1

Soit $\mathcal{O}'$ un ouvert $\subset \mathcal{O}$ comme indiqué sur la Figure 1. Lorsque $\varepsilon \to 0$ on a

$$(3.2) \qquad u_\varepsilon \to u \quad \text{dans} \quad L^2(\mathcal{O}) \text{ faible.}$$

$$\frac{\partial u_\varepsilon}{\partial x} \to \frac{\partial u}{\partial x}, \; \frac{\partial u_\varepsilon}{\partial y} \to \frac{\partial u}{\partial y} \quad \text{dans} \quad L^2(\mathcal{O}') \text{ faible}, \quad \forall \mathcal{O}'$$

où u est la solution unique de

$$(3.3) \qquad u \in L^2(\mathcal{O}), \; Du \in L^2(\mathcal{O})$$

$$(3.4) \qquad \begin{aligned} &Du + \alpha u - f \leq 0 \qquad u \leq 0, \\ &(Du + \alpha u - f) \; u = 0 \quad \text{dans} \quad \mathcal{O} \end{aligned}$$

$$(3.5) \qquad u|_{\Gamma^-} = 0 .$$

3.2.- Démonstration du théorème 3.1

On note $w = w_{\varepsilon\beta}$ la solution du problème non linéaire monotone

$$(3.6) \qquad \begin{cases} -\frac{\varepsilon^2}{2} \Delta w + Dw + \alpha w + \frac{1}{\beta} w^+ = f \quad \text{dans} \quad \mathcal{O} \\ w = 0 \quad \text{sur} \quad \Gamma . \end{cases}$$

On sait (LIONS [1]) que lorsque $\beta \to 0$ on a

(3.7) $$w \to u_\varepsilon \quad \text{dans} \quad H_0^1(\mathcal{O}).$$

On établit maintenant des estimations à priori sur w indépendantes de ε et β, d'où l'on déduit grâce à (3.7) des estimations sur u_ε indépendantes de ε. Les estimations suivantes sont classiques.

(3.8) $$|w| \leq C, \quad \varepsilon|\text{grad } w| \leq C, \quad |\frac{1}{\beta} w^+| \leq C .$$

<u>Lemme</u> 3.1

<u>Soit</u> $\varphi = \varphi(x,y)$ <u>une fonction régulière dans</u> $\bar{\mathcal{O}}$, <u>ayant son support dans</u> $\bar{\mathcal{O}}'$.

<u>On a</u>

(3.9) $$|\varphi\, Dw| \leq C .$$

<u>Démonstration</u>

On pose $g = f - \frac{w^+}{\beta}$. Multipliant (3.6) par $\varphi^2 Dw$ (1) on obtient (en désignant par $\frac{\partial}{\partial n}$ la dérivée normale prise vers l'extérieur de $\mathcal{O}$)

(3.10) $$- \frac{\varepsilon^2}{2} \int_\Gamma \frac{\partial w}{\partial n} \varphi^2 Dw \, d\Gamma + \frac{\varepsilon^2}{2} a_o(w,\varphi^2 Dw) + |\varphi Dw|^2 + \alpha(\varphi w,\varphi\, Dw) = (g,\varphi^2\, Dw)$$

où

$$a_o(v_1,v_2) = \int_{\mathcal{O}} \text{grad } v_1 \text{ grad } v_2 \, dx.$$

Mais

$$\frac{\varepsilon^2}{2} a_o(w,\varphi^2 Dw) = \frac{\varepsilon^2}{4} \sum_{i=1}^{2} \int_{\mathcal{O}} \varphi^2 D(\frac{\partial w}{\partial x_i})^2 dx_1\, dx_2 + O(1)$$

(d'après (3.8))

$$= \frac{\varepsilon^2}{4} \sum \int_{\Gamma^-} \varphi^2 (\frac{\partial w}{\partial x_i})^2 \cos(n,x) d\Gamma^- + O(1)$$

$$= - \frac{\varepsilon^2}{4} \int_{\Gamma^-} \varphi^2 (\frac{\partial w}{\partial x})^2 dy + O(1).$$

Le premier terme de (3.10) vaut $\frac{\varepsilon^2}{2} \int_{\Gamma^-} \varphi^2 (\frac{\partial w}{\partial x})^2 dy$ et par conséquent (3.10) donne

$$\frac{\varepsilon^2}{4} \int_{\Gamma^-} \varphi^2 (\frac{\partial w}{\partial x})^2 dy + |\varphi\, Dw| + O(1)$$

d'où (3.9).

---

(1) C'est une idée introduite par D. LASCAUX dans le cas des équations ; cf. l'exposé de la méthode de LASCAUX dans LIONS [2] .

Lemme 3.2

Soit $\psi(y)$ une fonction régulière à support compact dans $]0,1[$; on a

(3.11) $$|\psi \frac{\partial w}{\partial y}| \leq C .$$

Démonstration

On note $\frac{\partial}{\partial y} = \partial$. Appliquant ∂ aux deux membres de (3.6) il vient

(3.12) $$-\frac{\varepsilon^2}{2} \Delta\partial w + D\partial w + \alpha\partial w + \frac{1}{\beta} \partial w^+ = \partial f.$$

On prend le produit scalaire avec $\psi^2 \partial w$. On note que $\partial w = 0$ si $x_1 = 1$ ou $x_1 = 0$ et $\psi(0) = \psi(1) = 0$, d'où

(3.13) $$\frac{\varepsilon^2}{2} a_o(\partial w, \psi^2 \partial w) + \int_{\mathcal{O}} \frac{\psi^2}{2} D(\partial w)^2 \, dx_1 \, dx_2 + \alpha|\psi\partial w|^2 + \frac{1}{\beta}(\psi\partial w^+, \psi\partial w) = (\psi\partial f, \psi\partial w).$$

Mais

$$\frac{\varepsilon^2}{2} a_o(\partial w, \psi^2 \partial w) = \frac{\varepsilon^2}{2} \sum \int_{\mathcal{O}} \frac{\partial}{\partial x_i} (\partial w)(\psi^2 \frac{\partial}{\partial x_i} \partial w + 2\psi \frac{\partial \psi}{\partial x_i} \partial w) dx_1 \, dx_2 \geq$$

$$\geq \frac{\varepsilon^2}{2} \int_{\mathcal{O}} \psi^2 |\frac{\partial}{\partial x_i} \partial w|^2 \, dx_1 \, dx_2 - \frac{\varepsilon^2}{2} \sum \int_{\mathcal{O}} \psi^2 |\frac{\partial}{\partial x_i} \partial w|^2 \, dx_1 dx_2 - \frac{\varepsilon^2}{2} \int_{\mathcal{O}} (\psi')^2 (\partial w)^2 dx_1 dx_2$$

$$\geq O(1) \; ; \quad \int_{\mathcal{O}} \frac{\psi^2}{2} D(\partial w)^2 \, dx_1 \, dx_2 = 0 \quad ; \quad (\psi \, \partial w^+, \psi \, \partial w) = |\psi \, \partial w^+|^2$$

de sorte que (3.13) donne

$$\alpha|\psi \, \partial w|^2 \leq O(1) + |\psi \, \partial f| \; |\psi \, \partial w|$$

d'où (3.11). ∎

Des estimations (3.9) et (3.11) on déduit

(3.14) $$|\varphi \, Du_\varepsilon| + |\psi \, \partial u_\varepsilon| \leq C$$

et $$-\frac{\varepsilon^2}{2} \Delta u_\varepsilon + Du_\varepsilon + \alpha u_\varepsilon - f = -m_\varepsilon, \; m_\varepsilon \geq 0, \; |m_\varepsilon| \leq C, \; u_\varepsilon \leq 0, \; m_\varepsilon u_\varepsilon = 0 .$$

On peut extraire une sous-suite, encore notée u_ε, telle que

(3.15) $$u_\varepsilon \to u, \quad \varphi \, Du_\varepsilon \to \varphi \, Du, \quad \psi\partial u_\varepsilon \to \psi \, \partial u \;\text{ dans } L^2(\mathcal{O}) \text{ faible}$$

donc aussi $u_\varepsilon \to u$ dans $H^1(\mathcal{O}')$ faible et si $\mathcal{O}'$ est de frontière régulière on a

(3.16) $$u_\varepsilon \to u \text{ dans } L^2(\mathcal{O}) \; \underline{\text{fort}}.$$

On peut toujours supposer que

(3.17) $$m_\varepsilon \to m \text{ dans } L^2(\mathcal{O}) \text{ faible}.$$

On a $m_\varepsilon \geq 0$, $u \leq 0$ $Du+\alpha u - f = - m$ et grâce à (3.16) $mu = 0$.

Enfin comme $u_\varepsilon \to u$ dans $H^1(\mathcal{O}')$, il résulte des théorèmes de trace dans les espaces de Sobolev et du fait que $u_\varepsilon(0,x_2) = 0$ que $u=0$ sur Γ_-.

L'unicité se démontre comme pour le théorème 1.1.

Remarque 3.1

La structure particulière de l'ouvert n'intervient que pour le Lemme 3.2.

Remarque 3.2

Les résultats précédents, s'étendent à la situation suivante pour laquelle les méthodes probabilistes semblent inutilisables :

$$\frac{\varepsilon^2}{2}\Delta^2 u_\varepsilon + Du_\varepsilon + \alpha u_\varepsilon - f \leq 0 \tag{3.18}$$

$$u_\varepsilon \leq 0\ ,\quad (\frac{\varepsilon^2}{2}\Delta^2 u_\varepsilon + Du_\varepsilon + \alpha u_\varepsilon - f)u_\varepsilon = 0 \quad \text{dans } \mathcal{O}$$

$$u_\varepsilon = 0,\ \frac{\partial u_\varepsilon}{\partial n} = 0 \quad \text{sur } \Gamma. \tag{3.19}$$

Remarque 3.3

Le théorème 1.1 comme le théorème 3.1 s'étendent <u>au cas d'évolution</u> :

$$-\frac{\partial u_\varepsilon}{\partial t} - \frac{\varepsilon^2}{2}\Delta u_\varepsilon - \sum_i g_i \frac{\partial u_\varepsilon}{\partial x_i} - f \leq 0, \tag{3.20}$$

$$u_\varepsilon \leq 0,$$

$$u_\varepsilon\left[-\frac{\partial u_\varepsilon}{\partial t} - \frac{\varepsilon^2}{2}\Delta u_\varepsilon - \sum_i g_i \frac{\partial u_\varepsilon}{\partial x_i} - f\right] = 0 \quad \text{dans } \mathcal{O} \times]0,T[\ ,$$

$$u_\varepsilon = 0 \quad \text{sur} \quad \Gamma \times]0,T[, \tag{3.21}$$

$$u_\varepsilon(x,T) = 0 \quad \text{sur } \mathcal{O}. \tag{3.22}$$

BIBLIOGRAPHIE

A. BENSOUSSAN et J.L. LIONS [1] Note C.R.A.S. Paris (1973) t. 276, Série A, pp. 1411-1415, et Problèmes de temps d'arrêt optimal et inéquations variationnelles paraboliques, Applicable Analysis, 1974.

A. BENSOUSSAN et J.L. LIONS [2] Sur quelques questions liées au contrôle optimal. A la mémoire du Professeur PETROWSKI. Juin 1973. Ouspechi Mat. Nauk.

A. BENSOUSSAN et J.L. LION [3] Note C.R.A.S. Paris (1973), t. 276, Série A, pp. 1189-1192,
Note C.R.A.S. Paris, (1973), t. 276, série A, pp. 1333-1338.

H. BREZIS et G. STAMPACCHIA [1] Sur la régularité de la solution d'inéquations elliptiques, Bull. Soc. Math de France, 96 (1968), pp. 153-180.

A. FRIEDMAN [1] Stochastic games and variational inequalities. Archive fur Rational Mechanics and Analysis, 1973.

J.L. LIONS [1] <u>Quelques méthodes de résolution des problèmes aux limites non linéaires</u>, Dunod Paris 1969.

J.L. LIONS [2] Perturbations singulières dans les problèmes aux limites et en contrôle optimal. Lectures Notes in Math. Springer 1973.

J.L. LIONS et G. STAMPACCHIA [1] Variational inequalities. Comm. Pure Applied Math. XX (1967) pp. 493-519.

I.M. VISIK et L.A. LIOUSTERNIK [1] Dégénérescence régulière pour les équations différentielles linéaires avec un petit paramètre. Uspechi Mat. Nauk 12 (1957) 1-121.

METHODES DE RESOLUTION NUMERIQUE DES INEQUATIONS QUASI-VARIATIONNELLES

M. GOURSAT - S. MAURIN

IRIA - LABORIA

INTRODUCTION

De nombreux problèmes d'origine économique et mécanique, en particulier certains problèmes de contrôle impulsionnel, se ramènent à la résolution d'une inéquation quasi-variationnelle (I.Q.V.).

Dans cet article nous présentons un exemple de problème de gestion de stocks [BENSOUSSAN-LIONS [1]] et nous nous intéressons aux méthodes de résolution numérique de l'I.Q.V. associée: nous donnons la formulation en I.Q.V. et quelques résultats théoriques ; de ceux-ci nous tirons une méthode constructive en résolvant une suite d'inéquations variationnelles par un algorithme de relaxation. On introduit ensuite des méthodes de décomposition en montrant qu'elles s'adaptent aux I.Q.V.

Le plan adopté est le suivant :

I - LE PROBLEME PHYSIQUE

1.1. Politique de réapprovisionnement

1.2. Coût de gestion.

II - FORMULATION EN I.Q.V.

2.1. Commande optimale

2.2. Système d'inégalités

2.3. L'I.Q.V.

III - RESOLUTION NUMERIQUE I

3.1. Méthode de résolution

3.2. Résultats numériques.

IV - RESOLUTION NUMERIQUE II

4.1. Méthode de résolution

4.2. Résultats numériques.

BIBLIOGRAPHIE.

I - LE PROBLEME PHYSIQUE

Il s'agit de gérer un stock de n produits sur une période $[0,T]$; $x(x_1,\ldots,x_n)$ désigne le niveau du stock. La loi d'évolution du stock est une diffusion : $dx=\mu\, dt + \sigma\, dt_t$ $t\in[0,T]$ (μ et σ peuvent dépendre de t). Autrement dit les seules variations du stock sont dues à une demande ; si $D(t,s)$ représente la demande cumulée sur (t,s) on a

$$D(t,s) = \mu(s-t) + \sigma(b(s)-b(t)) \qquad s-t=\Delta t \tag{1.1}$$

$\Delta b=b(s)-b(t)$ est une variable aléatoire gaussienne de moyenne nulle et de matrice de covariance $\Delta t.I$.
La demande est donc modélisée par une partie déterministe et une perturbation gaussienne.

Les coûts de gestion sont :

- un coût de stockage et de rupture de stock

$$f(x)= \text{coût instantané} \begin{cases} \text{de stockage si } x>0 \\ \text{de rupture de stock si } x<0 \end{cases}$$

- un coût de commande.

il comprend un coût fixe de passation de commande et éventuellement un coût variable.

Exemple

Si on commande une quantité ξ on peut avoir un coût de commande $C(t,\xi)$ du type suivant :

$$C(t,\xi) = \begin{cases} 0 & \text{si } \xi=0 \\ k_1(t) + \sum\limits_i c_i(t)\,\xi_i & \text{si } 0 \leqslant \sum\limits_i \xi_i \leqslant Q_1 \\ k_2(t) + \sum\limits_i c'_i(t)\,\xi_i & \text{si } \sum\limits_i \xi_i > Q_1 \end{cases}$$

avec $k_1(t) > 0$, $k_2(t) > 0$.

Pour simplifier l'écriture nous prendrons dans ce qui suit $c(t,\xi)=k$.

1.1. Politique de réapprovisionnement et évolution de stock

A partir d'un stock de niveau x à l'instant t une politique de commande v_{xt} est la donnée d'une suite d'instants de commande θ^i_{xt} et de quantités commandées

ξ^i_{xt} à ces instants

(1.2) $v_{xt}=\left\{ \theta^1_{xt} , \xi^1_{xt} ;\ldots;\theta^i_{xt} , \xi^i_{xt};\ldots \right\}$.

Si $y_{xt}(s)$ désigne l'état du stock à l'instant s en partant de (x,t) et en suivant la politique v_{xt} on a :

$$(1.3)\quad \begin{cases} y_{xt}(t) = x \\ y_{xt}(s) = x-D(t,s) \quad s\in[t,\theta^1_{xt}[\\ y_{xt}(\theta^1_{xt}) = x-D(t,\theta^1_{xt})+ \xi^1_{xt} \end{cases}$$

puis on recommence en partant de $(y_{xt}(\theta^1_{xt}),\theta^1_{xt})$.

Remarque : Nous supposons ici que nous n'avons pas de délai de livraison.

1.2. Coût de gestion. Politique optimale

Prenons le schéma de la programmation dynamique du type décision-hasard et calculons le coût d'une transition élémentaire $(x,t) \rightarrow (y,t+\Delta t)$

en (x,t)
- . décision 1 : attente le coût est $f(x,t)\Delta t$
- . décision 2 : on commande une quantité ξ. Le coût est $k+f(x+\xi,t)\Delta t$

$J_{xt}(v_{xt})$ = coût associé à la politique v_{xt} sur [t,T] :

$$(1.3)\quad J_{xt}(v_{xt}) = E\left\{\sum_{i=1}^{\infty} \chi_{[t,T]} \quad (\theta^i_{xt}). k + \int_t^T f(y_{xt}(s),s)ds\right\}$$

avec $$\chi_{[t,T]} \quad (\theta^i_{xt}) = \begin{cases} 1 \text{ si } \theta^i_{xt} \leqslant T \\ 0 \text{ sinon.} \end{cases}$$

Si on introduit un taux d'actualisation α on obtient

$$(1.4)\quad J_{xt}(n_{xt}) = E\left\{\sum_{i=1}^{\infty} \chi_{[t,T]} (\theta^i_{xt})\text{-}k\ e^{-\alpha(\theta^i_{xt}-t)} + \int_t^T e^{-\alpha(s,t)} f(y_{xt}(s),s)ds\right\}.$$

Si $T=+\infty$ on doit avoir nécessairement $\alpha>0$ pour obtenir $J_{xt}(v_{xt})<+\infty$; (1.4) donne les coûts actualisés à l'instant t.

On note

$$(1.5)\quad \begin{cases} u(x,t) = \inf\limits_{v_{xt}} J_{xt}(v_{xt}) \\ v^*_{xt} \text{ telle que } u(x,t) = J_{xt}(v^*_{xt}) \end{cases}$$

v^*_{xt} est la politique optimale et son coût $u(x,t)$ le coût optimal de gestion.

II - FORMULATION MATHEMATIQUE : INEQUATION QUASI-VARIATIONNELLE

2.1. Commande optimale

Si on commande une quantité ξ en (x,t) avec un coût de commande k on a

$$u(x,t) = k+u(x+\xi,t).$$

La quantité optimale à commander ξ^* sera donc définie par

$$(2.1)\quad u(x+\xi^*,t) = \inf_{\xi\geqslant 0} u(x+\xi,t).$$

on note alors

$$(2.2)\quad Mu(x,t) = k + \inf_{\xi\geqslant 0} u(x+\xi,t).$$

2.2. Système d'inégalités

On montre dans BENSOUSSAN-LIONS [1] que $u(x,t)$ défini par (1.5) satisfait à :

$$(2.3)\quad \begin{cases} -\dfrac{\partial u}{\partial t} - \dfrac{\sigma^2}{2}\Delta u + \mu\dfrac{\partial u}{\partial x} + \alpha u - f \leqslant 0 \\ u - M(u) \leqslant 0 \\ -\left(\dfrac{\partial u}{\partial t} - \dfrac{\sigma^2}{2}\Delta u + \mu\dfrac{\partial u}{\partial x} + \alpha u - f\right)(u-M(u)) = 0 \end{cases}$$

avec la condition finale.

$$(2.4)\quad u(x,T) = 0$$

Remarque : $-\dfrac{\sigma^2}{2}\Delta u$ signifie $-\dfrac{1}{2}\operatorname{Tr}\sigma\sigma^*\Delta u$ et $\mu\dfrac{\partial u}{\partial x} = \sum\limits_i \mu_i\dfrac{\partial u}{\partial x_i}$.

Pour des raisons physiques le stock est borné, autrement dit $x \in \mathcal{O}$ ouvert borné de $\mathbb{R}^n$ de frontière Γ.

On rajoute les conditions aux limites suivantes

$$(2.5) \qquad \frac{\partial u}{\partial \nu} \leq 0 \qquad u - M(u) \leq 0 \qquad \frac{\partial u}{\partial \nu}(u - M(u)) = 0 \quad \text{sur } \Gamma.$$

2.3. L'inéquation quasi-variationnelle (I.Q.V.).

On définit pour $u, v \in H^1(\mathcal{O})$

$$(2.6) \qquad a(u,v) = \frac{\sigma^2}{2} \sum_i \int_{\mathcal{O}} \frac{\partial u}{\partial x_i} \frac{\partial v}{\partial x_i} dx + \sum_i \int_{\mathcal{O}} \mu_i \frac{\partial u}{\partial x_i} v \, dx + \alpha \int_{\mathcal{O}} u \, v \, dx$$

et pour $f, v \in L^2(\mathcal{O})$ $(f,v) = \int_{\mathcal{O}} f \, v \, dx.$

On cherche alors u solution de l'I.Q.V. :

$$(2.7) \qquad \begin{cases} -\left(\frac{\partial u}{\partial t}, v-u\right) + a(u,v-u) \geq (f,v-u) \\ \forall v \in H^1(\mathcal{O}) \text{ tel que } v \leq M(u) \\ u \leq M(u) \\ u(x,T) = 0 . \end{cases}$$

Pour le cas stationnaire c'est-à-dire avec $T=+\infty$, $\mu(t)=\mu$, $\sigma(t)=\sigma$, $f(x,t)=f(x)$ nous avons

$$(2.8) \qquad \begin{cases} a(u,v-u) \geq (f,v-u) \quad \forall v \leq M(u) \\ u \leq M(u). \end{cases}$$

III - RESOLUTION NUMERIQUE I

3.1. Méthode de résolution

a) Cas stationnaire

Nous rappelons ici un résultat essentiel pour la résolution : il s'agit d'un résultat d'existence de solution qui donne une méthode constructive :

dans le cas où $f \geq 0$, $f \in L^\infty(\Omega)$

on prend u^0= constante $\geq \frac{1}{\alpha}$ sup f

soit λ tel que

(3.1) $a(v,v) + \lambda\, |v|^2_{L^2} \geqslant c\|v\|^2_{H^1} \quad c>0 \quad \forall v\in H^1(\mathcal{O}).$

On définit u^n solution de l'inéquation variationnelle (I.V) :

$$(3.2)\quad \begin{cases} a(u^n,v-u^n) + \lambda(u^n,v-u^n) \geqslant (f+\lambda u^{n-1},v-u^n) \\ \forall v \leqslant M\, u^{n-2} \qquad u^n \leqslant M(u^{n-1}) \end{cases}$$

alors $u^n \to u$ dans $L^p(\mathcal{O})$ fort et dans $H^1(\mathcal{O})$ faible.

Ceci nous donne notre première méthode de résolution : nous allons résoudre une suite d'I.V.

Remarque : dans les exemples nous aurons généralement $\lambda=0$ et u^o solution du problème de Neumann :

$$a(u^o,v-u^o) \geqslant (f,v-u^o) \quad \forall v\in H^1(\mathcal{O}).$$

La discrétisation du problème est faite de manière standard par les différences finies et chaque I.V. est résolue par la méthode de relaxation avec projection (cf. GLOWINSKI-LIONS-TREMOLIERES [1] et CEA-GLOWINSKI [1]).

b) Cas évolutif

On se ramène à la résolution de problèmes stationnaires par discrétisation en temps de (2.7) :

$$(3.3)\quad \begin{cases} \left(\dfrac{u^k-u^{k-1}}{\Delta t}, v-u^k\right) + a(u^k,v-u^k) \geqslant (f,v-u^k) & \\ & k=1,\dots,N \\ \forall v \leqslant M\, u^k \qquad u^k \leqslant M\, u^k & \end{cases}$$

avec $u^o=0$ (correspondant à $u(x,T)=0$)

ce qui nous ramène à la résolution de N problèmes stationnaires $(N=\frac{T}{\Delta t})$.

3.2. Résultats numériques

Exemple 1 unidimensionnel évolutif

$\mathcal{O} =]-1,+3[\qquad T=2.$

pas de discrétisation $h=\frac{1}{40}$ en espace et $h_t=\frac{1}{50}$ en temps

$$\mu(t) = 5.(2{,}4+1{,}4 \sin \Pi t)$$

$$\sigma(t) = 0{,}3\ \mu(t)\ \sqrt{h_t}$$

$$f(x,t) = (1-0{,}125\ t)\ f(x) \text{ avec}$$

$$f(x) = \begin{cases} 8x & \text{si } x>0 \\ -60x & \text{si } x<0. \end{cases}$$

Coût de commande pour une quantité ξ : $k(\xi) = k_o+k_1\xi$ avec $k_o=k_1=0{,}3$ donc

$$M\ u(x,t) = k_o + \inf_{\xi\geq 0} \{k_1\xi+u(x+\xi,t)\}.$$

Test d'arrêt pour l'algorithme :

soit $\varphi(i)=u^n(i)$ $i=1,\ldots,N_\sigma$ la solution discrète de la $n^{\text{ième}}$ I.V. ($u^n \to u$ sol de l'I.Q.V.).

$\varphi^p(i)$ représente la $p^{\text{ième}}$ itération de relaxation pour obtenir $u^n(i)$.

Les deux tests d'arrêts sont

$$\sum_{i=1}^{N_\sigma} |\varphi^p(i)-\varphi^{p-1}(i)| < \varepsilon_1$$

$$\sum_{i=1}^{N_\sigma} |u^n(i)-u^{n-2}(i)| < \varepsilon_2$$

avec ici $\varepsilon_1=\varepsilon_2= .10^{-3}$

Le temps d'exécution est dans ce cas de 20s sur IBM 360-91. La figure 1 représente la politique obtenue c'est-à-dire :

C = zone de saturation des contraintes

$C = \{(x,t)\ ;\ u(x_1t) = M\ u(x,t)\}$; (s) = frontière de C

(S) = points où u est minimum c'est-à-dire que l'on a

$u(x,t)=k_o+k_1(y-x)+u(y,t)$ pour $(x,t)\in(S)$ et $(y,t)\in(S)$.

La politique optimale est la suivante : lorsque l'état est x en t si $(x,t)\in(s,S)$ on attend ; si x atteint (s) on commande $y-x$ avec $y\in(S)$.

La figure 2 représente le coût $u(x,o)$.

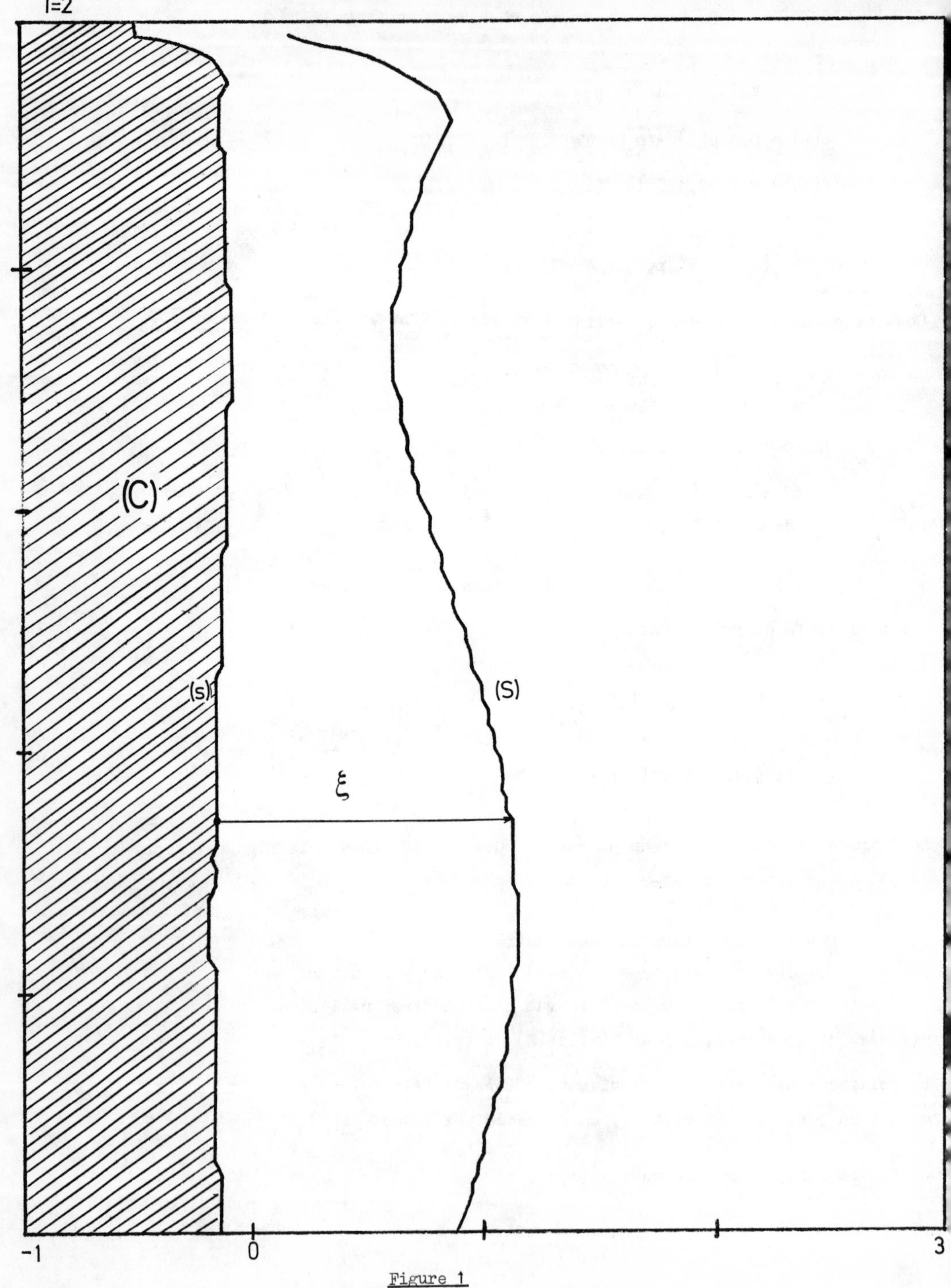

Figure 1

Remarque : nous ne donnons pas ici de résultats dans le cas stationnaire car le problème évolutif est mieux adapté aux applications. En effet, pour un problème stationnaire on résout le problème évolutif associé : on n'obtient évidemment pas le coût $u(x)$ (sauf si T est grand) mais la politique converge très rapidement vers la politique stationnaire.

La figure 3 représente la politique obtenue pour un problème analogue au précédent avec $\mu=8$ $\sigma=0{,}03\ \mu\sqrt{2}$ $f(x_1 t)=f(x)$ (comme dans l'exemple 1) la politique est stationnaire après 16 pas en temps.

Exemple 2 Cas bidimensionnel

$$\mathcal{O} =]-0{,}5\ ,\ +2[x]-0{,}5\ +\ 2[.$$

pas de discrétisation $h=(h_1,h_2)$ avec $h_1 = h_2 = \frac{1}{20}$.

Données :

$$\mu_1 = 12 \qquad \mu_2=8$$

$$\sigma_1=0{,}025\sqrt{2}\ \mu_1 \qquad \sigma_2=0{,}03\ \sqrt{2}\ \mu_2$$

$$M(u) = \operatorname{Inf}\ (M_1(u),M_2(u),M_3(u)) \quad \text{avec}$$

$$\begin{cases} M_1(u) = 0{,}7 + \displaystyle\operatorname*{Inf}_{\xi_1 \geqslant 0} \ u(x_1+\xi_1,x_2) \\ M_2(u) = 0{,}7 + \displaystyle\operatorname*{Inf}_{\xi_2 \geqslant 0} \ u(x_1,x_2+\xi_2) \\ M_3(u) = 1 + \displaystyle\operatorname*{Inf}_{\xi_1 \geqslant 0\ \ \xi_2 \geqslant 0} \ u(x_1+\xi_1,x_2+\xi_2) \end{cases}$$

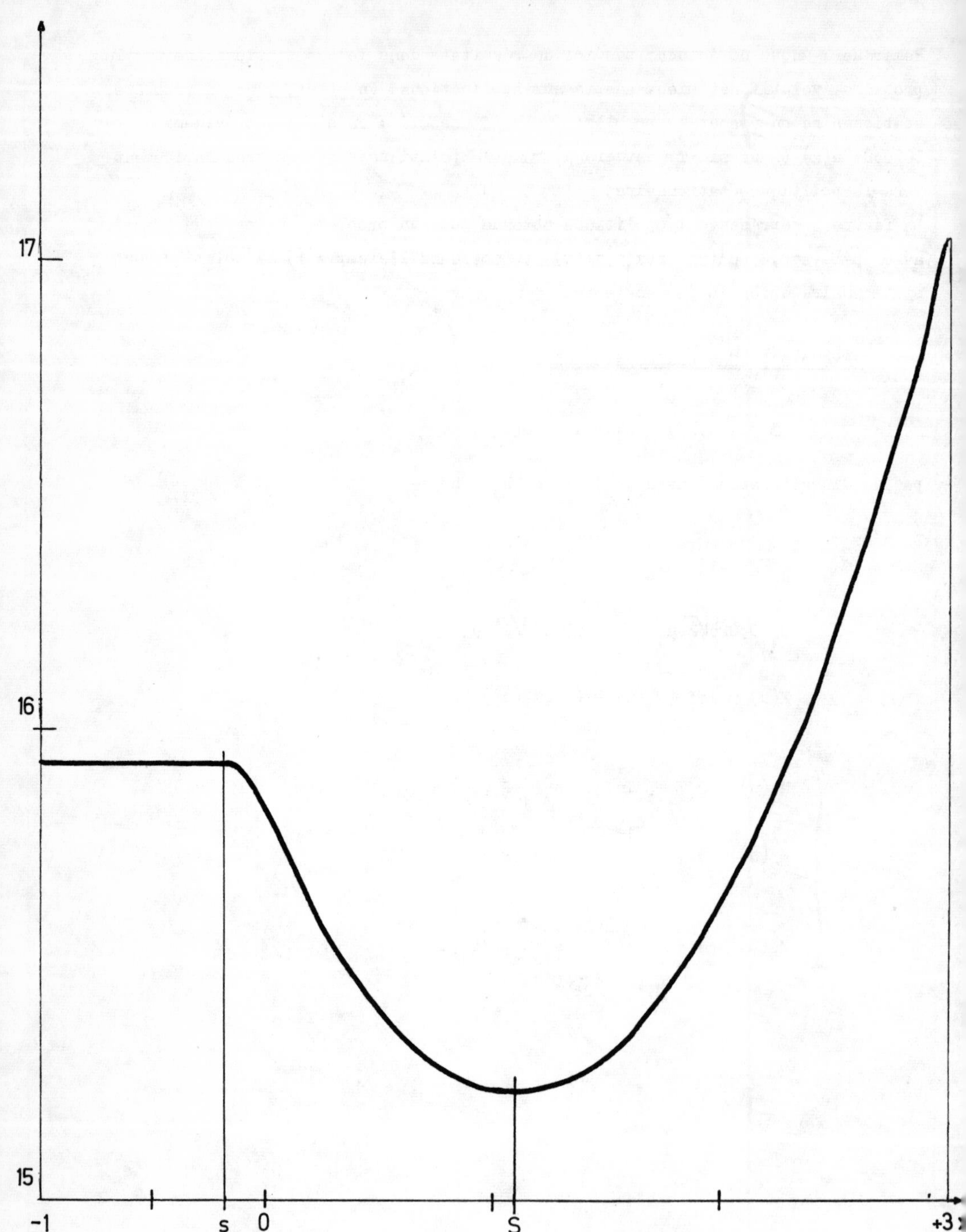

Figure 2

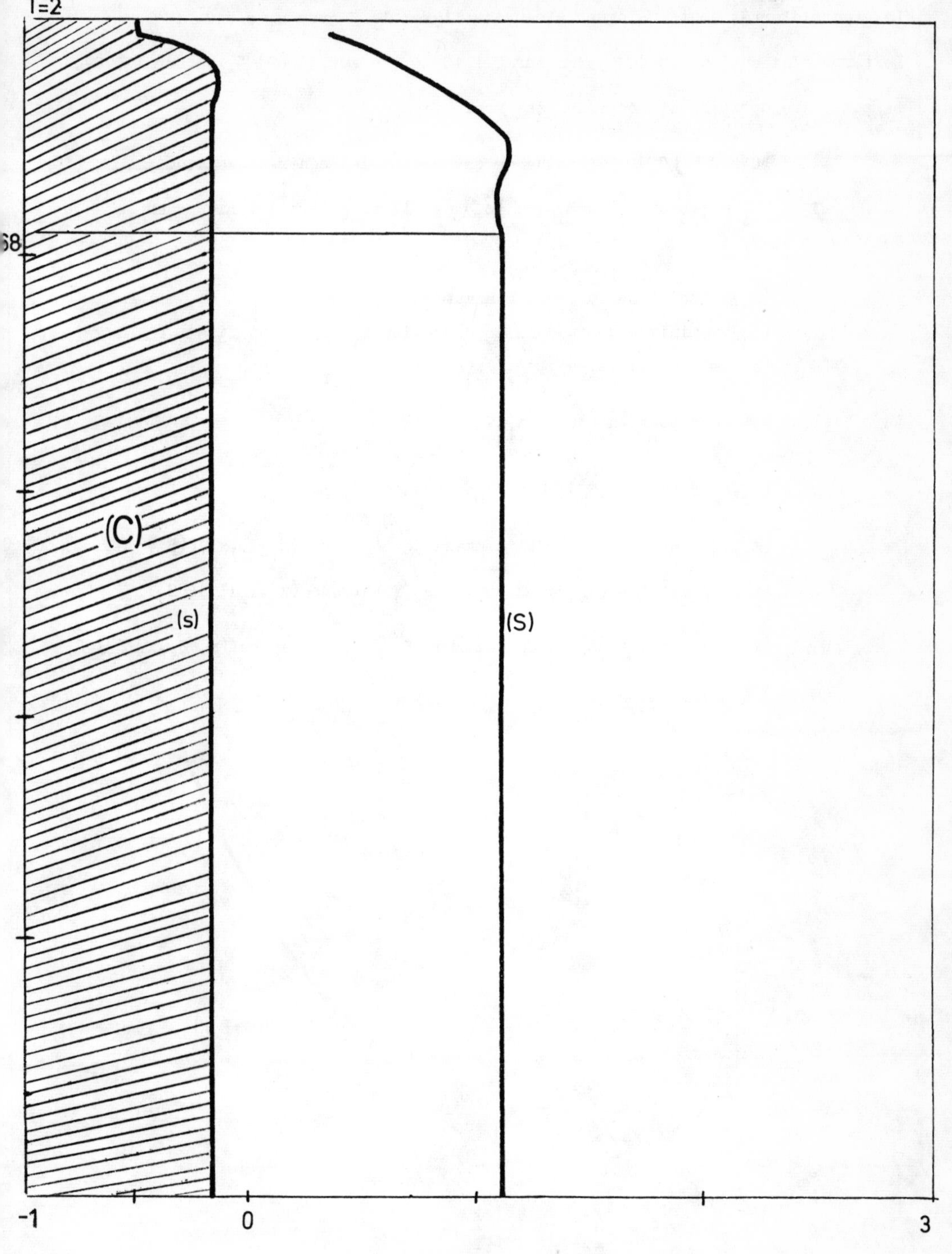

Figure 3

Les tests d'arrêt sont analogues à ceux de l'exemple 1 avec $\varepsilon_1=\varepsilon_2=10^{-2}$.

Le temps d'exécution est de 9,5 mn sur CII 10070 (1s sur 360-91 ≃ 15s sur 10070).

La politique optimale est représentée sur la figure 4 :

$C = C_1 \cup C_2 \cup C_3$ zone de saturation des contraintes avec

C_i = {points où $u=M_i u$} ; (s_i) est la frontière intérieure au domaine $\mathcal{O}$ de C_i ;

(S_3) = point où u est minimum

(S_1) = minimum de u dans la direction x_1, x_2 étant fixé

(S_2) = minimum de u pour x_1 fixé.

La politique est donc pour un stock (x_1,x_2)

$(x_1,x_2)\notin (s_1)\cup (S_2) \cup (S_3)$ on attend

(x_1,x_2) atteint (s_1) : on commande ξ_1 tel que $(x_1+\xi_1,x_2)\in(S_1)$

(x_1,x_2) atteint (s_2) : on commande ξ_2 tel que $(x_1,x_2+\xi_2)\in(S_2)$

(x_1,x_2) atteint (S) : on commande ξ_1 et ξ_2 tels que $(x_1+\xi_1,x_2+\xi_2)=S_3$

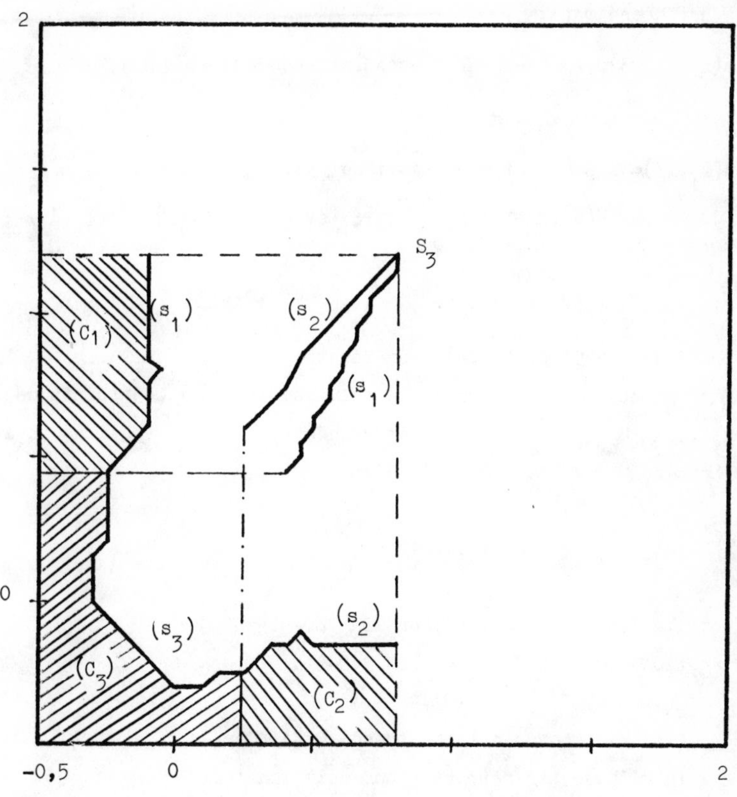

Figure 4

Exemple 3

On teste la méthode précédente par rapport à une solution de l'I.Q.V. stationnaire connue par avance.

On prend $\Omega=[-1,5,0] \times [-1,5,0]$

L'exemple tes est donné par :

$$u(x_1,x_2)=3+0,25\ [1-\cos\Pi\ x_1-\cos\Pi\ x_2-\cos\Pi\ x_1-\cos\Pi\ x_2]$$

pour $-1 \leqslant x_1 \leqslant 0$ et $-1 \leqslant x_2 \leqslant 0$

$$u(x_1,x_2)=3, \text{ pour } -1,5 \leqslant x_1 \leqslant -1 \text{ ou } -1,5 \leqslant x_2 \leqslant -1.$$

On prend M donné par $M(u)(x)=1+\inf\limits_{\xi\geqslant 0} u(x+\xi)$ avec $x=(x_1,x_2)$ $\xi=(\xi_1,\xi_2) \in \mathbb{R}^2$

$$\sigma_1^2 = \sigma_2^2 = 2,4/\Pi^2 \quad \mu_1 = \mu_2 = 2/\Pi \qquad \alpha=0,8.$$

h = 0,1 on prend donc 225 points de discrétisation.

Si $u_1^o, \ldots\ldots\ldots, u_{225}^o$ sont les valeurs de la solution test aux 225 points de discrétisation et $u_{1n}, \ldots, u_{n225}$ la solution approchée après n itérations on arrête l'algorithme lorsque

$$\sum_{i=1}^{225} |u_{ni}-u_i^o| \leqslant 225 \times 5 \times 10^{-3}.$$

Le temps de compilation est de 6s, 84 sur IBM 360-91

Le temps d'exécution est de 0s, 82 sur IBM 360-91.

Exemple 4 Cas tridimensionnel. On prend $\Omega=[-1.40]\times[-1.40]\times[-1.40]$. On teste la méthode par rapport à une solution test donnée par :

$$u(x_1,x_2,x_3) = 5+0,125\ [1-\cos\Pi\ x_1-\cos\Pi\ x_2-\cos\Pi\ x_3-\cos\Pi\ x_1\ \cos\Pi x_2$$
$$-\cos\Pi\ x_1\ \cos\Pi\ x_3-\cos\Pi\ x_2\ \cos\Pi\ x_3-\cos\Pi\ x_1\ \cos\Pi x_2\ \cos\Pi\ x_3]$$

si $-1 \leqslant x_1 \leqslant 0$ et $-1 \leqslant x_2 \leqslant 0$ et $-1 \leqslant x_3 \leqslant 0$.

$u(x_1,x_2,x_3) = 5,25$

si $-1,4 \leqslant x_1 \leqslant -1$ ou $-1,4 \leqslant x_2 \leqslant -1$ ou $-1,4 \leqslant x_3 \leqslant -1$.

$$\sigma_1^2 = \sigma_2^2 = \sigma_3^2 = 1,6/\Pi^2 \quad \mu = \mu_2 = \mu_3 = 1/\Pi \quad \alpha = 0,4$$

même opérateur M que dans l'exemple 3.

h pas de discrétisation = 0,1, on prend donc $2744=(14)^3$ points de discrétisation

on arrête l'algorithme lorsque

$$\sum_{i=1}^{2744} |u_i^n - u_i^o| \leqslant 2744 \times 5 \times 10^{-3}$$

avec les mêmes notations que dans l'exemple 3.

Le temps d'exécution est de 16,95s sur 360-91 IBM

Le temps de compilation est de 20s sur 360/91 IBM

IV - RESOLUTION NUMERIQUE II

On se propose dans ce paragraphe de montrer que la méthode de "splitting up" MARBCUK [1] YANENKO [1] TEMAM [1] BENSOUSSAN-LIONS-TEMAM [1] s'adapte aux problèmes d'I.Q.V. stationnaires. Il est en effet important de déterminer dans le cas des grandes dimensions soit le coût optimal soit la politique optimale de gestion.

4.1. Méthode de résolution

a) cadre théorique : Soit Ω ouvert de R^q borné "régulier"

soit $V_i = \{u \in \mathcal{L}^2(\Omega) \quad \frac{\partial u}{\partial x_i} \in \mathcal{L}^2(\Omega)\}$ pour $i=(1,\ldots,q)$

$$V = H^1(\Omega) = \bigcap_{i=1}^{q} V_i \quad H = \mathcal{L}^2(\Omega)$$

on note $\| \ \|_{V_i}$, $\| \ \|_{V}$, $\| \ \|_{\mathcal{L}^2}$, $\langle \ \rangle_{V_i}$, $\langle \ \rangle_{V}$, $\langle \ \rangle_{\mathcal{L}^2}$ les normes et les produits scalaires respectifs dans V_i, V et $\mathcal{L}^2$.

On a en identifiant H à son dual $V \hookrightarrow V_i \hookrightarrow H \hookrightarrow V_i' \hookrightarrow V'$ avec injections continues et denses.

Soient pour $i=(1,q)$ les q formes bilinéaires continues sur V_i respectivement

$$\forall u, v \in V_i \quad a_i(u,v) = \frac{\sigma_i^2}{2} \int_\Omega \frac{\partial u}{\partial x_i} \frac{\partial v}{\partial x_i} \, dx + \mu_i \int_\Omega \frac{\partial u}{\partial x_i} v \, dx + \alpha_i \int_\Omega u \, v . dx$$

On les suppose coercives c.a.d $\exists \ \beta_i > 0$ tels que

$$|a_i(u,u)| \geqslant \beta_i \|u\|_{V_i}^2 \qquad \forall u \in V_i$$

On a alors

$$a(u,v) = \sum_{i=1}^{q} a_i(u,v) \qquad \forall \ u, v \in V.$$

Soit A_i l'opérateur de V_i dans V_i' défini par $a_i(u,v) = \langle A_i u, v \rangle \ V_i' \times V_i \quad (\forall u, v \in V_i)$

Soit A l'opérateur de V dans V' défini par $a_i(u,v) = \langle Au, v \rangle \ V' \times V \quad \forall (u,v) \in V$

on a alors $A = \sum_{i=1}^{q} A_i$ de V dans V'

d'autre part si $a_i(u,v) = \langle g_i, v\rangle_{\mathcal{L}^2(\Omega)}$ on note A_i u l'élément de $\mathcal{L}^2(\Omega)$ tel que Ai u = g_i dans V'.

Pour plus de détails voir TEMAM [1] LIONS-MAGENES [1].

On suppose $f = \sum_{i=1}^{q} f_i$ $f_i \in \mathcal{L}^2(\Omega) \cap \mathcal{L}^\infty(\Omega)$ $f_i \geqslant 0$.

b) Algorithmes de décomposition

On se donne un paramètre Δt petit > 0.
On se donne également une valeur initiale u_o.

Cas d'une sur solution : on prend u_o satisfaisant à

$$a(u^o,v) \geqslant \langle f,v\rangle \quad \forall v \in V$$

il suffit de prendre u_o = constante $\geqslant \frac{\sup f}{\alpha}$.

Cas d'une sous solution : on prend $u_o = 0$.

Algorithme I : On résout les q équations suivantes :

$$\left\langle \frac{u^{1/q+1} - u_o}{\Delta t}, v\right\rangle_{\mathcal{L}^2(\Omega)} + a_1(u^{1/q+1}, v) = \langle f_1, v\rangle_{\mathcal{L}^2(\Omega)} \quad \forall v \in V_1$$

$$\left\langle \frac{u^{2/q+1} - u^{1/q+1}}{\Delta t}, v\right\rangle_{\mathcal{L}^2(\Omega)} + a_2(u^{2/q+1}, v) = \langle f_2, v\rangle_{\mathcal{L}^2(\Omega)} \quad \forall v \in V_2$$

$$\vdots$$

$$\left\langle \frac{u^{q/q+1} - u^{q-1/q+1}}{\Delta t}, v\right\rangle_{\mathcal{L}^2(\Omega)} + a_q(u^{q/q+1}, v) = \langle f_q, v\rangle_{\mathcal{L}^2(\Omega)} \quad \forall v \in V_q$$

puis on projette et on définit

$$u^1 = \inf(u^{q/q+1}, M(u^{q/q+1})) \in \mathcal{L}^2(\Omega)$$

par récurrence supposant connaître u^r $r \in \mathbb{M}$, alors on résout pour $i=(1,q)$ les équations suivantes

$$< \frac{u^{r+1/_{q+1}} - u^r}{\Delta t} , v>_{\mathcal{L}^2(\Omega)} + a_1(u^{r+1/_{q+1}},v) = < f_1, v>_{\mathcal{L}^2(\Omega)}$$

$$\forall\, v \in V_1$$

$$< \frac{u^{r+2/_{q+1}} - u^{r+1/_{q+1}}}{\Delta t} , v >_{\mathcal{L}^2(\Omega)} + a_2(u^{r/_{q+1}}, v) = <f_2,v >_{\mathcal{L}^2(\Omega)}$$

$$\forall\, v \in V_2$$

$$\vdots$$

$$< \frac{u^{r+q/_{q+1}} - u^{r+q-1/_{q+1}}}{\Delta t} , v >_{\mathcal{L}^2(\Omega)} + a_q(u^{r+q/_{q+1}}, v) = <f_q,v>_{\mathcal{L}^2(\Omega)}$$

$$\forall\, v \in V_q$$

et on définit u^{r+1} par

$$u^{r+1} = \inf(u^{r+q/_{q+1}}, M(u^{r+q/_{q+1}})) \quad \in \mathcal{L}^2(\Omega).$$

Remarques : 1) chacun de ces problèmes a une solution unique dans V_i TEMAM [1]

2) tous les $u^{r+i/_{q+1}}$ sont positifs $\forall\, r \in N$ $\forall i=(1,q)$.

Lemme 1 : Dans le cas d'une sur solution u_o

on a $u_o \geqslant u^{1+i/_{q+1}} \geqslant u^{2+i/_{q+1}} \dots \geqslant u^{r+i/_{q+1}} \dots \geqslant u^{r+1+i/_{q+1}} \geqslant 0$

pour chaque i la suite $(u^{r+i/_{q+1}})$ est décroissante positive dans $\mathcal{L}^2(\Omega)$.

Lemme 2 : Dans le cas d'une sous-solution u_o :

on a $u_o=0 \leqslant u^{1+i/_{q+1}} \leqslant u^{2+i/_{q+1}} \dots \leqslant u^{r+i/_{q+1}} \leqslant u^{r+1+i/_{q+1}} \dots \leqslant C$

pour chaque i la suite $(u^{r+i/_{q+1}})$ est croissante bornée dans $\mathcal{L}^2(\Omega)$

Lemme 3 : Les $u^{r+i/_{q+1}}$ tendent lorsque $r \to \infty$ vers $u_{\Delta t}^{i/_{q+1}}$ dans V_i faible et $\mathcal{L}^2(\Omega)$ fort pour $i=1$ à q et $u^r \to u_{\Delta t}^1$ dans $\mathcal{L}^2(\Omega)$ fort, où $u_{\Delta t}^{i/_{q+1}}$ sont solutions pour $i=(1,q+1)$ de l'inéquation QV "approchée" suivante

$$\left\{\begin{array}{l}
\sum_{i=1}^{q} A_i u_{\Delta t}^{i/q+1} \leq f \\
u_{1_{\Delta t}} \leq M(u_{\Delta t}^{q/q+1}) \\
(\sum_{i=1}^{q} A_i(u_{\Delta t}^{i/q+1}) - f)\,(u_{1_{\Delta t}} - M(u_{\Delta t}^{q/q+1})) = 0 \\
\frac{\partial u}{\partial \nu}{}_{A_i}\ \Delta t^{i/q+1} = 0 \quad i=1,q \\
\sum_{i=2}^{q+1} \| u_{\Delta t}^{i/q+1} - u_{\Delta t}^{i-1/q+1} \|^2_{\mathcal{L}^2(\Omega)} \leq C\ \Delta t.
\end{array}\right.$$

Remarque : 1) il est très probable et vérifié sur les exemples numériques mais non démontré que lorsque $\Delta t \to 0$ les $u_{\Delta t}^{i/q+1}$ tendent vers u solution de l'I.Q.V. stationnaire :

pour cela on peut s'aider de la Remarque suivante. Dans le cas non décomposé si on résout le problème suivant

$$\frac{u_{\Delta t}^{r+\frac{1}{2}} - u_{\Delta t}^{r}}{\Delta t} + A u_{\Delta t}^{r+\frac{1}{2}} = f \qquad r=0,1,2\ldots$$

$$u_{\Delta t}^{r+1} = \inf\,(u_{\Delta t}^{r+\frac{1}{2}}, M(u_{\Delta t}^{r+\frac{1}{2}})).$$

avec $u_o = 0$

On aboutit au mêmes résultats, de plus on peut montrer que si $\Delta t' < \Delta t$ alors

$$u_{\Delta t'}^{r} \leq u_{\Delta t}^{r} \quad \text{et} \quad u_{\Delta t'}^{r+\frac{1}{2}} \leq u_{\Delta t}^{r+\frac{1}{2}}$$

donc à la limite $\quad u_{\Delta t'}^{1} \leq u_{\Delta t}^{1} \quad u_{\Delta t'}^{\frac{1}{2}} \leq u_{\Delta t}^{\frac{1}{2}}$

avec respectivement $\quad a(u_{\Delta t'}^{\frac{1}{2}}, v-u_{\Delta t'}^{\frac{1}{2}}) \geq \langle f, v-u_{\Delta t'}^{\frac{1}{2}} \rangle$

$$u_{\Delta t'}^{1} \leq M(u_{\Delta t'}^{\frac{1}{2}})$$

$$v \leq M(u_{\Delta t'}^{\frac{1}{2}})$$

et $\quad a(u_{\Delta t}^{\frac{1}{2}}, v-u_{\Delta t}^{\frac{1}{2}}) \geq \langle f, v-u_{\Delta t}^{\frac{1}{2}} \rangle$

$$u_{\Delta t}^{1} \leq M(u_{\Delta t}^{\frac{1}{2}})$$

$$v \leq (u_{\Delta t}^{\frac{1}{2}})$$

en faisant tendre Δt vers zéro par valeurs décroissantes, $u^1_{\Delta t}$ décroit, alors il suffit de prendre

$$v \leqslant M(u) \Rightarrow v \leqslant M(u^{\frac{1}{2}}_{\Delta t}) \quad \forall \Delta t$$

et on peut passer à la limite dans les inégalités précédentes.

2) on trouvera la démonstration des Lemmes 1, 2, 3 dans un travail non publié de Bensoussan-Lions - Temam - Maurin.

<u>Algorithme II</u> : soit $M_i(u)(x_1 \ldots, x_i \ldots x_q) = 1 + \inf\limits_{\xi_i \geqslant 0} u(x_1 \ldots, x_i + \xi_i, \ldots x_q)$

On peut alors utiliser un algorithme en parallèle avec projection. On résout <u>une suite d'I.V.</u> pour $r=(0,\infty)$ $i=(1,q)$

$$a_i(u^{r+i/q+1}, v-u^{r+i/q+1}) + \left\langle \frac{u^{r+i/q+1} - u^r}{\Delta t}, v-u^{r+i/q+1} \right\rangle \geqslant \langle f_i, v-u^{r+i/q+1} \rangle$$

$$u^{r+i/q+1} \leqslant M_i(u^{r+i-1/q+1})$$

$$v \leqslant M_i(u^{r+i-1/q+1}) \quad v \in V_i$$

puis on projette

$$u^{r+1} = \inf \left(\sum_{i=1}^{q} \frac{1}{q} u^{r+i/q+1}, M \left(\sum_{i=1}^{q} \left(\frac{1}{q} u^{r+i/q+1} \right) \right) \right)$$

ou <u>une suite d'I.Q.V.</u>

pour $r=(0 \infty)$ $i=(1,q)$

$$a_i(u^{r+i/q+1}, v - u^{r+i/q+1}) + \left\langle \frac{u^{r+i/q+1} - u^r}{\Delta t}, v-u^{r+i/q+1} \right\rangle$$

$$\geqslant \langle f_i, v-u^{r+i/q+1} \rangle$$

$$u^{r+i/q+1} \leqslant M_i(u^{r+i/q+1})$$

$$v \leqslant M_i(u^{r+i-1/q+1}) \quad v \in V_i$$

puis on projette

$$u^{r+1} = \inf \left(\sum_{i=1}^{q} \frac{1}{q} u^{r+i/q+1}, M\left(\sum_{i=1}^{q} \frac{1}{q} (u^{r+i/q+1} \right) \right),$$

Algorithme III : On peut résoudre une suite d'I.V. sans projection en conservant l'opérateur M à chaque pas

$$a_i(u^{r+i/q+1}, v-u^{r+i/q+1}) + \langle \frac{u^{r+i/q+1} - u^{r+i-1/q+1}}{\Delta t}, v-u^{r+i/q+1} \rangle,$$

$$\geqslant \langle f_i, v-u^{r+i/q+1})$$

$$u^{r+i/q+1} \leqslant M(u^{r+i-1/q+1}) \qquad v \leqslant M(u^{r+i-1/q+1})$$

pour $r=0+\infty$, $i=1,q$

- ou une suite d'I.Q.V.

$$a_i(u^{r+i/q+1}, v-u^{r+i/q+1}) + \langle \frac{u^{r+i/q+1} - u^{r+i-1/q+1}}{\Delta t}, v-u^{r+i/q+1} \rangle$$

$$\geqslant \langle f_i, v-u^{r+i/q+1} \rangle$$

$$u^{r+i/q+1} \leqslant M(u^{r+i/q+1}) \quad v \leqslant M(u^{r+i/q+1})$$

pour $r=(0\ \infty)$ $i=(1,q)$.

On a pour ces 3 algorithmes les mêmes résultats que pour l'Algorithme I.

4.2. Résultats numériques

On a discrétisé de manière standard par les différences finies. Soit h le pas de discrétisation.

On peut remarquer qu'il n'y a aucun problème de convergence pour les problèmes discrétisés, mais le problème de la convergence des solutions discrétisees, vers la solution de l'I.Q.V. reste ouvert. On garde les mêmes notations que dans le cas non discrétisé.

Exemple 1 : On prend la même solution test que dans l'Exemple III du Paragraphe III et les mêmes données, on prend $\Delta t=\frac{1}{30}$, f donnée par le calcul direct.

Algorithme I : On résout par récurrence une suite de systèmes linéaires

$$(I+\Delta t Ai)u^{r+i/q+1} = \Delta t\ fi + u^{r+i-1/q+1}$$

la matrice $(I+\Delta tAi)$ est tridiagonale on résout le système par élimination YANENKO [1]. On prend $f_i=f/2$ $i=(1,2)$ avec les mêmes notations que dans l'exemple III du Paragraphe III, on arrête l'algorithme en r lorsque

$$\sum_{i=1}^{225} |u_i^r - u_o^i| \leqslant 225 \times 5 \times 10^{-3}$$

le temps de compilation est de 3s 50 sur 360-91 IBM

le temps d'exécution est de 1s 18 sur 360-91 IBM

Algorithme II : on résout pour chaque r et chaque i l'inéquation variationnelle suivante

$$a_i(u^{r+i/q+1}, v-u^{r+i/q+1}) + \left\langle \frac{u^{r+i/q+1} - u^r}{\Delta t}, v-u^{r+i/q+1} \right\rangle \geqslant \langle f_i, v-u^{r+i/q+1} \rangle$$

$$u^{r+i/q+1} \leqslant M_i(u^{r+i-1/q+1})$$

$$v \leqslant M_i(u^{r+i-1/q+1})$$

par Gauss-Seidel projeté (Méthode du N° III).

On arrête l'algorithme lorsque la différence de 2 itérées successives est inférieure en norme $\mathcal{L}^\infty$ à 10^{-1}.

On projette pour i=q+1 et on arrête l'algorithme en r avec le même test que l'Algorithme I.

Le temps de compilation est de 6s, 04

le temps d'exécution est de 3s,80 sur IBM 360-91.

Algorithme III : on remplace M_i par M dans l'algorithme précédent

Le temps de compilation est de 5s, 54

le temps d'exécution est de 3s, 10 sur IBM 360-91

pour la résolution de la suite d'I.V.

Exemple II : Cas tridimensionnel. On résout par l'Algorithme I, le même problème que dans l'Exemple 4 du Paragraphe III,

on prend $\Delta t = \frac{1}{20}$ $f_i = f/_3$ i=(1,2,3).

On résout les sytèmes pour chaque r et chaque i $(I+\Delta t\ Ai)\, u^{r+i/4} = \Delta t\ f_i + u^{r+i-1/4}$ de la même manière que dans l'Exemple I de ce Paragraphe.

On arrête l'algorithme en r lorsque

$$\sum_{i=1}^{2744} |u_1^r - u_i^o| \leqslant 2744 \times 5 \times 10^{-2}$$

Le temps de compilation est de 5s, 34

le temps d'exécution est de 15s, 21 sur IBM 360-91.

<u>Exemple III</u> : mêmes données que dans l'Exemple I de ce Paragraphe (bidimensionnel sauf f_1 et f_2 pris égaux a $f_1(x_1)= 4|x_1|$ $f_2(x_2)=4|x_2|$.

On s'intéresse à la politique optimale.

Par la <u>méthode du Paragraphe III</u> on obtient la politique stationnaire au bout de la résolution de 30 I.V.

Le temps de calcul est de 1s, 52 sur IBM 360-91. Voir Figure 1.

Par <u>la méthode de décomposition</u> (Algorithme I) avec $\Delta t=\frac{1}{30}$ on obtient la politique optimale au bout de 50 pas.

Le temps d'exécution est de 1s sur IBM 360-91. Voir Figure 2.

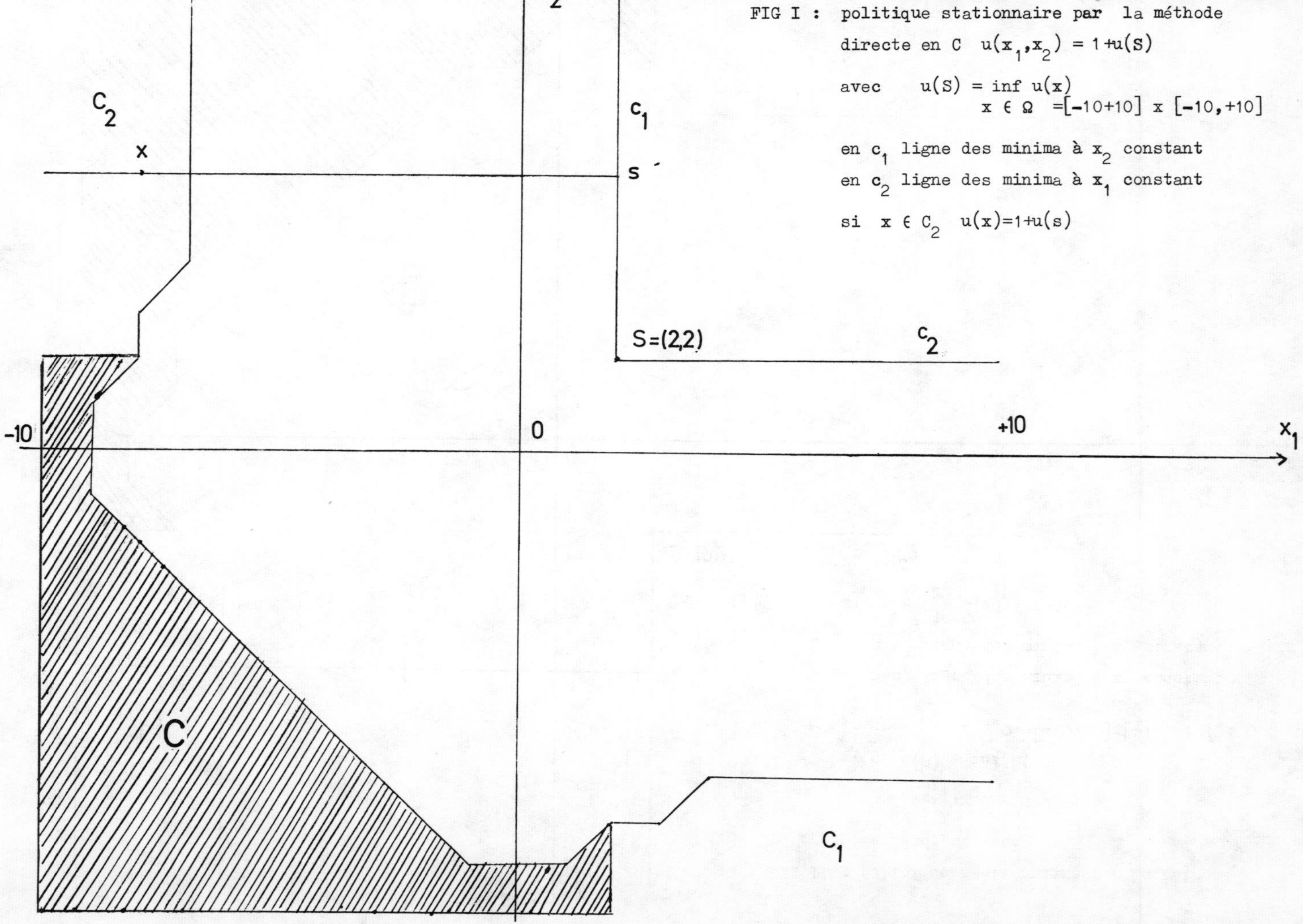

FIG I : politique stationnaire par la méthode directe en C $u(x_1,x_2) = 1+u(S)$

avec $u(S) = \inf_{x \in \Omega} u(x)$ $= [-10+10] \times [-10,+10]$

en c_1 ligne des minima à x_2 constant

en c_2 ligne des minima à x_1 constant

si $x \in C_2$ $u(x)=1+u(s)$

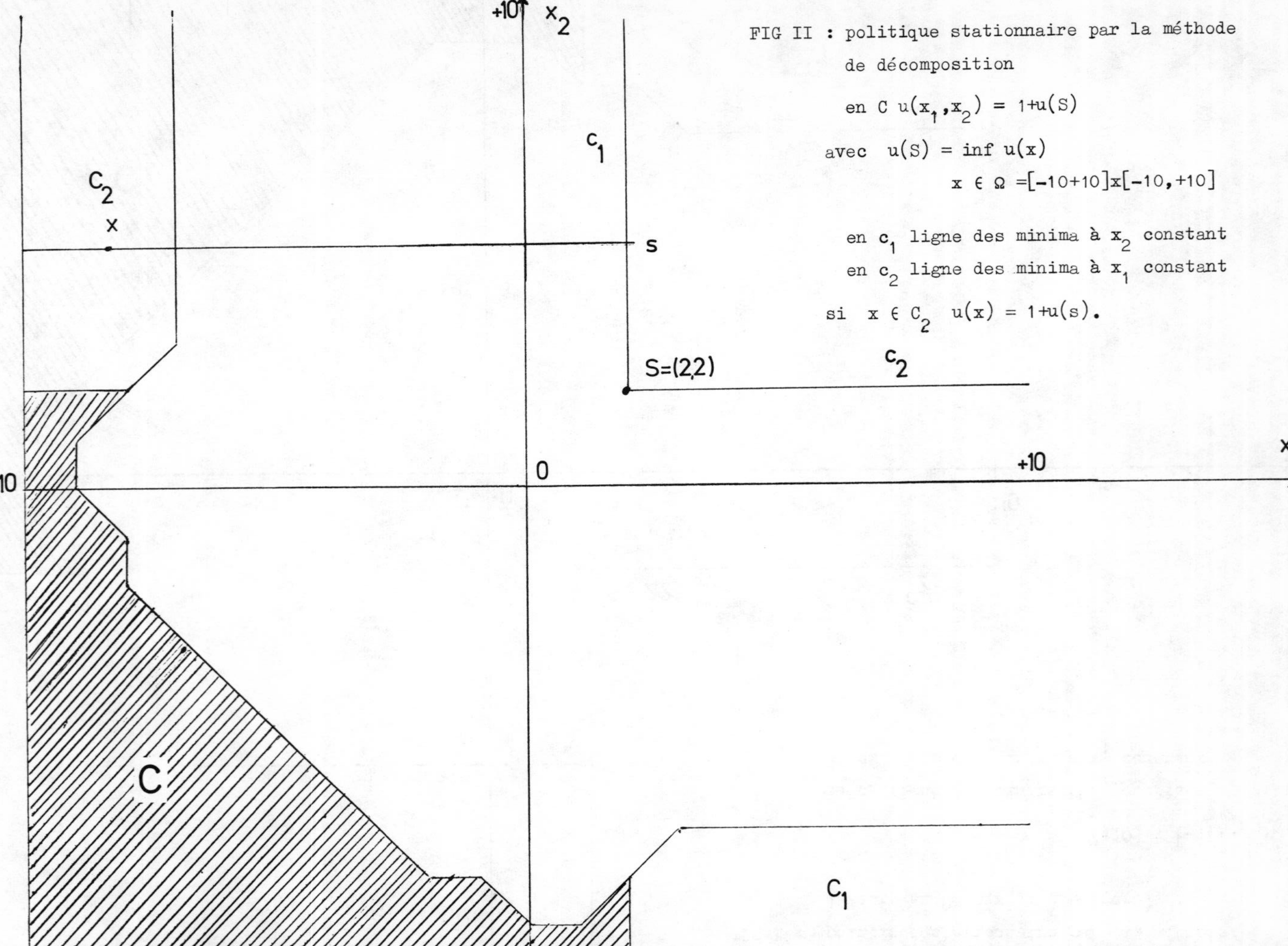

FIG II : politique stationnaire par la méthode de décomposition

en C $u(x_1,x_2) = 1+u(S)$

avec $u(S) = \inf u(x)$

$x \in \Omega = [-10+10] x [-10,+10]$

en c_1 ligne des minima à x_2 constant

en c_2 ligne des minima à x_1 constant

si $x \in C_2$ $u(x) = 1+u(s)$.

BIBLIOGRAPHIE

A. BENSOUSSAN-J.L. LIONS [1] CR Académie Sciences Paris

1) 276 (1973) pp. 1189- 193,

2) 276 (1973) pp. 1333-1337,

3) 278 (1974 pp. 675-679,

4) 278 (1974) pp. 747-751,

5) Ouvrage en préparation.

A. BENSOUSSAN-J.L. LIONS-R. TEMAM [1] Cahier IRIA n° 11 Juin 1972 p. 5-189.

J. CEA-R. GLOWINSKI [1] Méthodes numériques pour l'écoulement laminaire d'un fluide rigide viscoplastique incompressible. Int. J. Comp. Mach. Vol. 3 (1973), p. 225-255.

R. GLOWINSKI-J.L. LIONS-R. TREMOLIERES [1] Livre à paraître.

J.L. LIONS-E. MAGENES [1] Problèmes aux limites non homogènes Dunod Paris 1968 Tome I.

G.I. MARCHUK [1] Numerical methods in Meteorology. A. Colin Paris 1969.

R. TEMAM [1] Thèse. Paris 1967.

N.N. YANENKO [1] Méthodes à pas fractionnaires. A Colin 1968.

OPTIMISATION DE STRUCTURE
APPLICATION A LA MECANIQUE DES FLUIDES

O. PIRONNEAU

IRIA-LABORIA

1. INTRODUCTION

Depuis 1965 avec P.K.C. WANG [16], les techniques du contrôle optimal, développées pour le guidage des missiles sont appliquées aux systèmes physiques gouvernés par des équations aux dérivées partielles. Avec LIONS [5] le contrôle optimal des systèmes distribués a bénéficié des développements récents sur les équations aux dérivées partielles. Bien que cette branche du calcul des variations soit entrée, depuis plusieurs années déjà, dans sa phase opérationnelle, les applications à la mécanique des fluides restent très sporadiques. On peut penser que cela est dû à la non linéarité des équations de Navier-Stokes qui décrivent le phénomène physique ; mais on connaît maintenant de nombreux résultats sur l'existence, l'unicité et la régularité des solutions de ces équations (cf. LADYZHENSKAYA [3]). En fait la raison est ailleurs : c'est que dans la plupart des problèmes de contrôles que pose la mécanique des fluides, le contrôle (ou la variable d'optimisation) est un élément géométrique du domaine de définition des équations aux dérivées partielles. Je pense aux problèmes d'optimum design d'aéronautique et aux problèmes de transport (Vol optimal des avions à ailes battantes, nage ... cf. PIRONNEAU-KATZ [12]).

Nous sommes donc dans une branche très récente du contrôle optimal sur laquelle on a peu de résultat. On peut citer les travaux de BEGIS-GLOWINSKI [1], CEA, GIAN et MICHEL [2], MORICE [8], MURAT et SIMON [9], et de l'auteur [10 - 11]. Dans [10], on a montré qu'il était possible d'obtenir des conditions d'optimalité du premier ordre, pour ce type de problème, en calculant explicitement les termes du premier ordre de la variation du critère à optimiser en fonction de la distance normale entre la frontière optimale et une autre frontière admissible. Cette méthode permet d'une part, d'obtenir des conditions d'optimalité (i.e. une équation supplémentaire pour définir la surface), d'autre part, de construire une méthode de gradient pour résoudre le problème. Le but de cette étude était donc d'illustrer ces deux points. Cependant, comme le système différentiel et les conditions d'optimalité forment un système à frontière libre du type "problème de Stefan", on a choisi, pour illustrer la méthode, un problème qui ne dérivait pas d'un problème d'optimisation.

Le problème choisi est la détermination de la partie avant la plus aérodynamique d'un profil à partie arrière donnée et à volume donné ; ceci dans un écoulement gouverné par les équations de Navier-Stokes stationnaires dans les conditions d'application des méthodes de couche limite de Prandtl. Le problème est du type : trouver (φ,s) tel que $\Delta\varphi=o$ dans Ω_s $\varphi|_s=f$ $\frac{\partial\varphi}{\partial n}|_s=g$ où s est la frontière ou une partie de la frontière de Ω_s. Dans un premier temps, on remplace ce problème par un problème d'optimisation où une des conditions frontières est prise comme critère d'optimisation ; ce nouveau problème est alors résolu par la méthode décrite plus haut.

Les résultats numériques obtenus montrent que la méthode est bonne lorsque f et g sont des fonctions simples comme dans le cas du condensateur optimal, du meilleur profil en écoulement de Stokes ou du problème de Baiocchi, mais on rencontre des difficultés numériques sévères pour des problèmes plus compliqués comme celui que l'on s'est proposé de traiter où le rayon de courbure de s intervient explicitement dans le calcul de la direction de descente de l'algorithme. On doit alors concentrer son attention sur la construction d'une méthode itérative qui possède un degré de précision suffisant pour permettre à une rugosité apparue accidentellement de s'éliminer naturellement au cours du processus itératif.

2. POSITION DU PROBLEME.

On cherche un profil fermé Γ ayant un axe de symétrie Ox, une partie arrière Σ donnée et une partie avant s inconnue, tel que la solution φ^s du problème :

$$\left.\begin{array}{l} \Delta\varphi = o \quad \varphi|_s = \frac{3}{4}k\, s^{4/3}, \varphi|_\infty = u_o x, \left.\frac{\partial\varphi}{\partial n}\right|_\Sigma = o \\ \text{satisfasse à } \left.\frac{\partial\varphi}{\partial n}\right|_s = o \end{array}\right\} \qquad (1)$$

où k, u_o sont des constantes données et s est l'abscisse curviligne à partir de l'avant de s. Si la solution de ce problème existe alors le profil Γ est le plus aérodynamique de tous les profils d'égal volume et d'arrière Σ ; ceci en écoulement laminaire, incompressible, visqueux à grand nombre de Reynolds.

En effet, on a montré dans PIRONNEAU [11] que le profil Γ de volume donné qui a la plus petite résistance de trainée dans un écoulement gouverné par les équations de Navier-Stokes stationnaires devait satisfaire à l'équation :

$$\left\|\frac{\partial u}{\partial n}\right\|^2 + 2\,\frac{\partial u}{\partial n}\frac{\partial w}{\partial n} = \text{constante sur } \Gamma \qquad (2)$$

où u est la solution de

$$-\nu\Delta u + u\nabla u = \nabla p \qquad \nabla.u = o \qquad u|_\infty = u_o \qquad u|_\Gamma = o \qquad (3)$$

et w est la solution de

$$\nu\Delta w - \nabla u.w + u.\nabla w = -u.\nabla u + \nabla q \quad \nabla.w = o \quad w|_\Gamma = o \quad w|_\infty = o \qquad (4)$$

On sait depuis Prandtl que, en première approximation en 1/R, ($R=u_o x$(épaisseur de Γ)/ν), la solution de :

$$\text{rot } u = o,\ u|_\infty = u_o,\ u|_\Gamma = o \quad \nabla.u = o \qquad (5)$$

est solution de (3) sauf tout près de Γ où u est bien représenté par la solution de :

$$\left.\begin{array}{l} -\nu\dfrac{\partial^2 u_s}{\partial n^2} + u_s\dfrac{\partial u_s}{\partial s} + u_n\dfrac{\partial u_s}{\partial n} = U\dfrac{dU}{ds} \\ \dfrac{\partial u_s}{\partial s} + \dfrac{\partial u_n}{\partial n} = o \qquad u_s = u_n = o \text{ si } n = o,\ u_s = U \text{ si } n = +\infty \end{array}\right\} \qquad (6)$$

où s, n sont les coordonnées tangentielles et normales de Γ et U(s) est la trace sur Γ de la solution de (5).

Dans PIRONNEAU [11] on a montré que les techniques de Prandtl peuvent s'appliquer à (4) de telle sorte que, en première approximation, $w \equiv o$ sauf près de Γ où w_s est solution de

$$\left.\begin{aligned} &\nu \frac{\partial^3 w_s}{\partial n^3} + \frac{\partial^2 w_s}{\partial n \partial s} u_s + 2 \frac{\partial u_n}{\partial n} \frac{\partial w_s}{\partial n} + 2 \frac{\partial u_s}{\partial n} \frac{\partial w_s}{\partial s} = -\nu \frac{\partial^3 u_s}{\partial n^3} \\ &w_s = o \quad \text{si } n = o \quad w_s = \frac{\partial w_s}{\partial n} = o \quad \text{si } n = +\infty \end{aligned}\right\} \qquad (7)$$

En utilisant la transformation de Görtler [14], c'est à dire en posant :

$$\Phi = \int_o^s U(s)ds, \quad \xi = \Phi/u_o, \quad \eta = U.n/\sqrt{2\nu\Phi}, \quad u_s = U\frac{\partial f}{\partial \eta}$$

$$u_s + 2\, w_s = U\frac{\partial g}{\partial \eta}, \quad \beta(\xi) = 2\Phi U'(s)/U^2(s)$$

les équations (5) et (6) deviennent :

$$\left.\begin{aligned} &\frac{\partial^3 f}{\partial \eta^3} + f \frac{\partial^2 f}{\partial \eta^2} + \beta\left(1 - \left(\frac{\partial f}{\partial \eta}\right)^2\right) = 2\xi \left(\frac{\partial^2 f}{\partial \xi \partial \eta} \frac{\partial f}{\partial \eta} - \frac{\partial f}{\partial \xi} \frac{\partial^2 f}{\partial \eta^2}\right) \\ &f = \frac{\partial f}{\partial \eta} = o \quad \text{si } \eta = o \quad \frac{\partial f}{\partial \eta} = 1 \quad \text{si } \eta = \infty \end{aligned}\right\} \qquad (8)$$

$$\left.\begin{aligned} &\frac{\partial^3 h}{\partial \eta^3} - \frac{\partial^2 h}{\partial \eta^2}\left(f + 2\xi\frac{\partial f}{\partial \xi}\right) - \frac{\partial h}{\partial \eta}\left(\frac{\partial f}{\partial \eta} + 4\xi \frac{\partial^2 f}{\partial \xi \partial \eta}\right) + 2\,\beta\, h \frac{\partial^2 f}{\partial \eta^2} \\ &\qquad = -2\xi\left(\frac{\partial f}{\partial \eta} \frac{\partial^2 g}{\partial \xi \partial \eta} + 2 \frac{\partial^2 f}{\partial \eta^2}\frac{\partial g}{\partial \xi}\right) \end{aligned}\right\} \qquad (9)$$

$$h(o) = o \qquad h(\infty) = 1 \qquad \frac{\partial h}{\partial \eta}(\infty) = o$$

avec $\quad h = \dfrac{\partial g}{\partial \eta}$

On voit que (18) et (19) ont des solutions stationnaires si β est indépendant de ξ. De la définition de β on déduit $\beta = 2\Phi\Phi''/\Phi'^2$, soit encore :

$$\Phi = (R(s-s_o))^{\frac{2}{2-\beta}}, \quad U = (k(s-s_o))^{\beta/(2-\beta)}$$

et donc

$$\left.\begin{aligned} &\frac{\partial u_s}{\partial n} = \frac{1}{\sqrt{2}} f''(o)(k(s-s_o))^{\frac{2\beta-1}{2-\beta}} \\ &\frac{\partial u_s}{\partial n} + 2\frac{\partial w_s}{\partial n} = \frac{1}{\sqrt{2}} g''(o)(k(s-s_o))^{\frac{2\beta-1}{2-\beta}} \end{aligned}\right\} \qquad (10)$$

On peut donc satisfaire (2) en prenant $\beta = \frac{1}{2}$, ce qui implique :

$$U|_s = (k(s-s_o))^{1/3} \qquad (11)$$

Le problème (2)(3)(4) se réduit donc à (5)(11) qui peut encore s'écrire :

$$\Delta\varphi = o \qquad \varphi|_{\infty} = u_o x, \frac{\partial\varphi}{\partial n}\Big|_{\Gamma} = o \qquad \varphi|_{\Gamma} = \frac{3}{4}(k(s-s_o))^{4/3} \tag{12}$$

où φ est tel que $U = (\frac{\partial\varphi}{\partial y}, -\frac{\partial\varphi}{\partial x})$.

Comme $\frac{\partial\varphi}{\partial s}|_{\Gamma} = U|_{\Gamma}$, le système (12) ne peut avoir un profil fermé, convexe Γ comme solution car $U|_{\Gamma}$, l'écoulement autour de Γ, ne peut croître tout le long d'un profil fermé.

On sait d'ailleurs que, sauf pour la condition $\varphi|_{\infty} = u_o x$, le système (12) est satisfait si Γ est un cône d'angle 90°. On va donc chercher à ne satisfaire (11) que sur une partie S de Γ.
Ce type de problème est connu en mécanique des fluides sous le nom de "Problème inverse". LIGHTHILL [15] a donné quelques conditions nécéssaires pour l'existence et la résolution de ce type de problème.
Le problème (1) correspond donc bien à la recherche de l'avant le plus aérodynamique d'un profil arrière donné.
Le problème (1) est un problème à frontière libre sur lequel on ne connaît rien quant à l'existence d'une solution.

On se propose de construire la solution de (1), si elle existe, en résolvant :

$$\left.\begin{aligned} &\min_{S\in\mathcal{V}} \int_{\Omega_S} \nabla(\varphi-\psi)\nabla(\varphi-\psi)\,d\Omega \,\Big|\, \Delta\varphi = o,\ \varphi|_S = ks^{4/3},\ \frac{\partial\varphi}{\partial n}\Big|_{\Sigma} = o, \quad \varphi_{\infty} = u_o x \\ &\qquad\qquad \Delta\psi = o,\ \frac{\partial\psi}{\partial n}\Big|_S = o \quad \frac{\partial\psi}{\partial n}\Big|_{\Sigma} = o \quad \psi_{\infty} = u_o x \end{aligned}\right\} \tag{13}$$

Il est clair que si la solution de (1) existe alors c'est une solution de (13). Inversement, si la solution de (13) est telle que la valeur optimale du critère est nulle, alors c'est une solution de (1). Bien que le problème (1) ait été obtenu à partir d'un problème d'optimisation, les simplifications dans les équations de Navier-Stokes rendent l'étude de l'existence des solutions de (1) difficile. En revanche si la classe $\mathcal{V}$ est bien choisie, (13) a toujours une solution (cf. MURAT-SIMON [9]

3. OBTENTION DES CONDITIONS D'OPTIMALITE DES PROBLEMES DU TYPE (13).

Pour plus de clarté, on va chercher la solution de :

$$\left.\begin{aligned} &\min_{S\in\mathcal{V}} \int_{\Omega_S} \nabla(\varphi-\psi)\nabla(\varphi-\psi)\,d\Omega \,\Big|\, \Delta\varphi = o \text{ dans } \Omega_S,\ \varphi|_S = f|_S \\ &\qquad\qquad \Delta\psi = o \text{ dans } \Omega_S,\ \frac{\partial\psi}{\partial n}\Big|_S = g|_S \end{aligned}\right\} \tag{14}$$

où Ω_S est le domaine intérieur à S supposé être un contour fermé où $f \in H^2(\Omega)$,
$\Omega \supset \bigcup_{S \in \mathcal{V}} \Omega_S$; $g \in H^1(\Omega)$.

On suppose que les contours de $\mathcal{V}$ sont suffisamment réguliers pour que les systèmes différentiels de (14) aient, pour tout $S \in \mathcal{V}$, une solution unique continuement différentiable presque partout. (On peut montrer qu'il suffit que $S \in W^{2,\infty}$). D'autre part, on suppose que toutes les surfaces de $\mathcal{V}$ ont un axe de symétrie Ox et que $g(-y)=-g(y)$, de manière à assurer la condition $\int_S \frac{\partial \varphi}{\partial n} dS = o$, nécéssaire à l'existence des solutions du problème de Neumann.

Soit S une solution de (14) ; on suppose pour plus de simplicité, mais sans perte de généralité, que S est paramètrable par son abscisse curviligne s.

$$S = \{\xi(s) \mid s \in [o,L]\} \tag{15}$$

Soit E(S) la valeur du critère à optimiser dans (14). Soit S' une variation de S, définie par $\alpha : [o,L] \to R$, par :

$$S' = \{\xi(s) + \alpha(s)n(s) \mid s \in [o,L]\} \tag{16}$$

où n(s) est la normale extérieure à Ω_S au point $\xi(s)$.
On se propose de déterminer les termes du premier ordre en $\alpha(.)$ de :

$$\delta E = E(S') - E(S) \tag{17}$$

La méthode est exposée en détail dans PIRONNEAU [10]. On procède formellement :
si $\alpha \geqslant o$ et tel que S' soit symétrique par rapport à Ox :

$$\delta E = \int_{\Omega_{S'} - \Omega_S} \| \nabla(\varphi - \psi)\|^2 d\Omega + \int_{\Omega_S} \{\|\nabla(\varphi^{S'} - \psi^{S'})\|^2 - \|\nabla(\varphi^S - \psi^S)\|^2\} d\Omega \tag{18}$$

En utilisant la formule de la moyenne, la première intégrale devient :

$$\int_{\Omega_S} \alpha \|\nabla(\varphi-\psi)\|^2 d\Omega + o(\alpha) \tag{19}$$

soit, compte tenu des valeurs aux bornes de φ, ψ

$$\int_S \alpha \{ (\frac{\partial \varphi}{\partial n} - g)^2 + (\frac{\partial \psi}{\partial s} - \frac{\partial f}{\partial s})^2 \} dS + \sigma(\alpha) \tag{20}$$

En première approximation en $o(\varphi^{S'} - \varphi^S, \psi^{S'} - \psi^S)$, la deuxième intégrale de (18) est égale à : $2\int_{\Omega_S} \nabla(\varphi - \psi) . \nabla(\delta\varphi - \delta\psi)\, d\Omega$ (21)

On applique la formule de Green ; (21) devient :

$$-2\int_{\Omega_S} (\Delta(\varphi-\psi).\,\delta\varphi - (\varphi-\psi).\,\Delta\delta\psi)\,d\Omega + 2\int_S \Big[\Big(\frac{\partial\varphi}{\partial n} - \frac{\partial\psi}{\partial n}\Big)\,\delta\varphi - (\varphi-\psi)\frac{\partial\delta\psi}{\partial n}\Big]dS \quad (22)$$

Compte tenu de la définition de φ et ψ, (20), (22), (18) entraînent que :

$$\delta E = \int_S \alpha\Big[\Big(\frac{\partial\varphi}{\partial n}-g\Big)^2+\Big(\frac{\partial\psi}{\partial s} - \frac{\partial f}{\partial s}\Big)^2\Big]dS + 2\int_S \Big[\Big(\frac{\partial\varphi}{\partial n} - g\Big)\delta\varphi - (f-\psi)\frac{\partial\delta\psi}{\partial n}\Big]dS + \sigma(\alpha) \quad (23)$$

Ceci, car l'on peut montrer que $\varphi^{S'}$ et $\psi^{S'}$ tendent faiblement vers φ^S, ψ^S lorsque $\alpha \to o$ dans $C^\infty[o,L]$.

Un développement en série de Taylor au premier ordre montre que :

$$\delta\varphi = \varphi^{S'}(\xi(s)) - \varphi^S(\xi(s)) = \varphi^{S'}(\xi+\alpha n) - \alpha\frac{\partial\varphi}{\partial n}(\xi) - \varphi^S(\xi(s)) + \sigma(\alpha)$$

Or, par définition $\varphi^{S'}(\xi+\alpha n) - \varphi^S(\xi(s)) = f(\xi+\alpha n) - f(\xi)$, donc

$$\delta\varphi = \alpha\frac{\partial f}{\partial n} - \alpha\frac{\partial\varphi}{\partial n} + \sigma(\alpha) \quad (24)$$

De même

$$\delta\frac{\partial\psi}{\partial n} = \alpha\frac{\partial g}{\partial n} - \alpha\frac{\partial^2\psi}{\partial n^2} + \sigma(\alpha) \quad (25)$$

Donc

$$\delta E = \int_S \alpha\Big[\Big(\frac{\partial\varphi}{\partial n} - g\Big)^2 + \Big(\frac{\partial\psi}{\partial s} - \frac{\partial f}{\partial s}\Big)^2 - 2\Big(\frac{\partial\varphi}{\partial n} - g\Big)\Big(\frac{\partial\varphi}{\partial n} - \frac{\partial f}{\partial n}\Big) - 2(f-\psi)\Big(\frac{\partial g}{\partial n} - \frac{\partial^2\psi}{\partial n^2}\Big)\Big]dS + \sigma(\alpha) \quad (26)$$

Le même calcul étant valable pour $\alpha \leqslant o$, on montre facilement, à partir de (26), que la solution de (14) doit satisfaire à :

$$\Big(\frac{\partial\varphi}{\partial n} - g\Big)^2 + \Big(\frac{\partial\psi}{\partial s} - \frac{\partial f}{\partial s}\Big)^2 - 2\Big(\frac{\partial\varphi}{\partial n} - g\Big)\Big(\frac{\partial\varphi}{\partial n} - \frac{\partial f}{\partial n}\Big) - 2(f-\psi)\Big(\frac{\partial g}{\partial n} - \frac{\partial^2\psi}{\partial n^2}\Big) = o \text{ sur } S. \quad (27)$$

La condition (27) est l'analogue pour (14) de la condition $f'(Z)=o$ pour le problème $\min\{f(z) \mid z \in R^n\}$.

4. RESOLUTION DE (14) PAR UNE METHODE DE GRADIENT.

De (27) il ressort que si on pose :

$$\beta(s) = \Big(\frac{\partial\varphi}{\partial n} - g\Big)^2 + \Big(\frac{\partial\psi}{\partial s} - \frac{\partial f}{\partial s}\Big)^2 - 2\Big(\frac{\partial\varphi}{\partial n} - g\Big)\Big(\frac{\partial\varphi}{\partial n} - \frac{\partial f}{\partial n}\Big) - 2(f-\psi)\Big(\frac{\partial g}{\partial n} - \frac{\partial^2\psi}{\partial n^2}\Big) \quad (28)$$

$\alpha(s) = -\lambda\beta(s)$ avec $\lambda \geqslant o$ petit, alors $\delta E < o$ et donc S' est meilleur que S. On propose donc l'algorithme suivant :

<u>Pas o</u> Choisir $S_o = \{\xi_o(s) \mid s \in [o, L]\}$; poser i=o.

<u>Pas 1</u> Calculer φ^i, ψ^i solution de

$$\Delta\varphi = o \quad \varphi|_{S_i} = f|_{S_i} \; ; \Delta\psi = o \quad \frac{\partial\psi}{\partial n}|_{S_i} = g|_{S_i}$$

<u>Pas 2</u> Poser $S(\lambda) = \{\xi_i + \lambda\beta_i n_i \,|\, s \in [o,1]\}$ et calculer par dichotomie λ_i tel que :

$$\int_S -\frac{3}{4}\lambda_i \beta^2 dS \leqslant E(S(\lambda_i)) - E(S(o)) \leqslant -\int_S \frac{\lambda i}{4}\beta_i^2 dS \qquad (28)$$

<u>Pas 3</u> Poser $S_{i+1} = S(\lambda_i)$, $i = i+1$ et retourner en 1.

On notera que la méthode ci-dessus n'est autre que la méthode du gradient muni d'une recherche de pas à deux lignes (pour plus de détails, on renvoit à POLAK [13]). On notera aussi qu'il peut être plus intéressant de remplacer la recherche du pas par dichotomie par un développement de $E(S(\lambda))$ du 2ème ordre en λ fondé sur le développement de $\varphi-\psi$ en série entière de α donné en annexe. On remarque enfin que le pas 1 n'est pas indispensable puisque on a seulement besoin de $\frac{\partial\varphi}{\partial n}|_{S_i}$ et $\psi|_{S_i}$: d'où l'avantage de la méthode si le domaine est infini.

<u>Proposition</u>

Si tous les S_i sont suffisamment réguliers pour que φ^i, ψ^i existent, alors tout point d'accumulation de $\{S_i\}$, au sens de la convergence dans C^∞ des $\xi_i(.)$ satisfait à la condition (27).

<u>Démonstration</u>

D'après le modèle 1.3.9. de POLAK [13], il suffit de montrer que pour tout S ne satisfaisant pas à (27), il existe ε et δ tel que $E(S'')-E(S') \leqslant \delta < o$ pour tout S'' déduit de S' par une itération de l'algorithme et tout S' voisin de S par α avec $\|\alpha\|_{C^\infty} \leqslant \varepsilon$. De la convergence faible de $\varphi^{S'}$, $\psi^{S'}$ vers φ^S, ψ^S on déduit que E(S') est continu en α; de sorte que si S' et S sont voisins :

$$E(S'') - E(S') \leqslant \frac{1}{2}[E(S''') - E(S)] \leqslant -\frac{\lambda}{4}\int_S \beta^2 dS$$

où S''' est déduit de S par une itération de l'algorithme ; λ est le pas choisi et β est calculé sur S. Ce qui est bien la propriété cherchée car λ est borné inférieurement par $\lambda\min(S)$.

5. <u>PROGRAMMATION DE LA METHODE POUR RESOUDRE (1) VIA (13).</u>

Pour des raisons numériques dues à l'apparition de dérivées secondes dans (27), on a préféré résoudre :

$$\left.\begin{array}{l} \min\limits_{S \in \mathscr{V}} \{ \int_{\Omega_S} \|\nabla\varphi - D\,\psi\|^2 d\Omega \,|\, \Delta\varphi = o \quad \varphi|_S = \frac{3}{4}k\, s^{4/3} \quad \frac{\partial\varphi}{\partial n}|_\Sigma = o \quad \varphi|_\infty = u_o x \\ \qquad\qquad\qquad\qquad \Delta\psi = o \quad \psi|_S = o \quad \psi|_\Sigma = o \qquad \psi|_\infty = u_o y \end{array}\right\} \qquad (29)$$

où $D = (\frac{\partial}{\partial y}, -\frac{\partial}{\partial x})$.

Les lecteurs familiers avec les équations de l'écoulement potentiel reconnaitrons en φ la fonction potentielle du problème et en ψ la fonction de courant du problème. Le critère optimisé n'est autre que la différence entre l'écoulement potentiel autour de $S \cup \Sigma$(i.e. $D\psi$) et l'écoulement potentiel fictif autour de $S\cup\Sigma$ après avoir réparti sur S une distribution de singularité (source et puits) de manière à maintenir artificiellement $u_s = k\, s^{1/3}$ sur S.
Les conditions d'optimalité de (29) s'obtiennent de façon identique à la difficulté supplémentaire près que $s^{4/3}$ n'est pas une fonction différentiable dans Ω.
Lorsque l'on passe de S à S' par (16), $s^{4/3}$ devient : (R(s) rayon de courbure de S)

$$\frac{4}{3}\, s^{1/3} \int_o^s - \frac{\alpha(\sigma)}{R(\sigma)}\, d\sigma \; + \sigma(\alpha) \tag{30}$$

L'équation (24) doit donc être remplacée par :

$$\delta\varphi = -\alpha \frac{\partial\varphi}{\partial n} - k\, s^{1/3} \int_o^s \frac{\alpha}{R}\, d\sigma \; + \sigma(\alpha) \tag{31}$$

On obtient alors (α = o sur Σ) :

$$\delta E = \int_S \{ (\frac{\partial\varphi}{\partial n})^2 + (ks^{1/3} - \frac{\partial\psi}{\partial n})^2 + 2\, \frac{\partial\psi}{\partial n}(ks^{1/3} - \frac{\partial\psi}{\partial n}) - (\frac{\partial\varphi}{\partial n})^2 \}\, \alpha\, dS$$
$$+ 2 \int_S [\, Rs^{1/3} \frac{\partial\varphi}{\partial n} \int_o^s - \frac{\alpha}{R}\, d\sigma]\, dS + \sigma(\alpha)$$

Soit encore, après simplification et intégration par partie de la deuxième intégrale

$$\delta E = 2 \int_o^{s_1} -\alpha [(\frac{\partial\varphi}{\partial n})^2 + (\frac{\partial\psi}{\partial n})^2 - (ks^{1/3})^2 - \frac{2}{R} \int_{s_1}^s k\, s^{1/3} \frac{\partial\varphi}{\partial n}\, d\sigma] ds \tag{32}$$

où s_1 est l'abscisse curviligne du point de jonction de S et Σ. On note que le rayon de courbure R figure dans (32) ; d'où une complication numérique. Dans un premier temps, nous avons programmé l'algorithme en supposant que $\varphi|_{S_i} = \frac{3}{4} k s_o^{4/3}$ où s_o est l'abscisse curviligne de S_o ; la deuxième intégrale de (32) disparait et l'algorithme donne d'excellents résultats. Ce qui nous permet d'affirmer que si f,g dépendent de façon simple de S comme dans (14), alors la méthode est excellente.

Pour résoudre les problèmes mixtes de Dirichlet-Neumann, extérieurs de (13), on a utilisé l'adaptation de T.S. LUU[7] de la méthode des singularités discrètes. On rappelle que la méthode est fondée sur l'application de la formule de Green avec la fonction de Green du problème ; soit, dans notre cas : toute fonction harmonique à l'extérieur de $S\cup\Sigma$ et nulle à l'infini satisfait à :

$$2\pi\, \varphi(z) = \int_{S\cup\Sigma} \frac{\partial\varphi}{\partial n} \log|z - z(s)| - \varphi(z(s)) \frac{\partial}{\partial n_s} \log(z - z(s))] ds \tag{33}$$

Donc, si S et Σ sont donnés par un ensemble de segments d'extrèmités $\{z_i\}_{i=1}^N$, alors en tout point milieu z_j^c du segment $[z_j, z_{j+1}]$ $\frac{\partial\varphi}{\partial s}$ et $\frac{\partial\varphi}{\partial n}$ satisfont à :

$$\frac{\partial\varphi}{\partial s}(z_j^c) - i\,\frac{\partial\varphi}{\partial n}(z_j^c) = \sum_{1=1}^{N} \frac{q_1+i\gamma 1}{2\pi}\, e^{i(\beta j - \beta 1)} \log\left(\frac{z_j^c - z_1}{z_j^c - z_{1+1}}\right) \qquad (34)$$

où β_j est l'angle du segment $[z_j, z_{j+1}]$ avec l'axe Ox.

Pour calculer $\frac{\partial\varphi}{\partial n}\big|_S$ on a choisi q_1=o sur S et γ_1=o sur Σ. q_1 sur Σ et γ_1 sur S sont alors déterminés en résolvant la système linéaire constitué par la partie réelle de (34) avec $\frac{\partial\varphi}{\partial s}\big|_S = k\, s^{1/3} - u_\infty \cos\beta$. La partie imaginaire de (34) donne alors $\sin\beta + \frac{\partial\varphi}{\partial n}$ sur S. On calcule de même $\frac{\partial\varphi}{\partial n}\big|_S$ en choisissant q_1=o sur S et Σ.

La méthode est excellente mais elle ne permet pas de calculer $\frac{\partial\varphi}{\partial n}$, $\frac{\partial\psi}{\partial n}$ au point z_j mais seulement au point z_j^c. La construction de la surface S_{i+1} à partir de S_i pose donc un problème.

Dans un premier temps, on a pris pour $\frac{\partial\varphi}{\partial n}(z_j)$ la moyenne de $\frac{\partial\varphi}{\partial n}(z_1^c)$, 1=j, j+1, et la correction $\lambda\beta$ est portée sur la perpendiculaire à $[z_j^c, z_{j-1}^c]$, au point z_j. Mais l'algorithme obtenu n'est pas capable de corriger les rugosités qui peuvent apparaître sur S par suite des erreurs de tronquature. Il convient donc de prendre plus de points pour résoudre (30) que de sommets sur S, de manière à avoir une meilleurs précision sur $\frac{\partial\varphi}{\partial n}(z_j)$, $\frac{\partial\psi}{\partial n}(z_j)$. On a pris :

$$\Sigma = \{(x,y),\ (x-y) \mid y = \tfrac{1}{4}(x^2-1)(o.1x - 1.1),\ x \in [-1, o.46]\}$$

et

$$S_o = \{(x,y),(x,-y) \mid y = \tfrac{1}{4}(x^2-1)(o.1x-1-1)\quad x \in [o.46, 1]\}$$

avec 75 points de discrétisation.

Pour k=2.4, au bout de 9 itérations(1 mn IBM 360-91), on a obtenu la courbe de la figure 1, sur laquelle $\frac{\partial\psi}{\partial n} - k\, s^{1/3}$ est en valeur moyenne égal à o.o2 contre o.2 sur S_o. On notera, en revanche, que S_9 se raccorde mal avec Σ. Notre système possède en effet deux degrés de liberté supplémentaires que l'on n'a pas exploité ; d'abord la valeur de k, ensuite le point de raccordement de S sur Σ. Ce deuxième degré de liberté est trop difficile à utiliser mais, en l'abandonnant peut-être perdons nous l'existence d'une solution de (1). Quant à k, deux choix sont possibles : ou bien on optimise sur k aussi mais rien ne garantit le raccordement ou bien on choisit k de manière à avoir un bon raccordement et on itère. On a donc choisi k=1.8 (contre 2.4 plus haut). S_o est alors déjà une solution à 10%. Au bout de 6 itérations, on a obtenu S_6 qui satisfait la condition $\varphi|_S = ks^{1/3}$ à 1/100.

CONCLUSION

Les résultats sont donc de deux natures : d'une part on a montré que les problèmes à frontière libre pouvaient se résoudre en les remplaçant par un problème de contrôle à frontière variable ; si les fonctions f,g sont simples, la méthode marche bien - peut-être serait-il intéressant de tester cette méthode sur des problèmes justiciables des méthodes d'inéquations variationnelles (Problème de Stefan, Problème de la digue de Baiocchi) - d'autre part on a construit une solution de (1) à 1/100.

Ceci va nous permettre de construire le profil le plus aérodynamique à volume donné en intégrant les équations de la couche limite sur ε seulement.

Enfin, on a remarqué au paragraphe 5 qu'il était important de ne pas choisir n'importe quelle méthode, construire S_{i+1} à partir de Si ; un résultat qui reflète la complexité des problèmes d'optimisation de structure.

REFERENCES

[1] BEGIS D., GLOWINSKI R. : Application de la méthode des éléments finis à la résolution d'un problème de domaine optimal. Colloque IRIA, Décembre 1973.

[2] CEA J., GIAN A., MICHEL J. : Quelques résultats sur l'identification de domaine. Calcolo.

[3] LADYZHENSKAYA O. : The mathematical theory of viscous imcompressible flow. 1963.

[4] LATTES R., LIONS J.L. : The method of Quasi-reversibility. Elsevier, 1969.

[5] LIONS J.L. : Contrôle des systèmes gouvernés par des équations aux dérivées partielles. Dunod, 1968.

[6] LIONS J.L., MAGENES E. : Problèmes aux limites non homogènes. Dunod, Vol. 1. Paris 1967.

[7] LUU T.S. : Sur la technique des singularités en hydro et aérodynamique.
Colloque CNRS, 1971.

[8] MORICE P. : Une méthode d'optimisation de domaine appliquée au problème de la digue.
Ce symposium, 1974.

[9] MURAT R., SIMON J. : Quelques résultats sur le contrôle par un domaine géométrique.
Rapport N° 74003. Laboratoire d'Analyse Numérique. Université Paris VI. 1974.

[10] PIRONNEAU O. : On optimum profiles in Stokes flow.
J. Fluid. Mech. Vol. 59, pp. 117-128. 1973.

[11] PIRONNEAU O. : On optimum design in fluid mechanics.
J. Fluid. Mech. Vol. 64, pp. 97-111. 1974.

[12] PIRONNEAU O., KATZ D. : Optimal swimming of flagelated microorganisms.
A paraître dans J. Fluid. Mech.

[13] POLAK E. : Computational method in optimization.
Academic Press. 1971.

[14] ROSENHEAD L. : Laminar boundary Layers.
Oxford, 1966.

[15] THWAITES B. : Incompressible aerodynamics.
Oxford, 1960.

[16] WANG P.K.C. : Control of distributed parameter systems.
Academic Press. 1964.

ANNEXE

DEVELOPPEMENT EN SERIE DE LA SOLUTION D'UN PROBLEME DE DIRICHLET.

Soit $P = \{\xi(s) \mid s \in [o,1]^{n-1}\}$ une surface fermée régulière de R^n et $S = \{\xi(s)+\alpha(s)n(s) \mid s \in [o1]^{n-1}\}$ une variation de P.
n(s) est la normale extérieure de P au point $\xi(s)$; $\alpha \in L^\infty$.

Proposition

Soit φ la solution de $A\varphi = f$, $\varphi|_S = g$ dans $\Omega_S = \overset{o}{S}$. On suppose A elliptique d'ordre 2, coercif, et f dans $H^{m-2}(P)$, g dans $H^{m-1/2}(P)$, $m > \frac{n}{2} + k$.

Soit φ_o la solution de $A\varphi_o = f$ $\varphi_o|_P = g|_S$, soit φ_1 la solution de $A\varphi_1 = o$, $\varphi_1|_P = -\alpha\frac{\partial\varphi_o}{\partial n}|_P$ et plus généralement φ_i la solution de $A\varphi_i = o$, $\varphi_i|_P = -\sum_{j=o}^{i-1} \alpha^{i-j}\frac{\partial^{i-j}\varphi_j}{\partial n^{i-j}}$ dans ΩP.

Alors $\quad \varphi = \varphi_o + \varphi_1 + \ldots + \varphi_{k-i} + O_{k-i}(\alpha) \quad$ dans $\Omega_P \cap \Omega_S$ (1)

et $\quad \lim_{\alpha\to o} \left\| \frac{O_{k-i}(\alpha)}{\alpha^{k-i}} \right\|_{H^{1+m-i}(P)} = o \quad i=o, 1, \ldots, k.$ (2)

Démonstration

D'après le corollaire 9-1 de LIONS-MAGENES [6] et le théorème des traces $\varphi_i \in C^{m-i}(\overset{o}{P})$. On peut donc faire des développement de Taylors presque partout dans P jusqu'à l'ordre m-i. Si on prolonge φ_i dans $C^{m-i}(\Omega_S \cup \Omega_P)$ (on peut le faire, cf. LATTES-LIONS [4]) on a alors :

$$\varphi_i|_S = \varphi_i|_P + \frac{\partial\varphi_i}{\partial n}\alpha + \ldots + \frac{\partial^{k-i}}{\partial n^{k-i}}\varphi_i\,\alpha^{m-i} + \sigma(\alpha^{m-i}) \quad (3)$$

En utilisant (3) et la définition de φ_o, on obtient :

$$\varphi - \varphi_o|_S = -\alpha\frac{\partial\varphi_o}{\partial n}\Big|_P + \sigma(\alpha) \quad (4)$$

donc $\quad \|\varphi - \varphi_o\|_{H^{1+m}(\Omega_S)} \leq 2K \left\|\frac{\partial\varphi_o}{\partial n}\right\|_{H^{1/2+m}(S)} \|\alpha\|_{L^\infty[o,1]}$

où $K = \|A^{-1}\|$

On a aussi

$$\left\|\frac{1}{\alpha}(\varphi - \varphi_o - \varphi_1)\right\|_{H^m(\Omega_S)} \leq 2K\left(\left\|\frac{\partial^2\varphi_o}{\partial n^2}\right\|_{H^{m-1/2}(S)} \|\alpha\| + \left\|\frac{\partial\varphi_1}{\partial n}\right\|\right)$$

car de même :

$$\varphi - \varphi_o - \varphi_1|_S = -\frac{\alpha^2}{2}\frac{\partial^2\varphi_o}{\partial n^2}\Big|_P + o(\alpha^2) - \alpha\frac{\partial\varphi_1}{\partial n}\Big|_P$$

Or, de part la définition de $\varphi_1, \|\frac{\partial\varphi_1}{\partial n}\|_{H^{m-1/2}(S)} \to o$ lorsque $\alpha \to o$. On démontre donc facilement la proposition par récurrence.

Application

Calcul de $E(S(\lambda))$ au 3ème ordre en λ de manière à calculer le pas optimal dans l'algorithme du paragraphe 4.

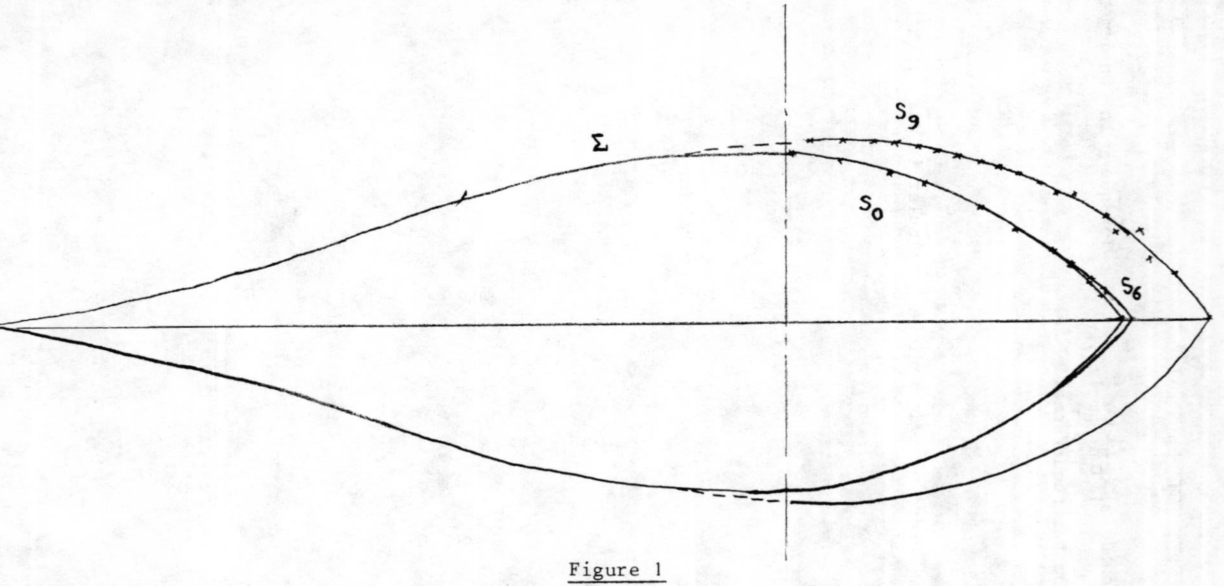

Figure 1

L'algorithme de gradient est utilisé en partant de S_o pour construire la solution de (1), c'est à dire le profil le plus aérodynamique à arrière Σ donné et à volume donné. Au bout de 9 itérations, on a trouvé S_9. La valeur de k choisie n'a pas donné un bon raccordement. On choisit alors k=1.8. Après 6 itérations, on obtient S_6 qui est solution du problème à 1/100 près.

REMARQUES SUR LES INEQUATIONS QUASI-VARIATIONNELLES

J.L. Joly, Université de Bordeaux I,
U. Mosco, Università di Roma (+)

Soit

(i) E un Banach réflexif réel, de dual E'

(ii) C une partie non-vide de E.

On se donne

(iii) une multiapplication Q de C dans E, à valeurs sous-ensembles convexes fermés non-vides de E,

(iv) une application L de E dans E', monotone, i.e.,

$$\langle Lu - Lv, u - v \rangle \geq 0 \qquad \forall\, u, v \in E ,$$

faiblement continue sur les sous-espaces de E de dimension finie, coercive sur C dans E, i.e., il existe $v_0 \in C$ tel que

$$\frac{\langle Lv, v - v_0 \rangle}{\|v\|_E^2} \to +\infty \quad \text{si } \|v\|_E \to \infty,\ v \in C .$$

Etant donné en outre $g \in E'$, on veut résoudre l'*inéquation quasi-variationnelle* (*i.q.v.*) [2] suivante:

$$(*) \quad \begin{cases} u \in Q(u), \\ \langle Lu, u - z \rangle \leq \langle g, u - z \rangle & \forall\, z \in Q(u) . \end{cases}$$

Si l'on introduit la *sélection variationnelle* $S : C \to 2^E$ de la multiapplication Q, définie par

(v) S(u) est, pour chaque $u \in C$, l'ensemble des solutions v de l'inéquation variationnelle (i.v.)

$$\begin{cases} v \in Q(u), \\ \langle Lv, v - z \rangle \leq \langle g, v - z \rangle & \forall\, z \in Q(u) \end{cases}$$

alors, évidemment, les solutions u de l'i.q.v. (*) sont rien autre que les *points fixes* de S,

$$u \in S(u) .$$

(+) Et GNAFA, Comitato per la Matematica del CNR.

On peut aussi rechercher des solutions u de l'i.q.v. (*) qui satisfassent une condition de " régularité abstraite"

(**) $u \in C_o$

où C_o est un sous-ensemble d'un Banach E_o qui a une injection continue, $\hookrightarrow$, dans E, avec $C_o \hookrightarrow C$.

Une propriété essentielle de continuité de la multiapplication Q par rapport à l'application $A = L - g$, qui permet d'obtenir l'existence de points fixes dans C_o de la sélection variationnelle S, est celle donnée dans la définition suivante.

Definition 1 Soit Q une multiapplication de C dans E, A une application de E dans E', $E_o \hookrightarrow E$, et $C_o \hookrightarrow C$ une partie non-vide de E.

On dit que Q est A-*continue sur* C_o *dans* E_o *faible*, si la propriété suivante est vérifiée:

Pour toute suite $(u_k, v_k) \in C_o \times C_o$ *vérifiant*

$$\begin{cases} v_k \in Q(u_k) \\ \langle Av_k, v_k - z \rangle \leq 0 \qquad \forall z \in Q(u_k) , \end{cases}$$

qui converge vers $(u, v) \in E_o \times E_o$ *dans* $E_o \times E_o$ *faible, on a*

(α) $v \in Q(u)$,

et

(β) *quel que soit* $w \in Q(u)$, *il existe des* $w_k \in Q(u_k)$ *tels que* $\lim \langle Av_k , w_k - w \rangle = 0$. ¤

Comme cas particulier d'un théorème d'existence et "régularité abstraite" de solutions $u \in C_o$ d'une inégalité du type

$$u \in C_1 \text{ t.q. } \varphi(u,u) + f(u,z) \leq \varphi(u,z) \qquad \forall z \in C_2 ,$$

où $\varphi : C_1 \times C_2 \to (-\infty, +\infty]$ et $f : C_2 \times C_2 \to (-\infty, +\infty)$, avec $C_o \subset C_1 \subset C_2$ parties non-vides d'un e.v. E, v. [6] , on obtient le théorème suivant

Théorème 1 *Avec les données (i)....(iv) et la notation (v), un vecteur* $g \in E'$ *étant aussi donné, on suppose que* E_o *est un Banach réflexif réel à injection continue* $\hookrightarrow$ *dans* E *et qu'il existe une partie convexe fermée non-vide* C_o *de* E_o, $C_o \hookrightarrow C$, *telle que les hypothèses suivantes soient vérifiées:*

(1) C_o *est stable par* S *et* SC_o *est borné dans* E_o ,

(2) Q *est A- continue sur* C_o *dans* E_o *faible, où* $A = L - g$.

Alors, il existe une solution u *du probléme* (*) (* *). ¤

<u>Corollaire</u> *Le théorème est ancore vrai si l'on remplace l'hypothèse (1) par*

(3) C_o *est stable par* S *et il existe un vecteur*

$$w_o \in \bigcap_{v \in C_o} Q(v)$$

tel que

$$\langle Lv, v - w_o \rangle / \|v\|^2_{E_o} \to +\infty \quad si \quad \|v\|_{E_o} \to \infty , \; v \in C_o . \; ¤$$

<u>Remarques</u>

(a) Si Q est *constante*, i.e. $Q(u) \equiv Q_o$ convexe fermé non vide de l'espace E, alors Q est A-continue sur C_o faible quelle que soit l'application A. Dans ce cas, le Théorème 1 se réduit, avec $E_o = E = C$ et $C_o = Q_o$, à un théorème d'existence de Browder [4] et Hartman-Stampacchia [5] pour l'i.v.

$$u \in Q_o \; : \; \langle Lu, u - z \rangle \leq \langle g, u - z \rangle \quad \forall \, z \in Q_o .$$

(b) Si A = 0, donc $S(u) \equiv Q(u)$, la multiapplication Q est A-continue dans E_o faible selon la Définition 1 si et seulement si elle est s.c.s dans E_o faible. Dans ce cas, le Théorème 1 est essentiellement une réformulation du théorème de point fixe de Kakutani et Ky Fan (ou de Schauder lorsque Q est univoque), v. aussi la remarque (c) suivante. Signalons d'autre part que nous n'avons pas obtenu de démostrations directes du Théorème 1 qui n'utilisent pas le théorème de Kakutani et Ky Fan, et pourtant tous les deux théorèmes sont fondés sur le lemme de Knaster-Kuratowski-Mazurkiewicz, v. [3] , [6] .

(c) Dans le Théorème 1 on peut remplacer l'espace "E_o faible" avec un espace localement convexe séparé E_o quelconque: on suppose alors C_o *compact* dans E_o et dans la définition de A-continuité on considère des filtres à la place des suites.

(d) Si A est *bornée*, i.e. A(B) est borné dans E' pour tout borné B dans E, alors une condition *suffisante* pour la A-continuité de Q sur C_o faible est la suivante:

(4) Q *est s.c.s dans* $E_o \times E_o$ *faible,* et en outre Q *est s.c.i dans* $E_o \times E$ *au sens suivant: pour toute suite* $u_k \in C$ *convergente vers* $u \in C$ *dans* E_o *faible et pour tout* $w \in Q(u)$, *il existe des* $w_k \in Q(u_k)$ *qui convergent vers* w *dans* E *fort.*

(e) La condition suffisante (4), comme (5) ci-dessous, ne dépend pas de l'application A. Toutefois, on peut montrer avec des exemples triviaux, et même avec E = R, que cette condition peut être violée et pourtant le probléme (*) admet des solutions. Ceci semble indiquer que la condition de continuité qu'il faut en général demander à Q doit être formulée par rapport à l'application A qui intervient dans le problème. Le Théorème 1 montre aussi que toute estimation "a priori" de la sélection variationnelle S (c'est à dire, toute estimation "a priori" des solutions de l'i.v. dans (v)), dans un espace E_o qui contient un convexe C_o laissé invariant par S, permet d'affaiblir les propriétés de continuité demandées à la multiapplication Q. Au moins en principe, on peut donc se donner le cas que pour une multiapplication Q donnée on puisse démontrer l'existence de solutions *régulières*, lorsque on dispose d'estimations "a priori" convenables de S(u) sous des hypothèses de régularité sur Q, L et g, sans qu'il soit possible de démontrer directement l'existence de solutions pour des données non régulières, v. aussi à cet égard l' Ex. 5,(35).

(f) Dans le Théorème 1 l'hypothèse de A-continuité de la multiapplication Q peut être affaiblie en remplaçant la condition (β) dans la Définition 1 par la condition (β') suivante:

(β') *quel que soit* $w \in Q(u)$, *il existe des vecteurs* w_k, *qui appartiennent à la fermeture de* $Q(u_k)$ *dans le dual* F'_A *de* F_A, *tels que*

$$\lim_k \langle Av_k , w_k - w \rangle_{F_A} = 0 ,$$

où F_A *est un Banach reflexif à injection continue dans* E' *et image dense contenant* AC_o, *et* $\langle \cdot,\cdot \rangle_{F_A}$ *dénote la dualité entre* F_A *et* F_A'.

(g) Signalons enfin , pour terminer ces remarques avant de passer aux exemples, une condition suffisante pour la A-continuité de Q sur C_o faible, lorsque $A = L - g$, avec $g \in E'$ quelconque, L est continu de E fort à E' *fort* et l'injection $E_o \subsetneq E$ est *compacte* (un operateur L comme dans (iv) est toujours continu de E fort à E' faible [8]):

(5) Q *est s.c.s. dans* E *fort*, et en outre, *pour toute suite* $u_k \in C$ *convergeante vers* $u \in C$ *dans* E *fort et pour tout* $w \in Q(u)$, *il existe des* $w_k \in Q(u_k)$ *qui convergent vers* w *dans* E *faible.* ¤

Exemples

Soit Ω un ouvert borné de R^N, de frontière Γ suffisamment régulière, cfr. [11].

Soit L un opérateur différentiel linéaire du deuxième ordre sous forme de divergence

$$L = - \sum_{i,j}^{1,N} \frac{\partial}{\partial x_i} (a_{ij}(x) \frac{\partial}{\partial x_j}) + c(x) \tag{6}$$

avec des coefficients $a_{ij} \in L^\infty(\Omega)$ qui vérifient une condition d'ellipticité uniforme dans Ω

$$\sum_{i,j}^{1,N} a_{ij}(x)\, \xi_i\, \xi_j \geq \alpha_o |\xi|^2 , \quad \xi \in R^N, \text{ x p.p. dans } \Omega , \tag{7}$$

où α_o constante > O .

Soit $c(x) \in L^\infty(\Omega)$, $c(x) \geq c_o > 0$ p.p. dans Ω . La forme bilinéaire

$$a(u , v) = \int_\Omega \Big(\sum_{i,j}^{1,N} a_{ij}(x) \frac{\partial u}{\partial x_i} \frac{\partial v}{\partial x_j} + c(x) u\, v \Big) dx \tag{8}$$

est continue et coercive sur l'espace de Sobolev $H^1(\Omega)$ et au moyen de l'identité

$$\langle Lu, v \rangle = a(u, v) \qquad u,v \in H^1(\Omega)$$

on définit L comme un opérateur linéaire continu coercif de $H^1(\Omega)$ dans son dual.

On va considérer d'abord des i.q.v. du type

$$(9)\quad \begin{cases} u \in Q(u) \\ a(u,u-w) \leq (g,u-w) \quad , \quad \forall\, w \in Q(u) \quad , \end{cases}$$

où $g \in L^2(\Omega)$ est donnée et Q est une multiapplication définie sur une partie C de $H^1(\Omega)$ à valeurs sous-ensembles convexes fermés non-vides de $H^1(\Omega)$.

On sait que la solution $v \in H^1(\Omega)$ de l'i.v.

$$(10)\quad \begin{cases} v \in Q(u) \\ a(v,v-w) \leq (g,v-w) \quad \forall\, w \in Q(u) \quad , \end{cases}$$

est *unique*, donc, la sélection variationnelle

$$v = Su \quad , \quad u \in C \quad ,$$

est maintenant une application univoque de C dans $H^1(\Omega)$.

On dénote dans la suite par u_N la solution du problème de Neumann

$$(11)\quad u_N \in H^1(\Omega) \quad , \; a(u_N\,, w) = (g,w) \quad \forall\, w \in H^1(\Omega).$$

Exemple 1

Soit

$$M : L^2(\Omega) \to L^2(\Omega)$$

qui vérifie les conditions (12) et (13) suivantes:

(12) M est *décroissant* p.p., i.e., $Mu' \leq Mu''$ p.p. dans Ω si $u'' \leq u'$ p.p. dans Ω ,

(13) $M : H^1(\Omega) \to H^1(\Omega)$ et M est *compact*, i.e., Mu_k converge fortement vers Mu dans $H^1(\Omega)$ si u_k converge faiblement vers u dans $H^1(\Omega)$.

Soit pour tout $v \in H^1(\Omega)$,

$$(14)\quad Q(v) = \{w \in H^1(\Omega) : w \leq Mv \text{ p.p. dans } \Omega\}.$$

Proposition 1 *Soit* M *vérifiant* (12) *et* (13), $g \in L^2(\Omega)$. *Alors le problème* (9) , *avec* Q *donnée par* (14), *a une solution* $u \in H^1(\Omega)$ *et* $Su_N \leq u \leq u_N$ *p.p. dans* Ω . ⌑

Dém. On utilise le Corollaire du Th. 1, avec $C = E = H^1(\Omega) = E_o$ et

$$C_o = \{v \in H^1(\Omega) : v \leq u_N \text{ p.p. dans } \Omega\} .$$

La propriété (3) est une conséquence de la décroissance de M et des théorèmes de comparaison pour les solutions des i.v., qui donnent aussi $Su_N \leq Sv$ p.p. pour tout $v \in C_o$. La propriété (2) -sous la forme suffisante (4)- est une conséquence immédiate de la hypothèse (13). ¤

Remarque Un opérateur M qui a les propriétés démandées dans la Proposition 1 est, par example

(15) $$Mu(x) = k(x) - (\text{mes } \Omega)^{-1} \int_\Omega u\, dx \quad ,$$

où $k \in H^1(\Omega)$. Pour une étude plus détaillée des i.q.v. associées à cet opérateur v. [1] . ¤

Exemple 2 (dû à la collaboration avec G. Geymonat)

Soient h, λ et ρ des fonctions données sur le bord Γ de Ω, qui appartiennent à l'espace des traces $H^{1/2}(\Gamma)$.

Pour tout v dans l'espace

$$H^1_L(\Omega) = \{v \in H^1(\Omega) : Lv \in L^2(\Omega)\} \quad ,$$

la dérivée conormale de v associée à l'opérateur L

$$\frac{\partial v}{\partial \nu} = \sum_{i,j}^{1,N} a_{ij} \frac{\partial v}{\partial x_i} \cos[\nu, x_j] \quad ,$$

ν normale extérieure à Ω, soit bien définie comme élément du dual $H^{-1/2}(\Gamma)$ de $H^{1/2}(\Gamma)$, avec

$$\left\|\frac{\partial v}{\partial \nu}\right\|_{H^{-1/2}(\Gamma)} \leq c \quad \|v\|_{H^1_L(\Omega)} \quad , \quad c > 0 \quad ,$$

où

(16) $$\|v\|_{H^1_L(\Omega)} = \left(\|v\|^2_{H^1(\Omega)} + \|Lv\|^2_{L^2(\Omega)} \right)^{1/2} \quad ,$$

(pour des théorèmes de traces de ce type v. [10] ch. 2).

Pour tout $v \in H^1_L(\Omega)$ on peut donc considérer le convexe fermé non-vide $Q(v)$ de $H^1(\Omega)$ defini par

(17) $$Q(v) = \{w \in H^1(\Omega) : w \geq h - \lambda \left(\rho, \frac{\partial v}{\partial \nu}\right)_\Gamma \text{ p.p. sur } \Gamma\}$$

où $(\cdot,\cdot)_\Gamma$ dénote le crochet de dualité entre $H^{1/2}(\Gamma)$ et $H^{-1/2}(\Gamma)$. Lorsque on considère des fonctions w de $H^1(\Omega)$ sur le bord Γ de Ω, on sous-entend le passage à la trace $w|_\Gamma$, qui appartient à l'espace $H^{1/2}(\Gamma)$.

<u>Proposition 2</u> *Soit* $\lambda \geq 0$ *et* $\rho \geq 0$ p.p. *sur* Γ. *Pour toute fonction* $g \in L^2(\Omega)$, *le problème* (9) , *avec* Q *donnée par* (17) *a une solution* $u \in H^1_L(\Omega)$. ¤

<u>Dém.</u> On applique le Corollaire du Th. 1, avec $E = H^1(\Omega)$, $C = H^1_L(\Omega)$, $E_O = H^1_L(\Omega)$ muni de la norme (16), et

$$C_o = \{v \in H^1_L(\Omega) : Lv = g \;,\; \frac{\partial v}{\partial \nu} \text{ mesure} \geq 0\}.$$

On vérifie facilement que C_o est un convexe fermé dans E_o, non-vide car $u_N \in C_o$.

C_o est stable par S : en effet, si $u \in C_o$ et $v = Su$ est la solution de l'i.v. (10), par le moyen de la formule de Green pour l'opérateur L on trouve que

$$Lv = g \text{ dans } \Omega$$

et $\frac{\partial v}{\partial \nu}$ distribution positive dans $H^{-1/2}(\Gamma)$, donc, $v \in C_o$.

En outre

$$\bigcap_{v \in C_o} Q(v) \neq \emptyset \quad ,$$

puisque, e.g., la fonction $\tilde{h} \in H^1(\Omega)$ telle que $\tilde{h} = h$ sur Γ et $L\tilde{h} = 0$, est dans $Q(v)$ pour tout $v \in C_o$, à cause des hypothèses $\lambda \geq 0$, $\rho \geq 0$ et de la positivité de $\frac{\partial v}{\partial \nu}$.

Enfin, comme

$$\|v\|^2_{H^1_L(\Omega)} = \|v\|^2_{H^1(\Omega)} + \|g\|^2_{L^2(\Omega)}$$

pour tout $v \in C_o$, la coercivité de $a(u, v)$ dans $H^1(\Omega)$ entraine la coercivité de $a(u, v)$ sur C_o dans $E_o = H^1_L(\Omega)$, i.e.

$$a(v,v)/\|v\|^2_{H^1_L(\Omega)} \to +\infty \qquad \text{si} \quad \|v\|_{H^1_L(\Omega)} \to \infty \quad , \quad v \in C_o .$$

L'hypothèse (3) du Corollaire est donc vérifiée.

Enfin, la propriété (2), sous la forme suffisante (4), est une conséquence immédiate des propriétés de continuité des traces $v \to v|_\Gamma$ et $v \to \frac{\partial v}{\partial \nu}$. En particulier, on peut vérifier la condition de s.c.i. dans (4) avec

$$w_k = w + \tilde{\lambda}[(\rho, \frac{\partial u}{\partial \nu})_\Gamma - (\rho, \frac{\partial u_k}{\partial \nu})_\Gamma],$$

où $\tilde{\lambda} \in H^1(\Omega)$, $\tilde{\lambda} = \lambda$ sur Γ : $w_k \in Q(u_k)$ et $w_k \to w$ fortement dans $H^1(\Omega)$. ¤

o O o

On va considérer dans la suite des exemples d'i.q.v. avec des données de Dirichlet sur le bord.

L'identité

$$\langle Lv, w \rangle = \sum_{i,j}^{1,N} \int_\Omega a_{ij}(x)\, v_{x_i}\, w_{x_j}\, dx, \quad v \in H^1(\Omega),\ w \in H^1_o(\Omega)$$

définit

$$L = -\sum_{i,j}^{1,N} \frac{\partial}{\partial x_i}\left(a_{ij}(x) \frac{\partial}{\partial x_j}\right), \tag{18}$$

où les coefficients a_{ij} vérifient (7), comme un opérateur linéaire continu de $H^1(\Omega)$ dans le dual $H^{-1}(\Omega)$ de $H^1_o(\Omega)$, coercif sur $H^1_o(\Omega)$ à cause de (7).

On considère des multiapplications Q d'une partie C de $H^1_o(\Omega)$ à valeurs sous-ensembles convexes fermés non-vides $Q(u)$ de $H^1_o(\Omega)$, et l'i.q.v.

$$\begin{cases} u \in Q(u) \\ \langle Lu, u - w\rangle \le (g, u - w) & \forall\, w \in Q(u) \end{cases} \tag{19}$$

où $g \in L^2(\Omega)$ est une fonction donnée.

La sélection variationnelle S est maintenant l'application (univoque) de C dans $H^1_o(\Omega)$ qui associe à $u \in C$ la solution

$$v = Su$$

de l'i.v.:

$$\begin{cases} v \in Q(u) \\ \langle Lv, v - w\rangle \le (g, v - w) & \forall\, w \in Q(u) \end{cases} \tag{20}$$

On dénote dans la suite par u_D la solution du problème de

Dirichlet

$$\begin{cases} u_D \in H^1_O(\Omega) \\ \langle Lu_D, w\rangle = (g, w) \qquad \forall\, w \in H^1_O(\Omega). \end{cases}$$

Exemple 3

Soit $M : L^2(\Omega) \to L^2(\Omega)$ vérifiant (12) et (13) comme dans l'Ex. 1 et, en outre, tel que

(21) $Mu \geq O$ p.p. sur Γ ,
pour toute $u \in H^1_O(\Omega)$, $u \leq O$ p.p. dans Ω .

Pour toute fonction v dans

(22) $C = \{v \in H^1_O(\Omega) : v \leq O$ p.p. dans $\Omega\}$,

soit

(23) $Q(v) = \{w \in H^1_O(\Omega) : w \leq Mv$ p.p. dans $\Omega\}$,

qui est un convexe fermé dans $H^1_O(\Omega)$, non-vide à cause de la condition (21).

Proposition 3 *Soit* Q *donnée par* (23) , *l'opérateur* M *vérifiant* (12), (13) *et* (21). *Pour toute fonction* $g \in L^2(\Omega)$, $g \leq O$ p.p. *dans* Ω, *le problème* (19) *a alors une solution* $u \in H^1_O(\Omega)$, *telle que* $Su_D \leq u \leq u_D$ p.p. *dans* Ω . ¤

Dém. On applique le Corollaire du Th. 1, avec $E = E_O = H^1_O(\Omega)$, C donné par (22), et

$$C_O = \{v \in H^1_O(\Omega) : v \leq u_D \text{ p.p. dans } \Omega\} .$$

Comme $g \leq o$ p.p. dans Ω , le principe de maximum nous donne $u_D \leq O$ p.p. dans Ω , donc $C_O \subset C$.

On vérifie comme dans l'Ex. 1, avec $H^1_O(\Omega)$ à la place de $H^1(\Omega)$ et u_D à la place de u_N, que l'hypothèse (3) est satisfaite et $Su_D \leq Sv$ pour tout $v \in C_O$.

La propriété (2), sous la forme suffisante (4), est encore une conséquence de la compacité de M; en particulier, on peut vérifier la condition de s.c.i. dans (4) avec

$$w_k = \inf\{w, Mu_k\} :$$

$w_k \in H^1_O(\Omega)$ puisque $Mu_k \geq O$ sur Γ , donc $w_k \in Q(u_k)$ et w_k converge fortement vers w dans $H^1_O(\Omega)$. ¤

Remarque L'opérateur (15), avec $k \in H^1(\Omega)$ tel que $k \geq 0$ p.p. sur Γ, a les propriétés demandées à M dans la Proposition 3.¤

Exemple 4

On se donne une fonction

$$g \in L^p(\Omega), \quad \text{où} \quad p > \frac{N}{2}, \quad p \geq 2$$

et un opérateur

$$M : L^2(\Omega) \to L^2(\Omega)$$

qui vérifient les hypothèses (25) et (26) suivantes, où E_o est l'espace défini par

(24) $$E_o = \{v \in H_o^1(\Omega) : Lv \in L^p(\Omega)\},$$

muni de la norme

$$\|v\|_{E_o} = (\|v\|^2_{H^1(\Omega)} + \|Lv\|^2_{L^p(\Omega)})^{1/2} :$$

(25) M est continu de E_o faible dans $L^2(\Omega)$ fort,

(26) Il existe $\nu = \nu_g \in L^p(\Omega)$ telle que pour toute fonction $v \in E_o$ qui vérifie la condition

(26') $$g \geq Lv \geq \inf\{g, \nu_g\} \text{ p.p. dans } \Omega$$

on a

$Mv \in H^{2,p}(\Omega)$, $L(Mv) \in L^p(\Omega)$,

$Mv \geq 0$ p.p. sur Γ et

$L(Mv) \geq \inf\{g, \nu_g\}$ p.p. dans Ω.

Comme exemples d'un opérateur M qui a les propriétés (25) et (26) signalons:

a) L'opérateur (15), où $k \in H^{2,p}(\Omega)$, $Lk \in L^p(\Omega)$ et $k \geq 0$ p.p. sur Γ : la condition (26) est satisfaite pour toute $g \in L^p(\Omega)$ telle que

$$g \leq 0 \text{ p.p. dans } \Omega,$$

lorsque on prend $\nu = Lk$;

b) Encore l'opérateur (15), mais avec le signe + à la place du signe $-$, et k est comme dans le cas précédent et vérifie en outre la condition

$$Lk \geq 0 \text{ p.p. dans } \Omega \quad :$$

la condition (26) est maintenant satisfaite pour toute

$$g \geq 0 \text{ p.p. dans } \Omega \quad ,$$

avec $\nu = 0$;

c) L'opérateur

$$(27) \qquad Mu(x) = k(x) + \frac{1}{\text{mes } \Omega} \int_\Omega |\text{grad } u| \, dx \; ,$$

où k est comme dans le cas (a): la condition (26) est satisfaite pour toute $g \in L^p(\Omega)$, avec $\nu = Lk$.

Remarque Tous ces opérateurs vérifient la propriété de continuité (25) d'une façon triviale. Dans l'Exemple 5 ci-dessous on considère des opérateurs M qui n'ont pas la propriété de compacité demandée dans les Ex. 1, 2, 3, mais qui vérifient la condition (25). Remarquons encore que l'opérateur (27) n'a pas des propriétés de croissance ou décroissance (v. aussi l'Ex. 2). ¤

Proposition 4 *Soit* $g \in L^p(\Omega)$, $p \geq 2$, $p > \frac{N}{2}$, *et* M *qui vérifie les hypothèses* (25) *et* (26). *Pour tout* $v \in H^1_0(\Omega)$, *soit* $Q(v)$ *le convexe* (23). *Alors, le problème* (19) *a une solution* $u \in H^1_0(\Omega)$, *telle que* $Lu \in L^p(\Omega)$. *En outre,*

$$g \geq Lu \geq \inf \{g, \nu_g\} \quad \text{p.p. } \textit{dans } \Omega$$

pour toute fonction $\nu_g \in L^p(\Omega)$ *telle que* (26) *soit verifiée.* ¤

Corollaire 1 *Soit* Ω *de classe* C^2. *Alors,* u *est continue et vérifie une condition d'Hölder sur* $\overline{\Omega}$. ¤

Corollaire 2 *Soit* Ω *de classe* C^2, $a_{ij} \in C^1(\overline{\Omega})$. *Alors,* $u \in H^1_0(\Omega) \cap H^{2,p}(\Omega)$ *et*

$$\|u\|_{H^{2,p}(\Omega)} \leq c(\|g\|_{L^p(\Omega)} + \|\nu_g\|_{L^p(\Omega)}) \quad , \quad c > 0 \; .$$

En outre, si $p > N$, *alors* $u \in C^1(\overline{\Omega})$ *et les derivées* u_{x_i} *vérifient une condition d'Hölder sur* $\overline{\Omega}$, *d'ordre* $\alpha = 1 - \frac{N}{p}$. ¤

Les corollaires de la Prop. 4 sont une conséquence directe de résultats bien connus de régularité pour le problème de Dirichlet.

Démonstration de la Proposition 4 On vérifie les hypothèses du Théorème 1, avec $E = H_0^1(\Omega)$, E_0 donné par (24), et

$$C_0 = C = \{v \in E_0 \text{ tel que } (26') \text{ soit vérifiée}\} \quad .$$

Pour tout $v \in C$, $Q(v)$ est un sous-ensemble convexe fermé non-vide de $H_0^1(\Omega)$, car $Mv \in H^1(\Omega)$ et $Mv \geq 0$ p.p. sur Γ , à cause de l'hypothèse (26).

En outre, C_0 est un convexe fermé et borné dans E_0 et C_0 est non-vide: en effet, $u_D \in C_0$. On va vérifier maintenant que C_0 est stable par S.

Soit donc $u \in C_0$ et $v = Su$ la solution de l'inéquation variationnelle (20). En conséquence de (26), on sait que $Mu \in H^{2,p}(\Omega)$ et donc, comme $p > \frac{N}{2}$ et $p \geq 2$, par le théorème d'immersion de Sobolev, on a que

$$Mu \in H^1(\Omega) \cap C^0(\overline{\Omega}) \quad ;$$

en outre, $L(Mu) \in L^p(\Omega)$ et $Mu \geq 0$ p.p. sur Γ . On peut donc appliquer l'inégalité de Lewy-Stampacchia [9] sous la forme donnée dans [11] , et on obtient l'estimation

$$g \geq Lv \geq \inf \{g, L(Mu)\} \text{ , p.p. dans } \Omega \text{ .}$$

Mais, encore par l'hypothèse (26),

$$L(Mu) \geq \inf \{g, \nu\} \text{ p.p. dans } \Omega \text{ ,}$$

donc v vérifie l'inégalité

$$g \geq Lv \geq \inf \{g, \nu\} \text{ p.p. dans } \Omega \text{ ,}$$

ce qui donne, en particulier, $Lv \in L^p(\Omega)$. Alors $v \in C_0$, ce qu'il fallait démontrer.

On vérifie maintenant l'hypothèse (2). La propriété (α) de la Déf. 1 est une conséquence de l'hypothèse de continuité (25) sur M. Pour vérifier la propriété (β), on prend

$$w_k = \inf \{w, Mu_k\} \quad .$$

Comme $w \in H_0^1(\Omega)$ et $u_k \in C_0$, on a $Mu_k \geq 0$ p.p. sur Γ , et donc $w_k \in H_0^1(\Omega)$ et $w_k \leq Mu_k$ p.p. dans Ω , c'est à dire, $w_k \in Q(u_k)$. En outre, comme u_k converge vers u faiblement dans E_0, à cause de (25) Mu_k converge fortement vers Mu dans $L^2(\Omega)$, donc w_k converge aussi fortement dans $L^2(\Omega)$ vers $\inf\{w, Mu\} = w$. D'autre part, Lu_k est borné dans $L^p(\Omega)$ et $p \geq 2$, donc

$$\lim_k (Lu_k, w_k - w) - (g, w_k - w) = 0 \ . \ \square$$

Exemple 5

Soit L un opérateur différentiel linéaire du second ordre , du type (18), uniformement elliptique dans Ω . On suppose que les coefficients de L soient dans $C^1(\overline{\Omega})$ et que Γ soit de classe C^2.

Pour tout vecteur

$$u = (u_i)_{1,\ell} \in [H_0^1(\Omega)]^\ell$$

soit

(28) $\qquad Q_i(u) = \{v_i \in H_0^1(\Omega) : v_i \leq M_i(u) \text{ p.p. dans } \Omega\}$,

où

(29) $\qquad M_i : [H_0^1(\Omega)]^\ell \to L^2(\Omega)$

sont des applications données, $i = 1, \ldots, \ell$.

On se donne encore des fonctions $g_1, \ldots, g_\ell$ dans $L^2(\Omega)$, et on considère le problème de trouver un *point de Nash* $u = (u_i)_{1,\ell}$ solution du système d'inéquations quasi-variationnelles

$$(30) \quad \begin{cases} u_i \in Q_i(u) \\ \langle Lu_i, u_i - w_i \rangle \leq (g_i, u_i - w_i) \quad \forall\, w_i \in Q_i(u),\ i = 1, \ldots, \ell . \end{cases}$$

On fait l'hypothèse

(31) $\qquad g_i \in L^p(\Omega)$, $\quad i = 1, \ldots, \ell$

où $p > \frac{N}{2}$, $p \geq 2$ et on dénote par E_i une copie du sous-espace (24) de $H_0^1(\Omega)$

(32) $\qquad E_i = \{v_i \in H_0^1(\Omega) : L v_i \in L^p(\Omega)\}$,

normé par

$$\|v_i\|_{E_i} = (\|v_i\|^2_{H^1(\Omega)} + \|L\, v_i\|^2_{L^p(\Omega)})^{1/2} \quad ;$$

à cause des hypothèses de régularité sur Ω et les coefficients de L , on a

$$E_i \subset H^{2,p}(\Omega) \; .$$

On suppose que chaque M_i vérifie des hypothèses analogues à celles de l'exemple précédent, à savoir:

(33) M_i est continu de $\prod_i^{1,\ell} E_i$ faible dans $L^2(\Omega)$ fort,

(34) Il existe $\nu_i = \nu_i(g_i) \in L^p(\Omega)$, telle que pour toute fonction $v = (v_i)_{1,} \in \prod_i^{1,\ell} E_i$ qui vérifie

$$g_i \geq L\, v_i \geq \inf\{\nu_i, g_i\} \quad \text{p.p. dans } \Omega \quad ,$$

on a $M_i v \in H^{2,p}(\Omega)$ et en outre

$M_i v \geq 0$ p.p. sur Γ , et

$L\,(M_i v) \geq \inf\{\nu_i, g_i\}$ p.p. dans Ω ,

$i = 1, \ldots\ldots, \ell$.

<u>Un exemple</u>: Soit

(35) $$M_i(u) = \beta_i - \sum_{j \neq i} u_j \qquad i,j = 1, \ldots\ldots, \ell$$

où $\beta_1, \ldots\ldots, \beta_\ell$ sont des fonctions données, avec

(36) $\beta_i \in H^{2,p}(\Omega)$, $\beta_i \geq 0$ p.p. sur Γ, $i = 1, \ldots\ldots, \ell$

Les hypothèses (33) et (34) sont satisfaites, en choisissant dans (34),

$$\nu_i = L\, \beta_i - \sum_{j \neq i} g_j \; , \qquad i,j = 1, \ldots\ldots, \ell$$

<u>Proposition 5</u> *On se donne* g_i *et* M_i, $i = 1, \ldots, \ell$ *qui vérifient* (31), (33) *et* (34), *avec* $p > \frac{N}{2}$, $p \geq 2$. *Alors le problème* (30)

admet une solution $u = (u_1, \ldots\ldots, u_\ell)$ *avec*

$$u_i \in H_0^1(\Omega) \cap H^{2,p}(\Omega) \quad , \quad i = 1, \ldots\ldots, \ell \, .$$

En outre,

$$g_i \geq L u_i \geq \inf \{ g_i, \nu_i \} \text{ p.p. } dans \ \Omega ,$$

pour toute fonction $\nu_i \in L^p(\Omega)$ *telle que* (34) *soit vérifiée.*¤

Corollaire *Si* $p > N$, *alors* $u_i \in C^1(\overline{\Omega})$ $\forall i$, *et les dérivées premières de* u_i *vérifient une condition d'Hölder sur* $\overline{\Omega}$, *d'ordre* $\alpha = 1 - \frac{N}{p}$. ¤

Dém. de la Proposition 5 On écrit le problème (30) sous la forme équivalente (19) dans l'espace produit

$$E = [H_0^1(\Omega)]^\ell \quad ,$$

avec $L : E \to E'$ donné par

$$\langle Lu, v\rangle = \sum_i^{1,\ell} \langle Lu_i, v_i\rangle \quad , \forall\, u = (u_i)_{1,\ell} \ , \ v = (v_i)_{1,\ell} \in E \ ,$$

$g = (g_i)_{1,\ell}$ et

$$Q(v) = \prod_i^{1,\ell} Q_i(v) \quad \text{pour tout} \quad v \in E \ ,$$

et on se réduit à appliquer le Théorème 1 comme dans la démonstration de la Prop. 4, avec

$$E_0 = \prod_i^{1,\ell} E_i \quad . \ ¤$$

Exemple 6

Avec les mêmes notations de l'Ex. 5, on considère le problème (30), où maintenant $Q_i(u)$ est défini par

(37) $$Q_i(u) = \{v_i \in H_0^1(\Omega) \ : \ v_i \leq \beta_i + \inf_{j \neq i}(u_j) \text{ p.p. dans } \Omega\}.$$

Si on ajoute l'hypothèse

(38) $$L\,\beta_i + L\,\beta_j \geq 0 \text{ p.p. dans } \Omega \ , \qquad \forall\, i \neq j \ ,$$

il est possible de définir un convexe fermé non-vide borné dans l'espace

$$E_o = \prod_i^{1,\ell} E_i \ , \ E_i \text{ défini par (32)} \ ,$$

qui est stable par l'*itérée* S^2 de S.

On ne peut pas donc appliquer directement le Théorème 1 dans ce cas. Toutefois, en utilisant les propriétés de croissance de S , on peut également démontrer la proposition suivante:

<u>Proposition 6</u> *On suppose que les fonctions* β_i *vérifient* (36) *et* (38) *et les fonctions* g_i *vérifient* (31) , $i = 1 \ldots\ldots \ell$, *avec* $p > \frac{N}{2}$, $p \geq 2$. *Alors le problème* (30) , *avec* Q_i *donnée par* (37), *a une solution* $u = (u_1, \ldots\ldots\ldots, u_\ell)$ *telle que*

$$u_i \in H_o^1(\Omega) \cap H^{2,p}(\Omega) \qquad \forall\, i \ .$$

De plus, si u_D^i *est, pour chaque* i , *la solution du problème de Dirichlet*

$$u_D^i \in H_o(\Omega) \quad , \quad L\, u_D^i = g_i \quad ,$$

alors la suite d'itérés $S^k u_D^i$ *converge, en décroissant, dans* $H^{2,p}(\Omega)$ *faible vers* u_{max}^i , $u_{max} = (u_{max}^1, \ldots\ldots\ldots, u_{max}^\ell)$ *étant la solution maximale du problème* (30). ⌑

Il est aussi possible de définir une suite croissante d'itérés convergeante faiblement dans $H^{2,p}(\Omega)$ vers une solution minimale du problème.

Pour d'autres détails, on renvoie à [7].

BIBLIOGRAPHIE

[1] A. Bensoussan, Journées d'Analyse Convexe à St. Pierre de Chartreuse, 1974.

[2] A. Bensoussan, J.L. Lions, Contrôle impulsionnel et inéquations quasi-variationnelles, Comptes rendus, 1973 et 1974.

[3] H. Brezis, L. Nirenberg, G. Stampacchia , A remark on Ky Fan's minimax principle, Bollettino U.M.I. (4) 6 (1972), 293-300.

[4] F. Browder, Non linear monotone operators and convex sets in Banach spaces, Bull. AMS 71 (1965), 780-785.

[5] P. Hartman, G. Stampacchia, On some non linear elliptic differential functional equations, Acta Math. 115 (1966), 271-310.

[6] J.L. Joly, U. Mosco, Sur les inéquations quasi-variationnelles, C.R. 1974.

[7] J.L. Joly, U. Mosco, Existence et régularité des solutions de certaines inéquations quasi-variationnelles, à paraître.

[8] T. Kato, Demicontinuity, hemicontinuity and monotonicity, Bull. AMS 70 (1964), 548-550; *idem*, Part II, *ibid.* 73 (1967), 886-889.

[9] H. Lewy, G. Stampacchia, On the smoothness of superharmonics which solve a minimum problem, J. Analyse Math. 23 (1970) 227-236.

[10] J.L. Lions, E. Magenes, Problèmes aux limites non homogènes et applications, vol. 1, Dunod éd., Paris 1968.

[11] U. Mosco, G.M. Troianiello, On the smoothness of solutions of unilateral Dirichlet problems, Bollettino U.M.I. (4) 8 (1973), 57-67.

PERTURBATIONS SINGULIERES dans un PROBLEME de CONTROLE OPTIMAL intervenant en BIOMATHEMATIQUE

C.M. BRAUNER[1] et P. PENEL[2]

=-=-=-=-=-=-=-=

I.- INTRODUCTION.-

1.1. - Le problème biochimique.-

Nous considérons ici une expérience réalisée au Laboratoire de technologie enzymatique de l'Université Technologique de Compiègne (voir KERNEVEZ [1], THOMAS [2] , KERNEVEZ et THOMAS [3] , et la bibliographie de ces travaux[3].

Une membrane enzymatique artificielle sépare deux compartiments. Chaque compartiment contient du substrat S , qui diffuse dans la membrane, et réagit avec l'enzyme E suivant la loi :

$$E + S \underset{k_{-1}}{\overset{k_1}{\rightleftarrows}} ES \overset{k_2}{\rightarrow} E + P \tag{1.1}$$

En fait, la réaction a lieu en présence d'un inhibiteur compétitif I , qui gêne la formation du complexe ES en formant avec le substrat un autre complexe EI .

En jouant sur la concentration d'inhibiteur au bord de la membrane, on contrôlera la réaction.

1.2. - Les équations d'état.-

La loi d'action de masse permet d'écrire les équations (sans dimension) suivantes :

(1) Laboratoire de Mathématiques et Informatique, Ecole Centrale de LYON, 69130 ECULLY (France)

(2) Département de Mathématique, Centre Scientifique et Polytechnique, Place du 8 mai 1945, 93206 Saint-Denis (France).

(3) Nous remercions J.P. KERNEVEZ et D. THOMAS de nous avoir communiqué ce problème.

(1.2) $$\frac{E_o}{K_M} \cdot \frac{\partial e}{\partial t} + \sigma(1+\frac{k_{-1}}{k_2})(es+e+z-1) + \sigma \frac{k_{-1}}{k_2}(ei-z) = O$$

(1.3) $$\frac{E_o}{K_M} \cdot \frac{\partial z}{\partial t} - \sigma \frac{k_{-1}}{k_2}(ei-z) = O$$

(1.4) $$\frac{\partial s}{\partial t} - \frac{\partial^2 s}{\partial x^2} + \sigma(1+\frac{k_{-1}}{k_2})es + \sigma \frac{k_{-1}}{k_2}(e+z-1) = O$$

(1.5) $$\frac{\partial i}{\partial t} - \frac{\partial^2 i}{\partial t} + \sigma(1+\frac{k_{-1}}{k_2})(ei-z) = O$$

où e,z,s,i représentent respectivement les concentrations de l'enzyme E , du complexe EI , du substrat S , et de l'inhibiteur I (ES a été éliminé). $K_M = \frac{k_{-1}+k_2}{k_1}$ est la constante de Michaëlis, E_o étant la concentration totale en sites actifs. Le modèle membranaire est caractérisé par le nombre σ .

Les conditions aux limites sont :

(1.6) $s(O,t) = \alpha \quad s(1,t) = \beta$, α et β donnés $> O$

(1.7) $i(O,t) = i(1,t) = v(t)$, $O \leq v(t) \leq M$.

Les concentrations initiales sont nulles pour i,s,z, alors que l'enzyme est présent à $t=O$:

(1.8) $s(x,O) = i(x,O) = O$

(1.9) $e(x,O) = 1$, $z(x,O) = O$

1.3. - L'hypothèse de l'état quasi-stationnaire .-

Selon l'hypothèse de BRIGGS et HALDAINE (1925), $\frac{\partial e}{\partial t}$ et $\frac{\partial z}{\partial t}$ sont très petits pendant la plus grande partie de la réaction, puisque l'enzyme est présent en faible quantité dans la membrane. Cette hypothèse - non satisfaisante sur le plan mathématique - conduit aux équations suivantes :

(1.10) $$\frac{\partial s}{\partial t} - \frac{\partial^2 s}{\partial x^2} + \sigma \frac{s}{1+i+s} = O$$

(1.11) $$\frac{\partial i}{\partial t} - \frac{\partial^2 i}{\partial x^2} = 0$$

Sans le terme de diffusion, l'équation (1.10) est connue sous le nom de <u>loi de Michaëlis-Menten (1913)</u> .

<u>En fait, c'est le rapport</u> $\frac{E_o}{K_M}$ <u>qui est petit</u> (une remarque analogue a été faite par HEINEKEN et al. [7] dans un cadre d'équations différentielles).

Simultanément, le rapport $\frac{k_{-1}}{k_2}$ peut être grand. C'est le cas que nous considérons ici . D'autres situations sont étudiées dans [15][16] .

1.4.- <u>Le problème de perturbation singulière</u>.-

Posant $\frac{k_{-1}}{k_2} = \frac{1}{\epsilon}$ et $\frac{E_o}{K_M} = \rho\epsilon$, les équations (1.2).....(1.5) deviennent :

(1.12) $$\rho\epsilon^2 \frac{\partial e}{\partial t} + \sigma(1+\epsilon)(es+e+z-1) + \sigma(ei-z) = 0$$

(1.13) $$\rho\epsilon^2 \frac{\partial z}{\partial t} - \sigma(ei-z) = 0$$

(1.14) $$\epsilon\left(\frac{\partial s}{\partial t} - \frac{\partial^2 s}{\partial x^2}\right) + \sigma(1+\epsilon)es + \sigma(e+z-1) = 0$$

(1.15) $$\epsilon\left(\frac{\partial i}{\partial t} - \frac{\partial^2 i}{\partial x^2}\right) + \sigma(1+\epsilon)(ei-z) = 0$$

Par combinaison linéaire, on obtient les nouvelles équations ([4])

(1.16) $$\frac{\partial s}{\partial t} - \frac{\partial^2 s}{\partial x^2} - \sigma(e+z-1) - \rho\epsilon\left(\frac{\partial e}{\partial t} + \frac{\partial z}{\partial t}\right) = 0$$

(1.17) $$\frac{\partial i}{\partial t} - \frac{\partial^2 i}{\partial x^2} + \rho\epsilon(1+\epsilon)\frac{\partial z}{\partial t} = 0$$

et donc (1.10)(1.11) apparaissent comme limites de (1.14).....(1.17) quand $\epsilon \to 0$. On verra au N° 3 que le comportement de e et z est <u>singulier</u> , celui de s et i restant régulier .

1.5- <u>Contrôle optimal</u>.-

La fonction $v(t)$, qui représente la concentration de l'inhibiteur au bord,

(4) Cette recombinaison des équations nous a été suggérée par le Professeur J.L.LIONS

est le contrôle. On cherche à réaliser un certain profil de substrat dans la membrane au cours du temps, en observant, par exemple, la concentration de substrat au milieu de la membrane (voir [13][14]) .

Naturellement, il est plus facile de contrôler le système (1.10)(1.11) (voir KERNEVEZ [1][4] et [13]) . C'est pourquoi il est intéressant de rechercher des résultats de convergence sur l'état et le contrôle quand $\epsilon \to 0$, ce qui est l'objet des Numéros 3 et 4 .

Au N° 5, on cherche, en vue des applications numériques, à écrire un système "adjoint" . On met alors en évidence deux états-adjoints différents, l'un étant associé à (1.12).....(1.15), l'autre à (1.14).....(1.17) : on constate que seul le second converge vers un état-adjoint associé au problème limite (1.10)(1.11) .

1.6- Notations.-

Pour simplifier l'écriture des équations, on utilisera les symboles suivants

(1.18) $$P = \frac{\partial}{\partial t} - \frac{\partial^2}{\partial x^2}$$

(1.19) $$D = \frac{\partial}{\partial t} .$$

On désignera par Ω l'ouvert $]0,1[$, par Q le rectangle $\Omega \times]0,1[$.

On notera en général $|.|$ la norme de l'espace $L^2(\Omega)$, et par $\|.\|$ celle de $L^2(Q)$, les normes des autres espaces étant précisées .

2.- RESULTATS D'EXISTENCE ET DE REGULARITE.-

2.1. - Le problème (P_ϵ) .-

Dans ce paragraphe, ϵ est un nombre > 0 fixé . En utilisant les notations du N°1.6, on considère dans Q le système non-linéaire suivant :

(2.1) $\rho\epsilon^2 De_\epsilon + \sigma(1+\epsilon)(e_\epsilon s_\epsilon + e_\epsilon + z_\epsilon - 1) + \sigma(e_\epsilon i_\epsilon - z_\epsilon) = 0$

(2.2) $\rho\epsilon^2 Dz_\epsilon - \sigma(e_\epsilon i_\epsilon - z_\epsilon) = 0$

(2.3) $\epsilon Ps_\epsilon + \sigma(1+\epsilon)\, e_\epsilon s_\epsilon + \sigma(e_\epsilon + z_\epsilon - 1) = 0$

(2.4) $\epsilon Pi_\epsilon + \sigma(1+\epsilon)(e_\epsilon i_\epsilon - z_\epsilon) = 0$

avec les conditions initiales et aux limites :

(2.5) $s_\epsilon(0,t) = \alpha \qquad s_\epsilon(1,t) = \beta$

(2.6) $i_\epsilon(0,t) = i_\epsilon(1,t) = v(t)$

(2.7) $s_\epsilon(x,0) = i_\epsilon(x,0) = 0$

(2.8) $z_\epsilon(x,0) = 0 \quad , \; e_\epsilon(x,0) = 1$

σ, ρ, α et β sont des nombres > 0 connus. v est une fonction mesurable donnée, telle que :

(2.9) $0 \le v(t) \le M \quad$ p.p.

Nous avons montré dans [13][14] l'existence et l'unicité d'une solution <u>positive</u> de (2.1).....(2.8)(5), à l'aide d'une méthode de semi-discrétisation par rapport au temps. Si $v \equiv 0$, le système se simplifie : il ne reste que deux équations qui ont été étudiées par KERNEVEZ et JOLY [1][5][6] à l'aide de la méthode des suites croissantes de TARTAR.

Nous avons obtenu le théorème suivant :

THEOREME 2.2. - <u>La fonction v vérifiant (2.9), le système (2.1).....(2.8) admet une solution positive unique</u> $(e_\epsilon, z_\epsilon, s_\epsilon, i_\epsilon)$ <u>dans</u>

(5) En fait, il existe quelques différences mineures entre (2.1).....(2.2) et le système étudié dans [13][14], mais ces différences ne modifient en rien les démonstrations.

$$\{H^1(O,T\ ;\ L^2(\Omega))\} \times L^2(O,T\ ;\ H^1(\Omega)) \times H^{1/2,1/4}(Q) \quad (6)$$

De plus, $O \leq e_\epsilon + z_\epsilon \leq 1$ p.p. et $O \leq s_\epsilon \leq \sup(\alpha,\beta)$, $O \leq i_\epsilon \leq M$ p.p.[7] .

Au cours de la démonstration de ce théorème (cf. [14] p.101 à 116) , nous avons décomposé s_ϵ et i_ϵ en :

(2.10) $\quad s_\epsilon = Y_\epsilon + Y \quad , \quad i_\epsilon = j_\epsilon + I$

où Y et I sont respectivement solutions dans $L^2(O,T\ ;\ H^1(\Omega))$ et dans $H^{1/2,1/4}(Q)$ de :

(2.11) $\quad \left| \begin{array}{l} PY = O \\ Y(O,t) = \alpha \quad , Y(1,t) = \beta \\ Y(x,O) = O \end{array} \right.$

(2.12) $\quad \left| \begin{array}{l} PI = O \\ I(O,t) = I(1,t) = v(t) \\ I(x,O) = O \end{array} \right.$

Les fonctions y_ϵ et j_ϵ vérifient alors les équations :

(2.13) $\quad \epsilon P y_\epsilon + \sigma(1+\epsilon)(e_\epsilon y_\epsilon + e_\epsilon Y) + \sigma(e_\epsilon + z_\epsilon - 1) = O$

(2.14) $\quad \epsilon P j_\epsilon + \sigma(1+\epsilon)(e_\epsilon j_\epsilon + e_\epsilon I - z_\epsilon) = O$

toutes les conditions initiales et aux limites étant nulles.

y_ϵ et j_ϵ sont dans l'espace Φ_O défini par :

(2.15) $\quad \Phi_O = \{\varphi \mid \varphi \in H^{2,1}(Q)\ ,\ \varphi(x,O) = \varphi(O,t) = \varphi(1,t) = O\}$ [8]

Cette décomposition de s_ϵ et i_ϵ nous sera très utile au cours des démonstra-

(6) Pour la définition de ces espaces, voir LIONS-MAGENES [8] t.2.
(7) Pour ce dernier résultat, du type " principe du maximum", voir [15] .
(8) Φ_O est muni de la topologie induite par celle de $H^{2,1}(Q)$. Notons que $\Phi_O \subset C^O(\bar{Q})$

tions ultérieures.

2.2. - Une hypothèse de régularité sur v. -

Nous supposerons désormais que :

(2.16) $v \in H^{1/4}(O,T)$ (9)

ce qui va nous donner de nouveaux résultats de régularité sur e, z et i.

THEOREME 2.2. - La fonction v vérifiant (2.9) et (2.16), la solution $(e_\epsilon, z_\epsilon, s_\epsilon, i_\epsilon)$ du système (2.1).....(2.8) est dans $\{H^1(O,T\ ;\ H^1(\Omega))\}^2 \times \{L^2(O,T\ ;\ H^1(\Omega))\}^2$ (9).

Démonstration

D'après un théorème de LIONS-MAGENES [8] t.2, J est dans $L^2(O,T\ ;\ H^1(\Omega))$, donc aussi i_ϵ.

Nous allons dériver formellement (2.1) et (2.2) pour écrire les équations qui sont vérifiées par $\frac{\partial e_\epsilon}{\partial x}$ et $\frac{\partial z_\epsilon}{\partial x}$.

On voit alors que $\frac{\partial e_\epsilon}{\partial x}$ et $\frac{\partial z_\epsilon}{\partial x}$ sont solutions dans $H^1(O,T\ ;\ L^2(\Omega))$ de :

(2.17) $$\rho\epsilon^2 D\left(\frac{\partial e_\epsilon}{\partial x}\right) + \sigma\frac{\partial e_\epsilon}{\partial x}\left[(1+\epsilon)(1+s_\epsilon) + i_\epsilon\right] + \sigma\epsilon\frac{\partial z_\epsilon}{\partial x} + \sigma e_\epsilon\left[(1+\epsilon)\frac{\partial s_\epsilon}{\partial x} + \frac{\partial i_\epsilon}{\partial x}\right] = O,$$

(2.18) $$\frac{\partial e_\epsilon}{\partial x}(x,O) = O$$

(2.19) $$\rho\epsilon^2 D\left(\frac{\partial z_\epsilon}{\partial x}\right) + \sigma\frac{\partial z_\epsilon}{\partial x} - \sigma\frac{\partial e_\epsilon}{\partial x} i_\epsilon - \sigma e_\epsilon\frac{\partial i_\epsilon}{\partial x} = O$$

(2.20) $$\frac{\partial z_\epsilon}{\partial x}(x,O) = O\ .$$

Ce raisonnement formel peut être justifié de la façon suivante : on régularise

(9) La fonction v n'est pas nécessairement continue (Cf. LIONS-MAGENES [8] t.1), e_ϵ et z_ϵ sont dans $C^0(\bar{\Omega})$, $s_\epsilon(t)$ et $i_\epsilon(t)$ sont dans $C^0(\bar{\Omega})$ p.p., ces résultats correspondant bien à la situation physique.

α , β , v , en les approchant dans $H^{1/4}(O,T)$ par des suites α_n , β_n , v_n d'éléments de $\mathcal{D}(]O,T[)$. On définit alors une suite $(e_\epsilon^n , z_\epsilon^n , s_\epsilon^n , i_\epsilon^n)$ de solutions de (2.1)..... (2.8) qui sont dans $\left\{C^\infty(\bar{Q})\right\}^4$, donc pour lesquelles la dérivation de (2.1)(2.2) a un sens. On établit facilement des estimations sur $\frac{\partial e_\epsilon^n}{\partial x}$ et $\frac{\partial z_\epsilon^n}{\partial x}$ indépendantes de n , à l'aide du lemme de GRONWALL , puis on passe à la limite de manière classique (voir [15] pour une démonstration détaillée).

2.3. - Le problème (P_o).-

Nous désignons ainsi le problème "modèle" étudié par KERNEVEZ dans [1][4], et par les auteurs dans [13]:

$$Ps_o + \sigma \frac{s_o}{1+i_o+s_o} = O \quad (10) \tag{2.21}$$

$$Pi_o = O \tag{2.22}$$

$$s_o(O,t) = \alpha \ , \ s_o(1,t) = \beta \tag{2.23}$$

$$i_o(O,t) = i_o(1,t) = v(t) \tag{2.24}$$

$$s_o(x,O) = i_o(x,O) = O \tag{2.25}$$

A partir des résultats donnés dans [1][13], on déduit facilement le théorème suivant :

THEOREME 2.3. - <u>La fonction v vérifiant (2.9) et (2.16), le système (2.21)....</u> <u>(2.24) admet une solution positive unique dans</u> $\left\{L^2(O,T ; H^1(\Omega))\right\}^2$. <u>De plus</u> , $O \le s_o \le \sup(\alpha,\beta)$ et $O \le i_o \le M$ p.p.

<u>Définissons</u> maintenant e_o et z_o par :

$$e_o = \frac{1}{1+i_o+s_o} \tag{2.26}$$

$$z_o = e_o i_o \ . \tag{2.27}$$

---

(10) Naturellement, on peut décomposer s_o en y_o+Y .

THEOREME 2.4. - e_o et z_o sont dans $L^2(O,T\,;\,H^1(\Omega)) \cap H^1(O,T\,;\,L^2_{loc}(\Omega))$.

Démonstration :

$\frac{\partial e_o}{\partial x}$ et $\frac{\partial z_o}{\partial x}$ sont dans $L^2(Q)$, puisque

(2.28) $$\frac{\partial e_o}{\partial x} = - \frac{1}{(1+i_o+s_o)^2}\left(\frac{\partial i_o}{\partial x} + \frac{\partial s_o}{\partial x}\right)$$

(2.29) $$\frac{\partial z_o}{\partial x} = \frac{\partial e_o}{\partial x} i_o + e_o \frac{\partial i_o}{\partial x}$$

Par contre, De_o et Dz_o ne sont pas dans $L^2(Q)$:

(2.30) $$De_o = - \frac{1}{(1+i_o+s_o)^2}\,(Di_o + Ds_o)$$

(2.31) $$Dz_o = De_o i_o + e_o Di_o \quad .$$

En effet, Di_o et Ds_o sont seulement dans $L^2(O,T\,;\,L^2_{loc}(\Omega))$.

Soit $\varphi \in \mathscr{D}(\Omega)$. Formons l'équation vérifiée par $s_o\varphi$

(2.32) $$P(s_o\varphi) = - \frac{\sigma s_o \varphi}{1+i_o+s_o} - s_o \frac{\partial^2 \varphi}{\partial x^2} - 2\frac{\partial s_o}{\partial x}\frac{\partial \varphi}{\partial x} \in L^2(Q)$$

(2.33) $$s_o\varphi(O,t) = s_o\varphi(1,t) = s_o\varphi(x,O) = O$$

d'où :

(2.34) $$Ds_o \,.\, \varphi = D(s_o\varphi) \in L^2(Q)$$

3. - CONVERGENCE DE $(e_\epsilon, z_\epsilon, s_\epsilon, i_\epsilon)$ VERS (e_o, z_o, s_o, i_o) .-

3.1. - Orientation.-

Dans tout ce N° , la fonction v est fixée dans $H^{1/4}(O,T)$ (avec (2.9)).

Mais les résultats s'étendront au cas où v est un contrôle optimal (N°4).

Nous allons montrer que $(e_\epsilon, z_\epsilon, s_\epsilon, i_\epsilon) \to (e_o, z_o, s_o, i_o)$ dans

$\left\{L^2(O,T\,;\,H^1(\Omega))\right\}^4$ fort quand $\epsilon \to O$. Mais alors que le comportement de s_ϵ et

i_ϵ est régulier, e_ϵ et z_ϵ ont au contraire un comportement singulier, puisque e_o et z_o sont seulement dans $H^1(O,T\,;L^2_{loc}(\Omega)) \cap L^2(O,T\,;H^1(\Omega))$ tandis que e_ϵ et z_ϵ sont dans $H^1(Q)$ (et même dans $H^1(O,T\,;H^1(\Omega))$.

Il apparaît donc une couche singulière au voisinage du bord de la membrane ([11]), couche que l'on peut étudier en régularisant e_o et z_o, et en introduisant des correcteurs internes, comme dans LIONS [11] .

Par contre, il n'apparait pas de couche limite au sens usuel, car $e_o(x,O) = 1$ et $z_o(x,O) = O$.

Nous nous contentons de donner ici les résultats de convergence, en renvoyant à [15][16] pour l'introduction de correcteurs et de développements asymptotiques, selon les techniques de LIONS [11] et d'ECKHAUS [12] .

3.2. - Le problème $(\hat{P}_\epsilon)$

En recombinant les équations (2.1).....(2.4) , on obtient deux nouvelles équations :

(3.1) $$Ps_\epsilon - \sigma(e_\epsilon+z_\epsilon-1) - \rho\epsilon(De_\epsilon+Dz_\epsilon) = O$$

(3.2) $$Pi_\epsilon + \rho\epsilon(1+\epsilon)Dz_\epsilon = O$$

En ajoutant les équations (2.3) et (2.4) :

(3.3) $$\epsilon Ps_\epsilon + \sigma(1+\epsilon)e_\epsilon s_\epsilon + \sigma(e_\epsilon+z_\epsilon-1) = O$$

(3.4) $$\epsilon Pi_\epsilon + \sigma(1+\epsilon)(e_\epsilon i_\epsilon - z_\epsilon) = O$$

et les conditions initiales et aux limites (2.5).....(2.8) , on obtient un nouveau problème $(\hat{P}_\epsilon)$ équivalent au problème initial (P_ϵ) .

Naturellement, on obtient un système analogue sur les "parties homogènes" y_ϵ

(11) Si v est plus régulière, la couche est concentrée au voisinage des "coins" $(x = O, t=O)$ et $(x=1, t=1)$.

j_ϵ introduites en (2.10).

(3.5) $Py_\epsilon - \sigma(e_\epsilon + z_\epsilon - 1) - \rho\epsilon(De_\epsilon + Dz_\epsilon) = 0$

(3.6) $Pj_\epsilon + \rho\epsilon(1+\epsilon)Dz_\epsilon = 0$

(3.7) $\epsilon Py_\epsilon + \sigma(1+\epsilon)(e_\epsilon y_\epsilon + e_\epsilon Y) + \sigma(e_\epsilon + z_\epsilon - 1) = 0$

(3.8) $\epsilon Pj_\epsilon + \sigma(1+\epsilon)(e_\epsilon j_\epsilon + e_\epsilon I - z_\epsilon) = 0$

toutes les conditions initiales et aux limites étant nulles.

3.3. - Les premières estimations indépendantes de ϵ .-

Nous considérons une suite $\epsilon \to 0$. Tout d'abord, la suite $(e_\epsilon, z_\epsilon, s_\epsilon, i_\epsilon)$ est bornée dans $\{L^\infty(Q)\}^4$ grâce au théorème 2.1 :

(3.9) $0 \le e_\epsilon \le 1$, $0 \le z_\epsilon \le 1$

(3.10) $0 \le s_\epsilon \le \sup(\alpha, \beta)$, $0 \le i_\epsilon \le M$ p.p.

ce qui donne également des estimations dans $L^\infty(Q)$ pour y_ϵ et j_ϵ .

En particulier, e_ϵ , z_ϵ , y_ϵ, j_ϵ étant dans $C^0(\bar{Q})$,

(3.11) $e_\epsilon(T)$, $z_\epsilon(T)$, $y_\epsilon(T)$, $j_\epsilon(T)$ sont bornés dans $L^\infty(\Omega)$

Enfin, s_ϵ et i_ϵ étant dans $C^0([0,T]; H^1(\Omega))$,

(3.12) $s_\epsilon(T)$ et $i_\epsilon(T)$ sont bornés dans $L^2(\Omega)$.

En utilisant (3.9) et (3.10) on obtient les premières estimations sur les dérivées :

De (3.7)(3.8), on tire :

(3.13) ϵPy_ϵ et ϵPj_ϵ sont bornés dans $L^2(Q)$.

D'où, P étant un isomorphisme de $L^2(Q)$ sur Φ_0 (voir (2.15)),

(3.14) ϵDy_ϵ et ϵDj_ϵ sont bornés dans $L^2(Q)$.

D'autre part, on déduit de (2.1)(2.2) que

(3.15) $\epsilon^2 De_\epsilon$ et $\epsilon^2 Dz_\epsilon$ sont bornés dans $L^2(Q)$.

3.4. - Estimations dans $L^2(O,T\ ;\ H^1(\Omega))$.-

LEMME 3.1. - Quand $\epsilon \to O$, la suite $(e_\epsilon, z_\epsilon, z_\epsilon, i_\epsilon)$ reste bornée dans

$$\left\{ L^2(O,T\ ;\ H^1(\Omega)) \right\}^4 .$$

Démonstration :

Multiplions l'équation (3.6) par j_ϵ , et intégrons sur Q (les notations sont celles du N°1.6).

$$(3.16) \qquad \tfrac{1}{2}\,|j_\epsilon(T)|^2 + \|\frac{\partial j_\epsilon}{\partial x}\|^2 = -\rho\epsilon(1+\epsilon)\int_\Omega z_\epsilon(T)\,j_\epsilon(T)dx + \rho(1+\epsilon)\int_Q z_\epsilon\,.\,\epsilon Dj_\epsilon$$

En utilisant (3.11) et (3.14) , et en effectuant un travail analogue sur (3.5), on voit que :

$$(3.17) \qquad j_\epsilon \text{ et } y_\epsilon \text{ sont bornés dans } L^2(O,T\ ;\ H^1(\Omega))$$

et par conséquence i_ϵ et s_ϵ sont bornés dans le même espace.

Rapportons nous maintenant aux équations (2.17).....(2.20) ; multiplions (2. 17) par $\frac{\partial e_\epsilon}{\partial x}$, et intégrons (notons $m_\epsilon = \frac{\partial e_\epsilon}{\partial x}$, $n_\epsilon = \frac{\partial z_\epsilon}{\partial x}$):

$$(3.18) \qquad \rho\epsilon^2|m_\epsilon(T)|^2 + 2\sigma\int_Q m_\epsilon^2\left[(1+\epsilon)(1+s)_\epsilon + i_\epsilon\right]dx\,dt$$
$$+ 2\sigma\int_Q m_\epsilon.\,\epsilon n_\epsilon dxdt + 2\sigma\int_Q m_\epsilon e_\epsilon\left[(1+\epsilon)\frac{\partial s_\epsilon}{\partial x} + \frac{\partial i_\epsilon}{\partial x}\right]dxdt = O$$

Grâce à la positivité de s_ϵ et i_ϵ ;

$$(3.19) \qquad 2\sigma\|m_\epsilon\|^2 \le \sigma\|m_\epsilon\|^2 + \sigma\|\epsilon n_\epsilon\|^2 + \frac{\sigma}{4}\|m_\epsilon\|^2 + C_1 .$$

D'autre part, multiplions (2.19) par $\epsilon^2\,\frac{\partial z_\epsilon}{\partial x} = \epsilon^2 n_\epsilon$, et intégrons sur Q :

$$(3.20) \qquad \rho\epsilon^4|n_\epsilon(T)|^2 + 2\sigma\|\epsilon n_\epsilon\|^2 - \sigma\int_Q \epsilon i_\epsilon\,.\,m_\epsilon.\,\epsilon n_\epsilon dx\,dt$$
$$- \sigma\epsilon\int_Q e_\epsilon\frac{\partial i_\epsilon}{\partial x}.\,\epsilon n_\epsilon dx\,dt = O$$

$$(3.21) \qquad 2\sigma\|\epsilon n_\epsilon\|^2 \le \sigma\epsilon M(\|m_\epsilon\|^2 + \|\epsilon n_\epsilon\|^2) + \frac{\sigma}{4}\|\epsilon n_\epsilon\|^2 + C_2 .$$

Ajoutons les inégalités (3.19) et (3.21) . Il vient :

$$(3.22) \qquad \frac{\sigma(1-\epsilon M)}{4}(\|m_\epsilon\|^2 + \|\epsilon n_\epsilon\|^2) \le C_1 + C_2 .$$

Donc, quand $\epsilon \to 0$,

(3.23) $m_\epsilon = \frac{\partial e_\epsilon}{\partial x}$ est borné dans $L^2(Q)$.

En multipliant alors (2.19) par $\frac{\partial z_\epsilon}{\partial x}$, on a alors

(3.24) $\frac{\partial z_\epsilon}{\partial x}$ est borné dans $L^2(Q)$

ce qui achève la démonstration du lemme 3.1.

3.5 - Introduction de puissances fractionnaires.-

Comme dans la démonstration précédente, on va utiliser de manière essentielle le fait que dans l'équation(2.1) z_ϵ n'intervient qu'avec un coefficient ϵ , ce qui crée un décalage d'un "ϵ" entre (2.1) et (2.2) .

LEMME 3.2. - $\epsilon^{3/2} De_\epsilon$ et $\epsilon^{3/2+1/4} Dz_\epsilon$ sont bornés dans $L^2(Q)$.

COROLLAIRE 3.1. -(12) $\epsilon^{1/2+1/4} Pj_\epsilon$ et $\epsilon^{1/2+1/4} Py_\epsilon$ sont bornés dans $L^2(Q)$.

Démonstration du lemme 3.2.

On multiplie l'équation (2.1) par ϵDe_ϵ , et on intègre sur Q :

(3.25) $$\rho\epsilon^3 \| De_\epsilon \|^2 + \frac{\sigma\epsilon}{2}(1+\epsilon)|e_\epsilon(T)-1|^2 = -\sigma \int_Q z_\epsilon \cdot \epsilon^2 De_\epsilon dxdt$$

$$- \sigma\epsilon \int_Q e_\epsilon De_\epsilon \left[(1+\epsilon)s_\epsilon + i_\epsilon\right] dxdt \ .$$

Mais :

(3.26) $$- \sigma\epsilon \int_Q e_\epsilon De_\epsilon \left[(1+\epsilon)s_\epsilon + i_\epsilon\right] dxdt$$

$$= -\frac{\sigma\epsilon}{2}\int_\Omega e_\epsilon^2(T)\left[(1+\epsilon)s_\epsilon(T) + i_\epsilon(T)\right]dx + \sigma\epsilon \int_Q e_\epsilon^2 \left[(1+\epsilon)\, dY + DI\right] dxdt$$

$$+ \sigma \int_Q e_\epsilon^2 \left[(1+\epsilon)\, \epsilon Dy_\epsilon + \epsilon Dj_\epsilon\right] dxdt.$$

(12) Le corollaire se déduit immédiatement du lemme grâce aux équations (3.5)(3.6) .

Grâce à (3.11)(3.12)(3.14) , et en remarquant que e_ϵ^2 est borné dans $L^2(0,T;H^1$ on sait borner le second membre de (3.26) ; en utilisant encore (3.15) , le second membre de (3.25) est donc borné, d'où :

(3.27) $$\epsilon^3 \|De_\epsilon\|^2 \leq \text{Cste}$$

ce qui démontre la première moitié du lemme 3.2.

Multiplions maintenant l'équation (2.2) par $\epsilon^{3/2} Dz_\epsilon$, et intégrons sur Q :

(3.28) $$\rho\epsilon^{3+1/2}\|Dz_\epsilon\|^2 + \sigma\frac{\epsilon^{3/2}}{2}|z_\epsilon(T)|^2 = \sigma\epsilon^{3/2}\int_\Omega e_\epsilon(T)i_\epsilon(T)z_\epsilon(T)dx$$
$$- \sigma\int_Q \epsilon^{3/2} De_\epsilon i_\epsilon z_\epsilon \, dxdt .$$

On conclut à l'aide de (3.11)(3.12) et de (3.27) que

(3.29) $$\epsilon^{3+1/2}\|Dz_\epsilon\|^2 \leq \text{Cste}$$

et le lemme est démontré.

3.6. - Extraction de sous-suites convergentes.-

On est maintenant en mesure d'extraire des sous-suites convergentes qui seront notées comme les suites initiales (les suites entières convergeront puisque le problème (P_0) admet une solution unique).

LEMME 3.3. - Quand $\epsilon \to 0$,

(3.30) $$j_\epsilon \to j_0 = 0 \text{ dans } L^2(0,T\,;H^{1-\eta}(\Omega)) \text{ fort } (\eta > 0)$$

(3.31) $$y_\epsilon \to y_0 \text{ dans } L^2(0,T\,;H^{1-\eta}(\Omega)) \text{ fort .}$$

Démonstration :

Les suites j_ϵ et z_ϵ étant bornées dans $L^2(0,T\,;H^1(\Omega))$, a fortiori :

(3.32) $$j_\epsilon + \rho\epsilon(1+\epsilon)z_\epsilon \text{ est borné dans } L^2(0,T\,;H^1(\Omega)) .$$

D'autre part, on déduit de (3.5) que :

(3.33) $$D(j_\epsilon + \rho\epsilon(1+\epsilon)z_\epsilon) = \frac{\partial^2 j_\epsilon}{\partial x^2} \text{ est borné dans } L^2(0,T\,;H^{-1}(\Omega)).$$

D'après un théorème de compacité (voir par exemple LIONS [10]):

(3.34) $\quad j_\epsilon + \rho\epsilon(1+\epsilon)z_\epsilon \rightarrow j_o$ dans $L^2(O,T\ ;\ H^{1-\eta}(\Omega))$fort , $\eta > O$

et il est facile de voir que $j_o = O$. Comme $\epsilon z_\epsilon \rightarrow O$ dans $L^2(O,T\ ;\ H^1(\Omega))$, $j_\epsilon \rightarrow O$ dans $L^2(O,T\ ;\ H^{1-\eta}(\Omega))$ fort.

De la même manière, on a (3.31), en déduisant de (3.5) que :

(3.35) $\quad D(y_\epsilon - \rho\epsilon(e_\epsilon + z_\epsilon)) = \dfrac{\partial^2 y_\epsilon}{\partial x^2} + \sigma(e_\epsilon + z_\epsilon - 1)$ est borné dans $L^2(Q)$.

LEMME 3.4.- <u>Quand $\epsilon \rightarrow O$:</u>

(3.36) $\quad e_\epsilon \rightarrow e_o$ dans $L^2(Q)$ fort

(3.37) $\quad z_\epsilon \rightarrow z_o$ dans $L^2(Q)$ fort

<u>démonstration</u> :

On combine les équations (3.7) et (3.8) pour obtenir :

(3.38) $$e_\epsilon = \frac{\sigma(1+\epsilon z_\epsilon) - \epsilon P y_\epsilon - \epsilon P j_\epsilon}{\sigma + \sigma(1+\epsilon)(j_\epsilon + I + y_\epsilon + Y)} \ .$$

Sachant que $\epsilon^{1/2+1/4} P j_\epsilon$ et $\epsilon^{1/2+1/4} P y_\epsilon$ restent bornés dans $L^2(Q)$, (3.36) est évident grâce au lemme 3.3.

Ensuite on obtient (3.37) en calculant z_ϵ :

(3.39) $$z_\epsilon = \frac{\epsilon}{\sigma(1+\epsilon)} P j_\epsilon - e_\epsilon j_\epsilon - e_\epsilon I \ .$$

LEMME 3.5. : <u>Quand $\epsilon \rightarrow O$</u> ,

(3.40) $\quad j_\epsilon \rightarrow O$ dans $L^2(O,T : H^1(\Omega))$ fort

(3.41) $\quad y_\epsilon \rightarrow O$ dans $L^2(O,T : H^1(\Omega))$ fort .

COROLLAIRE 3.2. :

(3.42) $\quad i_\epsilon \rightarrow i_o$ dans $L^2(O,T\ ;\ H^1(\Omega))$ fort

(3.43) $\quad s_\epsilon \rightarrow s_o$ dans $L^2(O,T\ ;\ H^1(\Omega))$ fort .

<u>Démonstration de lemme 3.5 :</u>

Notons que :

(3.44) $$Py_o - \sigma(e_o+z_o-1) = 0$$

et formons la différence entre (3.5) et (3.44)

(3.45) $$P(y_\epsilon-y_o) - \sigma(e_\epsilon-e_o) - \sigma(z_\epsilon-z_o) = \rho\epsilon D(e_\epsilon+z_\epsilon) \ .$$

On multiplie (3.45) par $y_\epsilon-y_o$, et on intègre sur Q .

Seule la convergence du terme $\int_Q \epsilon D(e_\epsilon+z_\epsilon)(y_\epsilon-y_o)\,dxdt$ n'est pas immédiate :

(3.46) $$\int_Q \epsilon D(e_\epsilon+z_\epsilon)(y_\epsilon-y_o)dxdt = \epsilon\int_\Omega (e_\epsilon(T)+z_\epsilon(T)(y_\epsilon(T)-y_o(T))dx$$
$$- \int_Q (e_\epsilon+z_\epsilon).\epsilon Dy_\epsilon dxdt + \epsilon\int_Q (e_\epsilon+z_\epsilon)Dy_o dxdt \ .$$

On vérifie que toutes les intégrales du second membre tendent vers 0 , ce qui démontre (3.43) . On démontre (3.42) de la même manière.

LEMME 3.6. : <u>Quand $\epsilon \to 0$</u> ,

(3.47) $$e_\epsilon \to e_o \quad \text{dans} \quad L^2(0,T\ ;\ H^1(\Omega)) \quad \text{fort}$$

(3.48) $$z_\epsilon \to z_o \quad \text{dans} \quad L^2(0,T\ ;\ H^1(\Omega)) \quad \text{fort} \ .$$

<u>Démonstration :</u>

Il faut revenir aux équations (2.17).....(2.20), avec les notations du lemme 3.1

Multiplions (2.17) par $m_\epsilon = \frac{\partial e_\epsilon}{\partial x}$, intégrons sur Q , en faisant apparaître $m_\epsilon - m_o(m_o = \frac{\partial e_o}{\partial x})$:

(3.49) $$\rho\frac{\epsilon^2}{2}\ |m_\epsilon(T)|^2 + \sigma\int_Q (m_\epsilon - m_o)^2\left[(1+\epsilon)(1+s_\epsilon)+i_\epsilon\right]dxdt$$
$$= \sigma\int_Q (m_o^2 - 2m_\epsilon m_o)\left[(1+\epsilon)(1+s_\epsilon)+i_\epsilon\right]dxdt$$
$$- \sigma\int_Q \epsilon n_\epsilon m_\epsilon\, dxdt - \sigma\int_Q e_\epsilon m_\epsilon\left[(1+\epsilon)\frac{\partial s_\epsilon}{\partial x} + \frac{\partial i_\epsilon}{\partial x}\right]dxdt \ .$$

Il est facile de voir que le membre de droite converge vers

$$- \sigma\int_Q m_o^2(1+i_o+s_o)dxdt \ - \sigma\int_Q e_o m_o(\frac{\partial i_o}{\partial x} + \frac{\partial s_o}{\partial x})\,dxdt$$

qui est nul d'après (2.28).

On vérifie ensuite sur (2.19) que $\frac{\partial z_\epsilon}{\partial x} \to \frac{\partial z_o}{\partial x}$ dans $L^2(Q)$ fort .

3.7.- Synthèse des résultats .-

Nous résumons tous les résultats obtenus dans un théorème :

THEOREME 3.1. - Quand $\epsilon \to O$, la fonction v étant fixée dans $H^{1/4}(O,T)$ et vérifiant (2.9), la solution $(e_\epsilon , z_\epsilon , s_\epsilon , i_\epsilon)$ des problèmes équivalents (P_ϵ) et $(\hat{P}_\epsilon)$ converge dans $\{L^2(O,T ; H^1(\Omega))\}^4$ fort vers la solution (e_o , z_o , s_o , i_o) du problème (P_o) .

4.- CONTROLE DU SYSTEME . CONVERGENCE D'UNE SUITE DE CONTROLES OPTIMAUX.-

Le contrôle optimal du problème (P_ϵ) (au sens de LIONS [9]) , a été étudié par les auteurs dans [13][14] . Le contrôle du problème (P_o) , beaucoup plus facile à réaliser numériquement , a fait l'objet de plusieurs expérimentations (voir KERNEVEZ [1] [4] , ainsi que [13]).

4.1. - Contrôle du problème (P_ϵ).-

Le contrôle est la fonction v . On définit l'ensemble des contrôles admissibles.

$$\mathcal{U}_{ad} = \left\{ v \mid O \le v(t) \le M \text{ p.p. } , \ |D^{1/4} v| \le C \right\} \tag{4.1}$$

et on prend comme observation la fonction $s_\epsilon(\frac{1}{2},t)$ qui représente la concentration du substrat au milieu de la membrane, et qui est un élément de $L^2(O,T)$.

Un élément s_d "désiré" étant donné dans $L^2(O,T)$, on introduit la fonction coût $J_\epsilon(v)$:

$$(4.2) \qquad J_\epsilon(v) = \int_O^T (s_\epsilon(v\,;\tfrac{1}{2}) - s_d)^2\,dt \;.$$

Un contrôle optimal u_ϵ est un élement de $\mathcal{U}_{ad}$ qui réalise le minimum de J_ϵ.

THEOREME 4.1. - ϵ <u>étant un nombre</u> $> O$ <u>fixé, il existe au moins un contrôle optimal</u> u_ϵ.

4.2. - Contrôle du problème (P_o).-

L'observation est maintenant la fonction $s_o(\frac{1}{2},t)$ qui est aussi dans $L^2(O,T)$. On introduit la fonction coût $J_o(v)$:

$$(4.3) \qquad J_o(v) = \int_O^T (s_o(v\,;\tfrac{1}{2}) - s_d)^2 dt \;.$$

Un contrôle optimal u_o est un élément de $\mathcal{U}_{ad}$ qui réalise le minimum de J_o.

THEOREME 4.2. - <u>Il existe au moins un contrôle optimal</u> u_o.

4.3. - Convergence d'une suite de contrôles optimaux.-

A chaque $\epsilon > O$, on associe <u>un</u> contrôle optimal u_ϵ du problème (P_ϵ). On s'intéresse au comportement de la suite u_ϵ quand $\epsilon \to O$.

THEOREME 4.3. - <u>Quand</u> $\epsilon \to O$, <u>on peut extraire d'une suite de contrôles optimaux</u> u_ϵ <u>une sous-suite</u> $u_{\epsilon'}$ <u>qui converge dans</u> $L^2(O,T)$ <u>vers un contrôle optimal</u> u_o <u>du problème</u> (P_o). <u>Simultanément</u>, $(e_{\epsilon'}(u_{\epsilon'}),\, z_{\epsilon'}(u_{\epsilon'}),\, s_{\epsilon'}(u_{\epsilon'})\,;\, i_{\epsilon'}(u_{\epsilon'}) \to (e_o(u_o), z_o(u_o), s_o), i_o(u_o))$ <u>dans</u> $\{L^2(Q)\}^2 \times L^2(O,T\,;\,H^1(\Omega)) \times L^2(O,T\,;\,H^{1-\eta}(\Omega))$ <u>fort</u>.

Démonstration :

A chaque u_ϵ, on associe la fonction $I_\epsilon(u_\epsilon)$ définie par :

$$(4.4)\quad \begin{cases} PI_\epsilon = O \\ I_\epsilon(O,t) = I_\epsilon(I,t) = u_\epsilon(t) \\ I_\epsilon(x,O) = O \end{cases}$$

La suite u_ϵ étant bornée dans $H^{1/4}(O,T)$,

(4.5) $\quad I_\epsilon(u_\epsilon)$ est bornée dans $L^2(O,T\ ;\ H^1(\Omega))$.

On peut extraire une sous-suite $u_{\epsilon'}$, telle que

(4.6) $\quad u_{\epsilon'} \to u_o$ dans $L^2(O,T)$ fort

(4.7) $\quad I_{\epsilon'}(u_{\epsilon'}) \to i_o$ dans $L^2(O,T\ ;\ H^{1-\eta}(\Omega))$ fort .

Les lemmes 3.1. à 3.5. sont encore vrais pour les suites $e_{\epsilon'}(u_{\epsilon'})$, $z_{\epsilon'}(u_{\epsilon'})$, $y_{\epsilon'}(u_{\epsilon'})$, $s_{\epsilon'}(u_{\epsilon'})$, $j_{\epsilon'}(u_{\epsilon'})$, $i_{\epsilon'}(u_{\epsilon'})$.

On a alors :

(4.8) $\quad s_{\epsilon'}(u_{\epsilon'}) \to s_o$ dans $L^2(O,T\ ;\ H^1(\Omega))$ fort

(4.9) $\quad i_{\epsilon'}(u_{\epsilon'}) = j_{\epsilon'}(u_{\epsilon'}) + I_{\epsilon'} \to i_o$ dans $L^2(O,T\ ;\ H^{1-\eta}(\Omega))$ fort

(4.10) $\quad e_{\epsilon'}(u_{\epsilon'}) \to e_o$ dans $L^2(Q)$ fort

(4.11) $\quad z_{\epsilon'}(u_{\epsilon'}) \to z_o$ dans $L^2(Q)$ fort

et on vérifie facilement que $s_o = s_o(u_o), i_o = i_o(u_o)$, etc.....

Il reste à montrer que u_o est un contrôle optimal du problème (P_o) ; pour cela, il suffit de passer à la limite quand $\epsilon' \to O$ dans l'inégalité :

$$(4.12)\quad \int_O^T (s_{\epsilon'}(\tfrac{1}{2},u_{\epsilon'}) - s_d)^2\,dt \le \int_O^T (s_{\epsilon'}(\tfrac{1}{2},v) - s_d)^2\,dt$$

où v est une élement quelconque fixé de $\mathcal{U}ad$.

5.- DEFINITION DES ETATS-ADJOINTS.-

Pour les applications numériques, il est important de savoir calculer la valeur du gradient de la fonction coût en un contrôle v fixé.

Dans ce but, on introduit en général un état-adjoint, qui permet de donner une expression simple du gradient.

Nous allons écrire les problèmes (P_ϵ) et $(\hat{P}_\epsilon)$ sous une forme plus abstraite qui va nous permettre en utilisant des résultats de différentiabilité démontrés dans [13] [14], de mettre en évidence deux états-adjoints différents.

5.1. - Résultats de différentiabilité relatifs au problème (P_ϵ)

Soit v un élément de $\mathcal{U}$ad fixé tel que :

(5.1) il existe $\gamma > O$ t.q. $v(t) \geq \gamma$ p.p. [13]

$I = I(v)$ désigne la solution de (2.12). Ecrivons le problème (P_ϵ) sous la forme générale :

(5.2) $$F_\epsilon(e_\epsilon, z_\epsilon, y_\epsilon, j_\epsilon ; I) = O$$

Pour simplifier l'écriture, adoptons les notations suivantes :

(5.3) $$E = H^1(O,T ; L^2(\Omega))$$

(5.4) $$D_O = \left\{ \varphi | \varphi \in E, \varphi(x,O) = O \right\}$$

(5.5) $$\mathbb{L}^2 = \left\{ L^2(Q) \right\}^4 .$$

F_ϵ est alors une application de $E^2 \times \Phi_O^2 \times L^\infty(Q)$ dans $\mathbb{L}^2$. Notons aussi :

(5.6) $$\omega_\epsilon = (e_\epsilon, z_\epsilon, y_\epsilon, j_\epsilon) .$$

On définit $\frac{\partial F_\epsilon}{\partial \omega}(\omega_\epsilon, I)$ comme étant l'application linéaire

(5.7) $$\frac{\partial F_\epsilon}{\partial \omega}(\omega_\epsilon, I) . \delta\omega = \frac{\partial F_\omega}{\partial e}(\omega_\epsilon, I).\delta e + \frac{\partial F_\epsilon}{\partial z}(\omega_\epsilon, I).\delta z + \frac{\partial F_\epsilon}{\partial y}(\omega_\epsilon, I).\delta y + \frac{\partial F_\epsilon}{\partial j}(\omega_\epsilon, I).\delta j .$$

C'est un élément de $\mathcal{L}(E^2 \times \Phi_O^2 ; \mathbb{L}^2)$. Plus précisement :

(13) Cette restriction à l'ensemble des contrôles admissibles est due à des considérations de différentiabilité. Simultanément, il faut supposer $i_\epsilon(x,O) = \gamma_O > O$ (voir [13] [14]).

THEOREME 5.1. - $\frac{\partial F_\epsilon}{\partial \omega}(\omega_\epsilon, I)$ est un isomorphisme de $D_o^2 \times \Phi_o^2$ sur $\mathbb{L}^2$.

Il est facile de voir que cet isomorphisme est représenté par la matrice d'opérateurs $A_\epsilon = A_\epsilon(v)$:

$$(5.8)\quad \begin{bmatrix} \rho\epsilon^2 D + \sigma(1+\epsilon)(1+s_\epsilon) + \sigma i_\epsilon & \sigma\epsilon & \sigma(1+\epsilon)e_\epsilon & \sigma e_\epsilon \\ -\sigma i_\epsilon & \rho\epsilon^2 D + \sigma & O & -\sigma e_\epsilon \\ \sigma(1+\epsilon)s_\epsilon + \sigma & \sigma & \epsilon P + \sigma(1+\epsilon)e_\epsilon & O \\ \sigma(1+\epsilon)i_\epsilon & -\sigma(1+\epsilon) & O & \epsilon P + \sigma(1+\epsilon)e_\epsilon \end{bmatrix}$$

Le lemme suivant est trivial :

LEMME 5.1. - Soit M un isomorphisme de $\mathbb{L}^2$ sur $\mathbb{L}^2$. Alors $M \circ A_\epsilon$ est un isomorphisme de $D_o^2 \times \Phi_o^2$ sur $\mathbb{L}^2$.

5.2. - Définition d'un état-adjoint.-

Tout d'abord, notons que l'application

$$(5.9)\qquad \Psi = (\Psi_i)_{i=1,4} \rightarrow 2\int_O^T (s_\epsilon(v\,;\tfrac{1}{2}) - s_d)\Psi_3(\tfrac{1}{2})dt$$

est en particulier une forme linéaire sur $D_o^2 \times \Phi_o^2$.

Par transposition de l'isomorphisme $M \circ A_\epsilon$, on peut donner alors la définition suivante d'un état-adjoint $p_\epsilon(M,v)$:

DEFINITION 5.1. - Soit M un isomorphisme de $\mathbb{L}^2$ sur $\mathbb{L}^2$. $p_\epsilon(M,v)$ est l'unique élément de $\mathbb{L}^2$ qui vérifie :

$$(5.10)\qquad ((p_\epsilon(M,v), M \circ A_\epsilon(v)\Psi)) = 2\int_O^T (s_\epsilon(v;\tfrac{1}{2}) - sd)\,\Psi_3(\tfrac{1}{2})dt$$

$$\forall\, \Psi = (\Psi_i)_{i=1,4} \in D_o^2 \times \Phi_o^2 .\ ^{(14)}$$

(14) Les doubles parenthèses désignent le produit scalaire dans $\mathbb{L}^2$.

Naturellement, il existe une infinité d'états-adjoints, chacun d'eux étant associé à un isomorphisme M . Cependant, deux choix de M sont naturels.

5.3.- L'état-adjoint associé au problème (P_ϵ) .-

Si on choisit pour M l'application identité, on trouve l'état-adjoint "canonique" associé au problème (P_ϵ) . $p_\epsilon = p_\epsilon(Id,v) = (p_i)_{i=1,4}$ est alors solution faible de : (15)

$$(5.11) \qquad -\rho\epsilon^2 \frac{\partial p_1}{\partial t} + \sigma(1+\epsilon)(1+s_\epsilon)p_1 + \sigma i_\epsilon p_1 - \sigma i_\epsilon p_2 + \sigma(1+\epsilon)s_\epsilon p_3 + \sigma p_3 + \sigma(1+\epsilon)i_\epsilon p_4 = 0$$

$$(5.12) \qquad -\rho\epsilon^2 \frac{\partial p_2}{\partial t} + \sigma p_2 + \sigma\epsilon p_1 + \sigma p_3 - \sigma(1+\epsilon)p_4 = 0$$

$$(5.13) \qquad \epsilon\left(-\frac{\partial p_3}{\partial t} - \frac{\partial^2 p_3}{\partial x^2}\right) + \sigma(1+\epsilon)e_\epsilon p_3 + \sigma(1+\epsilon)e_\epsilon p_1 = s_\epsilon - s_d \otimes \delta_{\frac{1}{2}}$$

$$(5.14) \qquad \epsilon\left(-\frac{\partial p_4}{\partial t} - \frac{\partial^2 p_4}{\partial x^2}\right) + \sigma(1+\epsilon)e_\epsilon p_4 + \sigma e_\epsilon p_1 - \sigma e_\epsilon p_2 = 0$$

Avec les conditions finales et aux limites :

$$(5.15) \qquad p_i(x,T) = 0 \quad , i=1 \text{ à } 4$$

$$(5.16) \qquad p_i(0,t) = p_i(1,t) = 0 \quad , i = 3 \text{ et } 4 .$$

$\epsilon > 0$ étant fixé, on a donc définit un état-adjoint p_ϵ dans $\mathbb{L}^2$. Mais on ne sait rien dire à priori du comportement de la suite p_ϵ ainsi construite quand $\epsilon \to 0$. Le système (5.11) (5.16) entre dans la catégorie des problèmes "raides" au sens de LIONS [11], et est étudié en tant que tel dans [15][16].

5.4.- L'état-adjoint associé au problème $(\hat{P}_\epsilon)$

Posons :

(5.17) $$M_\epsilon = \begin{bmatrix} -\frac{1}{\epsilon} & -\frac{1}{\epsilon} & \frac{1}{\epsilon} & 0 \\ 0 & \frac{1+\epsilon}{\epsilon} & 0 & \frac{1}{\epsilon} \\ 0 & 0 & 1 & 0 \\ 0 & 0 & 0 & 1 \end{bmatrix}$$

M_ϵ n'est autre que la matrice qui transforme le problème (P_ϵ) en le problème $(\hat{P}_\epsilon)$ (voir le N°. 3.2). M_ϵ est évidemment (à $\epsilon > 0$ fixé) un isomorphisme de $\mathbb{L}^2$ sur $\mathbb{L}^2$, et il est loisible de prendre $M = M_\epsilon$ dans (5.10).

Remarquons auparavant que

(5.18) $$\hat{A}_\epsilon = M_\epsilon \circ A_\epsilon$$

est la matrice d'opérateurs associée à $\frac{\partial \hat{F}_\epsilon}{\partial \omega}(\omega_\epsilon\, , I)$, le problème $(\hat{P}_\epsilon)$ étant écrit sous la forme :

(5.19) $$\hat{F}_\epsilon(e_\epsilon\, , z_\epsilon\, , y_\epsilon\, , j_\epsilon\, ; I) = 0 \; .$$

On calcule facilement $\hat{A}_\epsilon$:

(5.20) $$\hat{A}_\epsilon = \begin{bmatrix} -\rho\epsilon D - \sigma & -\rho\epsilon D - \sigma & P & 0 \\ 0 & \rho\epsilon(1+\epsilon)D & 0 & P \\ \sigma(1+\epsilon)s_\epsilon + \sigma & \sigma & \epsilon P + \sigma(1+\epsilon)e_\epsilon & 0 \\ \sigma(1+\epsilon)i_\epsilon & -\sigma(1+\epsilon) & 0 & \epsilon P + \sigma(1+\epsilon)e_\epsilon \end{bmatrix}$$

Notons maintenant $\hat{p}_\epsilon = p_\epsilon(M_\epsilon\, , v)$ la solution de (5.10) avec $M = M_\epsilon$.

(5.21) $$((\hat{p}_\epsilon\, , \hat{A}_\epsilon\, \psi)) = 2 \int_0^T (s_\epsilon(v\, ; \tfrac{1}{2}) - s_d)\psi_3(\tfrac{1}{2})dt \quad \forall \psi = (\psi_i)_{i=1,4} \in D_0^2 \times \Phi_0^2$$

$\hat{p}_\epsilon$ est alors l'état-adjoint associé au problème $(\hat{P}_\epsilon)$, $\hat{p}_\epsilon$ et p_ϵ étant liés par la relation :

(5.22) $$p_\epsilon = M_\epsilon^* \, \hat{p}_\epsilon$$

On vérifie facilement que $\hat{p}_\epsilon$ est solution faible de :

$$(5.23) \qquad -\rho\epsilon \frac{\partial \hat{p}_1}{\partial t} + \sigma \hat{p}_1 - \sigma(1+\epsilon)s_\epsilon \hat{p}_3 - \sigma\hat{p}_3 - \sigma(1+\epsilon)i_\epsilon \hat{p}_4 = 0$$

$$(5.24) \qquad -\rho\epsilon \frac{\partial \hat{p}_1}{\partial t} + \sigma\hat{p}_1 + \rho\epsilon(1+\epsilon)\frac{\partial \hat{p}_2}{\partial t} - \sigma\hat{p}_3 + \sigma(1+\epsilon)\hat{p}_4 = 0$$

$$(5.25) \qquad -\frac{\partial \hat{p}_1}{\partial t} - \frac{\partial^2 \hat{p}_1}{\partial x^2} + \epsilon\left(-\frac{\partial \hat{p}_3}{\partial t} - \frac{\partial^2 \hat{p}_3}{\partial x^2}\right) + \sigma(1+\epsilon)e_\epsilon \hat{p}_3 = (s_\epsilon - s_d) \otimes \delta_{\frac{1}{2}}$$

$$(5.26) \qquad -\frac{\partial \hat{p}_2}{\partial t} - \frac{\partial^2 \hat{p}_2}{\partial x^2} + \epsilon\left(-\frac{\partial \hat{p}_4}{\partial t} - \frac{\partial^2 \hat{p}_4}{\partial x^2}\right) + \sigma(1+\epsilon)e_\epsilon \hat{p}_4 = 0$$

avec les conditions finales et aux limites nulles sur tous les $\hat{p}_i$ (comparer avec (5.15) (5.16)).

5.6. - Convergence de la suite $\hat{p}_\epsilon$

Nous allons faire tendre formellement ϵ vers 0, dans (5.23).....(5.26).

Il vient à la limite :

$$(5.27) \qquad p_1^o - (1+s_o)p_3^o - i_o p_4^o = 0$$

$$(5.28) \qquad p_1^o + p_4^o - p_3^o = 0$$

$$(5.29) \qquad -\frac{\partial p_1^o}{\partial t} - \frac{\partial^2 p_1^o}{\partial x^2} + \sigma e_o p_3^o = (s_o - s_d) \otimes \delta_{\frac{1}{2}}$$

$$(5.30) \qquad -\frac{\partial p_2^o}{\partial t} - \frac{\partial^2 p_2^o}{\partial x^2} + \sigma e_o p_4^o = 0 \ .$$

En multipliant (5.28) par i_o, on voit que :

$$(5.31) \qquad p_3^o = p_1^o \frac{1+i_o}{1+i_o+s_o}$$

donc

$$(5.32) \qquad -\frac{\partial p_1^o}{\partial t} - \frac{\partial^2 p_1^o}{\partial x^2} + \sigma \frac{1+i_o}{(1+i_o+s_o)^2} p_1^o = (s_o - s_d) \otimes \delta_{\frac{1}{2}}$$

$$(5.33) \qquad -\frac{\partial p_2^o}{\partial t} - \frac{\partial^2 p_2^o}{\partial x^2} = \frac{\sigma p_1^o s_o}{(1+i_o+s_o)^2}$$

avec les conditions finales et aux limites :

$$(5.34) \qquad p_1^o(x,T) = p_2^o(x,T) = 0$$

$$(5.35) \qquad p_1^o(0,t) = p_1^o(1,t) = p_2^o(0,t) = p_2^o(1,t) = 0$$

On retrouve ainsi l'état-adjoint "canonique" associé au problème (P_o) dans [1][4][13] .

Nous renvoyons à [15] [16] pour une étude détaillée du système (5.23)..... (5.26) , ainsi que pour d'autres exemples de systèmes adjoints singuliers.

En conclusion, nous pensons que le choix du "bon état-adjoint" est d'une grande importance pour les applications numériques. La situation étudiée ici semble se présenter dans de nombreux domaines : Naturellement, toute la difficulté est dans le choix de la "matrice de recombinaison" M_ϵ .

REFERENCES BIBLIOGRAPHIQUES

[1] J.P. KERNEVEZ, - Thèse N° A.O.7246, Paris(1972).

[2] D. THOMAS, - Thèse N° A.O. 5407, Paris (1971).

[3] J.P. KERNEVEZ et D.THOMAS, - Numérical analysis of immobilized enzyme systems, Rapport de Recherche N°28, IRIA(1973).

[4] J.P. KERNEVEZ, - Controle of the flux of substrate entering an enzymatic membrane by an inhibition at the boundary, J. Optimization Theory Appl. , vol.12 N°1, 1973 .

[5] G. JOLY, - Thèse de 3ème cycle, Paris (1974).

[6] G. JOLY et J.P. KERNEVEZ, - Identification of kinetic parameters in biochemical systems, J.A.M.O. (à paraitre).

[7] F.G. HEINEKEN , H. M.TSUCHIYA, R.ARIS, - On the mathematical status of the pseudo-steady state hypothesis of biochemical kinetics, Mathematical Biosciences 1,95-113(1967).

[8] J.L. LIONS et E. MAGENES, - Problème aux limites non homogènes et applications, tomes 1 et 2, Dunod, Paris (1968).

[9] J.L. LIONS, - Contrôle optimal de systèmes gouvernés par des équations aux dérivées partielles, Dunod, Paris (1968).

[10] J.L. LIONS, - Quelques méthodes de résolution de problèmes aux limites non linéaires, Dunod, Paris(1969).

[11] J.L. LIONS, - Perturbations singulières dans les problèmes aux limites et en contrôle optimal, Springer-Verlag (1973).

[12] W. ECKHAUS, - Matched asymptotic expensions and singular perturbations, North-Holland (1973).

[13] C.M. BRAUNER et P. PENEL, - Thèse de 3ème cycle, Université de Paris-Sud (Orsay), 1972.

[14] C.M. BRAUNER et P. PENEL, - Un problème de contrôle optimal non linéaire en Biomathématique, Annali dell' Università di Ferrara, vol. XVIII, IV.7 (1973).

[15] C. M. BRAUNER, - Thèse (à paraître).

[16] C.M. BRAUNER et P. PENEL , - (à paraître).

THEORY AND APPLICATIONS OF SELF-TUNING REGULATORS

K. J. Åström

Lund Institute of Technology, Lund, Sweden

1. INTRODUCTION.

Linear stochastic control theory gives the potential to formulate and solve regulation problems for industrial processes in a fairly realistic manner. This has also been demonstrated in several applications, [1]. The use of the theory does, however, require mathematical models of the process and its disturbances. Models of the process dynamics can sometimes be obtained from physical laws. Modeling of the disturbances will almost always require experimental data from the process. To apply the theory it is thus necessary to perform plant experiments and to make a system identification.

Since the dynamic characteristics of the process and the disturbances may change with time, it is necessary to repeat the identification regularly to maintain the quality of the regulation. This imposes heavy restrictions on the operation of the system and requires persons with significant theoretical skills to keep the system regulating well.

From a practical point of view it is therefore highly desirable to investigate the possibilities to obtain control algorithms which can adapt to changes in the dynamics of the process and the disturbances. There are many ways to formulate control problems which lead to such algorithms. The main difficulty is to pose a problem which reflects the practical problem sufficiently well and whose solution gives a regulator of reasonable complexity.

---

†This work has been supported by the Swedish Board of Technical Development under contract 733546.

The changes in the process and its environment are very slow in many cases. A possible formulation is then to consider the problem of controlling a system with constant but unknown parameters. Such a problem can be solved at least for linear stochastic systems. Examples of such solutions are found in [2], [5]. The solutions obtained are not practical because it is necessary to introduce the conditional probability density of the parameters as a state. Even the solution of very simple problems will require computations which far exceed the capacity of computers available today.

It can then be attempted to analyse more modest problems. For systems with constant but unknown parameters one possibility is to construct control algorithms, which do not require knowledge of the system parameters and which converge to the optimal regulator that could be designed if the system parameters were known. Such algorithms are called <u>self-tuning</u> or <u>self-optimizing</u> regulators. The approach to such regulators is partly heuristic in the sense that it is necessary to develop enough insight into the problem to propose candidates for the algorithms. Once the algorithms are obtained there are, however, interesting and important problems of analysing them. Mathematically the problem changes from an optimization problem to an analysis problem.

There are many possibilities to generate the desired control algorithms. One possibility is to analyse the properties of solutions to simple optimal control problems that can be generated numerically. See [2]. Another possibility is to exploit on-line identification methods. Such a scheme based on least squares identification was proposed in [10]. Similar algorithms have also been investigated in [15] and [17].

This paper surveys the properties of a class of self-tuning regulators. In Section 2 a simple example is used to explain and motivate the algorithm. Some properties of the algorithm are analysed in Section 3. Limitations of the algorithm are discussed in Section 4 where some extensions also are given. Practical applications of the algorithm to control of industrial processes in paper and steel industry are briefly discussed in Section 5.

2. AN ALGORITHM.

A self-tuning control algorithm will now be given. The main ideas are illustrated using a simple example. Consider the system

$$y(t+1) + a\, y(t) = b\, u(t) + e(t+1) + c\, e(t) \qquad (2.1)$$

where u is the input, y the output and {e(t)} a sequence of independent, equally distributed, random variables. The number c is assumed to be less than one. Let the criterion be to minimize the variance of the output i.e.

$$\min V = \min Ey^2 = \min E \frac{1}{t} \sum_{k=1}^{t} y^2(k) \qquad (2.2)$$

It is easy to show [1] that the control law

$$u(t) = \frac{a - c}{b}\, y(t) \qquad (2.3)$$

is a minimum variance strategy, and that the output of the system (2.1) with the feedback (2.3) becomes

$$y(t) = e(t) \qquad (2.4)$$

Notice that the control law (2.3), which represents a proportional regulator, can be characterized by one parameter only.

A self-tuning regulator for the system (2.1) can be described as follows:

ALGORITHM (Self-TUning REgulator).

Step 1. (Parameter Estimation).

At each time t, fit the parameter α in the model

$$\hat{y}(k+1) + \alpha\, y(k) = u(k), \quad k = 1,\dots,t-1 \qquad (2.5)$$

by least squares, i.e. such that the criterion

$$\sum_{k=1}^{t} \varepsilon^2(k) \qquad (2.6)$$

where

$$\varepsilon(k) = y(k) - \hat{y}(k) \qquad (2.7)$$

is minimal. The estimated obtained is denoted α_t to indicate that it is a function of time.

Step 2. (Control).

At each time t, choose the control

$$u(t) = \alpha_t y(t) \tag{2.8}$$

where α_t is the estimate obtained in step 1.

□

Motivation.

There are several ways to arrive at the algorithm STURE given above. It can be obtained as a solution to a stochastic control problem based on the assumption that the problem can be separated into one identification problem and one control problem, which are solved separately. The algorithm is thus not a dual control in the sense of Feldbaum. The algorithm can also be obtained from the model reference principle. See e.g. [11].

Extensions.

The algorithm given for the simple example can be extended in many ways. A generalization to systems of n:th order is given in [3] and a multivariable version is given in [16].

3. ANALYSIS.

The properties of a closed loop system controlled by a self-tuning regulator will now be discussed. Since the closed loop system is nonlinear, timevarying and stochastic, the analysis is not trivial. In this section the major results will be stated for the simple example. For details we refer to the references [3], [12] and [13].

It is fairly obvious that the regulator will perform well if it is applied to a system (2.1) with $b = 1$ and $c = 0$, because in this case the least squares estimate α_t will be an unbiased estimate of a. The regulator (2.8) will thus converge to a minimum variance regulator if the parameter estimate α_t converges. It is surprising, however, that the regulator will also converge to the minimum variance regulator if $c \neq 0$ as will be demonstrated below.

The least squares estimate is given by the normal equation

$$\frac{1}{t}\sum_{k=1}^{t} y(k+1)y(k) + \alpha_{t+1}\frac{1}{t}\sum_{k=1}^{t} y^2(k) = \frac{1}{t}\sum_{k=1}^{t} y(k)u(k)$$

Assuming that the estimate α_t converges towards a value which gives a stable closed loop system, then it is straightforward to show that

$$\frac{1}{t}\sum_{k=1}^{t} (\alpha_{t+1} - \alpha_k)y^2(k) \rightarrow 0$$

Thus the closed loop system has the property

$$\lim_{t\to\infty} \frac{1}{t}\sum_{k=1}^{t} y(t+1)y(t) = 0 \qquad (3.1)$$

Furthermore, assuming that the system to be controlled is governed by (2.1), the output of the closed loop system obtained in the limit is given by

$$y(t) + [a - \alpha b]y(t) = e(t) + c\, e(t-1) \qquad (3.2)$$

The covariance of $\{y(t)\}$ at lag 1 is then given by

$$E\, y(t+1)y(t) = -f(\alpha) = \frac{(c-a+\alpha b)(1-ac+\alpha bc)}{1 - (a-\alpha b)^2} \qquad (3.3)$$

The condition (3.1) gives

$$f(\alpha) = 0$$

This is a second order equation for α which has the solutions

$$\alpha = \alpha_1 = \frac{a - c}{b} \qquad (3.4)$$

$$\alpha = \alpha_2 = \frac{a - 1/c}{b} \qquad (3.5)$$

The corresponding poles of the closed loop system are $\lambda_1 = c$ and $\lambda_2 = 1/c$ respectively. Since c was assumed less than one, only the value α_1 corresponds to a stable closed loop system. Notice that α_1 corresponds to the gain of the minimum variance regulator (2.3).

Hence, if the parameter estimate α_t converges to a value which gives a stable closed loop system, then the closed loop system obtained must be such that (2.7) holds. This means that the algorithm can be thought of as a regulator which attempts to bring the covariance of the output at lag one, i.e. $r_y(1)$, to zero in the same way as an integrating regulator brings the integral of the control error to zero.

If the system to be controlled is actually governed by (2.1), then the self-tuning regulator will converge to a minimum variance regulator if it converges at all. These properties of the self-tuning regulator are easy to extend to arbitrary n:th order systems. This is done in [4]. Simulations of the regulator are given in [4], [15] and [18].

As was already pointed out by Kalman [10], the regulator has strong stabilizing properties; because if the output assumes large values, then the terms containing u and y in (2.1) will dominate the stochastic terms, and the estimate α will become close to a/b. This will bring the closed loop poles close to the origin. Hence, if the values of the output becomes very large, they will quickly be reduced to the level of the disturbances again. A rigorous analysis is provided in [13].

The convergence of the algorithm is of course a key problem. This problem has been analysed in [12], [13], where necessary and sufficient conditions are given. The differential equation

$$\frac{d\alpha}{d\tau} = f(\alpha) \tag{3.6}$$

where the function f is defined by (3.3), plays a crucial role in the stability analysis. It follows from the previous discussion that $\alpha = \alpha_1$ is a stationary solution. It is shown in [12] that the parameter estimates will in a certain sense be close to the trajectories of (3.6). One condition required for convergence of the estimates is that the solution $\alpha = \alpha_1$ is a stable solution to (3.6). In the particular example the estimate will converge if $0 < b < 2$, [13].

Since the inputs to the system are generated by feedback from the output, it may conceivably happen that there will be many values of α which give the same values to the loss function. To understand the behaviour of the self-tuning algorithm it is easy to show that α is identifiable. Conditions for identifiability of systems under closed loop experimental conditions are discussed in [9].

4. LIMITATIONS AND EXTENSIONS.

Consider a system described by the n:th order difference equation

$$A(q^{-1})\,y(t) = B(q^{-1})\,u(t) + C(q^{-1})\,e(t) \qquad (4.1)$$

where $A(q^{-1})$, $B(q^{-1})$ and $C(q^{-1})$ are polynomials in the backward shift operator q^{-1} i.e.

$$A(q^{-1}) = 1 + a_1 q^{-1} + \ldots + a_n q^{-n}$$

$$B(q^{-1}) = \quad b_1 q^{-1} + \ldots + b_n q^{-n}$$

$$C(q^{-1}) = 1 + c_1 q^{-1} + \ldots + c_n q^{-n}$$

u is the input, y the output and $\{e(t)\}$ a sequence of uncorrelated, random variables. It has been found by simulation [18] that the self-tuning regulator will not necessarily converge when applied to the system (4.1) when the polynomial $B^*(x) = x^n B(x^{-1})$ has zeroes outside the unit circle, (non-minimum phase systems). There are also minimum phase systems for which the regulator does not converge,[13].

Several properties of the self-tuning regulator do also depend on a delicate balance of a bias in the least squares estimate and a modeling error. This balance is upset, if extra perturbation signals are introduced, or if the criterion (2.2) is changed e.g. to include weighting of the control signal. There are various modifications to the algorithm which can be introduced to overcome some of the difficulties. In [4] it was shown that the difficulties associated with non-minimum phase systems could sometimes be eliminated by modifying the control law.

In the algorithm given in section 2 no attempt is made to estimate the parameters of the polynomial C directly. These parameters will instead enter the procedure indirectly through the bias in the estimate of the parameter α of the model (2.5). In some cases e.g. when extra perturbation signals are introduced, or when the criterion includes penalty on the control actions,it is advantageous to estimate the parameters $c_1,\ldots,c_n$ too. This estimation problem is nonlinear. Approximative, linear estimation schemes may, however, sometimes be useful. One possibility is to substitute the estimation step of the algorithm by

Step 1 A. (Parameter Estimation)

At each time t the parameters of the model

$$\hat{y}(k+1) + A(q^{-1})\, y(k) = B(q^{-1})\, u(k) + C(q^{-1})\, \varepsilon(k) \qquad (4.2)$$

where ε is given by (2.7) and

$$A(q^{-1}) = \alpha_1 + \alpha_2 q^{-1} + \dots + \alpha_n q^{-n+1}$$

$$B(q^{-1}) = \beta_1 + \beta_2 q^{-1} + \dots + \beta_n q^{-n+1}$$

$$C(q^{-1}) = \gamma_1 + \gamma_2 q^{-1} + \dots + \gamma_n q^{-n+1}$$

are estimated by least squares i.e. in such a way that the criterion (2.6) is minimal.

Since the estimation problem is linear in the parameters, it is easily solved. Introduce the vectors

$$\theta = \mathrm{col}[\alpha_1,\dots,\alpha_n,\beta_1,\dots,\beta_n,\gamma_1,\dots,\gamma_n] \qquad (4.3)$$

$$\varphi(k) = [-y(k),\dots,-y(k-n+1),u(k),\dots,u(k-n+1),\varepsilon(k),\dots,\varepsilon(k-n+1)] \qquad (4.4)$$

The prediction $\hat{y}$ defined by (4.2) can then be written as

$$\hat{y}(k+1) = \theta\varphi(k) \qquad (4.5)$$

and the criterion (2.6) becomes

$$V(\theta) = \sum_{k=1}^{t} [y(k) - \theta\varphi(k-1)]^2 \qquad (4.6)$$

The value θ_t of θ which minimizes this function is given by the normal equations.

$$[\sum_{k=1}^{t} \varphi^T(k-1)\varphi(k-1)]\theta_t = \sum_{k=1}^{t} \varphi^T(k-1)y(k) \qquad (4.7)$$

since

$$y(k) = \hat{y}(k) + \varepsilon(k) = \varphi(k-1)\theta_k + \varepsilon_k$$

it follows from the normal equations that

$$\frac{1}{t}\sum_{k=1}^{t} \varphi^T(k-1)\varepsilon(k) + \frac{1}{t}\sum_{k=1}^{t} \varphi^T(k-1)\varphi(k-1)[\theta_t-\theta_k] = 0$$

If the parameter estimates converge to such values that the polynomial $x^n(1+C(x^n))$ is stable, then the second term converges to zero and we get

$$\lim \frac{1}{t}\sum_{k=1}^{t} \varphi^T(k-1)\varepsilon(k) = 0$$

or

$$\begin{aligned} &E \quad \varepsilon^o(t+\ell)y(t) = 0 && \ell = 1,\dots,n \\ &E \quad \varepsilon^o(t+\ell)u(t) = 0 && \ell = 1,\dots,n \qquad (4.8) \\ &E \quad \varepsilon^o(t+\ell)\varepsilon^o(t) = 0 && \ell = 1,\dots,n \end{aligned}$$

where ε^o is the residual calculated from the limiting estimate.

Hence, if the parameter estimates converge, then the limiting parameter estimates are characterized by the property that the covariances (4.8) must vanish. The conditions (4.8) are generalizations of (3.1).

Assuming that the estimation procedure is applied to a system described by (4.1), the conditions (4.8) give nonlinear equations for the possible limits of the estimates. It is straightforward to show that one solution is given by $\alpha_i = a_i$, $\beta_i = b_i$ and $\gamma_i = c_i$, $i = 1,2,\dots,n$.

For n = 1 it can also be shown that this is the only limiting point such that $x^n(1+C(x^{-1})$ is a stable polynomial if $(\Sigma_1 u^2(t))/t$ is stable. The problem of convergence is similar to the one for the simple algorithm.

The estimates obtained in this way are the same as those obtained by the method proposed in [20]. An alternative is to use a recursive version of the maximum likelihood method. When the parameter estimates are obtained, the control strategies can be determined in many different ways.

5. APPLICATIONS.

The algorithm outlined in Section 2 is easily implemented on a process computer. A program which handles systems of arbitrary order and includes tuning of feedforward parameters can be written using about 35 FORTRAN statements. Algorithms of this type have been successfully applied to control industrial processes. Applications to paper machine control are described in [6] and [8]. Control of an ore crusher is described in [7]. This application is of interest because of the extreme difficulty in modeling the process using physical principles only. It was implemented using tele-processing from a computer at Lund Institute of Technology to the plant in Kiruna covering a distance of about 1800 km.

In the applications the algorithm described in Section 2 is somewhat modified. An exponential discounting of past values is introduced in the criterion for estimation, (2.6), to allow for timevarying process characteristics. In the paper machine application [6] the regulator had 6 parameters which were updated using the algorithm (4 for the feedback and 2 for the feedforward). To control the ore crusher a regulator with 7 parameters was used.

The general conclusion that can be drawn from the applications is that self-tuning regulators can be useful for practical control problems. In paper machine applications it has been shown that the self-tuning regulator will perform just as well as a minimum variance regulator designed on the basis of plant experiment and system identification. The simulations shown in [3] are representative of what can be achieved in practice.

6. REFERENCES.

[1] Åström, K. J., "Computer Control of a Paper Machine - an Application of Linear Stochastic Control Theory", IBM J. Res. Dev. 11 (1967), 389-405.

[2] Åström, K. J. and Wittenmark, B., "Problems of Identification and Control", JMAA 34 (1971), 90-113.

[3] Åström, K. J. and Wittenmark, B., "On Self-Tuning Regulators", Automatica 9 (1973), 185-199.

[4] Åström, K. J. and Wittenmark, B., "Analysis of a Self-Tuning Regulator for Nonminimum Phase Systems", Submitted to IFAC Symposium

on Stochastic Control Theory, Budapest 1974.

[5] Bohlin, T., "Optimal Dual Control of a Simple Process with Unknown Gain", Report TP 18.196, IBM Systems Development Division Nordic Laboratory, Sweden 1969.

[6] Borisson, U. and Wittenmark, B., "An Industrial Application of a Self-Tuning Regulator", 4th IFAC Conference on Digital Computer Applications to Process Control, Zürich 1974.

[7] Borisson, U. and Syding, R., "Self-Tuning Control of an Ore Crusher", IFAC Symposium on Stochastic Control Theory, Budapest 1974.

[8] Cegrell, T. and Hedqvist, T., "Successful Adaptive Control of Paper Machines", 3rd IFAC Symposium on Identification and System Parameter Estimation, Hague 1973.

[9] Gustavsson, I. Ljung, L. and Söderström, T., "Identification of Linear, Multivariable Process Dynamics Using Closed Loop Estimation", Report 7401, Division of Automatic Control, Lund Institute of Technology, 1974.

[10] Kalman, R. E., Design of a Self-Optimizing Control System", Trans ASME 80 (1958) also in Oldenburger, R. (Ed.) Optimal Self-Optimizing Control MIT Press 1966, 440-449.

[11] Landau, I. D., "Model Reference Adaptive Systems - A Survey (MRAS) - What is possible and Why?", ASME J. of Dynamic Systems, Measurement and Control (1972), 119-132.

[12] Ljung, L., "Convergence of Recursive Stochastic Algorithms", Report 7403, Division of Automatic Control, Lund Institute of Technology, 1974.

[13] Ljung, L. and Wittenmark, B., "Asymptotic Properties of Self-Tuning Regulators", Report 7404, Division of Automatic Control, Lund Institute of Technology, 1974.

[14] Ljung, L., "Stochastic Convergence of Algorithms for Identification and Adaptive Control", Thesis for teknologie doktorsexamen, Lund Institute of Technology, Lund 1973.

[15] Peterka, V., Adaptive Digital Regulation of a Noisy System", 2nd IFAC Symposium on Identification and Process Parameter Estimation, Prague 1970, paper No. 6.2.

[16] Peterka, V. and Åström, K. J., "Control of Multivariable Systems with Unknown but Constant Parameters", 3rd IFAC Symposium on Identification and System Parameter Estimation, Hague 1973.

[17] Wieslander, J. and Wittenmark, B., "An Approach to Adaptive Control Using Real Time Identification, Automatica 7 (1971), 211-217.

[18] Wittenmark, B. "A Self-Tuning Regulator", Report 7311, Division of Automatic Control, Lund Institute of Technology, Lund, Sweden, April 1973.

[19] Wittenmark, B., "Self-Tuning Regulator", Thesis for teknologie doktorsexamen, Lund Institute of Technology, 1973 (also Report 7312, Division of Automatic Control, Lund Institute of Technology 1973.

[20] Young, P., "An Extension of the Instrumental Variable Method for Identification of A Noisy Dynamic Process", Report CN/70/1, University of Cambridge, Dept. of Eng.

SUPPLY AND DEMAND RELATIONSHIPS IN FISHERIES MANAGEMENT

by

Colin W. Clark

Department of Mathematics

The University of British Columbia

Vancouver, Canada

Introduction. This paper is concerned with the problem of dynamic optimization in fisheries management. We shall utilize a mathematical model of fishing essentially due to M. Schaefer (1957). Although Schaefer's model is a dynamic one, most economic analyses based upon it (including Schaefer's own analysis) have been purely static. In the first section of this paper, therefore, we review briefly a recent dynamic treatment of the Schaefer model due to the author (Clark, 1973).

An important assumption made both by Schaefer and Clark was that the price of fish is a constant, i.e. that demand is perfectly elastic. For the static case, it has been pointed out by Copes (1970) and Anderson (1973) that the presence of less than perfect demand elasticity can have serious effects on the theory, and consequently on the nature of optimal fishery regulation. On the other hand, it has been shown by Clark (1973) that the failure to consider the dynamic aspects of fishing also has serious theoretical and practical implications for optimal management. The purpose of the present paper is to study these two effects simultaneously.

From the mathematical viewpoint, we are led to a nonlinear problem in optimal control theory. Although the differential equations describing optimal trajectories of this problem take on a relatively simple algebraic form, they do not necessarily lead to simple solutions. Indeed these equations may typically possess multiple equilibria, and suffer from "structural instability" of the phase-plane topology, with corresponding discontinuities of optimal population levels as functions of price and other parameters.

The latter phenomenon has already been observed for the case of uncontrolled "common-property" fisheries, by Anderson (1973), and was alluded to by Schaefer. The usual static treatment of optimality in fishing, however, gives no hint that such phenomena might arise under dynamically optimal control.

1. The Dynamic Schaefer model

The following model will be presented here with very little justification. The reader should refer to the paper of Schaefer (1957), or to a forthcoming joint paper by Clark and Munro (1974) for further details regarding this model.

The model utilizes the one-dimensional state equation

$$\frac{dx}{dt} = f(x) - h(t), \quad 0 \leq t < \infty; \quad x(0) = x_0 \tag{1}$$

in which the symbols have the following interpretation. The state variable $x(t)$ represents the size of the fish population at time $t \geq 0$; the initial population is given by x_0. The control variable $h(t)$ represents the rate of harvest of fish, and is subject to the constraint

$$h(t) \geq 0 . \tag{2}$$

The given function $f(x)$ represents the natural net growth rate of the population in the absence of harvesting. This function is assumed to be smooth, and to have a graph of the form shown in Fig. 1. The values $x = \underline{K} \geq 0$ and $x = \bar{K} > 0$ are, respectively, unstable and stable natural population equilibria. In the simplest case, $\underline{K}$ vanishes, and $f(x) > 0$ for $0 < x < \bar{K}$. Our attention will be restricted to this case unless explicitly stated to the contrary.

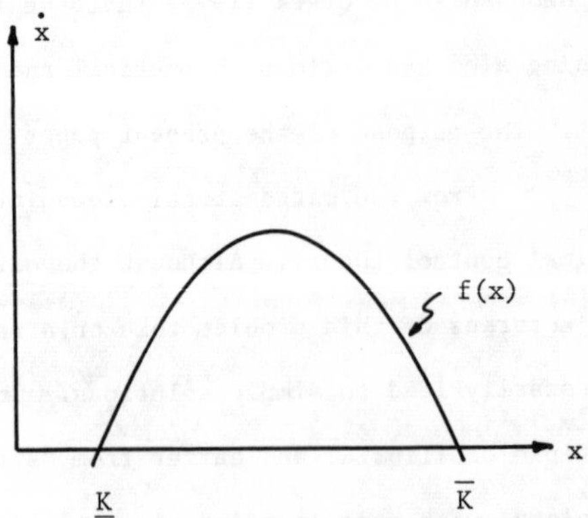

Figure 1. Graph of $f(x)$.

An obvious implication of Eq. (1) is that $f(x)$ also represents the sustainable yield as a function of the population level x . Standard examples of the growth function $f(x)$ include the Pearl-Verhulst logistic

$$f(x) = rx(1 - \frac{x}{K})$$

and the Gompertz law

$$f(x) = rx \ln \frac{K}{x} .$$

The harvesting of fish imposes on society a social opportunity cost $C(x,h)$, which will generally depend on the harvest rate h and also on the fish population level x. Harvesting also provides, through the consumption of fish, a certain social utility $U(h)$. The objective of fishery management is assumed to be the maximization of the present value of "net social utility"

$$\Pi(x,h) = U(h) - C(x,h) ,$$

that is, the maximization of

$$J = \int_0^\infty e^{-\delta t} \Pi(x,h)dt , \tag{3}$$

where $\delta > 0$ is the instantaneous rate of social time preference.

Formal manipulation using the maximum principle (Pontrjagin et al, 1962) leads easily to the following differential equation for $h(t)$:

$$\frac{dh}{dt} = \frac{1}{\Pi_{hh}} \{[\delta - f'(x)]\Pi_h - \Pi_x - \Pi_{xh} [f(x) - h]\} . \tag{4}$$

The solutions of the system (1), (4) are the trajectories of our problem (in (x,h) - space), among which must be selected the optimal trajectory, satisfying initial, terminal, and transversality conditions. This problem will be discussed in the following sections, but first we shall analyze the special case in which $\Pi(x,h)$ is a <u>linear</u> function of the control variable h.

Assume for the moment, therefore, that

$$\Pi(x,h) = ph - C(x)h \tag{5}$$

where $p = \text{const.}$ denotes the unit price of fish, and $C(x)$ the cost of a unit harvest when the population level is x. We adopt the reasonable assumptions

$$C(x) > 0; \quad C'(x) \leq 0 .$$

Equation (4), of course, does not apply to the linear problem. Instead we first determine the "singular" solution for which $\partial H/\partial h \equiv 0$ (where $H = e^{-\delta t}\Pi + \lambda\{f(x) - h\}$ is the Hamiltonian). For this we obtain the equation

$$p - C(x) = \frac{1}{\delta}\frac{d}{dx}(\{p - C(x)\} f(x)) . \qquad (6)$$

Thus, since Eq.(6) does not involve the time t, any singular solution is necessarily also an equilibrium solution: $x(t) \equiv x^*$ where x^* is a solution of Eq.(6).

The existence of a solution x^* to Eq.(6) is obviously assured provided that

$$C(K) < p < C(0^+) . \qquad (7)$$

These inequalities mean, respectively, that fishing is economically viable (price exceeds unit cost when $x = K$), and that extinction is not viable (unit cost exceeds price as $x \to 0$). Uniqueness of x^* occurs only under more restrictive conditions. For the specific model with

$$f(x) = rx(1 - \frac{x}{K}); \quad C(x) = \frac{c}{x} \qquad (8)$$

it is easy to verify that (6) does possess a unique solution $x^* > 0$ (provided $c < Kp$).

In the case that Eq.(6) possesses a unique solution, the overall optimal harvest policy $h^*(t)$ is easily deduced. Namely there is an initial "bang-bang" adjustment phase during wich the population is adjusted, as rapidly as possible, towards the optimal level x^*. If $x(0) < x^*$ then $h^*(t) = 0$ until $x(t)$ reaches x^*. If $x(0) > x^*$ and if there is no upper constraint on the control, then an impulse, or delta-function control is used to reduce x instantaneously to x^*. More realistically if it is assumed that

$$h(t) \leq h_{max}$$

then the optimal control $h^*(t)$ equals h_{max} until $x(t)$ reaches x^*. (There are some ambiguous cases if $h_{max} < f(x^*)$.)

It is clear from this description of the bang-bang adjustment phase that the linear model (5) is overly artificial. In practice the optimal rate of harvest during the development of an underexploited fishery will be determined by the level of demand, rather than by some fixed constraint on output $h(t)$. On the other hand, the complete closure of an overexploited fishery is unlikely to constitute the socially optimal method of recovery. The necessity of studying more general non-linear problems thus becomes apparent.

Before proceding to the nonlinear case, however, let us discuss some further aspects of the linear problem. First let us note the following "marginal" interpretation of Eq.(6). Consider, at the margin, the decision as to whether the existing population level x should be maintained by means of the sustained harvest $h(t) \equiv f(x)$, or whether (for example) it should be reduced by one unit. The net <u>gain</u> from a unit reduction equals $p - C(x)$, as on the left side of Eq.(6). On the other hand, since $\{p - C(x)\}f(x)$ represents the sutainable net revenue at the population level x, its derivative $\frac{d}{dx}(\{p - C(x)\}f(x))$ represents the <u>loss</u> in sustained revenue resulting from a unit reduction in x. The right side of (6) is the present value of this loss. Equation (6) simply asserts, then, that at the optimal population level $x = x^*$, the net immediate marginal gain must be the same as the present value of the net future marginal loss.

The two limiting cases $\delta \to 0$ and $\delta \to +\infty$ are of particular interest, since they correspond respectively to neglecting the immediate marginal gain, and neglecting the furture marginal loss. When $\delta = 0$, Eq.(6) becomes a prescription for maximizing sustainable economic yield $\{p - C(x)\}f(x)$, whereas when $\delta = +\infty$ it becomes a prescription for maximizing the immediate economic yield

$$\int_x^{x^0} \{p - C(z)\}dz .$$

When $0 < \delta < +\infty$, Eq.(6) determines the optimal "tradeoff" between these two extreme policies.

The theory of the competitive ("common-property") fishery (Gordon, 1954) predicts an equilibrium in which, as a result of the free entry of fishermen, the sustained economic rent $\{p - C(x)\}f(x)$ vanishes. Let x_∞ denote the corresponding population level:

$$p = C(x_\infty) . \tag{9}$$

Notice that, as shown by Scott (1955), this is the same level as in the limiting case $\delta = +\infty$ described above. Following Copes (1970), let us consider the competitive supply curve for this case. By definition, this curve shows the relationship between equilibrium output $Q = f(x_\infty)$ and price, p :

$$Q = f(C^{-1}(p)) . \tag{10}$$

Alternatively, let $TC(Q)$ denote the total cost of the sustained output level Q (per unit time):

$$TC(Q) = C(x)Q = C(f^{-1}(Q))\cdot Q$$

(where f^{-1} is multi-valued). Then Eq.(10) can be written as

$$p = C(f^{-1}(Q)) = \frac{TC(Q)}{Q} = AC(Q) \tag{10'}$$

where AC designates average cost. A typical such supply curve SS is shown in Fig. 2. As pointed out by Copes (1970), this supply curve inevitably possesses a "backwards-bending" segment, with the significant property that increases in price above a critical level p^* lead to a <u>decrease</u> in the level of output Q . This is usually referred to as (biological) <u>overfishing</u>.

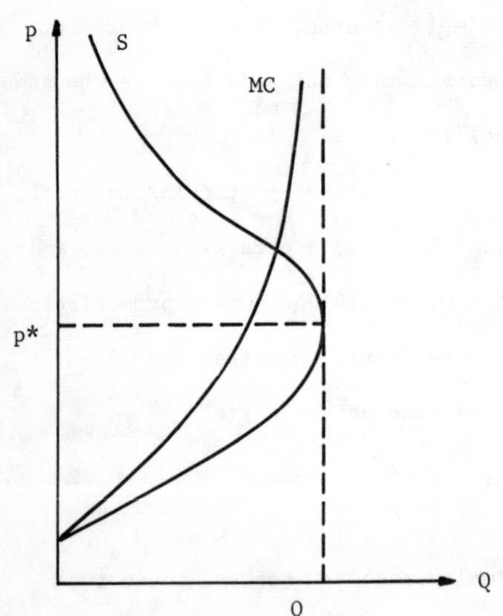

Figure 2. Competitive Supply Curve, and Marginal Cost Curve.

Copes also considered the marginal cost curve $MC = d(TC)/dQ$. This curve rises to a vertical asymptote at maximum sustainable yield $Q = Q_{max}$; the backwards-bending section of TC gives rise to an irrelevant negative marginal cost. It was observed that the usual condition for a welfare optimum, $p = MC$, would thus never lead to biological overfishing.

It should be noticed, however, that this analysis is entirely static, inasmuch as the curves TC, AC and MC are all based on an a priori assumption of constant output Q. Let us next consider the supply curve corresponding to the optimally controlled fishery; this curve again shows the sustained output $Q = f(x)$ corresponding to each price level p. In this case, x and p are related by Eq.(6), which we shall rewrite in the form

$$p = C(x) - \frac{C'(x)f(x)}{\delta - f'(x)} . \tag{11}$$

Also, Q and x satisfy

$$Q = f(x) . \tag{12}$$

Equations (11) and (12) together determine the "dynamic" supply curve S_δ (Fig. 3). Notice that, unless $\delta = 0$, this curve always possesses a backwards-bending segment, asymptotic to the vertical line $Q = Q_\delta$ where

$$Q_\delta = f(x_\delta); \quad f'(x_\delta) = \delta . \tag{13}$$

It is easy to verify that, for $\delta = +\infty$ the dynamic supply curve reduces to Copes' competitive supply curve (AC), while for $\delta = 0$ it is the same as Copes' marginal cost curve (MC).

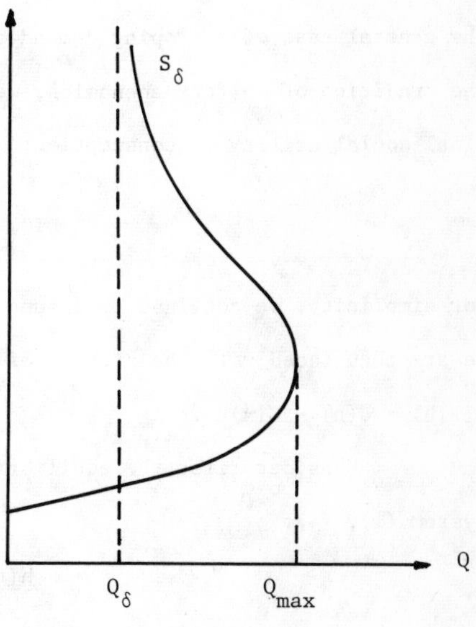

Figure 3. "Dynamic" Supply Curve.

Thus biological overfishing is not, as previously sometimes supposed, solely the result of common-property "inefficiency", but will in fact always occur under economic optimization $(\delta > 0)$ provided that the price p is sufficiently high. How serious will be the reduction in output Q depends on the ultimate level Q_δ, which is determined by the interaction between the biological growth rate $f(x)$ and the discount rate δ via Eq.(13). Notice, in fact, that if

$$\delta > f'(0) = \max f'(x)$$

then Eq.(13) has no solution x_δ. In this case the dynamic supply curve is asymptotic to the vertical axis $Q = 0$, and is thus essentially indistinguishable from the common-property supply curve studied by Copes.

For the case of whale populations, for example, $f'(0)$ is normally in the range from 5 - 10% per annum. It follows that common-property exploitation of whales will differ little in its outcome from profit-maximizing exploitation at normally risky discount rates (say 10 - 20% per annum; see Clark, 1973, 1974).

2. Inelastic demand functions.

In the previous section we treated price p as an exogenously determined parameter; i.e. the demand was assumed to be infinitely elastic. We now consider the general case of a sloping demand curve $p = D(Q)$, where $D'(Q) \leq 0$. Following the tradition of welfare economics, we suppose that demand can be equated with marginal social utility of consumption:

$$D(Q) = U'(Q) .$$

For simplicity, we continue to assume that costs of harvesting are linear in h. We are then faced with the problem of maximizing the objective functional (3), with $\Pi(x,h) = U(h) - C(x)h$.

Consider first the equilibrium solutions for the corresponding autonomous system (1), (4):

$$h(t) = f(x) ;$$

$$\{\delta - f'(x)\} \{U'(h) - C(x)\} = - C'(x)h .$$

These equations imply that

$$U'(f(x)) = D(f(x)) = C(x) - \frac{C'(x)f(x)}{\delta - f'(x)} . \tag{13}$$

Notice that, when $D(f(x)) = p = \text{const.}$, this reduces to the previous equilibrium result, Eq.(11). This means that the optimal equilibrium output $Q = f(x)$ can be characterized graphically as the abscissa of the point of intersection of the dynamic supply curve S_δ and the demand curve D (point P_1 in Fig.4).

Similarly the monopolist's optimization problem

$$\max \int_0^\infty e^{-\delta t}\{hD(h) - C(x)h\}dt$$

leads to the equilibrium condition

$$MR(f(x)) = C(x) - \frac{C'(x)f(x)}{\delta - f'(x)}, \qquad (14)$$

where $MR = (d/dQ)(QD(Q))$ denotes marginal revenue. In Figure 4 this corresponds to the point P_2 .

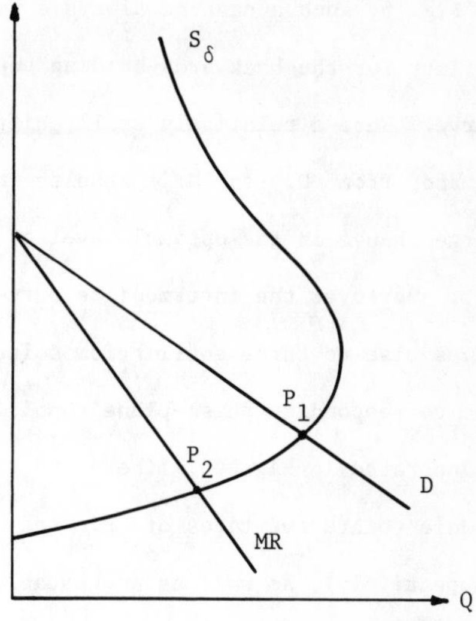

Figure 4. Supply and Demand.

It follows from this discussion that, as far as the equilibrium analysis of an optimally controlled fishery is concerned, the normal methods of supply and demand analysis can be utilized, S_δ playing the role of supply curve. The backwards-bending nature of S_δ, however, results in various phenomena not normally encountered in the theory of supply and demand. Under suitable circumstances, for example, the monopolistic owner may produce (<u>in equilibrium</u>) a greater output, at a lower price, than the social optimium (cf. Fig. 4 with the demand curve shifted considerably upwards). This seeming paradox is explained by the dynamics of the fishery: during the initial adjustment (nonequilibrium) stage, the optimal social output will exceed the monopoly level, ultimately driving the fishery to a level of biological "overexploitation."

Let us pass now to the more interesting question of dynamic stability of the system of equations (1) and (4). Suppose first that the supply and demand curves intersect in a uniquely determined point (Q^*,p^*), i.e. Eq. (13) possesses a unique solution $x = x^*$; thus $Q^* = f(x^*) = h^*$ and $p^* = D(Q^*)$. It turns out (the proof is the same as in Samuelson, 1965) that the equilibrium point (x^*,h^*) is a <u>saddle point</u> for the system (1), (4) - Fig 6a. The optimal trajectory (for the infinite time horizon) thus consists of one or the other of the separatrices converging to the equilibrium point - this corresponds to Samuelson's "asymptotic turnpike."

Consider next the case depicted in Fig. 5; such a case is always a possibility for the backwards-bending supply curve. Here a relatively small shift in demand, from D_1 to D_3, results in a large change in the optimal level of output. Moreover the intermediate curve D_2 gives rise to three equilibrium solutions. The corresponding phase-plane topology is illustrated in Fig. 6b. There are two saddle points (vestiges of P_1 and P_3 respectively), as well as an unstable centre. Similar cases of instability and multiple equilibria have been encountered in capital growth models; see Kurz (1968), Leviatan and Samuelson (1969), and Brock (1973).

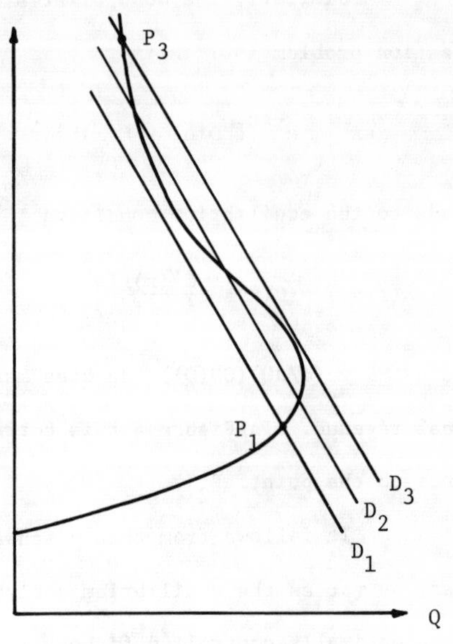

Figure 5. Instability

It can be seen that the point x_2 constitutes a "watershed" for turnpike solutions (cf. Leviatan and Samuelson, 1969) in the sense that if the initial population is above (below) x_2, then there is a unique turnpike leading to $x_3(x_1)$.

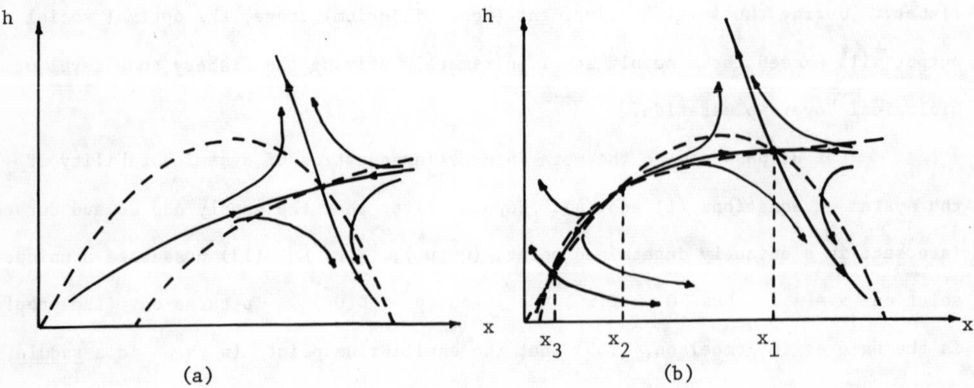

Figure 6. Stability and instability.

Of greater importance however, is the likelihood of severe instability in the optimal stock level resulting from relatively minor shifts in the demand curve.

In a famous paper that initiated the modern economic theory of the fishery, H.S. Gordon (1954) coined the phrase "bionomic equilibrium" to describe the outcome of a common-property fishery (see Eq. (9)). Perhaps there are cases where "bionomic instability" might be a more appropriate term.

References

Anderson, L.G., Optimum economic yield of a fishery, given a variable price of output, J. Fish. Res. Bd. Canada 30, 1973, 509-518.

Brock, W.A., Some results on the uniqueness of steady states in multisector models when future utilities are discounted, International Economic Review 14, (1973), 535-559.

Clark, C.W., The economics of overexploitation, Science 181 (No. 4100), Aug. 17, 1973, 630-634.

__________, Antarctic whaling: a model of joint production (to appear).

Clark, C.W., and Munro, G.R., The economics of fisheries and modern capital theory: a simplified approach (forthcoming).

Copes. P., The backward-bending supply curve of the fishing industry, Scot. J. Polit. Econ. 17, (1970), 69-77.

Gordon, H.S., Economic theory of a common-property resource: the fishery, J. Polit. Econ. 62, (1954), 124-142.

Kurz, M., Optimal economic growth and wealth effects, International Economic Review 9, (1968), 343-357.

Leviatan, N., and Samuelson, P., Notes on turnpikes - stable and unstable, Journal of Economic Theory 1, (1969), 454-475.

Pontrjagin, L.S., *et al.*, *The Mathematical Theory of Optimal Processes*, New York, Interscience, 1962.

Samuelson, P.A., The two-part golden rule deduced as the asymptotic turnpike of catenary motion, Western Econ. J. 6, (1967/68), 85-89.

Schaefer, M.B., Some consideration of population dynamics and economics in relation to the management of marine fisheries. J. Fish. Res. Bd. Canada 14, (1957), 669-681.

Scott, A.D., The fishery: the objectives of sole ownership, J.Polit. Econ. 63, (1955), 116-124.

COMMANDE STOCHASTIQUE D'UN SYSTEME DE STOCKAGE

G. BORNARD — Attaché de recherche au CNRS
J.F. CAVASSILAS — Maître-Assistant
Laboratoire d'Automatique de Grenoble

INTRODUCTION.

Dans les systèmes physico-chimiques complexes, l'interconnection des sous-systèmes fait généralement intervenir des éléments d'accumulation destinés à découpler partiellement ces sous-systèmes, en évitant que les perturbations se transmettent directement de l'un à l'autre, et à faciliter ainsi leur conduite simultanée. En effet les débits d'entrée et de sortie d'un bac peuvent être momentanément différents sans qu'il y ait solution de continuité dans le fonctionnement du système.

Cependant dans la pratique, la commande de ces éléments se réduit le plus souvent à une régulation linéaire du niveau autour d'un point de consigne fixe. Ce mode de commande ne permet sans doute pas d'exploiter au mieux les possibilités de découplage offertes. A la limite, une régulation "parfaite" du niveau rendrait même un bac équivalent à une ligne directe.

Une commande plus efficace peut être obtenue en remarquant que les impératifs à respecter sont les suivants :

- la quantité accumulée (ou le niveau du bac) doit être à tout instant comprise entre une borne supérieure et une borne inférieure (le bac ne doit ni déborder, ni se vider complétement).

- Entre ces deux limites le niveau peut par contre évoluer librement, et aucune valeur particulière n'est priviligiée.

Basée sur les deux points précédents, l'étude qui suit constitue une tentative d'utilisation rationnelle d'un élément d'accumulation pour le découplage des perturbations entre deux sous-systèmes.

Le système considéré est représenté sur la figure 1 : le bac est alimenté par le débit fluctuant d_1 (variable de perturbation) venant du sous-système 1, et alimente lui-même le sous-système 2 par le débit d_2 (variable d'action). On désire que le débit d_2 soit aussi peu fluctuant que possible, de façon à limiter les perturbations transmises du sous-système 1 au sous-système 2.

Dans ces conditions l'idéal serait de maintenir d_2 constant. Cet objectif étant évidemment impossible à satisfaire, nous avons choisi de modifier la valeur de d_2 le moins fréquemment possible, et de faire ainsi évoluer d_2 par paliers dont la longueur moyenne est la plus grande possible. Le critère de la commande envisagée consiste alors à choisir les instants de commutation et la nouvelle valeur de d_2 après chaque commutation de façon à rendre maximum l'espérance mathématique du temps qui s'écoule entre deux commutations.

Ce critère est justifié en particulier si cette espérance mathématique est grande devant le temps de réponse du sous-système 2.

Dans ce qui suit, deux approches sont proposées pour résoudre ce problème : la première, probabiliste, basée sur l'hypothèse gaussienne, passe par l'évaluation de la probabilité conditionnelle de sortie du domaine permis, à chaque période d'échantillonnage. La seconde, pour éviter ces calculs complexes, prend en compte moins d'information, et permet aussi de faire une hypothèse moins restrictive sur la nature des fluctuations. Des résultats de simulation sont présentés pour la seconde approche.

I - PREMIERE APPROCHE.

I-1. Hypothèses et notations.

La figure 2 représente le block-diagramme du système : le débit fluctuant d_1 est la somme de son espérance mathématique d_0 et d'une fluctuation résultant du passage dans un filtre linéaire d'un bruit blanc, gaussien, centré, stationnaire, d'écart type connu σ_η. Le débit d_2 vient se soustraire à d_1 à l'entrée de l'intégrateur dont la sortie est le niveau x_1 du bac.

Pour simplifier l'écriture on utilisera par la suite la variable $d=d_2-d_0$.

Le système sera représenté par une équation d'état discrète :

$$x(k) = A.x(k-1) + B.d(k-1) + C.\eta(k-1) \tag{1}$$

où l'élément x_1 du vecteur d'état x est le niveau du bac, et ou A B et C sont les matrices définissant les caractéristiques du bac et du bruit.

Remarquons que le bac lui-même étant représenté par un intégrateur, l'ordre du système est égal à celui du bruit plus 1.

I-2. Critère de commutation.

La stratégie de commande envisagée comporte, aux instants d'échantillonnage, deux étapes successives :

1 - Décision de modifier la valeur de d (si le système est arrivé dans une zone critique) ou de ne pas la modifier dans le cas contraire. La modification sera aussi appelée commutation.

2 - Choix de la nouvelle valeur du débit d si la commutation vient d'être décidée.

La première étape implique la définition d'un critère de commutation, susceptible de reconnaître une situation où la commutation s'impose ou est avantageuse. Un tel critère peut être par exemple basé sur la probabilité d'un dépassement des contraintes par le niveau, sur un horizon défini, compte tenu des mesures déjà faites. Un autre critère intéressant est celui utilisé dans la deuxième approche.

Ce critère de commutation définit dans l'espace d'état deux domaines : le domaine D à l'intérieur duquel la commutation n'est pas jugée opportune, et son complément où au contraire la commutation sera décidée.

Dans ce qui suit nous supposons que la connaissance du critère a permis de définir ce domaine D.

I-3. Calcul de la nouvelle valeur de débit d.

Lorsque la commutation a été décidée, la nouvelle valeur d° qui doit être affectée à d est celle qui rend maximum l'espérance mathématique TM du temps qui s'écoulera jusqu'à la prochaine commutation, compte tenu de l'état actuel du système.

Il n'est pas possible d'obtenir une expression analytique directe de d°. Cette valeur sera donc obtenue par maximisation numérique par rapport à d, de la fonction TM (d, x(kc)) (kc : instant de commutation actuel).

C'est donc TM que nous allons calculer maintenant en supposant connus le débit d, la densité de probabilité multidimensionnelle px (x,kc) de x(kc), ainsi que la densité de probabilité $p\eta(\eta)$ du bruit blanc.

Définissons les probabilités :

$$\varphi(k) = Pr.\{x(k) \notin D \cap x(i) \in D \,.\, \forall i \in [kc, k\text{-}1]\} \tag{2}$$

$$\phi(k) = Pr.\{x(i) \in D,\ \forall i \in [kc, k]\} \tag{3}$$

$$\psi(k) = Pr.\{x(k) \in D \;/\; x(i) \in D,\ \forall i \in [kc, k\text{-}1]\} \tag{4}$$

et de plus la densité de probabilité conditionnelle :

$$pc(x,k) = p\{x(k) = x \;/\; x(i) \in D,\ \forall i \in [kc, k\text{-}1]\} \tag{5}$$

Dans ces conditions la valeur de TM (exprimée en périodes d'échantillonnage) est donnée par la relation :

$$TM = \sum_{i=1}^{+\infty} \varphi(kc+i).(i-kc) \tag{6}$$

En utilisant en particulier la relation des probabilités conditionnelles

$$Pr.\{A \cap B\} = Pr.\{A/B\}.Pr.\{B\}$$

ainsi que l'équation d'état du système, on obtient pour l'évaluation de ρ, ϕ, ψ, et pc les relations récurrentes suivantes :

$$\phi(k) = \phi(k\text{-}1).\psi(k) \tag{7}$$

$$\varphi(k) = \phi(k\text{-}1).[1-\psi(k)] \tag{8}$$

$$\psi(k) = \int \!\ldots\! \int_D pc(x,k).dx \tag{9}$$

$$pc(x,k) = \frac{\int_{\eta_1}^{\eta_2} pc(\tilde{x}, k\text{-}1).p_\eta(\eta).d\eta}{\psi(k)} \tag{10}$$

avec η_1 et η_2 tels que $\tilde{x} \in D$ pour $\eta \in [\eta_1, \eta_2]$ (11)

et $\tilde{x} = A^{-1}.(x - B.d(k\text{-}1) - C.\eta)$ (12)

Cette dernière relation définit le sous-espace d'état où x(k-1) doit se trouver nécessairement pour pouvoir conduire à x(k) = x.

On connaît de plus les valeurs initiales ϕ(kc) et pc(x,kc) :

$$\phi(kc) = \int \!\ldots\! \int_D p_x(x, kc).dx \tag{13}$$

$$pc(x,kc) = p_x(x,kc) \tag{14}$$

Dans ces conditions nous disposons de tous les éléments pour calculer TM en fonction de d , donc pour calculer le débit optimal d° correspondant aux conditions initiales données.

I-3. Mise en oeuvre.

Si on suppose pouvoir résoudre en temps différé les équations (6) à (14), la mise en oeuvre en temps réel est simple. La décision de commutation et la valeur optimale d° du débit sont fonction de l'état x(k) du système. Si une estimation de cet état est disponible à chaque instant, il suffit d'avoir mémorisé les caractéristiques du domaine D et la valeur de d° pour un nombre de points suffisant de l'espace d'état. Le calcul de la commande consiste alors en une interpolation dans une table de données.

Par contre la résolution elle-même des équations (6) à (14) est très lourde. Du fait des équations (10) et (12), le calcul nécessite l'emploi successif, à chaque pas, de procédures numériques d'intégration très coûteuses en temps de calcul.

Pour pallier cet inconvénient, nous avons tenté d'employer une méthode plus simple : chaque élément du vecteur x a été quantifié. La densité pc(x) est alors remplacée par l'ensemble fini des probabilités pour x d'appartenir à chacun des domaines élémentaires, et les intégrales deviennent des sommes finies.

Cette procédure a donné de bons résultats dans le cas d'un système du premier ordre. Cependant, une précision suffisante n'a pas pu être obtenue pour les ordres supérieurs. En effet, la relation (12) ne fait pas de correspondre entre eux les domaines élémentaires aux instants k et k-1, ce qui pose un problème d'évaluation de pc($\hat{x}$, k-1) dans l'équation (10).

C'est une des raisons qui nous a conduits à développer la seconde approche présentée plus loin.

Des résultats partiels satisfaisants ont toutefois pu être obtenus, en particulier lorsque l'état initial est situé loin des contraintes. Quelques évolutions type sont présentées ci-dessous.

Les paramètres du système sont les suivants :

$$A = \begin{bmatrix} 1 & 0,907 \\ 0 & 0,819 \end{bmatrix} \qquad B = \begin{bmatrix} 1 \\ 0 \end{bmatrix} \qquad C = \begin{bmatrix} 0 \\ 0,2 \end{bmatrix}$$

$\sigma_\eta = 0,4$

$\sigma_x = 0$

Le critère de commutation, proche de celui cité en I-2., est défini pour les inégalités :

$$-1 < x_1 < 1$$
$$-1 < x_1 + 2.x_2 < 1$$

Les figures 3a, b, c montrent l'évolution des probabilités ψ, φ et ϕ pour $x(0) = \begin{bmatrix} 0,6 \\ 0 \end{bmatrix}$ et d=0. La stabilisation de la valeur de ψ traduit la mémoire finie du système. La figure 4 montre comment varie l'espérance mathématique TM en fonction du débit d.

II - DEUXIEME APPROCHE.

II-1. Présentation des notations du problème.

Les mesures du niveau et de la dérivée du niveau du liquide dans le réservoir, sont effectuées à des dates équidistantes dans le temps. A partir de chacune de ces époques, nous pouvons calculer l'évolution la plus probable m(t') du niveau puisque avec nos hypothèses, les fluctuations obéissent à la statistique d'un processus résultant du filtrage d'un bruit gaussien à corrélation microscopique par un filtre linéaire du second ordre amorti de plus de façon critique.

L'origine de t' est la date t à laquelle nous venons d'effectuer les dernières mesures du niveau et de sa dérivée. Nous supposons de plus que les fluctuations du niveau sont centrées à mi-hauteur du réservoir en l'absence de commande de débit et que par conséquent, ayant calculé et commandé un débit, nous sommes toujours à même de prévoir à chaque instant t la référence exacte R(t) des fluctuations, ce qui revient à dire qu'à chaque instant nous connaissons le niveau moyen du réservoir.

Pour un bruit x(t), centré, gaussien, du second ordre, la densité de probabilité conditionnelle $p\ (x(t')/x(t),\ \dot{x}(t))$ est donnée par :

$$\frac{1}{\sqrt{2\pi}\,.\,\sigma_1(t')} \cdot e^{-\frac{1}{2\sigma_1(t')^2}\left[x(t') - m(t')\right]^2} \tag{15}$$

où

$$m(t') = \frac{\Gamma(t')}{\Gamma(0)} \cdot x(t) + \frac{\alpha(t')}{\beta} \cdot \dot{x}(t) \tag{16}$$

$$\sigma_1^2(t') = \Gamma(0) - \frac{\alpha^2(t')}{\beta} - \frac{\Gamma(t')}{\Gamma(0)} \tag{17}$$

avec

$$\alpha(t') = -\frac{d}{dt'}\Big(\Gamma(t')\Big) \tag{18}$$

$$\beta = -\left[\frac{d^2}{dt'^2}\Big(\Gamma(t')\Big)\right]_{t'=0} \tag{19}$$

Ce calcul a été effectué simplement, en utilisant le théorème de Bayes et la formule bien connue :

$$E\left\{\frac{d^n}{dt^n}\,x(t+\tau).\frac{d^m}{dt^m}\,x(t)\right\} = (-1)^m\,\frac{d^{m+n}}{dt^{m+n}}\,\Gamma(\tau)$$

Dans le cas précis de notre problème, la prédiction du niveau, effectuable à la date t, sera donnée par :

$$m(t') = R(t) + [\,x(t) - R(t)].\frac{\Gamma(t')}{\Gamma(0)} + \frac{\alpha(t')}{\beta}\cdot\dot{x}(t) \tag{20}$$

R(t) représente le niveau moyen du réservoir à la date t. La formule (20) suppose naturellement que toute commande ait cessé pour $t' > 0$.

II-2. <u>Les idées proposées pour le choix d'une commande de débit</u>.

Supposons qu'à la date t, à laquelle nous venons de relever les mesures x(t) et $\dot{x}(t)$, nous nous proposions d'effectuer une commutation en choisissant une commande "acceptable" de débit.

Les courbes $m(t')+\sigma_1(t')$ et $m(t') - \sigma_1(t')$, nous permettant d'apporter un élément de réponse à notre question. En effet si l'une d'entre elles (la courbe $m(t') + \sigma_1(t')$ pour fixer les idées) intercepte le niveau supérieur admissible du réservoir, il est prévisible qu'il existe un certain danger qui peut être pallié en partie en commandant un débit tel que le niveau moyen du réservoir tende à diminuer. D'un autre côté, une action trop violente tendra à vider le réservoir trop rapidement.

Nous optons pour la solution qui consisté à choisir le débit d de façon à ce que la courbe $m(t') + \sigma_1(t') - d.t'$ soit tangente au niveau supérieur admissible, ce qui veut dire que la courbe $m(t')-\sigma_1(t') - d.t'$ coupera nécessairement le niveau admissible inférieur et qu'il faudra envisager à nouveau une commutation avant cette date.

II-3. Les idées proposées pour le choix des dates de commutation.

Appelons t'' la date à laquelle le niveau moyen du liquide dans le réservoir atteindrait le niveau inférieur si l'on appliquait la commande **d** à partir de l'instant t. Il est clair que nous avons tout intérêt à ce que l'intervalle (t''-t) soit aussi grand que possible. D'autre part nous avons intérêt également, compte-tenu de nos désirs, à ce que l'intervalle de temps séparant t, de la date t_c de la dernière commutation précédent t, soit également aussi grand que possible. Ceci nous amène à optimiser la quantité

$$\Delta = [(t - t_c) + (t'' - t)]$$

Lorsque pour un t donné la quantité Δ devient maximum nous décidons, de prendre cette date comme instant de commutation et de prendre pour débit de commande le débit d, calculé comme nous l'avons déjà indiqué.

II-4. Simulation.

Le programme de calcul de la commande est donc établi sur les bases suivantes :

1 - On calcul à chaque instant t les expressions
$m(t') \pm \sigma_1(t')$

2 - Si quel que soit $t' \geqslant 0$, nous avons $0 < m(t') \pm \sigma_1(t') < H$
alors t, n'est pas un instant de commutation

3 - Si $m(t') + \sigma_1(t') \geqslant H$ ou bien si $m(t') - \sigma_1(t') \leqslant 0$ pour certaines valeurs de t', alors t peut être un instant de commutation. Pour trancher, il faut étudier la quantité $\frac{d\Delta}{dt}$; si $\frac{d\Delta}{dt} > 0$ alors t, n'est pas un instant de commutation.

Une simulation a été faite en prenant un bruit dont la fonction d'autocorrélation est donnée par :

$$(1 + a|\tau|).e^{-a|\tau|}$$

Ce bruit est obtenu en prélevant la tension aux bornes de la capacité d'un circuit oscillant à l'amortissement critique, excité par un générateur de bruit blanc gaussien ; le coefficient a=667 Hz ; le temps d'échantillonnage est ici de 600µs. Nous avons de plus prélevé simultanément les valeurs des dérivées de ce bruit aux bornes de la résistance. Ces échantillons ont été enregistrés sur disque et traités ensuite sur T 2000. Dans l'application effectuée, l'écart-type du bruit sur le niveau est pris égal à 1 et la hauteur du réservoir égale à 3.

Sur la figure 5 , la courbe continue I représente la fonction d'autocorrélation des fluctuations du niveau sans commande. La courbe II est obtenue en joignant les points successifs caractérisant les hauteurs atteints à l'intérieur du réservoir lorsque le processus est commandé. Les instants de commutation sont mentionnés. Entre ces instants le débit de commande est naturellement constant (III). L'unité de débit est prise égale à a.σ.

CONCLUSION.

L'étude qui précéde a été motivée par l'idée que les éléments d'accumulation présents dans un système complexe ont un rôle particulier à jouer dans la commande, en vue de limiter la transmission des perturbations dans les sous-systèmes couplés en série.

Nous avons proposé pour un cas simple, une stratégie de commande. Deux approches du problème ont été effectuées. La seconde est de mise en oeuvre simple, et des résultats de simulation satisfaisants ont été obtenus.

Les limites du travail présenté se situent au niveau de la caractérisation des perturbations agissant sur le système. Si la caractérisation statistique retenue paraît raisonnable pour les fluctuations à fréquence élevée, celle des fluctuations à basse fréquence est beaucoup plus incertaine.

Pour ces dernières il serait certainement intéressant d'introduire une prédiction basée sur des informations en provenance des autres parties du système global : mesures, prédictions basées sur des modèles déterministes... etc, ce qui aurait aussi pour effet de situer le problème à son vrai niveau : celui de la commande du système global.

LISTE DES SYMBOLES UTILISES.

- Première approche

$d_1(k)$: débit d'entrée du bac

$d_2(k)$: débit de sortie du bac

$d(k)$: pseudo variable d'action = d (k) - $d_1(k)$

$\eta(k)$: séquence non corrélée, gaussienne, centrée, stationnaire

$x(k)$: vecteur d'état du système

σ_η : écart-type de η

σ_x : écart-type de l'estimation initiale de x

A, B, C : matrices de paramètres de l'équation d'état discrète (1)

TM : espérance mathématique, estimée lors d'une commutation, du temps qui s'écoulera avant la procheine commutation

d° : valeur optimale recherchée pour d

D : domaine d'excursion alloué au vecteur d'état x

$\psi(k)$: probabilité de sortie du domaine D entre k-1 et k conditionnellement à une évolution comprise dans D à tous les instants antérieurs à k.

$\rho(k)$: probabilité de première sortie entre k-1 et k

$\phi(k)$: probabilité d'évolution comprise dans D jusqu'à k

pc(x,k) : densité de probabilité d'avoir x(k) = x conditionnellement à une évolution antérieure à l'intérieur de D.

k : instant courant

kc : dernier instant de commutation avant k.

- Deuxième approche

$\Gamma(\tau)$: fonction d'autocorrélation des fluctuations de niveau

σ : écart-type de ces fluctuations

t : l'instant de mesure des grandeurs niveau et dérivée du niveau

t' : date de prédiction dont l'origine est t

t_c : date de la dernière commutation qui précéde t

H : hauteur totale du réservoir

x(t) : niveau du réservoir à la date t

$\dot{x}(t)$: dérivée du niveau de réservoir à la date t

m(t') : valeur moyenne prédite du niveau

$\sigma_1(t')$: écart-type de la prédiction

d : éventuelle valeur du débit de commande si la décision de commuter est prise

t'' : date à laquelle le niveau moyen du liquide atteindrait les niveaux extrêmes après une éventuelle commande de débit à la date t.

Système étudié

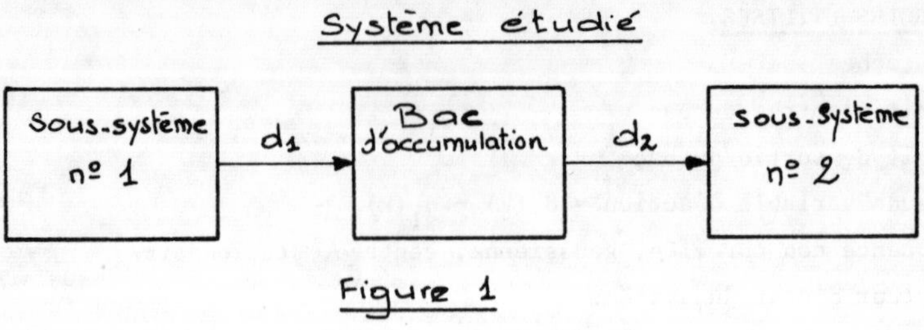

Figure 1

Block-diagramme

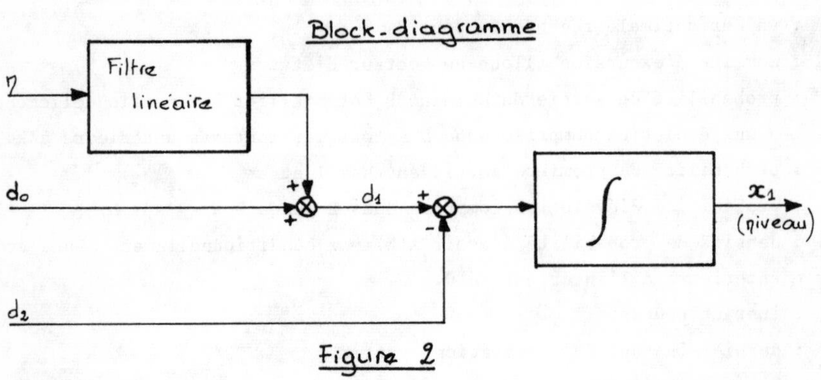

Figure 2

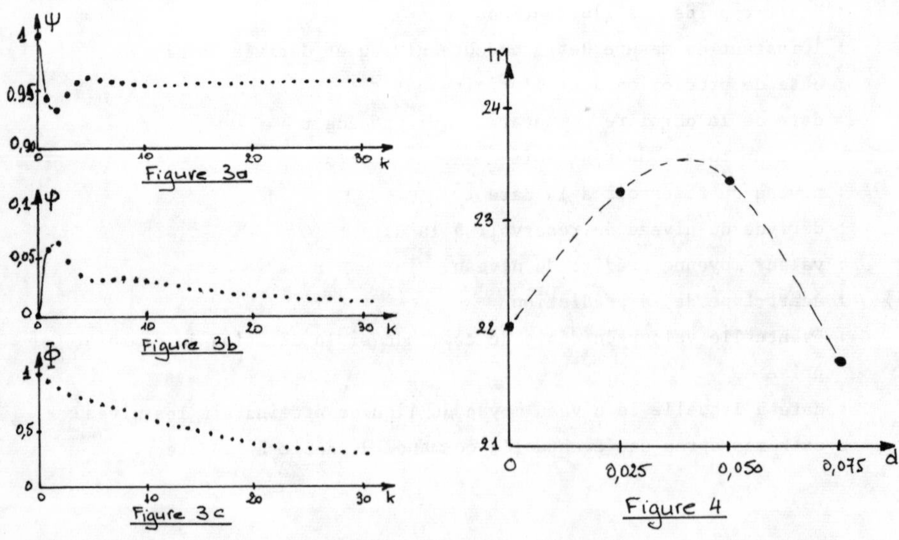

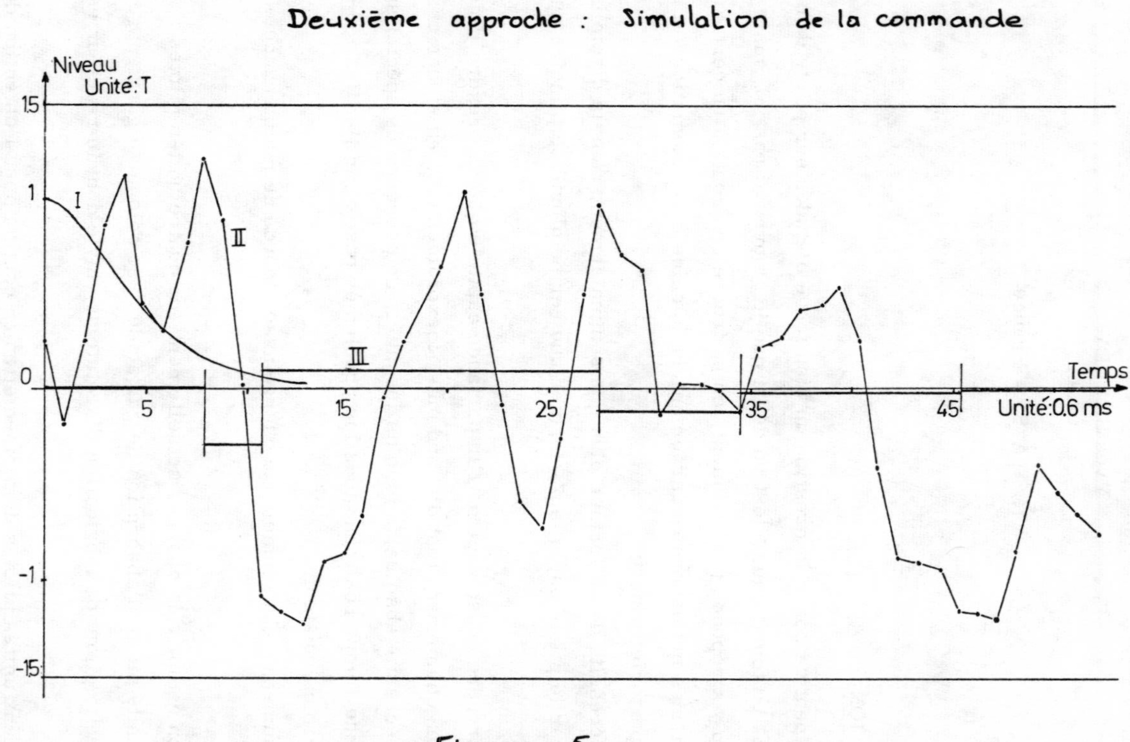
Deuxième approche : Simulation de la commande
Niveau
Unité: T
1.5
1
0
-1
-1.5
I
II
III
5
15
25
35
45
Temps
Unité: 0.6 ms

Figure 5

ETUDES D'AUTOMATIQUE SUR UNE UNITE PILOTE D'ABSORPTION ET SON MELANGEUR

par

Yves SEVELY
Professeur

L.A.A.S. - France

INTRODUCTION

On peut se poser la question du bien fondé d'études sur pilote dans un Laboratoire d'Automatique. C'est un débat que je n'ouvrirai pas ici, mais dès lors que l'on y a répondu positivement, on se trouve confronté à un certain nombre de problèmes qui débordent largement du domaine de l'automatique et que l'on peut répertorier ainsi :

- nécessité de se former plus que sommairement dans la discipline concernée par le pilote ou, mieux, former une équipe pluridisciplinaire autour de lui.

- prendre conscience de l'importance capitale que va prendre l'instrumentation, en équipement tout d'abord puis, inexorablement, en études ensuite.

- nécessité absolue, au moins jusqu'ici, de présence de spécialistes d'informatique-temps réel pour tous les problèmes posés par l'utilisation de mini ou micro calculateurs.

- admettre et assumer une charge assez lourde de maintenance et de gestion.

Mon équipe a la charge, actuellement, de quatre unités pilotes :

- unité pilote d'absorption de gaz acide avec colonne à garnissage

- unité pilote de fabrication d'acétate de vinyle dont le réacteur est en lit fluide

- deux unités pilotes de fermentation en continu pour l'obtention de protéines à partir, respectivement, d'hydrocarbures légers et de méthanol.

Nous nous intéresserons, dans cette communication, à l'unité pilote d'absorption [1] et au mélangeur [2] qui lui est associé.

DESCRIPTION SOMMAIRE DE L'UNITE (figure 1)

La partie centrale est la colonne d'absorption (hauteur : 2 m, diamètre : (10 cm) qui reçoit à la partie inférieure le mélange d'air et de gaz carbonique à épurer (dans l'industrie c'est par exemple le gaz naturel à débarasser du gaz carbonique et de l'anhydride sulfureux) et, à la partie supérieure, le liquide absorbant le CO_2 (mélange de diethanolamine et d'eau). Le liquide est récupéré en bas de colonne, débarrassé par chauffage de son CO_2 dans un dispositif de regénération et recyclé en tête de colonne. Le CO_2 libéré par l'amine est recyclé lui aussi, dans notre pilote, pour des raisons d'économie et mélangé à de l'air pour constituer le gaz à épurer, de débit et de concentration variables à notre gré (valeurs nominales ; débit gazeux 8 m^3/H, concentration en CO_2 de l'ordre de 20 à 30 %, débit liquide 150 l/H).

La colonne fonctionne, à une légère surpression près, à la pression atmosphérique et est garnie d'anneaux Raschig pour faciliter les échanges gaz-liquide (les unités industrielles sont à haute pression et comportent des colonnes à plateaux).

Le problème économique lié aux unités réelles est de minimiser le débit d'amine tout en assurant une concentration en gaz acides en tête de colonne impérativement inférieure à un certain seuil.

Les spécialistes du processus sont également intéressés par les problèmes suivants :

- suivre dans le temps la capacité d'absorption de l'amine
- peut-on améliorer la dépense énergétique liée à la regénération de l'amine en complétant l'injection en tête de colonne par une injection d'amine, peut être moins bien regénérée à une hauteur intermédiaire de la colonne, à déterminer
- déterminer la hauteur optimale des colonnes pour les futures installations.

Nous donnerons ci-dessous quelques exemples d'études d'application déjà effectuées ou en cours.

I - ETUDES LIEES AU MELANGEUR

Sans nous attarder sur cet aspect du problème nous signalerons la mise au point d'un programme permettant l'étalonnage automatique des divers débitmètres, depuis l'expérimentation (positionnement des vannes par le calculateur, attente du régime permanent, calcul du débit du compteur étalon à partir d'impulsions, scrutation des divers capteurs pour déterminer la composition du gaz, incrémentation de l'ouverture des vannes) jusqu'à l'établissement des relations explicites entre débits et tensions et le tracé automatique des courbes d'étalonnage.

L'importance de la composition du mélange air-CO_2 dans les indications du débitmètre (8 % de variation pour une concentration de CO_2 variant de 0 à 30 % (figure 2) nous a suggéré l'étude et la mise au point d'un débitmètre dont la lecture serait indépendante de la nature du gaz [3].

L'identification dynamique des analyseurs et des actionneurs a été faite par diverses méthodes déterministes et/ou stochastiques. Si nous prenons l'exemple d'une vanne pneumatique nous en avons estimé la structure par inspection des réponses indicielles (figure 3). Une non linéarité essentielle est ainsi apparue comme un frottement sec traduisible par un retour en h sign $(\dot{x})$ sur une transmittance (figure 4). Cette dernière a été identifiée comme un second ordre par la méthode du filtre auxiliaire [4-5] dont nous rappelons rapidement le principe.

Soit à identifier ω_n, ζ et K de l'équation $\ddot{y} + 2\zeta\omega_n\dot{y} + \omega_n^2 y = Ku$ où seuls u et y sont accessibles ; on les considère comme entrées de deux filtres identiques de transmittance $\frac{S}{R} = \frac{\alpha^3}{(p+\alpha)^3}$ $\left\{ \begin{array}{l} R = y \longrightarrow S = S^y \\ R = u \longrightarrow S = S^u \end{array} \right.$

On obtient à la sortie des filtres $s^u, \dot{s}^u, \ddot{s}^u$ et $s^y, \dot{s}^y, \ddot{s}^y$. On est donc capable, à un instant d'observation discret, à partir de la relation

$$[\ddot{s}^y]_k + 2\zeta\omega_n[\dot{s}^y]_k + \omega_n^2[s^y]_k = K[s^u]_k$$

de calculer les paramètres inconnus soit par la méthode des moindres carrés, soit par l'algorithme du filtrage linéaire de Kalman-Bucy

$$\underline{\theta}_{k+1} = \underline{\theta}_k$$
$$z_k = \underline{\theta}_k^T \underline{x}_k + \varepsilon_k$$

où $\underline{\theta}_k^T = [-2\zeta\omega_n \ \ -\omega_n^2 \ \ K]$, $z_k = [\ddot{s}^y]_k$, $\underline{x}_k^T = [\dot{s}^y \ \ s^y \ \ s^u]_k$

Les résultats obtenus par cette méthode ont été confirmés par la technique du filtrage non linéaire.

La vanne a alors été simulée sur calculateur analogique et les paramètres ajustés par la méthode du modèle (figure 5). La sortie de la vanne était en fait le déplacement de la tige du moteur pneumatique solidaire du clapet, enregistré par l'intermédiaire d'un capteur à variation d'inductance couplé à un pont de mesure.

La conduite du mélangeur (obtention des divers signaux désirés à l'entrée de la colonne) a été effectuée par calculateur numérique associé à un simulateur original. Le principe de ce dernier a déjà été exposé [2 , 9] . Nous le rappelons sur la figure 6.

Le problème d'automatique à résoudre est le suivant : en supposant que le processus de mélange soit statique (ce qui est physiquement raisonnable dans notre cas où le mélangeur est un T à chicanes) comment agir sur les débits d'entrée Q_1 (air) et Q_2 (air + CO_2 avec une concentration η de ce dernier) pour avoir en sortie un débit total Q et une concentration β . On est donc en présence d'un système à deux entrées (Q_1, Q_2), deux sorties (Q , β) et une perturbation mesurable (η).

$$Q = Q_1 + Q_2$$

$$\beta = \frac{Q_2 \, \eta}{Q_1 + Q_2}$$

On exige une non interaction entre Q et β.

Le principe retenu pour la synthèse globale a été :

- de nous affranchir, dans les transitoires, des capteurs physiques en utilisant les signaux du simulateur analogique (relations algébriques entre concentrations et débits)

- de réaliser des réponses "pile" (deadbeat response) sur chacune des entrées du mélangeur

- d'obtenir une faible sensibilité aux variations du gain des vannes en plaçant les pôles dominants de la transmittance du régulateur de débit sur les zéros d'un réseau correcteur introduit à cet effet.

La non interaction a été étudiée à partir des matrices de transfert (diagonalisation). Ce problème essentiellement non linéaire a été transformé en un problème linéaire à coefficiences variables dont le traitement est pris en compte par le calculateur numérique

$$\begin{bmatrix} dQ \\ d\beta \end{bmatrix} = \begin{bmatrix} 1 & 1 \\ -\frac{\beta_0}{Q_0} & \frac{\eta - \beta_0}{Q_0} \end{bmatrix} \cdot \begin{bmatrix} dQ_1 \\ dQ_2 \end{bmatrix}$$

L'étude a été menée en deux phases :

1° - mise en place d'une conduite en chaîne ouverte du mélangeur réel à partir d'un bloc synthétisé pour le simulateur (figure 7). On notera, sur cette figure, qu'ont été prises en compte, par le calculateur, les non linéarités des vannes.

2° - bouclage du mélangeur réel (figure 8) pour "contrer" les perturbations structurelles et extérieures (notons cependant que la principale perturbation, la concentration du gaz de recyclage, était déjà prise en compte dans la première phase).

La planche 1 donne des résultats de cette conduite pour des consignes simultanées en créneaux sur β et sinusoïdales sur Q.

La conduite du mélangeur est entièrement gérée par le calculateur numérique avec alarme lorsque l'on sort du domaine de fonctionnement permis en Q et β (figure 9) (domaine déterminé par les problèmes d'engorgement de la colonne et de précision des capteurs).

Dans le cas d'un fonctionnement normal, l'exécution de l'ensemble des tâches périodiques nécessaires à cette conduite a lieu toutes les secondes, elle dure 140 msec. La place approximative occupée en mémoire centrale pour l'ensemble des tâches est de 2 k mots de 12 bits.

Les diverses phases de cette étude ont été facilitées par un dispositif (hardware et software) permettant le tracé automatique du lieu des racines et l'étude de la sensibilité des systèmes synthétisés.

II - ETUDES LIEES A LA COLONNE D'ABSORPTION

II-1 - Modélisation - Identification

Il s'agit, rappelons-le, d'une colonne à garnissage (anneaux Raschig). Les équations à la hauteur z et au temps t sont obtenues en établissant un bilan matière dans le gaz et le liquide.

Nous avons retenu comme hypothèses simplificatrices de départ :

- homogénéité radiale (simplification habituelle en écoulement turbulent)
- écoulement piston (on néglige la diffusion axiale).

Notations :

W : "hold-up" liquide : moles de DEA/m^3 de colonne

U : "hold-up" gaz : moles de gaz/m^3 de colonne

X : concentration du CO_2 dans la DEA : moles de CO_2/moles de DEA

Y : concentration du CO_2 dans le gaz : moles de CO_2/moles de gaz

L : débit spécifique du liquide : moles de DEA/H/m^2

G_T: débit spécifique du gaz : moles de gaz/H/m^2

R : vitesse de transfert : moles de CO_2 transférées/H/m^2

a : surface spécifique de contact entre phases m^2/m^3 de colonne

S : section de la colonne.

$L(z+dz, t)$ ↓ — $z+dz$; z — ↓ $L(z,t)$

II-1-1 - Etablissement des équations

En écrivant le bilan de CO_2 dans le liquide, pour une tranche de hauteur dz pendant un temps dt :

CO_2 à la sortie	=	CO_2 à l'entrée	+	CO_2 transfére du gaz	-	CO_2 accumulé

$$\overline{LX}(z,t)\, S\, dt = \overline{LX}(z+dz,t)\, S\, dt + Ra\, S\, dz\, dt - \left[\overline{WX}(z,t) - \overline{WX}(z,t-dt)\right] S\, dz$$

En divisant par S dz dt et par passage à la limite on obtient

(1) $$\frac{\partial [WX]}{\partial t} = \frac{\partial [LX]}{\partial z} + Ra$$

Le bilan de CO_2 dans le gaz, pour cette même tranche, conduit à :

(2) $$\frac{\partial [UY]}{\partial t} = -\frac{\partial [G_T Y]}{\partial z} - Ra$$

Enfin, si l'on traduit le fait que le transfert de matière ne modifie pas le débit liquide :

(3) $$\frac{\partial L}{\partial z} = \frac{\partial W}{\partial t}$$

La prise en compte de la diffusion axiale revient à ajouter un terme de la forme $D_1 \frac{\partial^2 X}{\partial z^2}$ et $D_2 \frac{\partial^2 Y}{\partial z^2}$ dans les seconds membres des équations (1) et (2).

Ce système d'équations se simplifie

(1) + (3) → (1') $$W \frac{\partial X}{\partial t} = L \frac{\partial X}{\partial z} + Ra$$

Dans une colonne à garnissage on peut écrire

(4) $W \cong K_1 + K_2 L$

(5) $U \cong K_3 P$, P étant la pression, approximativement constante

d'où

(2) + (5) → (2') $$U \frac{\partial Y}{\partial t} = -\frac{\partial [G_T Y]}{\partial z} - Ra$$

Pour des raisons pratiques, si l'on mesure le débit d'air à l'entrée G_A et si on néglige l'air absorbé par la DEA, l'équation (2') devient

(2'') $$U \frac{\partial Y}{\partial t} = - G_A \frac{\partial}{\partial z}\left(\frac{Y}{1-Y}\right) - Ra$$

Enfin si l'on veut considérer le rapport molaire y (moles de CO_2/mole d'air) au lieu de la concentration molaire Y (moles de CO_2/mole de gaz), la deuxième équation s'écrit :

(2''') $$\frac{U}{(1+y)^2} \frac{\partial y}{\partial t} = - G_A \frac{\partial y}{\partial z} - Ra$$

Le terme Ra est une fonction inconnue de X et Y (ou y).

Le but de l'étude dans un premier temps est de calculer la loi de commande L (t) qui minimise $\int_{t_0}^{t_f} L(t)\,dt$ sous les contraintes $L(t) \leqslant L_{max}$, $\dot{L}(t) \leqslant \dot{L}_{max}$ et $Y(Z, t) \leqslant Y_{max}$ (Y (Z, t) : concentration en gaz acide en tête de colonne) pour des perturbations en débit G_T (0, t) et/ou en concentration Y (0, t) (ce que nous avons appelé Q et β dans la première partie de l'exposé).

Nos travaux se sont alors orientés dans deux directions :

1° - les perturbations et la commande s'effectuant aux limites, un modèle externe $Y(Z, t) = f\left[L(Z,t),\ G_A(0, t),\ Y(0, t)\right]$ est suffisant pour traiter le problème de l'optimisation.

2° - essayer d'identifier le terme de transfert Ra et faire la commande à partir du modèle aux dérivées partielles.

II-1-2 - Modèle externe

Le modèle statique a été obtenu à partir de résultats expérimentaux par la technique classique des régressions (moindres carrés). Un modèle adaptatif, les coefficients étant fonction du débit liquide, permet de couvrir l'ensemble du domaine avec la précision des capteurs :

$$\Delta Y(z,t) = -0{,}038\,\Delta L(t) + \left[0{,}96 - 0{,}11\,L(t)\right]\Delta Y(0,t) + \left[0{,}077 - 0{,}009\,L(t)\right]\Delta G_A(t) + \left[-0{,}045 + 0{,}002\,L(t)\right]\left[\Delta G_A(t)\right]^2$$

Le modèle dynamique a été obtenu par des réponses indicielles sur les divers sous systèmes.

Les transmittances peuvent être approchées par des expressions de la forme $e^{-\tau_i p}/1+T_i p$. Commeil était prévisible la dynamique relative à l'écoulement liquide (transmittance entre Δ L (0, p) et Δ Y (Z, p) est bien plus lente que celle de l'écoulement gazeux : cela posera des problèmes délicats pour la commande.

II-1-3 - Identification du modèle aux dérivées partielles

A partir des équations (1'), (2'''), (3) et (4) nous avons, dans le cadre d'un projet d'étudiants de dernière année INSA [7], procédé à une simulation hybride assez sommaire (couplage d'une TR 48 + DES 30 au calculateur numérique C 90-10) avec une expression linéaire de Ra: $Ra = AX + By$

La discrétisation a porté sur la variable temporelle. Les fonctions à mémoriser X_i (z), Y_i (z), W_i (z) et L_i (z) l'ont été à partir de 100 points avec un pas de discrétisation de 40 msec.

Dans un premier temps on résout les équations (1'), (3) et (4) en utilisant les valeurs obtenues au cycle précédent. Puis on résout (2''').

L'enchaînement des modes de fonctionnement de la calculatrice analogique se faisait à partir de la console logique DES 30, elle même pilotée par deux signaux u et v issus du calculateur numérique selon le schéma :

u	0	1	0
v	0	0	1
X	CI	OP	CI
y	H	CI	OP

CI : condition initiale
OP : fonctionnement
H : mémoire

Les paramètres inconnus A et B ont été obtenus par approximations successives manuelles.

Nous n'insisterons pas sur les résultats de ce travail exploratoire, l'étude devant être reprise, au Laboratoire, sur le calculateur hybride en cours de mise au point (EAI 680 + MITRA 15) :

1° - en tenant compte des termes diffusionnels

2° - en prenant un développement de Ra arrêté au second ordre

3° - en minimisant une distance entre y expérimental et y du modèle non seulement en régime permanent mais aussi en transitoire.

Il est clair que ce dernier objectif était des plus ambitieux car il pose de difficiles problèmes d'expérimentation.

II-1-4 - Dans le cadre des études générales d'instrumentation pour les unités pilotes chimiques, nous avons réalisé le couplage d'un spectrographe quadripolaire et d'un calculateur numérique en vue de l'analyse rapide, quantitative de mélanges gazeux et liquides [6].

En ce qui concerne les gaz, ici Y (z, t) nous avons pu descendre à une analyse par seconde et espérons augmenter encore un peu la fréquence d'échantillonnage.

Nous indiquons sommairement ci-après le principe de la mesure en supposant connus les spectres de masse des corps du mélange : on peut se reporter à la bibliothèque des spectres mais il est conseillé de reprendre cet étalonnage avec son propre appareil à la même pression d'introduction et avec la même énergie d'ionisation (nous donnons, figure 10, le spectre de masse du CO_2).

Les pics du spectre du mélange sont mis en mémoire tampon par l'intermédiaire d'amplificateurs de valeur crête et rangés en mémoire du calculateur sur interruption externe, ceci afin d'occuper le moins possible ce dernier.

Principe théorique de l'exploitation des spectres en vue de l'analyse quantitative

Supposons que l'on ait un mélange de n composés C_j de concentration x_j. Le spectre de masses de C_j donnant à la masse i une hauteur h_{ij}, le spectre du mélange donnera à cette même masse une hauteur :

$$y_i = \sum_{j=1}^{n} h_{ij} x_j = \underline{h}_i^T \underline{x} \qquad i = 1 \text{ à } m > n$$

Nous avons donc à calculer $\underline{x}$, vecteur des pourcentages relatifs, à partir de :

$\underline{y} = H \underline{x}$

$\underline{y}$: vecteur du spectre du mélange mesuré (dimension : m)

H : matrice des spectres des produits de référence (dimension mxn)

Etude de la précision

Nous avons étudié la précision des résultats dans deux cas :

H supposée comme sans erreur et H biaisée

a) H connue sans erreur :

nous avons $\underline{y} = H \underline{x} + \underline{e}$,

les composantes e_i étant supposées gaussiennes de variance σ^2 et indépendantes entre elles. On obtient x par minimisation de $\underline{e}^T \underline{e}$ ce qui donne

$$\hat{\underline{x}} = [H^T H]^{-1} H^T \underline{y}$$

Nous pouvons suivre l'évolution d'un mélange composé toujours des mêmes corps en calculant une fois pour toutes $[H^T H]^{-1} H^T$.

La variance d'erreur d'estimation sur x_i est P_{ii}, terme diagonal de la matrice $E[(\underline{x}-\hat{\underline{x}})(\underline{x}-\hat{\underline{x}})^T] = \sigma^2 [H^T H]^{-1} = [P_{ij}]$

On en déduit l'intervalle de confiance à deux écarts types pour x_i

$$x_i - 2\sqrt{P_{ii}} \leq x_i \leq x_i + 2\sqrt{P_{ii}}$$

L'erreur relative calculée est de l'ordre du millième (cf. tableau I).

b) H biaisée

Le calcul de la précision des résultats est basé sur la théorie du filtrage non linéaire.

L'équation de départ est $\underline{y} = \left[H + \delta H \right] \underline{x} + \underline{e}$

H représentant le biais sur H et $\underline{e}$ le bruit de mesure.

Nous utilisons la formule récurrente suivante :

$$y_{i+1} = \underline{h}^T_{i+1} \underline{x}_i + \underline{\delta h}^T_{i+1} \underline{x}_i + e_i$$

où y_{i+1} est la hauteur du spectre du mélange du pic de la masse i + 1

$\underline{h}_{i+1}$ vecteur des spectres des corps purs constitutifs du mélange à la masse i+1

$\underline{x}_i$ valeurs estimées des pourcentages des corps purs dans le mélange à partir de i premiers pics

L'exploitation de l'équation ci-dessus peut se faire que si l'on fait des hypothèses sur la nature du biais. Nous avons supposé : $\underline{\delta h}_{i+1} = \underline{\delta h}_i$ c'est-à-dire

$$\begin{bmatrix} \underline{x} \\ \underline{\delta h} \end{bmatrix}_{i+1} = \begin{bmatrix} \underline{x} \\ \underline{\delta h} \end{bmatrix}_i$$

Les résultats relatifs à un mélange d'air et de CO_2 sont donnés dans le tableau I. Malgré une meilleure précisions nous n'avons pas retenu cette méthode pour une utilisation en ligne compte tenu de la lourdeur des calculs.

Sur le plan pratique et moyennant certaines vérifications a posteriori on peut même suivre l'évolution de la concentration à partir du pic principal seulement et avec une précision acceptable.

Bouteille étalon % CO_2	% CO_2		Intervalle de confiance		$\frac{X}{X}$ %	
	moindres carrés	filtrage non linéaire	moindres carrés	filtrage non linéaire	moindres carrés	filtrage non linéaire
8, 9	8, 4	9, 1	0, 022	0, 044	0, 25	0, 49
17, 1	16, 2	16, 7	0, 027	0, 063	0, 15	0, 37
21, 3	20, 7	21	0, 025	0, 046	0, 12	0, 22
28, 6	28, 8	29, 4	0, 037	0, 089	0, 12	0, 31
60	56	57, 9	0	0, 077	0	0, 13

Tableau I

Résultats

A partir de l'appareil ainsi réalisé et du montage schématisé sur la figure 11, nous avons pu relever le profil statique Y (z, ∞) (figure 12) et dynamique Y (z, t) (figure 13) pour des échelons sur Y(0, t) avec L(t) = cte. Le tracé de cette surface caractéristique constitue, à notre connaissance, un résultat original en génie chimique.

Remarquons au passage que l'allure des courbes Y (z, t) ne permet pas de conserver l'hypothèse simplificatrice d'un écoulement piston : la modélisation doit tenir compte du terme diffusionnel.

II-1-5 - Etude d'un capteur permettant la mesure de X

C'est un problème important, sur le plan industriel, que de connaître l'évolution, dans le temps, de la capacité d'absorption de l'amine en CO_2 (et plus généralement en gaz acides). Sur le plan de la modélisation il serait intéressant également de pouvoir déterminer X (z, t).

Nous avons imaginé un dispositif, en cours d'expérimentation, qui ne permet pour l'instant que des mesures statiques ou, tout au moins, assez espacées.

Il ne nous est pas possible d'en donner ici le principe, un brevet étant en cours de dépôt.

II-2 - Commande - Optimisation

II-2-1 - Par programmation linéaire à partir du modèle externe

Cette approche par programmation linéaire, particulièrement simple lorsque l'on dispose du programme correspondant, nous a permis d'avoir une première idée des problèmes qui se posent.

En posant Y (Z, t) = y(t)

Y (0 , t) = u_1(t) perturbation connue

L (Z, t) = u_2(t) variable de commande

nous avons obtenu, comme modèle externe, à débit d'entrée constant :

$$y(p) = a \frac{e^{-4p}}{1+6p} u_1(p) - b \frac{e^{-3p}}{1+18p} u_2(p)$$

Problème : u_1 (t) étant connue (mesurable ou même prévisible) que doit être u_2 (t) qui minimise $J = \int_0^T u_2(t)\,dt$, sous la contrainte y (t) $\leq$ 0 (en supposant que l'on parte d'un régime permanent où y (t) = 0) avec $u_2(t) \leq u_{2max}$ = A puis avec $u_2(t) \leq u_{2max}$ et $\dot{u}_2(t) \leq \dot{u}_{2max}$.

En posant $x_1(p) = \frac{a}{1+6p} u_1(p)$ et $x_2(p) = \frac{b}{1+18p} u_2(p)$

nous obtenons, après discrétisation

$$
\begin{aligned}
x_1(k+1) &= f_1\, x_1(k) + g_1\, u_1(k) \\
x_2(k+1) &= f_2\, x_2(k) + g_2\, u_2(k) \\
y(k) &= x_1(k-k_1) - x_2(k-k_2) \leq 0
\end{aligned}
$$

Les résultats représentés sur la figure 14 correspondent à un échelon $u_1(t)$ appliqué à l'instant t = 0. Le critère à minimiser est $J = \sum_{n_A}^{n} u_2(k)$

n_A est l'instant où démarre la commande

n est la durée approximative du régime transitoire.

Pour que la solution soit possible la commande doit anticiper sur la perturbation.

II-2-2 - Commande et optimisation à partir du modèle à paramètres répartis

C'est certainement, du point de vue de l'automatique moderne l'approche la plus intéressante. Les études théoriques actuellement effectuées dans ce domaine au laboratoire et que nous résumons ci-dessous devraient pouvoir s'appliquer à notre problème.

1° - On considère un système global composé de N sous systèmes linéaires invariants interconnectés décrits par :

$$
\begin{cases}
\dfrac{\partial Y_i(x,t)}{\partial t} = M_i\left[Y_i(x,t)\right] + X_i(x,t) + B_i(x)\, U_i(x,t) \\
Y_i(x,0) = Y_{i0}(x) \\
L_i\left[Y_i(x',t)\right] = 0
\end{cases}
$$

$x \in \Omega \subset E^m$, $x' \in \delta\Omega$, $t \in [0,T]$, $i = 1$ à N

M_i, L_i opérateurs différentiels par rapport à x

$X_i = \sum_j C_{ij} Y_j$ représente l'interconnexion entre les sous systèmes

Problème : minimiser le critère séparable

$$J = \sum_i J_i = \sum_i \int_0^T \int_\Omega F_i\left[Y_i(x,t), U_i(x,t)\right] dx\, dt$$

En recherchant les variables de coordination qui conduisent à une forme séparable de l'hamiltonien associé, il est possible de définir des sous problèmes dont la résolution, après coordination par un niveau supérieur de commande, conduit à la solution globale cherchée.

L'utilisation des fonctions propres (ce qui implique, dans le cas de la colonne, que l'on retienne le terme diffusionnel, le modèle simplifié n'ayant pas de spectre discret) permet de se ramener à une coordination sur des fonctions du temps uniquement, ce qui constitue un gros avantage.

La décomposition servant de base à la mise en oeuvre de la méthode pourrait faire correspondre à chaque sous système une équation de la colonne.

2° - On considère le système décrit par

$$\frac{\partial y}{\partial t} = M[y(x,t)] + f(t)\,\delta(x - x^*)$$

M : opérateur différentiel parabolique par rapport à x

$$L[y(x',t)] = 0 \,, \quad x' \in \delta\Omega$$

$$y(0,t) = 0$$

On obtient le même point optimal $\hat{x}$ en minimisant $\int_0^T f^2(t)\,dt$, et en maximisant le volume atteignable. Ce point optimal est évidemment fonction des coefficients de l'opérateur M (en particulier quand le terme diffusionnel augmente le point $\hat{x}^*$ tend vers la limite x'.

A propos de ce problème on peut poser au théoricien ces questions, importantes du point de vue pratique.

a) quel est le modèle le plus concis compatible avec une certaine précision

b) ayant retenu un modèle à ν modes,

$$y(x,t) = \sum_{i=1}^{\nu} a_i(t)\,\Phi_i(x)$$

quelle est la meilleure place du point d'action x^*, cette place dépendant de ν.

REMERCIEMENTS

Programmé dans une session intitulée "Applications" cet exposé ne pouvait pas passer sous silence l'importance inéluctable que prend l'instrumentation dans toute étude sur pilote. Qu'il me soit permis de remercier ici le Service "Mesures, Capteurs et Instrumentation" du L.A.A.S. et son responsable Monsieur CLOT pour leur aide déterminante dans ce domaine. Les aspects théoriques sont envisagés dans le cadre de recherches inter-équipes, à l'intérieur du L.A.A.S.. C'est ainsi qu'il m'est agréable de citer :

- l'équipe "Aspects stochastiques de la commande automatique"
- l'équipe "Commande hiérarchisée"
- l'équipe "Modélisation et commande de processus thermiques considérés comme des systèmes à paramètres répartis"

et d'en remercier les responsables : MM. AGUILAR, GRATELOUP, TITLI, BABARY et leurs collaborateurs MM. ALENGRIN, CONTINENTE, PRADIN et AMOUROUX.

Je n'aurais garde de remercier mes propres collaborateurs : Madame PINGLOT, MM. POURCIEL, CANDELON et FADEL.

BIBLIOGRAPHIE

1 L. GIMENO

Identification d'une colonne d'absorption en vue de la commande du débit d'absorbant.
Thèse de Docteur Ingénieur 1972, L. A. A.S. - Toulouse

2 J.B. POURCIEL

Conduite non interactive d'un mélangeur de gaz par calculateur numérique et simulateur analogique.
Thèse de Spécialité (3ème Cycle) 1974, L.A.A.S. - Toulouse

3 J. POLIAK

Conception et réalisation de micro-débitmètres thermiques.
Thèse de Doctorat d'Université 1971, L. A. A.S. - Toulouse.

4 C. HERNANDEZ

Estimation linéaire des paramètres de systèmes dynamiques multidimensionnels et validation statistique du modèle. Applications à l'identification de processus.
Thèse de Spécialité (3ème Cycle) 1972, L.A.A.S. - Toulouse

5 P.C. YOUNG

An instrumental variable method for real time identification of noisy process.
Automatica, vol. 6, pp. 271-287, 1970.

6 D. PINGLOT

Couplage d'un spectromètre de masse quadripolaire à un calculateur numérique en temps réel : application à l'étude de processus chimiques.
Thèse de Docteur Ingénieur, 1974, L.A.A.S. - Toulouse

7 C. CERLES - J. FOISSEAU

Application des techniques de commande hiérarchisée à l'optimisation d'une colonne d'absorption.
Avant Projet INSA, dirigé par M. PRADIN, 1973

8 M. AMOUROUX

Sur la représentation des systèmes dynamiques linéaires à paramètres répartis et son application à l'étude de propriétés intrinsèques de ces systèmes.
Thèse de Spécialité (3ème Cycle), 1973, L. A. A.S. - Toulouse

9 J.B. POURCIEL - Y. SEVELY

Simulation originale hybride et commande dynamique d'un mélangeur comportant des actionneurs non linéaires.
Congrès International "Emploi des Calculateurs Electroniques en Génie Chimique", Paris, 24-28 Avril 1973.

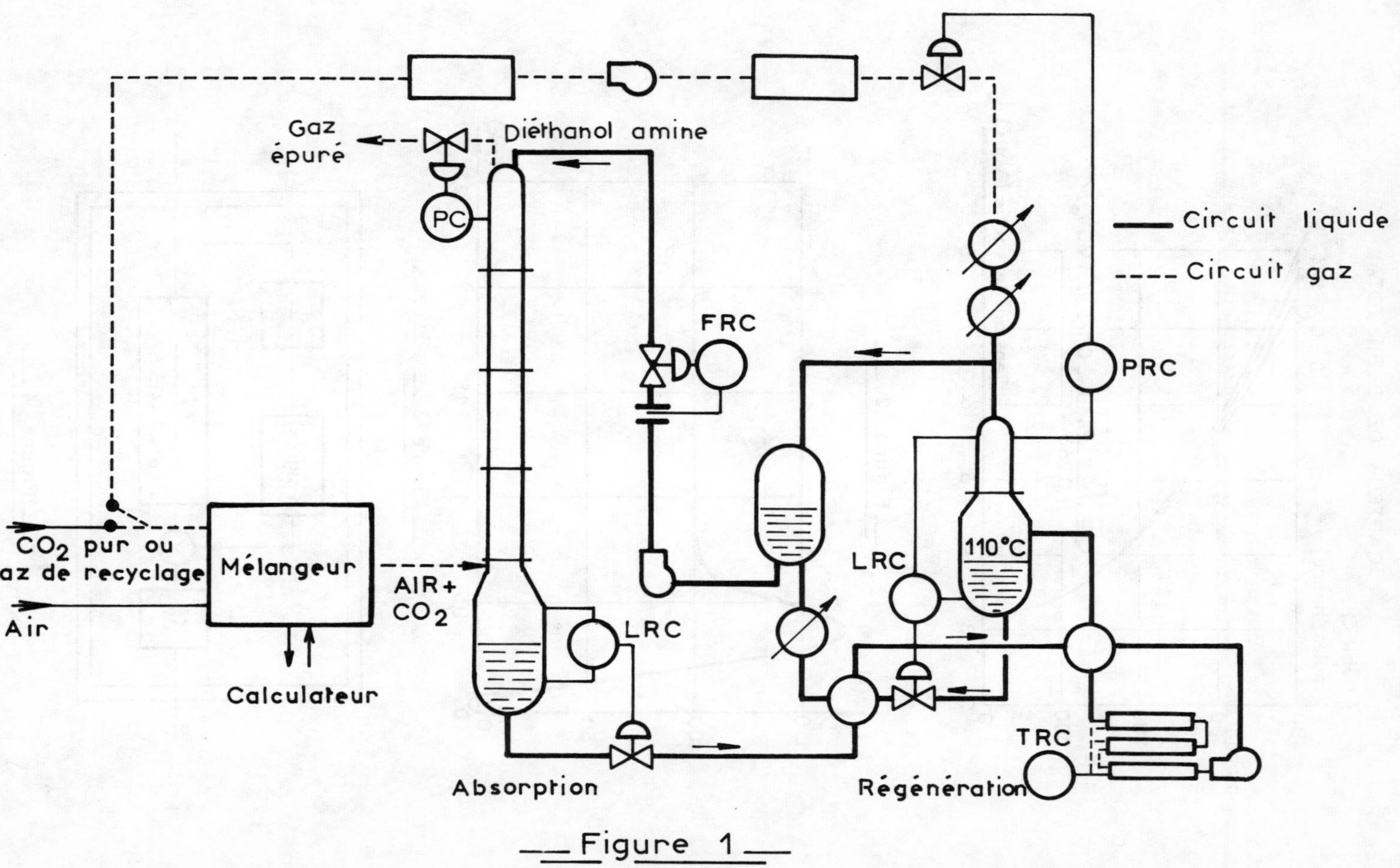
Gaz épuré
Diéthanol amine
PC
Circuit liquide
Circuit gaz
FRC
PRC
CO_2 pur ou gaz de recyclage
Mélangeur
Air
AIR + CO_2
Calculateur
LRC
110°C
LRC
TRC
Absorption
Régénération

Figure 1

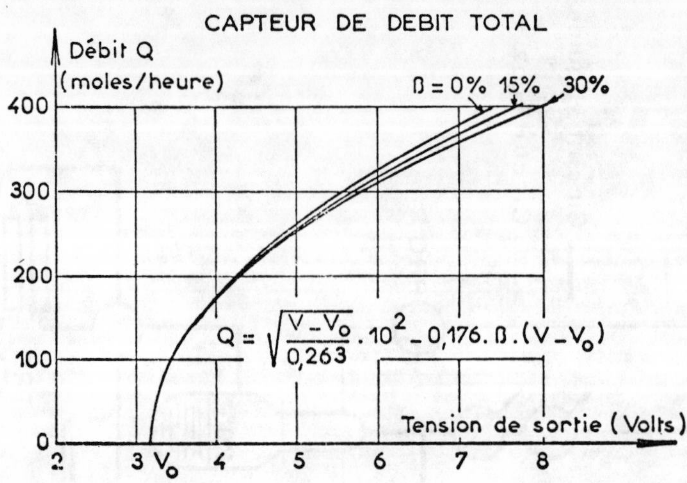

Figure 2

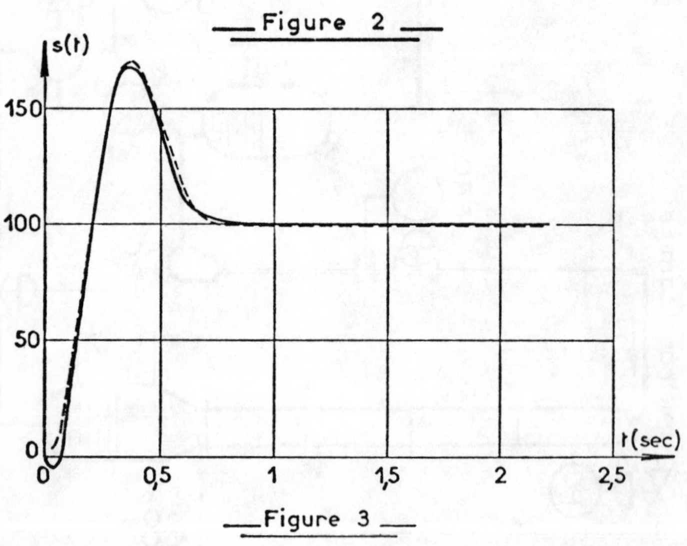

Figure 3

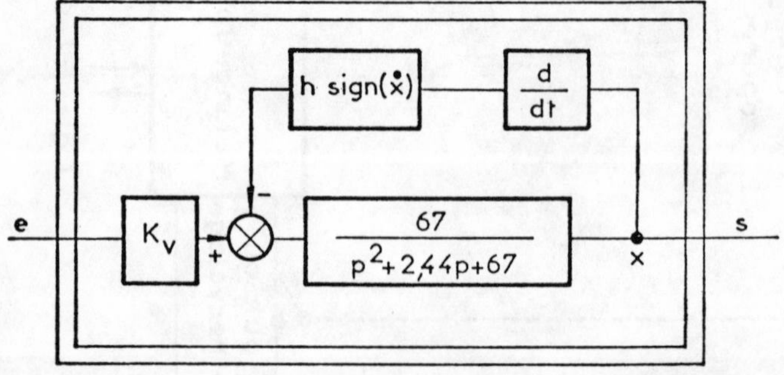

Figure 4

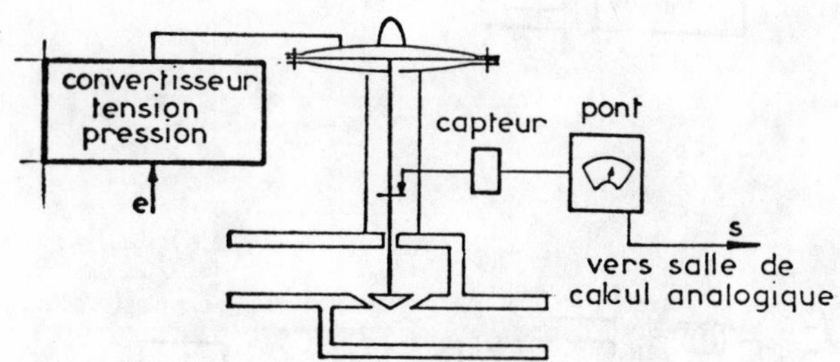
convertisseur tension pression
e
capteur
pont
s
vers salle de calcul analogique

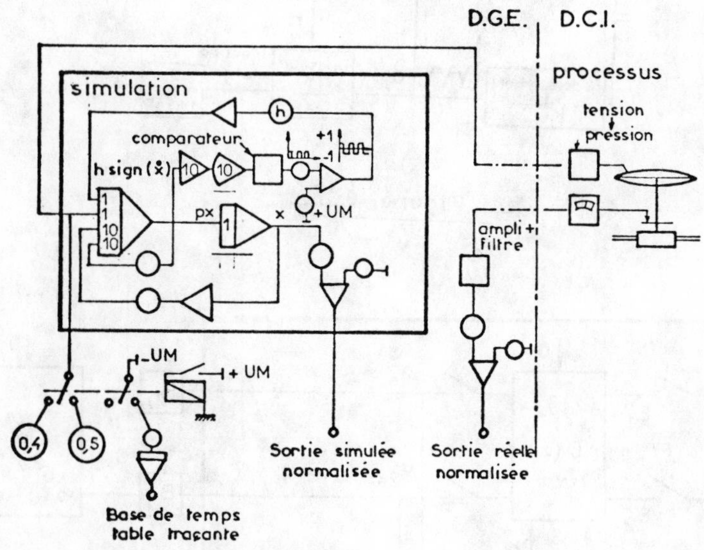
D.G.E.
D.C.I.
simulation
processus
h
comparateur
tension
pression
h sign ($\dot{x}$)
10
10
+1
-1
+UM
px
x
1
1
10
10
1
ampli + filtre
-UM
+ UM
0,4
0,5
Sortie simulée normalisée
Sortie réelle normalisée
Base de temps table traçante

Figure 5

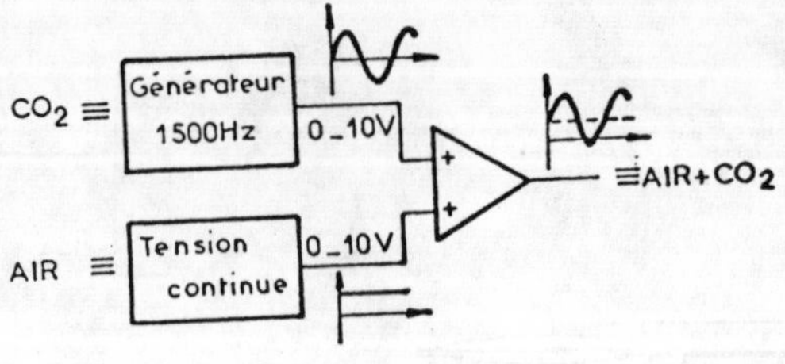

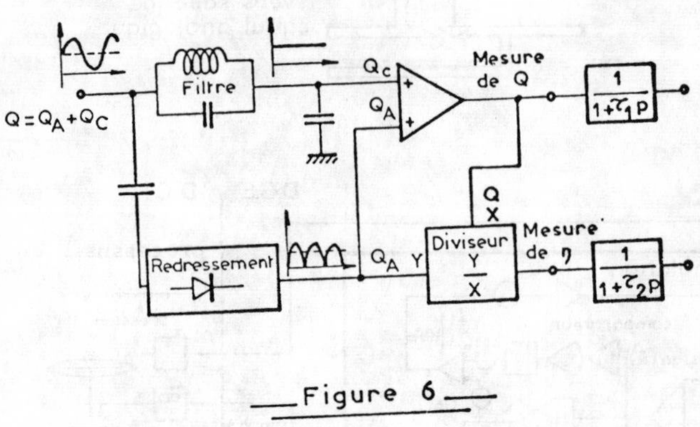

— Figure 6 —

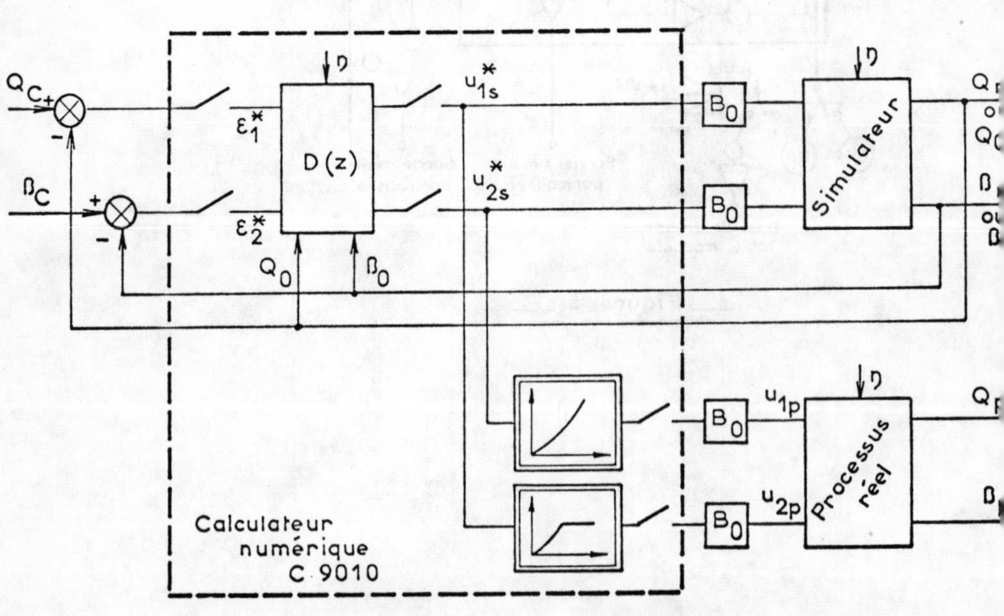

— Figure 7 —

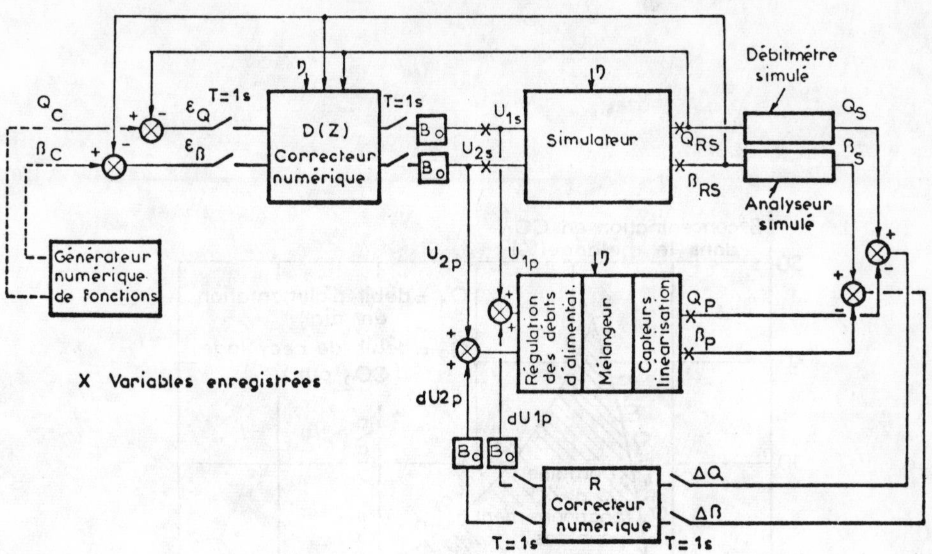

Figure 8

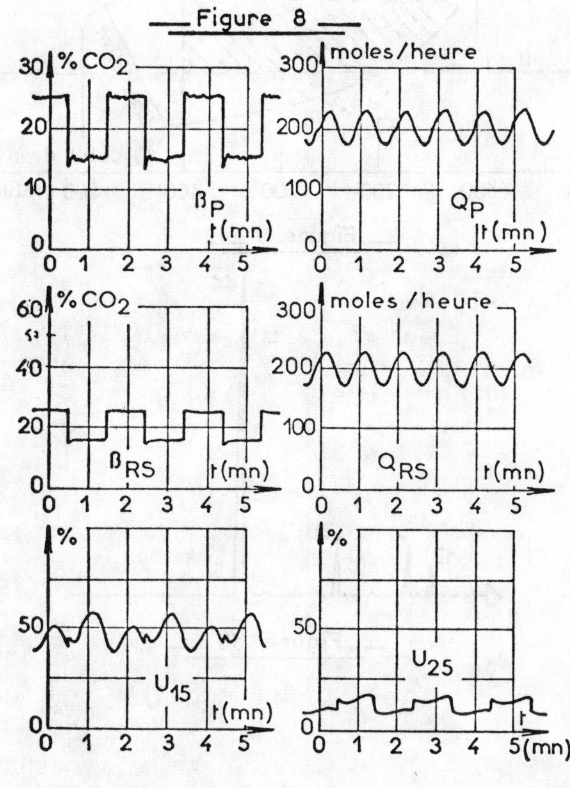

Planche 1

Enregistrement simultané de $\beta_P, Q_P, \beta_{RS}, Q_{RS}, U_{1S}, U_{2S}$ pour des consignes β_C en créneaux | 20% ± 5% | T = 1mn

Q_C sinusoïdale | 200 ± 30 moles/heure | T = 1 mn

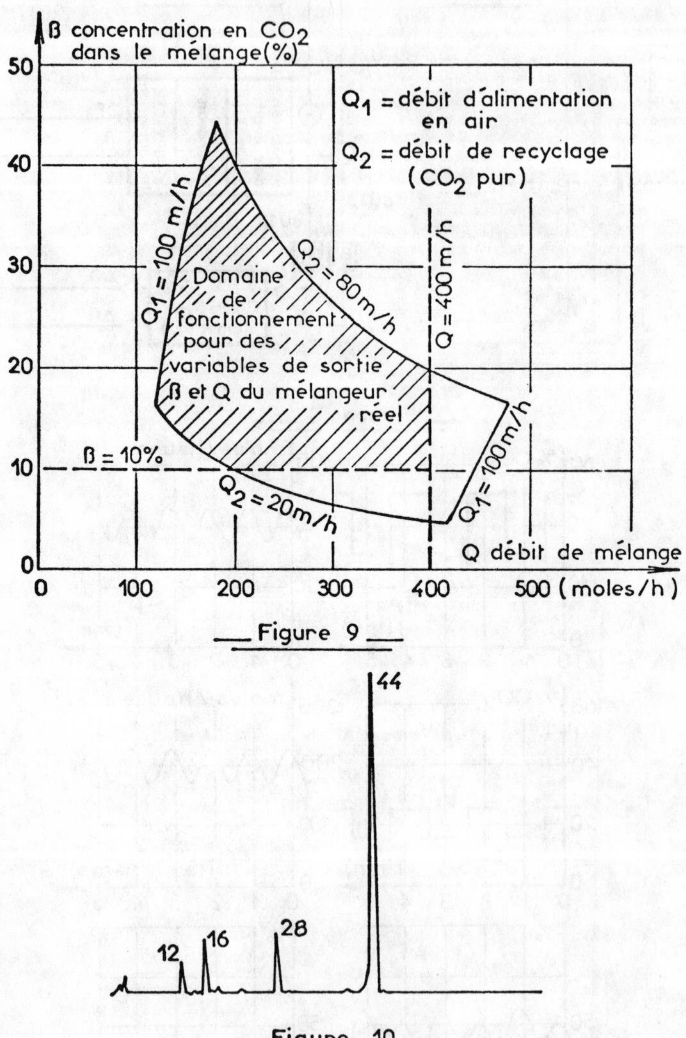

Figure 9

Figure 10

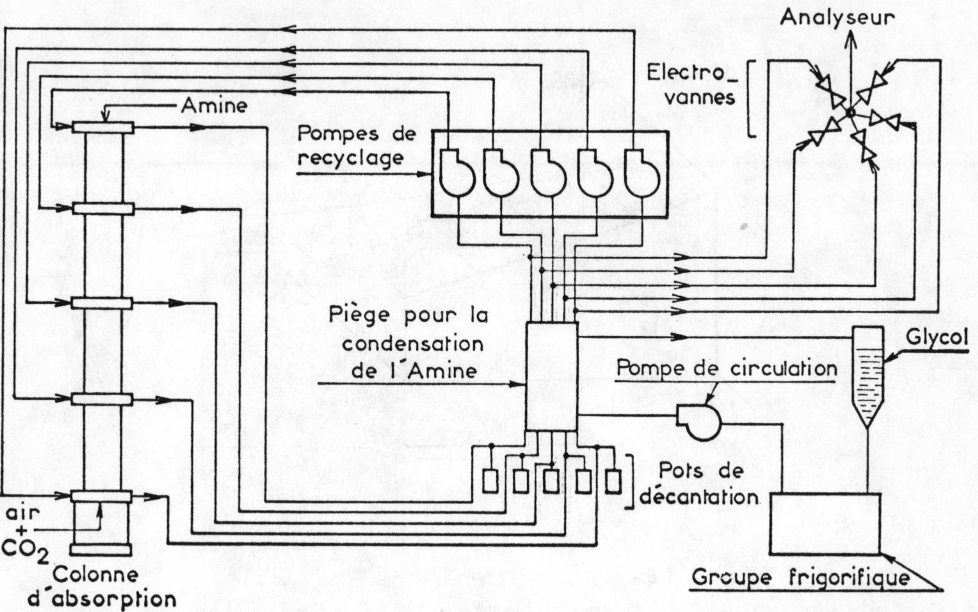

— Figure 11 —

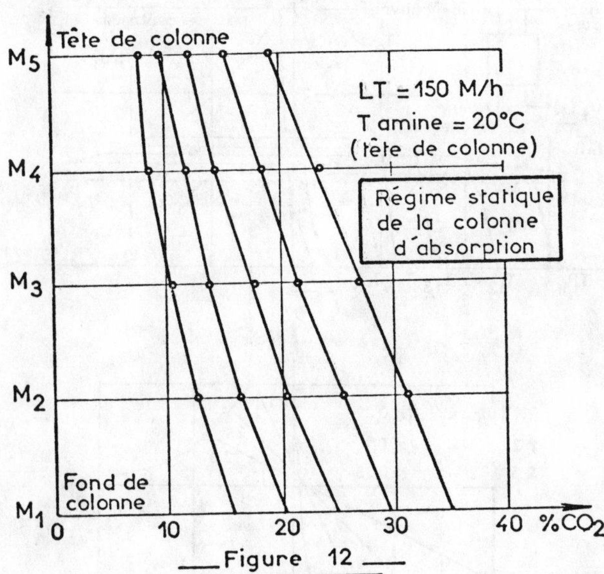

— Figure 12 —

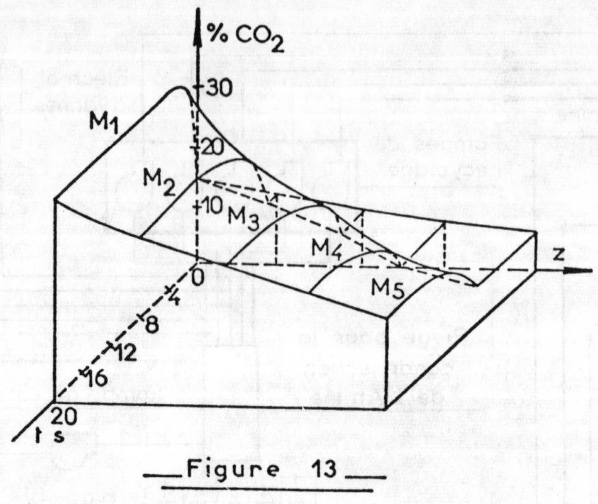

Figure 13

Fig. 14 see page 727

Commande optimale du débit d'amine avec contrainte

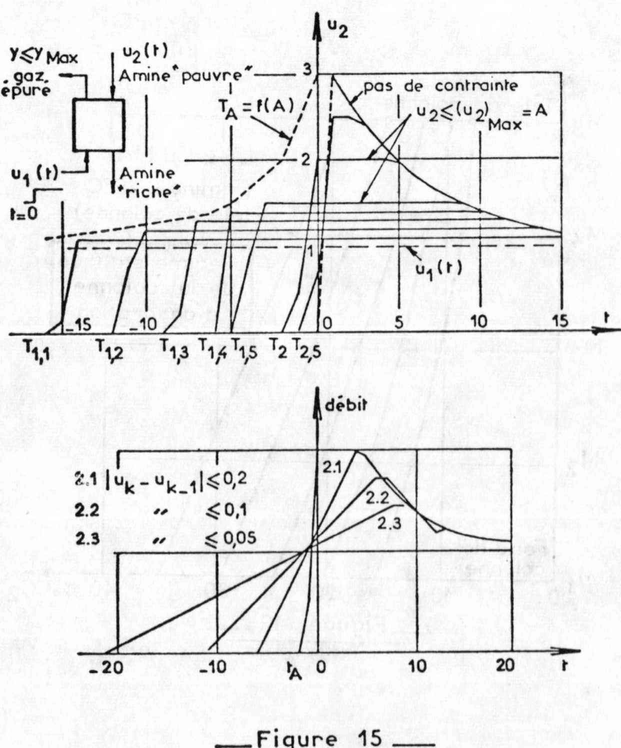

Figure 15

Commande optimale du débit d'amine avec contrainte.

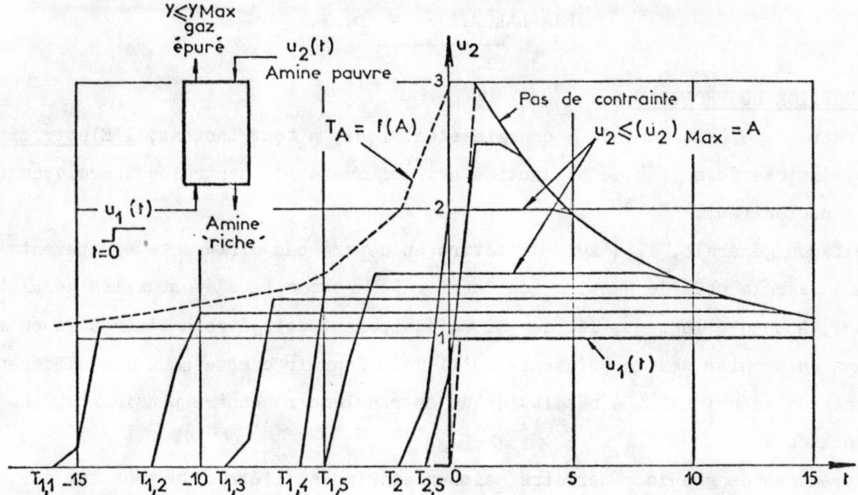

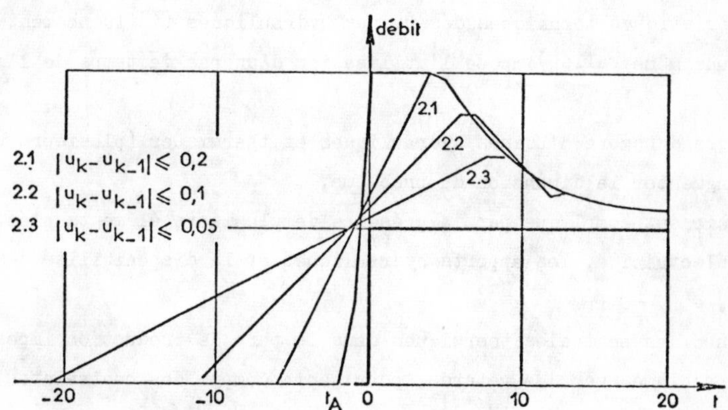

FIGURE 14

APPLICATION DU CONTROLE STOCHASTIQUE A LA GESTION DES CENTRALES THERMIQUES ET HYDRAULIQUES

C. LEGUAY - A. BRETON

IRIA-LABORIA - E. D. F.

I - POSITION DU PROBLEME.

Pour faire face à la demande d'énergie électrique, à tout instant, l'Electricité de France dispose de moyens de production hydraulique et de centrales thermiques (fossiles et nucléaires).

D'une façon générale, il s'agit de mettre en oeuvre ces différents équipements afin de minimiser le coût de gestion (ce dernier tenant compte d'éventuelles défaillances). Il faut, à chaque instant, décider si un équipement doit être à l'arrêt ou en marche, et dans ce dernier cas, la puissance qu'il doit fournir compte tenu des différentes contraintes techniques. La résolution de ce problème rencontre plusieurs types de difficultés :

- la présence de grands réservoirs saisonniers impose l'étude du problème sur un intervalle de temps de l'ordre de l'année,
- les variations de la demande d'électricité au cours d'une même journée ainsi que les caractéristiques techniques des usines hydrauliques (influencement des ouvrages au sein d'une même vallée) impose l'utilisation d'un pas de temps de l'ordre de l'heure,
- le très grand nombre d'usines hydrauliques et thermiques (plusieurs centaines) contribue à augmenter la dimension du problème,
- le problème est compliqué par la présence de plusieurs éléments aléatoires : la demande d'électricité, les apports hydrauliques et la disponibilité des centrales thermiques,
- la présence des centrales thermiques dans le parc de production impose une formulation du problème avec des retards, pour tenir compte des contraintes liées au démarrage des groupes thermiques.

Devant l'impossibilité pratique d'optimiser un tel problème (grande dimension et caractère stochastique) on procède intuitivement en combinant deux approches. On construit un premier modèle permettant d'aborder le caractère dynamique du problème. Ce modèle utilise une formulation axée essentiellement sur l'étude de l'influence des grands réservoirs saisonniers et fournit d'une part les quantités d'eau destockées chaque semaine par ces derniers et d'autre part, les différentes valeurs marginales de l'eau. En retenant les destockages précédents pour la première semaine (boucle ouverte adaptée, on peut faire appel à un second modèle étudiant sur une semaine l'utilisation, heure par heure, des équipements thermiques et hydrauliques. Il est même possible d'effectuer à ce niveau de légères retouches au programme hebdomadaire de destockage des grands réservoirs saisonniers en utilisant les valeurs marginales de l'eau précédentes.

Ce dernier modèle se présente donc sous la forme d'un problème d'optimisation stochastique de grande dimension avec des variables entières (marche ou arrêt des groupes) et des variables continues (possibilités de modulation). A l'heure actuelle, on ne sait pas traiter correctement ce problème et on se contente d'une approximation déterministe de celui-ci (consommation certaine, apports hydrauliques certains, disponibilité thermique certaine), mais en prenant par contre en considération la totalité des équipements et en tenant compte des principales contraintes techniques d'exploitation.

Cependant, on peut essayer de conserver au problème son caractère stochastique en se limitant à quelques centrales thermiques.

On obtient alors un problème simplifié, possédant la même structure que le problème réel, dans lequel il s'agit de placer, heure par heure, et sur une semaine, quelques centrales thermiques, en tenant compte des délais de démarrage et des possibilités de modulation de ces différents groupes. Il devient dès lors possible de cumuler les différents aléas (demande globale, production hydraulique à programme fixé en fonction des arrivées d'eau, disponibilité des autres équipements thermiques) pour construire le processus décrivant la demande à satisfaire pour ces quelques centrales thermiques. De plus, on peut envisager une représentation approximative des possibilités de production hydraulique, en utilisant les valeurs marginales de l'eau précédentes, pour construire le coût de la production supplémentaire d'énergie électrique d'origine hydraulique au voisinage d'un programme hydraulique déjà fixé.

C'est ce problème simplifié que nous nous proposons de résoudre.

II - FORMALISATION DES DONNEES PHYSIQUES.

2.1. La demande

La demande d'énergie électrique est une variable aléatoire (en abrégé V.A.) que nous représentons par une loi de diffusion, une telle modélisation étant très réaliste. (s,t) étant un intervalle de temps, on note $D(s,t)$ la demande cumulée entre les instants s et t ; on suppose :

$$(1.1) \qquad D(s,t) = \int_s^t \mu(\tau)d\tau + \int_s^t \sigma(\tau)db(\tau)$$

$\mu(t) \in \mathbb{R}$, $\mu(t) \geq 0$ $\forall t$, supposée connue dans l'intervalle $[0,T]$ sur lequel porte l'étude, représente la demande instantanée à l'instant t.

$$\int_s^t \mu(\tau)db(\tau)$$ est la partie déterministe de $D(s,t)$

$\sigma(t) \in R$ est une fonction connue, et :

$\int_s^t \sigma(\tau)db(\tau)$ est une v.a.* . gaussienne

. de moyenne nulle

. de variance $\int_s^t \sigma^2(\tau)d\tau$.

Ce terme est la partie aléatoire de $D(s,t)$.

2.2. Structure des coûts.

On distinguera :

2.2.1. Les coûts de production :

Thermique : si sur $(t,t+\Delta t)$ une centrale produit la puissance $p(t)$, alors le coût de production associé est : $p(t).C_p.\Delta t$ (C_p coût unitaire d'énergie thermique).

Hydraulique : Le coût de production d'une quantité $q(t)$ d'énergie hydraulique sur $(t,t+\Delta t)$, sachant que le niveau de stock hydraulique est x, sera $q(t).f(x,t).\Delta t$ $f(x,t) \geq 0$, fonction non nécessairement continue, représente le coût unitaire d'énergie hydraulique, relatif à un niveau de stock x.

2.2.2. Les coûts de lancement (d'unités thermiques)

Ces sont des coûts fixes. Le lancement d'une unité thermique entraîne un coût fixe K, correspondant à l'énergie qu'il faut fournir à cette unité pendant toute la durée du préchauffage.

On notera que l'arrêt d'une unité, instantané, pourrait être accompagné d'un coût d'arrêt, que l'on supposera ici nul.

2.3. Prise en compte d'autres phénomènes.

. On peut, sans changer la nature du problème, introduire un réapprovisionnement du stock d'eau ; on pourrait en particulier supposer que les apports suivent eux-mêmes une loi de diffusion.

. La taille du stock est limitée par X_{max} quantité maximale d'énergie hydraulique disponible sur $[0,T]$ et X_{min} quantité minimale, non nécessairement égale à zéro.

. Contraintes sur la capacité de turbinage : les turbines d'une centrale hydroélectrique sont utilisables sur une plage de puissance p_H vérifiant $0 \leq p_H \leq (p_H)_{max}$;

. Coût de défaillance dans le cas où le système thermique et le système hydraulique, en raison des contraintes de fonctionnement, ne peuvent satisfaire toute la demande,

* Pour la signification de l'intégrale stochastique $\int_s^t \sigma(\tau)db(\tau)$, on pourra consulter NEVEU [1] et BENSOUSSAN-LIONS [1].

l'on est appelé à produire la puissance défaillante par d'autres moyens que ceux envisagés dans l'étude (turbines à gaz, etc...). Nous pénaliserons toute défaillance par un coût proportionnel à la quantité d'énergie défaillante ; on note C_δ le coût unitaire de défaillance.

III - POLITIQUE OPTIMALE DE GESTION

3.1. Politique de décision.

Pour définir une politique de décision, on se place sur l'intervalle $[t,T] \subset [0,T]$, avec à l'instant t un stock x. A tout couple (t,x) on associe une politique de décision V_{tx} définie par une suite (dénombrable) d'instants de décision

$$\theta^1_{tx},\ \theta^2_{tx},\ \ldots\ldots\ \theta^n_{tx}\ \ldots.$$

vérifiant $\quad t \leq \theta^1_{tx} \leq \theta^2_{tx} \leq \ldots\ldots \leq \theta^n_{tx} \leq \theta^{n+1}_{tx} \ldots\ldots \leq T\ , \quad \forall n = 1,2,\ldots$

θ^n_{tx} est l'instant de la $n^{\text{ième}}$ décision postérieure à t sachant qu'à l'instant t le stock est x. Cette décision peut correspondre à un lancement ou à un arrêt suivant l'état thermique du système à l'instant t. Les θ^n_{tx} sont aléatoires (sauf dans le cas d'une demande déterministe) Ce sont des temps d'arrêt (cf. BENSOUSSAN-LIONS [1]).

3.2. Evolution du stock.

Nous envisageons ici le cas où le processus de décision est instantané, c'est-à-dire le cas où coïncident l'instant de lancement et l'instant de début de fonctionnement d'une unité thermique. Les phénomènes de retard sont pris en compte au § (3.4).
Nous formulons l'évolution du stock dans le cas d'une seule centrale thermique, afin de simplifier les notations.
On désigne par $y_{tx}(s)$ l'état du stock à l'instant $s \geq t$, sachant qu'à t le niveau de stock est x .
Donc $y_{tx}(t) = X$

. Si à t, l'unité thermique est arrêtée, il est clair que

$$y_{tx}(s) = x - D(t,s) \quad \text{sur} \quad [t,\theta^1_{tx}[$$

où θ^1_{tx} désigne l'instant de la première décision postérieure à t, c'est-à-dire un lancement. Ensuite

$$y_{tx}(s) = x - D(t,s) + p.(s - \theta^1_{tx})$$

$$= y_{tx}(\theta^1_{tx}) - D(\theta^1_{tx,s}) + p.(s-\theta^1_{tx}) \quad \text{sur} \ [\theta^1_{tx},\theta^2_{tx}[$$

où θ^2_{tx} est l'arrêt suivant et p la puissance nominale, supposée fixe ici, de l'unité thermique en fonctionnement.

. Si à t, l'unité thermique fonctionne, on aura :

$$y_{tx}(s) = x - D(t,s) + p.(s-t) \quad \text{sur} \quad [t,\theta^1_{tx}[$$

$$y_{tx}(s) = y_{tx}(\theta^1_{tx}) - D(\theta^1_{tx},s) \quad \text{sur} \quad [\theta^1_{tx},\theta^2_{tx}[$$

θ^1_{tx} étant cette fois un arrêt, et θ^2_{tx} un lancement.

Plus généralement,

$$(1.2)\quad \begin{cases} y_{tx}(s) = y_{tx}(\theta^{2n}_{tx}) - D(\theta^{2n}_{tx},s) \quad \text{sur} \quad [\theta^{2n}_{tx},\theta^{2n+1}_{tx}[\\ y_{tx}(s) = y_{tx}(\theta^{2n+1}_{tx}) - D(\theta^{2n+1}_{tx},s) + p.(s-\theta^{2n+1}_{tx}) \quad \text{sur} \quad [\theta^{2n+1}_{tx},\theta^{2n+2}_{tx}[\end{cases}$$

si à t, l'unité est arrêtée. θ^n_{tx} représente
- un lancement si n est impair
- un arrêt si n est pair

$$(1.3)\quad \begin{cases} y_{tx}(s) = y_{tx}(\theta^{2n}_{tx}) - D(\theta^{2n}_{tx},s) + p.(s-\theta^{2n}_{tx}) \quad \text{sur} \quad [\theta^{2n}_{tx},\theta^{2n+1}_{tx}[\\ y_{tx}(s) = y_{tx}(\theta^{2n+1}_{tx}) - D(\theta^{2n+1}_{tx,s}) \quad \text{sur} \quad [\theta^{2n+1}_{tx},\theta^{2n+2}_{tx}[\end{cases}$$

si à t, l'unité fonctionne. θ^n_{tx} représente
- un arrêt si n est impair
- un lancement si n est pair

N.B. Dans les 2 cas, on a pris la convention $\theta^o_{tx} = t$.

Remarque

Si la demande est une fonction continue de s, $y_{tx}(s)$ est également une fonction continue de s .

Si l'on considère $s \in [\theta^i_{tx},\theta^{i+1}_{tx}[$, et si l'on différentie les relations précédentes, on obtient la variation élémentaire du stock, pendant l'intervalle de temps ds,

$$dy_{tx}(s) = -\mu(s)ds - \sigma(s)dw(s) \quad \text{si l'unité thermique est arrêtée}$$

$$dy_{tx}(s) = [p-\mu(s)]ds - \sigma(s)dw(s) \quad \text{si elle fonctionne.}$$

En fait, le problème réel nous interdisant les variations positives de $y_{tx}(s)$, ceci correspondant au fait que :

"Toute énergie produite en trop est perdue",

nous avons été conduits à prendre :

(1.4)

$dy_{tx}(s) = -\mu(s)ds - \sigma(s)dw(s)$ si l'unité thermique est arrêtée
$dy_{tx}(s) = -[\mu(s)-p]^{+} - \sigma(s)dw(s)$ si elle fonctionne.

N.B. Ceci n'exclut pas totalement l'éventualité d'un $dy_{tx}(s) > 0$, mais la modélisation est satisfaisante dans le cas de perturbations gaussiennes pas trop grandes. Les relations écrites précédemment se généralisent sans difficulté au cas de 2 centrales.

3.3. Coût de production et politique optimale.

Notations

L'état du système, à l'instant t, est entièrement caractérisé par :

X le niveau du stock,

i indice de fonctionnement de la centrale (I)

j indice de fonctionnement de la centrale (II).

i et j sont des variables booléennes, dont la signification est :

$$i = \begin{cases} 0 & \text{si la centrale (I) est arrêtée} \\ 1 & \text{si la centrale (I) fonctionne.} \end{cases}$$

$$j = \begin{cases} 0 & \text{si la centrale (II) est arrêtée} \\ 1 & \text{si la centrale (II) fonctionne.} \end{cases}$$

On désignera par X_o le stock initial (énergie hydraulique maximale, disponible sur $[0,T]$)

X le niveau instantané.

. Coût de gestion associé à une politique V_{tx}

On suppose que pour les politiques V_{tx} choisies, l'évolution du stock est markovienne. Nous faisons appel au schéma habituel de la programmation dynamique (principe d'optimalité de BELLMAN) : le schéma du type "décision-hasard".

Etudions le cas d'une seule centrale thermique.

Si t est un instant de décision, avec évolution libre du système sur $[t,t+\Delta t[$, nous évaluons le coût d'une transition élémentaire :

$$(t,x) \quad \rightarrow \quad (t+\Delta t,\ x+\Delta x) \begin{cases} \text{Décision } 1 \\ \text{Décision } 2 \end{cases}$$

. Partons de l'état (t,x,0) (unité arrêtée).

Décision 1 : on lance l'unité thermique. Le coût de transition de t à $t+\Delta t$ est la somme : - du coût fixe de lancement, noté K ,

- du coût de production pendant l'intervalle Δt, soit

$$\underbrace{(\mu(t)-p)^{+}.f(x,t)\Delta t}_{\text{coût de production hydraulique}} + \underbrace{p.C_p . \Delta t}_{\text{coût de production thermique.}}$$

Décision 2 : on ne décide rien. Le coût de transition est $\mu(t).f(x,t).\Delta t$.

. Partons de l'état $(t,x,1)$ (unité en fonctionnement)

Décision 1 : on arrête l'unité thermique. Le coût de transition sur $[t,t+\Delta t[$ est

$$\mu(t).f(x,t).\Delta t$$

Décision 2 : on ne décide rien. Le coût de transition est

$$(\mu(t)-p)^{+}.f(x,t)\Delta t + p.C_p.\Delta t$$

En prenant l'espérance mathématique du coût sur toutes les transitions possibles, on obtient le coût global associé à la politique V_{tx}, sachant qu'à t on est parti de l'état (x,h) ($h=0$ ou 1) :

$$\mathbb{J}_{t,x,0}(V_{tx}) = \mathbb{E}\left\{\sum_{n=0}^{\infty} \chi_{[t,T]}(\theta^{2n+1}).K + \sum_{n=0}^{\infty}\int_{\theta^{2n}}^{\theta^{2n+1}} \chi_{[t,T]}^{(s)}\,\mu(s).f[y_{t,x}(s),s]ds + \right.$$

$$\left. + \sum_{n=0}^{\infty}\int_{\theta^{2n+1}}^{\theta^{2n+1}} [(\mu(s)-p)^{+}.f[y_{t,x}(s),s] + p.\,C_p]\,\chi_{[t,T]}(s)ds \right\}.$$

Et

$$\mathbb{J}_{t,x,1}(V_{tx}) = \mathbb{E}\left\{\sum_{n=1}^{\infty} \chi_{[t,T]}(\theta^{2n}).K + \sum_{n=0}^{\infty}\int_{\theta^{2n}}^{\theta^{2n+1}} [(\mu(s)-p)^{+}f[y_{t,x}(s),s]+p.C_p]\right.$$

$$\left.\chi_{[t,T]}(s)ds + \sum_{n=0}^{\infty}\int_{\theta^{2n+1}}^{\theta^{2n+1}} \mu(s).f[y_{t,x}(s),s]\chi_{[t,T]}(s)ds\right\}$$

où

$$\chi_{[t,T]}(s) = \begin{cases} 1 & \text{si } s < T \\ 0 & \text{sinon} \end{cases}$$

et où l'évolution du stock $y_{tx}(s)$ est donnée par (1.2) et (1.3).

On définit alors

$$(1.5)\qquad \begin{cases} u_o(t,x) = \inf\limits_{V_{tx}} \mathbb{J}_{t,x,0}(V_{tx}) \\ \\ u_1(t,x) = \inf\limits_{V_{tx}} \mathbb{J}_{t,x,1}(V_{tx}) \end{cases}$$

et la politique optimale V^{*}_{txk} associée à ces coûts optimaux. :

$$(1.6)\quad \begin{array}{ll} V^*_{t,x,0} & \text{est telle que } u_0(t,x) = J_{t,x,0}(V^*_{t,x,0}) \\ V^*_{t,x,1} & \text{est telle que } u_1(t,x) = J_{t,x,1}(V^*_{t,x,1}) \end{array}$$

Plus généralement, et de manière totalement analogue, on définit poûr le cas de 2 centrales

$u_{ij}(t,x)$ coût optimal de production de t à T, sachant qu'à l'instant t l'état du système est (x,i,j)

V^*_{txij} la politique optimale associée.

3.4. Prise en compte des phénomènes de retard.

τ désigne le délai de mise en fonctionnement d'une unité thermique (appelé "retard") : si à l'instant t l'on décide de lancer une centrale, son début de fonctionnement n'interviendra qu'à l'instant $t+\tau$, et entre t et $t+\tau$, nous laisserons le système évoluer librement.

N.B. La formulation du problème avec des décisions imbriquées n'a pas donné de résultats satisfaisants jusqu'à présent.

On désigne par $W^{kl}_{ij}(t,x)$ l'espérance du coût de production, sur $[t,T]$, sachant qu'à t l'on se trouve dans l'état (x,i,j), que l'on prend une décision de lancement qui nous amènera en $t+\tau$ dans l'état (y,k,l) et que toutes les décisions postérieures à $t+\tau$ sont optimales.

On obtient, dans le cas d'une seule centrale :

$$W^1_0(t,x) = \mathbb{E}\ \{ \int_t^{t+\tau} \mu(s) f[x- \int_t^s \mu(\xi)d\xi - \int_t^s \sigma(\xi)db(\xi),s]ds + u_1(t+\tau, x - \int_t^{t+\tau} \mu(s)ds - - \int_t^{t+\tau} \sigma(s)db(s))\}$$

ou encore :

$$(1.7)\quad W^1_0(t,x) = \int_t^{t+\tau} \mu(s)[\int_{\mathbb{R}} f(x+\eta,s)\Pi(s,\eta)d\eta]ds + \int_{\mathbb{R}} u_1(t+\tau,x+\eta)\Pi(t,\eta)d\eta$$

avec

$$\Pi(s,\eta) = \frac{1}{\sigma(s)\sqrt{2\Pi}} \exp\ [- \frac{1}{2\tau\ \sigma^2(s)} \quad (\eta + \tau\ \mu(s))^2\ ]$$

On généralise sans difficulté les résultats précédents et on obtient ainsi l'expression des $W^{kl}_{ij}(t,x)$, dans le cas de 2 centrales thermiques.

N.B. Nous verrons, dans l'écriture du modèle, que le calcul des $W^{kl}_{ij}(t,x)$ ne se fait pas directement, mais plutôt par résolution d'équations aux dérivées partielles.

IV - LE MODELE

Nous ne décrirons pas ici l'obtention des équations; on la trouvera sous forme complète dans LEGUAY [1].

Nous résumons néanmoins l'ensemble des décisions possibles pour chaque état thermique (i,j)

ETAT	DECISIONS
00	Attente ou Lancement de I Lancement de II Lancement de I et II
10	Attente ou Arrêt de I Lancement de II Arrêt de I et Lancement de II
01	Attente ou Arrêt de II Lancement de I Arrêt de II et Lancement de I
11	Attente ou Arrêt de I Arrêt de II Arrêt de I et II

On obtient alors le modèle suivant, constitué par <u>un système de 4 I.Q.V.</u> :

4.1. Le modèle dans le cas d'un contrôle impulsionnel.

$$\begin{cases} -\dfrac{\partial u_{oo}}{\partial t}(t,x) - \dfrac{\sigma^2(t)}{2}\dfrac{\partial^2 u_{oo}}{\partial x^2}(t,x) + \mu(t)\dfrac{\partial u_{oo}}{\partial x}(t,x) \leq \mu(t)f(x,t) \\ u_{oo}(t,x) \leq K_1 + W_{oo}^{10}(t,x) \\ u_{oo}(t,x) \leq K_2 + W_{oo}^{01}(t,x) \qquad \text{sur } t < T-\tau \\ u_{oo}(t,x) \leq K_1 + K_2 + W_{OO}^{11}(t,x) \\ \text{Produit nul des 4 conditions.} \end{cases}$$

$$\begin{cases} -\dfrac{\partial u_{10}}{\partial t}(t,x) - \dfrac{\sigma^2(t)}{2}\dfrac{\partial^2 u_{10}}{\partial x^2}(t,x) + (\mu(t)-p_1)^+\dfrac{\partial u_{10}}{\partial x}(t,x) \leq (\mu(t)-p_1)^+ f(x,t) + p_1 C_p^1 \\ u_{10}(t,x) \leq u_{oo}(t,x) \\ u_{10}(t,x) \leq K_2 + W_{10}^{11}(t,x) \qquad \text{sur } t < T-\tau \\ u_{10}(t,x) \leq K_2 + W_{10}^{01}(t,x) \\ \text{Produit nul des 4 conditions.} \end{cases}$$

$$\begin{cases} -\dfrac{\partial u_{o1}}{\partial t}(t,x) - \dfrac{\sigma^2(t)}{2}\dfrac{\partial^2 u_{o1}}{\partial x^2}(t,x) + (\mu(t)-p_2)^+\dfrac{\partial u_{01}}{\partial x}(t,x) \leq (\mu(t)-p_2)^+ f(x,t) + p_2 C_p^2 \\ u_{01}(t,x) \leq u_{oo}(t,x) \\ u_{01}(t,x) \leq K_1 + W_{01}^{11}(t,x) \\ u_{01}(t,x) \leq K_1 + W_{01}^{10}(t,x) \qquad \text{sur } t < T-\tau \\ \text{Produit nul des 4 conditions} \end{cases}$$

$$-\frac{\partial u_{11}}{\partial t}(t,x) - \frac{\sigma^2(t)}{2}\frac{\partial^2 u_{11}}{\partial x^2}(t,x) + (\mu(t)-p_1-p_2)^+\frac{\partial u_{11}}{\partial x}(t,x) \leq (\mu(t)-p_1-p_2)^+ f(x,t) + p_1 C_p^1 + p_2 C_p^2$$

$$\begin{cases} u_{11}(t,x) \leq u_{oo}(t,x) \\ u_{11}(t,x) \leq u_{10}(t,x) \qquad \text{sur } t \leq T \\ u_{11}(t,x) \leq u_{01}(t,x) \\ \text{Produit nul des 4 conditions.} \end{cases}$$

$$- \frac{\partial u_{oo}}{\partial t}(t,x) - \frac{\sigma^2(t)}{2} \frac{\partial^2 u_{oo}}{\partial x^2}(t,x) + \mu(t) \frac{\partial u_{oo}}{\partial x}(t,x) \leq \mu(t) f(x,t) \quad \text{sur} \quad [T-\tau,T]$$

$$\begin{cases} - \frac{\partial u_{10}}{\partial t}(t,x) - \frac{\sigma^2(t)}{2} \frac{\partial^2 u_{10}}{\partial t}(t,x) + (\mu(t)-p_1)^+ \frac{\partial u_{10}}{\partial t}(t,x) \leq (\mu(t)-p_1)^+ f(x,t) + p_1 C_p^1 \\ u_{10}(t,x) \leq u_{oo}(t,x) \qquad \text{sur} \quad [T-\tau,T] \end{cases}$$

Produit nul des 2 conditions

$$\begin{cases} - \frac{\partial u_{01}}{\partial t}(t,x) - \frac{\sigma^2(t)}{2} \frac{\partial^2 u_{01}}{\partial x^2}(t,x) + (\mu(t)-p_2)^+ \frac{\partial u_{01}}{\partial x}(t,x) \leq (\mu(t)-p_2)^+ f(x,t) + p_2 C_p^2 \\ u_{01}(t,x) \leq u_{oo}(t,x) \qquad \text{sur} \quad [T-\tau,T] \end{cases}$$

Produit nul des 2 conditions.

Et la condition finale, à l'instant T :

$$u_{oo}(T,x) = u_{10}(T,x) = u_{01}(T,X) = u_{11}(T,X) = 0, \qquad \forall X$$

. <u>Calcul des</u> W_{ij}^{kl}

On ne les obtient pas en calculant les intégrales du type (1.7), mais en résolvant backward les équations aux dérivées partielles :

$$\begin{cases} - \frac{\partial W_{ij}^{kl}}{\partial s}(s,x) - \frac{\sigma^2(s)}{2} \frac{\partial^2 W_{ij}^{kl}}{\partial x^2}(s,x) + \mu^*(s) \frac{\partial W_{ij}^{kl}}{\partial x}(s,x) = \mu^*(s) f(x,s) \\ W_{ij}^{kl}(t+\tau,x) = u_{kl}(t+\tau,x) \qquad \text{sur} \quad [t,t+\tau[\end{cases}$$

avec

$$\begin{vmatrix} \mu^*(s) = \mu(s) & \text{si I et II sont arrêtées sur } [t,t+\tau[\\ \mu^*(s) = (\mu(s)-p_1)^+ & \text{si I fonctionne} \\ \mu^*(s) = (\mu(s)-p_2)^+ & \text{si II fonctionne.} \end{vmatrix}$$

4.2. <u>Prise en compte des phénomènes de modulation.</u>

Dans le cas où la puissance de fonctionnement des centrales thermiques est modulable $p_{min} \leq p \leq p_{max}$, on obtient un problème dans lequel contrôles continu et impulsionnel interviennent simultanément.

On aboutit dans ce cas à un problème non linéaire.

Le modèle s'écrit de la même façon que précédemment, à condition de remplacer certaines inégalités d'évolution par :

$$- \frac{\partial u_{10}}{\partial t}(t,x) - \frac{\sigma^2(t)}{2} \frac{\partial u_{10}}{\partial x^2}(t,x) + \max_{p_1(t)} \{[\mu(t)-p_1(t)]^+(\frac{\partial u_{10}}{\partial x}(t,x)-f(x,t))-p_1(t)C_p^1\} \leq 0$$

$$- \frac{\partial u_{01}}{\partial t}(t,x) - \frac{\sigma^2(t)}{2} \frac{\partial u_{01}}{\partial x^2}(t,x) + \max_{p_2(t)} \{[\mu(t)-p_2(t)]^+(\frac{\partial u_{01}}{\partial x}(t,x)-f(x,t))-p_2(t)C_p^2\} \leq 0$$

$$- \frac{\partial u_{11}}{\partial t}(t,x) - \frac{\sigma^2(t)}{2} \frac{\partial u_{11}}{\partial x^2}(t,x) + \max_{p_1(t);p_2(t)} \{[\mu(t)-p_1(t)-p_2(t)]^+(\frac{\partial u_{11}}{\partial x}(t,x)-f(x,t)) -$$

$$- p_1(t)C_p^1 - p_2(t)C_p^2\} \leq 0$$

ainsi que les relations analogues, correspondant à la période $[T-\tau,T]$.

4.3. Conditions aux limites.

Les contraintes naturelles font que $X \in \Omega$ borné de R

$$\Omega = \{X \in R | X_{min} \leq X \leq X_{max}\}$$

Soit $\Gamma = \partial\Omega$. L'on a été amené à prendre comme conditions aux limites sur Γ

$$\left.\frac{\partial u_{ij}}{\partial n}\right|_{\Gamma} = 0 \qquad \text{(Neuman homogène)}$$

Il existe d'autres types de conditions, mais la condition de Neuman a une interprétation stochastique simple qui fait qu'elle est bien adaptée au problème (cf. BENSOUSSAN-LIONS [1]).

V - RESULTATS NUMERIQUES.

Nous présentons ici quelques résultats essentiels, obtenus grâce à la résolution du modèle, avec des données inspirées de celles de l'E.D.F.

La résolution donne, pour tout instant t et pour tout niveau de stock X, la valeur des fonctions de coût :

$$u_{11}(t,x),\ u_{01}(t,x),\ u_{10}(t,x),\ u_{00}(t,x)$$

ainsi que pour tout $t \in [0,T-\tau[$ et pour tout x, les termes de contraintes :

$$W_{00}^{10}(t,x),\ W_{00}^{01}(t,x),\ W_{00}^{11}(t,x),\ W_{10}^{11}(t,x),\ W_{01}^{11}(t,x)\ .$$

On déduit de tout cela la politique optimale, c'est-à-dire qu'à tout instant, et en fonction de l'état du système (stock x, état thermique (i,j)), on sait si l'on doit laisser le système évoluer librement, si l'on doit soit arrêter, soit lancer une ou deux centrales, soit en arrêter une et lancer l'autre.

Cette politique optimale peut être résumée dans des tables de décision, que nous ne figurerons pas ici. L'utilisation de ces tables permet d'effectuer des simulations. Nous mettrons ainsi en évidence la réponse du système à une demande donnée :

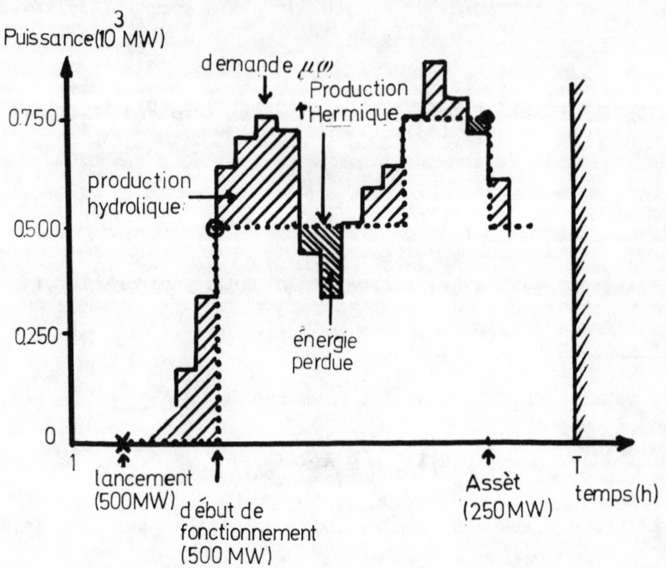

Nous allons donner quelques trajectoires simulées correspondant aux données suivantes*. Les puissances sont exprimées en megawatts (MW), les coûts en megafrancs (MF) ou en centimes par Kilowatt-heures (cts/kwh), les durées en heures (h).

L'étude porte sur une journée standard (24 h), avec un stock initial d'énergie hydraulique X_o=5 000 MWH.

Les puissances des deux centrales sont respectivement p_1=250 MW (2 tranches de 125 MW p_2=500 MW (2 tranches de 250 MW) et, dans le cas de la modulation $125 \leq p_1 \leq 250$ MW, $250 \leq p_2 \leq 500$ MW, le retard τ est de 5h.

Les coûts sont :

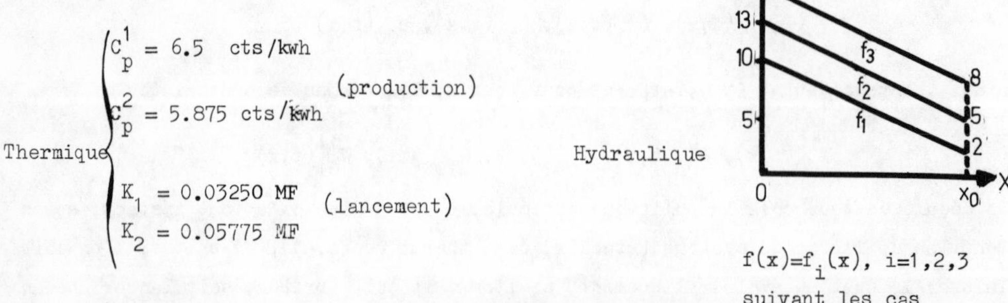

Thermique $\begin{cases} C_p^1 = 6.5 \text{ cts/kwh} \\ C_p^2 = 5.875 \text{ cts/kwh} \end{cases}$ (production)

$K_1 = 0.03250$ MF, $K_2 = 0.05775$ MF (lancement)

Hydraulique

$f(x)=f_i(x)$, i=1,2,3 suivant les cas

* Les résultats complets seront publiés dans LEGUAY [1].

Discrétisation $[0,T] = [0,1]$, pas de temps $\Delta t = \frac{1}{24}$ (1 heure)

$[0,X_o]=[0,5]$, pas d'espace $h=0.25$ (250 MWH)

$\Delta p_1 = 25$ MW, $\Delta p_2 = 50$ MW (dans le cas de la modulation)

5.1. Etude du cas impulsionnel (modèle linéaire)

5.1.a. Exemple test 1. Demande constante par morceaux.

La partie déterministe de la demande étant $\mu_1(t)$, la partie stochastique est donnée par

$$\sigma(t) = \sigma^* [\mu_1(t) + 1.5] \sqrt{\Delta t}$$

avec $\sigma^*=0.01$ (demande presque déterministe).

Nous avons simulé une trajectoire à partir de l'état $(X_o,0,0)$ (stock X_o, et les 2 centrales à l'arrêt). Cette trajectoire est représentée dans la Figure 1. On constate que la réponse du système thermique coïncide, autant que faire se peut, avec la partie déterministe de la demande.

5.1.b. Exemple test 2. Demande réelle.

La partie déterministe est $\mu_2(t)$, la partie stochastique $\sigma(t)=\sigma^*[\mu_2(t)+1.5] \sqrt{\Delta t}$ avec $\sigma^*=0.1$.

Nous avons, à partir des politiques optimales obtenues par la résolution du modèle simulé trois trajectoires, sachant qu'à $t=0$, on est dans l'état $(X_o,1,0)$; chacune des trajectoires (figure 2, figure 3, figure 4) correspond à un coût $f(x)$ égal à f_1, f_2, f_3 respectivement.

On notera l'influence importante du coût de production hydraulique, par rapport au coût de production thermique (analyse de sensibilité par rapport aux paramètres économiques).

5.2. Etude du cas impulsionnel et continu (modèle non linéaire)

5.2.a. Etude sur 1 journée standard.

La partie déterministe de la demande étant $\mu_2(t)$, et la partie stochastique

$$\sigma(t) = \sigma^*[\mu_2(t) + 1.5] \sqrt{\Delta t} \quad \text{avec} \quad \sigma^* = 0.1$$

Nous avons établi les tables de décision qui mentionnent la politique optimale (décisions de lancement, d'arrêt et puissances optimales de fonctionnement), et à partir de celles-ci nous avons simulé une trajectoire correspondant à $f(x) = f_3(x)$. C'est la figure 5.

5.2.b. Etude du problème complet.

L'E.D.F. s'intéresse essentiellement à la gestion sur une période d'une semaine, mais

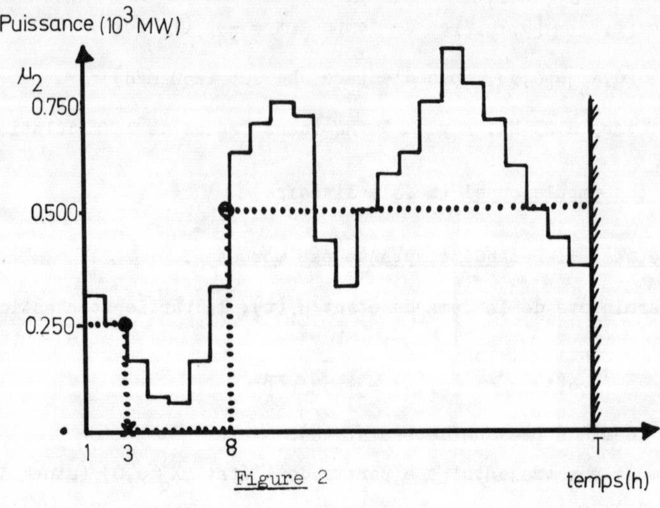

Figure 2

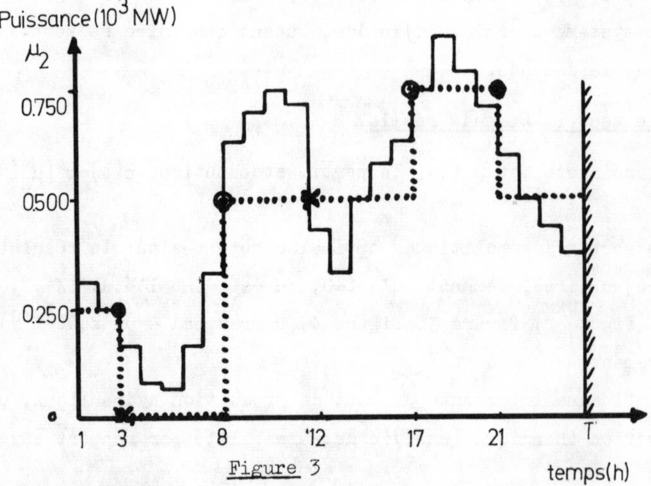

Figure 3

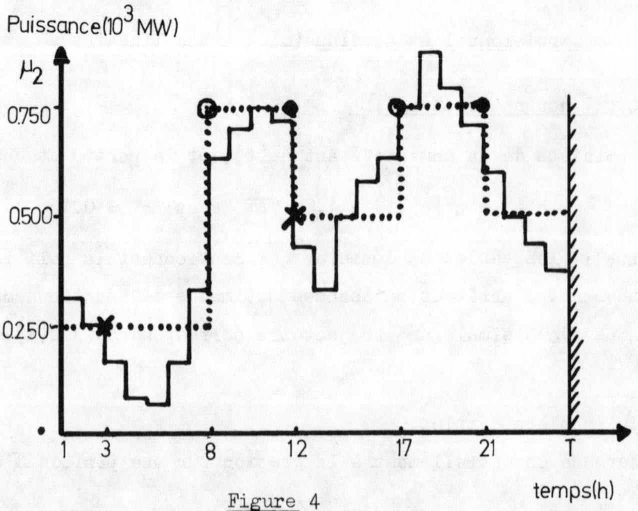

Figure 4

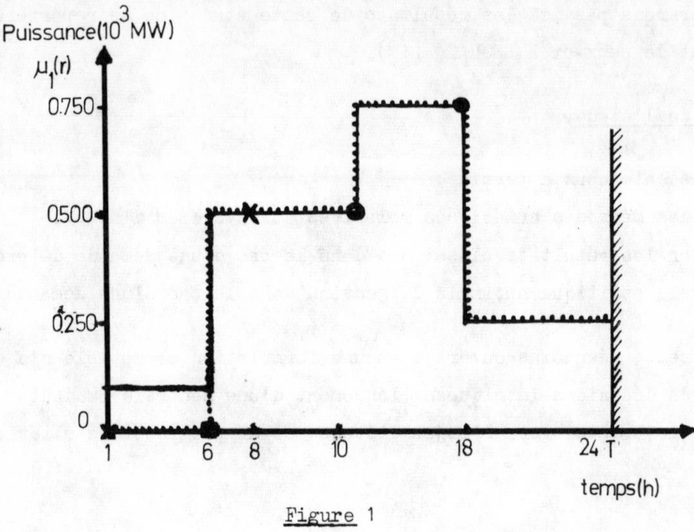

Figure 1

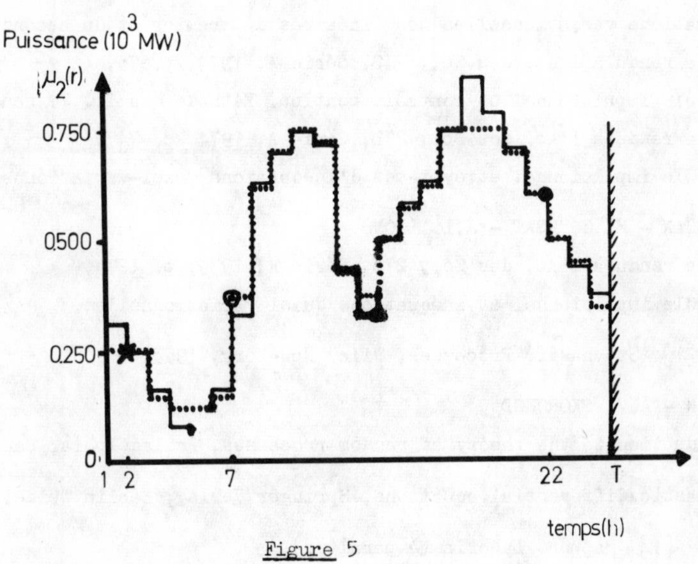

Figure 5

nous ne figurerons pas ici les résultats de cette étude (on se reportera, pour l'énoncé complet de ceux-ci à LEGUAY [1]).

Conclusions de l'étude.

L'étude précédente nous a permis :

- de tester des méthodes numériques relatives aux systèmes d'I.Q.V.
- de retrouver les résultats classiques dans le cas d'une demande déterministe.
- de définir la politique optimale de gestion dans le cas d'une demande aléatoire.

Un problème reste néanmoins ouvert : aucune formulation acceptable n'a été trouvée pour le cas de décisions imbriquées (lancement d'une centrale pendant le préchauffage de l'autre) ; ceci fera l'objet d'études et de publications ultérieures.

BIBLIOGRAPHIE

A. BENSOUSSAN - J.L. LIONS

[1] Ouvrage en préparation.

[2] Compte rendu à l'Ac. des Sc., 276, Série A, 1973, p1189,
Nouvelles formulations de problèmes de contrôle impulsionnel et application.

[3] Compte rendu à l'Ac. des Sc., 276, Série A, 1973, p1411,
Inéquations variationnelles non linéaires du premier et du second ordre.

[4] Compte rendu à l'Ac. des Sc., 278, Série A, 1974, p.675,
Contrôle impulsionnel et contrôle continu. Méthode des I.Q.V. non linéaires.

[5] Compte rendu à l'Ac. des Sc., 278, Série A, 1974, p. 747,
Contrôle impulsionnel et systèmes d'inéquations quasi-variationnelles.

A. BENSOUSSAN - M. GOURSAT - J.L. LIONS

[1] Compte rendu à l'Ac. des Sc., 276, Série A, 1973, p. 1279,
Contrôle impulsionnel et inéquations quasi variationnelles.

J.L. DOOB [1] Stochastic Processes, Wiley New-York, 1953.

I.I. GIHMAN - A.V. SKOROHOD

[1] Introduction to the theory of random processes, Philadelphia, Saunders 1969.

[2] Stochastic differential equations, Springer Verlag, Berlin Weidelberg, N.Y.1972.

M. GOURSAT [1] Rapport Laboria (à paraître).

C. LEGUAY [1] Thèse de Docteur-Ingénieur (à paraître).

S. MAURIN [1] Rapport Laboria (à paraître).

J. NEVEU [1] Cours de 3ème Cycle à la faculté des Sciences.

Automatic Sequential Clustering

of

Large Tables

by E. DIDAY

Summary

A method for clustering adapted to the processing of large data arrays is given ; this has been made possible by the economy of memory space. It is shown that one gets the same set of solutions either by the method described here or by the Dynamic Clusters Method. Some variants are given along with the criterion they optimize.

-=-

1 - Introduction

2 - Notations

3 - The SDCM algorithm

4 - Relations between the DCM and SDCM

4.1 - Reminders concerning the DCM

4.2 - Comparison between the solutions obtained with DCM and SDCM

4.3 - A condition sufficient to have R square in the sense of the DCM and the SDCM

4.4 - Comparative advantages of the DCM and the SDCM

5 - Conclusion

-=-

1 - Introduction

In any field where it is possible to define a set E in which the elements can be connected by a measure of similarity it is possible to find clusters; in other words, to try to detect classes of objects such that two objects of the same class are closer to each other than to objects belonging to other classes in the sense of the measure of simi-

larity.
Classifications may be conceived which are more or less suitable according to a given criterion.

In this paper, we will give methods which tend to optimize such criteri and which are also adapted to the processing of large tables requiring more than 100 K memory words.

. Many authors have taken interest in the hierarchical classification (cf. (2), (3), (6), (12), (13), (15), (18)). It is a method permitting to give partitions of E which are encased in one another, giving a hierarchy. This type of method runs into two difficulties :

a) The pruning of the tree into classes which allows to define types.
b) The necessity of storing a table with $\frac{N(N-1)}{2}$ arrays corresponding t the distances taken two by two of the elements of E.

. The minimum spanning tree (cf. (11) and (20)) permits an attractive approach to the problem in giving a tree whose vertices are the element of E and where the length of the edges represents the degree of similarity between the connected elements. This method allows to built a tree whose sum of the edges length is minimum. The visual representation of such a tree can be very enriching even though subjected to chain effect. To define types in the case of sets of large size, it is necessary to use cutting up techniques depending on thresholds determined before han

. One may conceive many methods which give solely a "good" partition of E. In our opinion, the most interesting ones are those which reduce the number of arbitrary parameters as much as possible, and which at the same time express the criterion being optimized ; the methods described in this paper require the knowledge of the number of classes of the sear ched partition, in other words, the degree of refinement desired by the user ; furthermore, a simple criterion allows to grasp the meaning of the partition which has been obtained. The so called "Dynamic Clusters Methods" are based on the choice of three functions :

- h is a function permitting to define the set E and a method of measure applied to this set (coding, similiraty measurements, a priori types, etc...).

- g a function of representation of any class of E ;

- f a function identifying any elements of E with the help of this representation.

The Dynamic Clusters Method (noted DCM - cf. (7) and (9)) consists in storing E in the central memory and in using alternativaly f and g until an equilibrium position is reached (corresponding to a local optimum of the criterion). The Sequentialized Dynamic Clusters Method, (SDCM) to which this paper is devoted, is based on the same principles than the DCM, but the storing in central unit of the data table associated to E is not necessary. This is of great advantage for large tables, there is not an important loss of time compared to the DCM.

The DCM and SDCM may be considered as a generalisation of a set of procedures called by some authors : "Iteration, relocation procedure" (Cormack (1971), Wishart(1971)), k-means by others (Mc Queen (1967), Watanabe (1972)). These procedures may be roughly described as follows.

One starts with k points choosen or estimated in a population, these k points will be the centers, all the points are assigned to the closest center according to one of the following two techniques :

a) The k centers may change when all the points are classified (Forgey (1965), Hall and Ball (1965), Jancey (1966)). The mean of each group is computed and gives a new point which is considered as the new center.

b) The k centers may move after the assignment of each point (Thorndike (1953), Mc Queen (1967), Beale (1969), Wishart (1971)). When a point is added to a group, the mean of the group is adjusted by taking in consideration this new point. This mean defines the new center.

Finally, the whole population is classified iteratively and several times according to techniques a) - Beale (1969), Wishart (1971) - and techniques b).

The procedure is ended when a number of iterations (predetermined) has occured, or when the process either comes to stability or oscillates.

Let us point out that classical anglo-saxon programs like ISODATA call out arbitrary tresholds of linkage and cutting out of the classes during the algorithm flow (cf. 16).

One may admit that a particular case of the DCM is a) when the function of representation g gives the centers of gravity.

In the SDCM one may use g as soon as all the individuals of a cluster are assigned. Even when g gives the centers of gravity, this method stays more general than b) and should save an appreciable amount of time by reducing the number of determinations of the centers.

The most important contributions of this paper are :

1) The introduction of the SDCM

2) The convergence properties of different variants with the associated criteria (meaningfull to the user)that the variants tend to optimize.

3) The demonstration of the equivalence of the set of solutions obtained with the DCM and the SDCM (therefore between a) and b)).

4) The description of the SDCM program which has already given a partition in 10 classes of a population of 5000 individuals characterized by 30 parameters in 7 min. on IBM 360-70. This partition could not have been achieved by the DCM because of the storage space required or else because of the number of exchanges that occur if the data are stored on auxillary storage units. However this partition corresponds well to the solution the DCM could have given.

2 - Notations

E is the set of the objects to be classified.

$\mathbb{L}_k = \{ L = (A_1, \ldots, A_k)/A_i \subset B \}$, B is a space depending on the variants.

$\mathbb{P}(E)$ is the set of the subsets of E.

$\mathbb{P}_k$ is the set of the E partition in k classes.

$P, Q \in \mathbb{P}_k$ will be noted $P = (P, \ldots, P_k)$ and $Q = (Q_1, \ldots, Q_k)$.

$D : E \times \mathbb{P}(B) \to \mathbb{R}^+$

$F : \mathbb{P}_k \times \mathbb{P}(E) \times \mathbb{L}_k \to \mathbb{P}_k$ is given by :

$F(Q,H,L) = P$ whereP is defined by the sequence $\{\pi_n\}$ of $\mathbb{P}_k$ as follows :

let us note $\pi_\ell = (\pi_{\ell 1}, \ldots, \pi_{\ell k})$ and $H = \{y_1, \ldots, y_n\}$.

let $\pi_\circ = Q$, the sequence π_n is defined by recurrence from $\pi_\circ$; once $\pi_{\ell-1}$ is given, one builts π_ℓ as follows : (let us suppose $y_\ell \in \pi_{\ell-1,j}$)

a) $\nexists i : D(y_\ell, A_i) \leqslant D(y_\ell, A_j)$ with j<i in case of equality ; one has then $\pi_\ell = \pi_{\ell-1}$

b) $\exists i : D(y_\ell, A_i) \leqslant D(y_\ell, A_j)$ with j>i in case of equality ; one has then $\pi_{\ell,i} = \pi_{\ell-1,i} \cup \{y_\ell\}$ $\pi_{\ell,j} = \pi_{\ell-1,j} - \{y_\ell\}$ and then $\pi_{\ell n} = \pi_{\ell-1,h}$ $\forall n$ such that $n \neq i$, $n \neq j$ and $n \in \{1,2\ldots,k\}$. Finally let us give $P = \pi_m$

R is a function which is defined upon $B \times]K] \times \mathbb{P}_k$ and has values in $\mathbb{R}^+$ (when $]K] = \{1,2,\ldots k\}$).

. Given k integers $n_1, \ldots, n_k$

$g^s : \mathbb{P}_k \rightarrow \mathbb{L}_k$ is such that $g^s(P) = L = (A_1,..,A_k)$ where $A_i = \{$the n_i elements of B that minimize $R(x,i,P)\}$.

. $W : V_k = \mathbb{L}_k \times \mathbb{P}_k \rightarrow \mathbb{R}^+$ is the criterion to be optimized, it can be written $W(v) = \sum_{i \in]k]} \sum_{x \in A_i} R(x,i,P)$ where $v = (L,P)$.

$$S' : S'(L,P) = W(v) \quad \forall v = (L,P)$$

3 - <u>The Basic algorithm of the SDCM</u>[1]

One starts from $P^{(\circ)}$ and one deduces $L^{(1)} = g^s(P^{(\circ)})$ and $P^{(1)} = F(P^{(\circ)}, H^{(\circ)}, L^{(1)})$.

Let us note $H^{(\circ)} H^{(1)},...,H^{(n)}$ a partition of E and let us also define the sequence $\{H^{(n)}\}$ in taking $H^{(n)} = H^{(r)}$ where r is the remainder of the division of N + 1 into n.

The SDCM is defined by the following sequence v_n :

one has $v_1 = (L^{(1)},P^{(1)})$ and $v_{n+1} = (L^{(n+1)},P^{(n+1)})$ is deduced from $v_n = (L^{(n)},P^{(n)})$ as follows :

$$P^{(n+1)} = F(P^{(n)},H^{(n)},L^{(n+1)}) \text{ and } L^{(n+1)} = g^s(P^{(n)}) \quad (2)$$

Let $u_n = W(v_n)$, $z_n = S'(L^{(n+1)},P^{(n)})$, $H^{(n)} = \{y_\circ^{(n)},...,y_m^{(n)}\}$, the sequence π_ℓ associated to $H^{(n)}$ will be noted $\pi_\ell^{(n)}$. To simplify, let us suppose that all the $H^{(n)}$ blocks are such that card $(H^{(n)}) = m$ fixed hence card $E = mN$ (all the following demonstrations are still valid if m is varying).

<u>Definition</u> :

We shall say that R is square in the SDCM sense if $\forall$ P,L,y

$$S'(L,F(P,y,L)) \leqslant S'(L,P)$$

<u>Theorem 1</u>

If R is square in the sense of the SDCM then the sequences u_n and z_n are convergent and decreasing and the sequence v_n converges.

<u>Demonstration</u>

Let $v_n = (L^{(n)},P^{(n)})$, one has $W(v_n) = S'(L^{(n)},P^{(n)})$ hence

$$u_n = \sum_{i=1}^{k} \sum_{x \in A_i^{(n)}} R(x,i,P^{(n)})$$

(1) SDCM means "Sequentialized Dynamic Cluster Method".

(2) $L^{(n+1)}$ differs from $L^{(n)}$ only if there exists at least one "improving element", his notion will be precised in [7].

As

$$z_n = S'(L^{(n+1)}, P^{(n)})$$

one has necessarily :

$$u_n \geqslant Z_n = \sum_{i=1}^{k} \sum_{x \in A_i^{(n+1)}} R(x, i, P^{(n)})$$

as $A_i^{(n+1)}$ is formed by elements of B that minimize $R(.,i,P^{(n)})$ and that $A_i^{(n)} \subset B$. As R is square one has :

$$S'(L^{(n+1)}, F(P^{(n)}, yo^{(n)}, L^{(n+1)})) \leqslant S'(L^{(n+1)}, P^{(n)})$$

in other words,

$$S'(L^{(n+1)}, \pi o^{(n)}) \leqslant S'(L^{(n+1)}, P^{(n)}) ;$$

one has likewise

$$S'(L^{(n+1)}, F(\pi_o^{(n)}, y_1^{(n)}, L^{(n+1)})) \leqslant S'(L^{(n+1)}, \pi_o^{(n)}),$$

or else

$$S'(L^{(n+1)}, \pi_1^{(n)}) \leqslant S'(L^{(n+1)}, \pi_o^{(n)}).$$

The same reasoning may be started over with $2^{(n)}, y_3^{(n)}$ until $y_m^{(n)}$ last element of $H^{(n)}$ hence

$$S'(L^{(n+1)}, \pi_m^{(n)}) \leqslant S'(L^{(n+1)}, \pi_{m-1}^{(n)}) \leqslant \ldots \leqslant S'(L^{(n+1)}, P^{(n)})$$

where

$$S'(L^{(n+1)}, P^{(n+1)}) \leqslant S'(L^{(n+1)}, P^{(n)})$$

this implies

$$u_{n+1} \leqslant Z_n$$

The sequence u_n is positive thus it converges and is decreasing. Futhermore it reaches its limits because it can take only a finite number of values.

The convergence of the sequence v_n can be demonstrated as follows :

let

$$u_M = \lim_{n \to \infty} u_n$$

there, exists :

$$v_M : u_M = w(v_M)$$

because u_n reaches its limit.

If an integer M_1 would exist such that $M_1 > M + N$ and $V_{M_1} \neq V_M$, two cases may occur because W is not necessarily injective :

a) $L^{(M_1)} \neq L^{(M)}$

b) $P^{(M_1)} \neq P^{(M)}$

Should case a) occur, one should find at least one improving element between the iteration M and M-1 hence $u_{M_1} < u_M$ which is impossible.

The case b) cannot occur because the kernels have not been modified during N successive iterations (N has been defined in 3).

Remark 1 :

When B is a function of $n \geqslant 1$ (the iteration number) and such that

$$B_n = \bigcup_{i=1}^{k} A_i^{(n)} \cup \{H^{(n)}\} \cup C$$

(where C may assume any value). The same result is obtained.

In fact $A_i^{(n+1)}$ is formed of the n_i elements of B_n which minimize $R(.,i,P^{(n)})$ and $A_i^{(n)} \subset B_n$; hence if improving elements exist in B_n one has necessarily $u_{n+1} < u_n$.

4 - Relations between the DCM and the SDCM

4.1 - Reminder concerning the DCM

R is defined as before :

A function f is introduced $f : \mathbb{L}_k \to \mathbb{P}_k$ such that $f(L) = P$ where

$$P_i = \{x \in E / D(x,A_i) \leqslant D(x,A_j)\}$$

if there is equality one assigns x to the class of smallest index.

The function $g : \mathbb{P}_k \to \mathbb{L}_k$ is given by $g(P) = L$ where $L = (A_1, \ldots, A_k)$ with $A_i = \{$the n_i elements of B which minimize $R(.,i,P)\}$.

S is a function $\mathbb{L}_k \times \mathbb{L}_k \to R+$ such that

$$S(L,M) = \sum_{i=1}^{k} \sum_{x \in A_i} R(x,i,f(M))$$

The criterion may then be written

$$W(v) = \sum_{i=1}^{k} \sum_{x \in A_i} R(x,i,P)$$

where $v = (L,P)$

4.2 - Comparison between the solutions obtained by the DCM and the SDCM

In order to simplify the statement of the following result, we shall say that an SDCM algorithm is associated with a DCM algorithm if the corresponding F, f, g^s and g functions are defined from the same D and R functions. Let us remind that a function R is said to be square in the sense of the DCM.

(cf (1), (2)) if $\forall\, L,M \in \mathbb{L}_k \; S(L,M) \leqslant S(M,M) \Longrightarrow S(L,L) \leqslant S(L,M)$ (2)

We have shown (cf [3]) that if R is square in the sense of the DCM, the basic DCM algorithm decreases the criterion W.

Theorem 2

Let $B \equiv E$. Having a basic DCM algorithm and its associated SDCM algorithm, if R is square in the sense of the SDCM and in the sense of the DCM, then these two algorithms decrease the same criterion and converge toward the same solutions.

Demonstration

R being square in the sense of the two methods, they both decrease the same criterion :

$$W(v) = \sum_{i=1}^{k} \sum_{x \in A_i} R(x,i,P) \quad \text{where} \quad V = (L,P) \quad \text{and} \quad L = (A_1,\ldots,A_k)$$

A last point should now be demonstrated namely that both algorithms converge necessarily on an unbiased element [(1)] (cf [8]) ; we are going to show that the same is true for the SDCM algorithm. $u_M = \lim_{n\to\infty} u_n$. Whatever $i \in \{1,2,\ldots,N\}$ (N is defined in 3) $L^{(M)} = L^{(M+i)}$ as a consequence of $u_M = u_{M+i}$, (if an improving element should exist, we would

(1) In other words an element $(L,P) \in L_k \times P_k$: $f(L) = P$ and $g(P) = L$

have $u_M < u_{M+i}$) ; As the $H^{(M+i)}$ constitutes an overlapping of E when i varies from 1 to N, one has $g(P^{(M)}) = L^{(M)}$.

By definition of F and because $L^{(M+i)} = L^{(M)}$ $\forall i \in 1,2,\ldots,N$ one has

$$\forall x \in \bigcup_{i=1}^{n} H^{(M+i)} = E \quad \text{and} \quad \forall x \in P_j^{(M+N)},$$

$$D(x, A_j^{(M)}) \leqslant D(x, A_q^{(M)})$$

(and $j \leqslant q$ in case of equality) because :

$$P^{(M+i+1)} = F(P^{(M+i)}, H^{(M+i)}, L^{(M+i+1)}) = F(P^{(M+i)}, H^{(M+i)}, L^{(M)})$$

Let $M_1 = M + N$ one has :

$$f(L^{(M_1)}) = P^{(M_1)}$$

as

$$g(P^{(M_1)}) = L^{(M_1)}$$

because M_1 is greater than M, $(L^{(M_1)}, P^{(M_1)})$ is really an unbiased element in the sense of the DCM.

Reciprocally any unbiased element v may be considered as a limit of the sequence $v_n = v$, this being true for both methods. Thus there is identity between the two spaces of the solutions obtained by the DCM and the SDCM.

Remark :

If the set B associated to the SDCM is function of n (the iteration number) and if B is such that $B_n = \bigcup_{j=1}^{k} A_j^{(n)} \cup \{H^{(n)}\}$ the same result is obtained.

As a matter of fact, let us come back to the previous demonstration ; one has $L^{(M)} = L^{(M+i)}$ because $v_M = v_{M+i}$, which means that no element of $A^{(M+i)}$ $\forall i \in \{0,1,\ldots N\}$ (hence no element of E) improves $L^{(M)}$, we consequently have $L^{(M)} = g(P^{(M)})$.

4.3 - Conditions sufficient to get R square in the sense of the DCM and SDCM

Proposition 1

In order to have R square simultaneously in the sense of the DCM and in the sense of the SDCM it is sufficient that the following condition be verified :

(1) $S'(L,P) = \sum_{i=1}^{k} \sum_{z\in P_i} \mu(z)\ D(z,A_i) \quad \forall L = (A_1,\ldots A_k) \in \mathbb{L}$

and $\quad P = (P_1,\ldots,P_k) \in \mathbb{P}$

Demonstration

Let us first show that (1) produces as a consequence R to be square in the sense of the DCM. If the relation $S'(L,P) = \sum_{i=1}^{k} \sum_{z\in P_i} D(z,A_i)$ is true $\forall P \in \mathbb{P}$ one has necessarily $\forall M \in \mathbb{L}_k$ such that :

$$f(M) = Q, S(L,M) = \sum_{i=1}^{k} \sum_{x\in A_i} R(x,i,Q) = S'(L,Q),$$

hence :

$$(2) \qquad S(L,M) = \sum_{i=1}^{k} \sum_{z\in Q_i} \mu(z)\ D(z,A_i) \quad \forall L,M \in \mathbb{L}.$$

this draws :

$$(3)\ S(L,L) = \sum_{i=1}^{k} \sum_{z\in Q_i} \mu(z)\ D(z,A_i) \text{ if } f(L) = Q' = (Q'_1,\ldots Q'_k)$$

Let us suppose that the same element $z \in E$ appears in (2) under the form $D(z,A_j)$ and in (3) under the form $D(z,A_i)$. As a result from the way Q' has been built up, one has necessary $D(z,A_i) \leqslant D(z,A_j)$. As each element $z \in E$ appears once and only once in (1) and (2) because Q and Q' are partitions of E, the same reasoning can be made $\forall\ z \in E$, hence $S(L,L) \leqslant S(L,M)$.

The function R is therefore square in the sense of the DCM.

It remains to be shown that the function R is square in the sense of the SDCM :

$$(1) \qquad S'(L,P) = \sum_{i=1}^{k} \sum_{z\in P_i} \mu(z)\ D(z,A_i) \quad \forall L \in \mathbb{L}_k \text{ and } P \in \mathbb{P}_k$$

$$(4) \qquad \Rightarrow S'(L,F(P,y,L)) = \sum_{i=1}^{k} \sum_{z\in P_i} \mu(z)\ D(z,A_i) \quad \forall y \in E.$$

where $P' = F(P,y,L)$.

Let us suppose that y belongs to P'_j, one knows that the two following cases may occur :

1) $\nexists i$: $D(y,A_i) \leqslant D(y,A_j)$ with $j < i$ in case of equality, then $F(P,y,L) \equiv P$

2) $\exists i : D(y,A_i) \leqslant D(y,A_j)$ with $j < i$ in case of equality, then by definition of $F(P,y,L)$, the expressions (1) and (4) are formed by identical terms, except for the term y which appears in (1) under the form $D(y,A_j)$ and in (4) under the form $D(y,A_i)$. In any case, one has :

$$S'(L,F(P,y,L)) \leqslant S'(L,P)$$

which proves that S' is really square in the sense of the SDCM.

4.4 - Comparative advantage of the two methods

The SDCM may be conceived in such a way as to require only once the storing of the whole population in the central unit at each iteration, the means being computed sequentially, block by block, while the DCM needs twice this loading in central unit at each iteration (once to compute the partition, a second time to calculate the kernels).

Furthermore for the DCM the Centerings are accomplished class by class, which is very unpractical when the data table is not stored in the central unit.

The drawback of the SDCM is the need of N-k (N being the number of parts H_i) additional and more expensive centerings at each iteration (the number of iterations before convergence seems to be quite the same for the two methods).

5 - Conclusion

The SDCM should already permit the processing of large tables. In a near future, the following directions should be fruitful :

a) The stochastic aspect, especially in the case of infinite tables, in order, for instance, to define cut-off tests,

b) The learning aspect by calling out labelled individuals during the flow of the algorithm,

c) The on line computing aspect (for keeping up to date nomenclature for instance),

d) The programming of different variants could perhaps improve the solutions obtained by changing rooted tree.

To classify allows man to give himself a concise representation of the universe (in agreement with a limited mind) and to locate any element according to this representation. It is obvious that man will have to confront continuously that representation with an objective reality because of new elements that have to be located. May be, philosophers, psychologists will tell us, some day, that h field function, g representation function and f location function allow the representation in machine of that dialectic and are constituting an approach to three implicit functions of knowledge.

-=-

Bibliography

[1] Beale (1969) "Euclidean cluster analysis" Bull. I.S.E., 43 Book 2, pp. 92-94 (London).

[2] Benzecri J.P. (1973) "Taxonomie - L'analyse des Données" (Dunod).

[3] Benzecri J.P. (1969) "Construction ascendante d'une classification hiérarchique" (L.S.M. I.S.U.P.).

[4] Benzecri J.P. (1970) "Algorithmes rapides d'agrégation" (Sup. Class.) (L.S.M. I.S.U.P.)

[5] Brianne J.P. (1972) "L'algorithme d'échange". Thèse de 3° cycle (L.S.M. I.S.U.P.)

[6] Cormack R.M. (1971) "A review of classification". The journal of the Royal Statistical Society serie A vol. 134, Part 3.

[7] Diday E. (1971) "Une nouvelle méthode en classification automatique et reconnaissance des formes : la méthode des nuées dynamiques". Revue de statistique appliquée. Vol. XIX n° 2.

[8] Diday E. (1972) "Optimisation en classification automatique et reconnaissance des formes" RAIRO vol. 3,p. 61 à 96.

[9] Diday E. (1973) "Introduction à l'analyse factorielle typologique". Rapport Laboria n° 27 (IRIA Rocquencourt (78)).

[10] Forgey E.W. (1965) "Cluster analysis of multivariate data" ARAS Biometric Society (WNRR) Riverside California USA.

[11] Gower J.C. and Ross G.J.S. (1969) "Minimum spanning trees and single-linkage cluster analysis". Appl. Stat. 18, 54-64.

[12] Jardine N and Sibson R. (1971) "Mathematical Taxonomy". J. Wiley and Sons ltd.

[13] Jambu M. (1972) "Techniques de classification automatique" Thèse de 3° cycle (L.S.M. I.S.U.P.).

[14] Jancey R.C. (1966) "Multidimensional group analysis" Aust. J. Bot. vol 14 p. 127.

[15] Lerman I.C. (1970) "Les bases de la classification automatique" Gauthiers Villars.

[16] Hall & Ball (1967) "A clustering techniques for summerizing multivariate data ". Behavioral Science vol. 12 n° 2.

Northouse R.A., From F.R. (1973) "Some results of Non-parametric clustering on large Date Problems". Proc. of the First Int. Joint Conf. on Pattern recognition.

[17] Mc. Queen J. (1967) "Some methods for classification and analysis of multivariable observations". 5th. Berkeley Symposium on Mathematical statistics and Probability vol. 1 n° 1 pp. 281-297.

[18] Roux M. (1968) "Un algorithme pour construire une hiérarchie particulière". Thèse de 3° cycle (L.S.M. I.S.U.P.).

[19] Thorndike R.L. (1953) "Who belongs in the family ?" Psychometrika, vol. 18 pp. 267-276.

[20] Zahn C.T. (1971) "Graph theoretical methods for detecting and describing gestalt clusters" I.E.E.E. Trans. and Comp. Vol C 20 n° 1.

Vol. 59: J. A. Hanson, Growth in Open Economics. IV, 127 pages. 1971. DM 16,–

Vol. 60: H. Hauptmann, Schätz- und Kontrolltheorie in stetigen dynamischen Wirtschaftsmodellen. V, 104 Seiten. 1971. DM 16,–

Vol. 61: K. H. F. Meyer, Wartesysteme mit variabler Bearbeitungsrate. VII, 314 Seiten. 1971. DM 24,–

Vol. 62: W. Krelle u. G. Gabisch unter Mitarbeit von J. Burgermeister, Wachstumstheorie. VII, 223 Seiten. 1972. DM 20,–

Vol. 63: J. Kohlas, Monte Carlo Simulation im Operations Research. VI, 162 Seiten. 1972. DM 16,–

Vol. 64: P. Gessner u. K. Spremann, Optimierung in Funktionenräumen. IV, 120 Seiten. 1972. DM 16,–

Vol. 65: W. Everling, Exercises in Computer Systems Analysis. VIII, 184 pages. 1972. DM 18,–

Vol. 66: F. Bauer, P. Garabedian and D. Korn, Supercritical Wing Sections. V, 211 pages. 1972. DM 20,–

Vol. 67: I. V. Girsanov, Lectures on Mathematical Theory of Extremum Problems. V, 136 pages. 1972. DM 16,–

Vol. 68: J. Loeckx, Computability and Decidability. An Introduction for Students of Computer Science. VI, 76 pages. 1972. DM 16,–

Vol. 69: S. Ashour, Sequencing Theory. V, 133 pages. 1972. DM 16,–

Vol. 70: J. P. Brown, The Economic Effects of Floods. Investigations of a Stochastic Model of Rational Investment Behavior in the Face of Floods. V, 87 pages. 1972. DM 16,–

Vol. 71: R. Henn und O. Opitz, Konsum- und Produktionstheorie II. V, 134 Seiten. 1972. DM 16,–

Vol. 72: T. P. Bagchi and J. G. C. Templeton, Numerical Methods in Markov Chains and Bulk Queues. XI, 89 pages. 1972. DM 16,–

Vol. 73: H. Kiendl, Suboptimale Regler mit abschnittweise linearer Struktur. VI, 146 Seiten. 1972. DM 16,–

Vol. 74: F. Pokropp, Aggregation von Produktionsfunktionen. VI, 107 Seiten. 1972. DM 16,–

Vol. 75: GI-Gesellschaft für Informatik e.V. Bericht Nr. 3. 1. Fachtagung über Programmiersprachen · München, 9–11, März 1971. Herausgegeben im Auftrag der Gesellschaft für Informatik von H. Langmaack und M. Paul. VII, 280 Seiten. 1972. DM 24,–

Vol. 76: G. Fandel, Optimale Entscheidung bei mehrfacher Zielsetzung. 121 Seiten. 1972. DM 16,–

Vol. 77: A. Auslender, Problemes de Minimax via l'Analyse Convexe et les Inégalités Variationelles: Théorie et Algorithmes. VII, 132 pages. 1972. DM 16,–

Vol. 78: GI-Gesellschaft für Informatik e.V. 2. Jahrestagung, Karlsruhe, 2.–4. Oktober 1972. Herausgegeben im Auftrag der Gesellschaft für Informatik von P. Deussen. XI, 576 Seiten. 1973. DM 36,–

Vol. 79: A. Berman, Cones, Matrices and Mathematical Programming. V, 96 pages. 1973. DM 16,–

Vol. 80: International Seminar on Trends in Mathematical Modelling, Venice, 13–18 December 1971. Edited by N. Hawkes. VI, 288 pages. 1973. DM 24,–

Vol. 81: Advanced Course on Software Engineering. Edited by F. L. Bauer. XII, 545 pages. 1973. DM 32,–

Vol. 82: R. Saeks, Resolution Space, Operators and Systems. X, 267 pages. 1973. DM 22,–

Vol. 83: NTG/GI-Gesellschaft für Informatik, Nachrichtentechnische Gesellschaft. Fachtagung „Cognitive Verfahren und Systeme", Hamburg, 11.–13. April 1973. Herausgegeben im Auftrag der NTG/GI von Th. Einsele, W. Giloi und H.-H. Nagel. VIII, 373 Seiten. 1973. DM 28,–

Vol. 84: A. V. Balakrishnan, Stochastic Differential System I. Filtering and Control. A Function Space Approach. V, 252 pages. 1973. DM 22,–

Vol. 85: T. Page, Economics of Involuntary Transfers: A Unified Approach to Pollution and Congestion Externalities. XI, 159 pages. 1973. DM 18,–

Vol. 86: Symposium on the Theory of Scheduling and Its Applications. Edited by S. E. Elmaghraby. VIII, 437 pages. 1973. DM 32,–

Vol. 87: G. F. Newell, Approximate Stochastic Behavior of n-Server Service Systems with Large n. VIII, 118 pages. 1973. DM 16,–

Vol. 88: H. Steckhan, Güterströme in Netzen. VII, 134 Seiten. 1973. DM 16,–

Vol. 89: J. P. Wallace and A. Sherret, Estimation of Product. Attributes and Their Importances. V, 94 pages. 1973. DM 16,–

Vol. 90: J.-F. Richard, Posterior and Predictive Densities for Simultaneous Equation Models. VI, 226 pages. 1973. DM 20,–

Vol. 91: Th. Marschak and R. Selten, General Equilibrium with Price-Making Firms. XI, 246 pages. 1974. DM 22,–

Vol. 92: E. Dierker, Topological Methods in Walrasian Economics. IV, 130 pages. 1974. DM 16,–

Vol. 93: 4th IFAC/IFIP International Conference on Digital Computer Applications to Process Control, Zürich/Switzerland, March 19–22, 1974. Edited by M. Mansour and W. Schaufelberger. XVIII, 544 pages. 1974. DM 36,–

Vol. 94: 4th IFAC/IFIP International Conference on Digital Computer Applications to Process Control, Zürich/Switzerland, March 19–22, 1974. Edited by M. Mansour and W. Schaufelberger. XVIII, 546 pages. 1974. DM 36,–

Vol. 95: M. Zeleny, Linear Multiobjective Programming. XII, 220 pages. 1974. DM 20,–

Vol. 96: O. Moeschlin, Zur Theorie von Neumannscher Wachstumsmodelle. XI, 115 Seiten. 1974. DM 16,–

Vol. 97: G. Schmidt, Über die Stabilität des einfachen Bedienungskanals. VII, 147 Seiten. 1974. DM 16,–

Vol. 98: Mathematical Methods in Queueing Theory. Proceedings of a Conference at Western Michigan University, May 10–12, 1973. Edited by A. B. Clarke. VII, 374 pages. 1974. DM 28,–

Vol. 99: Production Theory. Edited by W. Eichhorn, R. Henn, O. Opitz, and R. W. Shephard. VIII, 386 pages. 1974. DM 32,–

Vol. 100: B. S. Duran and P. L. Odell, Cluster Analysis. A survey. VI, 137 pages. 1974. DM 18,–

Vol. 101: W. M. Wonham, Linear Multivariable Control. A Geometric Approach. X, 344 pages. 1974. DM 30,–

Vol. 102: Analyse Convexe et Ses Applications. Comptes Rendus, Janvier 1974. Edited by J.-P. Aubin. IV, 244 pages. 1974. DM 25,–

Vol. 103: D. E. Boyce, A. Farhi, R. Weischedel, Optimal Subset Selection. Multiple Regression, Interdependence and Optimal Network Algorithms. XIII, 187 pages. 1974. DM 20,–

Vol. 104: S. Fujino, A Neo-Keynesian Theory of Inflation and Economic Growth. V, 96 pages. 1974. DM 18,–

Vol. 105: Optimal Control Theory and its Applications. Part I. Proceedings of the Fourteenth Biennual Seminar of the Canadian Mathematical Congress. University of Western Ontario, August 12–25, 1973. Edited by B. J. Kirby. VI, 425 pages. 1974. DM 35,–

Vol. 106: Optimal Control Theory and its Applications. Part II. Proceedings of the Fourteenth Biennial Seminar of the Canadian Mathematical Congress. University of Western Ontario, August 12-25, 1973. Edited by B. J. Kirby. VI, 403 pages. 1974. DM 35,–

Vol. 107: Control Theory, Numerical Methods and Computer Systems Modelling. International Symposium, Rocquencourt, June 17–21, 1974. Edited by A. Bensoussan and J. L. Lions. VIII, 757 pages. 1975. DM 53,–

GENERAL BOOKBINDING CO.
75 130BR 4 181 3 A 6000
QUALITY CONTROL MARK